Proceedings of the

SALT LAKE CITY MEETING

Proceedings of the

SALT LAKE CITY MEETING

Third Regular Meeting *(New Series)* of the Division of Particles and Fields of the American Physical Society

Hosted by the Physics Department,
University of Utah

Sheraton Triad Hotel and Towers,
Salt Lake City, Utah

January 14-17, 1987

Edited by

Carleton DeTar and James Ball

World Scientific
Singapore • New Jersey • Hong Kong

Published by

World Scientific Publishing Co. Pte. Ltd.
P.O. Box 128, Farrer Road, Singapore 9128

U.S.A. office: World Scientific Publishing Co., Inc.
687 Hartwell Street, Teaneck NJ 07666, USA

Library of Congress Cataloging-in-Publication data is available.

PROCEEDINGS OF THE SALT LAKE CITY MEETING

ISBN 9971-50-243-7
ISBN 9971-50-244-5 pbk

Printed in Singapore by General Printing and Publishing Services Pte. Ltd.

ACKNOWLEDGEMENTS

Sponsors

United States Department of Energy
National Science Foundation
University of Utah
American Physical Society

Program Advisory Committee

S. Adler
P. Bloch
C. DeTar (Chair)
G. Feldman
V. Fitch
B. Gittelman
S. Glashow
L. Littenberg
L. Pondrom
B. Pope
C. Quigg
D. Schramm
J. Steinberger
F. Wilczek
S. Wojcicki

Parallel Session Conveners

S. Tye
J. Dorfan
M. Strovink
G. Kane
E. Kolb
D. Sinclair
P. Nelson
P. Ramond
P. MacKenzie
D. Soper
D. Hitlin
T. Kirk
M. Breidenbach
R. Talman

Local Organizing Committee

C. DeTar (Chair)
J. Ball
E. Loh

Conference Secretariat

Nancy Parker
Chris McDonald
Shari Houston
Holli White
Cindy Allen

Proceedings Coordinator

Chris McDonald

Photographic Credits

Conference Black and White Snapshots

L. Casjens
Z. Peacock
J. Ball

Color Photographs and Plates

Dust cover	Barrier Canyon Pictographs	James Ball
Inside front cover	Dive of the Buckskin	James Ball
Inside back cover	Square Tower Group, Hovenweep National Monument	Laurel Casjens

Chapter Title Pages

Plenary Sessions	Sunset Point, Bryce Canyon (Conference Poster)	Laurel Casjens

Parallel Sessions

I	Jacob Hamlin Arch, Coyote Gulch	Laurel Casjens
II	Delicate Arch, Arches Nat'l Park	Carleton DeTar
III	Mount Superior, Alta, Utah	James Ball
IV	Calf Creek Falls	James Ball
V	Antelope Island, Great Salt Lake	Laurel Casjens
VI	Navajo Loop, Bryce Canyon	James Ball
VII	Terrace, Alpenglow Lodge, Alta	Maris Abolins

PREFACE

The third regular meeting (new series) of the Division of Particles and Fields of the American Physical Society, hosted by the University of Utah, was held January 14–17, 1987, in Salt Lake City, Utah, at the Salt Lake Sheraton Hotel and Towers. Since the 23rd International High Energy Physics Conference was held at Berkeley in the summer of 1986, an interval of sixteen months had been allowed to elapse since the previous divisional meeting at Eugene, Oregon, in June 1985. Although academic year commitments and the brief interval since the Berkeley conference were expected to hold attendance down, the attraction of Utah's famous snow and skiing opportunities and the enthusiasm of a large youthful membership drew an attendance of 330 with about 50 accompanying guests. It is remarkable that about 170 of the participants gave one or more talks.

The meeting was patterned after the previous successful Santa Fe and Eugene conferences. Seven parallel sessions covering major topical divisions of the field were constructed by two conveners each. With only a couple of exceptions, all contributed papers were assigned to a parallel session and were allotted a minimum of fifteen minutes presentation time. The conveners scheduled their parallel session, and were given a license to grant more time for selected contributions and to invite several mini-review speakers, some of whom had not originally made a contribution. The mini-review talks helped flesh out the subject covered only minimally by a series of short contributions. The conveners were selected with the advice and assistance of the Program Advisory Committee. It goes without saying that the success of this system depends heavily upon the good judgment, initiative and sense of fair play of each of the conveners. This meeting was particularly well served by its conveners, a number of whom had had experience with the system at Eugene and Santa Fe.

Plenary session talks were arranged with the advice of the Program Advisory Committee to provide good topical coverage of the most rapidly developing fields of current interest to the divisional membership. Although a rapporteur system was not used, some effort was made to alert the speakers to the relevant noteworthy contributions in the parallel sessions.

Because of the extraordinarily large number of talks (180) at this meeting, these Proceedings include only the invited talks solicited by the conveners in the parallel sessions, a few of the contributed papers selected by the conveners, and the talks in the plenary sessions. Only a few of the speakers were unable to submit contributions.

Since the main January attraction in Salt Lake City, other than an occasional physics conference, is the snow on the nearby slopes, it was a relatively simple procedure to provide an afternoon of skiing for conference participants and their guests. It was not as easy to provide a fresh order of Utah's famous powder snow on the designated afternoon, but novice and expert alike enjoyed a somewhat chilly but sunny half-day excursion into our beautiful Wasatch mountain canyons and a chance to ski downhill or cross-country. Those looking for deep powder will have to make a return visit.

The meeting will probably go down in history for featuring an after-dinner speech that struck a surprisingly controversial note among a few members of the audience. Professor James Clayton, Provost of the University of Utah and Professor of Economic History spoke about the "local", i.e. Mormon culture, and attempted to dispel some of the popular myths that surround it. Although he took a scholarly and objective approach, some in the audience assumed that his mission was to gloss over some of the more controversial political positions and attitudes of the Mormon Church. In the question period, Clayton was then placed in the awkward role of explaining and justifying church doctrine, some of which he, no doubt, privately took exception to. He survived with remarkable outward calm. Nonetheless, the talk did stimulate subsequent discussions.

Among the other highlights of the meeting were an introductory address by Norman H. Bangerter, Governor of the State of Utah, an evening public lecture on the supercollider by Roy Schwitters, which was attended by a modest but enthusiastic audience of high school students, reporters, and other interested people, and a concert by the Utah Symphony Orchestra in the gleamingly modern Salt Lake Symphony Hall. A contribution by Frank Avignone, reported the same day in the *New York Times*, described a surprising result from an experiment observing double beta decay (see these Proceedings), and created a flurry of excitement and a packed parallel session.

The effort and dedication of a large number of people made this meeting a success. In particular, the conference secretariat, Nancy Parker, Chris McDonald, Shari Houston, Holly White, and Cindy Allen kept the wheels turning prior to and during the conference. The conference database program was developed by John Ming, and provided a key to conference management. The expert logistical support of Dr. Lynn Higgs, Richard Ingebretsen, and Ziggy Peacock was essential. Janet Ball volunteered hours of her time planning and arranging excursions, developing information about local restaurants and other attractions, and providing travel agency assistance to conference attendees. And it goes without saying that the army of postdocs and graduate students who ran projectors, moved equipment, and manned tables was vitally important. Chris McDonald is especially to be thanked for her efficient and patient assistance in compiling these Proceedings.

Salt Lake City, Utah
May 1987

Carleton DeTar
James Ball

TABLE OF CONTENTS

PART I: PLENARY SESSION TALKS

SUPERSTRINGS – A PROGRESS REPORT

John H. Schwarz
Division of Physics, Math. and Astronomy
California Institute of Technology
Pasadena, California 91125 U.S.A.

ABSTRACT

General properties of superstrings and their use for unification are briefly reviewed. The structure of the perturbation expansion and attempts to construct nonperturbative formulations are then sketched.

1. Introduction

Superstring theory is an approach to the unification of fundamental particles and forces that shows much promise. Many of you have heard me give a survey of this subject at the Berkeley Conference in July[1).]While I must necessarily cover much of the same ground here, I will at least try to express it somewhat differently. Also, some more recent developments will be mentioned. For a more detailed survey of the subject, the reader is referred to the books listed in ref. 2.

Our starting point is the postulate that elementary particles are one-dimensional curves, called strings, rather than points as in conventional quantum field theory. Strings were first considered around 1970 as an approach to hadronic physics. At that time one supposed that their typical size was about one fermi. Now that strings are being proposed for unification, the natural scales are given by the fundamental constants $\hbar$, c, and G, where the latter is Newton's constant. In terms of these one can define the Planck length and the Planck mass

$$L_P = \left(\frac{\hbar G}{c^3}\right)^{1/2} \sim 1.6 \times 10^{-33} cm \qquad (1a)$$

$$M_P = \left(\frac{\hbar c}{G}\right)^{1/2} \sim 2.2 \times 10^{-5} gm \sim 1.2 \times 10^{19} \mathrm{GeV}/c^2. \qquad (1b)$$

Thus at these scales the effects of "stringiness" should be important, whereas for much larger distances or lower energies, an ordinary quantum field theory can be used as an effective theory.

String theory is a subject replete with miracles, which tend to enchant theorists who start working in it, as recent history demonstrates. Let me mention three of the most striking ones:

Miracle #1: A few string theories exist, which seem to be consistent with standard requirements such as quantum mechanics, relativity, causality, etc. This is significant because it is very nontrivial to construct consistent interacting relativistic quantum theories for extended objects. In fact, I would conjecture (with some confidence) that there are no such theories based on fundamental objects of more than one dimension.

Miracle #2: Each of the string theories that has been found contains gravity. In other words, the spectrum contains a massless spin-two excitation that interacts at low energies in accordance with the dictates of general covariance. This means that an effective low-energy action would be of the standard Einsteinian form plus an infinite sequence of corrections that are suppressed by powers of E/m_P. The fact that string theories are consistent quantum theories necessarily containing gravity is in striking contrast to the situation for point-particle theories. I believe it is fair to say that consistent quantum field theories based on point particles cannot accommodate gravity.

Miracle #3: The string theories that have been found seem very well-suited to explain many general features of the standard model. Specifically, the occurrence of a chiral theory based on specific groups and representations with a small number of families can naturally be achieved. Also, the general structure of the theories seems to accommodate (or even predict) many other popular ideas such as low-energy supersymmetry, axions, grand unification, and so forth. In short, while we are far from demonstrating that a particular theory is the correct one, all the ingredients that are required for a realistic "theory of everything" (as the popular media like to put it) seem to be in place.

2. Classification of string theories

Strings can occur in only two distinct topologies, called open and closed. As shown in fig.1, open strings have free ends, whereas closed strings have the topology of a circle. In each case, the strings may or may not carry an intrinsic orientation. If they do, we speak of oriented strings, which may be represented with arrows as shown in fig. 2. All theories that contain open strings necessarily contain closed strings as well. The reason is that the same basic interaction that allows two open strings to join ends and give one open string, also allows a single open string to join its ends and give a closed string. In such theories the closed strings are oriented if and only if the open strings are. Theories that contain closed strings need not contain open strings, however.

Fig.1 : Strings can be "open" with free ends as shown in (a) or "closed" with the topology of a circle as shown in (b).

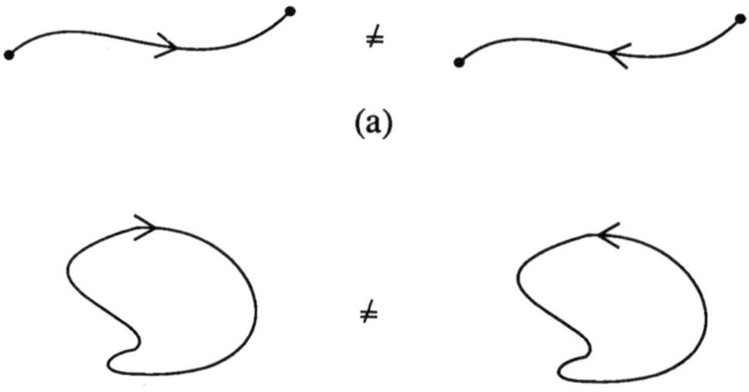

Fig. 2 : Strings can have an intrinsic orientation in which case two strings with opposite orientations (represented by arrows in the figure) that are otherwise identical are distinguished. This is shown for open strings in (a) and closed strings in (b).

Associated with any string theory ground state there is a two-dimensional quantum field theory that describes the dynamics of the strings. In other words, it is formulated on the two-dimensional world-sheet that is swept out by the motion of the string through space-time. The possible two-dimensional theories that one needs to consider are severely limited by the fundamental requirement of conformal invariance. I cannot go into the details here, but the bottom line is that only conformally invariant two-dimensional theories can be consistently interpreted as string theories. Two-dimensional field theories can be supersymmetric. In fact, the fundamentally distinct possibilities for string theories can be characterized by the number of local supersymmetries that the conformally invariant two-dimensional action possesses.

Let me begin the discussion of the classification of string theories by listing the theories that contain open strings. Conformally invariant string actions can be constructed with 0, 1, 2, or 4 local supersymmetries, so there are potential theories for each of these cases. However, in each case the requirement that there be no anomalies in the conformal invariance due to quantum effects determines a unique critical dimension D_c, which can be interpreted as the largest space-time dimension D for which the theory can be defined. It is possible to have $D < D_c$ if some modes are used to describe compactified dimensions or internal degrees of freedom.

Cancellation of the conformal anomalies gives the following critical dimensions:

$$\begin{aligned} N &= 0 \qquad & D_c &= 26 \\ N &= 1 \qquad & D_c &= 10 \\ N &= 2 \qquad & D_c &= 2 \\ N &= 4 \qquad & D_c &= -2. \end{aligned}$$

The last case does not make any physical sense at all, and the $N = 2$ theory cannot account for $D = 4$. Therefore we can restrict our considerations to $N = 0$ and $N = 1$.

The open string theory with $N = 0$ and $D_c = 26$, often referred to as the Veneziano model, was developed in the late 1960's. By attaching suitable "charges" to the ends of the open strings it can accommodate any classical group as a local gauge symmetry. The choices $SO(n)$ or $Sp(n)$ correspond to theories of unoriented strings, whereas $U(n)$ corresponds to oriented strings. These theories have several serious problems. First of all, their spectra do not contain any fermions; they are purely bosonic theories. Secondly, in flat 26-dimensional Minkowski space the classical spectrum contains a tachyon. A ground state that allows tachyonic excitations cannot be stable as a quantum theory. Either there is another stable vacuum (as in a Higgs theory) or the theory does not exist. A possible answer is suggested by a recent work of Orland[3]), which studied the $U(n)$ theory for large n. It concluded that for $n \to \infty$ the theory is free, and suggested that this might be the case in general. I find that conjecture quite attractive.

Let us turn now to the open string theory with one local world-sheet supersymmetry and $D_c = 10$. This theory, originally constructed in 1971[4]), was shown to have space-time supersymmetry as well, when properly interpreted[5]). At the classical level it can accommodate the same gauge groups as the Veneziano model. However, these theories have chiral fermions, and therefore one-loop effects can give rise to anomalies that break the gauge symmetry rendering the theory inconsistent. This turns out to be the case for every choice of group, except $SO(32)$[6]). Thus we are left with a unique possibility, called the $SO(32)$ type I superstring theory. It does not contain any adjustable dimensionless parameters. It is the only string theory containing open strings that is a candidate for a realistic theory. This theory has unoriented open and closed strings. In flat ten-dimensional Minkowski space it is supersymmetric, and the spectrum of string excitations is free from tachyons.

Let us now consider closed-string theories. In these theories the normal modes consist of left-moving and right-moving waves, in other words excitations that travel in one direction or the other on the string. The reason for this is very simple. The two-dimensional wave equation

$$\left(\frac{\partial^2}{\partial\tau^2} - \frac{\partial^2}{\partial\sigma^2}\right) X(\sigma, \tau) = 0 \tag{2}$$

has the general solution

$$X(\sigma, \tau) = X_R(\tau - \sigma) + X_L(\tau + \sigma), \tag{3}$$

where X_R and X_L describe right-moving and left-moving modes, respectively. By a Wick rotation $\tau \to i\tau$ it is possible to Euclideanize the world sheet, which can then be regarded as a Riemann surface and studied using methods of complex analysis. In particular, X_R becomes a holomorphic function (i.e., it depends only on the complex coordinate z) and X_L becomes an antiholomorphic function (it depends only on $\bar{z}$).

The classification of open-string theories that we described is immediately applicable to the study of closed strings. The reason for this is that left- or right-moving modes of closed strings are described by essentially the same mathematics as the modes of open strings (standing waves). Therefore we can associate a closed-string theory to any pair of open-string theories by the rule

$$\text{CLOSED} \sim (\text{OPEN}\#1)_{\text{L}} \otimes (\text{OPEN}\#2)_{\text{R}}.$$

If we want $D_c \geq 4$ (so that $D = 4$ is possible) there are only three candidates for the open-string theories: 26, 10_L, 10_R. The subscripts L and R refer to the fact that the $N = 1$, $D_c = 10$ superstring has an intrinsic handedness (the chirality of the supercharge). This distinction was not relevant previously because the two theories are equivalent. It does become relevant when we pair theories and consider the relative chirality of the two factors. Taking this into account there are four cases to be considered, as shown in the table.

Table. Four possible closed-string theories

Description	Name	D
$26 \otimes 26$	Shapiro–Virasoro	≤ 26
$26 \otimes 10_L$	heterotic	≤ 10
$10_L \otimes 10_R$	Type IIA	≤ 10
$10_L \otimes 10_L$	Type IIB	≤ 10

Just as was the case for the Veneziano model, the Shapiro–Virasoro model[7] has a tachyon and no fermions. It is tempting to conjecture once again that when properly understood it will turn out to be a free theory.

The heterotic theory[8] can be interpreted in $D = 10 - n$ dimensions by compactifying $16+n$ left-moving and n right-moving dimensions. The mass squared of a state gets a contribution M_L^2 from left-moving modes and M_R^2 from right-moving modes so that

$$M^2 = M_L^2 + M_R^2. \tag{4}$$

Moreover, the requirement that the physics of the closed string not depend on the choice of origin of the σ coordinate that is used to parametrize it implies that $M_L^2 = M_R^2$. Since $M_R^2 \geq 0$, this has the important consequence of ensuring that $M^2 \geq 0$, *i.e.*, that there are no tachyons in the spectrum. The heterotic string theory and the type II superstring theories[9] are based on oriented closed strings only.

The type IIA theory has complete left-right symmetry in all its states and couplings. Thus it is a "vectorlike" theory that is unable to account for the left-right asymmetry (parity violation) of the standard model. The type IIB theory on the other hand, is "chiral" (left-right asymmetric).

To summarize this section, there are three chiral string theories that are potentially realistic. They are the SO(32) type I superstring, the type IIB superstring, and the heterotic string. Of these the heterotic theory is generally considered the most promising for phenomenology, but the possibilities for constructing ground states are not understood well enough to say that the other two possibilities are excluded. In fact, there have been recent suggestions for constructing interesting compactifications of the type I theory[10)] and of the type II theory[11)]. Each of these could lead to structures as realistic as those provided by compactifications of the heterotic string[12)]. In fact, the type II theory tends to give a smaller gauge group, which may be an advantage.

3. Perturbation theory vacua

We have narrowed our search to just three theories, each of which is completely free from adjustable dimensionless parameters. However, to know a theory is to know the equations and not the solutions of the equations. Ideally the solution of lowest energy, called the "vacuum" or "ground state" of the theory, would be unique and would possess the symmetries observed in nature such as $D = 4$ Poincaré invariance and $SU(3)_c \times U(1)_{\rm EM}$. At our present level of understanding, we can only look for solutions to the classical string equations. There appear to be an enormous number that have unbroken supersymmetry and zero energy. (Zero energy means zero cosmological constant, which is the desired result. The real challenge is to obtain this result for a solution with spontaneously broken supersymmetry.) The properties of these classical solutions can be studied in perturbation theory, where, for the most part, they remain consistent (perhaps with small corrections) to every order. This vacuum degeneracy problem is a real threat to our most ambitious goal – a parameter-free explanation of the particle physics data. This requires not only a unique theory, but a unique vacuum as well. The hope (which is mostly based on wishful thinking) is that when nonperturbative effects are taken into account, the theory will resolve the degeneracy problem. This could happen either by splitting the levels or by rendering all but one of the solutions inconsistent. Since I cannot say more about this, let me briefly say a little about classical solutions.

The simplest cases are solutions that have ten-dimensional Poincaré symmetry. They are obviously unrealistic, but useful to consider as a warm-up exercise. Also, it will be a real challenge to find a nonperturbative effect that rules them out. For the $SO(32)$ type I theory there is a unique vacuum of this type. It has unbroken $SO(32)$ gauge symmetry and $N = 1$ supersymmetry (in the ten-dimensional sense). The vacuum of type II theory with $D = 10$ Poincaré invariance has no gauge group, but a larger ($N = 2$) supersymmetry. There are two possibilities, because the two supersymmetries can have either opposite or the same chirality (IIA and IIB). It is probably better to regard these as two different ground states rather than two different theories. The heterotic theory has three classical vacua with $D = 10$ Poincaré invariance that are tachyon-free. They are obtained by compactifying sixteen left-moving dimensions in ways that are restricted by the requirement of modular invariance. The three possibilities have $E_8 \times E_8$ or $SO(32)$ gauge symmetry with $N = 1$ supersymmetry [8] or $SO(16) \times SO(16)$ gauge symmetry with no supersymmetry[13)]. There are additional modular-invariant compactifications that do have tachyons[14)].

If one only asks for unbroken Poincaré symmetry in $D < 10$ dimensions, there are very many possibilities. Those examples that are known probably represent only a small fraction of what is possible. Even that is much more than I can explain here, however. Let me just describe one interesting class of compactifications of the heterotic string due to Narain[15)].

To define a heterotic string ground state with $D = 10 - n$, let us compactify $16 + n$ left-movers and n right-movers in a $(16 + 2n)$-dimensional torus. This basically means that each dimension forms a circle, although the circles are not required to be orthogonal to one another. The momentum quantum number for a periodic coordinate is quantized in integer multiples of a fundamental unit. Thus the compactified dimensions have discrete momenta $(k_1, k_2, \ldots, k_{16+n}\,;\, l_1, l_2 \ldots l_n)$ associated with them. The set of all allowed momenta form a $(16+2n)$-dimensional lattice of points, which Narain showed must be self dual and even in order to have modular invariance. In carrying out the construction it is necessary to use a "Lorentzian" metric. Thus, for example, $(k\,;\,l)^2 \equiv k^2 - l^2$. The statement that the lattice is even means that this is an even integer for all allowed $(k\,;\,l)$. The set of all such self-dual lattices that are equivalent is described by $(16 + n)n$ parameters and can be put in correspondence with the coset space

$$\frac{SO(16+n,n)}{SO(16+n)\times SO(n)}. \tag{5}$$

Many different gauge symmetries can be obtained depending on which point of this space is selected. They all have rank $16+2n$, however.

The construction described above is only the tip of the iceberg. There are many other possibilities based on Calabi-Yau spaces[12),16)] orbifolds[17)], and so forth.

4. The perturbation expansion

Given a solution to the classical equations of a string theory, one can examine perturbative expansions of scattering amplitudes expressed as a sum of Feynman diagrams. In practice, most analyses to date have been carried out for a classical background having $D=10$ or $D=26$ Poincaré symmetry. This is not essential, however, and the results are qualitatively similar in other cases.

Since strings sweep out two-dimensional surfaces in space-time, the Feynman diagrams are two-dimensional surfaces. Just as in point-particle theories, where the diagrams are given by interconnected lines, the diagrams of string theory can be classified by their topology. The classification is especially simple in the case of the type II and heterotic string theories, which contain oriented closed strings only. For these theories the relevant topologies are closed orientable manifolds, which (after a continuation to Euclidean space) can be regarded as Riemann surfaces. Such surfaces are classified topologically by a single integer g, known as the genus, which is the number of holes or handles, as shown in fig. 3.

The amplitude corresponding to a particular diagram is given by an integral that sums contributions from all inequivalent geometries of the particular topology under consideration. Because of the reparametrization and conformal invariances of the underlying two-dimensional field theory that describes the dynamics of the world sheet, the number of inequivalent geometries is severely restricted. In fact, conformally inequivalent geometries are characterized by a finite number of parameters (known as Teichmüller parameters). There is no parameter at all for $g=0$, which means that every genus zero two-dimensional surface is conformally equivalent to a unit sphere. For $g=1$, there is one complex parameter, and for $g\geq 2$ there are $3g-3$ complex parameters. Different choices of the Teichmüller parameters can actually describe equivalent geometries infinitely often. The mappings that relate these equivalent descriptions form a group known as the modular

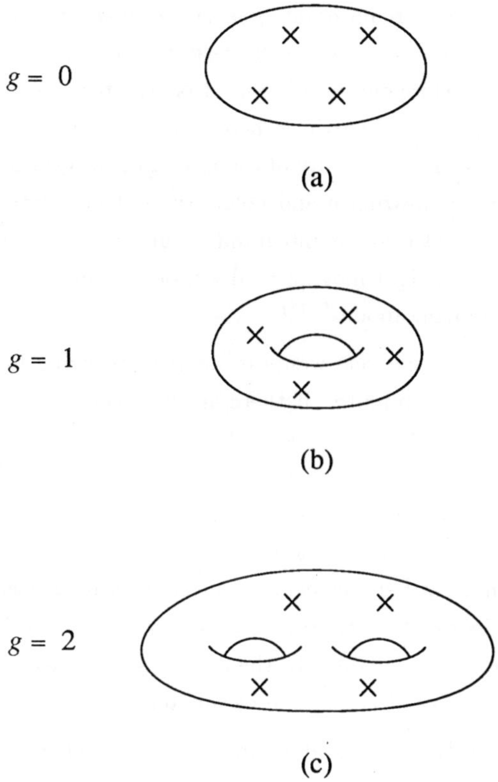

Fig. 3 : Closed orientable surfaces are classified by their genus. The genus zero case, with the topology of a sphere, is shown in (a). The genus one surface, with the topology of a torus, is shown in (b). The genus two surface, with two holes or handles, is shown in (c).

group or the mapping class group. In order to only count inequivalent configurations once each, it is necessary to restrict the integration to a fundamental region of Teichmüller space, called moduli space. It is necessary that the amplitude be independent of the choice of a fundamental region, in other words that it be modular invariant, in order for the physical amplitude to be well-defined. This important requirement was alluded to earlier.

In addition to integrating over all conformally inequivalent geometries, it is also necessary to keep track of the positions on the surface at which the external particles are attached. This requires integrating over n additional complex parameters, one for each external particle. In the cases of genus zero or one, the surface has isometries, and therefore only $n-3$ or $n-1$ parameters occur. Altogether, the g-loop diagram with n external particles involves an integral with $2(3g-3+n)$ real dimensions in all cases. The structure of the integral is largely determined by the requirement of modular invariance and holomorphic factorization. The latter is the nontrivial requirement that the integrand should factorize into a holomorphic function, associated with right-moving modes times an antiholomorphic function associated with left-moving modes[1),18)].

An obvious question that one wants to answer for string loop amplitudes is whether they are finite or infinite. If there are divergences the next question is whether there is a physically sensible way of dealing with them. The analysis has not been completed yet and so we may still be in for some surprises. However, it appears likely that the following is true: In general, there are divergences which have a sensible physical interpretation and can be dealt with in an unambiguous way. The situation is rather better than in conventional renormalizable field theories. In such theories the renormalized theory often contains dimensionless parameters that are not calculable. In the case of string theory, the divergences do not imply the existence of adjustable parameters.

Let me describe two classes of diagrams with divergences. In the first case, shown in fig. 4, the Riemann surface pinches off into two pieces of arbitrary genus. One of them contains external particle A and the other contains the rest of the external particles. Corresponding to the corner of the integration over moduli space associated with this picture, the amplitude has pieces of the form

$$\sum_X T_{n-1,X} \quad \frac{1}{m_X^2 - m_A^2} \quad T_{X,A}. \tag{6}$$

The sum over intermediate states X includes the particle A for which the propagator factor diverges. Therefore the result is infinite unless the self energy $T_{A,A}$ should happen to vanish. A divergence of this type naturally occurs in any field theory. It simply reflects the fact that particle A has been inserted with its classical mass value, which gets shifted by quantum corrections. The feature of string theory that is different is that there is no natural way to separate off the "one-particle irreducible parts" of diagrams. In any case, these divergences are expected and understood. They do not represent a fundamental problem. They have been discussed in several recent papers[19)].

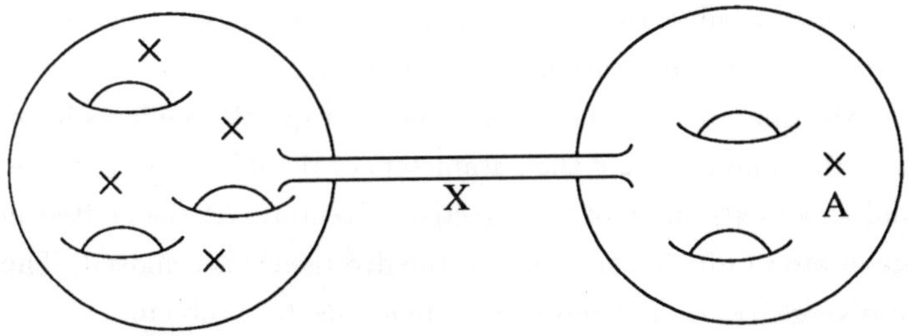

Fig. 4 : Loop amplitudes have divergences arising from the boundary of moduli space in which the surface pinches off in such a way that one external particle (A) is on one side and the rest are on the other.

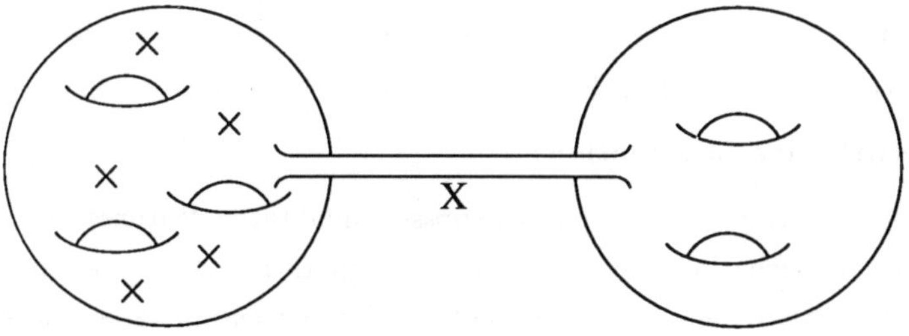

Fig. 5 : Loop amplitudes have divergences associated with "dilaton tadpoles" in which the surface pinches off in such a way that all the external particles are on one side.

The second class of divergences corresponds to the world sheet pinching off into two pieces with all the external particles attached to one of the pieces as shown in fig. 5. This gives a contribution of the form

$$\sum_X T_{n,X} \; \frac{1}{m_X^2} \; T_X. \tag{7}$$

The intermediate particles X must carry vacuum quantum numbers in this case. For all known classical vacua the spectrum of string excitations contains one or more massless scalars, called dilatons. The contribution when X is a massless dilaton D is divergent unless the dilaton tadpole T_D vanishes at the corresponding order of the perturbation expansion. In other words, the effective action has a term linear in the dilaton field. This is a signal that quantum effects have shifted the position of the minimum and the parameters of the dilaton have to be shifted to correspond to an extremum of the effective potential. At the shifted position, the complete dilaton tadpole vanishes and the divergence is removed. Thus, once again, these divergences do not represent a fundamental problem.

There is one circumstance in which the two types of divergences described above do not occur. Namely, if one starts with a supersymmetric ground state and restricts the external particles to be ones that are massless, the diagrams are presumably all finite. The reason is that certain "nonrenormalization" theorems imply that the massless self-energy diagrams $T_{A,A}$ and dilaton tadpoles T_D vanish when supersymmetry is unbroken. Of course, for a realistic vacuum supersymmetry must be broken, but this may depend on nonperturbative effects.

5. Nonperturbative formulations

The complete string theory is likely to possess qualitative features that are not present in the perturbation expansion (such as supersymmetry breaking). Also, we fervently hope that the great plethora of candidate vacuum states that exist in the perturbative framework can be eliminated by nonperturbative effects. This is essential if superstring theory is to have significant predictive power. One could imagine that some sort of string-theory instantons (which are beyond our present understanding) would play an important role in the analysis. In order to study questions such as these, a suitable nonperturbative formulation of string theory is required. (Historically, string theory has been defined by the perturbation series.) Such a formulation would presumably be based on more profound fundamental principles. The situation is somewhat analogous to knowing the perturbation expansion of S-matrix elements in general relativity without knowing the Einstein–Hilbert action. In general relativity one identifies different ground states by solving the field equations. This information is very hard to recover from an expansion of the S matrix in flat space-time.

Many proposals have been made in the last couple of years for a nonperturbative formulation of string theory. The conservative approach is to do what one does for point particles – *i.e.*, construct a field theory. More radical proposals suggest doing something quite different.

String theories are gauge theories *par excellence*. They have infinitely more gauge invariances than any theories considered before. Therefore all the subtle issues that are characteristic of gauge theories occur in full grandeur, which makes the subject somewhat difficult.

One approach is to avoid all the subtleties of gauge theories by constructing the field theory directly in a gauge-fixed form in which all unphysical gauge degrees of freedom have been eliminated. The most convenient choice is a light-cone gauge, which breaks manifest Lorentz symmetry. This has been carried out some time ago for bosonic strings[20)], later for superstrings[21)], and most recently for heterotic strings[22)]. The light-cone field theory exists for all string theories and seems to be correct. There are disadvantages, however. It is formulated in a specific background geometry (flat Minkowski space in practice), whereas a background-independent formulation would be preferable. This reflects the fact that the gauge choice obscures the underlying geometric principles that are probably required for a deeper understanding.

A covariant and gauge-invariant version of string field theory has been developed independently by groups in CERN[23)] and Kyoto[24)]. The two formalisms may differ in certain details, but they look very similar to an outside observer. In particular, they both require associating an unphysical length parameter to each string field in the action. It is then demonstrated that physical amplitudes do not depend on these parameters. I am concerned that these parameters may give rise to an infinite factor in loop amplitudes. These formulations are probably correct at tree level, however, and they can be implemented for every string theory.

The third class of string field theories are those based on a proposal of Witten[25)]. I gave a brief description of Witten's string field theory in my Berkeley talk[1)], which I will not repeat here. I should remark that references[23),25)] all build on the construction of the string BRST operator of Kato and Ogawa[26)] and subsequent work by Siegel, Zwiebach, Banks and Peskin[27)]. Witten's field theory is very simple and elegant, although it is necessary to do some rather elaborate analysis to prove that it is correct[28),29)]. Moreover, in a remarkable development, it has been reformulated in a background independent way[30)], which requires more careful scrutiny[31)], but also seems to be correct. Recently the quantization of the theory has been carefully investigated[32)] and shown to give rise to the simple Feynman rules that had been anticipated in ref. 29.

The Witten field theory as reformulated in ref. 30 seems to be as simple and beautiful as one could want. The problem is that it only exists for theories containing open strings. Thus it exists for the SO(32) type I superstring, but not for the type II or heterotic theories. One bizarre feature of the theory is that it is formulated entirely in terms of the open-string field even though the theory contains closed strings as well. This still needs to be understood better.

Let me now turn briefly to some more radical proposals for a fundamental non-perturbative formulation of string theories. One, due to Friedan and Shenker[33)], proposes to abandon the physical picture of strings propagating in space-time and base the theory on more abstract principles. The fact that string backgrounds correspond to conformally invariant two-dimensional field theories is regarded as fundamental. As we have discussed briefly, these are labeled by the points of moduli space. These are described by $3g - 3$ complex parameters for genus g (with $g > 1$), but Friedan and Shenker propose considering an infinite-dimensional generalization (roughly corresponding to infinite genus surfaces) called "universal moduli space." They then argue that the essential elements of string theory can be expressed in terms of the analytic geometry of this space. To be perfectly honest, I do not understand this proposal well enough to assess its chances for success.

An even more recent radical proposal has been put forward by Bowick and Rajeev[34)]. On the face of it, their suggestion seems quite different from that of Friedan–Shenker, but it may turn out to be equivalent. I find it very appealing because it seems to implement in a specific way, ideas that many of us have argued for some time should appear in a string extension of general relativity. For example, the space of all loops $X(\sigma)$ (called loop space) replaces the usual space-time manifold. The fundamental geometric principles are then formulated in terms of this infinite-dimensional space. At distances long compared to the Planck length, the size of the loops becomes important and the geometry reduces to ordinary Riemannian geometry.

An essential feature in the Bowick–Rajeev construction, which has no counterpart in ordinary space-time, is that a loop $X(\sigma)$ needs to be described with a parameter σ defined on a circle, but the loop is unaltered by any single-valued map of the circle onto itself $\sigma \to f(\sigma)$. One is thus led to consider the group of "diffeomorphisms of the circle" (modulo rigid translations) called diff S^1/S^1. This space admits a two-parameter class of Kähler metrics that gives rise to a unique Ricci curvature. By incorporating a suitable number of space-time dimensions (26 for bosonic strings) in a bundle over this space one is able to cancel off the Ricci curvature. This is apparently necessary in order that the vacuum be well-defined without a Berry phase ambiguity[35]. Even though this approach is still in its infancy, I find this result very tantalizing and am anxious to see where it leads.

6. Conclusion

Much interesting work has been done in string theory during the last two years. Thanks to the concerted efforts of many clever people, the pace of progress is very rapid. Nothing that has been learned makes us less optimistic than before that the ultimate goal of achieving an understanding of the fundamental principles of elementary particle physics is possible. Still, it is clear that we need to learn much more about the fundamentals of string theory, the role of nonperturbative effects, and the mechanisms by which a vacuum configuration is determined. It would be dumb luck if significant contact with experimental results were made prior to achieving such understandings. This might require another 10-20 years.

References

1. J. H. Schwarz, "Superstrings," to be published in the proceedings of the XXIII International Conference on High Energy Physics.

2. "Superstrings," in two vols., ed. J. H. Schwarz (World Scientific 1985);

   "Unified String Theories," eds. M. Green and D.Gross (World Scientific 1986);

   M. B. Green, J. H. Schwarz, and E. Witten, "Superstring Theory," in two vols., (Cambridge Univ. Press 1987).

3. P. Orland, Nucl. Phys. **B278** (1986) 790.

4. P. Ramond, Phys. Rev. **D3** (1971) 2415;

   A. Neveu and J. H. Schwarz, Nucl. Phys. **B31** (1971) 86; Phys. Rev. **D4** (1971) 1108;

   A. Neveu, J. H. Schwarz, and C. B. Thorn, Phys. Lett. **35B** (1971) 529;

   C. B. Thorn, Phys. Rev. **D4** (1971) 1112.

5. F. Gliozzi, J. Scherk, and D. Olive, Nucl. Phys. **B122** (1977) 253;

   M. B. Green and J. H. Schwarz, Nucl. Phys. **B181** (1981) 502.

6. M. B. Green and J. H. Schwarz, Phys. Lett. **149B** (1984) 117; Nucl. Phys. **B255** (1985) 93.

7. J. Shapiro, Phys. Lett. **33B** (1970) 361;

M. Virasoro, Phys. Rev. **177** (1969) 2309.

8. D. J. Gross, J. A. Harvey, E. Martinec, and R. Rohm, Phys. Rev. Lett. **54** (1985) 502; Nucl. Phys. **B256** (1985) 253; **B267** (1986) 75.

9. M. B. Green and J. H. Schwarz, Phys. Lett. **109B** (1982) 444.

10. J. A. Harvey and J. Minahan, "Open strings on orbifolds," Princeton preprint PUPT-1031 (Nov. 1986).

11. H. Kawai, D. Lewellen, and S.-H. Tye, "Four-dimensional type II strings and their extensions: type III strings," Cornell preprint CLNS 87/760 (Jan. 1987);

R. Blume, L. Dolan, and P. Goddard "A new method of incorporating symmetry into superstring theory," Rockefeller University preprint RU/B_1/187 (Dec. 1986).

12. P. Candelas, G. Horowitz, A. Strominger, and E. Witten, Nucl. Phys. **B258** (1985) 46.

13. L. Dixon and J. A. Harvey, Nucl. Phys. **B274** (1986) 93;

L. Alvarez-Gaumé, P. Ginsparg, G. Moore and C. Vafa, Phys. Lett. **171B** (1986) 155.

14. N. Seiberg and E. Witten, Nucl. Phys. **B276** (1986) 272.

15. K. Narain, Phys. Lett. **169B** (1986) 41;

K. Narain, M. H. Sarmadi, and E. Witten, Nucl. Phys. **B279** (1987) 369.

16. P. Candelas, talk given at the ITP Superstring Conference (Dec. 1986).

17. L. Dixon, J. Harvey, C. Vafa, and E. Witten, Nucl. Phys. **B261** (1985) 651; **B274** (1986) 285.

18. A. Belavin and V. Knizhnik, Phys. Lett. **168B** (1986) 201.

19. S. Weinberg, Phys. Lett. **156B** (1985) 309;

W. Fischler and L. Susskind, Phys. Lett. **171B** (1986) 336; **173B** (1986) 262;

N. Seiberg, "Anomalous dimensions and mass renormalization in string theory," Weizmann Institute preprint WIS/186/61/Nov-Ph.

20. M. Kaku and K. Kikkawa, Phys. Rev. **D10** (1974) 1110, 1823;

E. Cremmer and J.-L. Gervais, Nucl. Phys. **B76** (1974) 209; **B90** (1975) 410.

21. M. B. Green and J. H. Schwarz, Nucl. Phys. **B218** (1983) 43; **B243** (1984) 475;

M. B. Green, J. H. Schwarz, and L. Brink, Nucl. Phys. **B219** (1983) 437.

22. D. J. Gross and V. Periwal "Heterotic string light-cone field theory," NSF-ITP-86-121 (Sept. 1986).

23. A. Neveu and P. West, Nucl. Phys. **B278** (1986) 601;

Y. Kazama, A. Neveu, H. Nicolai, and P. West Nucl. Phys. **B278** (1986) 833 and references therein.

24. H. Hata, K. Itoh, T. Kugo, H. Kunitomo and K. Ogawa, Phys. Lett. **172B** (1986) 186, 195; **175B** (1986) 138 and Kyoto Univ. preprints.

25. E. Witten, Nucl. Phys, **B268** (1986) 253; **B276** (1986) 291.

26. M. Kato and K. Ogawa, Nucl. Phys. **B212** (1983) 443.

27. W. Siegel, Phys. Lett. **151B** (1985) 391, 396;

W. Siegel and B. Zwiebach, Nucl. Phys. **B263** (1986) 105;

T. Banks and M. Peskin, Nucl. Phys. **B264** (1986) 513.

28. D. Gross and A. Jevicki, "Operator formulation of interacting string field theory," Princeton preprint (May 1986);

part II: ITP preprint NSF-ITP-86-119 (Sept. 1986);

E. Cremmer, C. Thorn, and A. Schwimmer, "The vertex function in Witten's formulation of string field theory," Ecole Normale preprints LPTENS-86-14;

S. Samuel, Phys. Lett. **181B** (1986) 255.

29. S. Giddings, Nucl. Phys. **B278** (1986) 242;

S. Giddings and E. Martinec, Nucl. Phys. **B278** (1986) 91;

S. Giddings, E. Martinec, and E. Witten, "Modular invariance in string field theory," Princeton preprint (May 1986).

30. G. Horowitz, T. Lykken, R. Rohm, and A. Strominger, Phys. Rev. Lett. **57** (1986) 283.

31. G. Horowitz and A. Strominger, "Translations as inner derivations and associativity anomalies in open string field theory," U.C. Santa Barbara preprint UCSB-TH-86-34;

A. Strominger, "Point splitting regularization of classical string field theory," IAS preprint (1986).

32. M. Bochicchio, "Gauge fixing for the field theory of the bosonic string," Rome preprint no. 527 (Sept. 1986);

"String field theory in the Siegel gauge," Princeton Univ. preprint PUPT-1028 (Nov. 1986);

C. Thorn, "Perturbation theory for quantized string fields," IAS preprint (1986).

33. D. Friedan and S. Shenker, Nucl. Phys. **B281** (1987) 509.

34. M. J. Bowick and S. G. Rajeev, "String theory as the Kähler geometry of loop space," MIT preprint CTP # 1414 (Nov. 1986).

35. N. Warner, private communication.

SUPERCOMPUTATIONS IN QCD

Doug Toussaint

Department of Physics
University of California, San Diego
La Jolla, California 92093
USA

INTRODUCTION

The last few years have seen a tremendous increase in the use of large computers by particle theorists. In this talk I will give a brief and incomplete summary of what we are doing. The three major questions to be addressed are: why are we computing, what have we learned, and how should we continue? This is more an introduction than a review, so I will not attempt to give complete coverage of the field or complete references to the topics I mention.

Currently large scale computation in particle theory means numerical simulation of Euclidean quantum field theories. This activity is often called "Monte Carlo" simulation, since it usually involves use of random numbers. (The microcanonical ensemble methods are an exception.) However, Monte Carlo simulation is not the only theoretical approach to QCD which could require large scale computations. For example, one might want to do large scale variational calculations in QCD, or one might want to evaluate weak coupling or strong coupling series to high orders, or perhaps use some method as yet uninvented. Pursued seriously, any of these activities would place even greater demands on computer hardware and software than Monte Carlo simulation. Indeed, the reason that we do Monte Carlo simulation is that it is the easiest technique known, but let us keep in mind when estimating our computational needs that this is not all we will want to do in the future.

In QCD the Lagrangian to be studied is

$$L = -\frac{1}{4} F^{a}_{\mu v} F^{uv^a} + \sum_{\text{flavors}} \overline{\psi} (i \not{D} - m_f) \psi \quad . \tag{1}$$

In Euclidean space, or imaginary time, the Feynman path integral expression for the ground state expectation value of some operator O is

$$\langle O \rangle = \frac{\int [dA \; d\bar{\psi} \, d\psi] \, O \, e^{-\int L_E d^4x}}{\int [dA \; d\bar{\psi} \, d\psi] \, e^{-\int L_E d^4x}} \tag{2}$$

If L_E is real, as is the case for the gluonic part of L in Euclidean space, then the expression can be regarded as an expectation value in four dimensional classical statistical mechanics, where the Boltzmann factor $e^{-\int L_E(\{A\})d^4x}$ is the probability to find the system in the configuration $\{A\}$. ("Configuration $\{A\}$" means specifying the value of A^{μ} at every point. Inside the path integral A^{μ} is a number rather than an operator.) If we can find an algorithm which generates configurations $\{A\}$ with probability proportional to $e^{-\int L_E(\{A\})d^4x}$, then we can simply average O over a large enough set of such configurations to find $\langle O \rangle$.

To fit this system into a computer we must approximate it by a finite number of variables. This is accomplished by defining the variables on the sites or links of a lattice, and by restricting the size of the system. Roughly speaking, we replace derivatives in the action by differences. As was discovered by Wilson,[1] this can be done in such a way as to maintain exact gauge invariance, and in the simulations we generally do not have to worry about fixing the gauge. From a physical point of view, putting the system on a lattice of finite size restricts the range of length scales we see – the lattice spacing a gives an ultraviolet cutoff and the box size L gives an infrared cutoff.

Many methods can be used to produce configurations with probability e^{-S}. Among them are updatings of a single variable by the Metropolis, heatbath, or other algorithms,[2] integration of the Langevin equation,[3] or molecular dynamics integration of the equations of motion combined with heatbath updatings of the quadratic momenta.[4] All of these methods share the feature that they do not simply lay down a configuration with the desired probability. Instead they give a rule for generating a configuration from the previous one in such a way that as you continue the process the probability of being in any configuration approaches e^{-S}. Unfortunately this results in successive configurations that are strongly correlated. Since $\langle O \rangle$ is obtained by averaging over a number of samples, the accuracy of our estimate of $\langle O \rangle$ is proportional to $1/\sqrt{N}$, where N is the number of *statistically independent* samples. Since successive configurations are correlated, this is far less than the actual number of configurations generated.

It is not hard to understand why Monte Carlo simulations require large scale computing. First of all we need big lattices. If we are trying to simulate phenomena with some characteristic length scale ξ (say ξ is the size of a hadron), then we need the volume of our universe to be large: $L >> \xi$, and we need the lattice spacing, or the granularity of spacetime, to be small: $\xi >> a$. Combining these: $L >>>> a$. Since we are in four dimensions, the number of variables to be dealt with is $\sim (L/a)^4$. Also, as ξ/a grows we encounter a phenomenon called critical slowing down. This means that it takes more steps in the algorithm to get a statistically independent configuration. For local updating algorithms, we typically find that the number of updatings of each variable required to get an independent configuration increases as $(\xi/a)^2$. Putting together the need for many independent samples, the need for large lattices in four dimensions, and critical slowing down, we find that the amount of computing required increases very rapidly as we increase the quality of the simulation. (Just for fun, suppose that we choose ξ to be the geometric mean of L and a: $L/a = \xi^2$. Then, putting in a factor of $(\xi/a)^2$ for critical slowing down, we find that the time required to generate an independent configuration might scale as $(L/a)^4 \, \xi^2 \approx \xi^{10}$!)

The problem becomes even worse when we include fermions. In this case the weight in the path integral becomes

$$e^{-\int L_{gauge} - \bar{\psi}(i \not{D} - m)\psi} \quad . \tag{3}$$

The problem is that $\bar{\psi}$ and ψ are not ordinary numbers, they are anticommuting numbers, and we cannot interpret this expression as a probability. To proceed, we use the fact that the fermion action is quadratic. On the lattice, where the variables are indexed by i and j, the path integral has the form

$$Z = \int [dA \; d\bar{\psi}_i \, d\psi_j] \; e^{-S(A) - \bar{\psi}_i M_{ij}(A)\psi_j} \tag{4}$$

where $M_{ij}(A)$ is a matrix whose size is the volume of the lattice and whose entries are functions of the bosonic variables. Now the fermion integral can be done formally, and we find

$$\begin{aligned} Z &= \int [dA] \det M(A) \, e^{-S(A)} \\ &= \int [dA] \; e^{-S(A) + Tr \ln M(A)} \end{aligned} \tag{5}$$

At least this is a number, but it has the problem that it is nonlocal – the change in the Boltzmann weight as I change one variable depends on every other variable in the system. Clearly we can't just evaluate the determinant – $M(A)$ is an $(L/a)^4$ by $(L/a)^4$ matrix! There are a number of algorithms for taking this into account, among them pseudofermions,[5] the Langevin equation,[3] and refreshed molecular dynamics.[4] All of these methods require many times more computation than the simulations without fermions.

There are also problems of principle here, which I will not discuss – no one knows how to put chiral fermions on the lattice.[6]

APPLICATIONS OF SIMULATION

What can be done with Monte Carlo simulation? The possibility which has attracted the most attention is the computation of the masses and other properties of hadrons from first principles. Presumably if we simulate the QCD Lagrangian on a large enough lattice with a big enough computer, we will get any property of hadrons we desire. The question is whether when we have done this, presumably obtaining the real world numbers, we can be said to understand low energy QCD. In other words, does the ability to calculate something give insight into the physics? I don't think that the answer is completely yes, but I do think this computation would be a very valuable piece of the puzzle. Perhaps the greatest utility of simulation from this point of view is our ability to play with the theory. We can adjust the number of quarks or their masses and study the effects, indeed we can even butcher QCD into some other theory. We can also study in a very clean system phenomena that are difficult in the laboratory. An example of this is high temperature QCD, where high temperature means arounds 200 MeV or $10^{12}\,^\circ$K. We can also study other theories, whether they appear in nature or not. Even the study of quantum electrodynamics on a lattice has been remarkably interesting.[7]

The first application of Monte Carlo simulation to QCD was an exploration of the phase structure of lattice QCD.[8] This is worth mentioning here, because this is the best evidence we have that the asymptotic freedom known from perturbation theory continues smoothly into confinement at low energies.

Much effort has gone into the computation of hadron masses from numerical simulations. The idea here is that the propagator for a zero momentum particle which goes like e^{-imt} in Minkowski space goes like e^{-mt} in Euclidean space (imaginary time). Therefore we just need to measure the expectation value of an operator which creates the particle at some time and annihilates it at some other time.

$$\langle O_p(t_1)\, O_p(t_2)\rangle \approx e^{-m_p\,|t_1-t_2|} \qquad (6)$$

where the expectation value is computed as equation 2 and O_p is some operator, e.g.

$$\begin{aligned} \bar{\psi}\psi &\rightarrow \text{meson} \\ \psi\psi\psi &\rightarrow \text{baryon} \\ (F^{\mu\nu})^2 &\rightarrow \text{glueball} \end{aligned} \tag{7}$$

(When the anticommuting variables $\bar{\psi}$ and ψ are formally integrated out, as in eq. 5, a $\bar{\psi}(x)\psi(y)$ in the operator in front of the weight gives an $M^{-1}(y,x)$, so one must find elements of this inverse matrix.) The actual techniques used are much more sophisticated that can be described here, but this is the general idea. Most of the work on hadron masses to date has been done in the quenched approximation,[9] where the fermion determinant is simply ignored. In perturbation theory, this amounts to neglecting graphs with internal quark loops (see Figure 1). Obviously, this approximation is popular because it is easy. Today the standards for this sort of computation are $L/a \approx 16$ or so and $\beta = 6/g^2 \approx 6$. As a sample of results, Figure 2 shows work of Bowler et al.[10] Many things are right about this graph: the pion mass is going to zero as the square root of the quark mass, as expected for an approximate Goldstone boson. (Even though the determinant has been banished, m_q still appears in $M^{-1}(x,y)$ used in computing the propagators.) The nucleon and ρ appear to have finite masses as $m_q \rightarrow 0$, and two different operators creating the nucleon give the same mass. However, the ratio of nucleon mass to the ρ mass does not agree with the real world, as shown in Figure 3. This seems to be a general feature of these simulations, but the degree of agreement among different simulations is not as good as one would expect.[11] From the quenched results alone, it is not clear whether the "incorrect" mass ratios are due to the neglect of dynamical fermions or the practical limitations of the simulations. It is clear that we

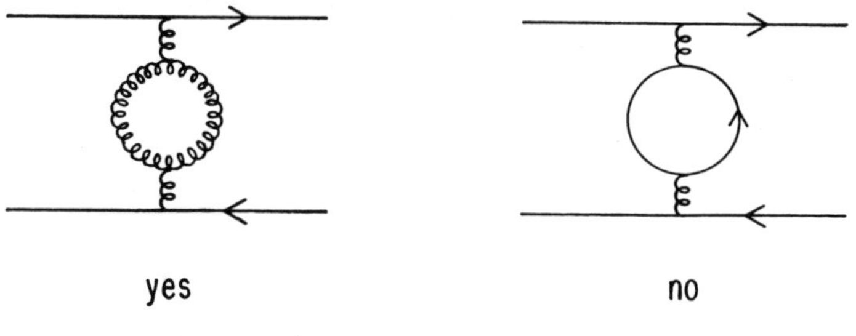

Figure 1. The quenched approximation.

need larger physical volumes (larger $L/5$), smaller lattice spacing (larger ξ/a), smaller quark masses, and more statistics to do this right.

Recently primitive calculations of the hadron spectrum with dynamical fermions have begun. Table 1 contains a few parameters of three of these works. Obviously there is a long way to go here. About the only conclusion that can be drawn here is that the fact that m_N/m_ρ in the quenched calculation is not that of the real world is more likely due to technical problems (lattice size, etc.) than the neglect of fermions.

A particularly simple application of QCD simulations in the quenched approximation is the computation of the potential between (infinitely) heavy quarks. Figure 4 is work of deForcrand and Stack.[15] The line on this figure is a fit to a "Coulomb plus linear" potential. Of course one may criticize a potential for infinite mass classical quarks in a world without dynamical quarks as unphysical, but this is still remarkable numerical evidence that QCD really behaves the way we expect it to. Incidentally, this computation was done on a 24^3 by 48 lattice, and required about 48 million words of storage. In addition to masses, numerical simulation can in principle be used to compute hadron wave functions[16] or hadronic matrix elements appearing in weak interactions, or even in proton decay in grand unified theories.[17] These applications are in an early stage of development and are much too technical to be summarized here.

A particularly clean application of numerical simulations is the study of QCD at high temperatures.[18] By now it is well established that QCD without fermions undergoes a first order deconfinement transition at high temperature. If we take $\sqrt{K} = 400$ MeV as the string tension, this transition occurs somewhere around 250 MeV and has a large latent heat, on the order of 3 Gev/fm^3 (this last number has large systematic errors).[19]

Table 1. Some recent mass spectrum calculations with dynamical fermions. I tabulate the size of the lattice, the number of flavors of degenerate quarks and the type of quark action, the number of lattice configurations on which propagators were measured, and the quoted value of the nucleon to rho mass ratio.

Ref.	Size	# Flavors	Formulation	# Configs	m_N/m_ρ
12	$6^3 \times 24$	2	Kogut-Susskind	500	1.5
13	$12^3 \times 3.8$	3	Kogut-Susskind	9	1.5
14	$9^3 \times 18$	2	Wilson	30	1.8 (1.3)

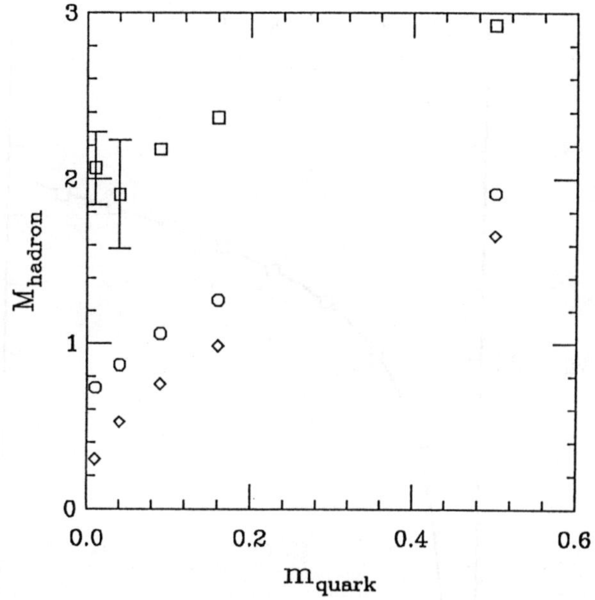

Figure 2. Quenched approximation results of Bowler et al. (ref. 10) on a $16^3 \times 24$ lattice. The diamonds, octagons, and squares represent the pion, rho, and nucleon respectively.

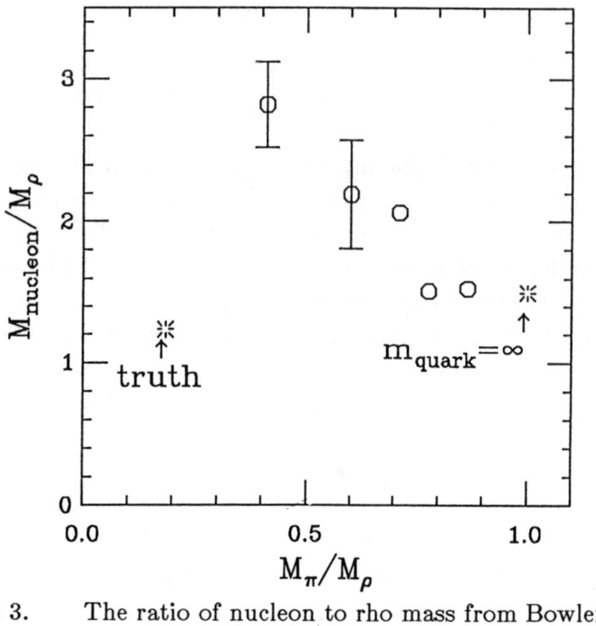

Figure 3. The ratio of nucleon to rho mass from Bowler et al.

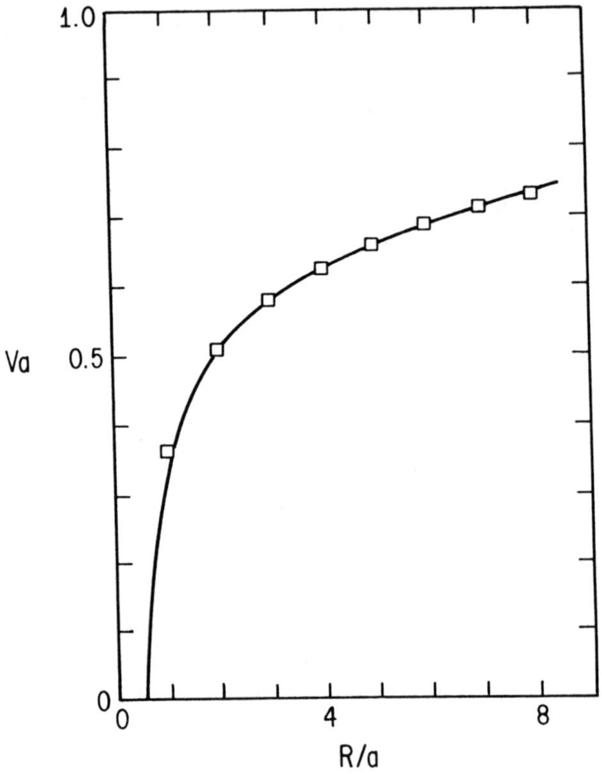

Figure 4. The QCD heavy quark potential in the quenched approximation, measured by de Forcrand and Stack.

Again, people are now beginning to include dynamical fermions in their simulations, and a surprise has emerged.[20] Figure 5 shows a phase diagram as a function of temperature and the mass of the light quarks. For infinite quark mass we have the pure gauge theory with a first order transition. As the quarks are "turned on" by lowering their mass the first order transition disappears at a fairly large quark mass, somewhere around $2 * T_c$. Then, at an order of magnitude smaller quark mass another transition appears. Since the creation of quark-antiquark pairs to break strings destroys confinement as defined in the pure gauge theory, this transition for small quark mass is probably best described as a restoration of chiral symmetry. (Incidentally, for two flavors of light quark it is not yet clear whether the transition is first order extending out to finite quark mass, or is a continuous transition existing only at exactly zero mass.)

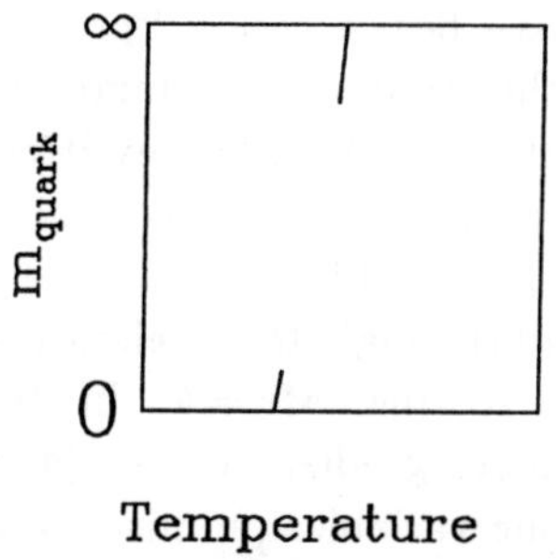

Figure 5. Phase diagram for QCD with four flavors of light quarks.

To complete this grab bag of applications, I would like to mention studies of the high energy behavior of the Weinberg-Salam model that have appeared in the last year.[21] The issue here is whether the cutoff in the theory can be removed for any value of the coupling other than zero. Here the lattice provides an explicit nonperturbative cutoff. To take the cutoff away, or take the lattice spacing to zero, we must adjust the coupling constants at the scale of the lattice to a critical point, a point where ξ/a is infinite. The possible long distance theories are then those which lie in the "relevant" directions from this critical point. This Monte Carlo simulation is used to search for these critical points. Among the a priori possibilities are 1) the existence of a nontrivial critical point with all the continuum coupling constants relevant in which case the theory seems fully consistent, 2) a critical point with too few relevant parameters, in which case the parameters of the continuum theory must satisfy relations, perhaps allowing computation of the Higgs mass, or 3) no suitable critical point, in which case the Weinberg-Salam model must be modified at high energies. This last possibility appears the most likely, but all the dust has not yet settled.

TECHNOLOGY

Although all these results are encouraging, it is clear that we are not yet doing accurate computations of hadron properties. To a large extent we are limited by technology. We need improvements in algorithms, in machines, and in software. Beyond noting their existence, I will not discuss improvements in software since the needs here are in no way special to theoretical QCD.

Slow but steady progress is being made in algorithms. Recent progress is summarized by Peter LePage in these proceedings.[22] The big problem is the inclusion of dynamical fermions, and some exciting things are happening here. In the last year we have begun seeing "production" results from the Langevin equation[3] such as the study of high temperature QCD by Fukugita and Ukawa mentioned earlier.[20] Another exciting development is the application to QCD of molecular dynamics methods with refreshing of the quadratic momenta.[41] I would also like to note the application of the well known technique of "preconditioning" to accelerate the conjugate gradient algorithm, which arises in solving $Mx = b$ in fermionic simulations.[23] Preconditioning of the conjugate gradient or similar acceleration techniques for the evolution of the bosonic variables promise to increase the efficiency of our algorithms by a significant factor.[3] However, the best methods of acceleration or the actual gain to be expected from the methods that have been suggested are not yet known.

The limited computing power of today's machines is probably the greatest obstacle to this industry. I will therefore survey briefly the machines available to us today and the prospects for improvement. We can distinguish three major types of machines used in these simulations: supercomputer centers such as those funded by NSF and DOE, dedicated machines with varying degrees of specialization, and custom built machines.

By far the greatest number of projects are done at supercomputer centers. Of the three types of machine, these clearly are the most flexible and the easiest to use (although we could wish for much more in ease of use!). Also, the supercomputers have larger memories than any of the dedicated machines in use today. For example the $24^3 \times 48$ lattice of de Forcrand mentioned earlier could not have been done today except on a supercomputer. The problem with this approach is its high cost – it costs tens of millions of dollars per year to run a supercomputer center. This means that the amount of time available at such centers will necessarily be limited.

A number of groups have used more or less dedicated machines for QCD simulations. Perhaps the first of these was the distributed array processor (DAP), a two dimensional array of very simple processors used very successfully in QCD. Incidentally, this was the machine used to generate the quenched approximation hadron masses I showed earlier.[10] The "cosmic cube" at Caltech[24] was motivated in part by QCD (and was used in reference 13 to generate configurations). This machine is notable for the part it played in spawning a collection of commercially produced hypercubes. Array processors, such as the FPS 164 and the ST100, are also in use for these simulations. The advantage of these dedicated machines is their low cost. The cost of buying and operating an array processor is an order of magnitude less than the cost of maintaining equivalent computing power at a supercomputer center. This means that a research group can reasonably hope to run one 24 hours a day, or 7000 hours a year. These machines are quite powerful. For example, in the refreshed molecular dynamics simulation an ST100 array processor ran at 2/3 the speed of a Cray XMP processor (1 processor!).[25] The disadvantages of this class of machines are that they tend to have limited memory and that programming them requires a lot of specialized knowledge and extra effort. However, this difficulty is mitigated by the fact that lattice Monte Carlo codes are simple in comparison to other computations using comparable resources.

Finally, there are a number of ambitious efforts in progress to build very large machines specifically for QCD Monte Carlo. Among these are a series of machines at Columbia[26] the GF11 at IBM,[27] the APE machine in Italy[28] and the STAC processor at Boston University.[29] The first of these machines at Columbia, a 16 processor version, has been running QCD simulations for over a year now. Obviously even more effort is required to develop the software for a custom machine than for any commercially available machine. In addition designing and building one is a tremendous effort. Finally, once it is built it is your problem to maintain it. The payoff is that you can contemplate building a very fast machine with very large memory – 16 Gigaflops in the case of Columbia version 3, or 11 Gigaflops for the GF11.

One that is often said in discussing these machines is that we choose their architecture to fit the problem. However, the various efforts have very different architectures, so this criterion must be ambiguous at best. There are some common features. All of these efforts are multiple processor and multiple memory machines. This is motivated by the availability of reasonably fast floating point chips of a few megaflops each, and the availability of cheap memory which is much slower than the desired access rate. The only way to achieve the desired speed cheaply seems to be to use many processing elements and many memories in parallel. In one sense lattice gauge theory simulations are particularly well suited to parallel architectures since we usually wish to do identical things to a large number of lattice sites and one usually imagines assigning each processor to update a separate section of the lattice. This is in contrast to a mode of parallelism which is very successful in things like analysis of events in high energy experiments, where each processor can be assigned to work on a separate event. In the case of event analysis only minimal communication is required among the processors, but in the case of a lattice gauge theory simulation the amount of communication is rather large. This means that the biggest issue in the design of a custom machine is the interconnection scheme among the processors. The various QCD projects are taking very different approaches to this problem. (If the memory requirements of your problem are modest enough, the most effective architecture is just a line of independent processors, such as 3081E's,[30] each working on an independent sample (see Figure 6a). Thus it is more the desire for larger shared memories than for faster processing which motivates these efforts.)

The Columbia machines are connected in a two dimensional mesh as sketched in Figure 6b, where each processor can access three memories. This machine is MIMD (multiple instruction multiple data), but host synchronization and *careful* programming must be used to prevent conflicting memory accesses. The STAC processor caricatured in Figure 6c is also intended to be a two dimensional mesh of processors, using the more obvious square lattice. Each STAC processor is vectorized with a natural vector length of eight and is intended to be used to update eight lattice sites simultaneously. In contrast the APE machine (Figure 6d) is a one dimensional array of processors and memories connected by a "barrel shifter", which can make a cyclic permutation of the correspondence between processors and memories. Notice that this allows equally fast access by a processor to any part of memory. The APE is a SIMD (single instruction multiple data) machine - all processors execute the same instruction in lockstep, while updating different parts of the lattice. A four processor version of the APE is in operation, with a sixteen processor version under construction. The most imaginative architecture is the GF11, in which a switching network can connect the 566 processors in any permutation (Figure 6e). This is also an SIMD machine. Here each processor has a local memory, but if you must access another node's memory all locations in the lattice can be accessed with equal speed. We see from the variety of designs that the problem constrains the architecture only in the most general sense.

There are a few general comments that can be made about designing a custom processor. The most commonly quoted measure of capacity for these and other machines is the "raw megaflops", or number of floating point units multiplied by their clock rate. However, this is only one measure of performance. I think that the memory bandwidth is a more important parameter for these machines. Of course, every machine is going to be limited by something, and with the availability of cheap floating point chips it is probably sounder design to be bottlenecked by memory bandwidth than by floating point capacity.

Unfortunately the effective memory bandwidth is much harder to quantify than the "raw megaflops". Register files and cache memories are used to decrease the actual bandwidth needed in the large memories, and their effect is complicated. Worse, architectural limitations may make it difficult to access the data you want – it might be in a memory inaccessible to the processor. This means that you must consider not only the total bandwidth needed but also where the data you want is stored and how easy it is to get there.

The commonly used benchmark for these machines is their speed on a Metropolis updating in the pure gauge (no quarks) theory. This can be misleading, since if you have enough register file or cache to hold a few matrices, the computation of the matrix products needed to compute the change in action require only about one memory access for every eight floating point operations. However, the time consuming step in the real problem is the solution of $Mx = b$, and with the conjugate gradient method for Kogut-Susskind fermions this requires about one memory access for every three floating point operations. More memory accesses are needed because in multiplying matrices times vectors you can't reuse numbers as much as when multiplying matrices by matrices. (For conjugate gradient with Wilson fermions the number is different again: about .15 reads per flop. This differs because you get to multiply four Dirac components of a colored spinor by the same color matrix.) For the Metropolis updating of the bosonic theory, all the accesses are local in the sense that the updating of each site needs only variables at nearest neighbor and next nearest neighbor sites. Thus a mesh of processors, or any design with short range connectivity, can work well. Again, with fermions the problem is the solution of $Mx = b$. With the conjugate gradient method the memory accesses are mostly local, but twice in each iteration one must compute a scalar which is a sum over all lattice sites and broadcast the result to all processors. The overhead on this can be painful. If Fourier acceleration or a similar method is to be used, then many memory accesses will be long range and a large locally connected mesh of processors might have trouble.

CONCLUSIONS

I think that it is clear that large scale computations can make important contributions to theoretical particle physics, in particular to low energy QCD. However, only a few things have really been done right at this point. We badly need more computing power and much more careful work. I have listed three classes of machines in use for these simulations, each with advantages and disadvantages. All three classes are useful in QCD; the best choice depends on the problem.

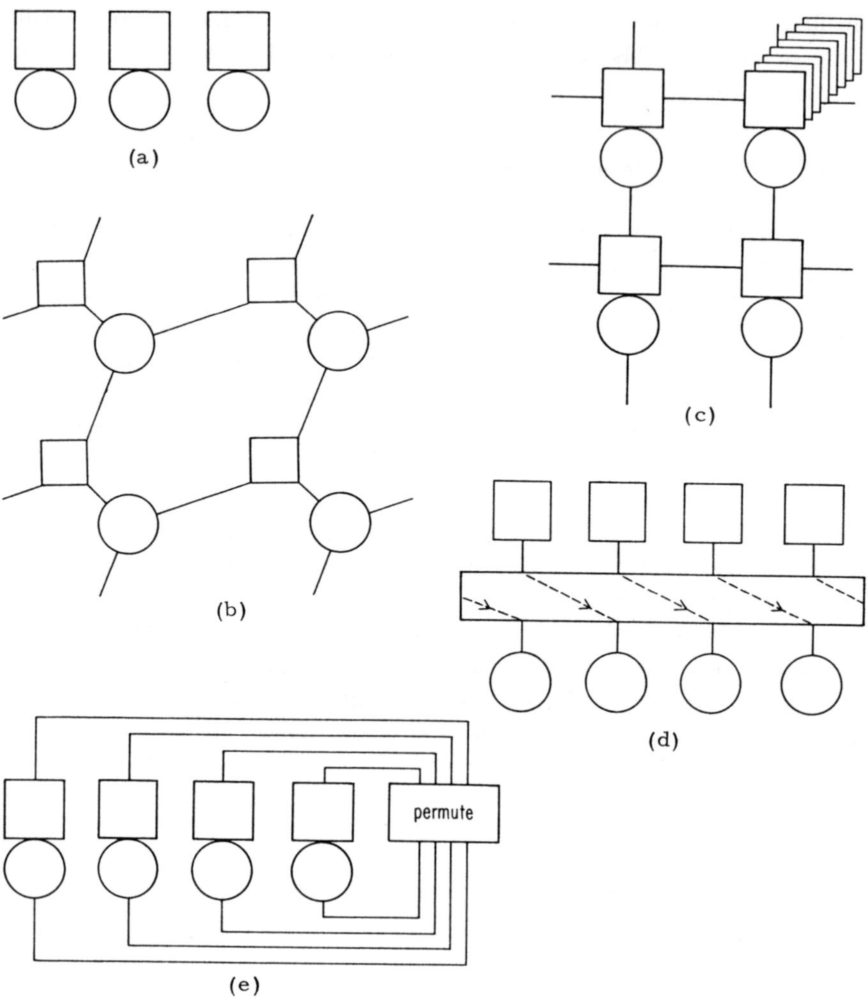

Figure 6. Architectures for customized QCD machines. The squares represent processors and the circles memories.

a) Trivial parallelism, or parallelism in parameter space. Each machine runs a separate lattice.

b) The Columbia architecture.[26]

c) The STAC architecture.[29] Each processor can be considered as eight processors working on eight different lattice sites.

d) The APE processor.[28] e) The GF11.[27]

REFERENCES

1) K.G. Wilson, Phys. Rev. **D14**, 2445 (1974).

2) Metropolis et al., J. Chem. Phys. **21**, 1087 (1953); for an introductory review see K. Binder, in "Phase Transitions and Critical Phenomena", C. Domb and M.S. Green, eds. (Academic, London, 1972) or J.M. Hammersley and D.C. Handscombe "Monte Carlo Methods" (Wiley, New York, 1964).

3) G. Parisi and Wu Yongshi, Sci. Sin. **24**, 483 (1981); G. Batrouni et al., Phys. Rev. **D32**, 2736 (1985).

4) S. Duane, Nuc. Phys. **B257** **[FS14]**, 652 (1985). S. Duane and J.B. Kogut, Phys. Rev. Lett. **55**, 2774 (1985); Nuc. Phys. **B253**, 398,(1986).

5) F. Fucito, E. Marinari, G. Parisi and C. Rebbi, Nuc. Phys. **B180**, 369 (1981).

6) H.B. Nielson and M. Ninomiya, Nuc. Phys. **B185, 20 (1981), B193**, 173 (1981) and an extensive subsequent literature under the general name of the "doubling problem".

7) H.G. Evertz, T. Jersak, T. Neuhaus and P.M. Zerwas, Nuc. Phys. **B251** **[FS13]**, 279 (1985); J. Jersak, T. Neuhaus, P.M. Zerwas, Nuc. Phys. **B251**, 299 (1985) and many references therein.

8) M. Creutz, Phys. Rev. Lett. **45**, 313 (1980), Phys. Rev. D **21**, 2308 (1980).

9) H. Hamber and G. Parisi, Phys. Rev. Lett. **47**, 1792 (1981). E. Marinari, G. Parisi and C. Rebbi, Phys. Rev. Lett. **47**, 1795 (1981). D.H. Weingarten, Phys. Lett. **109B**, 57 (1982).

10) K.C. Bowler, C.B. Chalmers, R.D. Kenway, G.S. Pawley and D. Roweth, Edinburgh preprint 86/269 (1986).

11) D. Barkai, K.J.M. Moriarty and C. Rebbi, Phys. Lett. **156B**, 385 (1985); H. Hamber, Phys. Lett. **178B**, 277 (1986).

12) A. Billoire and E. Marinari, preprint ROM2F-86/017

13) F. Fucito, K.J.M. Moriarty, C. Rebbi and S. Solomon, Phys. Lett **B172**, 235 (1986)

14) M. Fukugita, Y. Oyanagi and A. Ukawa, Phys. rev. Lett. **57**, 953 (1986).

15) P. de Forcrand, J. Stat. Phys. **43**, 1077 (1986).

16) B. Velikson and D. Weingarten, Nucl. Phys. **B249**, 433 (1985); S.A. Gottlieb, in "Advances in Lattice Gauge Theory", D.W. Duke and J.F. Owens, eds. (World Scientific, Singapore, 1985).

17) C. Bernard et al., Phys. Rev. Lett. **55**, 2770 (1985); N. Cabibbo, G. Martinelli and R. Petronzio, Nucl. Phys. **B244**, 381 (1984); R.C. Brower, M.B. Gavela, R. Gupta and G. Maturana, Phys. Rev. Lett. **53**, 1318 (1984); R.C. Brower et al., Comput. Phys. Comm. **38**, 9 (1985); C. Bernard, T. Draper, G. Hocksey and A. Soni, preprint UCLA-86TEP/48.

18) J. Kuti, J. Polonyi and K. Szlachanyi, Phys. Lett. **98B**, 199 (1981). L. McLerran and B. Svetitsky, Phys. Lett. **98B**, 195 (1981); Phys. Rev. **D24**, 450 (1981).

19) J. Engels, F. Karsch, I. Montvay and H. Satz, Phys. Lett. **101B**, 89 (1981), Nucl. Phys. **B205**, 545 (1982); F. Karsch, Nucl. Phys. **B205** 285 (1982); T. Celik, J. Engels and H. Satz, Phys. Lett. **129B**, 323 (1983); N.H. Christ and A. Terrano, Phys. Rev. Lett. **56**, 111 (1985); S.A. Gottlieb et al., Phys. Rev. Lett. **55**, 1958 (1985), Phys. Lett. B, to be published.

20) J. Polonyi, H.W. Wyld, J.B. Kogut, J. Shigemitsu and D.K. Sinclair, Phys. Rev. Lett. **53**, 644 (1984); Fukugita and Ukawa, Phys. rev. Lett. **57**, 503 (1986); J.B. Kogut, E. Kovacs and D.K. Sinclair, presented by D.K. Sinclair at the Brookhaven conference on lattice gauge theory (1986). S.A. Gottlieb et al., in preparation.

21) P. Hasenfratz and J. Nager, presented at the Brookhaven conference on lattice gauge theory (1986); D.J.E. Callaway and R. Petronzio, preprint CERN TH.45111/86 (1986); R. Shrock, presented at the Brookhaven conference on lattice gauge theory (1986); J. Jersak, C.B. Lang, T. Neuhaus and G. Vones, Phys. Rev. D **32**, 2761 (1985).

22) P. Lepage, these proceedings.

23) Y. Oyanagi, to be published.

24) C. Seitz, Journal of VLSI and Computer Systems, Vol. 1, No. 3 (1984).

25) D.K. Sinclair, presented at the Brookhaven conference on lattice gauge theory (1986).

26) N. Christ and A. Terrano, IEEE Transactions on Computers, **C-33**, 344 (1984).

27) J. Beetem, M. Denneau and D. Weingarten, J. Stat. Phys. **43**, 1171 (1986).

28) E. Marinari, in "Lattice gauge theory", B. Bunk and K.H. Mutter, eds. (Plenum,1986); private communication.

29) R. Giles, private communication.

30) P.F. Kunz et al., preprint SLAC 3069, CERN-DD-83-03 (1983).

CP AND OTHER TESTS OF THE STANDARD MODEL

B. Winstein

Enrico Fermi Institute and the Department of Physics
The University of Chicago

INTRODUCTION

In this lecture, we will review recent progress in testing the Standard Model of the electroweak interactions particularly with respect to CP non-conservation. We will begin with a brief discussion of those experiments at Brookhaven National Laboratory which are addressing the issue of highly suppressed or even forbidden decay modes of the charged and neutral Kaon. Then we will treat the phenomena of mixing in heavy flavor systems and discuss the possible evidence for the phenomena presented by the UA1 collaboration. Finally, we will treat the phenomena of CP non-conservation and present the preliminary result from the FNAL experiment measuring ϵ'/ϵ.

RARE KAON DECAY EXPERIMENTS[1]

The Standard Model of the electroweak interactions has been highly successful in accomodating the vast array of phenomena in its domain. Nevertheless it is incomplete in that many of its parameters must be input from experiment. One very attractive way of looking for the inevitable breakdown of the model is to do a high sensitivity search for processes which it strictly forbids. Of course what one looks for has to be guided by one's intuition on where breakdowns are likely to show up as such experiments are by nature usually very demanding both with respect to human and financial resources. The experiments to be discussed here are all either in a test mode or soon will be.

It is well known that the three generations of quarks mix whereas the leptons do not. The associated "mixing angles" are a set of parameters that need to be input into the model and, as long as the charged current mixing matrix is Unitary, essentially any set of angles would be allowed in the model. We will mention the pattern of such angles later, but for now it should be pointed out that one might guess that this obvious asymmetry between the quarks and the leptons might be restored at some level at perhaps a higher energy scale. This is the essential motivation for the searches for the decay modes of the Kaon which do not conserve lepton number. The goal with respect to branching ratio sensitivity for these experiments is largely governed by what the current limit is in that the desired mode may be in fact absolutely forbidden or it may lie just at the current limit; it is difficult to get precise theoretical guidance.

The goal of AGS E780[2] is a measurement of the branching ratio for $K_L \to \mu e$ with a sensitivity at the 10^{-11} level. Since the current level is about 10^{-8}, this represents a major improvement. The significance of the improvement is very clear should a signal be seen; if not, one sets limits on new phenomena within the context of the given model, be it lepton compositness, supersymmetry, or technicolor. The major background for this decay is $K_L \to \pi e \nu$ where the pion decays in such a way as to fake the sought for mode. One needs good resolution and particle identification as well. In E780, electron identification is accomplished by means of an atmospheric hydrogen Cerenkov counter as well as lead glass. Muon identification is accomplished by means of a range array of scintillators (see Figure 1). One serious limitation of such experiments is the instantaneous rate in the apparatus from the incident beam - the Kaon decays within the beam as well as the significant number of accompanying neutrons and any halo particles around the beam itself. The incident beam for E780 will be in the neighborhood of 10^8 K_L per pulse and although there is a hole through the entire detector, such a flux will clearly tax the apparatus. In particular, one must be concerned about any radiation damage that the lead glass might suffer.

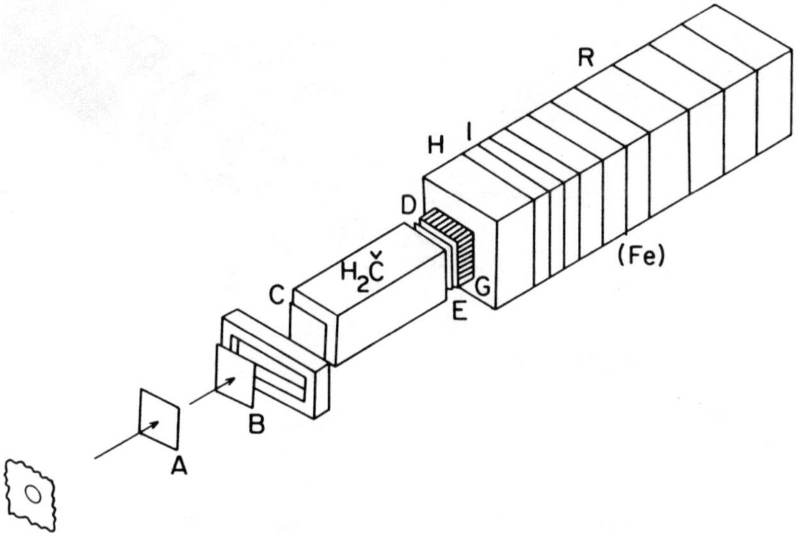

Figure 1. The Yale-BNL $K_L \to \mu e$ spectrometer, BNL E780.

The goal for E791[3] is to have a sensitivity again for $K_L \to \mu e$ in the range of 10^{-12}, about an order of magnitude better than that for E780. The apparatus for this effort is shown in Figure 2. To reach its goal, the experiment needs to be able to run at an intensity of about 5×10^8 incident Kaons per pulse. Again lead glass and a gas Cerenkov counter are used for electron identification and a range measurement is made on the muons. To achieve the added rejection against background, the charged particle momenta are measured twice in this experiment. At this level of sensitivity, about 10^4 $K_L \to 2\mu$ should be seen and in a future effort, the experimenters hope to measure the polarization of the positive muon with an error of about 15%. A signal at this level would signal a sizable non-standard model CP violation as either the Superweak or KM model would predict a polarization at the level of ϵ, or about 2×10^{-3}.

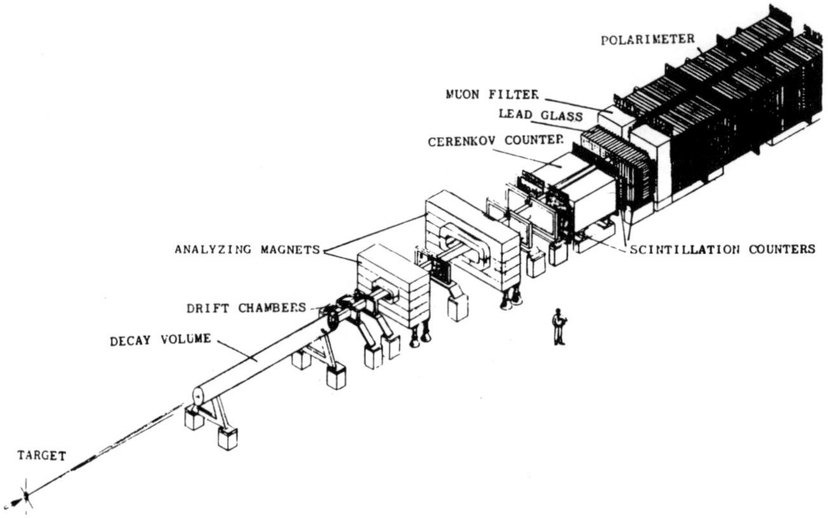

Figure 2. The spectrometer for BNL E791.

The E777 collaboration,[4] is looking for the mode $K^+ \to \pi\mu e$ with a sensitivity in the range of 10^{-11} and their spectrometer is shown in Figure 3. This search also requires a very intense beam and excellent particle identification but there are perhaps advantages from a purely experimental point of view in the rejection of background. In particular, the incident particle has a known momentum and three charged particles are observed which must form a good vertex which lies in the region of the incident beam.

The experiments described so far will also have sensitivities to other Kaon decay modes as well. The future of such searches, in the event that no signals are seen in the first round of these experiments, will largely depend upon the experience gained. It may always be desirable to increase the sensitivity significantly and in principal this would be possible at a "Kaon Factory," the feasibility for which is being studied at several laboratories. The experience gained in these efforts will also be valuable towards other very high rate experiments, even for those at the SSC. The instantaneous rates in these efforts are very significant where more than one Kaon decay within the resolving time of the detectors is a significant source of background. Additionally, the triggering resembles that for large collider experiments with various levels of sophisticated microprocessors operating.

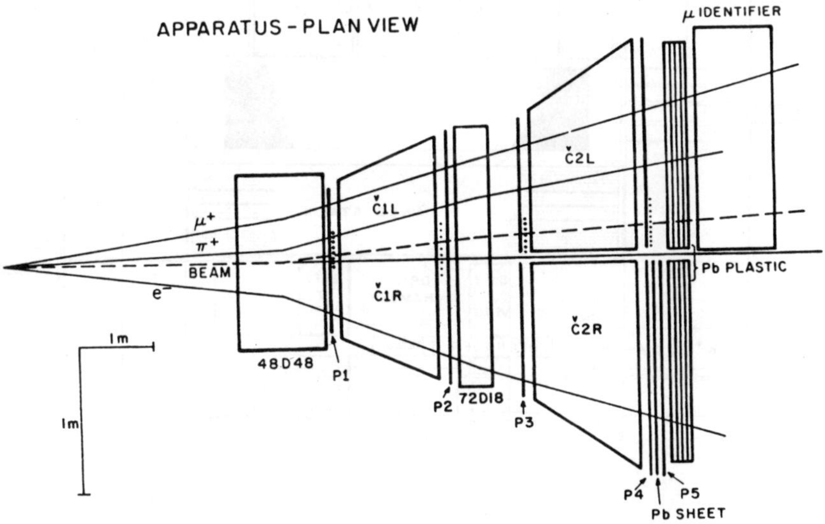

Figure 3. The spectrometer for BNL E777.

Experiment E787,[5] is looking for the mode $K^+ \to \pi^+$ + "nothing" with a sensitivity of about 10^{-10}, a major improvement over existing limits. "Nothing," which of course means a weakly interacting particle or particles, could be a new particle (for example an axion,[6] familon,[7] or Higgs-like particle[8]) which would result in a monochromatic π^+: the experiment consists in a 4π detector surrounding a stopped K^+ beam. Or it could be two weakly interacting particles and in fact a measurable signal is expected from $\nu\bar{\nu}$ pairs. The apparatus is shown in Figure 4 where the great attention paid to rejecting the dominant $K^+ \to \pi^+\pi^0$ mode can be seen. The π^+ is identified by its decay to the muon and the muon decay is observed as well.

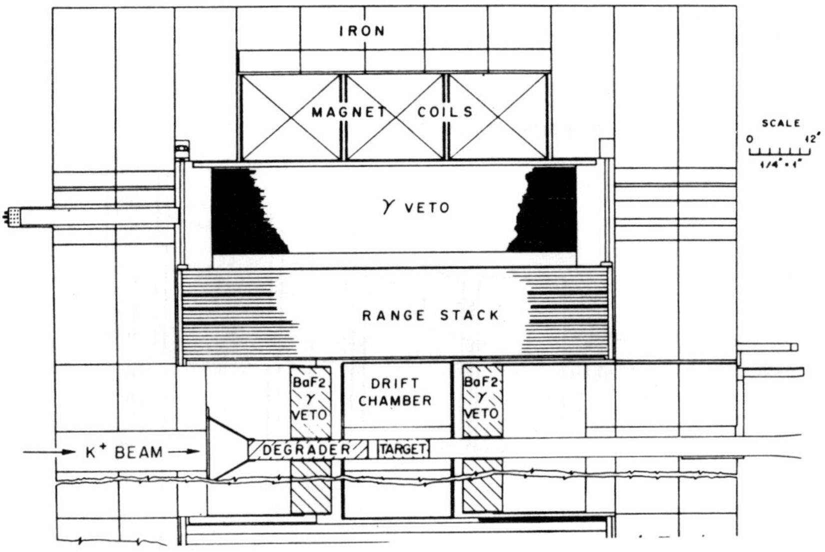

Figure 4. The detector for BNL E787.

The diagrams which will contribute to this mode are shown in Figure 5 and the expected branching ratio is in the range of a few $\times 10^{-11}$ per neutrino type. The branching ratio is highly GIM suppressed but that suppression is weakened[9] as the top quark mass is increased. Thus a signal may even be seen in this first round experiment and in any case there is a large window for possible new physics to contribute to this mode.

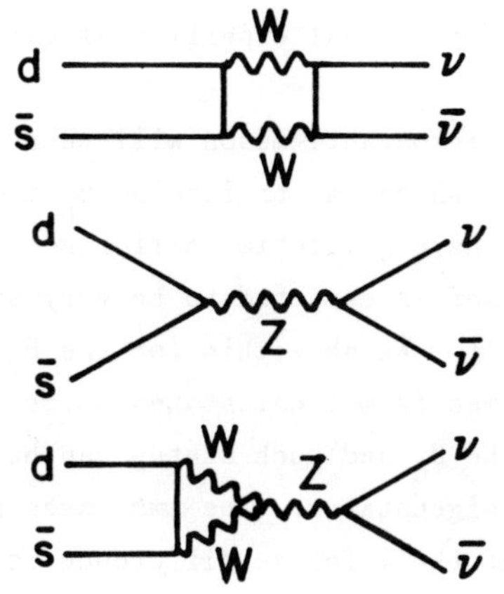

Figure 5. Standard Model contributions to $K^+ \rightarrow \pi^+\nu\bar{\nu}$.

MIXING IN HEAVY FLAVOR SYSTEMS

Based upon the phenomena in the $K^0\bar{K}^0$ system, one would expect mixing and possible CP non-conservation in neutral systems of heavy flavors. The B^0 system is the most promising for the observation of such effects and this follows from elementary consideration of the mixing angles involved.

We will adopt the Wolfenstein[10] parameterization of the charged current (Kobayashi-Maskawa) coupling matrix where the elements are expressed as powers of the Cabibbo angle, here denoted by λ, and the diagonal elements are approximately equal to one:

$$V = \begin{pmatrix} 1 - \lambda^2/2 & \lambda & \lambda^3 e^{i\delta} \\ -\lambda & 1 - \lambda^2/2 & \lambda^2 \\ \lambda^3(1 - e^{-i\delta}) & -\lambda^2 & 1 \end{pmatrix}$$

The parameter δ allows for CP non-conservation. We stress of course that this representation is useful for getting rough estimates for processes but that the actual experimental values are somewhat different.

A given system of meson-anti-meson will then have two eigenstates of the weak interaction which can be labeled B_H and B_L for heavy and light. These two will have a lifetime difference as well as a mass difference but the former is expected to be very small in comparison with the lifetime itself. We show this for the B_d system. The lifetime difference comes from final states which are common decay modes for the B_d and the $\bar{B}_d$ and such states can be either CP eigenstates or CP non-eigenstates. The two cases are shown in Figure 6 where it is seen that the lifetime difference is governed by λ^6 so that, with respect to the lifetime itself, it is suppressed by two powers of the Cabibbo angle. Recall that for the $K_L K_S$ system, the lifetime difference is comparable to the (K_S) lifetime.

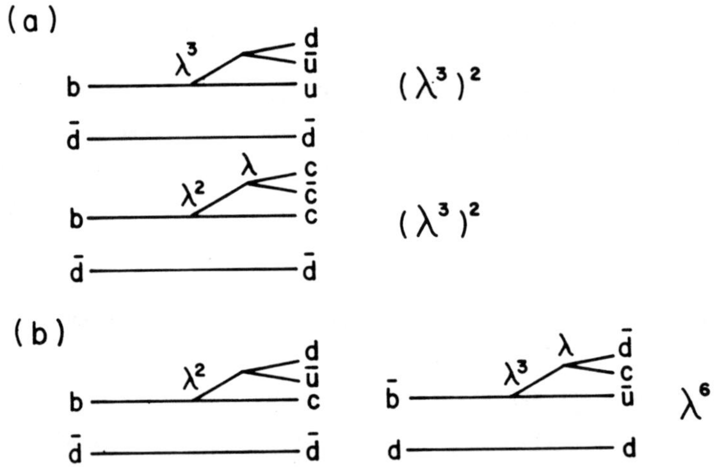

Figure 6. Diagrams contributing to $\Delta\Gamma$ for the B_d system.
(a) Intermediate CP eigenstate
(b) Intermediate non CP eigenstate.

The mass difference can be estimated by similarly considering the relevant "box" diagrams. In Figure 7, we indicate the degree of suppression that the relevant diagrams have for mixing in the three particle/antiparticle systems and it is seen that it is only for the B_s^o system that the mass difference is appreciable.

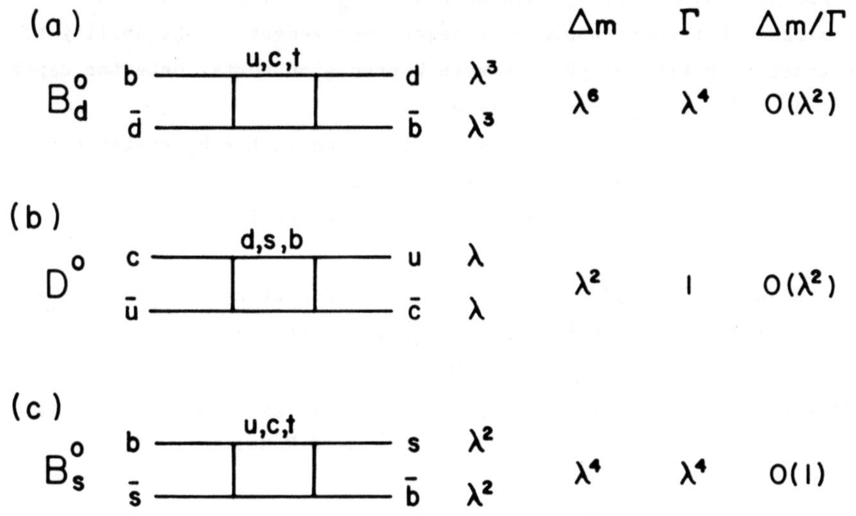

Figure 7. Diagrams contributing to Δm for three heavy quark systems; (a) B_d^0, (b) D^0, (c) B_S^0. Indicated in the figure are the dependences on the Cabibbo angle, here parameterized by λ, of each "flavor changing" line of the diagram.

In fact, the UA1 group has claimed[11] to have observed such mixing already. The relevant signature is the appearance of same-sign charged leptons, either electrons or muons, coming from the associated production of B mesons. Various topological cuts are applied to the data to reduce background but what remains is a sizable contamination which is thought to be due to cascade decays. For the ratio of same sign to opposite sign dileptons, the group reports

$$R = (N_{++} + N_{--})/N_{+-} = 0.46(7)(5)$$

where the first error (the uncertainty on the last digit in parentheses) is statistical and the second systematic. The "expected" signal, where several event simulations have been done, is about equal to 0.24 so that the group reports that "the observed R value is consistent with full B_s^o mixing and no B_d^o mixing." It should be remarked that there would be a major improvement in the ability to extract this kind of physics with a precision vertex detector capable of identifying secondary vertices.

A recent limit[12] on the size of mixing in the B_d system has been given by the CLEO group; they report:

$$R < 0.24 \text{ (90\% confidence level).}$$

[Note added: The ARGUS collaboration is reporting[13] a positive signal for mixing within the B_d system. They report that

$$R = 0.23(7)(3)$$

If confirmed, this would imply[14] a rather large value for the top quark mass in the framework of the Standard Model.]

FNAL EXPERIMENT ON DIRECT CP VIOLATION

The Fermilab experiment determining the ratio of ϵ'/ϵ via a precision measurement of the ratio of the two observable CP non-conserving parameters characterizing the two pion decays, η_{+-}/η_{00}, has done a preliminary analysis on some test data and a new result is available. At present, the two best values are the following:

$$|\eta_{+-}/\eta_{oo}|^2 = 0.972(35) \quad \text{(reference 15)}$$

and

$$|\eta_{+-}/\eta_{oo}|^2 = 1.010(49) \quad \text{(reference 16)}$$

The members of this collaboration[17] are listed below:

L. Gibbons, K. Nishikawa, V. Papadimitriou, R. Patterson, Y. Wah, B. Winstein, R. Winston, M. Woods, H. Yamamoto, The University of Chicago; E. Swallow, Elmhurst College; G. Bock, R. Coleman, B. Hsuing, K. Stanfield, R. Stefanski, T. Yamanaka, Fermilab; G. Gollin, G. Grazer, M. Karlson, J. Okamitsu, Princeton University; J. Brisson, P. Debu, P. Jarry, R. Daudin, B. Peyaud, R. Turlay, B. Village, CEN, Saclay.

The experiment had a test run for about six weeks during the summer of 1985 and the data taking run will be in the Spring/Summer of 1987.

For this experiment, a new K_L beam was built with attention to minimizing secondary sources of low energy neutrons and photons. The detector employed the same lead-glass as was used earlier[15] (although arranged differently) together with a new drift chamber spectrometer giving good resolution and rate characteristics. The magnet aperture was significantly enlarged for greater acceptance and a significant number of photon veto's were employed to enable the rejection of the copious background of $K_L \to 3\pi^0$. Finally, the regenerator which is used to provide K_S's was "active" in the sense that it contained 24 anti-coincidence counters for rejection of inelastic events. The apparatus is shown in elevation view in Figure 8; the lead-glass (the blocks are 58mm by 58mm by 20rl) arrangement is shown in Figure 9. As can be seen, two closely-spaced beams originating from the same production target (approximately 180m upstream of the lead-glass array) are employed: one of them strikes the regenerator so that K_L and K_S decays can be observed simultaneously. By moving the regenerator, the roles of the two beams are interchanged every accelerator pulse (once per minute) so that many of the systematic errors in the normalization are effectively eliminated.

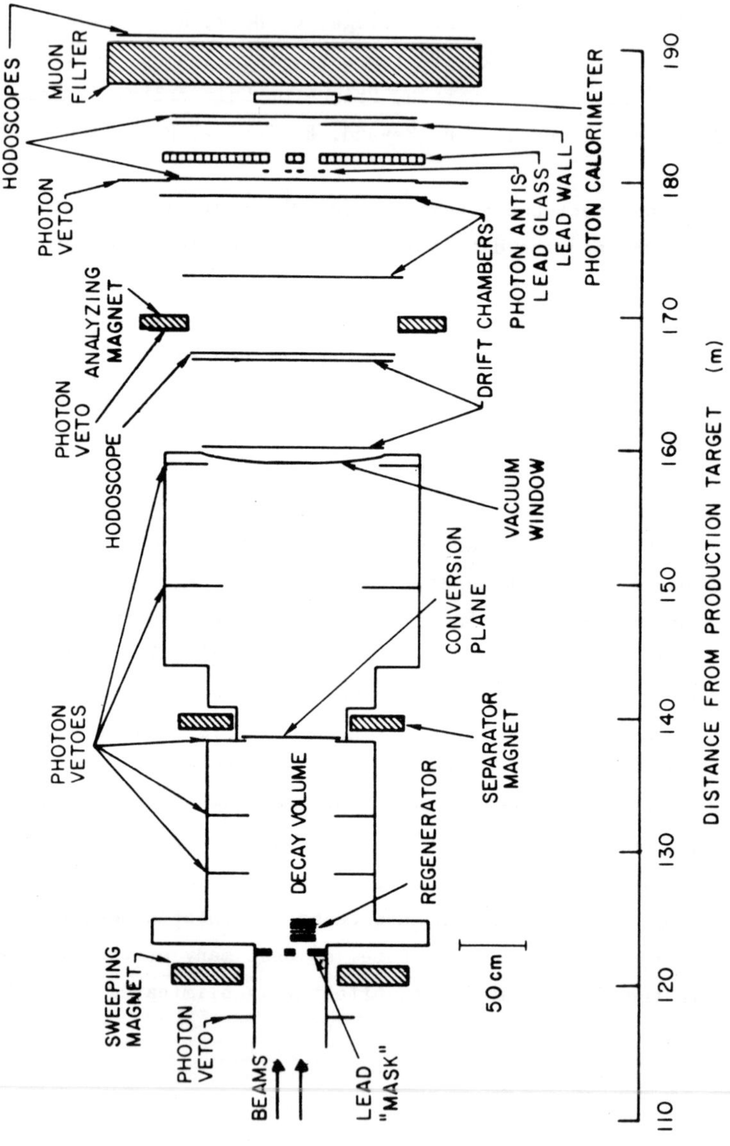

Figure 8. Elevation view of the detector for FNAL E731.

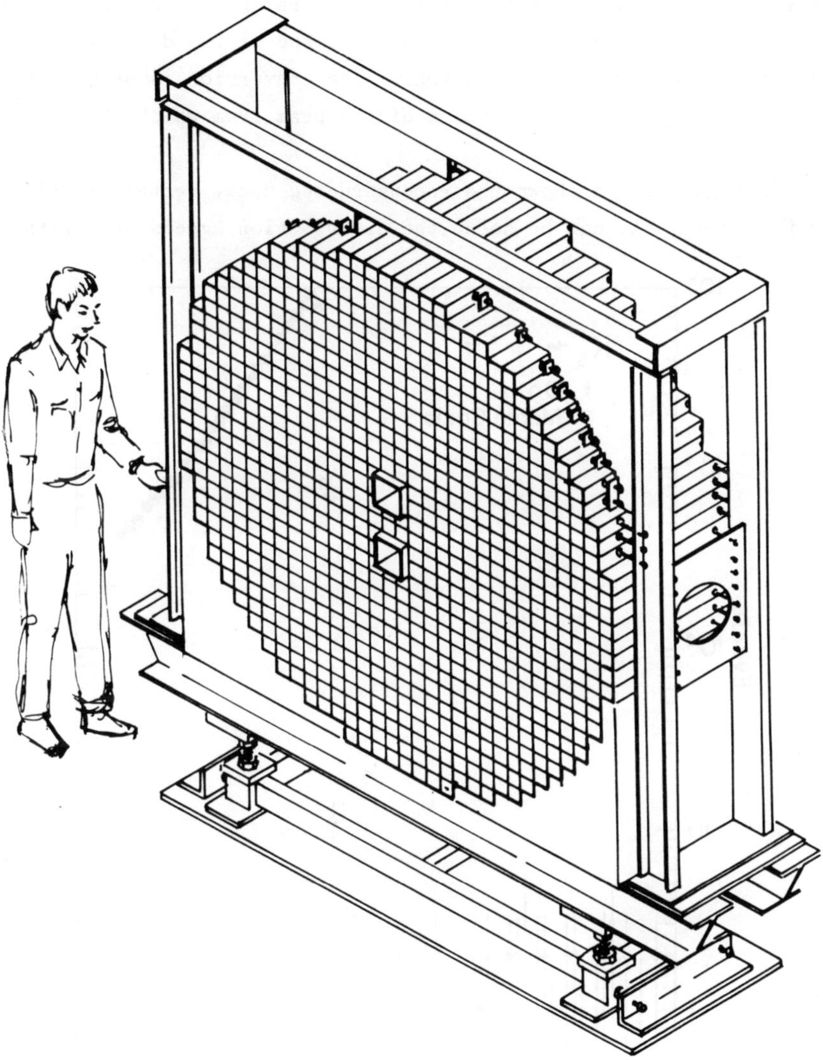

Figure 9. Schematic of the lead-glass array for FNAL E731.

The experiment is run in two distinct modes: charged and neutral. Aside from some changes in the trigger, the only important change to the hardware is that for the neutral mode running, an 0.1rl converter is introduced at the downstream end of the decay volume in order to convert one of the photons. The conversion is necessary so that an unambiguous determination of the beam from which the Kaon decayed can be made.

In Figure 10 is shown the reconstructed decay vertex distribution for $K_S \to \pi^+\pi^-$. As can be seen, the distribution agrees well with the

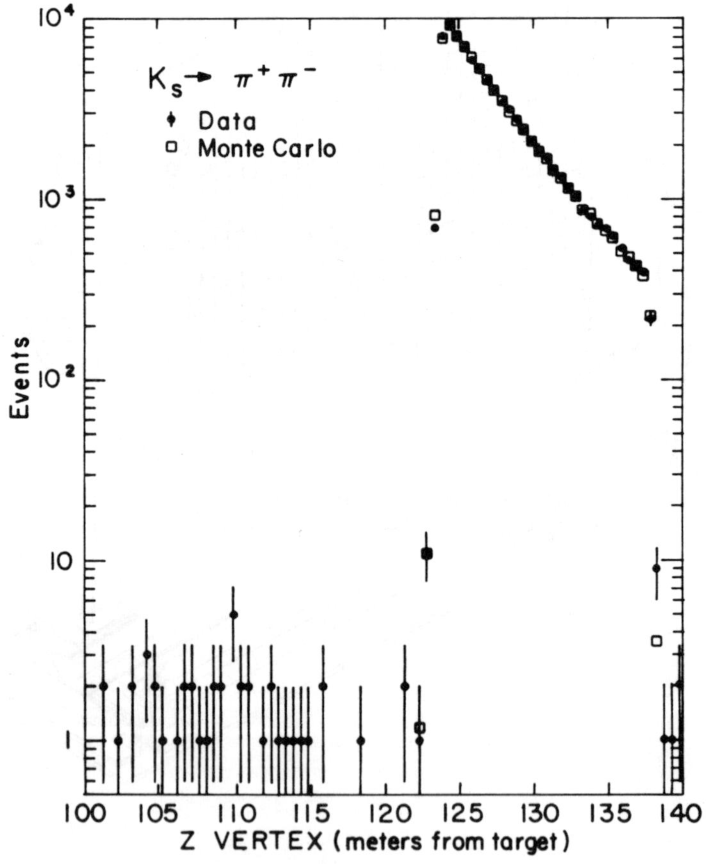

Figure 10. Decay vertex distributions for $K_S \to \pi^+\pi^-$.

Monte-Carlo prediction and the background is negligible. It is important that the acceptance in decay vertex be well understood because the decay distributions for the K_L's and K_S's are different.

Another potential problem is that due to incoherent regeneration in the regenerator which will not be present for the decays in vacuum. Figure 11 shows the reconstructed P_T^2 distribution for the charged mode decays and it can be seen that there is only a very small residual background within the resolution of the coherent peak which

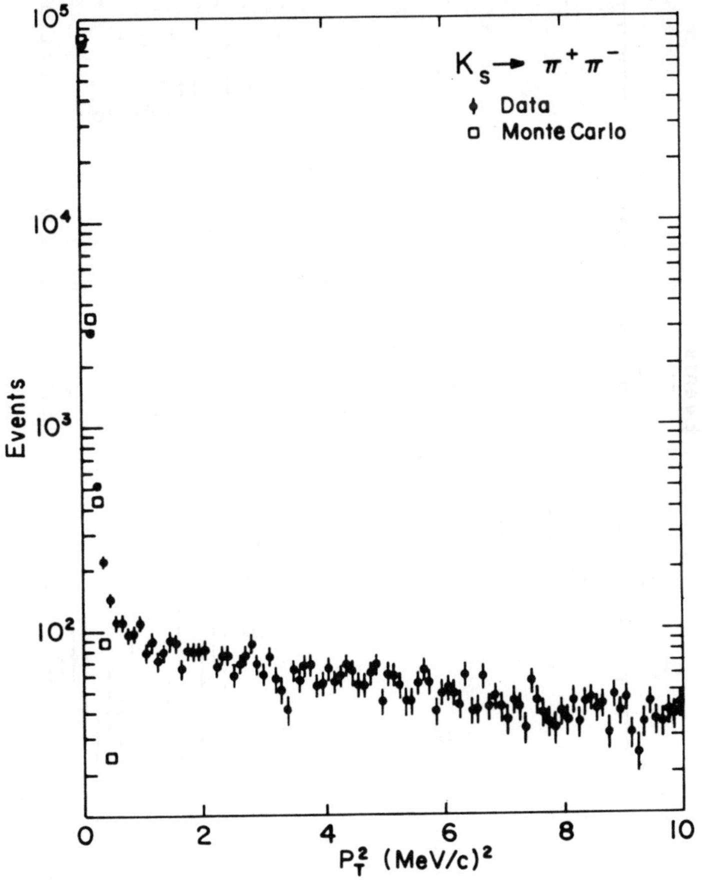

Figure 11. P_T^2 distributions for $K_S \rightarrow \pi^+\pi^-$.

is easily subtracted with little uncertainty. In addition, the shape of the incoherent signal is well understood itself, arising both from diffraction off the individual nuclei within the regenerator and from processes with particle production or nuclear breakup. The latter have been considerably reduced with the "active" regenerator.

Figure 12 shows the $\pi\pi$ invariant mass distribution for those decays from the vacuum beam. The shape is well reproduced by the Monte Carlo but a background is present. This is predominantly due to

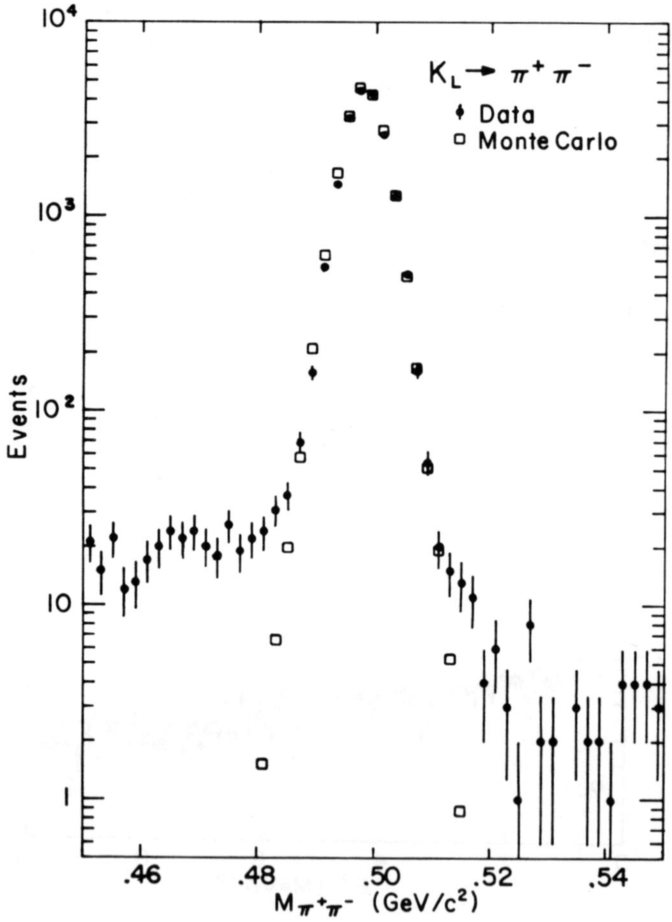

Figure 12. Mass distributions for $K_L \rightarrow \pi^+\pi^-$.

$K\mu_3$ decays where the muon is misidentified as a pion as a result of an interaction in the muon filter (3m of iron). This background is about 1% and presently is known within an uncertainty of about 25% of itself.

The reconstruction of events in the neutral mode is not as straightforward. For the charged mode, the energy scale can be as easily checked (and, as is usually the case, adjusted) with the value of K^0 and the Λ mass. For the neutral mode, the correct pairing of the four photons is done on the basis of consistent vertices for the two π^0's in the event; the vertex is taken to be a properly weighted average assuming the π^0 mass so that the resultant 4-gamma invariant mass is not sensitive to the overall energy scale. The scale can be checked with the distribution of the lead-glass energy for the electron pair compared to its measured momentum and also by a study of the positions of the "edges" of the distributions, namely the upstream edge for the regenerated events and the downstream edge for the vacuum events. By means of such studies, the energy scale in the neutral mode with respect to that in the charged is known with an uncertainty of about 0.2%. The rms energy resolution for electrons can be represented roughly by

$$\Delta E/E = 1\% + 5\%/\sqrt{E}$$

and the position resolution varies from about 2mm to about 5mm depending upon energy and where within a block the electron (or photon) hits. A great deal of attention is given to the calibration and monitoring of the lead glass array. Every week, a special run is taken where several hundred thousand electron pairs are used for calibration and the gains can be tracked by means of a light flasher system to about 0.5%. In addition, it is important to correct for non-linearities and non-Gaussian response of the glass. Other all neutral modes have been very useful ($3\pi^0$ and 2γ). The K_S signal is again very clean with negligible background under the mass peak. There is a larger correction necessary for the remnants of diffraction and inelastic regeneration under the coherent peak which is significantly broader than in the charged mode. However, the uncertainty is not significant in that the shape of the background is well known from the charged mode data.

In Figure 13 is shown the 2π invariant mass distribution for the $K_L \to 2\pi^0$ event sample. There is a background of about 2% which is due to remnant $3\pi^0$ events where photons are either lost or fuse with other photons in the lead-glass. Also shown in the figure is a simulation of the 3π background and the shape is in good agreement with the

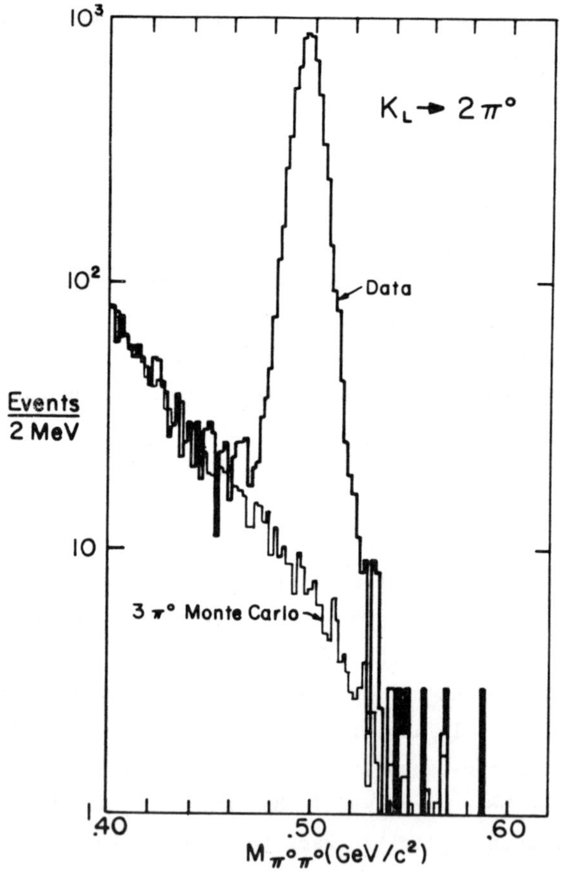

Figure 13. Mass distribution for $K_L \to 2\pi^0$ with a simulation of the background from $K_L \to 3\pi^0$ superimposed.

observed background. In the Table below, we list the final event sample for each mode together with the percentage of background and the uncertainty in the last digit of that percentage. Also shown are the background percentages and uncertainties from Ref. 15.

MODE	EVENTS	BACKGROUND[%]	BACKGROUND[%], REF. 15
$K_S \to \pi^+\pi^-$	109104	0.20(2)	1.68(8)
$K_L \to \pi^+\pi^-$	27178	1.2(3)	3.0(2)
$K_S \to \pi^0\pi^0$	21541	3.2(4)	12.7(4)
$K_L \to \pi^0\pi^0$	6738	2.0(3)	7.6(5)

The above numbers are corrected for acceptance and then both the regeneration amplitude and its momentum dependence can be calculated together with a value for ϵ'/ϵ. Checks can be made with respect to previous determinations of regeneration in this energy range; the agreement is reasonable and further studies are in progress.

In extracting the value for ϵ'/ϵ from the data, we have taken the phase of ϵ' (which is very close to that of ϵ) from determinations of the π-π scattering phase shifts in the appropriate isospin channels. The result is:

$$\epsilon'/\epsilon = 0.0035(30)(20)$$

where the first error is statistical and the second systematic. The systematic error arises from four sources and these are at present estimated as shown in the following table:

SOURCE	SYSTEMATIC ERROR
Uncertainty in backgrounds	0.0010
Uncertainty in neutral mode energy scale	0.0005
Uncertainty in charged mode acceptance	0.0012
Uncertainty in neutral mode acceptance	0.0012
TOTAL	0.0020

A comparison, then, of this experiment with those of Ref. 15 and Ref. 16 is shown in the following Table:

DETERMINATION	ϵ'/ϵ (x 10^4)	STAT. ERROR(x 10^4)	SYST. ERROR(x 10^4)
Ref. 16	17	70	42
Ref. 15	-46	53	24
This test run (preliminary)	35	30	20

In Figure 14 these results are plotted together with three other earlier determinations. We see that this result is consistent with earlier measurements, with zero, and with the KM[21] model.

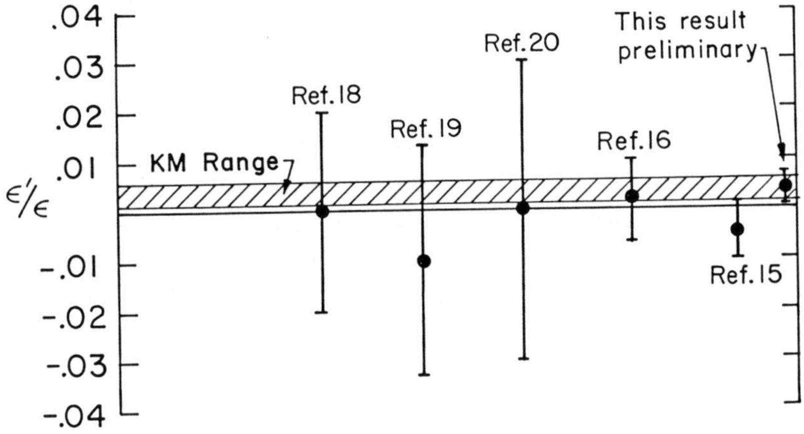

Figure 14. Comparison of FNAL E731 preliminary result with other recent results and with Standard Model predictions.

As the systematic errors for these experiments are pretty well decoupled, it is perhaps not unreasonable to add all of the errors in quadrature. The resulting world average value then becomes:

$$\epsilon'/\epsilon = 0.0015(29) \qquad \text{(world average)}$$

FUTURE IMPROVEMENTS

The systematic error on this result from the FNAL experiment should come down as the analysis is completed. What is more important is that the group has learned a great deal and expects to further reduce both the statistical and systematic error in the data taking run which will commence in late spring of 1987. To this end, the experiment will be run at higher beam intensity (up to 3×10^{12} protons on target per 20 sec spill) and with a better trigger which results principally from a hardware cluster finder which can in a small number of microseconds after each trigger determine how many clusters are present in the lead-glass. The time to record an event has been substantially reduced with the conversion to Fastbus electronics. Steps have been taken to further reduce the backgrounds in the neutral mode and as well an improved calibration procedure will be employed. A micro VAX will allow more sophisticated on-line monitoring. A result with an error of 0.001 is expected.

We conclude by noting that in the next few years it is planned that there will be three new results with errors in the range of 0.001: this experiment, the CERN[22] fixed target experiment which already has its data in hand, and the LEAR[23] experiment which is now in the construction phase. For these three efforts, the systematic errors are sufficiently dissimilar that universal agreement will be very gratifying. Should the result still be consistent with zero, for example, one would have to contemplate an even more ambitious effort. As difficult as this might be, it is very important that another manifestation of CP non-conservation be observed and, if the Standard Model is the correct description of the phenomena, it might still be easier to improve the measurement on ϵ'/ϵ by another substantial factor than to observe the phenomena in the heavy quark systems directly.

REFERENCES AND FOOTNOTES

1. This program has recently been summarized in the excellent review article of L. Littenberg, BNL #38348.
2. AGS E780, BNL-Yale collaboration.
3. AGS E791, UCLA-LANL-Pennsylvania-Stanford-Temple and William and Mary collaboration.
4. AGS E777, BNL-SIN-Yale-Washington collaboration.
5. AGS E787, BNL-Princeton-TRIUMF collaboration.
6. S. Weinberg, Phys. Rev. Lett. 40, 223 (1978); F. Wilczek, Phys. Rev. Lett. 40, 279 (1978).
7. F. Wilczek, Phys. Rev. Lett. 49, 1549 (1982).
8. M.B. Wise, Phys. Lett. 103B, 121 (1981).
9. F.J. Gilman and J.S. Hagelin, Phys. Lett. 133B, 443 (1983).
10. L. Wolfenstein, Phys. Rev. Lett. 51, 1945 (1984).
11. K. Eggert, Physikalisches Institut, Technische Hochschule, Aachen Preprint 86/12.
12. S. Stone, private communication; submitted for publication.
13. H. Albrecht et al., DESY preprint, March 1987.
14. J. Ellis, J.S. Hagelin, and S. Rudaz, CERN-TH.4679/87.
15. R.H. Bernstein et al., Phys. Rev. Lett. 54, 1631 (1985) (Chicago-Saclay collaboration.
16. J.K. Black et al., Phys. Rev. Lett. 54, 1628 (1985) (BNL-Yale collaboration).
17. FNAL E731, Chicago-Elmhurst-Fermilab-Princeton-Saclay collaboration.
18. M. Holder et al., Phys. Lett. 40B, 141 (1972).
19. M. Banner et al., Phys. Rev. Lett. 28, 1597 (1972).
20. J.H. Christenson et al., Phys. Rev. Lett. 43, 1209 (1979).
21. See L. Wolfenstein, Ann. Rev. Nucl. Part. Sci. 36, 137 (1986) and J.L. Rosner, these proceedings, for discussions of the current predictions for ϵ'/ϵ based upon the KM model.
22. CERN NA31, CERN-Dortmund-Edinburgh-Maintz-Orsay-Pisa-Siegen collaboration.
23. CERN PS195, Basile-CERN-Demokritos-Fribourg-Liverpool-Saclay-Stockholm-Thessaloniki-SIN-ETH collaboration.

QUARK MIXING AND CP VIOLATION

Jonathan L. Rosner

Enrico Fermi Institute and Department of Physics
University of Chicago, Chicago, IL 60637

ABSTRACT

Some aspects of CP violation are discussed within the context of the three-generation Cabibbo-Kobayashi-Maskawa model of quark mixing. Specific systems treated include those involving light quarks, charmed hadrons, and B mesons. Brief remarks are made on recent developments for quark mass matrices, and on extensions beyond the three-generation model.

1. INTRODUCTION

The phenomenon of CP violation[1] has continued to elude a deep theoretical understanding for more than 20 years.[2,3] The existence of at least three generations of quarks and leptons implies that, unless expressly forbidden, CP-violating weak couplings will arise as a result of complex values of the fermion mass matrices,[4] when one transforms the quarks from weak eigenstates to mass eigenstates.[5] Much is known about this Cabibbo/Kobayashi-Maskawa (KM) transformation. The purpose of this review is to highlight theoretical and experimental work that can tell in the next few years whether CP violation really can be described within the KM framework. We will stress the importance of improved lattice QCD estimates, detection of $b \to u$ transitions, discovery of the t quark, and refinement of measurements of the ratio ε'/ε in CP violating decays $K_L \to 2\pi$. At the same time, we will dwell briefly on other aspects of the KM matrix, its possible origins, and extensions.

For phenomena involving CP violation, theorists are only able to provide rough guidelines in most instances. Details of hadron physics are still not understood quantitatively enough; other processes (besides those involving CP violation) help to test our control of these details.

We begin in Sec. 2 by reviewing some CP violation signatures and theories, if only to focus attention on the few that will be discussed. Knowledge regarding KM parameters is summarized in Sec. 3. Light-quark systems of interest here include hyperons (Sec. 4) and kaons (Sec. 5). Charmed hadrons are treated in Sec. 6. A program for studying CP violation in B meson systems is outlined in Sec. 7. Some comments on parametrizations of mixing and mass matrices occupy Sec. 8. Prospects for extending our view beyond the three-generation KM model are noted in Sec. 9. Sec. 10 concludes.

2. CP VIOLATION SIGNATURES AND THEORIES

2.1 Signatures.

We list some signatures for CP violation and sources of recent information. Those with an asterisk will be discussed in this talk.

*2.1.1. $K_L \to 2\pi$. Ratios of CP violating to CP conserving decay amplitudes for both charged and neutral pions may be measured, leading to information on complex parameters η_{+-} and η_{00}, or ε and ε':

$$\frac{<\pi^+\pi^-|H_W|K_L>}{<\pi^+\pi^-|H_W|K_S>} \equiv \eta_{+-} = |\eta_{+-}|e^{i\phi_{+-}} = \varepsilon + \varepsilon' \ , \tag{2.1}$$

$$\frac{<\pi^0\pi^0|H_W|K_L>}{<\pi^0\pi^0|H_W|K_S>} \equiv \eta_{00} = |\eta_{00}|e^{i\phi_{00}} = \varepsilon - 2\varepsilon' \ . \tag{2.2}$$

Values listed in the compilation of Ref. 6 include

$$|\eta_{+-}| = (2.275 \pm 0.021) \times 10^{-3} \tag{2.3}$$

$$\phi_{+-} = (44.6 \pm 1.2)^0 \tag{2.4}$$

$$\phi_{00} = (54.5 \pm 5.3)^0 \ , \tag{2.5}$$

while recent attention[7–9] has focused on improved measurements of $|\eta_{00}/\eta_{+-}|$, leading to a world average[9] of

$$|\eta_{00}/\eta_{+-}| = 0.9955 \pm 0.0087 \ , \tag{2.6}$$

implying a very small value of ε'/ε:

$$\varepsilon'/\varepsilon = 0.0015 \pm 0.0029 \ , \tag{2.7}$$

if ε'/ε is nearly real, as expected theoretically.[2,3]

2.1.2. *Charge asymmetry in K_{l3} decay* is sensitive to Re ε and is consistent with the values (2.3) - (2.6).[6]

2.1.3. $K_S \to 3\pi$. The predominant decay amplitude for $K_S \to \pi^+\pi^-\pi^0$ is expected to be S-wave with regard to all the pions, and hence would be CP-violating. The parameter

$$\eta_{+-0} \equiv \frac{<\pi^+\pi^-\pi^0|H_W|K_S>}{<\pi^+\pi^-\pi^0|H_W|K_L>} = |\eta_{+-0}|e^{i\phi_{+-0}} \tag{2.8}$$

should be of order η_{+-} in many models of CP violation, but at present all that is known is[10]

$$|\eta_{+-0}| = 0.023 \pm 0.029 \ , \tag{2.9}$$

$$\phi_{+-0} = (202 \pm 159)^0 \ . \tag{2.10}$$

Substantial improvements in the accuracy of these measurements are expected.

2.1.4. *A neutron electric dipole moment* would signal violation of time-reversal invariance. Present experimental limits

$$d_N = \begin{cases} (-2.0 \pm 1.0) \times 10^{-25}\ e\cdot cm & (Ref.\ 11) \quad (2.11) \\ (-1.8 \pm 2.9) \times 10^{-25}\ e\cdot cm & (Ref.\ 12) \quad (2.12) \end{cases}$$

already are close to challenging models in which CP violation is due to Higgs boson exchange.[13]

2.1.5. *Transverse polarizations in decays*, i.e., nonzero expectation values of $\sigma\cdot(\mathbf{p}_1 \times \mathbf{p}_2)$, are T-violating observables, under circumstances in which final-state interactions are negligible. When these interactions are calculable, useful bounds can sometimes be placed, as in K_{l3} decays.[14]

2.1.6. *Triple products of momenta*, $(\mathbf{p}_1 \times \mathbf{p}_2)\cdot\mathbf{p}_3$, can acquire nonzero expectation values through final-state interactions or T-violation. Suggestions for employing such observables as T-violation tests in $(e^+e^-$ or $\bar{p}p) \to$ jets or in $\bar{p}p \to \bar{\Lambda}\Lambda \to \pi^+\bar{p}\pi^-p$ have been made.[15,16]

2.1.7. *Charm decays* appear to be unpromising as sources of information on CP violation.[17] We shall comment briefly on some prospects.

*2.1.8. *B decays* are expected to display mixing phenomena analogous to those in neutral K decays. We shall outline a program (see also Ref. 17) which could lead to the observation of CP violation in B decays if the KM model is the major source of the observed CP violation in kaon physics.

2.2. Theories

We shall discuss only the KM model$^{(*)}$ in any detail, but will make brief remarks on others (†) regarding the possible involvement of a 20 *TeV* mass scale in observed CP-violating phenomena.

† 2.2.1. *The superweak model*[18] of CP violation invokes a $\Delta S = 2$ interaction to mix K^0 and $\bar{K}^0$; no CP violation occurs in direct $\Delta S = 1$ decays, and the parameter ε' in $K_L \to 2\pi$ decays is expected to vanish. The neutron electric dipole moment is expected to be very small, $d_n < 10^{-30}\ e\cdot cm$, in this model.[19]

2.2.2. *Models involving Higgs couplings.* It has been proposed that the CKM matrix is real, but that complex vacuum expectation values of Higgs fields are responsible for CP violation.[20] In such models ε'/ε is expected to be $O(10^{-2})$, [19] and the neutron electric dipole moment should be $10^{-24} - 10^{-25}\ e\cdot cm$. [13,19] Ref. 21 contains a description of recent constraints on this model.

*2.2.3. *The Kobayashi-Maskawa (KM) model* envisions CP violation as arising from a complex (unitary) transformation matrix between mass eigenstates and weak eigenstates of quarks. We shall discuss in some detail its predictions for ε'/ε, which may be within the range ($\gtrsim 10^{-3}$) of experimental study. The electric dipole moment of the neutron is expected to be 10^{-31} to $10^{-32}\ e\cdot cm$ in this model.[19] The value of $\varepsilon \simeq \eta_{+-}$ may, in fact, provide the most stringent test of the model, as we shall see.

† 2.2.4. *Right-handed W's*, if they exist, probe new aspects of the quark mass matrix. In particular, CP violation can arise with only two generations of quarks.[22] Values of ε'/ε (10^{-2} to 10^{-3}) and d_n (10^{-25} to 10^{-26}) large enough to detect in the near future have been quoted.[19,23]

3. THE KM MATRIX

3.1. Notation and Conventions.

The interaction of quarks with a charged current may be written in the form

$$L_{int} = g(\overline{U}_L \gamma_\mu V D_L) W^\mu + h.c. \quad , \tag{3.1}$$

where U and D are three-component vectors of Dirac fields,

$$U = \begin{bmatrix} u \\ c \\ t \end{bmatrix} \quad ; \quad D = \begin{bmatrix} d \\ s \\ b \end{bmatrix} \quad , \tag{3.2}$$

and V is the unitary ($V^+V = 1$) KM matrix:

$$V = \begin{bmatrix} V_{ud} & V_{us} & V_{ub} \\ V_{cd} & V_{cs} & V_{cb} \\ V_{td} & V_{ts} & V_{tb} \end{bmatrix} \quad . \tag{3.3}$$

In Fig. 1 we depict typical KM elements associated with weak $\Delta Q = \pm 1$ transitions. An "O" will be used to denote a $d \to u$ or $\overline{u} \to \overline{d}$ transition, while an X will denote a $u \to d$ or $\overline{d} \to \overline{u}$ transition.

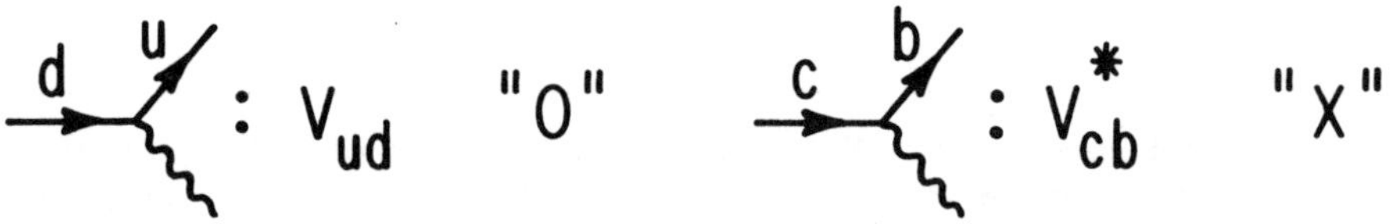

Fig. 1. Conventions for KM elements.

CP-violating observables are invariant with respect to redefinition of phases of quark fields. As a result, all such observables are composed of products of four elements in a 2×2 submatrix of the KM matrix, such that one "X" and one "O" lies in each row and each column.[24] An example is

$$\begin{bmatrix} X & O \\ O & X \end{bmatrix} = V_{ud}^* V_{us} V_{cd} V_{cs}^* \quad . \tag{3.4}$$

In Sec. 8 we shall discuss a parametrization[25] of the KM matrix (and its generalization to more than three generations) such that all elements on the diagonal and all just above the diagonal are real ("R"), so V may be written

$$V = \begin{bmatrix} R & R & ? \\ \surd & R & R \\ \surd & \surd & R \end{bmatrix} \tag{3.5}$$

The element marked "?" has a phase which must be specified; those below the diagonal (marked " √ ") are specified both in magnitude and in phase in terms of those on and above the diagonal by the unitarity condition $V^+V = 1$. We shall adopt the convention (3.5).

3.2. Magnitudes and Phases.

3.2.1. *By comparing nuclear beta-decays and the muon lifetime*, taking careful account of radiative corrections,[26] one finds

$$V_{ud} = 0.9744 \pm 0.0010 \tag{3.6}$$

In a two-generation model parametrized by a single angle θ_c,[5] one would have $V_{ud} = \cos\theta_c$.

3.2.2. *Analysis of hyperon*[27] *and K_{e3}*[28] *decays* leads to

$$V_{us} = 0.220 \pm 0.002 \quad . \tag{3.7}$$

In the two-generation model,[5] one would have $V_{us} = \sin\theta_c$.

3.2.3. *Charmed particle decays* yield the estimate[29]

$$V_{cs} = 0.95 \pm 0.14 \quad , \tag{3.8}$$

where in a two-generation model we would expect $V_{cs} = \cos\theta_c$.

3.2.4. *Charm production by neutrinos* has been employed to extract[29]

$$|V_{cd}| = 0.207 \pm 0.024 \quad ; \tag{3.9}$$

$V_{cd} = -\sin\theta_c$ in a two-generation model. Some questions have been raised[30,31] regarding the model-dependence of this estimate; an alternative method,[32] using three Cabibbo-suppressed ($c \to d$) semileptonic D decays, implies a value consistent with (3.9) but with $\approx 35\%$ errors.

3.2.5. *The b lifetime and semileptonic branching ratio* allow one to conclude[31,33] (if $|V_{bu}/V_{bc}| << 1$) that

$$V_{cb} = 0.048\,(1 \pm 0.2) \quad . \tag{3.10}$$

3.2.6. *Limits on $b \to u$ transitions* yield model-dependent conclusions on the magnitude of $|V_{bu}|$; a rough estimate (see Ref. 34 for ranges of limits) is

$$|V_{ub}/V_{cb}| \lesssim 0.2 \tag{3.11}$$

3.2.7. *The parameter ε in $K \to 2\pi$* can be explained within the KM framework only if V_{ub} has a large imaginary part, as we shall see in Sec. 5.

4. HYPERONS

4.1 Σ^- Beta Decay.

After many years of doubt, the sign of g_1/f_1 in $\Sigma^- \to n e^- \bar{\nu}_e$ now is found to agree with the predictions of the Cabibbo theory.[35–37] The small errors quoted in Eq. (3.7) stem in part from this happy situation. The new experiment[36,37] also has provided the first measurement of the form factor g_2 ("axial weak magnetism"), where the $\Sigma^- \to n$ weak current is proportional to

$$\langle n | J_\mu | \Sigma^- \rangle = \bar{u}(n)[f_1(q^2)\gamma_\mu + f_2(q^2)\sigma_{\mu\nu}\frac{q^\nu}{M_\Sigma} + g_1(q^2)\gamma_\mu\gamma_5 + g_2(q^2)\sigma_{\mu\nu}\frac{q^\nu}{M_\Sigma}\gamma_5]u(\Sigma) \quad , \tag{4.1}$$

and we have neglected terms whose effect vanishes for vanishing lepton mass. Here q is the 4-momentum transfer between Σ^- and n. It is found that[37] that

$$g_1(0) = 0.20 \pm 0.08 \quad , \tag{4.2a}$$

$$g_2(0) = -\,0.56 \pm 0.37 \quad , \tag{4.2b}$$

with large error correlations between the two values, since what is measured is essentially $g_1^{eff} = g_1 - 0.237\, g_2$. Most quark models[38] indeed predict a large value of $g_2/g_1 \approx -2$. The magnitude of g_2 affects the quality of $SU(3)$ fits, which tend to prefer a larger value of $g_1 \approx 0.28$, [36] and hence a smaller $|g_2|$. It would be interesting to see if $|g_2/g_1|$ is large in other hyperon beta-decays, such as $\Sigma^\pm \to \Lambda e^\pm \nu_e$ and $\Xi^- \to \Lambda e^- \overline{\nu}_e$.

4.2 Radiative Hyperon Decays.

Considerable experimental progress has been made in the past year or two in the study of these processes.[39,40] The CERN hyperon beam group has published the branching ratio[41]

$$B(\Lambda \to n\gamma) = (1.02 \pm 0.33) \times 10^{-3} \quad ; \tag{4.3}$$

results on $\Xi^- \to \Sigma^-\gamma$ are expected soon. An experiment at Fermilab[39,42] finds

$$B(\Xi^0 \to \Lambda\gamma) = (1.3 \pm 0.2) \times 10^{-3} \quad . \tag{4.4}$$

The processes $\Xi^- \to \Sigma^-\gamma$ and $\Omega^- \to \Xi^-\gamma$ are particularly interesting[43] in that they are sensitive to the $s \to d\gamma$ transition; since negative hyperons have no u quarks, the processes $su \to ud$ cannot contribute to their radiative decays. We shall see in Sec. 5 that it is important for interpreting kaon decays to understand the role of $s \to d\gamma$ transitions.

5. KAONS

5.1. CP Violation in $K_L \to 2\pi$.

In the absence of weak decays, the neutral K^0 and $\overline{K}^0$ are expected (if CPT is valid) to be degenerate states. When the weak interactions are turned on both K^0 and $\overline{K}^0$ can couple to two pions. This coupling can split the degeneracy of the two states. An analogy is provided by the two excitations of a drum-head illustrated in Fig. 2.[44]

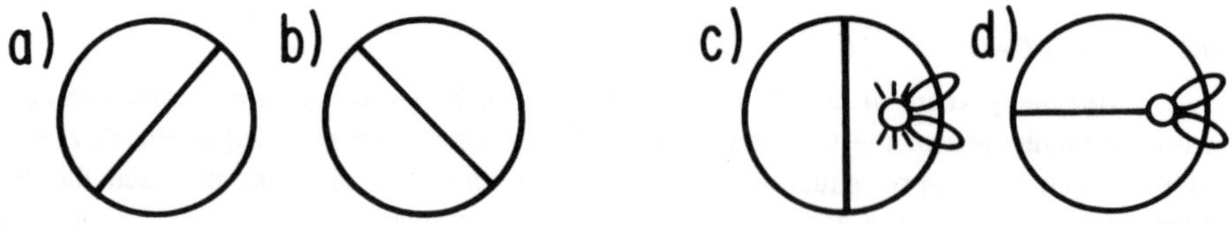

Fig. 2. Modes of a drum-head. a), b) degenerate modes; c), d): degeneracy lifted.

In Figs. 2a, 2b, any two modes whose nodes are orthogonal to one another may be used as a basis. In Figs. 2c, 2d, the degeneracy has been lifted by a fly which alights on the drum. One mode will couple strongly to the fly; the other will decouple from it. Fig. 2c is analogous to the combination ($\approx K_S$) of K^0 and $\overline{K}^0$ which couples to 2π; Fig. 2d is analogous to the combination ($\approx K_L$) which decouples from it.

The two-pion state with $J = 0$ has positive CP. If CP is a valid symmetry, the states of definite mass and lifetime are then those of definite CP:

$$K_S = (K^0 - \overline{K}^0)/\sqrt{2} \quad , \tag{5.1}$$

$$K_L = (K^0 + \overline{K}^0)/\sqrt{2} \quad . \tag{5.2}$$

If CP is violated but CPT holds, the deviation from this parametrization may be characterized[2,3] by a single complex parameter ε:

$$K_S = \frac{1}{\sqrt{2(1+|\varepsilon|^2)}} \{(1+\varepsilon)K^0 - (1-\varepsilon)\overline{K}^0\} \quad , \tag{5.3}$$

$$K_L = \frac{1}{\sqrt{2(1+|\varepsilon|^2)}} \{(1+\varepsilon)K^0 + (1-\varepsilon)\overline{K}^0\} \quad . \tag{5.4}$$

The CP-violating parameter ε is the same one which appears in $K_L \to 2\pi$ decays in Eqs. (2.1) and (2.2); mixing alone between K^0 and $\overline{K}^0$ cannot produce a difference between the charged-pion and neutral-pion amplitude ratios. Direct CP violation in the $I = 2$ decay amplitude for $K_L \to 2\pi$ is required in order that these ratios differ. Thus the parameter ε′ in Eqs. (2.1) and (2.2) is a signal of such direct CP violation.

5.2 Sources of ε and ε′.

The parameter ε, describing the mixing of K^0 and $\overline{K}^0$, receives short-distance contributions in the KM model from the graphs of Fig. 3.

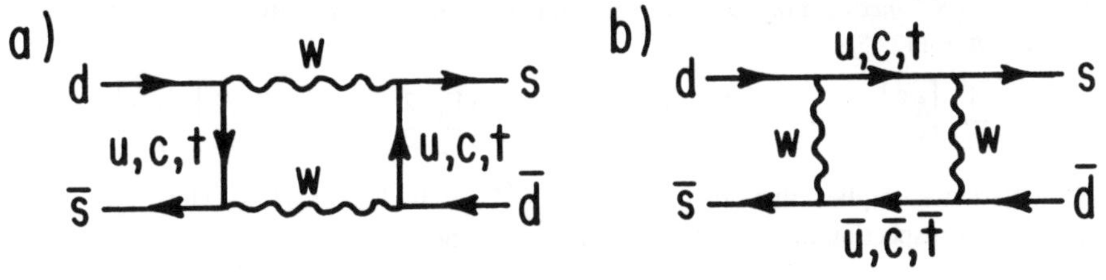

Fig. 3. Graphs describing contributions to ε in the KM model.

"Direct" decays also can contribute to ε, as long as they provide equal contributions to η_{+-} and η_{00}.

The parameter ε′, in contrast to ε, will be nonvanishing only in the presence of CP-violating direct decays. In the KM model, an important contribution to these decays[45] is the $s \to d$ transition with the mediation of one or more gluons, commonly known as the "penguin" graph (Fig. 4).

5.3. Ingredients in Estimates.

5.3.1. *Estimates of* ε involve, among other things, the following parameters: a) The use of PCAC allows one to connect the $J^P = 0^-$ $d\bar{s}$ quark state to the hadron K^0 via the constant f_K. b) A parameter B describes the value of the matrix elements $\langle\overline{K}^0|(s\bar{d})^2|K^0\rangle$ relative to that which would follow from insertion of the vacuum intermediate state in all

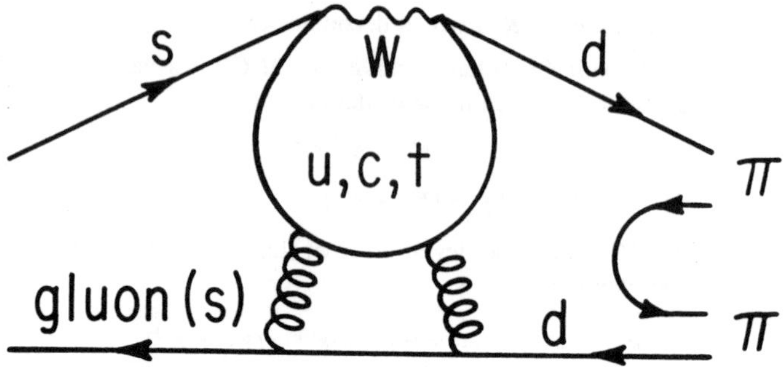

Fig. 4. The "penguin" graph contribution to the $K \rightarrow 2\pi$ decay.

possible ways. This matrix element is the major source of uncertainty coming from hadron physics in evaluating ε. Progress in its evaluation is described in Ref. 46. c) The properties of the CP-violating part of H_W at short distances may be estimated on the basis of KM parameters and QCD.

5.3.2. *Estimates of* ε′ involve all of the above quantities, as well as: d) the short-distance properties of the penguin operator, and e) matrix element of the penguin operator. It is these last which render the estimates of ε′ indeterminate to at least a factor of three.[47] We shall comment presently on a further source of a possible reduction in ε′ as a result of electromagnetic effects.[48]

5.4. Relative Phases of ε and ε′.

5.4.1. $K^0 - \overline{K}^0$ *mixing* may be described via mass and decay matrices $M = M^+$, $\Gamma = \Gamma^+$ such that ($\hbar = 1$)

$$i \frac{\partial}{\partial t} \begin{bmatrix} K^0 \\ \overline{K}^0 \end{bmatrix} = [M - i \frac{\Gamma}{2}] \begin{bmatrix} K^0 \\ \overline{K}^0 \end{bmatrix} = \begin{bmatrix} M_{11} - i\Gamma_{11}/2 & M_{12} - i\Gamma_{12}/2 \\ M_{21} - i\Gamma_{21}/2 & M_{22} - i\Gamma_{22}/2 \end{bmatrix} \begin{bmatrix} K^0 \\ \overline{K}^0 \end{bmatrix} . \quad (5.5)$$

CPT invariance implies that $M_{11} = M_{22}$, $\Gamma_{11} = \Gamma_{22}$. If $|\text{Im}\, M_{12}| >> |\text{Im}\, \Gamma_{12}|$, as one expects from inspection of unitarity relations,[49] one finds

$$arg\ \varepsilon \approx \tan^{-1} (\frac{-2\Delta M}{\Delta\Gamma}) \simeq 45^0 \ , \quad (5.6)$$

where[6]

$$\Delta M = m_L - m_S = 0.477\ \Gamma_S \ , \quad (5.7)$$

$$\Delta\Gamma = \Gamma_L - \Gamma_S = -\ 0.998\ \Gamma_S \ . \quad (5.8)$$

5.4.2 *Decay amplitudes* may be labeled by the isospin $I = 0, 2$ of the final dipion state. They will contain an explicit final state phase δ_I, and an intrinsic weak part A_I. The phases of A_I from various contributions are shown in Fig. 5.

The net $I = 0$ and $I = 2$ amplitudes arising from the sources shown in Fig. 5 are summarized in Fig. 6. There is a relative phase between the $I = 0$ and $I = 2$ amplitudes. This relative phase is a signal of direct CP violation. In a convention[50] in which A_0 is real, one finds

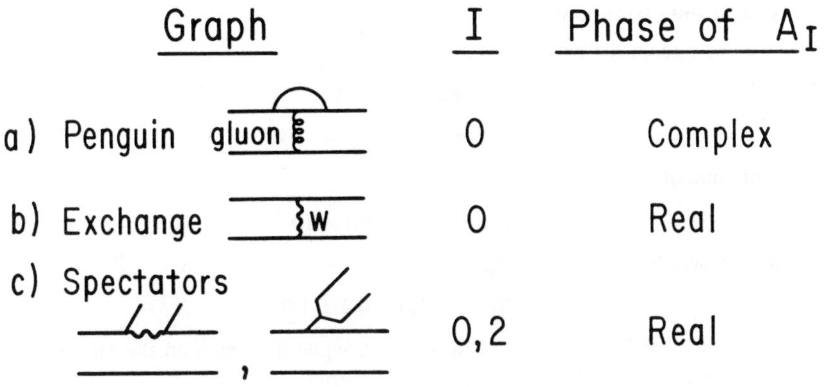

Fig. 5. Graphs contributing to $K_L \to 2\pi$ decay amplitudes $A_I e^{i\delta_I}$ and corresponding phases of A_I in the 6-quark KM model.

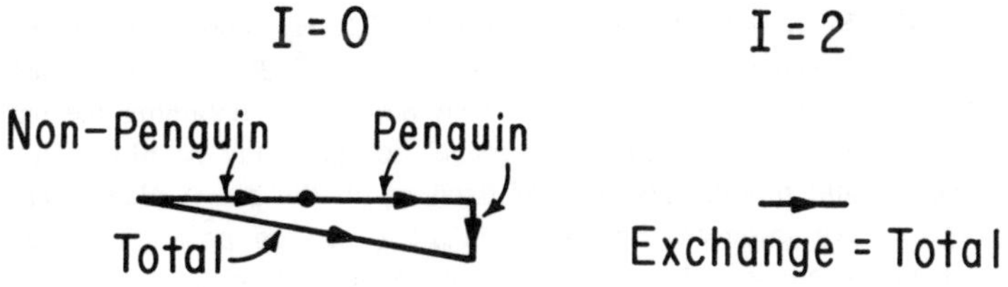

Fig. 6. Contributions to $I = 0$ and $I = 2$ amplitudes A_I in $K_L \to 2\pi$.

$$\varepsilon' = \frac{i}{\sqrt{2}} \frac{\text{Im}\, A_2}{A_0} e^{i(\delta_2 - \delta_0)} \quad . \tag{5.9}$$

Since $\delta_2 - \delta_0 \approx - \pi/4$, one concludes that *arg* ε' must *also* be around 45^0, as is *arg* ε.[2] The fact that ε'/ε is expected to be nearly real means that measurements[7–9] of $|\eta_{00}/\eta_{+-}|$ should be very sensitive to ε'/ε. However, this conclusion rests on the assumption of CPT invariance, which we shall relax presently.

One of the electromagnetic effects proposed in Ref. 48 includes the exchange of the photon as well as a gluon in Fig. 5a. As a result, the (complex) penguin amplitude now can contribute to $I = 2$ as well as $I = 0$ decays. [Without the photon, the $s \to d$ transition is of course purely $\Delta I = 1/2$.] The sign of the imaginary part of this contribution to A_2 has been found to be *opposite to* that of the imaginary part of the gluonic penguin contribution to a_0. The phase difference between A_2 and A_0 thus is *increased*, leading to an *increase* in the predicted value of ε'/ε. A compensating *decrease* arises from isospin impurities due to u–d quark mass differences.[48] The net effect is to introduce a 25% uncertainty in ε'/ε which could go in either direction.

5.5. Constraints from ε and ε′.

We shall adopt the parametrization[51]

$$V_{cb} = 0.048A \quad , \tag{5.10}$$

$$V_{ub} = 0.011A\ (\rho - i\eta)\ . \tag{5.11}$$

The b lifetime of[52]

$$\tau_b = (1.16 \pm 0.16) \times 10^{-12}\ s \tag{5.12}$$

and average semileptonic branching ratio of

$$B_{SL} = (11.0 \pm 0.2 \pm 1.0)\% \tag{5.13}$$

imply[53] $A = 1 \pm 0.2$, as mentioned in Sec. 3.2, while the limits on the ratio of $b \to u$ to $b \to c$ transitions, implying $|V_{ub}/V_{cb}| \lesssim 0.2$, restrict

$$\rho^2 + \eta^2 \lesssim 3/4\ . \tag{5.14}$$

5.5.1. *The box diagram* of Fig. 3 contributes to ε in the form[54]

$$|\varepsilon| = 3.1\times10^{-3}\, A^2\, \eta\, B\,[1 + \frac{1}{2}\, A^2(1-\rho)(\frac{m_t}{42\ GeV})^2]\ , \tag{5.15}$$

for $m_t \simeq 42\ GeV$. With present values of $A = 1 \pm 0.2$, m_t (none), η (the weak upper bound (5.14)), and B (taken as < 0.66, given estimates[55,56] $B \simeq 1/3$ which are probably uncertain[57] to at least a factor of 2), there is no problem in obtaining the experimental value of $|\varepsilon| \simeq 2.27\times10^{-3}$. The "most permissive" solution may be illustrated by letting $A = 1.2$, $B = 0.66$, and plotting contours of fixed m_t in the $\rho - \eta$ plane. The result is shown in Fig. 7. An improved bound on $|V_{ub}|$ (i.e., on $\rho^2 + \eta^2$) could require a (moderately) heavy t quark. These bounds pose no problem yet, since no t quark evidence exists.[58] We desperately need measurements of m_t and $|V_{ub}|$!

5.5.2. *The penguin diagram* implies a value of $|\varepsilon'/\varepsilon|$ now estimated[54] to be

$$\varepsilon'/\varepsilon = (3\ to\ 10)\ A^2\, \eta \times 10^{-3}\ . \tag{5.16}$$

The experimental value (2.7) of ε′/ε, which we may quote approximately as

$$\varepsilon'/\varepsilon \lesssim 0.005\ (1\sigma)\ , \tag{5.17}$$

only implies $\eta \lesssim 1.16$ for $A = 1.2$ on the basis of Eq. (5.16), which is not a severe constraint. Values of $\varepsilon'/\varepsilon < 0.003$, within the grasp of present experiments,[9,59] could add valuable new information in this context.

5.6. CPT Tests.

The possibility of a difference between ϕ_{00} [Eq. (2.5)] and ϕ_{+-} [Eq. (2.4)] exists at the 2σ level:

$$\phi_{00} - \phi_{+-} = 9.9 \pm 5.4^0 \tag{5.18}$$

Since $\eta_{+-} - \eta_{00} = 3\varepsilon'$, if $|\eta_{00}/\eta_{+-}| \simeq 1$, the phase difference (5.18) would imply

$$\varepsilon'/\varepsilon = (5.7 \pm 3.1) \times 10^{-2}\, e^{-i\pi/2}\ , \tag{5.19}$$

a relative phase incompatible with CPT invariance.[60,61] Several experimental tests are immediately suggested by this possibility.

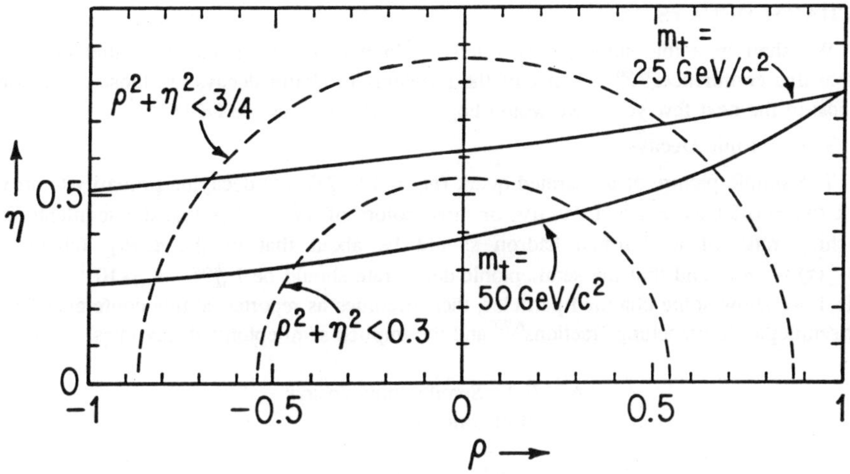

Fig. 7. Contours of fixed top-quark mass m_t (solid curves) implied by Eq. (5.15) for $A = 1.2$, $B = 0.66$. The dashed semicircles denote the bounds $\rho^2 + \eta^2 \leq 3/4$ [Eq. (5.14)] and an improved (hypothetical) bound $\rho^2 + \eta^2 \leq 0.3$.

5.6.1. *An improved measurement of* $\phi_{00} - \phi_{+-}$ is needed, and is planned.[62] Experiments[63] to measure this difference are very challenging.

5.6.2. *The rate asymmetry in* $K^\pm \to \pi^\pm\pi^0$ is expected to be[61]

$$\frac{\Gamma(K^+ \to \pi^+\pi^0) - \Gamma(K^- \to \pi^-\pi^0)}{\Gamma(K^+ \to \pi^+\pi^0) + \Gamma(K^- \to \pi^-\pi^0)} = (9.5 \pm 5.5) \times 10^{-3} \tag{5.20}$$

taking the phase difference (5.18) seriously; present data give $(4 \pm 6) \times 10^{-3}$ for this asymmetry.

5.6.3. *The ratio* $(\tau^+ - \tau^-)/\tau^+$, where $\tau^\pm \equiv \tau(K^\pm)$, is expected to be[61] $(4 \pm 2.3) \times 10^{-3}$ on the basis of (5.18); present data imply $(1.1 \pm 0.9) \times 10^{-3}$.

5.6.4. *Comparisons of* $K^0 \to l^\pm\nu\,\pi^\mp$ *and* $\bar{K}^0 \to l^\mp\nu\pi^\pm$ as a function of decay times can yield much useful information on CPT invariance.[60]

5.6.5. *A direct measurement of* ε' is possible in the reaction[64,65]

$$e^+e^- \to \phi \to K_S(\hat{z})K_L(-\hat{z}) - K_L(\hat{z})K_S(-\hat{z}) \ . \tag{5.21}$$

The final state is expected to be antisymmetric in K_S and K_L, even if CP or CPT is violated, as long as C is conserved in the strong and electromagnetic interactions. Suppose the kaon moving along $+\hat{z}$ decays to $\pi^+\pi^-$ at proper time t_1, while that moving along $-\hat{z}$ decays to $\pi^0\pi^0$ at proper time t_2. By studying asymmetries in $\Delta t = t_1 - t_2$, one acquires a direct measurement of ε'; no such asymmetries would be present if ε' vanished. It appears one needs at least 10^8 ϕ's to detect a magnitude of ε'/ε corresponding to the CPT-violating value (5.19), or at least 10^9 ϕ's to have an impact on measurements of a real ε'/ε down to the 10^{-3} level.

6. CHARM DECAYS

We shall treat this subject only briefly. Much more detail may be found in other talks at this conference.[34,66,67] Many of the problems in charm decays are those of hadron physics; in the next few years, we would hope for help from lattice QCD.[46,68]

6.1. Semileptonic Decays.

The simple picture of a charmed quark ($m_c = 1.6\ GeV/c^2$) decaying primarily to an s quark ($m_s = 0.5\ GeV/c^2$) and $e\nu$, $\mu\nu$, or three colors of $u\bar{d}$ implies that the semileptonic branching ratio of a charmed hadron should be about that of the τ, B_{SL} (charm) $\approx B_{SL}(\tau) \approx 18\%$, and that the semileptonic decay rate should be $\Gamma_{SL}^{theor} \approx 1.8\times10^{11}\ s^{-1}$. In Table 1 we show some charmed particles, their lifetimes as reported at this conference,[66,69] their semileptonic branching fractions[6,70] and the implied semileptonic decay rates.

TABLE 1. Semileptonic decays of charmed hadrons

Charmed particle	Lifetime ($10^{-13}s$)	B_{SL} (%)	Γ_{SL} ($10^{11}s^{-1}$)
$D^0(c\bar{u})$	$4.42^{+0.15}_{-0.14}$	7.5 ± 1.2	1.70 ± 0.28
$D^+(c\bar{d})$	$10.29^{+0.44}_{-0.40}$	17.0 ± 2.0	1.65 ± 0.21
$\Lambda_c(cud)$	$1.9^{+0.4}_{-0.3}$	4.5 ± 1.7	2.4 ± 1.0

All the semileptonic decay rates are compatible with the simple free-quark decay picture. It is in the nonleptonic decays that important differences occur.

6.2. Effects in Nonleptonic Charm Decays.

The nonleptonic decays of charmed quarks in a hadron are affected by what other quarks that hadron contains. A review of these effects may be found in Refs. 66, 67 and 71. Other recent treatments include those of Refs. 72-74.

6.2.1. *An overall QCD enhancement* of the operator for $c \rightarrow su\bar{d}$ which is flavor-antisymmetric in s and u is expected.[75] This effect raises all nonleptonic decay rates, lowering the predicted value of B_{SL} for charmed particles to $\approx 15\%$.[71]

6.2.2. *Quark interference* via the Pauli principle in the process $c\bar{d} \rightarrow su\bar{d}\bar{d}$ is expected to decrease the rate for the two-body process $D^+ \rightarrow \bar{K}^0\pi^+$.[76] If this result can be generalized to other D^+ decays (such as to $\bar{K}^{*0}\pi^+$, $\bar{K}^0\rho^+$, and genuine $\geq$ 3-body final states), one estimates[71] $B_{SL}(D^+) =$ 19%. However, the role of this interference effect has not been demonstrated for multi-body D^+ decays.

6.2.3. *Dropping* $1/N_c$ *terms* in an expansion in the number N_c of quark colors[72] lowers $B_{SL}(D^0)$ to 12%.[71] Since N_c is three and not infinity, this prescription remains at present of phenomenological rather than fundamental nature. We look to theories of hadrons (perhaps lattice QCD) for its justification.

6.2.4. *W exchange processes* are easily capable of lowering the Λ_c^+ lifetime to the observed value[77] via the subprocess $cd \rightarrow su$. This subprocess also affects the lifetime of $\Xi_c^0 = csd$ but not of $\Xi_c^+ = csu$, which may explain why the Ξ_c^+ and not the Ξ_c^0 is seen in experiments[78] selecting for a finite track length of the decaying baryon. The role of W exchange in D^0 decay (via the subprocess $c\bar{u} \rightarrow s\bar{d}$) is less clear; it requires the $c\bar{u}$ not to be in a $J^P = 0^-$ state, in order to avoid the familiar helicity suppression associated with the V-A charged-current interaction. Other values of J^P for $c\bar{u}$ could of course be present if gluons play an important role in the D^0 wave function. Estimates involving single-gluon emission[79] are not encouraging, leading to $B_{SL}(D^0) > 10\%$.[71] But why should single-gluon emission be the whole story?

6.2.5. *Final state interactions* undoubtedly play a role in charmed particle decays.[67,74,80–82] The $I = 1/2$ and $I = 3/2$ $\bar{K}\pi$ amplitudes in $D^{0,+} \rightarrow \bar{K}\pi$ cannot be relatively real.[81] Indeed, one expects this from Regge pole theory; the $I = 1/2$ $\bar{K}\pi$ elastic channel is non-exotic, so the part due to ρ and f_0 exchange should be imaginary, while the $I = 3/2$ $\bar{K}\pi$ channel is exotic, so its contribution from ρ and f_0 exchange should be real.[83] Both amplitudes also have a contribution from Pomeron exchange, which one expects to be imaginary.

6.2.6. *The role of multi-body final states* affects many recent theoretical conclusions[73] about charmed particle decays. Recent results of Fermilab Experiment No. E-691,[66,84] such as the observation of the modes $D_s^+ \rightarrow K\bar{K}\pi$ (nonresonant) and $\phi 3\pi$, indicate (at least to the present author) the importance of such final states. The importance of such states will also hinge on whether, as has been suggested,[34,85] the D^0 and D^+ branching ratios quoted by Mark III[70] are high by some 20%.

6.3. $D^0 - \bar{D}^0$ Mixing.

Conventional estimates of $D^0 - \bar{D}^0$ mixing are very small, corresponding to $\Delta m/\Gamma = 10^{-3}$ to 10^{-1}.[54] The lower value is thought to be more likely.[86,87] The only (feeble) evidence for this mixing at present comes from the reaction[70]

$$e^+e^- \rightarrow \psi(3770) \rightarrow D^0\bar{D}^0 \rightarrow K^\pm K^\pm + \ldots, \tag{6.1}$$

in which the Mark III collaboration detects three same sign pairs, to be compared with 162 opposite-sign pairs. Thus

$$2r \equiv \frac{\text{rate for same sign pairs}}{\text{rate for opposite sign pairs}} \tag{6.2}$$

is quoted to be about two percent, with an error of ± 60%. In a study of same-sign dimuons coming from $\pi \rightarrow$ charm, Experiment E-615 at Fermilab[88] quotes $r \lesssim 1/2\%$. It is estimated[70] that particle misidentification contributes only 0.4 ± 0.2 event to the signal. We would expect the ratio for double-Cabibbo-suppressed decays $D^0 \rightarrow K^+ + \ldots$, governed by the quark transition $c \rightarrow du\bar{s}$, to the Cabibbo-favored ones (mediated by $c \rightarrow su\bar{d}$) to be about

$$2r = \frac{2\cos^4\theta_c \sin^4\theta_c}{\cos^8\theta_c} \approx 0.5\% \tag{6.3}$$

in the process (6.1). Adding this to the background from particle misidentification, it is not clear there is evidence for mixing. Bose statistics imply that for *identical* final states (e.g., $K^-\pi^+$) r must vanish unless mixing is present.[86,89] So far, the observed final states in the Mark III experiment are not identical. The experiment to watch in this context is Fermilab

E-691,[66,84] which will be able to separate double-Cabibbo-suppressed decays and mixing effects from one another by their different dependence on decay times. [Note added: preliminary results from this experiment now appear to favor the interpretation of wrong-sign events as due to double-Cabibbo-suppressed decays.[90]]

7. CP VIOLATION IN B SYSTEMS

The greatest hope for observing CP violation outside the kaon system is that involving B mesons, at least within the KM framework. Many authors have studied these possibilities in detail;[3,91–94] our purpose is to give an overview of the effort needed to observe CP violation in B systems.

7.1. An Experimental Program.

B decay channels associated with large rates generally are characterized by small CP-violating asymmetries, while the largest asymmetries usually are expected for the rarest decays. We outline a program which takes account of this limitation.

7.1.1. *See B's.* A good tag for the inclusive production of B's is needed. One suggestion[95] for hadronic reactions has been the use of the inclusive decay $B \to \psi + X$, whose branching ratio is some 1/2%. One then expects

$$\sigma(pp \to B + ... \to \psi + ... \to l^+l^-) \sim 1\ pb \tag{7.1}$$

for 800 GeV protons on a fixed target. A first priority is to measure this cross section. In e^+e^- interactions above $B\bar{B}$ threshold, certain energies may be most suitable for specific channels, such as $B_u\bar{B}_u$ or $B_d\bar{B}_d$ production (probably the $\Upsilon[4S]$), $B\bar{B}^*$ production, or $B_s\bar{B}_s$ production (possibly the $\Upsilon[5S]$ at $E_{cm} = 10.86\ GeV$).[97] Scanning at CESR above $B\bar{B}$ threshold will be required to learn these "magic energies". The optimum energies for $F\bar{F}$ (actually, $F\bar{F}^*$) production in e^+e^--annihilations are just now being mapped out.[66,70]

7.1.2. *Measure branching ratios.* Low-multiplicity final states (such as $D\pi$, ψK, . . .) are expected to occur with relative rates of about 10^{-3}.

7.1.3. *Measure the relative production of B_u, B_d, and B_s.* An educated guess based on experience with the fragmentation of lighter quarks is

$$B_u : B_d : B_s = 2 : 2 : 1\ . \tag{7.2}$$

These fractions may be very sensitive to the production process. The estimate (7.2), if it applies at all, should be valid only far above $b\bar{b}$ threshold.

7.1.4. *See $B_d - \bar{B}_d$ and $B_s - \bar{B}_s$ mixing.* Conservative estimates are $\Delta m/\Gamma \sim 0.1$ for $B_d - \bar{B}_d$, O(1) for $B_s - \bar{B}_s$. Somewhat more optimistic estimates are quoted in Ref. 54 and will be discussed below.

7.1.5. *See the $b \to u$ transition and the t quark.* As discussed in Sec. 5, both pieces of information are needed to verify that the KM framework is the major source of CP violation for kaons.

7.1.6. *See CP violation in the B system.* As we shall see below, one will need at least 10^7 B's, luck, or both.

7.1.7. *Measure time-dependent CP violation.* If one has been fortunate enough to see CP violation at all in B systems, time-dependent effects can be quite large.[94] For example, asymmetries can be enhanced by suitable cuts on proper decay times.

7.2. B − $\overline{\text{B}}$ Mixing.

A recent proposal[98] for high-statistics studies of B's contains a useful summary of expectations for $B_d - \overline{B}_d$ and $B_s - \overline{B}_s$ mixing. These results can be reduced to the predictions

$$\frac{\Delta m_{B_d}}{\Gamma_B} = 0.035\,\left(\frac{m_t}{40\ GeV/c^2}\right)^2 [(1-\rho)^2 + \eta^2]\ , \tag{7.3}$$

$$\frac{\Delta m_{B_s}}{\Gamma_B} = 1.5\,\left(\frac{m_t}{40\ GeV/c^2}\right)^2\ , \tag{7.4}$$

where the dominant diagrams contributing to the mixing involve t quark loops, and are shown in Fig. 8.

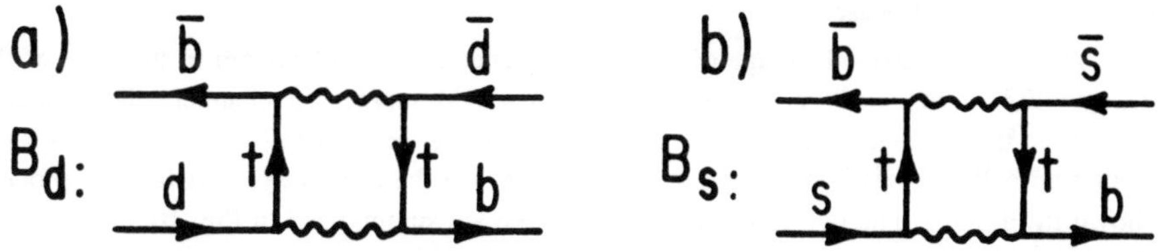

Fig. 8. Graphs giving dominant contributions to (a) $B_d - \overline{B}_d$, (b) $B_s - \overline{B}_s$ mixing.

To show the assumptions implicit in these relations, we should multiply Eq. (7.3) by the factor

$$(f_{B_d}/100\ MeV)^2 \times \{|V_{td}|^2/[(0.011)^2\{(1-\rho)^2 + \eta^2\}]\}B_{B_d}(\tau_B/1.16 \times 10^{-12}s),$$

and (7.4) by

$$(f_{B_s}/150\ MeV)^2\ |V_{ts}/0.048|^2 B_{B_s}\ (\tau_B/1.16 \times 10^{-12}s).$$

Here f_{B_d} and f_{B_s} are pseudoscalar meson decay constants; B_{B_d} and B_{B_s} express the ratio of matrix elements to vacuum saturation values in the manner of the parameter B in Sec. 5, and ρ and η are parameters noted in Eq. (5.11) for describing V_{ub}. Eqs. (7.3) and (7.4) imply that:

7.2.1. *The UA1 collaboration should be able to observe $B_s - \overline{B}_s$ mixing,*[99,100] even if the t quark is too heavy for it to detect.[58]

7.2.2. *Smaller values of* $\rho^2 + \eta^2$ *imply larger* $(1-\rho)^2 + \eta^2$. The $b \to u$ and $t \to d$ transitions cannot both vanish within a three-generation model, given what we know about KM matrix elements.

7.2.3. *In* $e^+e^- \to B^0\overline{B}^0$, where the primary lepton from B^0 is usually l^+ and from $\overline{B}^0$ is usually l^-, one will sometimes see primary l^-l^- or l^+l^+ pairs. (By primary we mean the leptons coming from $\overline{b} \to \overline{c}l^+\nu_l$ or $b \to cl^-\overline{\nu}_l$; energy cuts are needed to distinguish such leptons from those arising in charm decays). The ratio of same-sign to opposite-sign primary leptons will be (for very small $\Delta\Gamma$) [89]

$$\frac{N(l^-l^-) + N(l^+l^+)}{N(l^+l^-)} = \frac{(\Delta m/\Gamma)^2}{2 + (\Delta m/\Gamma)^2}\ , \tag{7.5}$$

which approaches 1 for large $\Delta m/\Gamma$.

7.3 . CP Asymmetries for B's.

In order to see a CP asymmetry between the rates for B to decay to a final state f and $\bar{B}$ to decay to $\bar{f}$, one needs a decay amplitude of the form[17]

$$<f|B> = A_1 e^{i\delta_1} + A_2 e^{i\delta_2} , \quad (7.6)$$

where A_1 and A_2 are intrinsic decay amplitudes (as in Sec. 5.4.2) and δ_1, δ_2 are final-state phase shifts. Then

$$<\bar{f}|\bar{B}> = A_1^* e^{i\delta_1} + A_2^* e^{i\delta_2} , \quad (7.7)$$

$$|<\bar{f}|\bar{B}>|^2 - |<f|B>|^2 \sim \mathrm{Im}(A_1 A_2^*)\sin(\delta_1 - \delta_2) . \quad (7.8)$$

Observation of an asymmetry requires both a relative phase in weak amplitudes and different final state phases.

One way to take advantage of this circumstance may be illustrated in the case of the decay[91] $B_d(\equiv \bar{b}d) \to D^+\pi^-$. Note that this is a KM-suppressed channel. This process occurs via the graph of Fig. 9a. Another route for B_d to reach the $D^+\pi^-$ final state is for it to mix with $\bar{B}_d(\equiv b\bar{d})$, which then decays to $D^+\pi^-$ via the (relatively) KM-favored process shown in Fig. 9b. The two routes to the final state are summarized in Fig. 10.[94]

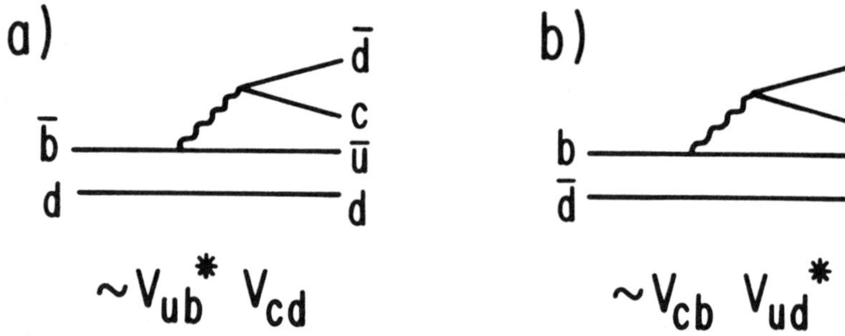

Fig. 9. Graphs denoting decays of (a) B_d and (b) $\bar{B}_d$ to $D^+\pi^-$.

7.4. Required statistics for CP asymmetries.

Suppose N_+ B's are seen to decay to a final state f, while N_- $\bar{B}$'s decay to $\bar{f}$. Define the total number of B's $N \equiv N_+ + N_-$ and the asymmetry $A \equiv (N_+ - N_-)/(N_+ + N_-)$. Then in order to see a nonzero value of A at the n-standard-deviation level, it may be shown that one needs

$$N \geq n^2(1 - A^2)/A^2 \quad (7.9)$$

events. For a process with branching ratio B_f (averaged over $B \to f$ and $\bar{B} \to \bar{f}$) we then require a total number of B's

$$N_{tot} \geq n^2(1 - A^2)/B_f A^2 . \quad (7.10)$$

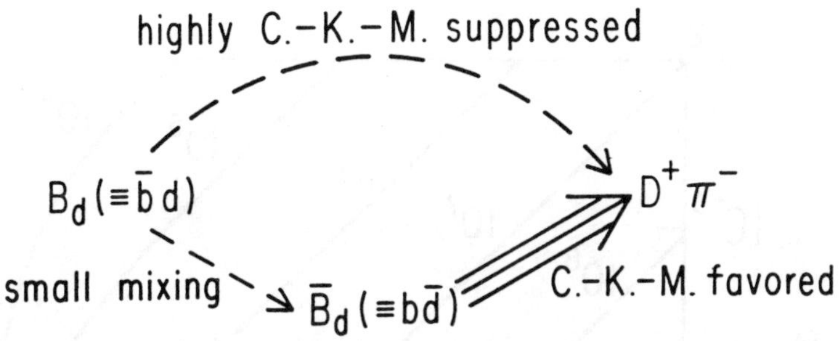

Fig. 10. Two interfering amplitudes for $B_d \to D^+\pi^-$ leading to differences between the rate for $B_d(t{=}0) \to D^+\pi^-$ and that for $\bar{B}_d(t{=}0) \to D^-\pi^+$.

In Fig. 11 we plot contours of the relation (7.10) for $n = 3$, i.e., for a three-standard-deviation asymmetry. We also indicate expected asymmetries and branching ratios for some processes.[94] The branching ratios contain estimated detection efficiencies. One sees that at least 10^6 B_s's or 10^7 B_d's are needed to study CP-violating asymmetries in any process. Some decay channels, such as B_s, $B_d \to D^+\pi^-$ and $B_d \to \pi^+\pi^-$, K^+K^- are more favorable than others for displaying the expected asymmetries.

It will not be easy to observe CP violation in B systems, if our guesses based on the KM model are correct. However, lest one be daunted by the prospect of detecting 10^8 B's, recall that kaon physics also began with the examination of individual events. The best one can do is to make a start. The program appears feasible both in e^+e^- [98] and in hadron[95,101] machines.

8. MIXING AND MASS MATRICES

In this section we give examples of efforts to parametrize the KM matrix (and its generalizations to more than three generations), and to understand why its elements have their particular values.

8.1. Parametrizations of the KM Matrix.

For n generations of quarks, the unitary $n \times n$ KM matrix V may be expressed as $V = \exp(iH)$, where H is Hermitian. It is, perhaps, easier to see that H can be expressed in terms of n^2 real parameters. Consequently, there are n^2 independent real parameters in V.

Because there are $2n$ quarks, there are $2n-1$ relative phases in V that do not have physical consequences. There are $n(n-1)/2$ angles for an $n \times n$ rotation matrix. Consequently, the number n_ϕ of physically significant phases in V is $n_\phi = n^2 - (2n-1) - n(n-1)/2$, or

$$n_\phi = (n-1)(n-2)/2 \ . \qquad (8.1)$$

For $n = 2$, there is one angle (the Cabibbo angle θ_c) and no phase. For $n = (3,4)$ there are (3,6) angles and (1,3) phases.

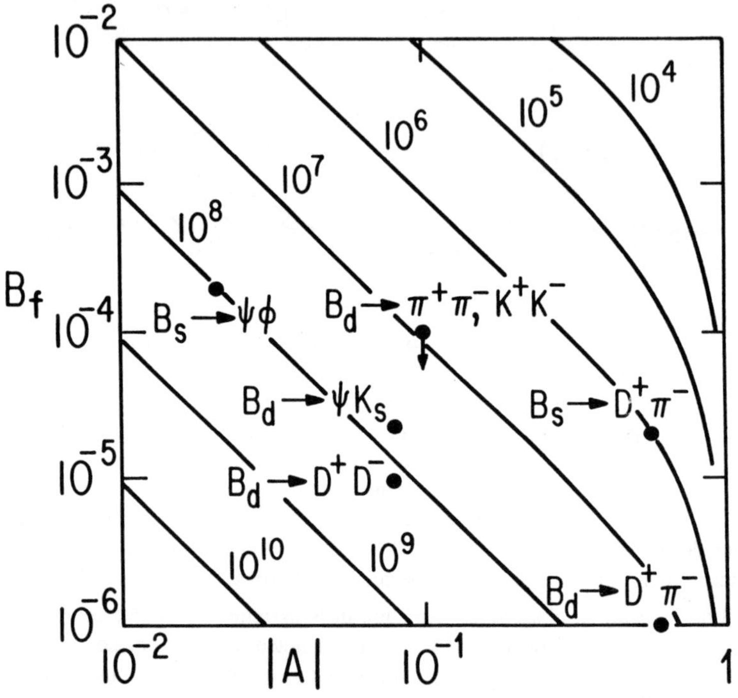

Fig. 11. Contours of total numbers of B's as function of asymmetry A and branching ratio B_f required to observe a 3σ asymmetry [$n = 3$ in Eq. (7.10)]. Branching ratios include assumed detection efficiencies.

While many parametrizations[102] of V have appeared in the literature, we cite just one[25] as most easily recalled. We use the freedom of choosing $2n - 1 = n + (n - 1)$ relative quark phases to make the n diagonal elements and $n - 1$ additional elements of the KM matrix real. The obvious $n - 1$ additional elements are those adjacent to (either above or below) the diagonal. Thus, we take V of the form

$$V = \begin{bmatrix} R & R & & \\ & R & R & \\ & & R & R \\ & & & R \end{bmatrix} \tag{8.2}$$

where "R" stands for "real". The remaining $(n - 1)(n - 2)/2$ elements above the entries marked "R" are assigned the $(n - 1)(n - 2)/2$ independent phases. The phases and magnitudes of the elements below the diagonal are given in terms of the others by unitarity. The imaginary parts of "plaquettes" (products of entries in 2×2 submatrices of the form (3.4)) with no elements below the diagonal constitute independent CP-violating observables. For the 3×3 matrix, there is only one such independent observable. In the 3×3 case, this observable is proportional to the determinant of the commutator of the U and D quark mass matrices.[103] The imaginary parts of plaquettes below the diagonal may be related to those of plaquettes and values of $|V|$ above the diagonal, and so one can express all CP-violating observables in terms of the n_ϕ phases and values of $|V|$. In the three-generation KM case, the fact that V_{ub} appears to have a large imaginary part (required for the observed $|\varepsilon|$ unless the t quark is very massive) suggests that in some sense CP violation is "maximal".[104] Opinions do differ,[102,103] however, on the question of "maximal" CP violation; a conclusive answer would require that we understand the origin of both the magnitudes and the phases of the KM elements.

8.2. A Guess for KM Elements.

If one can predict the (nondiagonal) quark mass matrices, one knows the KM elements, since these arise as a result of the diagonalization procedure. Many guesses have been made for the form of mass matrices;[105–107] we present only one example. If these are of the form[107]

$$M^U = \begin{bmatrix} 0 & i\sqrt{UC} & 0 \\ i\sqrt{UC} & 0 & \sqrt{CT} \\ 0 & \sqrt{CT} & T \end{bmatrix} ; \; M^D = \begin{bmatrix} 0 & \sqrt{DS} & 0 \\ \sqrt{DS} & 0 & \sqrt{SB} \\ 0 & \sqrt{SB} & B \end{bmatrix} , \tag{8.3}$$

where we use the shorthand $U \equiv m_u$, etc., one finds

$$V_{us} = \sqrt{(D/S) + (U/C)} \simeq 0.23 \quad , \tag{8.4}$$

$$V_{cb} = \sqrt{S/B} - \sqrt{C/T} \quad , \tag{8.5}$$

$$\frac{|V_{ub}|}{V_{cb}} = \sqrt{\frac{U}{C} + \frac{(D/S)(S/B)^3}{V_{cb}^2}} \quad . \tag{8.6}$$

Using $S/D = 19.6 \pm 1.6$, $C/U = 264.7^{+124.2}_{-67.7}$, $B/S = 30.3^{+10.1}_{-5.2}$, and $C(1\ GeV) = 1.35 \pm 0.05\ GeV$, one obtains $|V_{ub}|/V_{cb} = 0.067^{+0.017}_{-0.015}$, or a ratio $\Gamma(b \to ue\nu)/\Gamma(b \to ce\nu)$ which is about 1/10 the present upper bound.[34] The range $|V_{cb}| = 0.04 - 0.06$ corresponds via (8.5) to a prediction $m_t(1\ GeV) = 50 - 150\ GeV$.

9. BEYOND THE KM MODEL

9.1 A Fourth Generation?

9.1.1. *A very heavy t'* could be seen in its decay to W+...[108] Detailed studies of this signature are in progress.[109] Indeed, if the t is heavy enough, *it* could decay to $W + b$.

9.1.2. *A possible b' at 23 GeV* has been suggested[110] on the basis of $e^+e^- \to \mu^+$+... events with low thrust, seen only at energies above 44 GeV.[111,112] Such an interpretation will be easily confirmed or rejected at TRISTAN. We show in Fig. 12 some quarkonium levels expected for a 23 GeV b' quark in a QCD-inspired potential.[113] The present PETRA energy scan is not sensitive enough to exclude this level structure.[114]

9.2. Multi-TeV Scales

9.2.1. *Right-handed W's,* when exchanged in loop diagrams like Fig. 3, can lead to CP violation in a 2-generation model.[22,23] The relevant mass scale for understanding $\varepsilon = O(M_{W_L}^2/M_{W_R}^2)$ on this basis is probably $M(W_R) \geq 10\ TeV$,[115] since W_R's are much more effeetive than W_L's in generating contributions to ε.[116]

9.2.2. *New degrees of freedom* in quantum mechanics have been proposed[117] as a source of an effective superweak CP violation, arising to second order in perturbation theory. Just

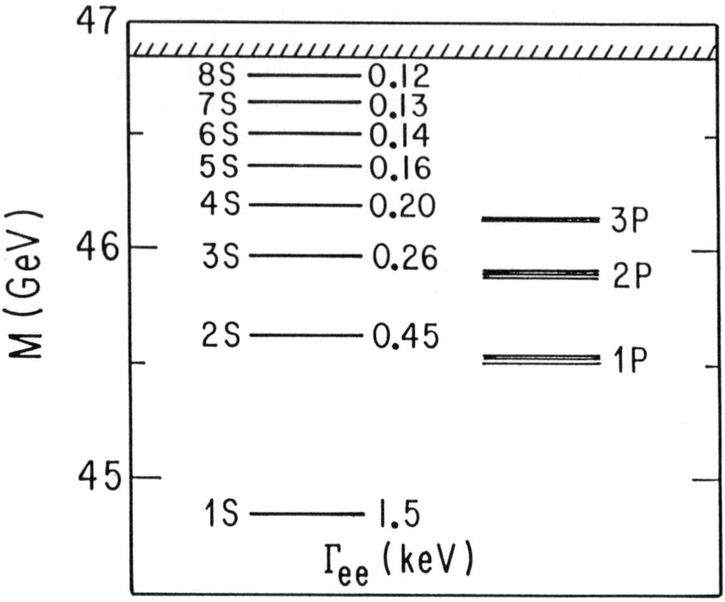

Fig. 12. Some quarkonium levels expected for a 23 *GeV* b' quark. The horizontal line at the top denotes flavor threshold.

as for right-handed W's, the relevant mass scale for new physics is around 20 *TeV*.

9.3 Neutral Leptons: Windows on $\geq 10^9$ *GeV*.

Neutrinos are much lighter than other fermions. If the electron neutrino has any mass at all, it is probably less than a few tens of *eV*. One mechanism for generating a small neutrino mass m_D is via Dirac (m_D) and large Majorana (M) terms, leading to[118] $m_u \simeq m^2/M$. If $m_\nu \lesssim 1$ *eV* and $m_D \sim 1$ *GeV*, then $M \geq 10^9$ *GeV*.

9.3.1. *Neutrino oscillations* are one possible explanation of anomalous electrons observed at CERN[119] and Brookhaven.[120]

9.3.2. *Spontaneously broken lepton number*[121] can lead to Nambu-Goldstone bosons coupling to neutrinos, detectable in principle in studies of double beta decay.[122]

10. CONCLUSIONS

The three-generation Cabibbo-Kobayashi-Maskawa framework for quark mixing passes all experimental tests, for the present. Its utility for describing CP violation in kaons depends on a) finding the t quark, b) measuring $|V_{ub}|$, and c) refining $|\varepsilon'/\varepsilon|$ to $\pm$ 0.001. Many prospective experiments, such as measurement of the phase of η_{00}, studies of $\bar{p}p \to K^{\pm}\pi^{\mp}\overset{(-)}{K}{}^0$, initial investigations of B mesons and nearly null results for CP violation in charmed mesons, will test the KM model further.

We have sidestepped the big question: where do quark masses and KM phases come from?

ACKNOWLEDGMENTS

Special thanks are due to Isard Dunietz for enjoyable collaboration on many of the topics mentioned here, and for helpful advice. I wish to thank also J. D. Bjorken, L. L. Chau, G. B. Gelmini, J. Hauser, D. Hitlin, B. Holstein, C. Quigg, R. G. Sachs, M. Sokoloff, S. Stone, B. Winstein and M. Witherell for helpful discussions, and G. B. Word for pointing out an error in the original version of Fig. 7.

This work was supported in part by the United States Department of Energy under Contract No. DE-AC02-82ER 40073.

REFERENCES

1. J. H. Christenson, J. W. Cronin, V. L. Fitch, and R. Turlay, Phys. Rev. Lett. **13**, 138 (1964).
2. Early reviews and discussions are given by T. T. Wu and C. N. Yang, Phys. Rev. Lett. **13**, 380 (1964); T. D. Lee and C. S. Wu, Ann. Rev. Nucl. Sci. **16**, 511 (1966); J. S. Bell and J. Steinberger, *Proc. Oxford Int. Conf. on Elementary Particles, 19/25 Sept., 1965*, edited by T. R. Walsh *et al.*, Rutherford High Energy Lab., Chilton, England, p. 195; K. Kleinknecht, Ann. Rev. Nucl. Sci. **26**, 1 (1976); V. Fitch, Rev. Mod. Phys. **53**, 367 (1981); J. W. Cronin, Rev. Mod. Phys. **53**, 373 (1981) and Acta Physica Polonica **B15**, 419, 721 (1984); Ling-Lie Chau, Phys. Rep. **95**, 1 (1983); V. V. Barmin et al., Nucl. Phys. **B247**, 293 (1984); Andrzej J. Buras, in *Proc. of the Workshop on the Future of Intermediate Energy Physics in Europe, Freiburg, W. Germany, 1984*, edited by H. Koch and F. Scheck (Kernforschungzentrum Karlsruhe, Karlsruhe, Germany, 1984), p. 53.
3. For later reviews consult Ling-Lie Chau, in *Meson 50 (Proc. Kyoto Int. Symposium on the Jubilee of the Meson Theory), Kyoto, Japan, Aug. 15-17, 1985*, edited by M. Bando, R. Kawabe, and N. Nakanishi, Prog. Theor. Phys. Suppl. **85**, 147 (1985); L. Wolfenstein, Comm. Nucl. Part. Phys. **14**, 135 (1985); I. Bigi and A. I. Sanda, Comm. Nucl. Part. Phys. **14**, 149 (1985); John F. Donoghue and Barry R. Holstein, Phys. Rep. **131**, 319 (1985); L. Wolfenstein, Ann. Rev. Nucl. Part. Sci. **36**, 137 (1986); I. Bigi and A. Sanda, SLAC report SLAC-PUB-3949, May, 1986, submitted to Nucl. Phys. B; Barry R. Holstein, presented at 2nd Int. Conf. on Intersections of Particle and Nuclear Physics, Lake Louise, Canada, May 1986, to be published; John F. Donoghue, presented at XXIII Int. Conf. on High Energy Physics, Berkeley, Calif., July, 1986, to be published; A. I. Sanda, presented at 6th Int. Conf. on Physics in Collision, Chicago, Sept. 3-5, 1986, to be published.
4. M. Kobayashi and T. Maskawa, Prog. Theor. Phys. **49**, 652 (1973).
5. N. Cabibbo, Phys. Rev. Lett. **10**, 531 (1963).
6. M. Aguilar-Benitez *et al.*, (Particle Data Group), Phys. Lett. **170B**, 1 (1986).
7. R. H. Bernstein *et al.*, Phys. Rev. Lett. **54**, 1631 (1985).
8. J. K. Black *et al.*, Phys. Rev. Lett. **54**, 1628 (1985).
9. Bruce Winstein, plenary talk, this conference.
10. Fermilab E-621, G. Thomson, Spokesman (seminar at U. of Chicago, May 19, 1986).
11. V. M. Lobashev and A. P. Serebrov, J. de Physique **C3**, Supp. **45**, 11 (1984), as quoted in Ref. 13.
12. J. M. Pendlebury *et al.*, preprint NBS-PUB-711-1986, as quoted in Ref. 13.
13. I. Bigi and A. Sanda, SLAC preprint SLAC-PUB-4157, December, 1986.
14. J. Sandweiss, *et al.*, Phys. Rev. Lett. **30**, 1002 (1972).
15. J. Donoghue and G. Valencia, Phys. Rev. Lett. **58**, 451 (1987); J. Donoghue, B. Holstein, and G. Valencia, Phys. Lett. **178B**, 319 (1986).
16. It is not clear at present whether the jet signature can be freed from final state interaction effects. We are grateful to R. G. Sachs for a discussion of this point.

17. I. Bigi and A. Sanda, SLAC report SLAC-PUB-3949, May, 1986, subm. to Nucl. Phys. B.
18. L. Wolfenstein, Phys. Rev. Lett. **13,** 562 (1964).
19. See, e.g., Table 3 of L. Wolfenstein, Ann. Rev. Nucl. Part. Sci. **36**, 137 (1986).
20. S. Weinberg, Phys. Rev. Lett. **37,** 657 (1976)
21. Jiang Liu and Lincoln Wolfenstein, Carnegie-Mellon Univ. preprint CMU-HEP-86-21, 1986, subm. to Nucl. Phys. B.
22. J. C. Pati and A. Salam, Phys. Rev. Lett. **31**, 661 (1973); Phys. Rev. **D10**, 275 (1974); R. N. Mohapatra and J. C. Pati, ibid. **11**, 566 (1975); **11**, 2558 (1975).
23. G. Ecker and W. Grimus, Phys. Lett. **153B**, 279 (1985); Nucl. Phys. **B258**, 328 (1985).
24. O. W. Greenberg, Phys. Rev. **D32**, 1841 (1985); C. Jarlskog, Phys. Rev. Lett. **55,** 1039 (1985); Z. Phys. **C29**, 491 (1985); I. Dunietz, O. W. Greenberg, and Dan-di Wu, Phys. Rev. Lett. **55**, 2935 (1985); Dan-di Wu, Phys. Rev. **D33**, 860 (1986).
25. Isard Dunietz and J. D. Bjorken, manuscript in preparation.
26. A. Sirlin and R. Zucchini, Phys. Rev. Lett. **57,** 1994 (1986); A. Sirlin, New York University preprint NYU/TR11/86, 1986.
27. John F. Donoghue, Barry R. Holstein, and Stefan W. Klimt, Phys. Rev. **D35**, 934 (1986).
28. H. Leutwyler and M. Roos, Z. Phys. **C25**, 91 (1984).
29. William Marciano, presented at XXIII Int. Conf. on High Energy Physics, Berkeley, Calif., July, 1986, and abstract no. 167, this conference.
30. David Hitlin, comments in plenary session, this conference.
31. Sheldon Stone, Cornell Univ. preprint CLNS-86/753, October, 1986, presented at 6th Int. Conf. on Physics in Collision, Chicago, IL., Sept. 3-5, 1986.
32. David Hitlin, private communication; Mark III Collaboration, in preparation.
33. Lincoln Wolfenstein, Ref. 3.
34. Sheldon Stone, plenary talk, this conference.
35. S. Y. Hsueh *et al.*, Phys. Rev. Lett. **54,** 2399 (1985).
36. Roland Winston, presented at XXIII Int. Conf. on High Energy Physics, Berkeley, Calif., July, 1986.
37. Joseph T. Lach, Abstract no. 67, this conference.
38. J. O. Eeg, H. Hogaasen, and O. Lie-Svendsen, Z. Phys. **C31**, 443 (1986).
39. C. James, presented at XXIII Int. Conf. on High Energy Physics, Berkeley, Calif., July, 1986. This paper contains extensive references to earlier theoretical work.
40. M. Bourquin, presented at XXIII Int. Conf. on High Energy Physics, Berkeley, Calif., July, 1986.
41. S. F. Biagi *et al.*, Z. Phys. **C30**, 201 (1986).
42. C. James, thesis, Univ. of Minnesota, 1986; seminar at Fermilab, January, 1986.
43. See, e.g., E. Golowich and B. R. Holstein, Phys. Rev. **D26**, 182 (1982); A. N. Kamal and R. C. Verma, *ibid*, **D26**, 190 (1982); Lo Chong-Huah, *ibid*; **D26**, 199 (1982); M. D. Scadron and M. Visinescu, *ibid.*, **D28,** 1117 (1983); J. Eeg, Z. Phys. **C21,** 253 (1984).
44. See, e.g., Jon Mathews and R. L. Walker, *Mathematical Methods of Physics*, Second Ed., (W. A. Benjamin, Inc., Menlo Park, Calif., 1970), p. 294.
45. Frederick J. Gilman and Mark B. Wise, Phys. Lett. **83B,** 83 (1979); F. J. Gilman and J. S. Hagelin, Phys. Lett. **133B,** 443 (1983).
46. C. Bernard, Abstract no. 204, this conference.
47. B. Guberina and R. D. Peccei, Nucl. Phys. **B163**, 289 (1980); Branko Guberina, in *Flavour Mixing and CP Violation (Proc. of the Fifth Moriond Workshop, La Plagne, France, Jan.*

13-19, 1985), edited by J. Tran Thanh Van (Editions Frontières, Gif-sur-Yvette, France, 1985), p. 305.

48. John F. Donoghue, Eugene Golowich, Barry R. Holstein, and J. Trampetić, Phys. Lett. **179B**, 361 (1986) and erratum, in preparation; John F. Donoghue, Ref. 3. The sign of the electromagnetic penguin was first thought to *decrease* ε'/ε; this sign error has now been corrected. (B. Holstein, private communication.) For earlier discussions of electromagnetic penguins, see Buras, Ref. 2, and J. Bijnens and M. B. Wise, Phys. Lett. **137B**, 245 (1984).

49. Bell and Steinberger, Ref. 2.

50. Wu and Yang, Ref. 2.

51. L. Wolfenstein, Phys. Rev. Lett. **51**, 1945 (1983).

52. See, e.g., the compilation by Vera Lüth, Abstract no. 168, this conference; S. Stone, Ref. 34; M. Witherell, plenary talk, this conference.

53. See, e.g., L. Wolfenstein, Ref. 3; S. Stone, Ref. 31.

54. Lincoln Wolfenstein, Ann. Rev. Nucl. Part. Sci. **36**, 137 (1986).

55. J. F. Donoghue, E. Golowich, and B. R. Holstein, Phys. Lett. **119B,** 412 (1983).

56. A. Pich and E. de Rafael, Phys. Lett. **158B,** 477 (1985).

57. A. J. Buras, W. Slominski, and H. Steger, Nucl. Phys. **B238,** 529 (1984); **B245,** 369 (1984).

58. We were informed by D. Cline at the end of this talk that lower bounds on the mass of the *t* quark are to be expected soon from the UA1 collaboration.

59. CERN Experiment No. NA-31, H. Wahl, spokesman.

60. N. W. Tanner and R. H. Dalitz, Ann. Phys. (N.Y.) **171**, 463 (1986).

61. V. V. Barmin *et al.*, Ref. 2.

62. Fermilab Experiment No. E-731, Bruce Winstein, spokesman; Fermilab Experiment No. E-773, George Gollin, spokesman. The importance of measuring $\phi_{00} - \phi_{+-}$ as a CPT test has also been stressed by R. G. Sachs, Phys. Rev. **D33**, 3283 (1986).

63. J. H. Christenson *et al.*, Phys. Rev. Lett. **43,** 1209 (1979).

64. Isard Dunietz, Jay Hauser, and Jonathan L. Rosner, Enrico Fermi Institute report EFI 86-54, November, 1986, to be published in Phys. Rev. D.

65. G. Feldman, B. Winstein, M. Witherell, and H. Yamamoto, 1986 (unpublished).

66. M. Witherell, plenary talk, this conference.

67. Ling-Lie Chau, Abstract No. 58, this conference.

68. D. Toussaint, plenary talk, this conference.

69. Vera Lüth, Abstract no. 168, this conference.

70. R. Schindler (Mark III Collaboration), SLAC-PUB-4055, presented at XXIII Int. Conf. on High Energy Physics, Berkeley, Calif., July 1986; R. Schindler, SLAC-PUB-4135, 1986, to be published in *High Energy Electron-Positron Physics* (World Scientific, Singapore, 1987).

71. R. Rückl, DESY report no. 86-116, presented at XXIII Int. Conf. on High Energy Physics, Berkeley, Calif., July, 1986.

72. A. J. Buras, J.-M. Gérard, and R. Rückl, Nucl. Phys. **B268,** 16 (1986).

73. M. Baur, B. Stech, and M. Wirbel, Univs. of Heidelberg and Dortmund preprint HD-THEP-86-19/DO-THEP-86/21, November, 1986, to be published in Z. Phys. C.

74. Ling-Lie Chau and Hai-Yang Cheng, Univ. of Calif. (Davis) preprint UCD-86-32, 1986, and earlier references therein.

75. M. K. Gaillard and B. W. Lee, Phys. Rev. Lett. **33,** 108 (1974); G. Altarelli and L. Maiani, Phys. Lett. **52B,** 351 (1974); G. Altarelli, N. Cabibbo, and L. Maiani, Nucl. Phys. **B88,** 285 (1975); Phys. Lett. **57B,** 277 (1975); R. L. Kingsley, S. Treiman, F. Wilczek, and A. Zee,

Phys. Rev. **D11,** 1919 (1975); M. B. Einhorn and C. Quigg, Phys. Rev. **D12,** 2015 (1975); J. Ellis, M. K. Gaillard, and D. V. Nanopoulos, Nucl. Phys. **B100,** 313 (1975); C. Quigg, Z. Phys. **C4,** 55 (1980).

76. B. Guberina, S. Nussinov, R. D. Peccei, and R. Rückl, Phys. Lett. **89B,** 111 (1979); T. Kobayashi and N. Yamazaki, Prog. Theor. Phys. **65,** 775 (1981); R. D. Peccei and R. Rückl, Symp. on Special Topics in Gauge Field Theories, Ahrenshoop, E. Germany, Nov. 8-13, 1981 (Akad. Wiss., Berlin, 1981), p. 8; M. B. Voloshin and M. A. Shifman, Yad. Fiz. **41,** 187 (1985) [Sov. J. Nucl. Phys. **41,** 120 (1985)].

77. V. Barger, J. P. Leveille, and P. M. Stevenson, Phys. Rev. Lett. **44,** 226 (1980).

78. S. F. Biagi *et al.*, Phys. Lett. **122B,** 455 (1983); Phys. Lett. **150B,** 230 (1985); Z. Phys. **C28,** 175 (1985); P. Coteus *et al.*, Abstract No. 73, this meeting.

79. S. P. Rosen, Phys. Rev. Lett. **44,** 4 (1980); M. Bander, D. Silverman, and A. Soni, Phys. Rev. Lett. **44,** 7 (1980); H. Fritzsch and P. Minkowski, Phys. Lett. **90B,** 455 (1980); W. Bernreuther, O. Nachtmann, and B. Stech, Z. Phys. **C4,** 257 (1980).

80. M. K. Gaillard, B. W. Lee, and J. L. Rosner, Rev. Mod. Phys. **47,** 277 (1975).

81. A. N. Kamal, SLAC preprint SLAC-PUB-3443, 1984 (unpublished); A. N. Kamal and M. D. Scadron, Phys. Rev. **D32,** 1164 (1985).

82. John F. Donoghue and Barry R. Holstein, Phys. Rev. **D21,** 1334 (1980).

83. J. Rosner, in *Proc. 8th Annual Eastern Theoretical Physics Conference, Syracuse, N.Y., Oct. 10, 11, 1969,* edited by F. Rohrlich (Physics Department, Syracuse Univ., 1969), p. 231.

84. J. C. Anjos *et al.* (Fermilab E-691), Abstracts nos. 76, 81-84, this conference.

85. E. Thorndike, Abstract No. 180, this conference.

86. I. Bigi and A. Sanda, Phys. Lett. **171B,** 320 (1986).

87. L. Wolfenstein, Phys. Lett. **164B,** 170 (1985); John F. Donoghue, Eugene Golowich, Barry R. Holstein, and Josip Trampetić, Phys. Rev. **D33,** 179 (1986).

88. W. C. Louis *et al.*, Phys. Rev. Lett. **56,** 1027 (1986).

89. A. Pais and S. B. Treiman, Phys. Rev. **D12**, 2744 (1975); L. B. Okun', V. I. Zakharov, and B. M. Pontecorvo, Lett. Nuov. Cimento **13,** 218 (1975); R. L. Kingsley, Phys. Lett. **63B,** 329 (1976); Maurice Goldhaber and Jonathan L. Rosner, Phys. Rev. **D15,** 1254 (1977).

90. B. R. Kumar (E-691 Collaboration), presented at Lake Louise Winter Institute, Lake Louise, Canada, Feb. 16-21, 1987.

91. J. Ellis, M. K. Gaillard, D. V. Nanopoulos, and S. Rudaz, Nucl. Phys. **B131,** 285 (1976); A. B. Carter and A. I. Sanda, Phys. Rev. Lett. **45,** 952 (1980); Phys. Rev. **D23,** 1567 (1981); I. I. Bigi and A. I. Sanda, Nucl. Phys. **B193,** 85 (1981); J. S. Hagelin, *ibid.* **B193,** 123 (1981); Robert G. Sachs, Enrico Fermi Institute report EFI 85-22, February, 1985.

92. Dongsheng Du, Isard Dunietz, and Dan-di Wu, Phys. Rev. **D34,** 3414 (1986).

93. Dan-di Wu, Institute of High Energy Physics (Beijing) report BIHEP-TH-85-27, November, 1985.

94. Isard Dunietz and Jonathan L. Rosner, Phys. Rev. **D34,** 1404 (1986).

95. Fermilab proposal P-771 (May, 1986), B. Cox, spokesman.

96. M. S. Alam *et al.*, Phys. Rev. **D34,** 3279 (1986); H. Albrecht *et al.*, Phys. Lett. **162B**, 395 (1985).

97. S. Ono, N. A. Törnqvist, J. Lee-Franzini, and A. I. Sanda, Phys. Rev. Lett. **55,** 2938 (1985).

98. R. Eichler, *et al.*, Swiss Inst. for Nucl. Research preprint PR-86-13, Nov., 1986.

99. Don Summers (UA1 Collaboration), Abstract no. 166, this conference.

100. Daniel Froidevaux, plenary talk, this conference.

101. B. Cox, F. Gilman, and T. Gottschalk, summary report on Heavy Flavor Production, to be published in *Proceedings of the 1986 Summer Study on the Physics of the Superconducting Super Collider, Snowmass, Colo, June 23 - July 11, 1986*, edited by R. Donaldson and J. N. Marx.

102. L. Maiani, Phys. Lett. **62B,** 183 (1976) and in *Proc. Int. Symp. on Lepton and Photon Interactions at High Energies, Hamburg, 1977*, edited by F. Gutbrod (DESY, Hamburg, 1977), p. 867; L.-L. Chau and W.-Y. Keung, Phys. Rev. Lett. **53,** 1802 (1984); H. Fritzsch, Phys. Rev. **D32,** 3058 (1985); M. Gronau, R. Johnson, and J. Schechter, Phys. Rev. Lett. **54,** 2176 (1985); O. W. Greenberg (Ref. 24); Haim Harari and Miriam Leurer, Phys. Lett. **B181,** 123 (1986); H. Fritzsch and J. Plankl, Max-Planck-Institut preprint MPT-PAE/PTh 20/86, 1986; M. Gronau, Technion report no. PHYS-86-27, 1986, presented at International Symposium on Production and Decay of Heavy Flavours, Heidelberg, May, 1986.

103. C. Jarlskog, Ref. 24.

104. David Hochberg and Robert G. Sachs, Phys. Rev. **D27,** 606 (1983).

105. H. Fritzsch, Phys. Lett. **73B,** 317 (1978); Nucl. Phys. **B155,** 189 (1979).

106. B. Stech, Phys. Lett. **130B,** 189 (1983).

107. Michael Shin, Brown Univ. report HET-597, presented at XXIII Int. Conf. on High Energy Physics, Berkeley, Calif., 1986, and references therein.

108. J. L. Rosner, Phys. Lett. **146B,** 108 (1984).

109. S. C. Lee, Fermilab report, 1987, in preparation.

110. F. Cornet, *et al.*, Phys. Lett. **174B,** 224 (1986).

111. B. Adeva et al., Phys. Rev. **D34,** 681 (1986); Min Chen, Abstract no. 30, this conference.

112. M. Kuhlen, DESY report no. 86-052, 1986, presented at 21st Rencontre de Moriond, Les Arcs, France, Mar. 9-16, 1986.

113. Peter J. Moxhay and Jonathan L. Rosner, Phys. Rev. **D28,** 1132 (1983).

114. See, e.g., Ref. 111 and S. Komamiya, *Proc. Int. Symposium on Lepton and Photon Interactions at High Energy, Kyoto, Aug. 19-24, 1985*, edited by M. Konuma and K. Takahashi (Kyoto Univerity, Kyoto, 1986), p. 611.

115. See, e.g., Haim Harari and Miriam Leurer, Nucl. Phys. **B233,** 221 (1984).

116. G. Beall, M. Bander, and A. Soni, Phys. Rev. Lett. **48,** 848 (1982); F. Gilman and M. H. Reno, Phys. Rev. **D29,** 937 (1984).

117. Stephen L. Adler, Phys. Rev. Lett. **57,** 167 (1986); Phys. Rev. **D34,** 1871, 3934(E) (1986).

118. M. Gell-Mann, P. Ramond, and R. Slansky, in *Supergravity*, edited by D. Z. Freedman and P. van Nieuwenhuizen (North-Holland, Amsterdam, 1979), p. 315; T. Yanagida, in *Proc. of the Workshop on Unified Theory and Baryon Number in the Universe*, edited by O. Sawada and A. Sugamoto (KEK, Tsukuba, Japan, 1979); Rabindra N. Mohapatra and Goran Senjanović, Phys. Rev. Lett. **44,** 912 (1980); Phys. Rev. **D23,** 165 (1981).

119. G. Bernardi *et al.*, Phys. Lett. **B181,** 173 (1986).

120. Brookhaven National Laboratory Experiments No. 776, Wonyong Lee, Spokesman (unpublished) and No. 816, F. Vannucci, Spokesman (unpublished).

121. Y. Chikashige, R. N. Mohapatra, and R. D. Peccei, Phys. Lett. **98B**, 265 (1981); G. B. Gelmini and M. Roncadelli, Phys. Lett. **99B,** 411 (1981); Howard M. Georgi, Sheldon Lee Glashow, and Shmuel Nussinov, Nucl. Phys. **B193**, 297 (1981).

122. F. T. Avignone *et al.*, submitted to Phys. Rev. Lett., and this conference.

e^+e^- COLLIDER RESULTS

Sheldon Stone
Floyd R. Newman Laboratory of Nuclear Studies
Cornell Univeristy
Ithaca, N. Y. 14853

ABSTRACT

Predictions of the standard model are tested against results from e^+e^- machines. In particular, estimates from QCD are compared to measurements of R and quarkonium decays. Determinations of the standard model parameters V_{cb} and V_{ub} are discussed using data from B decay. Information on the ν_τ mass and the τ lifetime is presented.

1. INTRODUCTION

In this review, I compare the experimental results from e^+e^- colliders with predictions of the standard model.[1] I focus on: strong interactions and QCD, heavy quark decays, and τ lepton decays, unfortunately ommiting several other interesting areas.

2. STRONG INTERACTIONS and QCD

2.1 Measurments of α_s

In order to ascertain the viability of perturbative QCD, the "running" coupling constant $\alpha_s(Q)$ should be measured in many different ways. To define $\alpha_s(Q)$, both an expansion scheme and a scale for Q must be chosen. Many schemes have been suggested: MOM, MS, $\overline{\text{MS}}$, etc.[2] Brodsky, Lepage and Mackenzie (BLM) tie the choice of scale to the choice of scheme.[3]

The simplest experimental quantity to calculate in perturbative QCD is R,

$$R = \frac{\sigma(e^+e^- \to \text{hadrons})}{\sigma(e^+e^- \to \mu^+\mu^-)} .$$

In QCD,

$$R(Q^2) = 3\sum_q e_q^2\left[1 + \frac{\alpha_s(Q)}{\pi} + C_2\left(\frac{\alpha_s(Q)}{\pi}\right)^2\right]$$

where e_q are the individiual quark charges expressed as fraction of the electrons charge and C_2 is scheme dependent. In the $\overline{\text{MS}}$ scheme, C_2=1.4, while in the BLM scheme C_2=0.02 and the scale Q is equal to 0.71 times the center-of-mass energy.[2] At PEP and PETRA energy there are five active quark flavors and $R=3.67[1+\alpha_s/\pi+C_2(\alpha_s/\pi)^2]$. Experimentally this is very difficult to measure because systematic errors dominate and are notoriously difficult to estimate. Nevertheless, several groups have reported results in the 29-34 GeV energy region.[4,5] These are shown in Table I. The world average value for R then turns out to be 3.97±0.02±0.06.

Table I

Measurements of R

Experiment	Center-of-Mass Energy (GeV)	R
TASSO	34	4.05±0.06±0.19
JADE	34	3.99±0.04±0.10
MARK J	34	3.95±0.05±0.22
MARK II	29	3.90±0.05±0.25
MAC	29	3.96±0.03±0.09

I derive a world average value for α_s by combining the statistical and systematic errors in quadrature. In the $\overline{MS}$ scheme, α_s=0.23±0.04 at Q=31 GeV and in the BLM scheme α_s=0.26±0.05 at Q=.71*31= 22 GeV. Including corrections for Z^0 exchange and quark masses, lowers the α_s value to 0.17±0.03 in $\overline{MS}$.[6] Next, I compare α_s found by other processes with this value.

Many groups have sought to determine α_s by using shape of events produced in high energy e^+e^- annihilation. Here a significant fraction of events contain quark-antiquark-gluon final states, and this fraction is proportional to the size of α_s. Unfortunately, the value obtained depends on the models used to hadronize the quarks and gluons. Values anywhere from 0.12 to 0.20 have been reported.[3,6]

Υ decays have also be used to find α_s. Most of the hadronic decays of the Υ are believed to occur via three gluons, and shape analysis of Υ decay data show that this is valid. There is also a hadronic decay via an intermeadiate γ into quark-antiquark, but this can be subtracted off the total rate. Several Υ decay diagrams are shown in Fig. 1. The total decay width, Γ_{tot}, is found from measuring the branching ratio

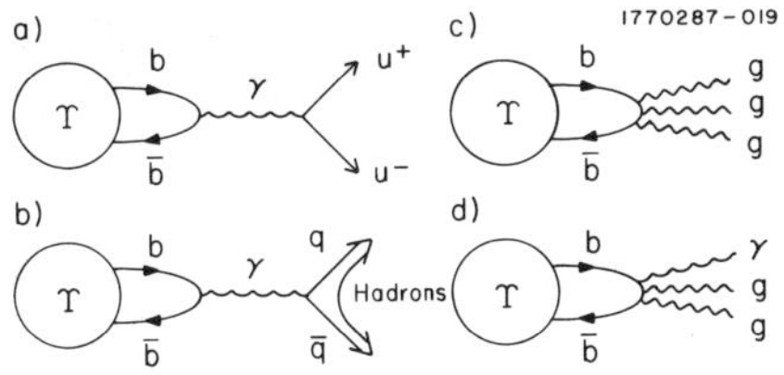

Fig. 1: Υ Decay Diagrams

into muon pairs $B_{\mu\mu}$ and the electronic width Γ_{ee}, since $\Gamma_{ee}=B_{\mu\mu}\Gamma_{tot}$, if we assume electron muon-universality. Γ_{ee} is found by integrating the total measured cross section for $e^+e^-\to\Upsilon$ over energy. (The width of the Υ peak is determined by the machine energy resolution.) Then radiative corrections must be performed to extract a value. Recently, new radiative calculations have been made by Kuraev and Fadin,[7] which differ from the old prescription of Jackson and Scharre.[8] Using the new radiative corrections, the values of Γ_{ee} and Γ_{tot} have been redetermined and are given in Table II.[9]

Table II

Υ Decay Widths

	$B_{\mu\mu}$	Γ_{ee}	Γ_{tot}
Υ(1S)	2.63±0.13	1.33±0.06	51± 3
Υ(2S)	1.6 ±0.4	0.60±0.04	37±10
Υ(3S)	1.6 ±0.4	0.43±0.03	27± 6

For the Υ(1S), $\Gamma_{ggg}=\Gamma_{tot}-(3+R)\Gamma_{ee}$, where the factor of three comes from the virtual photon decay into leptons and the factor of R comes from the virtual photon decay into quark-antiquark. This yields a value for $\Gamma_{ggg}(1S)=42\pm3$ KeV.

We can determine α_s by comparing the decay rate into three gluons with the decay rate into muon pairs, $\Gamma_{\mu\mu}$, since both rates are proportional to the square of the $b\bar{b}$ wave function at the origin,[10] $|\psi(0)|^2$. Since Γ_{ggg} is proportional to $|\psi(0)|^2\ \alpha_s^3$ and $\Gamma_{\mu\mu}$ is proportional to $|\psi(0)|^2\ \alpha_{QED}^2$, $|\psi(0)|^2$ will cancel in the ratio. Using the $\overline{MS}$ scheme for the perturbative expansion we have:

$$\frac{\Gamma_{ggg}}{\Gamma_{\mu\mu}} = C_o \frac{\alpha^3_{\overline{MS}}(Q^*)}{\alpha^2_{QED}} \left[1 - 14 \frac{\alpha_{\overline{MS}}(Q^*)}{\pi} \right] , \quad Q^* = 0.157M_\Upsilon ,$$

and $C_o= 10(\pi^2-9)/81\pi e_b^2$.

Notice that the correction term has a large negative coefficient (14) and the magnitude of this term is close to one. Other perturbative expansion schemes can have smaller correction terms. However, BLM claim that use of these schemes just hides the fact that the perturbative expansion of this system is subject to large higher order corrections.[2] This is similar to orthopositronium decay into three photons. So the theoretical uncertainties prevent a meaningful α_s determination from the rather nicely measured data.

The Υ also decays into a γ and two gluons [$\Upsilon\to\gamma gg$], shown in Fig. 1d. Here the large theoretical corrections cancel in the ratio:

$$\frac{\Gamma_{\gamma gg}}{\Gamma_{ggg}} = \frac{36e_b^2}{5} \frac{\alpha_{QED}}{\alpha_{\overline{MS}}(Q^*)} \left[1 - 2.2\frac{\alpha_{\overline{MS}}(Q^*)}{\pi} \right] , \quad Q^* = 1.5 \text{ GeV}.$$

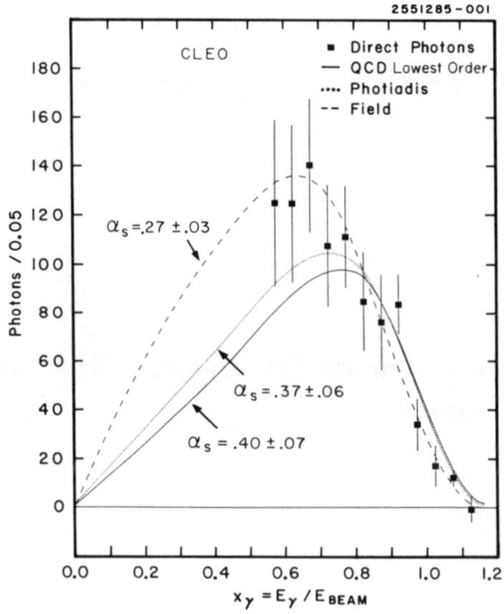

Fig. 2: γ spectrum from the Υ(1S) with fits to theoretical spectra

However, here the data are poor. The CLEO inclusive photon spectrum in shown in Fig. 2.[11] To find a rate, the data must be extrapolated into the low x_γ region and this requires a model of the spectral shape. Three such models exist[12] and make acceptable fits to the data, although the Field model fits best. The extracted α_s values are 0.27±0.03, 0.36±0.06 and 0.40±0.07 for the lowest order QCD, Photiadis and Field models, respectively. CUSB also has measurements, but their data fits the lowest order QCD shape and is inconsistent with the Field model.[13] They find $\alpha_s=0.20^{+0.12}_{-0.06}$. The experimental situation will get clarified in the future.

The R measurement provides the best determination of α_s. Other methods are inaccurate but consistent.

2.2 Radiative ψ decays

Another prediction of QCD is that two gluons can form a bound state colloquially called a "glueball". We examine the evidence for glueballs in radiative ψ decays.[14] For any final state X, we can compare γX with ωX and with ϕX. The decay diagrams for these various processes are shown in Fig. 3. The hope is that we can learn about the quark and gluon content of the X system, since the presence of the ω means that $u\bar{u}$ and $d\bar{d}$ quarks are present and the presence of the ϕ indicates $s\bar{s}$.

We first consider the K^+K^- system. The Mark III data are shown in

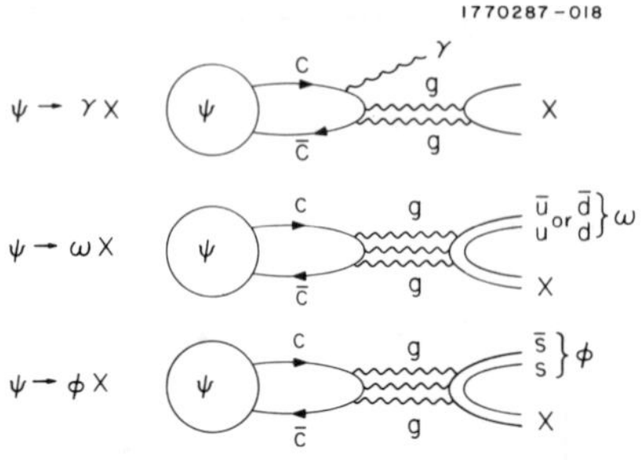

Fig. 3: Decay diagrams for $J/\psi\to\gamma X$, $J/\psi\to\omega X$, and $J/\psi\to\phi X$.

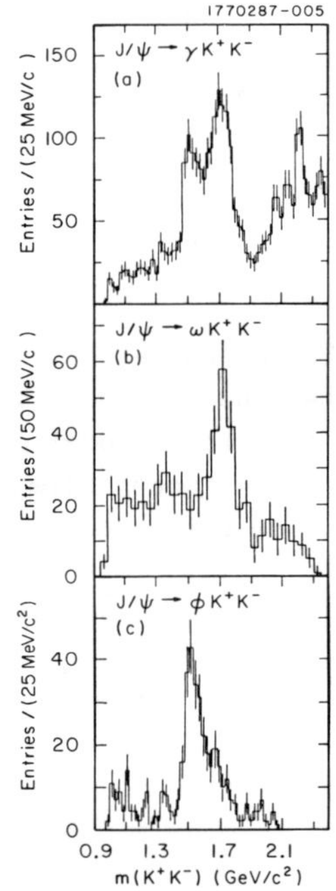

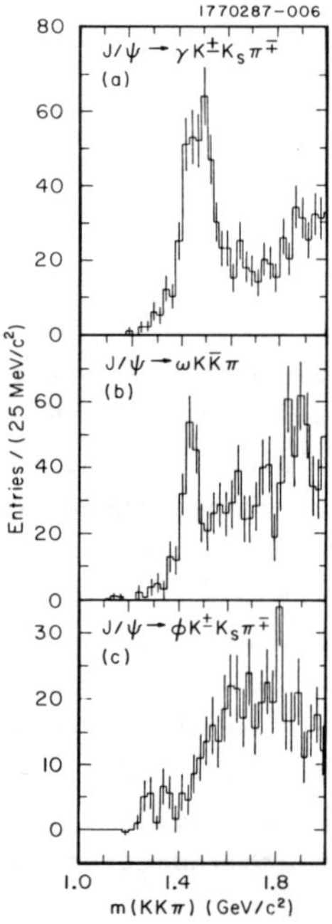

Fig. 4: Invariant K^+K^- mass in the reactions (a) $J/\psi\to\gamma K^+K^-$, (b) $J/\psi\to\omega K^+K^-$, and (c) $J/\psi\to\phi K^+K^-$

Fig. 5: Invariant $K\bar{K}\pi^-$ mass in the reactions (a) $J/\psi\to\gamma K^{\pm}K_S^0\pi^{\mp}$, (b) $J/\psi\to\omega K^{\pm}K_S^0\pi^{\mp}$ + $\omega K^+K_S^-\pi^0$, and (c) $J/\psi\to\phi K^{\pm}K_S^0\pi^{\mp}$

Fig. 4.[15] (There are also data of comparable quality from DM2).[16] Opposite a γ, the data show three distinct structures, the f′(1520), the θ(1720) and the ξ(2230).[17] The measured J^{PC} are 2^{++}, 2^{++} and (even)$^{++}$, respectively. The ξ width is found to be 22^{+18}_{-14}MeV. Opposite the ω, a large bump appears that is consistent in mass and width with the θ, but the spin-parity analysis has not been performed. Opposite the ϕ, a strong f′ signal is evident, but there is only a weak shoulder at the θ mass. Since the f′ seems to be the $q\bar{q}$ state which fills out the 2^{++} nonet, the θ is redundant. It, therefore, is a good glueball candidate. It is somewhat mysterious, however, that the θ, which decays into K^+K^-, couples much more strongly to the the ω, which has no strange quarks, than to the ϕ.

The narrow ξ is also seen by Mark III in the $K^o_sK^o_s$ channel, but is not seen by DM2 in either kaon decay channel. Quantitatively, Mark III measures the product branching fraction $B(\psi\to\gamma\xi)B(\xi\to K^+K^-)=4.2^{+1.9}_{-1.6}\times10^{-5}$, while DM2 limits this to $<1.2\times10^{-5}$ at 95% confidence level. If the ξ is real, the question remains as to what it is. Godfrey et al.[18] suggest that it is an L=3 resonance, where the angular momentum limits the possible decays modes so the width could be of the order of 50 MeV.

Next, I examine the $K^o_sK^+\pi^-$ system. The Mark III data are shown in Fig. 5.[19] There is a wide bump opposite the γ, with mass of 1459±5 MeV, width of 99±11 MeV and J^{PC} equal to 0^{+-}. Opposite the ω, there is a narrower bump with mass 1445^{+11}_{-21} MeV, width of 40^{+20}_{-16} MeV, and $J^P\neq0^-$. Nothing appears opposite the ϕ. Remember the E/ι puzzle.[20] Both resonances decay into $KK\pi$ at about the same mass. The E has been measured to be 1^{++} by most groups (although some 0^- measurements have been reported), while the ι is 0^{-+}.[21] The data here suggest that both the E and ι are being produced opposite a γ, while only the E is being produced opposite the ω. If this is the case, then the ι remains a good glueball candidate. Unfortunately, radiative ψ decays have led to a rich array of final states, which have served to confuse rather than elucidate light quark spectroscopy.[22] More data especially from Υ decays could lead in the future to a better understanding.[23]

Some new information on the E/ι question has been found by two experiments doing $\gamma\gamma$ collisions. Here the process is $e^+e^-\to e^+e^-\gamma\gamma$, where the virtual $\gamma\gamma$ collide and produce hadrons. The TPC group has separated their data[24] into two samples, one where both photons are nearly real and the other where one photon is highly virtual (γ^*). In this later sample, one of the final state electrons is located in a special tagging detector. For the two nearly real photons there is no signal, but for the $\gamma^*\gamma$ sample a large peak appears at 1420 MeV. See Fig. 6. The Mark II group has confirmed this signal (Fig. 6c).[25] A spin one particle cannot be made with two real photons, but can be made with $\gamma^*\gamma$. This implies that the spin-1 E is being observed in two photon colllisions. Note, that it is expected that glueballs cannot be made in two photon collisions; only $q\bar{q}$ states couple to two photons.

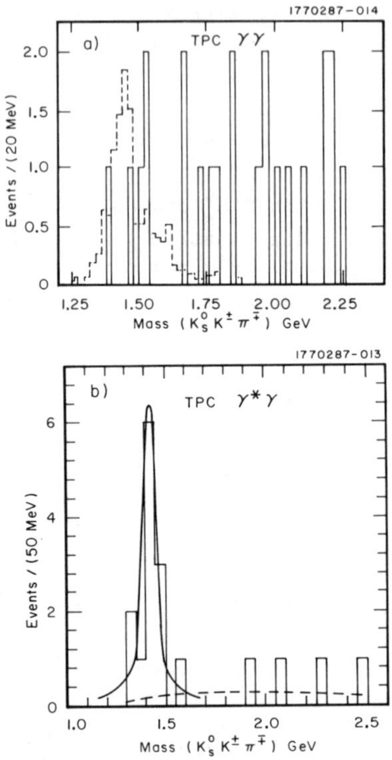

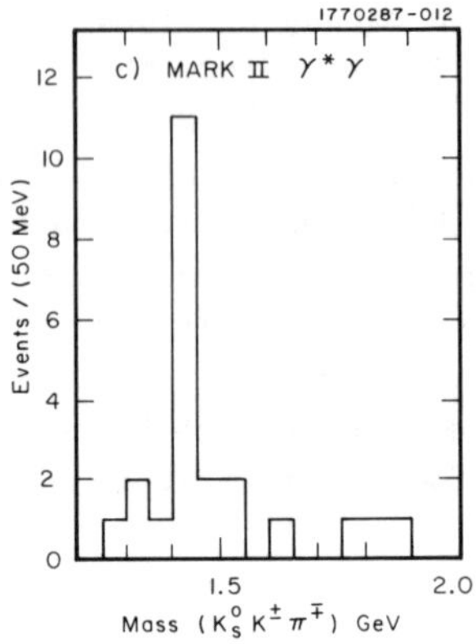

Fig. 6: $K^{\pm}K_S^0\pi^{\mp}$ mass spectrum from two-photon collisions (a) TPC data where both γ's are nearly real, the dashed histogram show the 95% c. ℓ. upper limit for ι production, (b) TPC data where one γ is virtual. The fit gives a mass value of 1.420±0.018 GeV, and (c) MARK II data where one γ virtual.

2.3 Quarkonium Potentials - Fine Structure Splitting

It is well known that both the ψ and Υ systems can be described by a common quark-antiquark potential V(r). In fact, several different functions work.[26] One of these is a QCD inspired potential,

$$V(r) = -4\alpha_s/(3r) + kr. \tag{1}$$

Here the first term comes from one gluon exchange, while the second comes from scalar confinement. The k is a constant which can be calculated in some models. However, these models, in the main, do not directly connect with QCD. They just show that a description in terms of $q\bar{q}$ is plausible.

Observed transitions among $c\bar{c}$ or $b\bar{b}$ states occur via ππ or single photons. Some of the possible transitions in the Υ system are shown in Fig. 7. The spin dependence of the potential can be investigated by measuring the mass splitings in the triplet P states. The observed states are the triplet S and triplet P levels (χ and χ′). The χ states were observed in γ transitions from the Υ(2S).[27] Recently, CUSB using a new BGO detector has remeasured the χ′ splittings by observing Υ(3S)→γχ′ decays.[28] Their data are shown in Fig. 8a. The three peaks

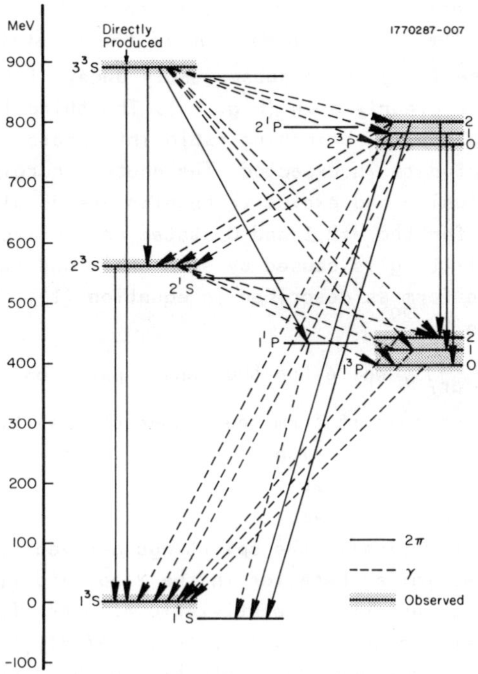

Fig. 7: Possible γ and $\pi\pi$ transistion in the Υ system.

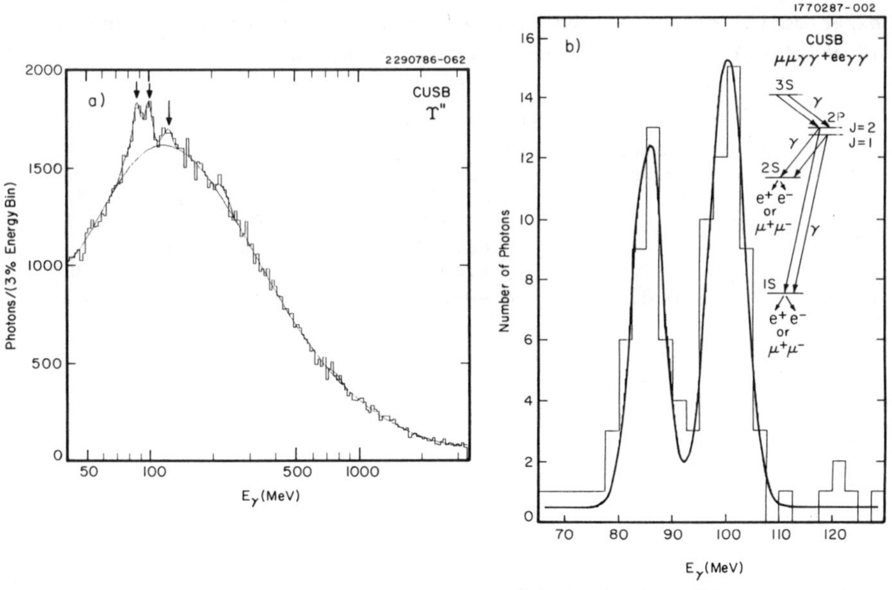

Fig. 8: γ energy spectra from the Υ'', (a) inclusive (b) energy of lowest energy γ for exclusive $\gamma\gamma\ell^+\ell^-$ events.

in photon energy indicated by arrows correspond to the three states with J=2,1 and 0. They also have data where the χ' state decays via a γ to the $\Upsilon(1S)$ or the $\Upsilon(2S)$ which subsequently decays into $\ell^+\ell^-$. Two of the three lines are clearly seen (Fig. 8b). The third line (J=0) is expected to have a smaller branching ratio into photons, so its absence with this amount of data is expected. The photon energies obtained by combining the inclusive and exclusive results are 85.9±0.5, 99.9±0.5 and 124.2±2.3 MeV for the J=2,1,and 0 states respectively.

This fine splitting is caused by spin-orbit and tensor forces. For a potential of the form as described in equation (1), the spin orbit splitting is given by:[29]

$a = \frac{1}{2M^2}\left\langle \frac{3}{r}\frac{dV}{dr} - \frac{1}{r}\frac{dS}{dr}\right\rangle$, while for the tensor part: $b = \left\langle \frac{1}{r}\frac{dV}{dr} - \frac{1}{r}\frac{d^2V}{dr^2}\right\rangle$

The χ state masses are split from the center-of-gravity $\bar{M}$ by:

$$M(\chi_2) = \bar{M} + a - {}^2/_5 b$$
$$M(\chi_1) = \bar{M} - a + 2b$$
$$M(\chi_o) = \bar{M} - 2a - 4b.$$

Fig. 9. shows diagramatically the splittings derived for the χ' system.[30] We have similar behavior in the $\chi(b\bar{b})$ and $\chi(c\bar{c})$ systems, but the charm system has uncertain relativistic effects. It is convienent to define a parameter r where $r = \{M(\chi_2)-M(\chi_1)\}/\{M(\chi_1)-M(\chi_o)\}$. A value of r equal to 0.8 corresponds to a potential with no scalar confinement (i. e. a Coloumb like potential). In table III, I compare the experimental determination of r with a typical theoretical prediction (Buchmüller)[31] for the three different χ systems.

Table III

r values in the χ systems

	Theory	Experiment	Group	Ref.
$\chi(b\bar{b})$	0.73	0.69±0.04	ARGUS	27
		0.61±0.10	Crystal Ball	27
		0.71±0.08	CUSB	27
		0.68±0.03	World Average	
$\chi'(b\bar{b})$	0.72	0.57±0.07	CUSB	28
χ (cc)	0.61	0.49±0.02	Crystal Ball	32

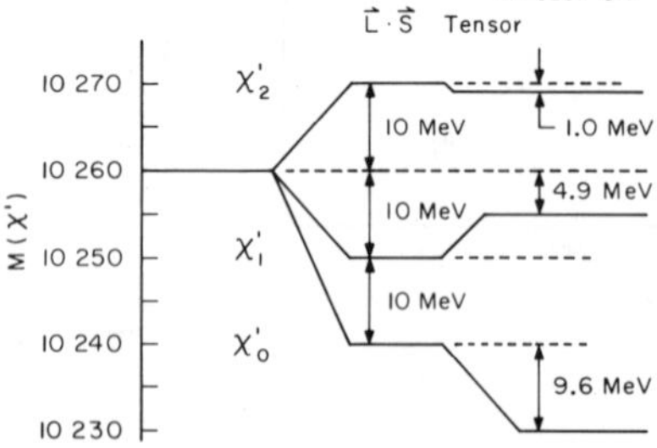

Fig. 9: Masses of the χ' states.

Why does r change so much between the χ and χ' systems? This is not expected in the Buchmüller model. Perhaps there is more confinement because the χ' is larger radially than the χ system. The CUSB data are preliminary and they have recently more than doubled their data sample, so we will soon see if this trend in the data is confirmed.

2.4 $\pi\pi$ transitions

These are very important in ψ', Υ' and Υ'' decays. For example: $B(\psi'\to\pi\pi\psi)=50\%$.[21] CLEO has new measurements of the $\pi^+\pi^-$ transition from the Υ''.[33] The missing mass squared opposite candidate $\pi^+\pi^-$ pairs is shown in Fig. 10a. In Fig. 10b the data sample is restricted to events with four charged tracks including a $\ell^+\ell^-$ pair. In the latter case three different transitions are observed. A hint for the $\Upsilon(3S)\to\pi^+\pi^-\Upsilon(1^1P_1)$ is also seen in the inclusive spectrum. (The 1^1P_1 is forbidden to decay into $\ell^+\ell^-$). This is only a 2.5 standard deviation effect, but the data sample has recently been almost tripled and we anxiously await the analysis.

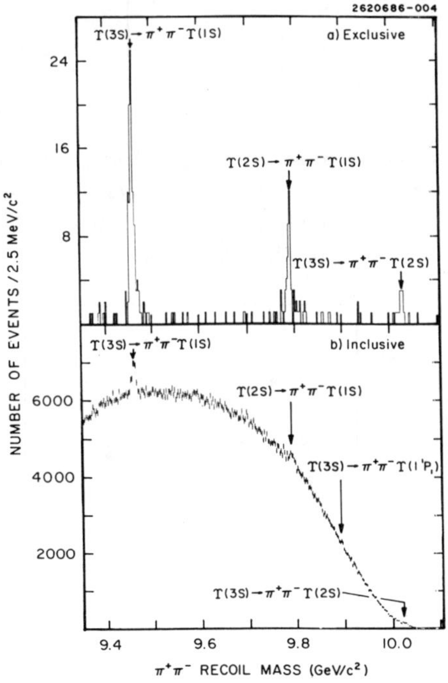

Fig. 10: Missing mass squared opposite $\pi^+\pi^-$ for (a) $\Upsilon''\to\pi^+\pi^-\ell^+\ell^-$ events, (b) inclusive $\Upsilon''\to\pi^+\pi^-X$ events.

The $\pi^+\pi^-$ mass spectrum has been looked at in these transitions. In Fig. 11a the square of the $\pi\pi$ invariant mass is shown for $\psi'\to\pi\pi\psi$ decays.[34] It peaks at high mass and is inconsistent with phase space. The shape of this spectrum is consistent with a model of Yan which has one free parameter.[35] Fig. 11b has a plot of the $\pi^+\pi^-$ invariant mass for $\Upsilon'\to\pi^+\pi^-\Upsilon$ from CLEO.[36] The spectrum has a similar shape to that from the ψ' decay. In contrast to this nice agreement, the $\pi^+\pi^-$ mass spectrum for $\Upsilon''\to\pi^+\pi^-\Upsilon$ shows a peaking at low mass (Fig. 12). One way

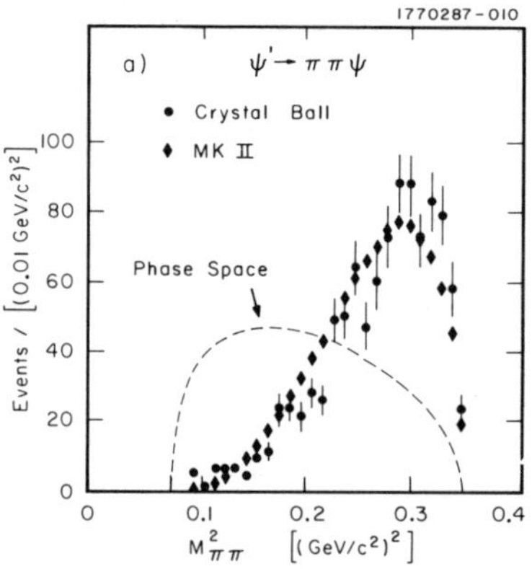

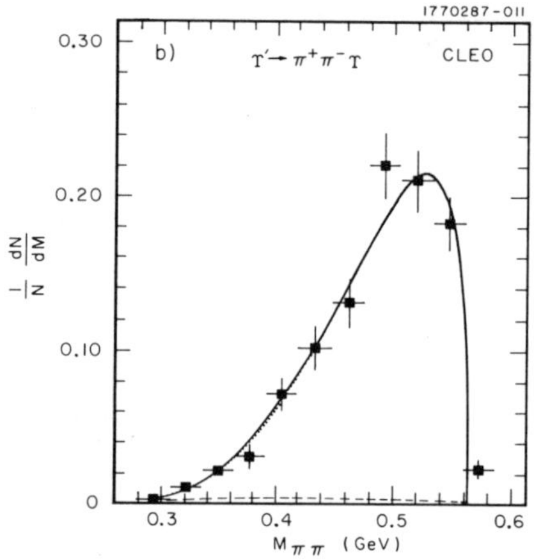

Fig. 11: Invariant mass squared of $\pi\pi$ (a) for $\psi'\to\pi\pi\psi$ decays, (b) invariant mass of $\pi\pi$ for $\Upsilon'\to\pi^+\pi^-\Upsilon$ decays. The dotted curve is shape given by Yan and the dashed curve is a Breit Wigner (see text). The solid curve is the sum of the other two.

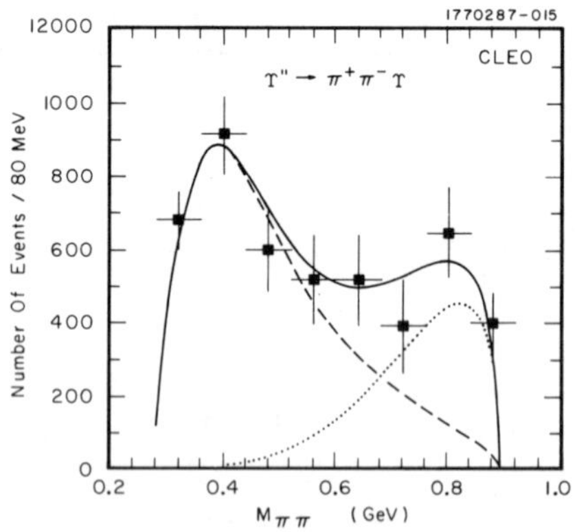

Fig. 12: Invariant mass of $\pi\pi$ for $\Upsilon''\to\pi^+\pi^-\Upsilon$ decays. The dotted curve is shape given by Yan and the dashed curve is a Breit Wigner (see text). The solid curve is the sum of the other two.

for these mass spectra to differ is for an angular momentum greater than zero wave to be present in the dipion system.[37] This wave can couple differently to the ϒ′′ transition because there is more phase space availible. Note that C conservation forbids the dipion system from being in a C= -1 state. Thus the 1^{--} ρ dominated spin wave is missing. A simultaneous fit to both the ϒ′′ and ϒ′ data, with a Breit-Wigner resonance shape (suitably suppressed by threshold) and a Yan shape for the background gives a good fit to both spectra (curves on the figures) and gives a mass of 300 MeV and width of 300 MeV for the Breit-Wigner. Is this a manifestation of a low mass glueball? A spin analysis is in progress.

3. HEAVY QUARK DECAYS

3.1 Introduction

In the "standard" electroweak model[1] there are three generations of left handed quark doublets and right handed singlets (see Fig. 13).

$$\begin{pmatrix} u \\ d' \end{pmatrix} \qquad \begin{pmatrix} c \\ s' \end{pmatrix} \qquad \begin{pmatrix} t \\ b' \end{pmatrix} \begin{matrix} \text{charge} = 2/3 \\ \text{charge} = -1/3 \end{matrix}$$

Fig. 13: Left handed quark doublets

The charge -1/3 quarks decay because they are mixed. Kobayashi and Maskawa noticed that a three generation model with quark mixing would provide a natural explanation of CP violation in the K^o system.[38] The mixed quarks, denoted by primes, are related to the flavor eigenstates by the Kobayashi-Maskawa (KM) matrix, as follows:

$$\begin{pmatrix} d' \\ s' \\ b' \end{pmatrix} = \begin{pmatrix} V_{ud} & V_{us} & V_{ub} \\ V_{cd} & V_{cs} & V_{cb} \\ V_{td} & V_{ts} & V_{tb} \end{pmatrix} \begin{pmatrix} d \\ s \\ b \end{pmatrix} .$$

We would like to be able to explain the phenomena seen in K, D and B meson decays in terms of s, c and b quark decays. I will concentrate here on B decays.[39] There are a number of interesting questions to answer. 1) The possible quark level diagrams for B decay are shown in Fig. 14. Which of these are important? Theoretically, the spectator graphs are expected to dominate.[40] 2) What are the decay rates of the b quark to the c quark and to the u quark. 3) Is there any evidence for the non-standard model B decays?

3.2 Flavor Changing Neutral Currents

One type of non-standard model decay occurs via flavor changing neutral currents (FCNC), as depicted in Fig. 15. These have not been

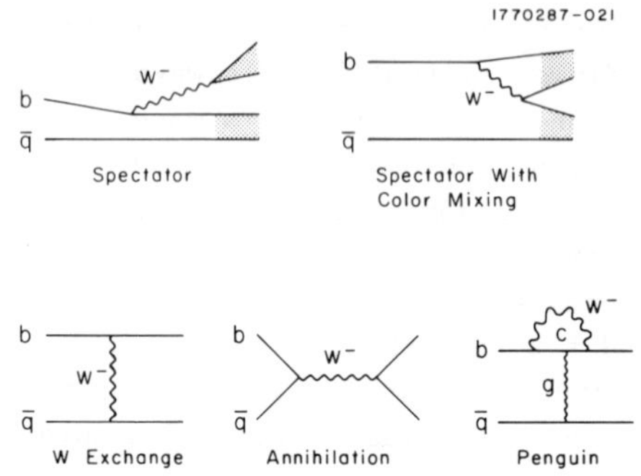

Fig. 14: Possible B decay diagrams.

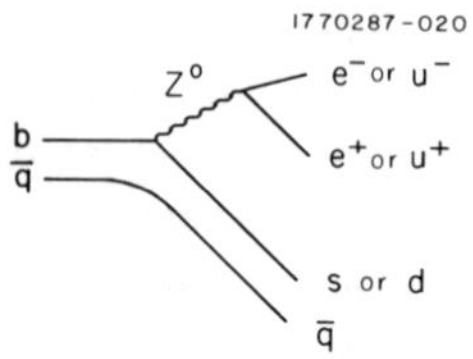

Fig. 15: Diagram for B decay via a Flavor Changing Neutral Current.

observed. The best limit in b decay is from CLEO which finds $B(B\to X\ell^+\ell^-)/B(B\to All) < 0.13\%$ at 90% confidence level.[41] This limit is sufficiently strong to exclude models where the b-quark is in a left handed singlet and decays via its mixture with s and d.[42] Other non-standard models predict FCNC but do not specify the rate.[43]

3.3 V_{cb}

V_{cb} is extracted from mesurements of the B lifetime.[44] The B's are produced at PEP and PETRA. The measured lifetimes shown in Fig. 16 are obtained from an unknown mixture of B^o, B^-, B_s and B-baryons. The world average lifetime is 1.17±0.14 psec. Further analysis requires the assumption that the lifetimes of the above mentioned species are almost equal. The only information on relative B lifetimes comes from CLEO who restrict $0.4<\tau_{B^0}/\tau_{B^-}<2.0$ at 90% confidence level.[45] The KM elements are best extracted through the equation for the semileptonic width:[46]

$\Gamma_{SL}=\Gamma_o(0.45|V_{cb}|^2 + |V_{ub}|^2)$, where $\Gamma_o=G^2m_b^5/192\pi^3$ and 0.45 is a phase space factor. Since the semileptonic branching ratio is related

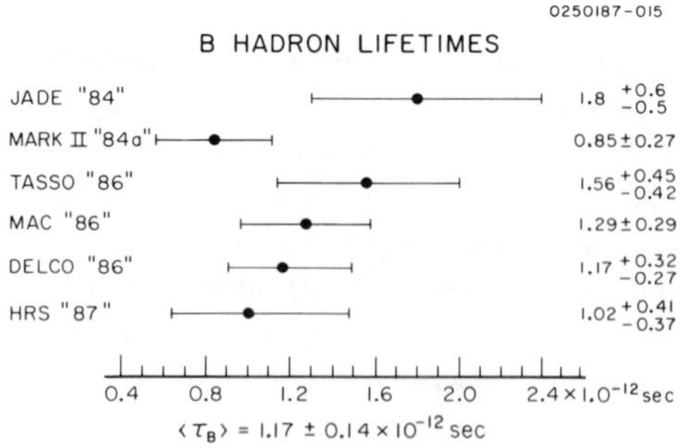

Fig. 16: Measurements of the B hadron lifetime.

to the semileptonic width by $B_{SL}=\Gamma_{SL}/\Gamma_b=\Gamma_{SL}\tau_b$, we can use the CLEO measured semileptonic B branching ratio of 11.0±1.0% in conjunction with the world average τ_b measurement to provide numerical constraints on V_{cb} versus V_{ub}.[47] We still need to supply a value for the b-quark mass (m_b). It is not clear what an appropriate value is. I use 4.95±0.20 GeV with great trepidation.[48] Fig. 17 shows the allowed region for V_{ub} versus V_{cb}.

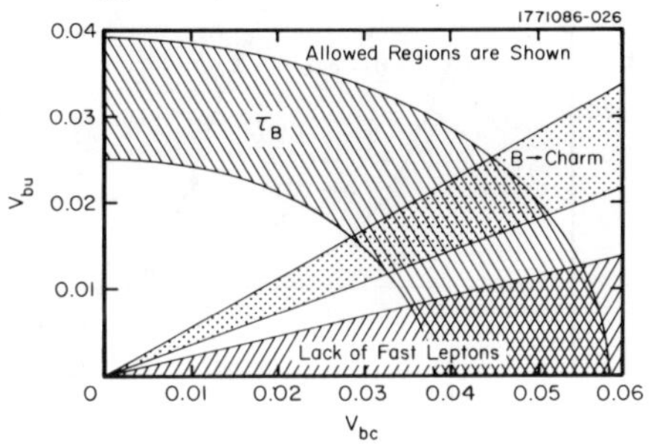

Fig. 17: Allowed regions for $|V_{ub}|$ versus $|V_{cb}|$.

3.4 V_{ub}

3.4.1 Lepton momentum spectrum in semileptonic B decay

In semileptonic B meson decay, $b\to W^-$(c or u) and the W^- materializes as a $e^-\bar{\nu}_e$ or $\mu^-\bar{\nu}_\mu$ pair. These decays can occur only through the spectator graph on Fig. 14. Since the c-quark is heavier than the u-quark, the lepton momentum spectrum spectrum should extend to higher momenta for $b\to u$ transitions than for $b\to c$ transitions. This naive notion is the basis for several detailed models of the shape of the lepton spectrum.[49] In order to extract a ratio of the coupling strengths the data are fit to a particular model for $b\to c\ell\nu$, $b\to u\ell\nu$ and for the cascade decay chain $b\to c\to s\ell\nu$, where the lepton spectra are smeared by the relatively slow B motion since, at the Υ(4S) resonance, the B mesons are produced almost at rest.[50] The CLEO e^- spectrum is shown in Fig. 18.[51] The data are adequately fit without a $b\to u\ell\nu$ component. Upper limits on V_{ub}/V_{cb} range from 0.11 using the Altarelli et al. model to 0.21 using the model of Grinstein et al. at 90% c. ℓ. Instead of fitting the entire spectrum, CLEO has also just used the measured lepton cross-section between 2.4 and 2.6 GeV. This is above the maximum momentum allowed for a $b\to c\ell\nu$ transition (i. e. $B\to D\ell\nu$). As can be seen in Fig. 18, the cross-section is consistent with zero. Upper limits then can be determined just from using models of $b\to u\ell\nu$. The upper limits found range from 0.13 (Altarelli) to 0.23 (Grinstein) again at 90% c. ℓ. Using the largest upper limit, the range on the V_{ub} versus V_{cb} plot is restricted as shown in Fig. 17.

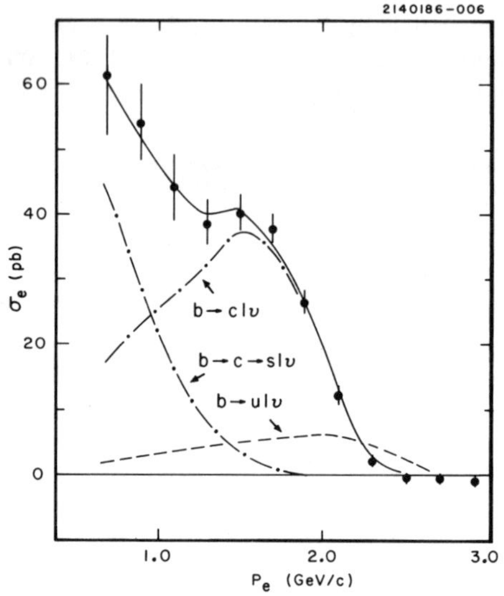

Fig. 18: Electron momentum spectrum from B decay.

3.4.2 Search for charmless exclusive B decays

As the amount of b→u allowed is model dependent we look for other methods of limiting or finding evidence for b→u. In the first such method exclusive charmless B decays are hunted.

The first mode looked for is $B^- \to \rho^0 \ell^- \bar{\nu}$. CLEO requires a ρ^0 opposite a fast lepton with momentum between 2.4 and 2.6 GeV.[52] They obtain a limit on the branching fraction $B(B^- \to \rho^0 \ell^- \bar{\nu})/B(B \to All) < 2.5 \times 10^{-4}$ when the lepton is in the restricted range mentioned above. The fraction of leptons in this momemtum range is model dependent. Grinstein et al. claim its 9%,[49] and Wirbel et al.[53] predict 12%. The upper limits on V_{ub}/V_{cb} range from 0.15 using the Wirbel et al. model to 0.26 using the model of Grinstein et al. at 90% c. ℓ.

Secondly, CLEO[50] and ARGUS[54] have looked for several modes including $B^- \to \rho^0 \pi^-$ and $B^0 \to \pi^+ \pi^-$ where upper limits of 0.02 and 0.03%, respectively, have been set. It is unclear how to translate these into upper limits on V_{ub}/V_{cb}, but one method has been suggested by Bauer and Stech.[55]

3.4.3 Inclusive B decays into charm

The last method used to try and find b→u transitions, relies on finding the amount of inclusive charm in B decay; what is left is presumed to be b→u. The total amount of charm present could exceed one per B decay, since the virtual W^- could materialize as a $\bar{c}s$ pair. If b→c were 100% we would have 1.15 charm particles per B decay. Evidence for the $W^-\to\bar{c}s$ transition is given by the observation of ψ and D_s (also known as F) mesons in B decay by both the CLEO and ARGUS groups.[56] These particles can be produced by the diagrams shown in Fig. 19. CLEO finds that 64±22% of the D_s's have momenta that make it likely that they are produced in a two-body decay $B\to DD_s$. ψ production is only at the 1% level, and D_s production is consistent with being less than 15%, depending on the value of the $D_s\to\phi\pi$ branching ratio. For the following discussion, we assume that all the D_s and ψ production comes from the $W^-\to\bar{c}s$ and that D^0, D^+ and charmed baryons must account for the remainder of the b→c decays.

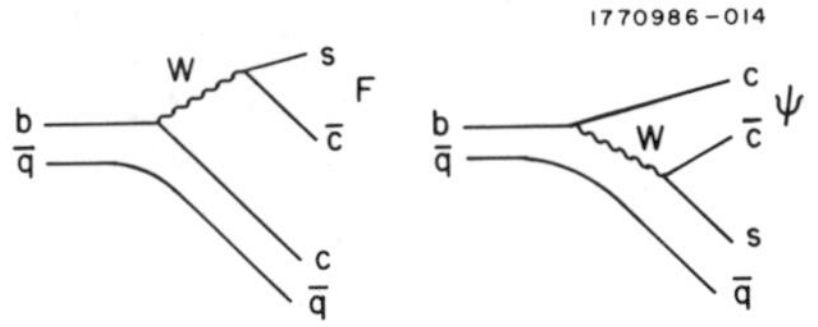

Fig. 19: Mechanisms for D_s and ψ production from B decay.

The signals for D^0, D^+ and D^{*+} mesons from both the Υ(4S) and the continuum below from the CLEO experiment are shown in Fig. 20.[57] D^0's are detected when they decay into $K^-\pi^+$, D^+'s are detected when they decay into $K^-\pi^+\pi^+$, and D^{*+} are detected in the π^+D^0 channel when the D^0 decays to $K^-\pi^+$. The D^{*+} are analyzed because they provide an important consistency check; if D^*'s are present, we expect more D^0's than D^+'s. ARGUS has a preliminary analysis of these decays.[54] The CLEO and ARGUS results are summarized in Table III. The Mark III D branching ratios, $B(D^0\to K^-\pi^+)=5.6\pm0.4\pm0.3\%$ and $B(D^+\to K\pi^+\pi^+) = 11.6\pm1.4\pm0.7\%$, are used.[58]

Table III

B to D Branching Fractions

	CLEO	ARGUS
$Br(B\to D^0X)$	0.39±0.05±0.04	0.50±0.07±0.08
$Br(B\to D^+X)$	0.17±0.04±0.02	0.23±0.08±0.05
Sum	0.56±0.06±0.05	0.73±0.11±0.09

For CLEO the first error is their statistical and systematic error added in quadrature and the second error is due to the quoted error on the D branching ratios. For ARGUS the first error is their statistical error and the second error the overall systematic error.

The weighted average sum is 0.60±0.07 and this is clearly far from 1.0. It is possible that charmed baryons are produced in B decay. CLEO

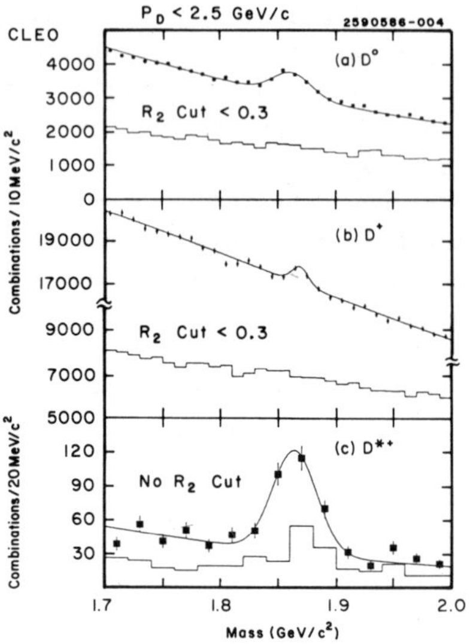

Fig. 20: Invariant Mass plots for D candidate momenta below 2.5 GeV/c. $\Upsilon(4S)$ data are shown by solid points, while the scaled continuum data are shown by histograms. The solid lines are fits to the data with a Monte Carlo based signal shape plus background polynomial. (a) $K^-\pi^+$ mass spectrum, $R_2<0.3$ (b) $K^-\pi^+\pi^+$ mass spectrum, $R_2<0.3$ (c) $K^-\pi^+$ mass spectrum cut on $K\pi\pi-K\pi$ mass difference, no R_2 cut.

has measured $B(B\to\Lambda X)=4.2\pm0.7\%$ and $B(B\to pX)=5.6\pm1.6\%$.[59] If all the Λ's come from charmed baryon decay, then the decay process is $\bar{B}\to\Lambda_c\bar{N}X$, where $\bar{N}$ denotes either a $\bar{p}$ or $\bar{n}$ and is needed to conserve baryon number. For data taken on the $\Upsilon(4S)$ there is also a B in the event $[\Upsilon(4S)\to\bar{B}B]$ and 22% of the time the B decays semileptonically, $B\to X\ell^+\nu$. Ensuring that the lepton comes from the primary B decay by requiring it to be at high momentum, CLEO has looked for $\Lambda\ell^+$ events and $\Lambda\ell^-$ events (see Fig. 21). The strong signal in the $\Lambda\ell^+$ sample (26.6±7.9 events) versus the absence of events (3.3±4.8) in the $\Lambda\ell^-$ sample indicates that charmed baryons are source of baryons in B decay. Alternative models, for example $\bar{B}\to\Lambda\bar{\Lambda}X$, predict equal numbers of $\Lambda\ell^+$ and $\Lambda\ell^-$. Assuming all baryons in B decay come from charmed baryons, CLEO extracts a (preliminary) branching ratio of $B(\bar{B}\to\text{charmed baryons} + X)=\sim8\%$.[59] Combining this new result with the inclusive D results, gives $V_{ub}/V_{cb}=0.46\pm0.10$ in contradiction to the results from the lepton momentum spectrum (see Fig. 17).

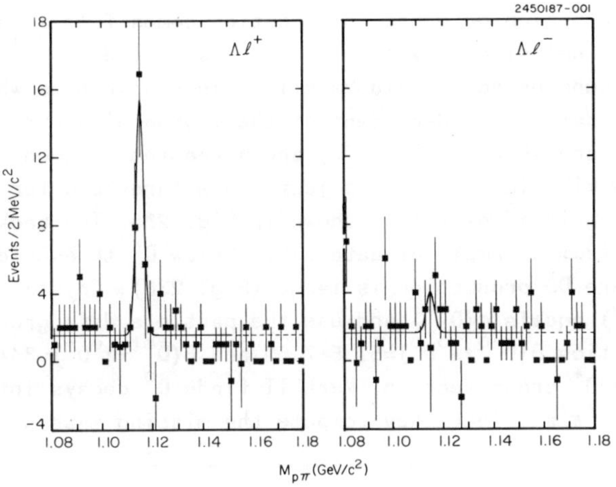

Fig. 21: Invariant mass of $p\pi^-$ candidates for events with (a) an identified e^+ or μ^+ or (b) an identified e^- or μ^-. The p and π^- candidates are required to form a vertex away from the rest of the tracks and the p candidate is required to have an ionization consistent with that expected for a p.

There are several possible explainations of this low D rate. First, the Υ(4S) may not decay entirely into $\bar{B}B$. However, the charged particle momentum spectrum from the Υ(4S) vanishes at half the beam energy excluding this possibility.[47,50] Secondly, Penguin diagrams (see Fig. 14) could contribute to B decay. Then $b \to s$ directly without a charmed particle. This possibility is excluded by the CLEO measurement of D^o's in semileptonic B decay of 0.35±0.09.[57] Since Penguin's can't occur in semileptonic decay, the low D^o rate is evidence that Penguins aren't important in hadronic decay. We are left with two other possibilities: 1) $b \to u$ is large and the models of the lepton momentum spectrum are wrong and 2) the Mark III branching ratios are too big.

Next I will discuss two tests of the Mark III branching ratios. Theoretically, the expected R value for charm is 4/10 when four quark flavors are active. Several groups have measured $\sigma_D/(.4{*}\sigma_{tot})$ using the Mark III branching ratios. CLEO,[57] ARGUS,[60] and HRS[61] find that this ratio is 46±6, 55±10 and 46±8%, respectively. It is possible that D_s and Λ_C production account for the remaining half of charm. Note, however, if $B(D_s^+ \to \phi\pi^+)$=4%, then D_s production accounts for only 5.6% of contiuum charm and charmed baryon production would have to be ~45%. Use of the Mark II D branching ratios[62] would give $\sigma_D/(.4{*}\sigma_{tot})$=86±8%.

New information on D branching ratios comes from a comparison of Crystal Ball inclusive D^* yields[63] with Mark II data.[64] The Mark II yields are found using the old Mark II branching ratios while the Crystal Ball data are independent of the D branching ratios. The measured R_{D^*} are shown in Fig. 22, and agreement is good. Crystal Ball measures low kinetic energy π^0's just above threshold for D^* production. Their π^0 signal is shown in Fig. 23a. The background is ascertained from a sample of data taken below D^* threshold at the $\psi(3772)$, where $D\bar{D}$ production is large (Fig. 23b). Crystal Ball then assume $\sigma(D^{*0})$ equals $\sigma(D^{*+})$ and use the particle data group[21] branching ratios $B(D^{*0}\to\pi^0 D^0)=51.5\pm7.6\%$ and $B(D^{*+}\to\pi^0 D^+)=34\pm7\%$ in order to extract a D^* cross section. Mark II finds D^0 decays into $K^-\pi^+$ and D^+ decays into $K^-\pi^+\pi^+$. Then they compute the missing mass squared opposite

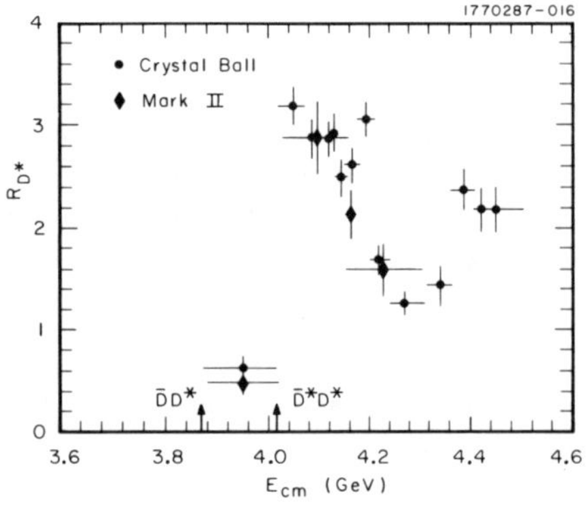

Fig. 22: R_{D^*} for Crystal Ball (circles) and Mark II data (diamonds).

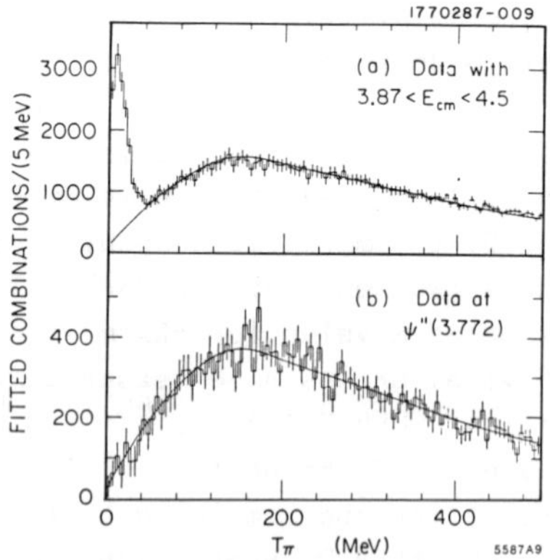

Fig. 23: Distribution of π^0 kinetic energies (a) for data with $3.87<E_{cm}<4.5$ GeV and (b) for data taken at the $\psi''(3772)$. The shape of the smooth curves is derived from the $\psi''(3772)$ sample.

the detected D. The distribution is shown in Fig. 24. It is consistent with D being produced in three final states $D\bar{D}$, $D\bar{D}^*$ or $D^*\bar{D}^*$. From this distribution Mark II finds the fraction of $D\bar{D}^*$ and $D^*\bar{D}^*$ and the D^* cross sections.

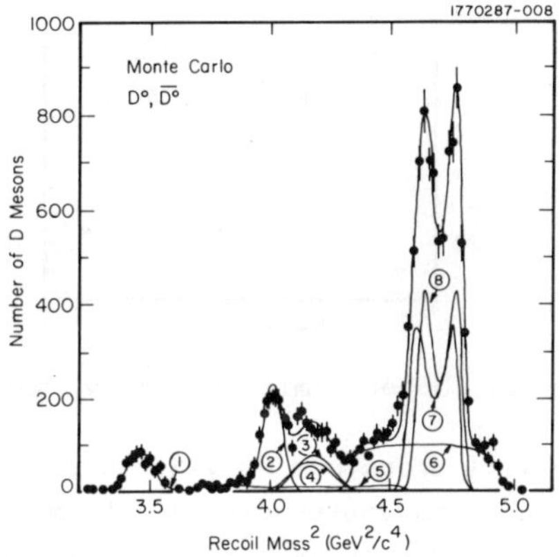

Fig. 24: The Monte Carlo produced recoil mass squared spectrum produced opposite a detected D^o at 4.16 GeV. The upper cuve is a fit to the spectrum. The number curves show contributions from various processes (see ref. 64).

4. τ DECAYS

I will discuss tests of the standard electroweak model in the lepton sector using the heavist known lepton, the τ. It has been studied only at e^+e^- collliders.

4.1 Upper limits on ν_τ mass

If neutrinos are mixed as the quarks are, they must have mass.[65)] Searches for ν_e mass have not been conclusive and are pressing the 30 ev level.[66)] The ν_μ mass upper limit is 0.25 MeV at 90% c. ℓ. Note, that in some models the ν mass would scale as the mass squared of its lepton partner. Thus, the ν_τ could show a mass at the 100 MeV level and still be compatible with the measured limits on ν_e and ν_μ masses. The production reaction for τ's is $e^+e^-\to\tau^+\tau^-$. They then decay: $\tau^-\to X^-\nu_\tau$. To measure the ν_τ mass, the energy or mass spectrum of the X^- is investigated. No positive evidence has been found for a ν_τ mass.[67)] The groups involved choose to present 95% c. ℓ. upper limits (see Fig. 25).

4.2 Measurement of the Michel parameter in τ decay

If τ decays behave as we expect from the standard model, then the

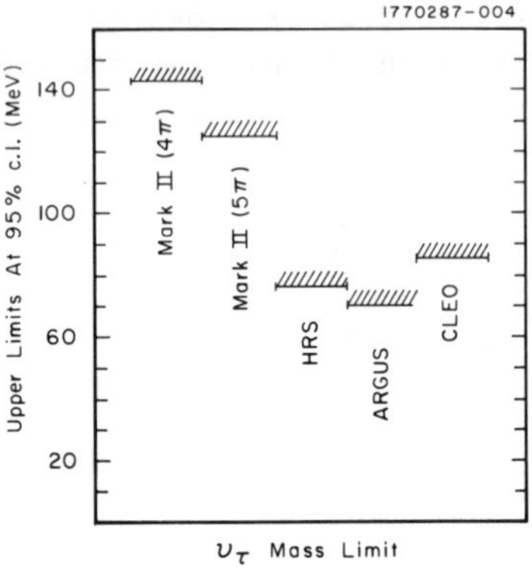

Fig. 25: Upper limits on the ν_τ mass.

weak matrix elements will transform as V-A, just as every other weak decay process. The Michel parameter can be determined by measuring the lepton energy spectrum in the decay $\tau^- \to \ell^- \bar{\nu}_\ell \nu_\tau$, where the ℓ^- can be a e^- or μ^-. For V-A the expected value of the Michel parameter is 0.75. Recall, that in muon decay the measured value is 0.752±0.003.[20] There are two relevant measurements in τ decay. DELCO[68] found 0.72±0.15 and CLEO[69] found 0.71±0.09±0.03. Clearly, there is a long way to go before τ's are measured with comparible accuracy to μ's.

4.3 Measurement of the τ lifetime

The muon decay width is given by:[47] $\Gamma(\mu^- \to \nu_\mu e \bar{\nu}_e) = G^2 m_\mu^5 / 192\pi^3$ ·(radiative corrections)·(phase space corrections).
The theoretical width for $\tau^- \to \nu_\tau e^- \bar{\nu}_e$, obtained by replacing the muon mass by the τ mass and making suitable corrections in the other terms, is 1.595 psec^{-1}.[70] This theoretical prediction is equal to the ratio of the branching fraction for $\tau^- \to \nu_\tau e^- \bar{\nu}_e$ and the τ lifetime (τ_τ). The world average value for $B(\tau^- \to \nu_\tau e^- \bar{\nu}_e)$=17.70±0.47%.[71] The similar rate $B(\tau^- \to \nu_\tau \mu^- \bar{\nu}_\mu)$ has been measured as 17.99±0.49%.[71] Phase space should suppress the muon decay relative to the electron decay by 3%. The measured ratio is 1.016±.038 consistent with the expectation of 0.97. We can view this as a test of electron-muon universality in τ decay. Using both the electron and muon measurements, and assuming universality, I derive $B(\tau^- \to \nu_\tau e^- \bar{\nu}_e)$=18.12±0.34%.

A summary of τ lifetime measurements is given in Fig. 26.[72] The world average is 0.296±0.010 psec, using the usual procedure of adding statistical and systematic errors in quadrature. Then taking the appropriate ratio $\tau_\tau / B(\tau^- \to \nu_\tau e \bar{\nu}_e)$=0.296±0.10psec/0.1812±0.0034 =1.63±0.04 psec, in good agreement with the prediction of 1.595 psec.

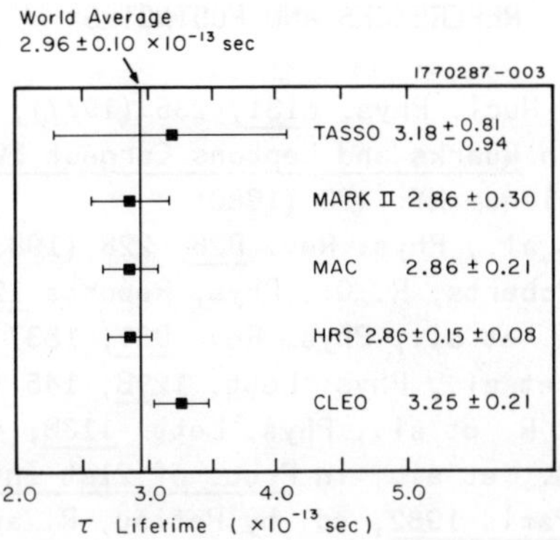

Fig. 26: Measurements of the τ lifetime.

Thus τ decays appear to obey the predictions of the standard model. Several exotic τ decays have been looked for and not found. Examples are: $B(\tau\to e\gamma)<3.2\times10^{-4}$,[73] $B(\tau\to\mu\mu\mu)<2.9\times10^{-5}$ and $B(\tau\to\mu^-\pi^+\pi^-)<4\times10^{-5}$.[74]

5. CONCLUSIONS

No violations of the standard model have been found in quarkonia, B decay or τ decay. α_s has been found using precise measurements of R from PEP and PETRA. Other data are not in conflict. V_{cb} has been found using measurements of the B lifetime. There is a conflict in the limit set on V_{ub}/V_{cb} by using the lepton momentum spectrum and explicitly enumerating the charmed particles. The resolution of this conflict may lie in the values of the D branching ratios. τ decay measurements are becoming more precise and don't show any anomalies. With Tristan, SLC, LEP and BEC joining the already existing machines, we expect that e^+e^- colliders will continue to make important measurements and will perhaps provide crucial new insights into the standard model or beyond.

6. ACKNOWLEDGMENTS

I thank P. Lepage, J. Rosner, G. Gilchriese, B. Gittelman and H. Tye, for useful discussions. This work was supported by the National Science Foundation.

REFERENCES AND FOOTNOTES

1 Ellis, J. et al., Nucl. Phys. B131, 285 (1977); Gaillard, M. K., and Maiani, L., in Quarks and Leptons Cargése 1979, ed. Lévy, M. et al. Plenum NY 433 (1980).

2 Brodsky, S. J. et al., Phys. Rev. D28, 228 (1983).

3 Duke, D. W. and Roberts, R. G., Phys. Reports 120, 276 (1985).

4 MAC: Fernandez, E. et al., Phys. Rev. D31, 1537 (1985); JADE: Bartel, W. et al., Phys. Lett. 129B, 145 (1983); TASSO: Brandelik, R. et al., Phys. Lett. 113B, 499 (1982); MARK J: Burger, J., et al., in Proc. of 21st Int. Conf. on High Energy Phyiscs, Paris 1982, ed. by Petiau, P. and Porneuf M. J., Phys. (Paris) Collog. 43, C3 (1982); MARK II: Hollebeek, R. et al., ibid.

5 The CELLO collaboration has presented results at W=44GeV. They find $\alpha_s=0.18\pm^{0.05}_{0.08}$. See "Measurement of the Total Hadronic Cross Section in e^+e^- Annihilation between 14 and 47 Gev", contributed paper to XXIII Int. Conf. on High Energy Physics, Berkeley (1986).

6 Bartel, W., in these proceedings.

7 Kuraev, E. A. and Fadin, V. S., Sov. J. Nucl. Phys. 41, 466 (1985); see also Baru, E. S. et al., Z. Phys. 30, 551 (1986).

8 Jackson, J.D. and Scharre, D.L., Nucl. Inst.& Meth. 128, 13 (1975).

9 Cooper, S., "Spectroscopy of Light and Heavy Quarks", invited paper to XXIII Int. Conf. on High Energy Physics, Berkeley (1986), SLAC-PUB-4139.

10 Quigg, D. and Rosner, J., Phys. Reports 56C, 167 (1979).

11 Csorna, S. E. et al., Phys. Rev. Lett. 56, 1222 (1986).

12 Koller, K. and Walsh, T., Nucl. Phys. B140, 449 (1979); Field, R. D., Phys. Lett. 133B, 248 (1983); Photiadis D. M., Phys. Lett. 164B, 160 (1985).

13 Schamberger, R. D. et al., Phys. Lett. 138B, 225 (1984).

14 Ronan, M. T., et al., Phys. Rev. Lett. 44, 367 (1980); Scharre, D. L., et al., Phys. Rev. D23, 43 (1981); Abrams, G. S. et al., Phys. Rev. Lett. 44, 114 (1980).

15 Coffman, D. M. et al., "Radiative Decays of the J/ψ into $\gamma\pi^+\pi^-$ and γK^+K^-", SLAC-PUB-3720 (1986); Baltrusaitis, R. M. et al., Phys. Rev. Lett. 56, 107 (1986).

16 Jean-Marie, B., "DM2 Results on Hadronic and Radiative J/ψ Decays", contributed paper to XXIII Int. Conf. on High Energy Physics, Berkeley (1986), LAL 86-33.

17 I have avoided using the new Particle Data Group names in fear of confusing both myself and the audience. In the new nomenclature f′is $f_2(1525)$ and θ is $f_2(1720)$.

18 Godfrey, S. et al., Phys. Lett. 141B, 439 (1984).
19 Mallik, U., Les Arcs, March 1986, SLAC-PUB-4946.
20 The new Particle Data Group names for the E/ι depend on the spin-parity assignment. The 1^{++} state (perhaps the E?) is called the $f_1(1420)$ and the 0^{-+} state (perhaps the ι?) is called the $\eta(1440)$.
21 Particle Data Group, Phys. Lett. 170B, 1 (1986).
22 There are more states produced in radiative ψ decay than I have discussed. See Königsman, K., Phys. Reports 139, 243 (1986).
23 Bean, A. et al., Phys. Rev. D34, 905 (1986).
24 Eisner, A. M., "Resonance Formation in $\gamma\gamma$ Collisions: Recent Results from the TPC/2γ Collaboration", UC-IIRPA-86-01 (1986).
25 Private communication from G. Gidal.
26 Berkelman, K., Rep. Prog. Phys. 49, 1 (1986).
27 ARGUS: Albrecht, H. et al., Phys. Lett. 160B, 331 (1985);
CLEO: Haas, P. et al., Phys. Rev. Lett. 52, 799 (1984);
CUSB: Klopfenstein, C. et al., Phys. Rev. Lett. 51, 160 (1983);
Crystal Ball: Nernst, R. et al., Phys. Rev. Lett. 51, 2195 (1985).
28 Lovelock, D. M. et al., "Fine Structure of the χ_b' States", and Lee-Franzini, J. et al., "Observation of the $\Upsilon'' \to \chi_b' \gamma \to (\Upsilon')\Upsilon\gamma\gamma \to \mu^+\mu^- (e^+e^-)\gamma\gamma$ and the Hadronic Widths of the χ_b' States", papers submitted to this conference.
29 Rosner, J. L., "Heavy Quark Spectroscopy", in in Proceedings of the 1985 Int. Symposium on Lepton and Photon Interactions at High Energies, ed. Konuma, M., and Takahashi, K., Kyoto Univ. 448(1985).
30 Königsmann, K., "Production, Spectroscopy and Decays of Heavy Quark Bound States", DESY 86-136 (1986).
31 Buchmüller, W., Phys. Lett. 112B, 479 (1982).
32 Gaiser, J., thesis, Stanford Univ. SLAC-255 (1982).
33 Bowcock, T. et al., Phys. Rev. Lett. 58, 307 (1987).
34 Oreglia, M. et al., Phys. Rev. Lett. 45, 959 (1980).
35 Yan, T.-M., Phys. Rev. D22, 1652 (1980).
36 Besson, D. et al., Phys. Rev. D30, 1433 (1984); Gelphman, D. et al., Phys. Rev. D32, 2893 (1985).
37 Peskin, M. E., in Proc. of 11th SLAC Summer Inst. SLAC- 267 (1984).
38 Kobayashi, M. and Maskawa, T., Prog. of Theor. Phys. 49, 652(1973).
39 See talks of B. Winstein on K decays and M. Witherell on Charm decays, in these proceedings.
40 Leveille, J. P., "B-Decays", Univ. of Michigan preprint UMHE81-18 (1981). See Chau, L.-L., Prog. Theor. Phys. Sup. 85, 147 (1985).
41 Guida, J. et al., "Improved Upper Limit on Flavor-Changing-Neutral-Current Decays of the b Quark", CLEO CBX 87-2 (1987).
42 Kane, G. L. and Peskin, M. Nucl. Phys. B195, 29 (1982).
43 See for example: Ma, E., Phys. Rev. Lett. 57, 535 (1986).

44 JADE 84: Bartel, W., et al., DESY preprint DESY 86-001 (1986). MarK II 84A: Jaros, J., in Int. Conf. on Physics in Collision IV, Santa Cruz Cal. (1984). TASSO 86: Caldwell, A., Paper presented to XXIII Int. Conf. on High Energy Physics, Berkeley (1986). MAC 86: Ash, W. W., et al., Phys. Rev. Lett. 58, 640 (1987). DELCO 86: Klem, D. E., Ph.d. thesis, Stanford Univ. SLAC -0300 (1986). HRS 87: Blockus, D., et al., "Measurement of the Lifetime of Hadrons Containing b-Quarks", Argonne preprint ANL-HEP-86-137. See V. Lüth in these proceedings.

45 Bean, A. et al., Phys. Rev. Lett. 58, 183 (1987).

46 This is done more rigorously in Stone, S., "Experimental Determination of the KM Matrix", CLNS-86/753 (1986).

47 Stone, S., in Proceedings of the 1983 Int. Symposium on Lepton and Photon Interactions at High Energies, ed. Cassel D. G., and Kreinick, D., Cornell Univ. p203 (1983).

48 Gasser, J. and Leutwyler, H., Phys. Reports 87, 77 (1982).

49 Ali, A. and Yang, T. C. Phys. Lett. 65B 275 (1976); Altarelli G., et al., Nucl. Phys. B208, 365 (1982); Grinstein, B., Isgur, N., and Wise, M., Phys. Rev. Lett. 56, 298 (1986); Angelini, L. et al., Phys. Lett. 172B 447 (1986); Tye, S-H. H., and Trahern, G. "Semileptonic B Decays", CLEO CBX-85-45 Cornell Univ. (1985).

50 Bebek, C. et al., "Exclusive Decays and Masses of the B Mesons", Cornell preprint CLNS 86/742 (1987).

51 Behrends S. et al., "Semileptonic Decays of B Mesons", Cornell Preprint CLNS-86/738 (1986).

52 Private communication from N. Katayama.

53 Wirbel, W. et al., Z. Phys. C29, 637 (1985).

54 Weseler, S. "Experimental Review on B Meson Decays", in Proceedings of the Int. Symp. on Production and Decay of Heavy Hadrons, ed. by Schubert, K. R. and Waldi, R., DESY,Hamburg (1986).

55 Stech, B., "Lessons from Weak Decays of Heavy Quarks", Heidelberg preprint HD-THEP-86-7 (1986); Bauer, M., and Stech, B., Phys. Lett. 152B 380 (1985).

56 Haas, P., et al., Phys. Rev. Lett. 56, 2781 (1986), and Albrecht, H., et al., "Observation of Inclusive F Production in B Meson Decays", contributed paper to Berkeley Conf. (1986).

57 Bortoletto, D., et al., Phys. Rev. D35, 19 (1987).

58 Baltrusaitis, R. M., et al., Phys. Rev. Lett. 56, 2140 (1986).

59 Alam, M. S. et al., Phys. Rev. Lett. 51, 1143 (1983); Blucher, E. and Stone, S., "Footprints of Charmed Baryons from B Decay, CLEO collaboration internal report CBX-86-65 Cornell Univ. (1986).

60 Gringrich, D., contributed paper.

61 Abachi, S. et al., "Charm D Production in e^+e^- Annihilation at 29 Gev", Argonne preprint ANL-HEP-CP-86-75 (1986).

62 Schindler, R. H. et al., Phys. Rev. D24, 78 (1981).

63 Osterheld, A. et al., "Measurements of total hadronic and inclusive D^* cross sections in e^+e^- annihilations between 3.87 and 4.5 GeV", SLAC-PUB-4160 (1986).

64 Coles, M. W. et al., Phys. Rev. D26, 2190 (1982).

65 Vannucci, F., in Pro. of Int. Europhysics Conf. on HEP, ed. by Nitti, L., Bari, Italy (1985).

66 Bergkvist, E., in Proceedings of the 1985 Int. Symposium on Lepton and Photon Interactions at High Energies, ed. Konuma, M., and Takahashi, K., Kyoto Univ. 448(1985).

67 Bacino, W. et al., Phys. Rev. Lett. 42, 749 (1979); Matteuzzi, C. et al., Phys. Rev. Lett. 52, 1869 (1984) and Phys. Rev. D32, 800 (1985); Burchat, P. R. et al., Phys. Rev. Lett. 54, 2489 (1985); Csorna, S. E. et al., "Limit on the mass of the τ neutrino", CLNS 86/743 (1986).

68 Bacino, W. et al., Phys. Rev. Lett. 42, 749 (1979).

69 Behrends, S. et al., Phys. Rev. D32, 2468 (1985).

70 Gilman, F. and Rhie, S. H., Phys. Rev. D31, 1066 (1985).

71 Gilchriese, M., rapporteur talk at the XXIII Int. Conf. on High Energy Physics, Berkeley (1986).

72 MAC: Fernandez, E. et al., Phys. Rev. Lett. 54, 1624 (1985); MARK II: Jaros, J. A. et al., Phys. Rev. Lett. 51, 955 (1983); TASSO: Althoff, M. et al., Phys. Lett. 141B, 264 (1984); CLEO: Behrends, S. et al., CBX 86-70 (1986); HRS: private communication from D. Miller.

73 Crystal Ball Collab., "Search for Exotic τ Decays", submitted to XXIII Int. Conf. on High Energy Physics, Berkeley (1986).

74 Albrecht, H. et al., "Search for Lepton Number and Lepton Flavour Violation in τ Decays", DESY 86-137 (1986).

ADVANCED ACCELERATOR RESEARCH

R. H. Siemann*

Accelerator Division, Fermi National Accelerator Laboratory**
Batavia, Illinois 60510

ABSTRACT

Advanced accelerator research is aimed primarily at electron-positron collisions at the TeV energy scale. Constraints from particle physics and economics are reviewed and some examples of accelerator concepts are presented.

INTRODUCTION

An important component of progress in particle physics is the development of detectors and accelerators, the instruments of the science. Major accelerator developments in the past include:

1. Strong focusing[1,2] - the basis of the magnetic confinement of every modern synchrotron and storage ring;
2. High power klystrons[3] - the radio frequency (RF) power source of high energy electron linacs, storage rings, and synchrotrons;
3. Stochastic cooling[4] - key to the discovery of the weak intermediate bosons.

At the present time two technically innovative accelerators, the Stanford Linear Collider (SLC)[5] and the Tevatron,[6] are being commissioned, and the experiments at these accelerators may give insight into physics beyond the Standard Model. The Tevatron is based on superconducting magnets; this technology is ready for further exploitation and is the heart of the Superconducting Super Collider (SSC) design.[7] The linear collider concept[8,9] is being tested for the first time at the SLC. With linear colliders electron-positron collisions at energies comparable to the SSC may be possible in the future. This will depend on the success of the SLC and additional

---

* Permanent address: Newman Laboratory of Nuclear Studies, Cornell University, Ithaca, N. Y. 14853

** Operated by Universities Research Association, Inc under contract with the United Stated Department of Energy

accelerator research and development. The need for additional research comes from a combination of constraints from particle physics and economics. These constraints and examples of approaches designed to meet them are discussed in this paper. A more extensive review has been prepared for publication elsewhere.[10]

PARAMETERS AND CONSTRAINTS FROM PARTICLE PHYSICS

Center-of-mass energy and luminosity are the central parameters from particle physics. There is only general guidance to the energy at which new phenomena will occur. The S$\bar{p}$pS collider, the Tevatron, the SLC, and LEP may give better guidance in the future, but without this most accelerator designers are considering the problems associated with a beam energy, $\gamma m_e c^2$, of 0.5 TeV or greater. Because of the γ^{-2} dependence of the annihilation cross section, luminosities greater than 10^{33} $cm^{-2}sec^{-1}$ are needed.

In addition to the hard collisions leading to the production of elementary particles the beams, interact through their electromagnetic fields. One consequence of this is "disruption", one beam focusing the other. The luminosity is given by

$$L = \frac{N^2 \, f \, H_D}{4 \, \pi \, \sigma_x \, \sigma_y}$$

where N is the number of particles per beam bunch (assumed to be equal for the two beams), f is the collision frequency, H_D is a luminosity enhancement factor caused by disruption,[11,12] and σ_x and σ_y are the rms horizontal and vertical beam sizes without disruption.

A second consequence is "beamstrahlung", the radiation of photons in the beam-beam collision. The average fractional energy loss due to beamstrahlung has been calculated in the "classical" regime where the radiation spectrum is that of synchrotron radiation with the critical energy much less than the beam energy. The result is[13]

$$\delta_{cl}(\hbar\omega_c << \gamma) = 0.88 \, \frac{r_e^3 \, N^2 \, \gamma \, H_D}{\sigma_z \, (\sigma_y + \sigma_x)^2} \, .$$

In this equation r_e is the classical electron radius, and σ_z is the rms bunch length. When the transverse acceleration approaches that corresponding to a uniform magnetic field of 4.4×10^{13} Gauss/γ, the critical energy approaches the beam energy. The radiation spectrum is cut-off at the beam energy, and the beamstrahlung energy loss changes to that appropriate for the "quantum" regime[14]

$$\delta_q (\hbar\omega_c >> \gamma) \doteq 0.5 \left[\frac{\alpha^4 N^2 r_e \sigma_z H_D}{(\sigma_x + \sigma_y)^2 \gamma} \right]^{1/3}$$

where α is the fine structure constant. The approximations used in obtaining the above results and the transition from the classical to the quantum regime have been studied in a recent paper.[15]

Combining the above equations for luminosity and δ with an expression for the power of a single beam,

$$P_b = N \gamma m_e c^2 f ,$$

leads to scaling equations for linear colliders[16]

$$\frac{L^2 \gamma^3}{\delta_{cl}} = 7.2 \times 10^{-3} \frac{(\sigma_x + \sigma_y)^2 \sigma_z H_D P_b^2}{r_e^3 (m_e c^2)^2 \sigma_x^2 \sigma_y^2} ,$$

and

$$\frac{L^2 \gamma}{\delta_q^3} = 0.05 \frac{H_D (\sigma_x + \sigma_y)^2 P_b^2}{\alpha^4 r_e (m_e c^2)^2 \sigma_x^2 \sigma_y^2 \sigma_z} .$$

Given the luminosity, center-of-mass energy, and permissible energy spread, which fixes δ, there is a relation between beam sizes and beam power; this is shown in Figure 1.

The spot sizes are extremely small; to achieve these, low emittance particle sources and final focus systems with aberrations corrected to high order are needed. There is a trade-off between beam power and spot size; the minimum practical spot will determine the beam power and, consequently, the operating power of the accelerator complex.

With a short bunch beamstrahlung is in the quantum regime while with a long bunch it is in the classical regime. Usually the bunch length is about 1% of the RF wavelength, and the transition between regimes is for wavelengths in the range of 1 cm to 10 cm.

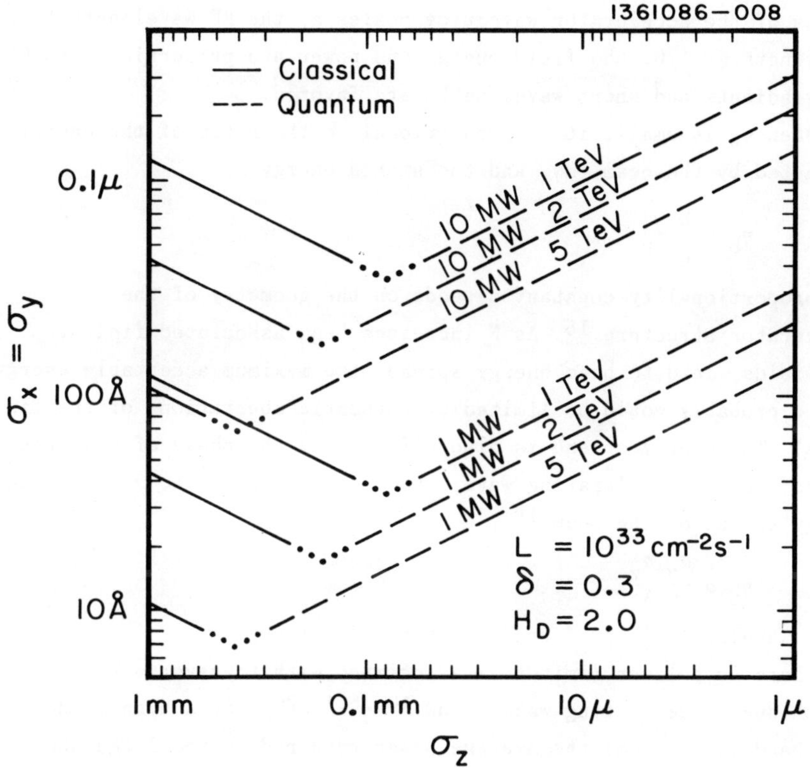

Figure 1: Linear collider scaling laws in the classical and quantum beamstrahlung regimes

ECONOMIC FACTORS

Economic factors and their influences on the choice of wavelength have been discussed by Palmer.[17] An example from this discussion showing the importance of short wavelength is below.

In a linear accelerator the group velocity is significantly less than the speed of light, and an appropriate picture is that the accelerator waveguide is filled with electromagnetic energy. After a filling time the beam enters and is accelerated. It extracts a fraction η_b of the energy, and the remaining energy is lost in the waveguide walls and external loads. The field energy and the average operating power are proportional to the square of a typical field, taken to be the gradient G, times the accelerator volume. Since the radius of the accelerator waveguide scales as the RF wavelength λ and the length as 1/G, the field energy and power are proportional to $G\lambda^2$. Low gradients and short wavelengths are favored.

When η_b is small, it is proportional to the ratio of the energy extracted by the beam, $N\gamma$, and the stored energy

$$\eta_b \propto \frac{N\gamma}{G\lambda^2} .$$

The proportionality constant depends on the geometry of the accelerator structure.[16] As N increases beam associated fields, wakefields, lead to beam energy spread; the maximum acceptable energy spread probably would be limited by chromatic aberrations of the final focus. The energy spread is also affected by the phase of beam with respect to the accelerating wave; choosing this phase to minimize the energy spread of the beam,[18]

$$\eta_b \propto \frac{\sigma_\gamma \sigma_z}{\gamma\lambda} .$$

Short wavelength is preferred for high efficiency.

There is a lower limit to λ associated with transverse emittance growth due to deflecting wakefields.[10,16,19,20] For a high gradient at S-band (λ = 10 cm) the average power required is too large and the energy extraction efficiency too low for use in a high energy linear collider. A wavelength λ = 1 cm and gradients in the 80 MeV/m to 500 MeV/m range are being considered in a number of preliminary designs as a reasonable compromise between accelerator length, wakefields, efficiency, and average power.

TWO-BEAM ACCELERATORS

The RF power requirements of a collider are considerable, and the capabilities and limitations of RF power sources, particularly peak power and efficiency, are central to collider design. For a fixed geometry accelerator, one in which all feature sizes scale as λ, the filling time per unit length is proportional to $\lambda^{3/2}$.[16] Given the stored energy dependence on λ, the peak power demand falls as $\lambda^{1/2}$. However, for a given microwave tube design scaled with wavelength the peak power capability falls as λ^2,[21] and the gap between achievement and requirement grows with decreasing λ.

In addition, higher overall efficiency is possible with higher peak power. η_b is one factor in the overall efficiency which is given by

$$\eta_{overall} = \eta_b \, \eta_s \, \eta_p \, .$$

η_s, the structure efficiency, accounts for energy lost during the filling time, and η_p is the power source efficiency. η_s can be increased by designing an accelerator structure with a high group velocity and a short filling time.[16,18] Therefore, both short wavelength and improving η_s increase the peak power demand, and new ideas for short wavelength power sources with high peak power, high average power, and high efficiency are needed. The "two-beam" accelerator is such an idea.[22]

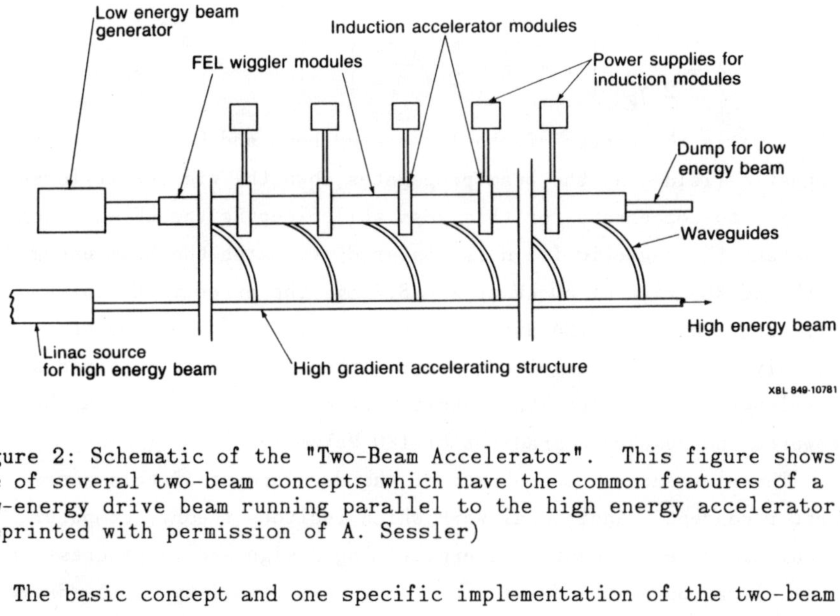

Figure 2: Schematic of the "Two-Beam Accelerator". This figure shows one of several two-beam concepts which have the common features of a low-energy drive beam running parallel to the high energy accelerator. (Reprinted with permission of A. Sessler)

The basic concept and one specific implementation of the two-beam accelerator is illustrated in Figure 2. A low-energy, high-current beam propagates parallel to the collider beam. Energy is extracted from the low-energy beam and guided to the high gradient accelerating structure; this energy is replaced by acceleration modules. Specific two-beam concepts have different energy extraction and reacceleration mechanisms. The advantage of the two-beam accelerator is that in comparison to individual RF power tubes a relatively high energy beam can be used in the power source; this results in high peak power.

In the original two-beam concept of Sessler,[22] which is illustrated in Figure 2, energy is extracted with a Free Electron Laser (FEL) and the beam is reaccelerated with an induction linac. The FEL power source has been tested experimentally, and the results have been published recently.[23] In this experiment, performed with the ETA accelerator at Lawrence Livermore National Laboratory, an 850 A, 3.5 MeV beam was sent through a 3 m long FEL wiggler. In contrast to an oscillator, the FEL power source is an amplifier producing RF with a fixed phase relative to a low level drive signal. The wiggler gives the electron beam a velocity component in the direction of the transverse electric field of the RF wave; this allows transfer of energy from the beam to the wave. The RF wavelength is related to the wiggler wavelength λ_w by

$$\lambda \doteq \frac{\lambda_w}{2\,\gamma_\ell^2} \left\{ 1 + \frac{1}{2} \left[\frac{e\,B_w\,\lambda_w}{2\,\pi\,m_e c} \right]^2 \right\}$$

where γ_ℓ is the energy of the low-energy beam and B_w is the wiggler magnetic field. As the beam propagates down the wiggler it losses energy; to compensate for this loss while keeping the RF wavelength constant the magnetic field is "tapered" to match the beam energy.[24] With and RF drive of 50 kW at $\lambda = 8.7$ mm the experiment achieved an efficiency of $\eta_p = 0.34$ and peak power of 1 GW.[23] This power is roughly an order of magnitude greater than previously obtained at this wavelength. Using the FEL a short section of accelerator has been powered and reached a gradient of 180 MeV/m.[25]

These measurements are part of the development of this accelerator concept. Additional work which includes theory, computer simulations, experiments and engineering design are in progress on: reacceleration of the drive beam, long term stability of the drive beam, microwave components, accelerator structure fabrication, and staging and relative phasing of the drive and high energy beams. Results are generally encouraging.[26]

A two-beam concept being investigated at CERN uses a long wavelength, superconducting linac powered by a CW klystron to accelerate the drive beam.[18] Superconducting RF allows efficient storage of energy between drive beam pulses. In this concept energy is transferred from the low-energy to the collider beam through transfer structures similar to the energy extraction cavities of a klystron. With this energy extraction mechanism the drive beam must be longitudinally bunched at the wavelength of the high gradient accelerator.

Other variants which are being considered are the "relativistic klystron" which employs an induction linac plus klystron output cavities and longitudinal bunching,[27,28] and the combination of superconducting RF with an FEL.[29] All four two-beam accelerators have attractive features; the optimum choice should emerge after study of a wide range of issues such as those mentioned above.

COLLECTIVE IMPLOSION ACCELERATOR

The collective implosion accelerator[30] is based on laser guiding of low energy electron beams, and is illustrated in Figure 3. It works as follows:

1. A high-current, low-energy electron beam (the charging beam) travels through a low density gas creating electron-ion pairs. The electrons are ejected by the radial electric field of the beam.
2. The charging beam has a sharp tail, and it leaves behind an ion column with a line-charge density approximately equal to that of the beam. The ion density can be controlled with gas pressure.
3. Before the ions can disperse a laser pulse photoionizes the gas outside the ion column. The photoelectrons flow towards the ion column producing fields which are accelerating and focusing for the high energy electron beam following closely behind the laser pulse.

The charging beam loses kinetic energy as it travels down the accelerator. The sum of the electrostatic field energy and kinetic energy of the electrons hitting the wall equals this loss. The relative fraction of these two pieces depends on the ion density. Estimates are that between one-quarter and one-half of the beam energy can be converted into electrostatic field energy;[30] this is a relatively high efficiency.

Technical requirements have been estimated by Briggs.[30] For a charging beam such as could be produced by the ATA accelerator at

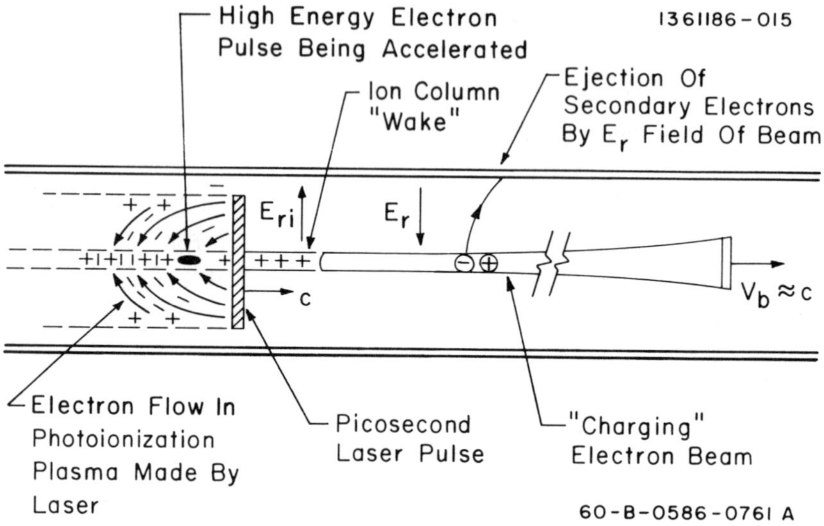

Figure 3: Schematic of the collective implosion accelerator. (Reprinted with permission of R. Briggs)

Livermore the estimated gradient is several hundred MeV/m. The photoionization laser power requirement is close to that available today. The stage length is set by diffusion of the charging beam tail and is expected to be several hundred meters. At the beginning of each stage the charging beam tail would be sharpened, and, if necessary, the laser beam amplified and the charging beam reaccelerated.

A disadvantage is that positron acceleration is not natural. The in-rush of photoelectrons is the initial transient of a plasma oscillation, and it may be possible to use a later phase of this oscillation for positrons.[31] Overall this idea is attractive and deserves further study.

GRAND DISRUPTION

Plasma accelerators have the possibility of gradients in excess of 1 GeV/m, and for this reason they have attracted considerable attention. In the "plasma beatwave accelerator"[32,33] an accelerating wave is produced with a laser with two output frequencies which differ by the plasma frequency, and in the "plasma wakefield accelerator"[34,35] the accelerating wave is produced by a low energy electron beam. "Grand disruption" is closely related;[36] a strong focusing lens is produced in a plasma. This example is chosen as illustrative of two points the first of which is the work on plasma accelerators.

The second is the value of exploring a variety of acceleration mechanisms. The particle physics and economic requirements for a linear collider are extremely demanding, and it is unclear that a speculative idea such as a plasma accelerator can meet these. However, with the trade-off between beam power and spot size a strong final focusing lens can have important impact without having to meet restrictive efficiency demands.

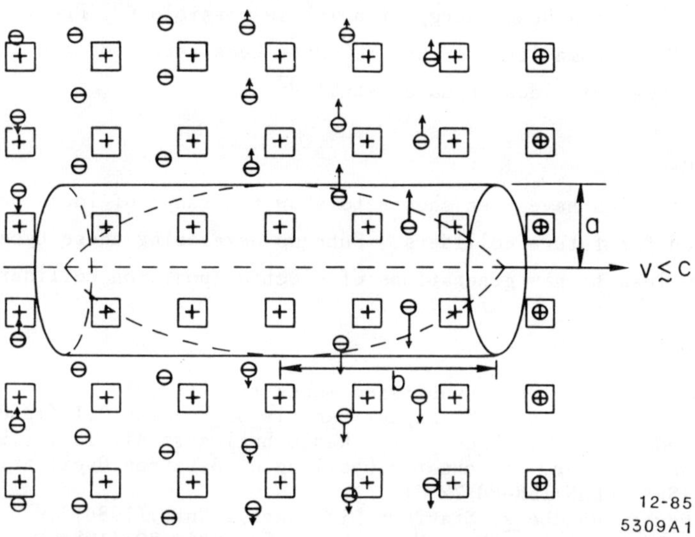

Figure 4: A "standard" electron bunch traversing a plasma. The radius and half-length are a and b, respectively. The parabolic curves indicate the longitudinal density distribution. For the plasma the squares represent ions and the circles electrons. (Reprinted with permission of P. Chen).

The basic idea is illustrated in Figure 4. An idealized electron beam of charge density

$$\rho (r, z) = \frac{3 N}{2 \pi a^2 b} \left(1 - \frac{r^2}{a^2} \right) \left(1 - \frac{(z + b)^2}{b^2} \right)$$

where N is the number of particles in the beam and a and b are defined in the figure. In the case of a long beam bunch and the plasma wavelength, λ_p, chosen such that $b >> \lambda_p/2\pi >> a$ there is a strong focusing force given by[36]

$$F_{perp} = \frac{3 e^2 N}{a^2 b} \left[1 - \frac{(z+b)^2}{b^2} \right] r .$$

The beam is focused, but the focusing force depends on the longitudinal coordinate, z and is a maximum at the midpoint of the bunch.

To be useful as a lens the longitudinal charge distribution must be tailored. With an thin bunch containing $\lambda_p/2\pi b$ of the main bunch charge a distance $\lambda_p/4$ ahead of the main bunch, the focusing is independent of z.[36] For a positron bunch the precursor would have to be $3\lambda_p/4$ ahead. The potential of such a lens is enormous. A focal length of 5 m at a beam energy of 5 TeV is possible.[36] Preliminary studies of a plasma lens for the SLC have been reported, and the plans are to pursue this idea in more detail.[37]

CONCLUSION

The examples have been chosen to show the range of ideas being considered for future colliders. Success developing these types of ideas may lead to new generations of electron-positron colliders.

REFERENCES

1. N. C. Christofilos, unpublished manuscript (1950).
2. E. D. Courant and H. S. Snyder, Annals of Physics **3**,1 (1958).
3. M. Chodorow et al, Proc. Inst. Radio Engineers **41**, 1584 (1953).
4. S. Van der Meer, "Stochastic Coolling of Betatron Oscillations in the ISR", CERN/ISR-PO/72-31.
5. SLC Design Handbook, Stanford Lin. Accel. Cntr.(1984).
6. H. T. Edwards, Ann. Rev. Nucl. Part. Sci. **35**, 605(1985).
7. Superconducting Super Collider Conceptual Design, SSC-SR-2020(1986).
8. M. Tigner, Nuovo Cimento **37**, 1228(1965).
9. U. Amaldi, Phys. Lett. **B61**, 313(1976).
10. R. H. Siemann, "Advanced Accelerator Concepts and Electron-Positron Linear Colliders", to be published in Ann. Rev. Nucl. Part. Sci..
11. R. Hollebeek, Nucl. Instrum. Methods **184**, 333(1981).
12. W. M. Fawley and E. P. Lee, UCID-18584, Lawrence Livermore National Laboratory(1980).
13. M. Bassetti and M. Gygi-Hanney, LEP Note 221, CERN(1980).
14. T. Himel and J. Siegrist, AIP Conf. Proc. **130**, 602 (1985).
15. R. Blankenbecler and S. D. Drell, submitted to Phys. Rev. D.
16. P. B. Wilson, SLAC-PUB-3674, SLAC(1985). An earlier version which contains some errors is P. B. Wilson, AIP Conf. Proc. **130**, 560(1985).

17. R. Palmer, Lecture presented at the 1985 Summer School on the Physics of Particle Accelerators.
18. Schnell, W. CERN-LEP-RF/86-14, CERN(1986).
19. K. L. F. Bane, IEEE Trans. Nucl. Sci. **NS-32**, 2389(1985).
20. V. Balakin et al, Proc. 12th Int. Conf. on High Energy Accel., pp. 119(1983)
21. V. L. Granatstein, private communication.
22. A. M. Sessler, AIP Conf. Proc. **91**, 154(1982).
23. T. J. Orzechowski et al, Phys. Rev. Lett. **57**, 2172(1986).
24. T. J. Orzechowski et al, IEEE J. Quant. Electron. **QE-21**, 831 (1985).
25. D. B. Hopkins, and R. W. Kuenning, IEEE Trans. Nucl. Sci. **NS-32**, 3476(1985).
26. J. Wurtele, AIP Conf. Proc. **156**, in press.
27. R. Marks, LBL-20918/UC-34A, Lawrence Berkeley Lab.(1985).
28. A. M. Sessler and S. S. Yu, "The Relativistic Klystron Two-Beam Accelerator" (1987).
29. U. Amaldi, and C. Pellegrini, CERN CLIC Note 16, CERN(1986).
30. R. J. Briggs, Phys. Rev. Lett. **54**, 2588(1985).
31. R. J. Briggs, Talk presented at the Symposium on Advanced Accelerator Concepts, Madison, Wis (1986).
32. T. Tajima and J. M. Dawson, Phys. Rev. Lett. **43**, 267(1979).
33. C. Joshi et al, Nature **311**, 525(1984).
34. R. D. Ruth et al, Part. Accel. **17**, 171(1985).
35. P.Chen et al, Phys. Rev. Lett. **54**, 693(1985).
36. P. Chen, Part. Accel. **20**, 171(1987).
37. D. B. Cline et al, "Transverse Focussing using Plasma Wake Fields", submitted to the 1987 Accelerator Conference.

STUDY OF ULTRA HIGH ENERGY COSMIC RAYS WITH THE FLY'S EYE

Pierre Sokolsky

Department of Physics, University of Utah
Salt Lake City, UT 84112 USA

ABSTRACT

The operation of the Fly's Eye cosmic ray observatory is described. This detector observes $\geq 10^{17}$ eV cosmic ray extensive air showers via air fluorescence. Results on the cosmic ray spectrum, composition and the ultra high energy neutrino flux, are presented. Plans for a next generation high resolution eye are described.

1. INTRODUCTION

The "Fly's Eye" is a cosmic ray observatory designed to detect ultra high energy cosmic rays ($E \geq 10^{17}$ eV) via air fluorescence.[1,2] It consists of two stations (F.E. I and F.E. II) seperated by 3.3 km. F.E. I consists of 67 62" mirrors, 880 associated photomultipliers and Winston light collecting funnels arranged in clusters of 12 or 14 tubes mounted in the focal place of each mirror. F.E. II consists of 36 such mirrors with 464 associated photomultipliers.

Even though the atmosphere is a poor scintillator ($\leq .1\%$ efficiency between 3000 and 4000 Å), the overwhelming amount of energy liberated by the large number of charged particles in an extensive air shower (EAS) ($n = 10^{7} \sim 10^{12}$) makes it possible to optically detect cosmic rays cosmic rays with energies of 10^{20} eV up to distances of 20 km from the detector. This large fiducial area is necessary to detect fluxes as low as 10^{-17}/cm^{2}-sec-sr.

Experimental measurements presently being carried out with this detector include the following: (1) cosmic ray spectrum between 10^{17} and 10^{20} eV, (2) cosmic ray anisotropy, (3) cosmic ray composition, (4) search for extragalactic neutrino flux, (5) measurement of

$\sigma_{p\text{-}air}^{inel}$, and (6) search for τ-like particles in the cosmic ray flux.

The Fly's Eye can also be used to detect lower energy showers (10^{14}-10^{16} eV) if their axis points towards the detector. In that case, the copious Cherenkov light produced allows a measurement of the shower direction and an estimate of the energy. This mode of operation is primarily useful for γ-ray point source searches and has recently been discussed in detail in reference[3].

Recently, the Fly's Eye group has been joined by a group from the University of Michigan[4] who propose to install a 1000 m² underground muon array around F.E. II. We are deploying a surface electron array to complement theirs. Somewhat further in the future, a much larger, high density surface electron array will be built by a group from the University of Chicago.[5] The surface array will measure (in the case of the Chicago array) EAS directions for $10^{13} \leq E \leq 10^{16}$ eV with $< 1^0$ error, while the underground muon array will tag photon induced showers since the e/μ ratio for hadronic and γ induced EAS is expected to be very different. F.E. II will observe the Cherenkov light from the shower and provide good energy calibration for the surface array. An improvement of several orders of magnitude is expected over present sensitivities to point γ-ray sources.

2. EVENT RECONSTRUCTION & SHOWER SIZE ANALYSIS

Figure 1 shows the geometry of an EAS as seen by the Fly's Eye.

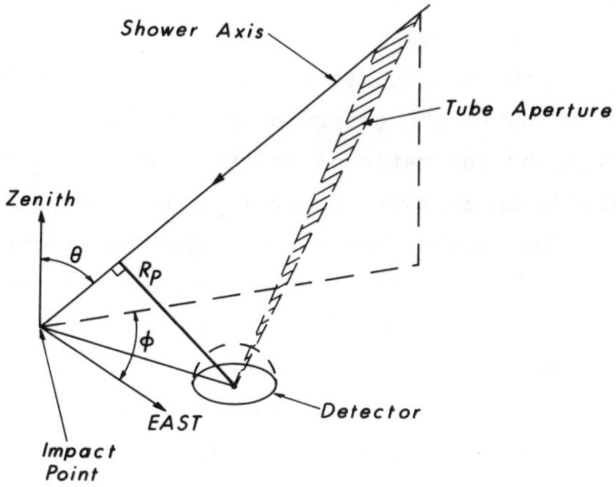

Fig. 1. Event and detector geometry

We obtain the location of the track in space by measuring four parameters: The direction vector of the plane in which the EAS lies, the impact parameter R_p, and ψ, the ground impact angle. The plane is found from the geometry of hit PMT's while the other two parameters are obtained from the relative tube time delays.

For events seen in stereo, there are several additional constraints on the geometry. The track must lie at the intersection of the two event planes defined by the two eyes, and relative timing is available with respect to both detectors.

Once the geometry of the event is known, we use the recorded PMT integrals to convert the light yield into shower size Ne as a function of atmospheric depth, through the following expression:

$Ne = \frac{4\pi}{Y\Delta\theta} \frac{1}{\epsilon A} R_p N_{pe} \exp(R/\lambda)$ where Ne = shower size at the observed location along trajectory, Y = fluorescent light yield (~5 photons/m/electron), ϵ = combined light collection efficiency and photoelectron conversion efficiency (= .17 ± 20%), A = effective light gathering area (1.7 m^2), λ = attenuation length of 3600 Å photons in air (~ 18.0 km), R = distance of EAS segment to detector, R_p = impact parameter of EAS to detector, and $\Delta\theta$ = field of view of PMT.

At viewing angles less than 20-30° Cherenkov light contributes to the photoelectron yield. We deal with this problem by either cutting out tracks viewed at small emission angles or by subtracting the Cherenkov light numerically.

The energy of the EAS is determined by fitting a function (typically a gaussian) to the observed size distribution, integrating it, and multiplying by the ratio of critical energy to the radiation length. This yields an average energy loss per charged particle of 2.18 MeV/gm/cm^2. This number has been checked by intergrating the energy loss over the energy distribution of electrons in an EAS as calculated by Hillas[6] which leads to an average energy loss per particle of 12.24 MeV/gm/cm^2.

Typical reconstruction errors are ± 2-4 degree for zenith angles, ± .1 km for R_p and ± 20% for energy. Systematic errors on the energy calibration are estimated to be 25%. The accuracy of the reconstruction depends on the number of PMT's that view an EAS. This means that although events with $R_p > 20$ km can trigger the system (if their energy is high enough), they will not be reconstructed with adequate accuracy.

The Fly's Eye has been upgraded recently by the addition of filters which pass light with $3200 < \lambda < 3800$ Å. This increases the signal to noise by a factor of 2 and results in a 3 fold increase in data rate. The average reconstructible trigger rate is ~ 30/hour.

3. THE COSMIC RAY SPECTRUM ABOVE 10^{17} eV

j(E), the differential cosmic ray spectrum, is related to N(E), the distribution of EAS observed by the Fly's Eye, through $j(E) = N(E)/tA\Omega(E)$ where $A\Omega(E)$ in km^2 sr is the efficiency corrected acceptance for events with energy between E and E + dE and t is the exposure time. $A\Omega(E)$ is calculated using Monte Carlo techniques, assuming an isotropic cosmic ray flux and an interaction length and shower profile parameterization consistent with the data. Fake events are generated and passed through the standard reconstruction and analysis programs. The calculation is checked by comparing the impact parameter and zenith angle distribution for fake event and data. Figure 2 shows the Fly's Eye acceptance, prior to the addition of filters, as a function of energy.

The data[7] (see Figure 3) shows that between 10^{17} and 10^{19} eV, the differential energy spectrum can be well represented as a power law with exponent $\alpha = 2.94$. Between 10^{19} and 5×10^{19} eV, the spectrum flattens with $\alpha = 2.42 \pm .27$. This flattenting has a 2.4 σ statistical significance. If this spectrum were to continue with this slope beyond 5×10^{19} eV, we expect 11 ± 5 events; we see one event. This data is consistent with recent calculations of the effect of interactions of the primary cosmic ray flux with 2.7° black body radiation (the Greisen-Zatsepin cut-off).[8] If this is correct, cosmic ray sources of the $>10^{19}$ eV flux are likely to be extragalactic since the mean free path for this interaction is ~ 6 Mpc. Recent theories suggest that the local supercluster of galaxies centered on Virgo, may be the source of this flux. However, the evidence is still statistically weak.

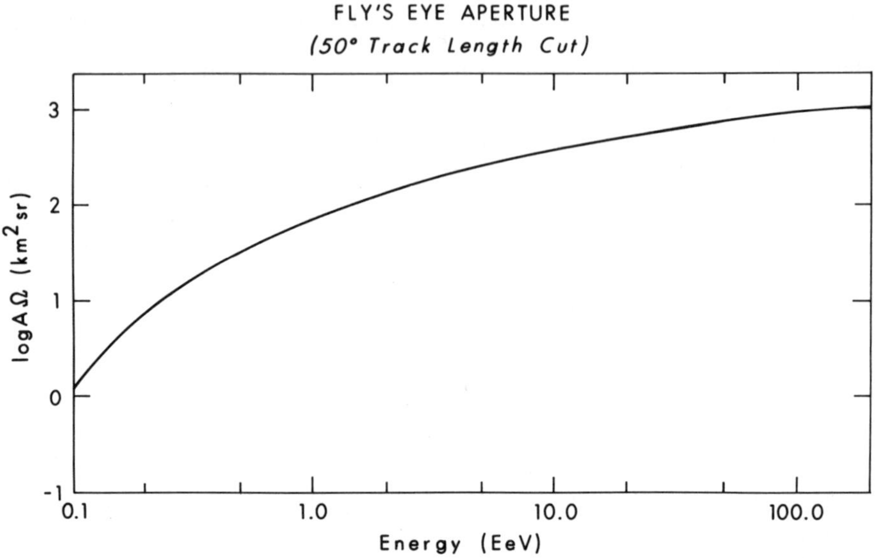

Fig. 2. Fly's Eye aperture (no filters).

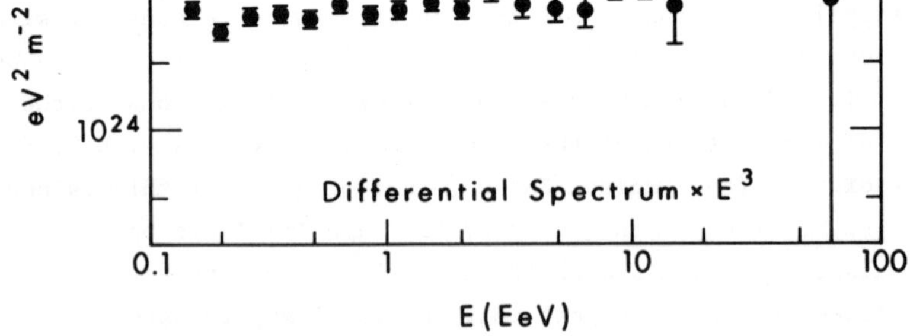

Fig. 3. Differential energy spectrum (no filters).

We have analyzed recent data taken with optical filters. Comparison of the ratio of data rates with and without filters can be used to test the adequacy of the acceptance Monte Carlo since $N(\text{filters})/N(\text{no filters}) = d(A\Omega)_{\text{filt}} t_{\text{filt}} / d(\Omega)_{\text{no filt}} t_{\text{no filt}}$ where N is the number of events as a funtion of energy and t is the accumulated running time for a given condition. Figure 4 shows that the calculation and the data are in good agreement. A preliminary analysis of the spectrum above 10^{18} eV shows that the new data is consistent with the old (see Figure 5).

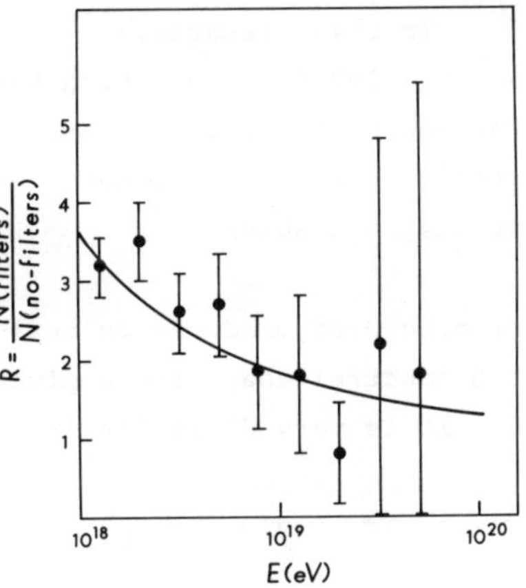

Fig. 4 Ratio of apertures (filters to no filters). Solid line is theoretical prediction.

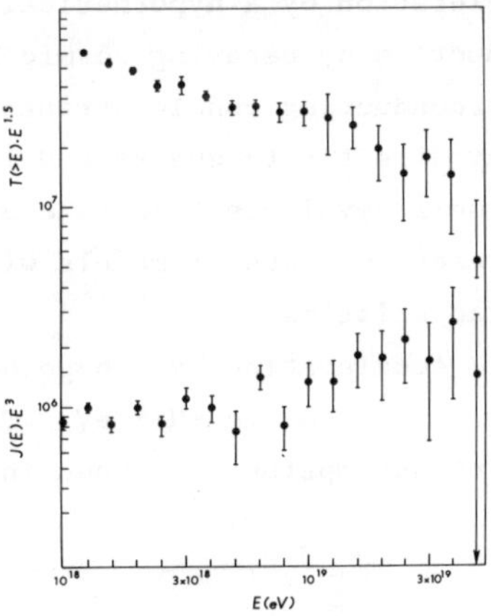

Fig. 5. Preliminary differential and integral spectrum using data from filter & no filter runs. Vertical scale has arbitrary units.

5. IMPLICATIONS OF SPECTRAL FLATTENING FOLLOWED BY CUT OFF.

Observation of structure in the Cosmic Ray spectrum at $>10^{19}$ eV can be used to infer whether the sources of such particles are galactic or extragalactic.

Galactic cosmic rays could have a composition ranging from pure protons to pure Fe, they should begin to show an anisotropy towards the galactic disk at the highest energies (unless they are highly enriched in Fe) and the observed spectral structure above 10^{19} eV must be a consequence of the nature of the acceleration mechanism.

Extragalactic cosmic rays arriving from, say, the Virgo supercluster must be protons, since photo spallation reactions on the 2.7° black body radiation will break-up any heavy nuclei. The cutoff is produced by the Greisen mechanism, i.e., threshold N^* photo production by protons interacting with the 2.7° radiation. If the cosmic ray sources are not uniformly destributed throughout the universe, an anisotropy will be apparent towards the source direction (certainly for $\leq$ 20 Mpc sources).

Until recently, there have been no candidate models for cosmic ray acceleration at $\geq 10^{19}$ eV. There are now two contending models--acceleration by a hypothetical galactic wind,[9] and particle production by decaying cosmic strings.[10] In this scenario, superconducting cosmic strings decay to $\sim 10^{15}$ GeV fermions which then decay directly to quarks and leptons. Although it may be too early in their development to take any model too seriously, these may represent classes of models with similar distinguishing characteristics.

Acceleration by a hypothetical galactic wind produces galactic cosmic rays out to 10^{20} eV. The expected spectral shape for a given set of assumptions is shown in Figure 6. It is very difficult in

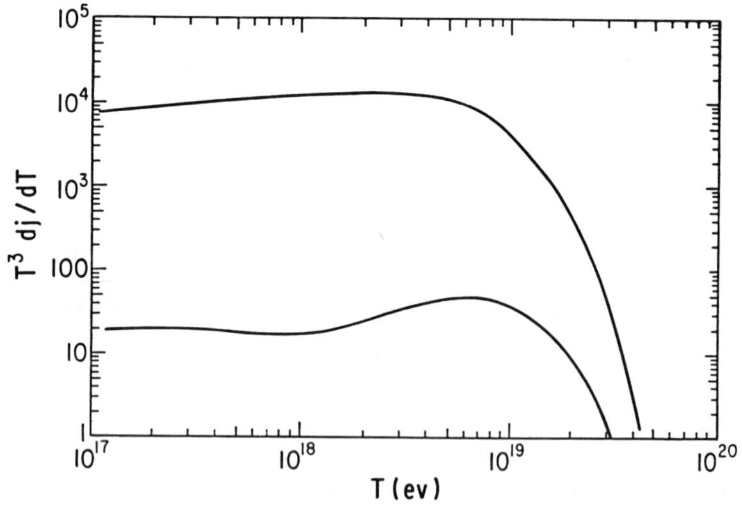

Fig. 6. Cosmic Ray spectrum for galactic wind acceleration model.[9]

this model to accelerate protons beyond 10^{19} eV and hence it would predict an Fe enriched flux near the cut-off. Decaying cosmic strings, on the other hand, are presumably extragalactic, have no heavy nucleon component, and have an associated high energy neutrino flux which may be considerably larger than the flux associated with pion decays from the Greisen mechanism. There may also be a "rebound" in the cosmic ray spectrum at energies considerably greater than the presently observed cut-off (see Fig. 7). It is clear that a

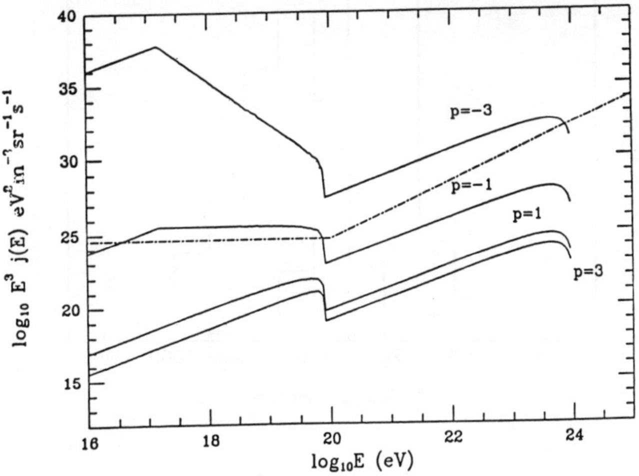

Fig. 7. Spectrum from decaying cosmic strings.[10] P parameter describes evolution of primordial B field.

careful study of the cosmic ray composition above 10^{19} eV will be of crucial importance in discriminating between possible acceleration mechanisms.

6. THE COSMIC RAY COMPOSITION AT $\sim 10^{17}$ eV.

The cosmic ray composition can be measured indirectly by studying the distribution of X_{max}, the depth of EAS shower maximum, in the atmosphere. For a beam of nuclei of atomic number A, the distribution of X_{max} is given by the convolution of the distribution of X_1, the interaction point of the nucleus, with shower development fluctuations. The distributions of X_1 for iron is expected to lie at shallower atmospheric depths than for protons and the shower development fluctuations are smaller for iron than for protons. Figure 8 shows the expected X_{max} distributions for iron and protons

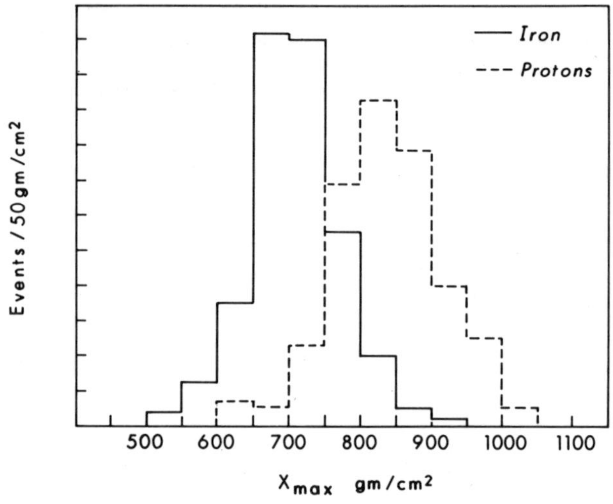

Fig. 8. Expected X_{max} distribution for iron and protons.

with the Fly's Eye acceptance folded in. Figure 9 shows the event distribution for a sample of events with <E> ~ 6 x 10^{17} e V and tight cuts on reconstruction errors. For a simplified two component model of the cosmic ray flux, the data is consistent with a 40% $\pm$ 20% proton composition.

Since geometrical biases can seriously affect this measurement, new stereo data will be crucial. Approximately 40% of tracks are now viewed in stero with much improved reconstruction systematics. The two eyes have been accumulating data since October 1986, and more precise data on composition up to ~ 5 x 10^{18} eV will soon be available.

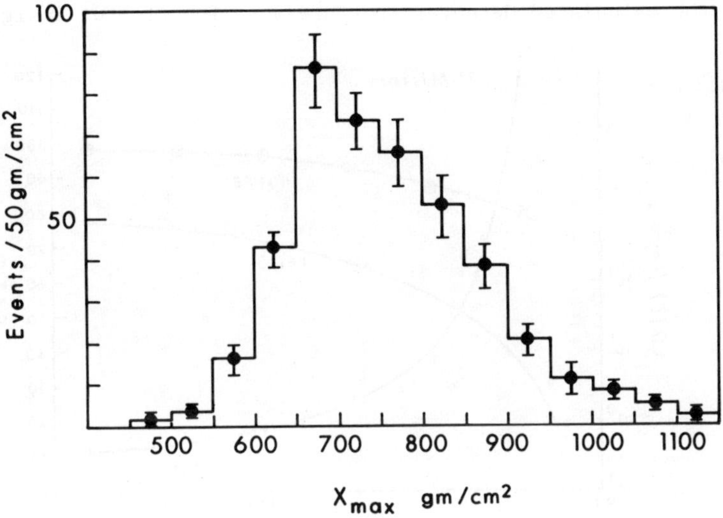

Fig. 9. X_{max} distribution (data).

7. HIGH RESOLUTION EYE.

It is clear that data above 10^{19} eV will be statistically limited with the present detector, and composition studies will remain rudimentary. We have, therefore, embarked on the design of the next generation Fly's Eye detector. The primary goal of this detector is to increase event rates at $\geq 10^{19}$ eV and to improve the X_{max} resolution at all energies for a more precise study of composition. This will be done by decreasing the phototube angular aperture from the present 5^0 to 1^0 and by increasing the mirror diameter from 5′ to between 6′ and 8′. Since the signal to noise of each tube is $\alpha\ (\frac{A}{\Delta\Omega})^{1/2}$ where A is the mirror area and $\Delta\Omega$ is the angular aperture, this should yield a factor of 10 signal to noise improvement over the present eye. Since the signal to noise determines how distant a given energy shower can be before it falls below the triggering thresholds, improvement in S/N implies a larger fiducial volume which in turn implies a greater data rate. Figure 10 shows the calculated detector aperture at a function of energy. A

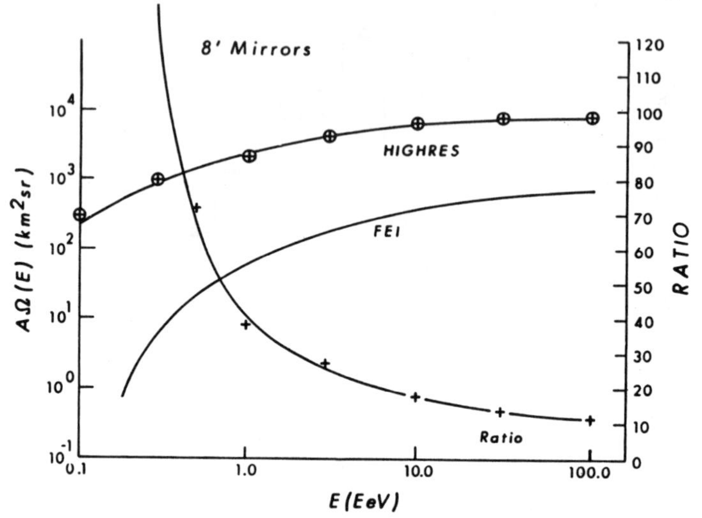

Fig.10. High Resolution Eye aperture & ratio to present Eye aperture.

factor of 10x improvement in data rate relative to the present eye above 10^{19} eV is clearly possible. This would yield ~200 events/year at $>10^{19}$ eV.

Finer sampling of the EAS due to reduced tube aperture also implies an improvement in X_{max} resolution from the present 80 gm/cm² to 15-20 gm/cm². Figure 11 shows a Monte Carlo calculation with full detector resolution of a heavy vs. light composition. Discrimination between these extremes appears possible even by simple inspection. It may also be possible to measure the small differences in fluctuations in individual shower widths expected for protons and Fe and hence have an additional independent method for measuring composition.

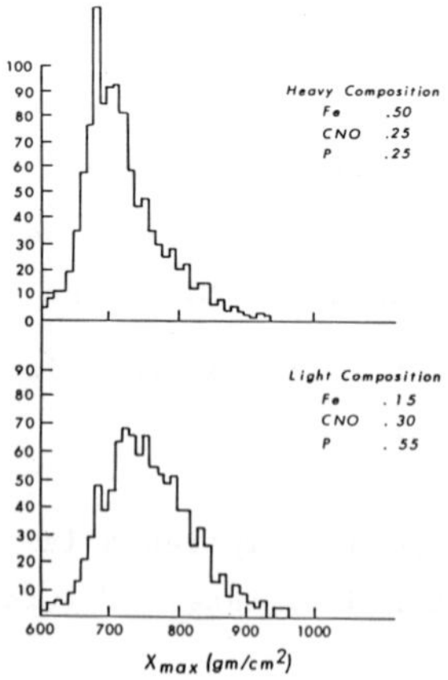

Fig. 11 Expected X_{max} distribution for light and heavy composition for High Resolution Eye.

8. SEARCH FOR UHE NEUTRINO FLUX.

A concomitant effect of the Greisen cut-off is the predicted existence of a diffuse neutrino flux. This flux has recently been calculated by Hill and Schramm[8] and is found to extend to 10^{19} eV, We note that high energy electrons produced by ν_e interaction will have their radiation length increased in the Earth's crust due to the LPM effect[11] for $E_e > 10^{17}$ eV. We can search for such a flux[12] by looking for upward EAS originating in ~100 m of the Earth's crust below the detector. No candidates have been found and we can set limits that are a factor of 50 or more above current predictions. Current cosmic string models are already impacted by these limits, however.[10] The High Resolution Eye should be able to improve on these limits by a factor of 10-50.

9. CONCLUSION

The present eyes have provided a coarse look at the UHE spectrum. We have now identified a specific area of interest, the observed structure in the $>10^{19}$ eV spectrum. To study this further and to distinguish between classes of competing models, a higher resolution detector is required. We are proceeding with the design and prototyping of such a detector. In the meantime, stereo data will provide a detailed look at the composition of cosmic rays in the 10^{17} to 10^{19} eV energy range.

10. REFERENCES

1. The members of the collaboration are: Baltrusaitis, R.M., Cassiday, G.L., Cooper, R., Dawson, R.R., Elbert, J.W., Fick, B.E., Liebing, D.F., Loh, E.C., Sokolsky, P., and Steck, D., all from the University of Utah.
2. Baltrusaitis, R.M., et al., Nucl.Inst and Meth. A240 (1985) 410.
3. Elbert, J.W., et al. in Proceeding of the 1986 Moriond Meeting, in press.

4. Chapman, J.W., Llope, W., Passens, A., Sinclair, D., Thornton, G., and Van der Velde, J.C., "Proposal to Complement the Fly's Eye with Subsurface Muon Detector", University of Michigan.
5. Cambell, M., Cronin,J., Gibbs, K., Krimm, H., Mascarenhas, N., Muller, D., Rosenberg, L.J.,Sauders, H., and Wiedenbeck, M., "A Proposal for a Large Surface Array at Fly's Eye II", Surico Fermi Institute, University of Chicago.
6. Hillas, A.M., J. Phys. G,8, (1982) 1461.
7. Baltrusaitis, R.M., et al., Phys. Rev. Lett. 54 (1985) 1875.
8. Hill, C.T., and Schramm, D.N., Phys. Rev. D, 31 (1984) 564, and references therein.
9. Jokipii, J.R., and Morfill, G., "Ultra-High-Energy Cosmic Rays in a Galactic Wind and its Termination Shock", Jokipii, J.R., and Morfill, G., Ap.J. to be published.
10. Hill, C.T., Schramm, D.N., and Walker, T.P., "Ultra-High Energy Cosmic Rays from Superconducting Cosmic Strings", Fermilab-Pub. 86/146-T, 1986.
11. Baltrusaitis, R.M., et al., Astrophys.J. 281 (1984) L9.
12. Baltrusaitis, R.M., et al., Phys. Rev. D. 31 (1985) 2192.

HEAVY QUARK PHYSICS

M.S. Witherell
Department of Physics
University of California
Santa Barbara, California 93106, USA

ABSTRACT

This is an experimental review of two areas in heavy quark physics: charm decays and production of heavy quarks. Within the area of charm decay, the topics covered are lifetimes, semileptonic decays, and D_s decay modes.

1. INTRODUCTION

The field of heavy quark physics has been an unusually active one in the last year. A large fraction of all the new experimental results presented at this conference fall under this classification. Fortunately, a number of the other speakers in the plenary sessions discussed some of these results, so I can confine my attention to topics not covered by them. I will therefore review the areas of charm decays and production of heavy quarks.

2. CHARM DECAYS

I will first review the general picture of charm decays that has developed over the last few years. There are four quark diagrams that describe the Cabibbo-allowed decays of charmed mesons, and they are shown in Fig. 1. The first two are both spectator diagrams in which the light quark is not one of

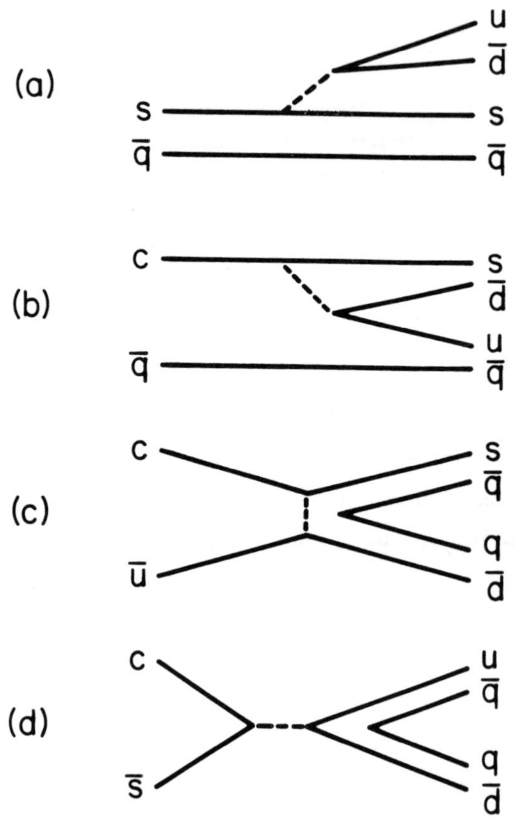

Fig. 1. The quark diagrams for Cabibbo-allowed decays of charmed mesons: (a) spectator diagram with automatic color alignment; (b) spectator diagram without automatic color alignment; (c) W exchange diagram in D^o decay; and (d) W annihilation diagram in D_s^+ decay. In all cases "q" means any one of the three light quarks, u, d, or s.

the four fermions at the weak vertex. Diagram (a) can be called "color-aligned" because the two $q\bar{q}$ pairs are necessarily color singlets. Diagram (b) would be color-suppressed without QCD corrections, because in the absence of gluons the $q\bar{q}$ pairs are not automatically color singlets. The last two diagrams in Fig. 1 are not spectator diagrams, for the light quark is involved at the weak vertex. Diagram (c) is the W-exchange diagram and contributes to D^o decay only. Diagram (d) is W-annihilation and will only produce a Cabibbo-allowed decay in the case of the D_s^+ (formerly the F^+).

An instructive starting point for discussing the physics of heavy quark decay is the ratio of the lifetimes of the D mesons. For some time we have known that the lifetime of the D^+ is much longer than that of the D^o. (The recent measurements give a ratio $\tau_D^+/\tau_D^o \simeq 2.4$). According to the simplest model, in which the charm quark decays with little effect from the light antiquark, this ratio would be unity. Even allowing for final state interactions, one would expect the ratio to be within 20% of this value. The first question to answer, therefore, is why is this ratio so large? Although a great deal of theoretical effort has gone into detailed answers to this question, I will simplify the discussion by listing two effects which could be quite large.

The most obvious candidate to explain the large lifetime ratio is a large contribution from the exchange and annihilation diagrams.[1] Such diagrams contribute to D^o, D_s^+, and Λ_c^+ decays, but are absent from D^+ decays except at the level of Cabibbo-suppressed decays. Thus the D^+ lifetime is quite naturally larger than the other three. These effects were discarded in initial considerations because in the D^o and D_s^+ they are reduced by helicity suppression.

The cleanest way to study the helicity suppression experimentally is in the decay $D^+ \to \mu^+\nu_\mu$. The decay width for this mode is

$$\Gamma(D^+ \to \mu^+\nu) = \tfrac{1}{8\pi}\, G^2\, f_D^2\, m_D\, m_\mu^2\, |Vcd|^2\, (1-(m_\mu/m_D)^2)^2,$$

where f_D is the D decay constant which measures the overlap of the quarks in the D^+ wave function. Although most calculations give values near $f_D = 150$ MeV,[2] it would take a value of around 700 MeV to dominate the D^o decays. Mark III has measured $B(D^+ \to \mu^+\nu_\mu)$ to be less that 8.4×10^{-4} at the 90% confidence level.[3] This corresponds to a limit of 340 MeV for f_D. In the perturbative calculations this limits the nonleptonic exchange and annihilation contribution to less than about 25% of the total D^o decay rate.

The difficulty in applying this result to the nonleptonic decay rate is that higher order corrections including more gluons could be much larger. We must turn to experiment for guidance, and look for decays which are forbidden in the spectator picture. The one such decay already observed is $D^o \to \overline{K}^o \phi$, which has a branching ratio of about 1.2%.[4] When one accounts for the small phase space and the probability of producing an $s\bar{s}$ pair from the vacuum, this decay rate shows no suppression relative to other pseudoscalar-vector modes. This may be due to large nonperturbative gluon corrections. On the other hand, some theorists have suggested that the large $\overline{K}^o \phi$ rate may be due to rescattering of such common final states as $K^- \rho^+$ and $K^{*-} \pi^+$.[5]

A more unambiguous signature of nonspectator effects is available in the D_s decays. It is clear from Figure 1 that all spectator decays of the D_s will have a strange quark in the final state. On the other hand, the annihilation diagram should be dominated by decays with no strange quarks. Thus if annihilation were dominant $B(D_s \to \rho\pi)$ should be larger than $B(D_s \to \phi\pi)$, while it is much smaller in the spectator picture. We do not yet have a sensitive measurement of or limit on this mode, but such a measurement should be possible.

The second possible cause of the large difference in D lifetimes is suppression of the D^+ decay rate by interference.[6] The two spectator diagrams in Fig. 1 lead to identical final states if the spectator quark is a $\bar{d}$. In calculations of either inclusive decay rates at the quark level or exclusive rates into individual final states, there is a negative interference. The net effect of this interference is to raise the D^+ lifetime by as much as a factor of 2-3, so it may account for much of the lifetime difference.[7] Such a large coherent effect is not necessarily maintained for nonresonant multibody decays, but since the total decay rate seems to be dominated by two-body decays, it should be possible to calculate the lifetime ratio with some reliability by summing up these modes.

Already the results on lifetimes and D branching ratios have advanced our picture of charm decays. There are a number of fundamental physics questions which can be given decisive experimental answers. The experimental information that will help us understand more about charm decay physics falls mostly into the areas of (1) precise lifteimes, including baryons; (2) exclusive semileptonic decays; (3) Cabibbo-suppressed and other difficult-to-measure decays of the D^o and D^+; (4) D_s decay modes, especially with no $s\bar{s}$; and (5) mixing and the related double-Cabibbo-suppressed decays. All of these difficult experimental goals are now within our grasp. In the past years we have received results on many of them. Much more will follow in the next year or two.

2.1 Lifetimes

Now that the D^+ and D^o lifetimes are known quite precisely, the confrontation of models with data will occur in the D_s and Λ_c lifetimes. In most models the D_s lifetime is predicted to be quite close to the D^o, and the Λ_c is significantly shorter. The exchange diagram for the Λ_c may well not be as small as for the D^o. There is no helicity suppression, and the wave function could be considerably different.

Besides their intrinsic interest there are other reasons to measure the lifetimes well. Models calculate ratios of decay widths for exclusive modes. We should know the lifetime ratios better than the branching ratios to compare with the models. In addition, many of our theoretical ideas about B decays are based on extrapolation from D decays. We must understand the source of the $D^+ - D^o$ lifetime difference to scale properly to the B system.

The experimental information on lifetimes was presented in parallel sessions, and was reviewed in more detail in V. Lüth's talk.[8] The dominant experiment in the D^o and D^+ lifetimes is E691.[9] Fig. 2 shows the mass spectra for three samples of D^os, and the associated decay time spectra. There are a total of 1360 decays with rather small background, and the corrections to the time distribution are small. The resulting lifetime is $.435 \pm .015 \pm .010$ psec. The D^+ lifetime measured by E691 on the basis of 969 D^+ decays is $1.06 \pm .05 \pm .03$ psec. Other experiments are in agreement with these two measurements, with errors ranging from about 12-16%.[8] The ratio of lifetimes from E691 is $2.44 \pm .14 \pm .08$. This agrees with the ratio of semileptonic branching ratios from Mark III, which is $2.3^{+0.5}_{-0.4} \pm 0.1$.[9]

There has been more recent progress in the measurement of the D_s^+ lifetime. Fig. 3 shows the mass spectra and decay time spectra for two decay modes from E691. On the basis of 99 D_s decays, the measured lifetime is $0.48^{+.06}_{-.05}$ $\pm .02$ ps.[10] A recent measurement from NA11/32 is $0.33^{+.10}_{-.06}$ ps. The world average[8] is $0.44^{+.05}_{-.04}$ ps, which is dominated by these two experiments. The ratio $\tau(D_s^+)/\tau(D^o)$ is therefore 1.0 ± 0.1, which agrees with most predictions.

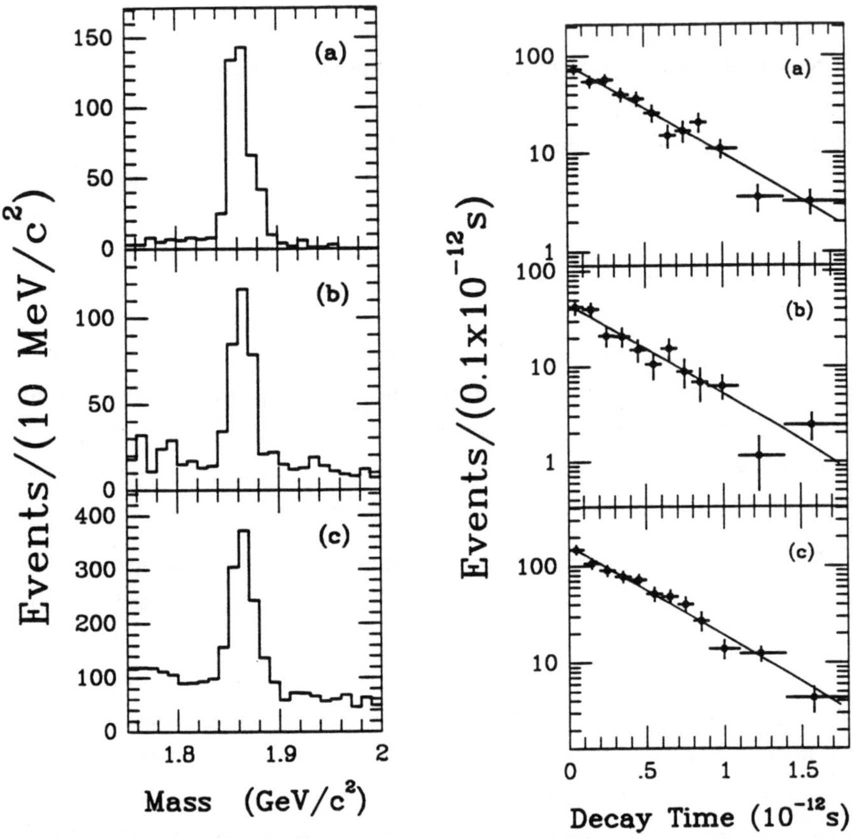

Fig. 2. The events used for the D^o lifetime measurement by E691. On the left are the mass spectra and on the right are the decay time spectra for the three samples: (a) $D^{*+} \to \pi^+ D^o$, $D^o \to K^-\pi^+$; (b) $D^{*+} \to \pi^+ D^o$, $D^o \to K^-\pi^+\pi^-\pi^+$; and (c) $D^o \to K^-\pi^+$, no D^{*+}.

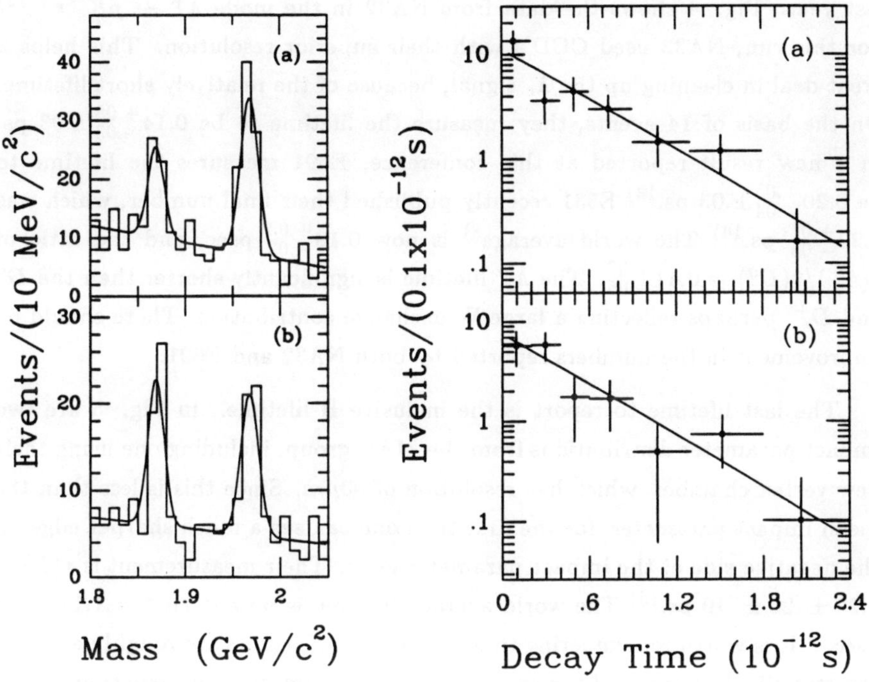

Fig. 3. The events used for the D_s^+ lifetime measurement by E691. On the left are the mass spectra, and on the right are the decay time spectra for the two samples: (a) $D_s^+ \to \phi\pi^+$, and (b) $D_s^+ \to \overline{K}^{*o}K^+$.

It is worth noting that the error on the D_s lifetime has been reduced from about 40% error in the 1986 summary by the Particle Data Group[11] to $\pm 10\%$ presently.

The measurement of the Λ_c^+ lifetime has also improved significantly in the last year. Fig. 4 shows the data from NA32 in the mode $\Lambda_c^+ \to pK^-\pi^+$.[12] For this run, NA32 used CCD's with their superior resolution. This helps a great deal in cleaning up the Λ_c signal, because of the relatively short lifetime. On the basis of 14 events, they measure the lifetime to be $0.14^{+.05}_{-.03} \pm .03$ ps. In a new result reported at this conference, E691 measures the lifetime to be $0.20^{+.05}_{-.04} \pm .03$ ps.[13] E531 recently published their final number, which was $0.20^{+.07}_{-.05}$ ps.[14] The world average[8] is now $0.18^{+.03}_{-.02}$ psec, and the ratio of $\tau(\Lambda_c^+)/\tau(D^o) = 0.41^{+.07}_{-.05}$. The Λ_c^+ lifetime is significantly shorter than the D^o and D_s^+, perhaps reflecting a large W exchange contribution. There should be improvement in the numbers reported by both NA32 and E691.

The last lifetime to report is the inclusive B lifetime. In Fig. 5 are two impact parameter distributions from the MAC group, including one using their new vertex chamber, which has resolution of $90\mu m$. Since this is less than the mean impact parameter, for the first time one can see a much sharper edge on the negative side of the impact parameter peak. Their measurement is $\tau(B) = 1.29 \pm .20 \pm .19$ ps.[15] The world average lifetime is now $< \tau_B >= 1.17 \pm .14$ psec.[8] It is worth remembering that the systematic errors are roughly equal to the statistical errors, and the systematic errors from different experiments are not independent. To improve the situation of the B lifetime further it will be necessary to separate B^o from B^+ events. The ramifications of the B lifetime are discussed in S. Stone's plenary talk at this conference.

2.2 Semileptonic Decays

The reason for giving special attention to the measurement of exclusive semileptonic decays is that they are easier to interpret. Only spectator diagrams can contribute, so there is no uncertainty due to annihilation or exchange diagrams. In addition there is no interference with the leptons in the final state.

To explain the exclusive decays, one must take into account the available hadronic states. The weak interaction leaves a low mass $s\bar{q}$ state which during hadronization becomes K, K^* or $K\pi$. From knowledge about the D and K

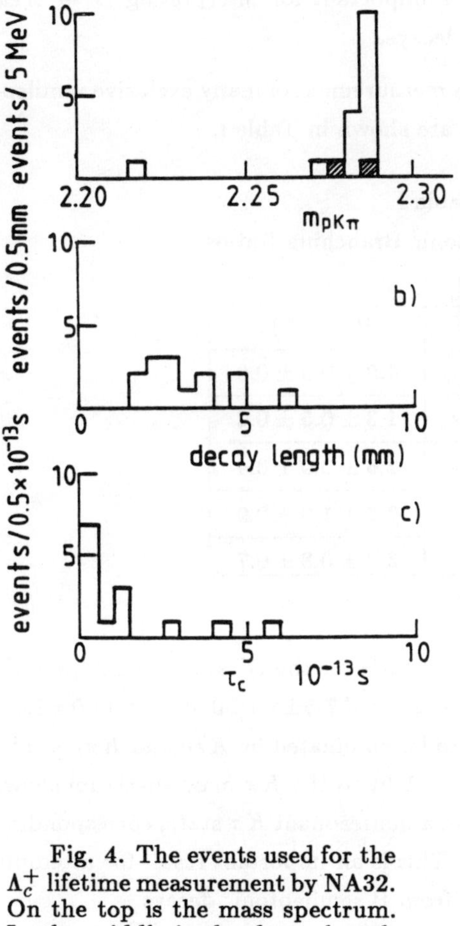

Fig. 4. The events used for the Λ_c^+ lifetime measurement by NA32. On the top is the mass spectrum. In the middle is the decay length spectrum. On the bottom is the decay time spectrum.

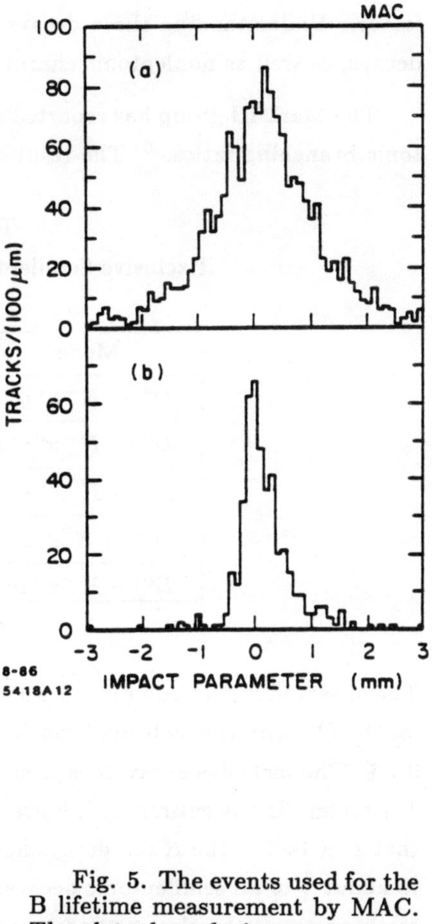

Fig. 5. The events used for the B lifetime measurement by MAC. The plots show the impact parameter distribution for events (a) without and (b) with the new vertex chamber.

wave functions, one should be able to predict the fractions into these states and the spectrum in momentum transfer, $t = m_{e\nu}^2$. The $D \rightarrow Ke\nu$ is a pure vector transition, with a decay rate proportional to $G^2|V_{cs}|^2 \; |f_+(t)|^2$. If one assumes $|V_{cs}| \simeq cos\theta_c$, then the absolute decay rate fixes the normalization of the form factor. Understanding these decays is important for interpreting $B \rightarrow Xe\nu$ decays, as well as nonleptonic charm decays.

The Mark III group has reported a measurement of many exclusive semileptonic branching ratios.[16)] The results are shown in Table I.

Table I

Exclusive Semileptonic Branching Ratios

Mode	B.R. (%)
$D^o \rightarrow K^- e^+ \nu$	$4.0 \pm 0.5 \pm 0.6$
$D^o \rightarrow K^- \pi^o e^+ \nu$	$1.3 \pm 0.5 \pm 0.2$
$D^o \rightarrow \overline{K}^o \pi^- e^+ \nu$	$2.6 \pm 0.9 \pm 0.4$
$D^+ \rightarrow \overline{K}^o e^+ \nu$	$7.2 \pm 1.9 \pm 0.9$
$D^o \rightarrow K^- \pi^+ \nu$	$3.9 \pm 0.8 \pm 0.7$

The observed modes add up to $7.9 \pm 1.1 \pm 0.7\%$ for the D^o and $13.1 \pm 2.2 \pm 1.4\%$ for the D^+ with the inclusive branching ratios of $7.5 \pm 1.1 \pm 0.4\%$ and $17.0 \pm 1.9 \pm 0.7\%$. The inclusive decay does seem to be dominated by $Ke\nu$ and $K\pi e\nu$. The $K\pi$ system is not entirely K^*, however. A fit to the $K\pi$ mass spectrum shows that $45 \pm 14\%$ of the $K\pi e\nu$ decays have a nonresonant $K\pi$ state, corresponding to about $\frac{1}{4}$ of all semileptonic decays. This is an important result for attempts to characterize the electron spectrum from B semileptonic decays.

2.3 D_s Decays

The lifetime is not the only area in which our knowledge about the D_s has increased recently. From three experiments we now have the beginning of detailed branching ratio measurements. The underlying physics issues can be seen in Fig. 1. Consider the two spectator diagrams for the case where the light antiquark in the initial state is $\overline{s}$. Diagram (a) in Fig. 1 is responsible for the decay $D_s^+ \to \phi\pi^+$, while diagram (b) describes the decay $D_s \to \overline{K}^{*o}K^+$. A comparison of these two modes, then, will measure the relative size of these two diagrams in D_s decays.

The W annihilation diagram leads also to the decay $D_s \to \overline{K}^{*o}K^+$, if the $q\overline{q}$ produced from the vacuum happens to be $s\overline{s}$. As mentioned earlier, the unambiguous signal of the existence of this diagram would be in modes without strangeness, such as $\rho\pi$ and $\omega\pi$. As in the case of the D^o and D^+, the ability to calculate lifetime ratios depends on the extent to which the total decay rate is dominated by the two-body modes. This is easiest to check in the mode $K^+K^-\pi^+$.

The ARGUS group has reported the observation of three decay modes of the D_s^+. Besides the $\phi\pi^+$ mode, which for every experiment is the easiest to observe, they published a D_s signal in the mode $\phi\pi^+\pi^-\pi^+$.[17)] Because of the difficult background problem, the signal was just visible, and it was less clear in more recent data. The $\overline{K}^{*o}K^+$ mode has been reported more recently.[18)]

At this conference Mark III reported preliminary results from their running at e^+e^- center-of-mass energy just above threshold for $D_s\overline{D}_s^*$ production.[19)] Their plots of the beam constrained mass are shown in Fig. 6. There are peaks at the D_s mass with modest statistics but low background in three modes: $\phi\pi^+$, $\overline{K}^oK^+$, and $\overline{K}^{*o}K^+$. This represents the only observation of the $\overline{K}^oK^+$ mode to date.

E691 has analyzed the $K^+K^-\pi^+$ mode used for the lifetime analysis, but with tighter vertex cuts which reduce the background even more. The mass plots for the modes $\phi\pi^+$, $\overline{K}^{*o}K^+$ and $K^+K^-\pi^+$ (nonresonant) are shown in Fig. 7. The backgrounds are low enough that even in the nonresonant channel the signal-to-background ratio is about 2. A Dalitz-plot analysis using the lifetime information to help separate signal from background was used to extract the three branching ratios. Finally, E691 also sees the $\phi\pi^+\pi^-\pi^+$ mode, as shown in Fig. 9. Although in all of the other plots the D^+ and D_s^+ are seen in roughly equal numbers, after vertex cuts there is no sign of the D^+ in this channel.

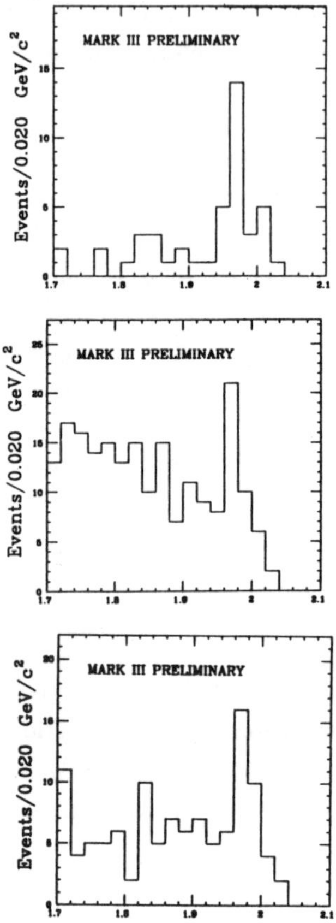

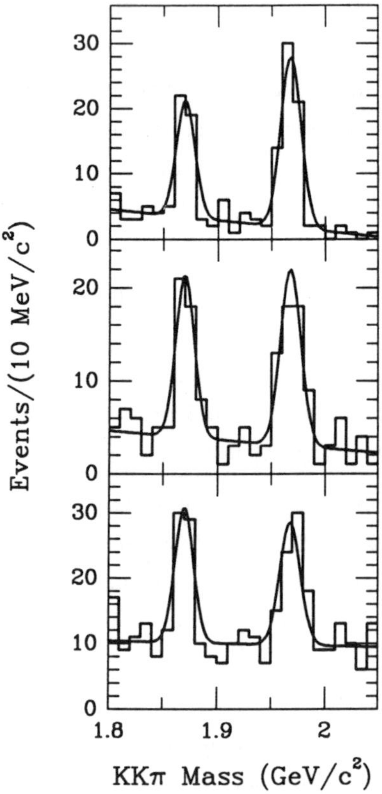

Fig. 6. The D_s^+ peaks from Mark III: Top, the $\phi\pi^+$ mass spectrum; Middle, the $\overline{K}^o K^+$ mass spectrum; Bottom, the $\overline{K}^{*o} K^+$ mass spectrum.

Fig. 7. The $K^+K^-\pi^+$ mass spectra from E691: Top, the $\phi\pi^+$ mass spectrum; Middle, the $\overline{K}^{*o} K^+$ mass spectrum; Bottom, the nonresonant $K^+K^-\pi^+$ mass spectrum. All show peaks at the D^+ and D_s^+.

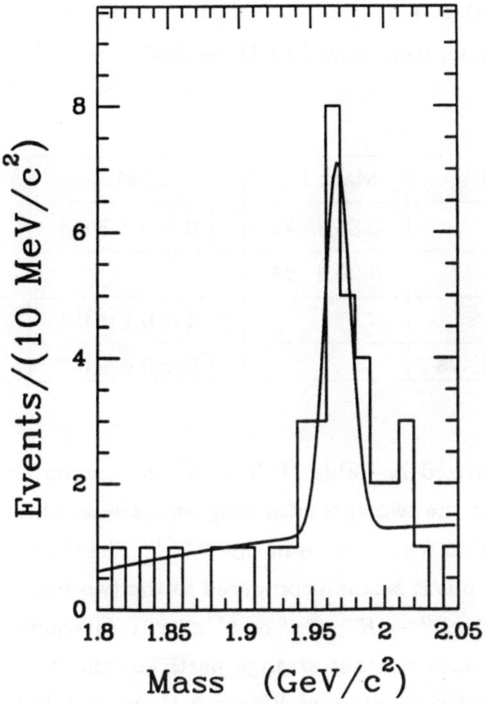

Fig. 8. The mass spectrum for the channel $\phi\pi^+\pi^-\pi^+$ from E691.

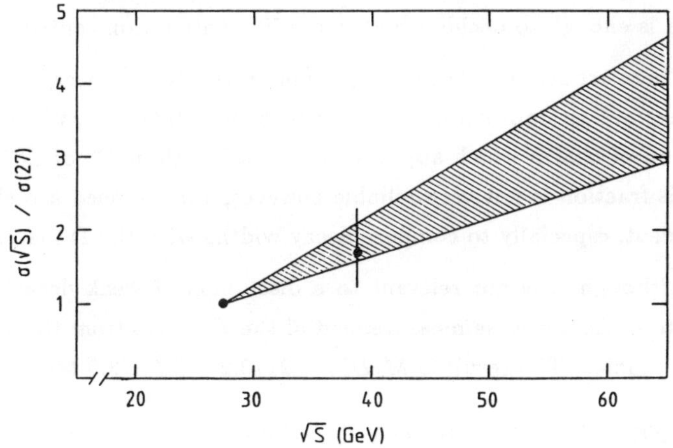

Fig. 9. The measurement by the LEBC Group of the total $D/\overline{D}$ cross section normalized to the value at $\sqrt{s} = 27.4$ GeV versus c.m. energy. The hatched band shows the result of fusion model calculations.

Table II

Relative D_s Branching Ratio, Normalized to $D_s \to \phi\pi^+$

D_s^+ Decay Mode	ARGUS	Mark III	E691
$\overline{K}^{*o}K^+$	$1.44 \pm .37$	$0.85 \pm .23$	$1.0 \pm 0.2 \pm 0.1$
K^oK^+		$0.90 \pm .25$	
$K^+K^-\pi^+$			$0.3 \pm 0.1 \pm 0.1$
$\phi\pi^+\pi^-\pi^+$	$1.11 \pm .37 \pm .28$		$1.0 \pm 0.4 \pm 0.3$

The branching ratios are summarized in Table II. The $\overline{K}^{*o}K^+$ mode is roughly equal to $\phi\pi^+$, confirming that the two spectator diagrams are roughly equal in the D_s system. This agrees with what is seen in the D^+.[20] The three-body decay $D_s^+ \to K^+K^-\pi^+$ is seen, but it is small compared to the two-body modes. This is similar to what is seen in $D^o \to K^-\pi^+\pi^o$ or $\overline{K}^o\pi^+\pi^-$. Although there is no evidence yet for decay modes without strange particles, the very small backgrounds in the $\overline{K}^{*o}K^+$ and $\phi\pi^+$ plots of Figure 8 show that the sensitivity for the decay $D_s \to \rho\pi^+$ will be a small fraction of the $\phi\pi^+$ mode, which is enough to decide whether the W annihilation contribution is large.

There are still no absolute branching ratios for D_s decays. All experiments can extract an approximate measurement for $B(D_s \to \phi\pi^+)$ by assuming the fraction of charm which appears as D_s^+ rather than D^o or D^+. The estimate of this fraction is not very reliable however, and we need a real absolute measurement, especially to compare decay widths with the D meson.

Although it is not relevant to a discussion of weak decays, Mark III was able to make a precise measurement of the D_s^* mass from the samples used for these studies. The result is $M(D_s^*) = 2110.2 \pm 2.1 \pm 3.6$ MeV.

3. PRODUCTION OF HEAVY QUARKS

3.1 Hadroproduction

The first order $[0(\alpha_s^2)]$ QCD diagrams for the hadroproduction of charm are $gg \to c\bar{c}$ and $q\bar{q} \to c\bar{c}$. The fusion model, which is based on calculations to this order, predicts the distributions well but underestimates the cross sections by a factor of 5-10. It is expected that this discrepancy is due to large higher order diagrams.

The average p_T in the data is about 1 GeV/c.[21] The dependence on longitudinal momentum varies from $(1 - x_F)^1$ for charmed particles containing a quark in common with the beam particle, to $(1 - x_F)^5$ for those which do not. The big puzzle in charm hadroproduction is the energy dependence of the charm cross section. The ratio of $\sigma(p + p \rightarrow D + \overline{D} + X)$ at $\sqrt{s} = 60$ GeV is at least 20 times that which is measured at 27 GeV. The LEBC group has recently measured the charm cross section at the Tevatron energy of $\sqrt{s} = 39$ GeV to compare with its SPC measurement at $\sqrt{s} = 27$ GeV.[22] The result, shown in Fig. 9, is $\sigma_{D\overline{D}}(\sqrt{s} = 39\ GeV)/\sigma_{D\overline{D}}(\sqrt{s} = 27\ GeV) = 1.7^{+0.6}_{-0.5}$. This agrees well with the fusion model prediction of 1.6-2.1. The cross section would have to rise by at least a factor of 12 between 39 GeV and 60 GeV to match the ISR results, whereas the fusion model predicts a factor of 2. The LEBC measurement is particularly useful because the systematics should be very similar for the two measurements. Also, there is no A-dependence correction necessary to compare to the ISR results. Finally, the LEBC experiments are not only sensitive to central production, so one cannot ascribe the discrepancy to large forward production.

Even if the perturbative calculations do not give the absolute charm cross sections, there is reason to expect that at the bottom quark mass, the perturbative picture is valid.[23] Unfortunately the data are not in a mature state. WA75 has observed one event in emulsion, corresponding to a $\sigma(b\overline{b})$ of about 5-10 nb.[24] WA78 has reported a signal in muon production from πU interactions, corresponding to a cross section of $4.5 \pm 1.5 \pm 1.5$ nb.[25] Both of these measurements are at $\sqrt{s} = 26$ GeV. UA1 reported at this conference dimuon signal due to bottom production.[26] The cross section is $1.2 \pm 0.1 \pm 0.2 \mu b$ for $p\overline{p}$ interactions at $\sqrt{s} = 600$ GeV, with the selection $p_T > 5\ GeV/c$ and $|\eta| < 2$.

All of these results rely on model-dependent corrections, although the results are in crude agreement with the perturbative calculations. The situation is reminiscent of the first observation of charm in hadroproduction some years ago. To test the model seriously, we need more data from experiments with larger acceptances. Then the experiments can test not only the cross sections, but also such detailed predictions as that the average p_T of the bottom particles should be about 5 GeV/c. Note that at the colliders the fraction of events with $b\bar{b}$ production is about 10^{-4}. With the background suppression now observed in charm experiments using silicon vertex detectors, collider experiments should be able to obtain very clean signals for B decays.

3.2 Photoproduction

The dominant diagram in charm photoproduction should be photon-gluon fusion, in which the photon interacts with a gluon from the nucleon to produce a $c\bar{c}$ pair. In this case the perturbative calculations should be more reliable (perhaps much more) than for charm hadroproduction. There is only one gluon distribution, which makes the calculation less sensitive to the distribution assumed. In addition, the higher order corrections are lower because the photon is in the initial state rather than a gluon.

With E691, there is an experiment with well over 10^4 charm events with low background, and very few cuts on kinematic variables beyond $x_F > 0.2$. For the first time we will be able to make extremely detailed comparison with the photon gluon fusion model, including the correlations between the charm pair. Some early results were presented at this conference.[27)] The $\overline{D}/D$ ratio is measured to be 1.02±.05, consistent with no contribution from $\overline{D}\Lambda_c$. This was not the case at lower energy. The average value of p_T^2 is $1.34 \pm .05 \pm .08\ GeV^2/c^2$, similar to that seen in hadroproduction. The cross section $\sigma(\gamma Be \rightarrow c\bar{c}) = 3.17 \pm 22 \mu b$/nucleus, with a normalization error of ±25%. Finally, the charm cross section has risen by a factor of 1.8 ± 0.5 between 100 and 200 GeV in photon energy.

All of these results agree with the predictions of the photon-gluon fusion model. It will be interesting to see if the model is able to predict the details of the large event sample, including the large double charm sample. In particular, it may be possible to separate the effects of the charm quark mass from those of the gluon structure functions.

4. CONCLUSIONS AND SUMMARY

A review of this type is necessarily a summary of a great deal of work. Rather than summarize it further, I will indulge myself by giving a brief wish list. It contains some of the experimental results that should shape heavy quark physics in the near future. It also allows me to mention interesting results that I did not have time to cover in detail.

In charm decays, measurements of the charm baryons to the level that the mesons are known would give much more information to confront the models. The D_s decay modes should be filled out in much the way that has already happened in the D system. Decay modes such as $\rho\pi$ and $\phi e\nu$ are of particular importance. Very sensitive studies of D mixing are interesting, especially in light of the tantalizing events from Mark III. Finally, there should be further work on the absolute branching ratios in the D system.[28)]

In spectroscopy, the $\ell = 1$ $c\overline{q}$ system looks to be the next area of progress. ARGUS has already seen a broad peak in the $D^*\pi$ mass spectrum which may be some combination of 2 $J^P = 1^+$ states and the 2^+ state.[29)] There are already models of the $c\overline{q}$ potential which predict the characteristics of these states. We need the mass and width of all 4 states. In addition, the charm strange baryons also will be the object of intense study. E-400 presented new results at this conference on the Ξ_c^+.[30)]

In production, it would be very useful to have a high statistics study of charm hadroproduction with few kinematic cuts in the selection process. A careful look at the A-dependence of the charm cross section would also help to clear up old problems, and might lead to new insights. Of course, in bottom production any new information is going to make a large impact.

We heard at this conference some early looks at mixing in the B_s and B_d.[31)] Measurements of the mixing parameters for each meson individually should be a high priority. The observation of $b \rightarrow u$ decays, rather than just an upper limit, would contribute the most to our understanding of the family-mixing matrix. Individual lifetime measurements for the B^+, B_d^o, and B_s^o may be not too far away. Of course the most important goal for the B system is to see enough decays to *expect* to see CP violation, assuming the standard phenomenology. Unfortunately this goal may lie outside the reach of any experiment now in progress or under construction.

Finally, I come to the top quark. The fact that there is nothing new to report is in itself important. One must now assume that we have no idea what the mass of the top quark is. Its unambiguous discovery would be a breakthrough, of course.

There has been a great deal of experimental progress in the last year in the area of heavy quarks. To see this, one needs only to look at the fraction of time in the experimental plenary talks at this conference devoted to heavy quark physics. There are a number of theoretical predictions that are well determined from quite fundamental considerations, and that can be given decisive experimental tests. To the traditional method of studying heavy quark decays in the clean e^+e^- system, we can add the technique of photo- and hadroproduction with high statistics and vertex detectors to clean up charm signals. The safest prediction one can make is that at the next DPF meeting there will be a similarly large number of new results in all areas of heavy quark physics.

REFERENCES

1. M. Bander, D. Silverman, and A. Soni, *Phys. Rev. Lett. 44*, 7 (1980); H. Fritzsch and P. Minkowski, *Phys. Lett. 90B*, 455 (1980).

2. V.A. Novikov *et al.*, *Phys. Rev. Lett. 38*, 626 (1972); V.S. Mathur and M.T. Yamawaki, *Phys. Rev. D29*, 2057 (1984); H. Krasemann, *Phys. Lett. 96B*, 397 (1980).

3. S. Wasserbaech, These Proceedings; R.H. Schindler, SLAC-PUB-4055.

4. H. Albrecht *et al.*, *Phys. Rev. Lett. 158B*, 525 (1985); R.M. Baltrusaitis *et al.*, *Phys. Rev. Lett. 56*, 2136 (1986); and C. Bebek *et al.*, *Phys. Rev. Lett. 56*, 1893 (1986).

5. J. Donoghue. *Phys. Rev.*, *D33*, 1516 (1986), but also see L.-L. Chau, in Proceedings of the International Symposium on Production and Decay at Heavy Hadrons, Heidelberg (1986) (hereafter referred to as Heidelberg Proceedings).

6. B. Guberina *et al*, *Phys. Lett. 89B*, 111 (1979); R.D. Peccei and R. Rückl, in Proceedings at the Symposium on Special Topics in Gauge Field Theories, Ahrenhoop, DDR (1981); and M.B. Voloshin and M.A. Shifman, *Yad. Fiz. 41*, 187 (1985).

7. M. Bauer and B. Stech, *Phys. Lett. 152B*, 380 (1985); M. Wirbel, in Heidelberg Proceedings.

8. V. Lüth, Heidelberg Proceedings and these Proceedings.

9. J. Anjos *et al.*, *Phys. Rev. Lett.*, *58*, 311 (1987).

10. J. Anjos *et al.*, Fermilab preprint, Fermilab-PUB-87/29-E, submitted to *Phys. Rev. Lett.* (1987).

11. Particle Data Group, Review of Particle Properties, *Phys. Lett. 170B*, 1 (1986).

12. S. Barlag *et al.*, *Phys. Lett. 184B*, 283 (1987).

13. L. Cremaldi, contributed talk at this conference.

14. N. Ushida *et al.*, *Phys. Rev. Lett. 56*, 1767 (1986).

15. W.W. Ash *et al.*, *Phys. Rev. Lett. 58*, 640 (1987).

16. D. Coffman, *Caltech Thesis* (1986); and D. Hitlin, Heidelberg Proceedings.

17. H. Albrecht *et al.*, *Phys. Lett. 153B*, 343 (1985).

18. H. Albrecht *et al.*, *Phys. Lett. 179B*, 398 (1986).

19. W. Toki, SLAC-PUB-4153 (1986) and S. Wasserbaech, These Proceedings.

20. R.M. Baltrusaitis *et al.*, *Phys. Rev. Lett. 55*, 150 (1985).

21. For a detailed summary of these results see S. Reucroft, Proceedings of the 6th International Conference on Physics in Collision, Chicago (1986).

22. R. Ammar *et al.*, *Phys. Lett.* *183B*, 110 (1987).

23. E.L. Berger, J.C. Collins, and D.E. Soper, Argonne preprint ANL-HEP-PR-86-80 (1986).

24. J.P. Albanese *et al.*, *Phys. Lett.* *158B*, 186 (1985).

25. M.G. Catanesi *et al.*, Berkeley Conf. Preprint, Berkeley (1986).

26. G. Arnison *et al.*, UA1 preprint (1986); D. Summers, These Proceedings.

27. M. Purohit, These Proceedings.

28. See E.H. Thorndike, These Proceedings, for a detailed discussion of the absolute branching ratios.

29. H. Albrecht *et al.*, *Phys. Rev. Lett.* *56*, 549 (1986).

30. P. Coteus, contributed talk at this conference.

31. D. Summers and B. Winstein, These Proceedings.

STRING THEORY IN 1987

EDWARD WITTEN*

Joseph Henry Laboratories
Princeton University
Princeton, New Jersey 08544

Let us first recall some facts about quantum field theory, in which the Feynman propagator is so central.

To Feynman, the propagator

$$\langle y | \frac{1}{\Box} | x \rangle \tag{1}$$

was a sum over physical paths from x to y. Indeed, one writes

$$\begin{aligned} \langle y | \frac{1}{\Box} | x \rangle &= \int_0^\infty d\tau \, \langle y | e^{-\tau \Box} | x \rangle \\ &= \int_0^\infty d\tau \int_{X(0)=x}^{X(\tau)=y} \mathcal{D}X \, exp - \frac{1}{2} \int_0^\tau d\tau' \left(\frac{dX}{d\tau'} \right)^2 \end{aligned} \tag{2}$$

In the last formula, τ can be interpreted as proper time, and $\mathcal{D}X$ is an integral over all paths that run from x to y in proper time τ.

* Research supported in part by NSF Grants PHY80-19754 and 86-16129.

So when we look at a Feynman diagram (figure (1)), we can think of it as a sum over actual processes in spacetime. In the figure, initial and final particles are emitted and absorbed at spacetime points x, y, z, and w. Interactions —joining or splitting of particles —occur at vertices marked a, b, c, d, e, f. A solid line in the diagram denotes a Feynman propagator, or in other words an integral over paths joining the initial and final points, according to the prescription of equation (2).

Now, there are basically two defects of quantum field theory.

(a) The infinities are a chronic nuisance, and above all make it impossible to include gravity.

(b) There are too many theories, infinitely many, in fact.

These two properties can be viewed as consequences of one basic feature of Feynman diagrams, namely the fact that a "graph" is not a manifold. The graph depicted in figure (1) is a manifold except for singularities at the interaction vertices marked a, b, c, d, e and f. It is because of these singularities that we call it a Feynman graph, not a Feynman manifold.

The infinities arise in the region of integration over a, b, c, d, e and f where these points (or suitable subsets thereof) are coincident in spacetime. This is a Lorentz invariant notion, since all observers can agree that a, b, c, d, e and f were exceptional moments in the history depicted by the graph.

The other consequence of the fact that a Feynman diagram is not a manifold is that there are too many quantum field theories. Even after choosing the integration measure of (2) for integrating over free motion of a particle, we are still entitled to include special factors at interaction vertices, in formulating our rules for evaluation of Feynman diagrams. This is sketched in figure (2). The option of including special factors at interaction vertices is not just a technicality; it is used, in a rather intricate way, in constructing the most beautiful field theories, such as Yang-Mills theory and general relativity. It is the reason that there are infinitely many possible field theories.

Suppose, though, that we wish to make a *physical model* of a Feynman diagram. We don't have mathematical lines at our disposal, so we use, perhaps, hollow tubes (figure (3)). Now we are discussing string theory.[†] The tube is the world sheet of a closed loop or string moving forward in time, just as the lines in a Feynman diagram are world lines of propagating point particles. The basic novelty of the string diagram depicted in the right of figure (3) is that it is a smooth two dimensional manifold with no analogue of the interaction vertices of the ordinary Feynman diagram. Thus, any small part of a string diagram looks locally like it describes a small piece of freely propagating string, and after we decide what we mean by propagation of a free string there is little more that can be done in building the interacting diagram.

From these simple remarks about string diagrams, it is reasonable to expect that string theories—if they make sense at all—will be free of infinities, and few in number.

The real surprise is that the resulting string theories are vastly richer and more *physical* than one could think. When one quantizes the motion of a free string, one is bound to find some spectrum of Lorentz multiplets with definite spin and mass. There is no obvious reason, though, that this should be a particularly interesting spectrum. The real treat is that when one quantizes the motion of a free string, one finds massless spin one and spin two fields with Yang-Mills and gravitational couplings. (The details depend on which string theory one considers.) One finds, in fact, not just gauge theories and gravity, but also their supersymmetric generalizations.

To be more blunt about it, when one quantizes the motion of a string, one finds theories that

---

† For reviews, see e.g. D.J. Gross and M.B. Green, *Unified String Theories* (World Scientific, 1986); M.B. Green, J.H. Schwarz, and E. Witten, *Superstring Theory* (Cambridge University Press, 1987). In this talk I will not attempt a systematic review of string theory or its history. A more general review of string theory has been given at this conference by Schwarz.

(a) are finite and few in number ($\lesssim 4-6$).

(b) have the most beautiful field theories as their low energy limits.

And one finds that the seemingly innocent step of replacing points by strings leads to theories with a remarkably rich mathematical structure, bringing us into the realm of Riemann surfaces, algebraic geometry, infinite dimensional Lie algebras, and more. In fact, in many areas, string theory brings us to the frontiers of mathematics.

There are three things about string theory which show that we should take it seriously. I will number them from 0 to 2 (not 1 to 3) and the first point on my list is indeed often passed over without care.

(0) The first point is simply that string theory is a consistent generalization of the known logical framework of physics. This is remarkable in itself, as that framework is very rigid. Its rigidity is indeed one of the main reasons that we've been able to make progress in theoretical physics. Experimental input has always been crucial, but our ability to interpret experimental data correctly has been greatly enhanced by a tight-knit logical framework —relativity, quantum mechanics, etc. —which severely constraints the ideas that theorists can try. History shows just a few cases when the then current logical frame work of physics was successfully deformed to something new. For instance, Einstein deformed Galilean relativity to special relativity; the physicists of the 1920's deformed classical mechanics to quantum mechanics. Such developments have become increasingly difficult as the logical structure of physics has become increasingly rigid. History gives few examples when physicists conceived of a logically consistent deformation of the basic structure of twentieth century physics which however proved to be wrong.

(1) String theory overcomes what has been for so long the central, most inaccessible problem of physics —to reconcile general relativity and quantum field theory. Insofar as there have been periods in the last few decades when this problem was not on the forefront of our minds, it was largely because it appeared to be too difficult. History shows us how fruitful it is to think about clashes between established

fundamental theories such as general relativity and quantum field theory. Many fundamental advances in the twentieth century were derived from attempts to overcome just such problems. Special relativity, for instance, was born from Einstein's efforts to reconcile electromagnetism and classical mechanics. General relativity was likewise born from Einstein's efforts to reconcile special relataivity with Newtonian gravity. And finally, quantum field theory was created by physicists who wished to reconcile special relativity with nonrelativistic quantum mechanics.

Thus, history shows that reconciling fundamental but incompatible theories is very often a way to make significant progress. What is more, there is another lesson. History shows that in dealing with such a fundamental problem, physicists usually flounder around for a long time without saying anything significant. Then, when a real idea emerges, it proves to be on the right track much more often than you would think. It is not the case that the right idea is usually preceded by N good but wrong ideas.

(2) Finally, we've learned in the last few years that the remarkably rich structure known as string theory is capable, at least in broad terms, of unifying the forces and giving a low energy world that looks like what we see. We can write a dictionary relating observed physics in four dimensions to properties of an underlying string theory vacuum, though there are many blank entries in this dictionary and many unsolved problems. The outstanding blank entry in the dictionary is undoubtedly the whole question of why the cosmological constant vanishes after supersymmetry breaking.

Let us now turn our attention to another theme. I would like to compare the development of two branches of physics.

Quantum mechanics started in 1900 with Planck's work on black body radiation. It was largely ignored for five years. Einstein in 1905 and then Debye in 1906-7 introduced new ideas about the photoelectric effect and the specific heat of solids. Still another period went by with quantum mechanics not yet at the forefront of the attention of most physicists. Then in 1913, Bohr made the first real try of a quantum model of an atom. After this, it finally was generally realized that quantum mechanics was something very important. There followed about eleven years of confusion about what quantum mechanics really was, until after 1924 Schrodinger and others straightened things out. It still took another quarter century before the proper understanding of QED.

The latter thought is sobering. Planck's black body formula was really a formula about electromagnetism in the quantum domain, but a proper understanding of QED took fifty years after Planck.

What about string theory? The Veneziano model in 1968 started a period of explosive growth that lasted until 1974. Then interest faded, because the original motivation of applying string theory to strong interactions faded with the emergence of QCD. In the early 1980's, the subject was revived, by Green and Schwarz especially, and it attracted broad attention after about 1984. In January, 1987, we are beginning what might be called the nineteenth year of string theory.

Nineteen years after Planck, it was clear that quantum mechanics was important, but not at all clear what quantum mechanics was. Nineteen years after the Veneziano model, we are in a similar situation.

In 1919, most physicists —even those working on quantum mechanics —greatly underestimated the scope of the intellectual departure that quantum mechanics would entail. This is, again, probably the case today. Indeed, it is part of our basic quandary as string theorists.

General relativity was invented in what we may call a "logical" way. Einstein began with the physical concepts —the equivalence principle, and general covariance. Then he formulated the mathematical framework, which was Riemannian geometry. Finally, he formulated the theory.

String theory had very different origins. It was invented almost by accident, certainly with no grand conception of what it is all about. It was developed through a long process of tinkering —and I must apologize for not attempting in this talk to review this history, with the many important contributions that were made.

Probably something like this is the only way string theory *could* have been invented in the twentieth century. We certainly haven't had the conception for it. It is in this spirit that it has been said that string theory is part of the physics of the twenty-first century that fell by chance into the twentieth century. It is our good fortune, as twentieth century physicists, that string theory was invented, giving us the chance to work on this marvelous structure, but the price we pay for the strange way that string theory was discovered is that we are working on a theory whose fundamental concepts are unknown to us.

As a presumed worthy successor of general relativity, string theory must be based on some suitable generalization of the principle of equivalence and Riemannian geometry. But we have only glimmers of this.

The limited understanding of what string theory is hampers us immensely in trying to use string theory to solve physical problems. And it means that we have scarcely begun to conceive the full impact of string theory on physics.

I cannot give you a good idea of what the logical setting for string theory may prove to be. To try to convey a flavor of the storm clouds that —as in 1919 —however over the physical landscape, I will mention two lines of thought often considered by workers in the field.

(a) It is not at all clear that spacetime points mean anything in string theory. This shows up, for instance, in the difficulty in defining local observables. It is very possible that a proper understanding of string theory will make the spacetime continuum melt away, just as the continuum of classical phase space melted away in 1924. That happened, recall, when the coordinates p and q of classical phase space ceased to commute:

$$[p, q] = -i\hbar$$

Space-time might likewise melt away if, e.g. continuous functions on spacetime are replaced by some noncommutative ring, or if everything is somehow generated from geometry of Riemann surfaces, with space-time as some effective concept.

(b) The transition from classical to quantum mechanics is remarkably smooth in the case of string theory. The classical theory involves certain computations on the Riemann sphere, while the quantum corrections involve precisely the same computations on a surface of higher genus (figure (5)). This exceptionally close relation between classical and quantum mechanics in the case of string theory leads some physicists to a radical view. The view is that in string theory, it is only the quantum theory that is truly natural; and the logical structure that we seek will emerge only in the quantum domain. If so, if the desired successor to the principle of equivalence were intrinsically quantum mechanical, then string thoery might well be said to have explained quantum mechanics.

My personal guess about these questions is that space-time will melt away but quantum mechanics will remain an enigma. But all of this remains to be seen.

I would like to conclude by briefly citing a few examples of physical ideas that came to fruition, and got experimental proof, only after many years. Since our subject is so closely tied to general relativity, I will choose examples from relativity and relativistic astrophysics.

In the 1930's, it was clear that old stars (of certain mass) should collapse to neutron stars. Due to the smallness and dimness of those stars, observing them seemed like science fiction, and for decades interest in them was slim. Then, in the late 1960's, radio astronomers were lucky, and discovered neutron stars in the form of pulsars.

Gravitational lenses were put forward by Einstein in the 1930's for illustrative purposes. Observing them seemed fanciful for forty years —until they were discovered in observations of quasars. Now they are emerging as a basic astrophysical tool.

Finally, I come to gravitational waves. They were one of the basic predictions of general relativity when it took its final form in 1915. For many decades it appeared that there was no hope of testing this fundamental prediction. That other bit of science fiction, neutron stars, had to come true before it became possible to use the binary pulsar to test the theory of gravitational waves. And sometimes experimental techniques develop farther than we think. For many years, direct laboratory detection of gravitational waves was not a serious prospect. But now there is a good chance that this will occur by the end of our century.

What is the moral of this story? To test string theory, we will probably have to be lucky. But in physics, there are many ways of being lucky.

FIGURE CAPTIONS

(1) A Feynman diagram. Particles are emitted or absorbed at space-time events x, y, z, w. Particles split or rejoin at space-time events a, b, c, d, e, f. Solid lines represent sums over physical paths of propagating particles.

(2) (a) A freely propagating particle;

(b) Interaction vertices at which special factors may be inserted.

(3) A Feynman diagram and its string counterpart.

(4) The world line of a point particle (left) and the world sheet of a string (right). Time is flowing vertically.

(5) The perturbative expansion of string theory involves a sum over Riemann surfaces of varying topology. The first corresponds to the classical approximation, while others are quantum corrections.

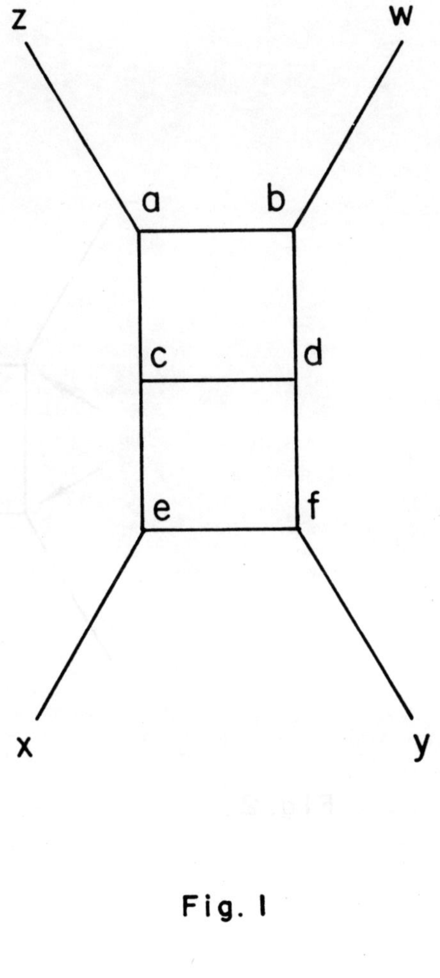

Fig. 1

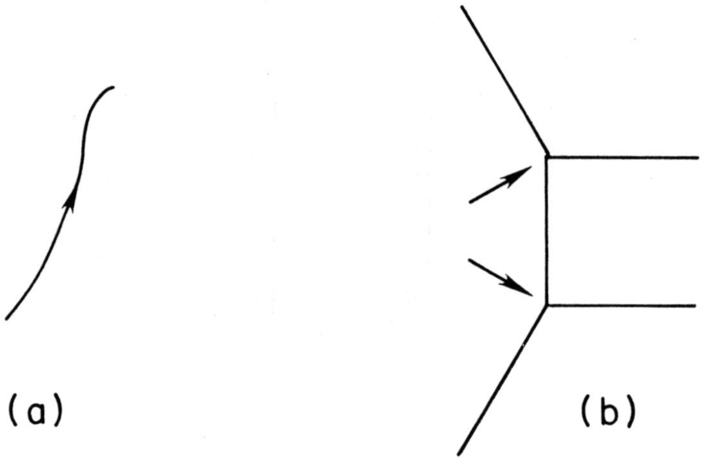

Fig. 2

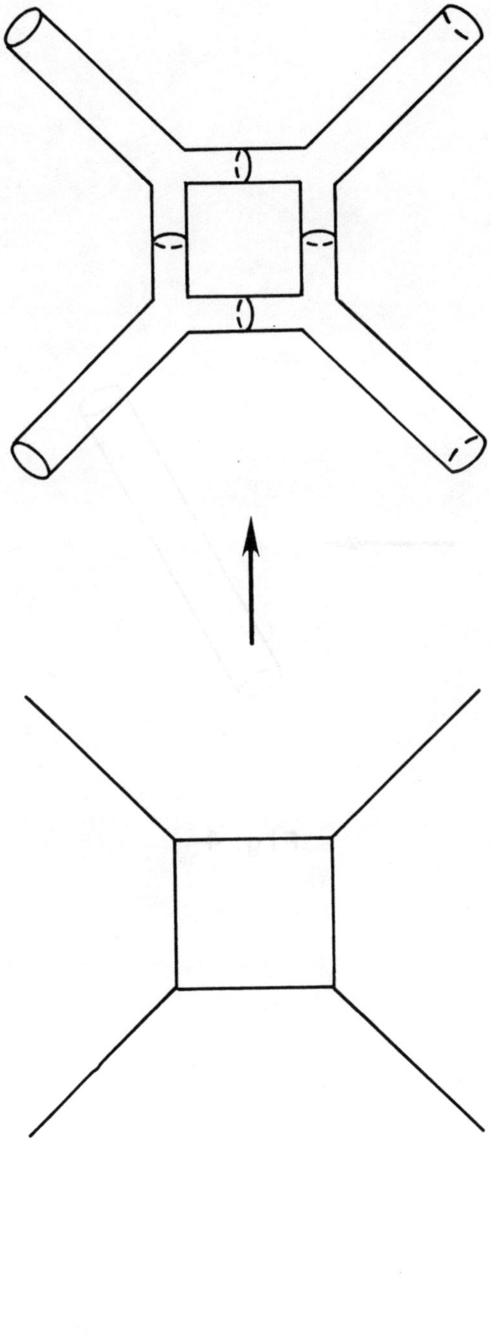

Fig. 3

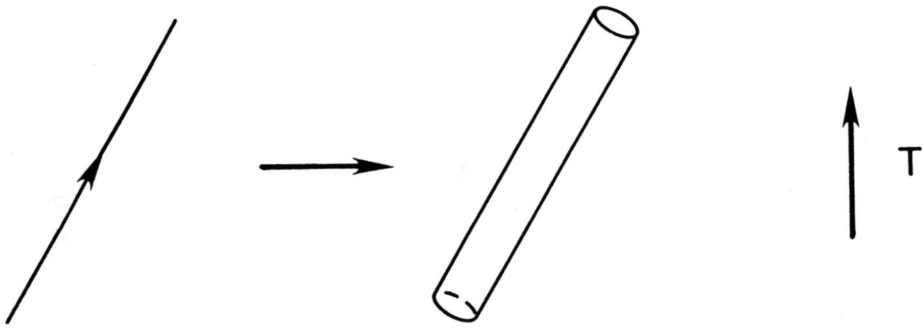

Fig. 4

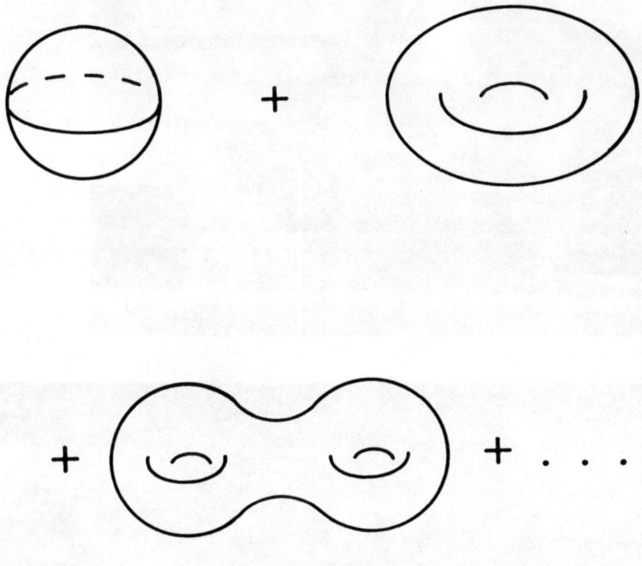

Fig. 5

PART II: PARALLEL SESSION TALKS

I. Electroweak Interactions 1

Conveners — S. Tye and J. Dorfan

CHARGED CURRENT UNITARITY AND EXTRA NEUTRAL GAUGE BOSONS

W.J. Marciano

Department of Physics
Brookhaven National Laboratory
Upton, New York 11973

and

A. Sirlin

Department of Physics
New York University
New York, N.Y. 10003

ABSTRACT

The experimental status of the KMC matrix is surveyed an shown to provide a precision test of the standard model at the level of its $O(\alpha)$ radiative corrections. Implications for new physics and constraints on extra neutral gauge bosons are described.

The three generation charged current quark mixing matrix, hereafter referred to as the KMC (Kobayashi-Maskawa-Cabibbo) matrix,

$$V = \begin{pmatrix} V_{ud} & V_{us} & V_{ub} \\ V_{cd} & V_{cs} & V_{cb} \\ V_{td} & V_{ts} & V_{tb} \end{pmatrix} \tag{1}$$

is constrained by unitarity to satisfy

$$\sum_i V^*_{ij} V_{ik} = \sum_i V^*_{ji} V_{ki} = \delta_{jk} \tag{2}$$

Those orthonormality conditions provide a powerful consistency check on the standard model. Any apparent deviation from unitarity which could not be accounted for by hadronic effects or electroweak radiative corrections[1] would be evidence for new physics beyond the standard model.

The present "best" values for directly measured KMC elements are given below along with some brief comments.

$|V_{ud}| = 0.9744 \pm 0.0010$: This value[2] is obtained from the average of eight precisely measured $0^+ \to 0^+$ nuclear superallowed beta decays after accounting for rather large $\sim 4\%$ electroweak radiative corrections.[1,2] An important development is a new calculation of the $Z\alpha^2$ Coulomb corrections which brings the Ft values from all 8 decays into good agreement with one another.[3] If instead of employing an average, one takes the most precisely measured individual nuclear decay, which happens to be $^{26}Al^m$, one finds[2]

$$|V_{ud}| = 0.9748 \pm 0.0010 \quad (\text{from } ^{26}Al^m \text{ alone}) \tag{3}$$

That decay has the smallest experimental uncertainty as well as the least variation in theoretical estimates of its nuclear matrix element correction (an error that spans the range of those estimates has been included[2,4]); so it should be quite reliable. We employ both values in our subsequent discussion.

$|V_{us}| = 0.220 \pm 0.002$: This value is obtained by averaging the K_{e3} determination 0.2196 ± 0.0023 by Leutwyler and Roos[5] with the hyperon decay value $0.220 \pm 0.001 \pm 0.003$ found by Donoghue, Holstein and Klimt[6] after taking SU(3) breaking into account. It would be nice if the experimental error in the K_{e3} determination, which is quite clean theoretically, could be further reduced.

$|V_{cd}| = 0.207 \pm 0.024$ and $|V_{cs}| = 0.95 \pm 0.15$: These values are extracted from CDHS dimuon data on charm production in deep-inelastic neutrino scattering.[7] They have been updated[8] by using larger $c \to \mu$ branching ratios and newer strange quark sea distributions. In the future, exclusive charm decays should provide somewhat better determinations of these parameters.

$|V_{ub}| < 0.012$ and $|V_{cb}| = 0.043^{+0.006}_{-0.008}$: These quantities are based on the average b lifetime of 1.1 picoseconds and $\Gamma(b\to u)/\Gamma(b\to c) < 0.2$. There is, however, still considerable disagreement concerning the theoretical analysis of those quantities. Further experimental and theoretical scrutiny is needed.

Assuming that the bound on $|V_{ub}|$ is correct and allowing for a 2σ variation in $|V_{cb}|$, one finds from three generation unitarity the indirect constraints

$$0.9984 \leq |V_{tb}| \leq 1 \quad (4a)$$

$$0 < |V_{td}| \leq 0.024 \quad (4b)$$

$$0 < |V_{ts}| \leq 0.054 \quad (4c)$$

$$|V^*_{ts}V_{td}| \leq 0.0013 \quad (4d)$$

The stringent constraint on $|V^*_{ts}V_{td}|$ in Eq. (4d) implies a suppression of virtual t quark loop effects in kaon decays such as $K^+ \to \pi^+ \nu\bar{\nu}$. Therefore, if a relatively large branching ratio is found in the about to start experiment at BNL ($\sim 10^{-9}$), it would most likely signal the existence of a fourth generation which automatically negates the strong constraint in Eq.(4d). (A weaker bound $|V^*_{ts}V_{td}| \leq 0.01$ can be obtained from $K_L \to \mu^+\mu^-$ decays.)

From the above values, one finds for the first row in V

$$|V_{ud}|^2 + |V_{us}|^2 + |V_{ub}|^2 = 0.9979 \pm 0.0021 \quad (5a)$$

$$= 0.9986 \pm 0.0021 \quad (^{26}\mathrm{Al}^m \text{ alone}) \quad (5b)$$

The good accord with unitarity must be viewed as a significant quantum loop triumph for the standard model, since without radiative corrections one would have obtained ~ 1.04 (an apparent violation of unitarity).

The result in Eq.(5) can be used to constrain supersymmetry[9], bound the scale of compositeness, limit fourth generation mixing[10], bound masses of additional neutral gauge bosons[11] etc. For example, Eq.(5a) limits potential 4th generation (ν_L, L, t', b') mixing to

$$|V_{ub'}| < 0.07 \quad (90\% \text{ C.L.}) \quad (6)$$

Unfortunately, that constraint is not very stringent when one compares it with the bound $|V_{ub}| < 0.012$. Indeed, unless the mixing pattern is peculiar[8], one would expect $|V_{ub'}| < |V_{ub}|$.

In the remainder of this talk, we will concentrate on mass bounds for additional neutral gauge bosons implied by the good agreement with unitarity in Eq.(5). The existence of such bosons would lead to additional $O(\alpha)$ radiative corrections to hadronic and muon beta-decays (see fig. 1). Such corrections would shift the $|V_{ui}|^2$ and the summation in Eq.(5) by a factor $1+\Delta$, where Δ is of $O(\alpha)$. (Muon decay normalizes the KMC matrix.) However, 3 generation unitarity limits the allowed magnitude of such a shift to (at 90% C.L.)

$$\Delta \geq -0.0006, \quad \Delta \leq 0.0048 \qquad \text{(from Eq.(5a))} \tag{7a}$$

$$\Delta \geq -0.0013, \quad \Delta \leq 0.0041 \qquad \text{(from Eq.(5b))} \tag{7b}$$

and thereby constrain the masses and couplings of such bosons. (We note that Eq.(7a) turns out to be much more constraining for the class of models we consider: so, bounds from both Eqs.(7a&7b) will be quoted in our subsequent discussion.)

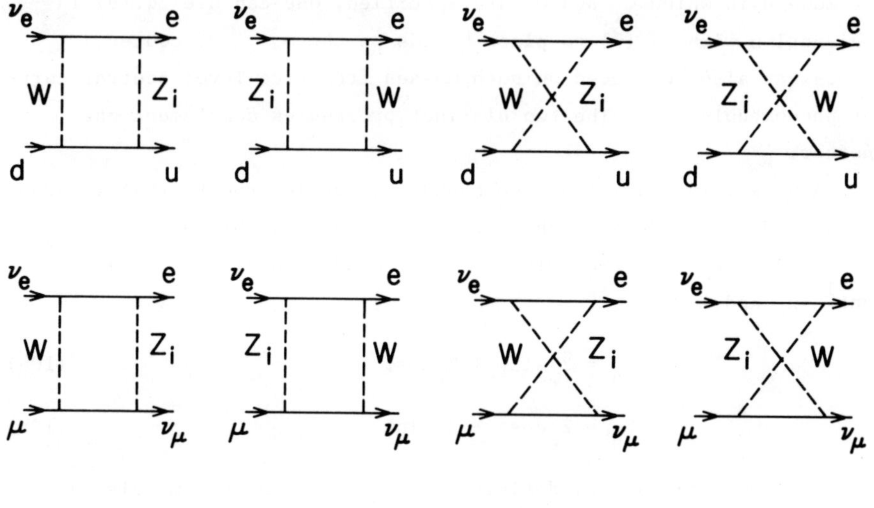

Fig. 1 Box diagram corrections to quark and muon beta decays involving Z_i bosons. For the case of V_{us} and V_{ub}, $d \to s$ or b in the above diagrams.

Given an interaction Lagrangian of N additional neutral gauge bosons Z_i, i=1,2...N with fermions

$$L_{int.} = \frac{-3g}{2\sqrt{10}} \sum_{i=1}^{N} C_i Z_i^{\mu}(x) J_{\mu}^{i}(x) \tag{8a}$$

$$J_{\mu}^{i}(x) = \sum_{f} (Q_{f_L}^{i} \bar{f}\gamma_{\mu}\gamma_{-}f + Q_{f_R}^{i} \bar{f}\gamma_{\mu}\gamma_{+}f) \tag{8b}$$

with $g = e/\sin\theta_W$, $\gamma_{\pm} = (1 \pm \gamma_5)/2$, and C_i & Q^i arbitrary, we find[11] from the diagrams in fig. 1 (assuming generation universality i.e. $Q_{e_L}^{i} = Q_{\mu_L}^{i}$, $Q_{d_L}^{i} = Q_{s_L}^{i}$ etc.) the general result

$$\Delta = \frac{-27\alpha}{40\pi\sin^2\theta_W} \sum_{i=1}^{N} |C_i|^2 Q_{e_L}^{i} (Q_{e_L}^{i} - Q_{d_L}^{i}) \frac{\ln x_i}{x_i - 1} \tag{9a}$$

$$x_i = m_{Z_i}^2 / m_W^2 \tag{9b}$$

For models in which C_i and Q^i are specified, one can use Eq.(9) in conjunction with Eq.(7) to place bounds on the m_{Z_i}. Of course, bounds can also be placed on such masses from tree level neutral current phenomenology.[12] The two distinct procedures complement one another.

A nice illustration of the bounds derivable from Eq.(9a) is provided by E_6 GUT models which contain $SU(3)_c \times SU(2)_L \times U(1) \times U(1)_\chi \times U(1)_\psi$. Those models predict the existence of two additional neutral gauge bosons[12] Z_η and $Z_{\eta'}$

$$Z_\eta = Z_\psi \sin\phi + Z_\chi \cos\phi \tag{10a}$$

$$Z_{\eta'} = Z_\psi \cos\phi - Z_\chi \sin\phi \tag{10b}$$

where $-\pi/2 < \phi \le \pi/2$ and by definition $m_{Z_\eta} \le m_{Z_{\eta'}}$. In such models, $|C_\eta|^2 \simeq |C_{\eta'}|^2 \simeq 5\tan^2\theta_W/3 \simeq 0.5$ and[11]

$$Q_{e_L}^{\eta} (Q_{e_L}^{\eta} - Q_{d_L}^{\eta}) = \frac{4}{3}\cos^2\phi (1 + (5/27)^{\frac{1}{2}}\tan\phi) \tag{11a}$$

$$Q_{e_L}^{\eta'} (Q_{e_L}^{\eta'} - Q_{d_L}^{\eta'}) = \frac{4}{3}\sin^2\phi (1 - (5/27)^{\frac{1}{2}}\cot\phi) \tag{11b}$$

We, therefore, obtain from Eqs.(9a) & (7) the constraints

$$\cos^2\phi\,(1+(\tfrac{5}{27})^{\frac{1}{2}}\tan\phi)\frac{\ell n\, x_\eta}{x_\eta - 1} + \sin^2\phi\,(1-(\tfrac{5}{27})^{\frac{1}{2}}\cot\phi)\frac{\ell n\, x_{\eta'}}{x_{\eta'} - 1} \begin{array}{l} \leq 0.13 \ (\text{Eq.7a}) \\ \leq 0.29 \ (\text{Eq.7b}) \end{array} \qquad (12)$$

The case $\phi = 0$ corresponds to effective SO(10) models for which our bound becomes

$$\begin{array}{lll} m_{Z_\eta} & \geq 413 \text{ GeV} & (\text{from Eq.(7a)}) \\ & \geq 235 \text{ GeV} & (\text{from Eq.(7b)}) \end{array} \qquad (13)$$

The second bound which comes from $^{26}Al^m$ is less restrictive but more reliable. Similar bounds can be obtained for a range of ϕ values.[11] For $m_{Z_\eta} \simeq m_{Z_{\eta'}}$, the bounds in Eq.(13) apply to all ϕ.

The above example illustrates the utility of KMC unitarity for probing new physics beyond the standard model. It would, of course, be nice if the uncertainties could be further reduced and the bound on Δ improved. Indeed, for $m_{Z'} \simeq 300$-500 GeV, Δ scales approximately as $m_{Z'}^{-1.2}$; so, a factor of 2 improvement in the Δ constraints would improve the implied mass bounds by almost a factor of 2. To do better, however, requires a reduction in the error on V_{ud}, since it dominates the uncertainty in Eq.(5). That error is primarily due to hadronic structure uncertainties coming from charged current axial-vector loop effects.[10] To reduce the error one needs better theoretical calculations for the heavy nuclear decays from which $|V_{ud}|$ is presently extracted. Alternatively, new high precision measurements involving simpler systems such as pion beta decay $\pi^+ \to \pi^0 e^+ \nu_e$ or neutron decay $n \to p e \bar{\nu}_e$ can be undertaken. In the latter case, precision measurements of both the lifetime and g_A/g_V must be carried out. The feasibility of such experiments should be carefully scrutinized and if possible, pushed as far as possible.

REFERENCES

1. Sirlin, A., Rev. Mod. Phys. 50, 573 (1978).
2. Sirlin, A., NYU preprint TR11/86.
3. Sirlin, A. and Zucchini, R., Phys. Rev. Lett. 57, 1994 (1986); Jaus, W. and Rasche, G., Universitat Zurich preprint (1986).
4. Towner, I.S. and Hardy, J.C., Phys. Lett. 73B, 20 (1978).
5. Leutwyler, H. and Roos, M., Z. Phys. C25, 91 (1984).
6. Donoghue, J., Holstein, B. and Klimt, S., Phys. Rev. D35, 934 (1987).
7. Abramowicz, H. et al., Z. Phys. C15, 19 (1982).
8. Marciano, W.J. and Parsa, Z., Ann. Rev. Nucl. Part. Sci. 36, 171 (1978).
9. Barbieri, R., Bouchiat, C., Georges, A. and leDoussal, Nucl. Phys. B269, 253 (1986).
10. Marciano, W.J. and Sirlin, A., Phys.Rev. Lett. 56, 22 (1986).
11. Marciano, W.J. and Sirlin, A., Phys. Rev. D35, 1672 (1987).
12. Durkin, L.S. and Langacker, P., Phys. Lett. 166B, 436 (1986); London, D. and Rosner, J., Phys. Rev. D34, 1530 (1986).

Physics of Quark Mixing and CP Noninvariance: The Uses of the Quark Diagram Scheme

Ling-Lie Chau

Department of Physics, University of California
Davis, CA 95616

ABSTRACT

The uses of the model-independent Quark Diagram Scheme are demonstrated in analyzing recently measured exlusive two-body decays of charm mesons. Predictions are made for future experimental measurements. Specific beauty particle nonleptonic decays are pointed out for experimental measurements to establish the nonzeroness of $|V_{ub}|$, and to measure $|V_{ub}/V_{cb}|$. Some interesting decays, which are from W-loop diagrams only and easy for experimental detections are pointed out. General properties of partial decay rate differences in charm, beauty and t particles are discussed, also using the Quark Diagram Scheme. Some interesting $B^{\pm}$ decays with all charged particles in the final states are given for the search of CP noninvariance in the decay amplitudes. Some highlights are pointed out in the event that higher than three generations of quarks exist.

TABLE OF CONTENTS

INTRODUCTION

It has been shown[1,2] that all meson decays can be expressed model-independently in terms of six quark diagrams (shown in Fig. I): $\mathcal{A}$, the external W-emission diagram; $\mathcal{B}$, the internal W-emission diagram; $\mathcal{C}$, the W-exchange diagram; $\mathcal{D}$, the W-annihilation diagram, these we call the W tree diagrams; $\mathcal{E}$, the horizontal W-loop diagram; $\mathcal{F}$, the vertical W-loop diagram. These quark diagrams are specific, well defined physical quantities. They are classified according to the topology of weak interactions with only the quark and intermediate bosons explicitly shown. All QCD strong-interaction effects are there though not explicitly shown. QCD effects can also be treated perturbatively, e.g., the one-gluon-exchange diagram of the $\mathcal{E}$ graph is the so called Penguin diagram. These diagrams keep their identities under all QCD effects. Only so such classification is useful. It has recently been successfully used[3] to analyze two-body charm decay data, to compare experimental data with various theoretical models to check their validity, and further to make predictions and to suggest new experimental measurements. See Section I.

Since the proposal of Kobayashi-Maskawa (KM) scheme of quark mixing for CP noninvariance[4] and the discovery of the beauty particles,[5] much progress has been made in measuring the quark mixing matrix in the case of three generations of quarks.[1] Besides the urgently needed observation of the t quark and its decays to measure V_{tb}, V_{ts}, V_{td}, presently the measurement of V_{ub} and the verification of its nonzeroness is most urgent. If $|V_{ub}|$ is too small or zero, higher than three generations of quarks will be needed if we still want to describe CP noninvariance through the quark mixing phenomena. Theoretically the neatest way to measure $|V_{ub}|$ is to observe $B_u \to \tau\nu_\tau$. Unfortunately the branching ratio of such decay is very small $Br(B_u \to \tau\nu_\tau) \lesssim 10^{-5}$, using the current bound on $|V_{ub}/V_{cb}|^2 < (\frac{1}{5})^2$. It is therefore important to find other decay channels which are nonzero only when $|V_{ub}|$ is not zero. Certainly the observation of charmless semileptonic decay $b \to \ell^- \overline{\nu}_\ell X_{c=0}$ inclusively or exclusively will establish $|V_{ub}| \neq 0$. Here I would like to point out some pure nonleptic decays which are non-zero only if $|V_{ub}| \neq 0$, and some ratios of nonleptonic decays that are proportional to $|V_{ub}/V_{cb}|$. Such a selection is possible based upon the general model-independent Quark-Diagram Scheme.

This scheme can also provide a model-independent way to pick out nonleptonic decays of beauty particles, which are nonzero only if $|V_{ub}|$ is nonzero, see Section II. The result is that the following decay channels are purely proportional to $|V_{ub}|$:

$B_{d,u} \to F^+ X_{c=0,s=0}$ or $D X_{c=0,s=1}$, which include exclusive channels like:

$B_u^+ \to F^+\pi^0,\ F^+\rho^0,\ F^+\pi^+\pi^-,\ F^+\phi,\ F^+K^+K^-,\ F^+\eta_8,\ F^+\eta_0,\ F^+\eta_c,$
F^+J/ψ, etc; $D^0K^+,\ D^+K^0,\ D^{*0}K^+,\ D^0K^{*+},\ D^{*0}K^{*+}$, etc.

$B_d^0 \to F^+\pi^-,\ F^+\rho^-,\ F^+\pi^-\pi^0$, *etc.*; $D^0K^0,\ D^{*0}K^0,\ D^0K^+\pi^-,\ D^{*0}K^{0*},\ D^{*0}K^{0*}$, *etc.*

$B_s \to F X_{c=0,s=-1}$, or $D X_{c=0,s=0}$, which include exclusive channels like:

$B_s^0 \to F^+K^-,\ F^+K^{*-},\ F^+\overline{K}^0\pi^-$, *etc.*; $D^0\pi^0,\ D^0\eta_8,\ D^0\eta_0, D^0\phi,\ D^{*0}\phi,\ D^{*0}\eta_8,$
$D^{*0}\eta_0,\ D^0\eta_c,\ D^0J/\psi$, *etc.*; $D^+\pi^-,\ D^+\rho^-,\ D^+\pi^-\pi^0,\ D^{*+}\pi^-,\ D^{*+}\rho^-$, *etc.*

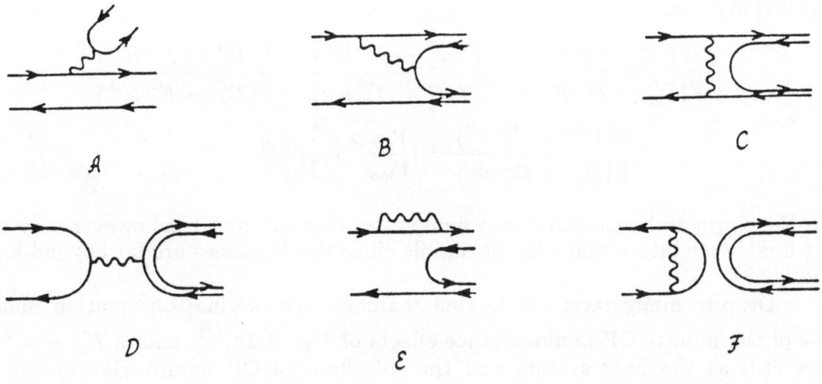

Fig. I. The six quark diagrams for a meson → two mesons.

They typically have a branching ratio of 10^{-4} within the reach of current experimental capability. Note that all these channels have a charm particle in the final states.

Recently bounds on charmless nonleptonic decays like $B^0 \to \pi^+\pi^-$ have been used as direct measurement of bound on $|V_{ub}|$. Caution must be given in such a procedure. Observation of charmless beauty particle decays does not necessarily mean $|V_{ub}| \neq 0$. The important point is that those charmless decays do not have to be zero if $|V_{ub}| = 0$ according to the general Quark Diagram Scheme since they have W-loop diagram contributions with the dominant coefficient V_{cb}. Only if we can establish experimentally that the W-loop-diagram contributions are negligibly small comparing to the tree-diagram contributions then charmless decays can give indications on $|V_{ub}|$. To find ways to study the contributions of the W-loop diagram, again the Quark-Diagram Scheme is useful. It can be shown that the following decays can have only W-loop-diagram contributions, even with the dominant coefficient $V_{cs}V^*_{cb}$:

$$B^0_d \to K^0\phi,\ K^{*0}\phi;\ B^0_s \to \overline{K}^0\phi,\ \overline{K}^{*0}\phi;\ \text{and}$$
$$B^0_{d,s} \to \phi\phi,\ K^0\overline{K}^0,\ K^{*0}\overline{K}^0,\ K^0\overline{K}^{*0},\ K^{*0}\overline{K}^{*0}.$$

Experimentally studying these decays can give direct measurements of the size of the W-loop diagrams.

Once $|V_{ub}|$ is established to be nonzero, we ought to determine its precise value. It is then worthwhile to measure the dynamically simplest $B \to \tau\nu_\tau$ in a dedicated experiment. Another way is to find other b particle decays which have the same Quark-Diagram amplitudes as listed above except with known mixing matrix elements. Then the ratios of such decays can give ratios of $|V_{ub}/V_{cb}|$. The idea here is similar to a previous well known case in charm decays[1,6] $\tilde{\Gamma}(D^+ \to \pi^0\pi^+)/\tilde{\Gamma}(D^+ \to \overline{K}^0\pi^+) = \frac{1}{2}|V_{cd}/V_{cs}|^2$. The ratios of decay widths which give

$|V_{ub}/V_{cb}|$ are, e.g.

$$\frac{2\tilde{\Gamma}(B_u^+ \to F^+\pi^0)}{\tilde{\Gamma}(B_d^0 \to D^-K^+)} = \frac{\tilde{\Gamma}(B_d^0 \to D^0K^0)}{\tilde{\Gamma}(B_d^0 \to \overline{D}^0K^0)} = \frac{\tilde{\Gamma}(B_s^0 \to F^+K^-)}{\tilde{\Gamma}(B_s^0 \to F^-K^+)}$$

$$= \frac{2\tilde{\Gamma}(B_s^0 \to D^0\pi^0)}{\tilde{\Gamma}(B_s^0 \to D^-\pi^+)} = |\frac{V_{cs}}{V_{us}}|^2 \cdot |\frac{V_{ub}}{V_{cb}}|^2 .$$

Here we ought to be careful about final-state interactions. It is however very likely that final-state interactions are negligible since the B masses are far beyond known resonances.

Despite many recent dedicated searches, the original observation made in 1964 of the minute CP noninvariance effects of $K_L \to 2\pi$,[(7)] and in $K_L \to \pi^\pm e^\mp \nu_e$ stays still as the only system and the sole kind of CP noninvariance. Yet CP noninvariance makes up one of the essential ingradiences in explaining the matter dominance of our universe.[(8)] Further, the origin of CP noninvariance seems to reside at the same reservoir of ignorance of our fundamental understanding of the now triumphant electroweak unified theories, i.e. the mass generating mechanism. It is natural that the searching for new kind of CP noninvariance in reactions other than the K_L system continue to be dedicated efforts at various laboratories, and to be the central theme in the discussions on possible searches in future experiments.

In weak decays basically there are two types of mechanisms making CP noninvariance: the mass-matrix type (also called superweak type),[9] originated in the particle-antiparticle mixing of neutral mesons, and are therefore limited only to neutral mesons; and the decay-amplitude type, originated directly in the weak decay amplitudes. The ϵ from $K_L \to 2\pi$ belongs to the mass-matrix type. $\epsilon' \equiv |\eta_{+-}/\eta_{00}|-1$ is of the decay amplitude type.[(10)] Unfortunately ϵ' is suppressed relative to ϵ due to the $\Delta I = \frac{1}{2}$ rule, i.e. $\Gamma(K^+ \to \pi^+\pi^0)/\Gamma(K^0 \to \pi^+\pi^-) \approx \frac{1}{670}$. Ironically such suppression was vital for the discovery of parity violation in the 1950's. So far ϵ' has still eluded experimental observation.[11] Dedicated experiments are being carried out to search for its possible presence.[12] And to avoid the $\Delta I = \frac{1}{2}$ rule suppression of decay-amplitude CP noninvariance ϵ', it was pointed out[13] that the time integrated partial decay rate difference in $K^0, \overline{K}^0 \to 2\gamma$ can be as large as ϵ. It is therefore important to do an experiment on this decay. In section II, we shall mainly concentrate on CP noninvariance of partial decay rate differences in heavy quark decays.

A general convenient way to search for such decay-amplitude CP noninvariance is to measure the partial decay rate differences between particle and its antiparticle. It was shown that the Quark Diagram Scheme can provide a model-independent general survey of whether decay-amplitude CP noninvariance can exist in a particular exclusive decays in the KM way of violating CP invariance, and such general survey has been carried out in Ref. (14), (see also Ref. (15)-(17)). Many channels of charm and beauty particle decays were shown to possibly have partial decay rate differences via basic interferences between the quark diagrams, belonging to the following major categories:[(14)] interferences among the tree graphs alone, between the tree graphs and the loop graphs, and between the loop graphs themselves. How big these partial decay rate differences are depends upon specific model calculations. A series of investigations has been carried out based on this Quark-Diagram Scheme,[18,19] using the currently available models. A general picture has emerged that the partial decay rate differences can be very large, 10's%, in beauty particle decays. Such large partial decay rate differences are very encouraging.

The decay of $B_d^0 \to K^+\pi^-$ calculated in Ref. (19) has been singled out in the physics studies for SSC due to its simplicity in detecting its final particles K^- and π^+, which are all charged. Here some clarifying discussions are in order to compare the advantages of searching partial decay rate differences in the charged $B^\pm$ decays over the neutral B° decays. It was pointed out in Ref. (19), the partial decay rate differences in neutral B^0 decays like $B_d^0 \to K^+\pi^-$, D^-F^+, etc., i.e. $B_d^0 \to$ total $s = 0$, $c = 0$ states, which are not CP self-conjugate state and only B^0 (not $\overline{B}^0$) can decay into it, are purely decay-amplitude type of CP noninvariance. There is no $B^0 \leftrightarrows \overline{B}^0$ mixing effects nor mass-matrix CP noninvariance involved. Therefore the interpretations of any observations of such partial decay rate differences are straight forward. There are neutral channels both B^0 and $\overline{B}^0$ can decay into, e.q. $B_d^0 \to \pi^+\rho^-$, $\pi^-\rho^+$, $\pi^+\pi^-$, D^+D^-, etc. Their time integrated partial decay rate differences are results of all three distinct effects: decay-amplitude CP noninvariance, mass-matrix CP noninvariance, and $B^0 \leftrightarrows \overline{B}^0$ mixing.[19,20] Therefore the interpretation of observed results of partial decay rate differences in these decays are more complicated. To untangle these three effects, two more other types of experiments must be supplemented. Here I shall confine my discussions on the decays in which partial decay rate differences are purely from the decay-amplitude CP noninvariance. The charged $B^\pm$ decays for studying partial decay rate differences have the further advantage over the neutral B^0 decays, for which we must tag experimentally in order to avoid the contamination from $B^0 \leftrightarrows \overline{B}^0$ mixing in the measurement of integrated partial decay rate differences.

In Section III, I elaborate these points and then point out a few more such beauty particle decays, which have large partial decay rate differences, yet simple for experimental observation, e.g.

$$B_u^\pm \to K^\pm\rho^0,\ \overline{K}^{*0}\pi^\pm,\ K^\pm\pi^+\pi^-,\ \text{etc.}$$

These decays are from similar kinds of quark diagrams as in $B_d^0 \to K^-\pi^+$, so we expect similar partial decay rate differences ($\sim 10's\%$) and branching ratios ($\sim 10^{-4}$ to 10^{-5}).[21]

I shall also give a simple explanation why the partial decay rate differences are naturally large in beauty particle decays, but rather small in charm decays. The partial decay rate differences in t particle decays can also be large, as in the beauty particle decays. Unfortunately the branching ratios of two or three particle decays of t particles are very much suppressed due to the weightiness of the t particles. To enhance the branching ratio, high-multiplicity final states must be studied. This certainly complicates the experimental search. In charm particle decays, the difficulty in the search for CP noninvariance of partial decay rate differences is quite different. In general, the branching ratios of two-body exclusive decays of charm particles are bigger than in beauty particle decays, but the partial decay rate differences are in general much smaller than in beauty particle decays.

In Section IV, some highlights will be pointed out in the quark mixing matrix, and CP noninvariance in the events that higher than three generations of quarks exist.

I. ANALYSIS OF EXCLUSIVE TWO-BODY DECAYS OF CHARM MESONS USING THE QUARK DIAGRAM SCHEME

Recently, many two-body decays of charm particles have been beautifully measured,[22–26] which we list in Tables I and II. Here we shall put the quark-diagram formalism to use, analyzing all existing charm two-body decay data and discussing their implications for various theoretical model calculations.

We begin with the PV decays because of the relative simplicity in presenting the discussion, though the data of PV decays are not yet as good as those for some of the PP decays. The simplicity in discussing the PV decays comes from the purity of the quark contents in ϕ and ω. Many PV decays are given by one type of amplitude, as shown in Table I: e.g., $F^+ \to \phi\pi^+$ ($\propto a'$), $D^+ \to \phi\pi^+$ ($\propto b'$), $D^0 \to \phi\bar{K}^0$ ($\propto \tilde{c}'$), respectively. Thus from the decay rates, using the known lifetime measurements,[6] we can determine their absolute values:

$$|a'| = (2.50 \pm 0.42) \times 10^{-6},\ |b'| = (3.67 \pm 0.51) \times 10^{-6},\ |\tilde{c}'| = (1.68 - 2.10) \times 10^{-6}. \tag{1.1}$$

The only theoretical assumption used here is that $|exp(i\delta^{\phi\pi})| = 1 = |exp(i\delta^{\phi\bar{\lambda}})|$. Note here that neither amplitude b' nor amplitude $\tilde{c}'$ is negligible, as preferred by some model calculations.

$D^+ \to \bar{K}^{*0}\pi^+[\propto (a' + b')]$ is an exotic channel which implies elastic and small $\delta_{3/2}^{\bar{\lambda}^*\pi}$. Therefore, the rate gives

$$|a' + b'| = (1.16 \pm 0.37) \times 10^{-6}. \tag{1.2}$$

Combining Eqs. (1) and (2), we obtain

$$a' = (2.50 \pm 0.42) \times 10^{-6}, \quad b' = -(3.67 \pm 0.51) \times 10^{-6}. \tag{1.3}$$

From $D^0 \to K^{*-}\pi^+$, $\bar{K}^{*0}\pi^0$ measurements we obtain the following two solutions.

$$(a' + c')/(a' + b') = 2.36 \pm 0.67 \to c' = -(5.25 \pm 0.39) \times 10^{-6},$$
$$\Delta_{\bar{K}^*\pi} = (52^{+30}_{-52})^\circ, \tag{1.4a}$$

$$(a' + c')/(a' + b') = -1.70 \pm 0.67 \to c' = -(0.53 \pm 0.39) \times 10^{-6},$$
$$\Delta_{\bar{K}^*\pi} = 180^\circ - (52^{+30}_{-52})^\circ, \tag{1.4b}$$

We note that the errors on $\Delta_{\bar{K}^*\pi}$ are so large that the data can be accommodated by real amplitudes without final-state interactions. The amplitude c' is quite different from $\tilde{c}'$ in Eq. (1.1) (Amplitudes with tildes involve strange-quark and -antiquark pair production.) To make definite conclusions, we need better measurements.

One nice prediction from this analysis is $B(D^0 \to \phi\pi^0) \simeq 0.21\%$, from $D^0 \to \phi\pi^0$ ($\propto b'$). This will be an important measurement if we are to test this scheme.

We next proceed to determine the unprimed amplitudes form $D \to p\bar{K}$ and $D^0 \to \omega\bar{K}$ decays. From Table I it follows that $D^+ \to \rho^+\bar{K}^0$, $D^0 \to \omega\bar{K}^0$ determine

$$|a+b| = (2.18 \pm 0.25) \times 10^{-6}, \quad |b+c| = (2.57 \pm 0.51) \times 10^{-6}. \tag{1.5}$$

From the measurements of $D^0 \to \rho^0\bar{K}^0, \rho^+K^-$, we find two solutions for $(a+c)/(a+b)$. Combining all those results we obtain the following three possible solutions for amplitudes a, b, and c: two for $\Delta_{\rho\bar{K}} = (24^{+25}_{-24})^\circ$,

$$a = (4.06 \pm 0.38) \times 10^{-6}, \ b = -(1.88 \pm 0.28) \times 10^{-6}, \ c = -(0.68 \pm 0.28) \times 10^{-6}; \tag{1.6a}$$

$$a = (1.50 \pm 0.38) \times 10^{-6}, \ b = (0.68 \pm 0.28) \times 10^{-6}, \ c = (1.89 \pm 0.28) \times 10^{-6}, \tag{1.6b}$$

and one for $\Delta_{\rho\bar{K}} = 180^\circ - (24^{+21}_{-24})^\circ$,

$$a = (1.41 \pm 0.38) \times 10^{-6}, \ b = (0.77 \pm 0.28) \times 10^{-6}, \ c = -(3.34 \pm 0.28) \times 10^{-6}, \tag{1.6c}$$

The measurement of $D^+ \to \bar{K}^{*0}K^+$, Table I, gives us $|a' - \bar{d} + \delta_e| = (2.70 \pm 0.61) \times 10^{-6}$.

Future measurements of $\bar{K}^{*0}\eta_8$ [$\propto$ $(b' + c' - 2\bar{c})$] and $\bar{K}^{*0}\eta_0$ [$\propto$ $(b' + c' + \bar{c})$], can help to determine amplitudes $\bar{c}$ and $(b' + c')$, and then c', since b' is known. From future measurements of $D^0 \to \bar{K}^{*0}K^0$, $\bar{K}^{*0}\bar{K}^0$ [$\propto$ $(c - c')$], and the known information on c' we can determine the amplitude c. Then we can check which solution of Eq. (6) will be picked, and thus determine a and b individually. From $F^+ \to \rho^+\pi^0$ [$\propto$ $(d - d')$], $F^+ \to \omega\pi^+$ [$\propto$ $(d + d')$], we can determine d and d'. We then know all the amplitudes a, b, c, d, and a', b', c', d', and their relative signs. The rest of the PV decays are predictable, up to SU(3) breaking and final-state interactions. These results must be conformed to by any theoretical calcutions.

Next we discuss the case of charm meson decay into two pseudoscalars, $P_c \to PP$. (Note that the amplitudes a to f here for PP decays have no relation to those for the PV decays. When needed for clarity, we use subscript PP to denote the distinction.) Here, the data are of great accuracy than for the $P_c \to PV$ case, Table II. From $D^+ \to \pi^+\bar{K}^0$, $D^0 \to K^-\pi^+$, $\bar{K}^0\pi^0$, we can conclude definitely that real amplitudes $(a, b, c,)_{PP}$ (without effects like final-state interactions included) cannot fit the data.[12,13] From $D^+ \to \pi^+\bar{K}^0$, we obtain

$$|a+b|_{PP} = (1.66 \pm 0.11) \times 10^{-6} \text{ GeV}. \tag{1.7}$$

Then from $D^0 \to K^-\pi^+$, $\bar{K}^0\pi^0$, we obtain the following two solutions:

$$\begin{array}{lcl} [(a+c)/(a+b)]_{PP} = 1.95 \pm 0.14, & \text{or} & [(a+c)/(a+b)]_{PP} = -1.28 \pm 0.14, \\ \Delta_{\bar{K}\pi} = (79^{+10}_{-14})^\circ; & & \Delta_{\bar{K}\pi} = 180 - (79^{+10}_{-14})^\circ. \end{array} \tag{1.8}$$

We want to caution about the interpretation of the phase shift obtained here. Its relation to the hadronic scattering phase shifts is complicated by the other competing channels, e.g., $\pi\pi\bar{K}$, $\pi\pi\pi\bar{K}$ (not including $\rho\bar{K}$, $\pi\bar{K}^*$, which do not communicate with $\bar{K}\pi$ through strong interactions).

Unlike the PV decays, it is much harder here to determine individual amplitudes, since none of the decays is given by a single amplitude. It is interesting to point out that the nonspectator-diagram amplitudes $\bar{c}$ and d can be measured in a model-independent way by observation of the following decay modes:

$$\frac{\Gamma(D^0 \to K^0\eta_8)}{\Gamma(D^0 \to K^0\eta_0)} = \frac{1}{2}\left|\frac{(b+c-2\bar{c})}{(b+c+\bar{c})}\right|^2_{PP}, \quad \frac{\Gamma(F^+ \to \eta_8\pi^+)}{(F^+ \to \eta_0\pi^+)} = 2\left|\frac{(a-d)}{(a+2d)}\right|^2_{PP}. \quad (1.9)$$

From the absolute rates of these decays we can determine $[(b+c), \bar{c}$; and $a, d]_{PP}$. Combining these with the solutions Eq. (1.7) and (1.8), we shall determine all amplitudes $(a, b, c,$ and $d)_{PP}$ and their relative signs.

Next we go to the mixing-matrix singly, suppressed measurement of $\Gamma(D^0 \to K^+K^-)/\Gamma(D^0 \to \pi^+\pi^-) \neq 1$. From Table II, we see that such differences can be attributed to the SU(3)-breaking effect of $(\delta e + 2\delta f)_{PP}$, which contributes with opposite sign to $D^0 \to K^+K^-$, $\pi^+\pi^-$ and/or to the final-state-interaction effect (e.g., $\delta_0^{\pi\pi}$ has a larger absorptive part than $\delta_0^{K\bar{K}}$). To clarify these mechanisms it is of paramount importance to measure $D^0 \to \pi^0\pi^0$ (see Table II), since the same unknowns, $(\delta e + 2\delta f)_{PP}$, $\delta_0^{\pi\pi}$, are present, but the rest of the amplitudes, $(b-c)_{PP}$, $(a+b)_{PP}$, are known [from Eqs. (1.7) and (1.8)].

With the known relation between η, η' and η_8, η_0 (here the mixing angle of -10° is used; in the analysis of future measurements of decays involving η', care should be taken to subtract any component in η' that is not η_0 or η_8), the current measurement of $D^0 \to \bar{K}^0\eta$, Table II, gives $[1.23b - 04.9c]_{PP} = (4.57 \pm 1.01) \times 10^{-6}$ GeV, if there is no SU(3) breaking, i.e., $\bar{c} = c$; or it gives $|b+c|_{PP} = (3.71 \pm 0.83) \times 10^{-6}$ GeV, if SU(3) breaking is maximal, i.e., $\bar{c} = 0$. Future measurements of $D^0 \to \bar{K}^0\eta_8$ [$\propto (b+c-2\bar{c})$], or $\bar{K}^0\eta'$, will help to determine amplitudes $[(b+c), \bar{c}$; thus $b, c]_{PP}$ individually when combined with the results of Eqs. (1.7) and (1.8). The measurement of $D^+ \to \bar{K}^0K^+$ gives $|a - \bar{d} + \delta e|_{PP} = (4.29 \pm 0.66) \times 10^{-6}$ GeV. Future measurments of $F^+ \to \pi^+\eta_8$ [$\propto (a-d)_{PP}$], and $\pi^+\eta_0$ [$\propto (a+2d)_{PP}$], can give $(a, d)_{PP}$; $\bar{\Gamma}(D^+ \to \bar{K}^0K^+)$ is predicted to be equal to $(s_1/c_1)^2 \times \bar{\Gamma}(F^+ \to \pi^+\eta_8)$ if there is no SU(3) breaking. The measurements of $D^0 \to K^0\bar{K}^0$, $\eta_0\eta_0$, which are nonzero only from SU(3) breaking, can give a direct modification of SU(3)-breaking effects. The long-predicted relation

$$\bar{\Gamma}(D^+ \to \pi^+\pi^0)/\bar{\Gamma}(D^+ \to \bar{K}^0\pi^+) = \frac{1}{2}|V_{cd}/V_{cs}|^2$$

should be checked by experiments.

II. QUARK MIXING MATRIX AND THE SEARCH FOR $|V_{ub}|$ AND $|V_{ub}/V_{cb}|$ USING THE QUARK - DIAGRAM SCHEME

First I give a brief account of the status of the KM matrix. Recent results show that the quark mixing matrix can be parameterized most conveniently in the following form[27]

$$V = \begin{bmatrix} 1 & 0 & 0 \\ 0 & c_y & s_y \\ 0 & -s_y & c_y \end{bmatrix} \begin{bmatrix} c_z & 0 & s_z e^{-i\phi} \\ 0 & 0 & 0 \\ -s_z e^{i\phi} & 0 & c_z \end{bmatrix} \begin{bmatrix} c_x & s_x & 0 \\ -s_x & c_x & 0 \\ 0 & 0 & 1 \end{bmatrix}$$

$$
\begin{array}{c} \\ = \end{array}
\begin{array}{c}
\begin{array}{ccc} \mathrm{d} & \mathrm{s} & \mathrm{b} \end{array} \\
\begin{bmatrix} c_x c_z & s_x c_z & s_z e^{-i\phi} \\ -s_x c_y - c_x s_y s_z e^{i\phi} & c_x c_y - s_x s_y s_z e^{i\phi} & s_y c_z \\ s_x s_y - c_x c_y s_z e^{i\phi} & -c_x s_y - s_x c_y s_z e^{i\phi} & c_y c_z \end{bmatrix}
\end{array}
\begin{array}{c} u \\ c \\ t \end{array}
$$

$$
\underset{s_i \approx 0}{\approx} \begin{bmatrix} 1 & s_x & s_z e^{-i\phi} \\ -s_x - s_y s_z e^{i\phi} & 1 - s_x s_y s_z e^{i\phi} & s_y \\ s_x s_y - s_z e^{i\phi} & -s_y - s_x s_z e^{i\phi} & 1 \end{bmatrix}. \tag{2.1}
$$

An important feature of the three-generation quark mixing matrix as first shown in Ref. (22) is that all CP noninvariance in decays of those three-generation quark particles are described by one universal phase-convention-independent parameter

$$
\begin{aligned}
X_{CP} &= V_{us} V_{ud}^* (V_{cs} V_{cd}^*)^*; \text{ for strange particle decays;} \\
&= V_{cs} V_{us}^* (V_{cd} V_{ud}^*)^*; \text{ for charm particle decays;} \\
&= V_{ub} V_{ud}^* (V_{cb} V_{cd}^*)^*, \\
&= V_{ub} V_{us}^* (V_{cb} V_{cs}^*)^*; \text{ for beauty particle decays;} \\
&= V_{tb} V_{cb}^* (V_{ts} V_{cs}^*)^*, \\
&= V_{tb} V_{ub}^* (V_{ts} V_{us}^*)^*; \text{ for t particle decays;} \\
&= s_x s_y s_z s_\phi c_x c_y c_z^2, \\
&\approx 10^{-5}, \quad \text{if } m_t \approx 50 \text{ GeV is used in fitting } \epsilon.
\end{aligned} \tag{2.2}
$$

Because of this property we can put the phase factor at the most convenient place, i.e. at where the matrix element is the smallest.[28] This has the advantage that for all practical purposes without involving CP noninvariance, the matrix can be considered to be real. Further in Section II I shall demonstrate that this parametrization is also the simplest in considering CP noninvariance of partial decay rate differences.

Another important feature of this parametrization is that it takes advantage of all the experimental information: each sine is directly related to one type of experiment[29]

$$ s_x \approx 0.22, \quad \text{determined from strange particle decays,} \tag{2.3a} $$

$$ s_y \approx 0.05, \quad \text{from b particle life time,} \tag{2.3b} $$

$$ s_z \lesssim 0.01, \quad \text{from bounds on } (b \to u)/(b \to c). \tag{2.3c} $$

This is the inherited convenience from the original Maiani parametrization,[30] the rotation order was found originally in Ref. (27) and with final form in the real angles agreeing with Maiani's. We pick the subscript x, y, z because that was the order of experimental measurements and easiest to remember. Such parametrization via sequences of rotations and put phases at the furthest corners can be generalized to cases with higher generating of quarks.[31,32] See Section IV.

In this parametrization we can see easily that if $V_{ub} = 0$, there is no phase factor for CP noninvariance, and the value of $|V_{ub}|$ needed for ϵ is not too far below the upper bound[33] on $|V_{ub}|$ from $b \to \ell^- X$. It is very important to settle the issue whether $|V_{ub}|$ is nonzero or not.

Theoretically the clearest way to look for V_{ub} is from $B \to \tau \overline{\nu}_\tau$, since it has the least complication dynamically. (Here we pick the heaviest of the leptons to avoid the helicity suppression in the light-lepton case, as $\pi \to e\nu_e$ is very suppressed compared to $\pi \to \mu\nu_\mu$.) Unfortunately, the branching ratio is very small:[34]

$$\Gamma(B_u \to \tau\nu_\tau) = |V_{ub}|^2 (G_F^2 f_B^2/8\pi) m_B m_t^2 (1 - m_t^2/m_B^2), \qquad (2.4a)$$

using

$$\Gamma(b \to cX) \approx |V_{cb}|^2 G_F^2 m_b^5/(192\pi^3), \qquad (2.4b),$$

we obtain

$$Br(B \to \tau\nu_\tau) = |V_{ub}/V_{cb}|^2 \times 2.1 \times 10^{-3} \lesssim (0.01/0.05)^2 \times 2.1 \times 10^{-3} \approx 8.4 \times 10^{-5}. \qquad (2.4c)$$

Therefore we need to find other ways to establish the nonzeroness of $|V_{ub}|$. The observation of charmless $B \to \ell^+ X_{c=0}$ inclusively or exclusively will establish $|V_{ub}| = 0$. Another way is to study those hadronic decay channels which is nonzero only when $|V_{ub}|$ is not zero. Here the model-independent description of Quark-Diagram Scheme can be very useful. From the Quark-Diagram Scheme, we find the following exclusive decays which are nonzero only if $|V_{ub}| \neq 0$:

$B_u^+ \to F^+ X_{c=0,s=0}$, including the following decay modes:

$F^+\pi^0 \times \sqrt{2}$, $F^+\eta_8 \times \sqrt{6}$, $F^+\eta_0 \times \sqrt{3}$,
$F^+\rho^0$, $F^+\pi^+\pi^-$, etc.: $\quad V_{cs}V_{ub}^* \, (\mathcal{A})$; (2.5a)
$F^+\phi$, $F^+K^+K^-$, etc.: $\quad V_{cs}V_{ub}^* (\mathcal{D})$; (2.5b)

$\to DX_{c=0,s=+1}$:
D^0K^+, D^0K^{*+}, $D^{*0}K^+$, $D^{*0}K^{*+}$, etc.: $\quad V_{cs}V_{ub}^* (\mathcal{B}+\mathcal{D})$; (2.5c)

$B_d^0 \to F^+ X_{c=0,s=0}$:
$F^+\pi^-$, $F^+\rho^-$, $F^+\pi^-\pi^0$, etc.: $\quad V_{cs}V_{ub}^* \, (\mathcal{A})$; (2.5d)

$\to DX_{c=0,s=+1}$:
D^0K^0, D^0K^{0*}, $D^0K^+\pi^-$, $D^{*0}K^0$, $D^{*0}K^{0*}$, etc.: $\quad V_{cs}V_{ub}^* \, (\mathcal{B})$; (2.5e)

$B_s^0 \to F^+ X_{c=0,s=-1}$:
F^+K^-, F^+K^{*-}, $F^+\overline{K}^0\pi^-$, etc.: $\quad V_{cs}V_{ub}^* (\mathcal{A}+\mathcal{C})$; (2.5f)

$\to DX_{c=0,s=o}$:
$D^0\pi^0 \times \sqrt{2}$, $D^0\eta_8 \times \sqrt{6}$, $D^0\eta_0 \times \sqrt{3}$,
$D^+\pi^-$, $D^+\rho^-$, $D^+\pi^-\pi^0$, D^0J/ψ, etc.: $\quad V_{cs}V_{ub}^* \, (\mathcal{C})$. (2.5g)

(For simplicity, here we use the same amplitude symbols for final states from different multiplets. But do remember that they are in general different.) Notice that all these decays have a net charm quantum number equal to one in the final state. Essentially they are all from graphs with $\overline{b} \to \overline{u}W^+$ and $W^+ \to c\overline{s}$, therefore it is always the combination $V_{cs}V_{ub}{}^*$ appearing in these decays. There are other decays involving $V_{cd}V_{ub}{}^*$, e.g. $D^+\pi^-$, $D^0\pi^0$ which are nonzero only if $|V_{ub}| \neq 0$, but their branching ratios are suppressed by a factor $|V_{cd}/V_{cs}|^2 \approx 0.05$ compared to the previous case.

Decays with $V_{ud}V_{ub}^*$ factors always can have other diagrams involving $V_{cd}V_{cb}^*$. For example the charmless decay $B_d^0 \to \pi^+\pi^-$ are given by the following quark amplitudes and quark mixing factors,

$$B_d^0 \to \pi^+\pi^- : V_{ud}V_{ub}^*(\mathcal{A} + \mathcal{C} + \mathcal{E}_{u-t}) + V_{cd}V_{cb}^*(\mathcal{E}_{c-t}). \qquad (2.6a)$$

We see that even when $|V_{ub}| = 0$, $B_d^0 \to \pi^+\pi^-$ may not be zero due to the W-loop diagrams.

So the search for $|V_{ub}|$ is not best in the charmless decays of beauty particles but in those specific beauty particle decays into one charm particle states Eqs. (2.5).

However if we can establish experimentally that the W-loop diagram contributions even with their large mixing matrix coefficient are negligible comparing to the tree-diagram contributions (in many cases with very small mixing matrix coefficients), then the charmless decays like $B_d^0 \to \pi^+\pi^-$ can give estimates on $|V_{ub}|$. To do such an analysis we need to measure the W-loop diagrams independently. From the Quark-Diagram Scheme, we find the following decays can have only W-loop contributions (even with the dominant coefficient $V_{cs}V_{cb}^*$),

$$B_d^0 \to K^0\overline{K}^0,\ K^{*0}\overline{K}^0,\ K^0\overline{K}^{*0},\ K^{*0}\overline{K}^{*0} : V_{ud}V_{ub}^*(\mathcal{E}_{u-t} + \mathcal{F}_{u-t}) + V_{cd}V_{cb}(\mathcal{E}_{c-t} + \mathcal{F}_{c-t});$$

$$B_d^0 \to K^0\phi,\ K^{*0}\phi : \ V_{ud}V_{ub}^*(\mathcal{E}_{u-t}) + V_{cd}V_{cb}^*(\mathcal{E}_{c-t});$$

$$B_d \to \phi\phi : \ V_{ud}V_{ub}^*(\mathcal{F}_{u-t}) + V_{cd}V_{cb}^*(\mathcal{F}_{c-t});$$

$$B_s^0 \to \overline{K}^0\phi : \ V_{ud}V_{ub}^*(\mathcal{E}_{u-t}) + V_{ud}V_{ub}^*(\mathcal{E}_{c-t});$$

$$B_s^0 \to \phi\phi,\ K^0\overline{K}^0,\ K^{*0}\overline{K}^0,\ K^0\overline{K}^{*0},\ K^{*0}\overline{K}^{*0} : V_{us}V_{ub}^*(\mathcal{E}_{u-t} + \mathcal{F}_{u-t}) + V_{cs}V_{cb}^*(\mathcal{E}_{c-t} + \mathcal{F}_{c-t}). \qquad (2.6b)$$

(Here $\mathcal{E}_{i-j} \equiv \mathcal{E}_i - \mathcal{E}_j$). Combined analysis of decays of Eq. (2.6b) and charmless beauty decays like Eq. (2.6a) can help to shed light on $|V_{ub}|$. It is interesting to note that $B_d \to \phi\phi$ comes only from the vertical W-loop diagram $\mathcal{F}$; and $B_d \to K^0\phi$, $K^{*0}\phi$, $B_s^0 \to \overline{K}^0\phi$, $\overline{K}^{*0}\phi$ comes only from the horizontal loop diagram $\mathcal{E}$.

We can also use ratios of nonleptonic decays to measure $|V_{ub}/V_{cb}|$. In the following we list some of the nonleptonic decay amplitudes which are proportional to V_{cb}, e.g.

$$B_d^0 \to D^-K^+ : \ V_{us}V_{cb}^*(\mathcal{A}); \quad (2.7a) \qquad B_s^0 \to F^-K^+ : \ V_{us}V_{cb}^*(\mathcal{A} + \mathcal{C}); \quad (2.7c)$$

$$\to \overline{D}^0K^0 : \ V_{us}V_{cb}^*(\mathcal{B}); \quad (2.7b) \qquad \to D^-\pi^+ : \ V_{us}V_{cb}^*(\mathcal{C}). \quad (2.7d)$$

Assuming that the final state interaction effects are small,[3,21] by taking the proper ratios between the decays in Eqs. (2.5) and in Eqs. (2.7), we obtain

$$\frac{2\tilde{\Gamma}(B_u^+ \to F^+\pi^0)}{\tilde{\Gamma}(B_d^0 \to D^-K^+)} = \frac{\tilde{\Gamma}(B_d^0 \to D^0K^0)}{\tilde{\Gamma}(B_d^0 \to \overline{D}^0K^0)} = \frac{\tilde{\Gamma}(B_s^0 \to F^+K^-)}{\tilde{\Gamma}(B_s^0 \to F^-K^+)}$$

$$= \frac{2\tilde{\Gamma}(B_s^0 \to D^0\pi^0)}{\tilde{\Gamma}(B_s^0 \to D^-\pi^+)} = |\frac{V_{cs}}{V_{us}}|^2 \cdot |\frac{V_{ub}}{V_{cb}}|^2 . \tag{2.8}$$

Here $\tilde{\Gamma}$ is the reduced width with unequal phase factors factored out. Thus measuring such ratios we can measure $|V_{ub}/V_{cb}|^2$.

III. PARTIAL DECAY RATE DIFFERENCES FORM THE QUARK-DIAGRAM SCHEME

The partial decay rate differences are defined as

$$\Delta \equiv \frac{|A|^2 - |\overline{A}|^2}{|A|^2 + |\overline{A}|^2}, \tag{3.1}$$

where A, $\overline{A}$ stand for decay amplitudes of particle and antiparticle into a specific exclusive final state and their corresponding antiparticle state. To map out which exclusive decay channels can possibly have CP noninvariance of partial decay rate differences, the general Quark Diagram Scheme has been indispensible.[3,14,18,19] It provides a model-independent way to find which channels can not have CP noninvariance and which channels can.

The best way to explain the scheme is to give an example: $B_u^+ \to K^+\rho^0$ is given in terms of the quark diagrams as follows,

$$A = V_{ub}V_{us}^*(\mathcal{A} + \mathcal{B} + \mathcal{D} + \mathcal{E}_{u-t}) + V_{cb}V_{cs}^*(\mathcal{E}_{c-t}) \equiv V_{ub}V_{us}^*A_1 + V_{cb}V_{cs}^*A_2, \tag{3.2a}$$

$$\overline{A} = V_{ub}^*V_{us}(\mathcal{A} + \mathcal{B} + \mathcal{D} + \mathcal{E}_{u-t}) + V_{cb}^*V_{cs}(\mathcal{E}_{c-t}) \equiv V_{ub}^*V_{us}A_1 + V_{cb}^*V_{cs}A_2. \tag{3.2b}$$

Note that the difference between particle and antiparticle decays are mainly in $V_{ij} \to V_{ij}^*$. Substituting Eqs. (3.2) in Eq. (3.1),

$$\Delta = \frac{-4Im[(V_{ub}V_{us}^*(V_{cb}V_{cs}^*)^*]ImA_1A_2^*}{|V_{ub}V_{us}^*A_1 + V_{cb}V_{cs}^*A_2|^2 + |V_{ub}^*V_{us}A_1 + V_{cb}^*V_{cs}A_2|^2},$$

$$= \frac{-4X_{cp}\ ImA_1A_2^*}{2|V_{ub}V_{us}^*A_1 + V_{cb}V_{cs}^*A_2|^2 + 4X_{cp}Im(A_1A_2^*)}. \tag{3.3}$$

Here $V_{ij}V_{k\ell}^*$ are not phase invariant quantities and are parametrization dependent. They become simplified when we use the parametrization of Eq. (2.1), which always put the imaginary part at the smallest place so that quite often the imaginary part can be dropped. For example $V_{cb}V_{cs}^* = s_y(1 - s_xs_ys_ze^{-i\phi}) \approx s_y$ to a very good approximation.

Also the expression $V_{ub}V_{us}^*A_1 + V_{cb}V_{cs}^*A_2$ holds for partial decay rate differences of many b particle decays, like $B_d^0 \to K^+\pi^-$, which we shall denote in general as $b \to s$. After some rearrangement and using Eqs. (2.1),(2.2) Eq. (3.3) becomes,

$$\Delta_{b\to s} \approx \frac{-4(X_{CP}/s_xs_ys_z)Im(B_1B_2^*)}{2|\eta_{bs}^uB_1 + \eta_{bs}^cB_2|^2 - \text{numerator}}, \tag{3.4}$$

where $B_1 \equiv s_xs_zA_1$, $B_2 \equiv s_yA_2$, $\eta_{bs}^u = e^{-i\phi}$, $\eta_{bs}^c = 1 - s_xs_ys_ze^{-i\phi}$. The important point here is that $X_{CP}/s_xs_ys_z \approx s_\phi$, which can be of order one. In cases when calculations give $|B_1| \sim |B_2|$ and $|\eta_{bs}^uB_1 + \eta_{bs}^cB_2|^2 \sim Im(B_1B_2^*)$, $\Delta_{b\to S}$ become fractional. This does happen quite often.

Similarly for another class of b decays, like $B_s^0 \to K^-\pi^+$, which we denote by $b \to d$

$$\Delta_{b\to d} \simeq \frac{-4(X_{CP}/s_x s_y s_z) Im(B_1' B_2'^*)}{2|\eta_{bd}^u B_1' + \eta_{bd}^c B_2'|^2 - \text{numerator}}, \tag{3.5}$$

where $B_1' \equiv s_z A_1'$, $B_2' \equiv -s_x s_y A_2'$, $\eta_{bd}^u = e^{-i\phi}$, $\eta_{bd}^c = 1 + (s_y s_z/s_x)e^{-i\phi}$. Again the same factor $X_{CP}/s_x s_y s_z \approx s_\phi$ appears, and model calculations can give $\Delta_{b\to d}$ close to one.

For the t quark decays we have the following two cases for the partial decay rate differences:

$$\Delta_{t\to c} = \frac{-4(X_{CP}/s_x^2 s_y^2) I_m(T_1 T_2^*)}{2|\eta_{tc}^d T_1 + \eta_{tc}^s T_2|^2 - \text{numerator}}, \tag{3.6a}$$

where $T_1 \equiv s_x^2 s_y A_1$, $T_2 \equiv -s_y A_2$, $\eta_{tc}^d = 1 - (s_z/s_x s_y)e^{i\phi}$, $\eta_{tc}^s = 1 + (s_x s_z/s_y)e^{i\phi}$. Again here $X_{CP}/s_x^2 s_y^2 \approx (s_y/s_x s_y)e^{i\phi} \approx s_\phi$, like in the b particle decays. Another case of the t particle decay is

$$\Delta_{t\to u} = \frac{-4(X_{CP}/s_x^2 s_y^2) I_m(T_1' T_2'^*)}{2|\eta_{tu}^d T_1 + \eta_{tc}^s T_2|^2 - \text{numerator}}, \tag{3.6b}$$

where $T_1' \equiv s_x s_y A_1'$, $T_2 \equiv -s_x s_y A_2'$, $\eta_{tu}^d = 1 - (s_z/s_x s_y)e^{i\phi}$, $\eta_{tu}^s = 1 + (s_x s_z/s_y)e^{i\phi}$. Here $X_{CP}/s_x^2 s_y^2 \approx s_\phi$ again like in the b decays. Here we see that the partial decay rates of the t particles can be appreciable like in the b particle decays.

For charm decays the partial decay rate difference has one form:

$$\Delta_c \approx \frac{-4(X_{CP}/s_x^2) Im(C_1 C_2^*)}{2|\eta_{cu}^d C_1 + \eta_{cu}^s C_2|^2 - \text{numerator}}, \tag{3.7}$$

where $C_1 = -s_x A_1$, $C_2 = s_x A_2$, $\eta_{cu}^d = 1 + (s_y s_z/s_x)e^{i\phi}$, $\eta_{cu}^s = 1 - s_x s_y s_z e^{i\phi}$. Here $X_{CP}/s_x^2 \approx 10^{-2} s_\phi$, which is 10^{-2} of that of the b and t particle decays. Except when there is exact cancellation between the real parts of C_1, C_2 the first term in the denominator dominates over the second term, and $\Delta_c << 1$. In model calculations usually $I_m(C_1 C_2^*)/|\eta_{cu}^d C_1 + \eta_{cu}^s C_2|^2 \approx 1$, thus $\Delta_c \lesssim 10^{-2}$.

To state it simply and intuitively, the reason for the possibilities of large partial decay rate differences Δ in the b and t particle decays is that Δ must come from mixing matrix suppressed diagrams. In the three generation case the numerator of Δ is proportional to a universal factor X_{cp}, therefore those decays with the smaller denominator in Δ win. The b and t decays have smaller denominators in Δ than charm decays since the mixing-matrix suppressions are much more severe in the b and t particle decays than in the charm particle decays.

Let's now look at the branching ratios. The widths (or lifetimes) of heavy-quark particle decays are usually calculated using the simple W-emission diagram, e.g.

$$\Gamma_B = |V_{cb}|^2 \; \frac{G_F^2 m_B^5}{192\pi^3}. \tag{3.8}$$

Let's compare this to the decay width of $B \to P_1 P_2$, two pseudoscalar final states, via a W-emission diagram

$$\Gamma_B \to P_1 P_2 = \frac{2m_B |\vec{p}_1| |V_{ij} V_{k\ell}^* (G_F/\sqrt{2}) f_{P_1} F^{BP_2} m_B^2|^2}{16\pi m_B^3} \approx |V_{ij} V_{kl}^*|^2 f_{P_1}^2 \frac{G_F^2 m_B^3}{32\pi}, \tag{3.9}$$

where $|\vec{p}_1| 2m_B = [m_B^2 - (m_1 + m_2)^2]^{\frac{1}{2}} \times [m_B^2 - (m_1 - m_2)^2]^{\frac{1}{2}}$, and f_{P_1} is the P_1 particle decaying constant and usually we take $f_{P_1} \approx f_B \approx f_\pi \approx 130$ MeV is taken, and F^{BP_2} is the form factor between B and P_2, usually we take $F^{BP_2} \approx 1$. Taking the ratio of Eq. (3.8)/Eq (3.9), we obtain the branching ratio.

$$\mathcal{A} \qquad B_r(B \to P_1P_2) = \Gamma_B \approx 6\pi^2 \frac{|V_{ij}V_{k\ell}^*|^2 f_{P_1}^2}{|V_{cb}|^2 m_B^2} \begin{cases} \approx 10^{-2}, \text{ for } V_{ij}V_{k\ell}^* = V_{cb}V_{cs}^*; \\ \approx 10^{-4} \sim 10^{-5}, \text{ for } V_{ij}V_{k\ell}^* = V_{ub}V_{us}^*. \end{cases} \tag{3.10}$$

In partial decay rate differences, the W-loop graphs $\mathcal{E}$, $\mathcal{F}$ have coefficients $V_{cb}V_{cs}^*$, but in the current model calculations, the loop graphs are much smaller such that $V_{cb}V_{cs}^*\mathcal{E} \approx V_{ub}V_{us}^* \mathcal{A}$. Therefore the branching ratio is rather small for the B decay. Actually it is the smallness of the decay width helping to make Δ_b large since $\Gamma_{B\to P_1P_2}$ do appear in the denominator of Δ_b. This actually is a nice trade off since the number of events needed is inversely proportional to the square of Δ and inversely proportional to the branching ratio, Br., i.e.

$$N \gtrsim \Delta^{-2} B_r^{-1}, \tag{3.11}$$

as was discussed in Ref. (18).

Another point we see from Eq. (3.10) is that the two-body-decays branching ratio is inversely proportional to the square of the mass of the decaying particle. Therefore the charm two-body-decays in general have larger branching ratio then beauty two-body decays, and the two body decays of the t quark particles will have very small branching ratios. All these explain why it is most favorable to look for partial decay rate differences in particle decays.

Here I shall list a few nice charged $B^\pm$ decay modes:

$$B_u^+ \to K^\pm \rho^0, \ \overline{K}^{*0}\pi^\pm, \ K^\pm\pi^+\pi^-, \text{ etc.} \tag{3.12}$$

which have similar quark-diagram structure as $B_d^0 \to K^+\pi^-$. Therefore it is very natural to expect that the decay modes of Eq. (3.12) have the similar partial decay rate differences of 10's%, and branching ratios of 10^{-4} to 10^{-5}. The following decay modes should also be good interesting,

$$B_u^+ \to K^+\phi, \ K^+K^+K^-; \ K^+w, \ K^+\pi^+\pi^-\pi^0, \text{ etc.} \tag{3.13}$$

though our current calculations are not as reliable for these channels as the ones given in Eq. (3.12).

IV. BEYOND THE THREE GENERATIONS OF QUARKS

When there are more than three generations of quarks, we can generalize the sequential rotations as given in Eq. (2.1), e.g. for four generations of quarks, the quark mixing matrix can be written as follows:

$$V_4 = \begin{bmatrix} 1 & 0 & 0 & 0 \\ 0 & 1 & 0 & 0 \\ 0 & 0 & c_u & s_u \\ 0 & 0 & -s_u & c_u \end{bmatrix} \begin{bmatrix} 1 & 0 & 0 & 0 \\ 0 & c_v & 0 & s_v e^{-i\phi_3} \\ 0 & 0 & 1 & 0 \\ 0 & -s_v e^{i\phi_3} & 0 & c_v \end{bmatrix} \begin{bmatrix} c_w & 0 & 0 & s_w e^{-i\phi_2} \\ 0 & 1 & 0 & 0 \\ 0 & 0 & 1 & 0 \\ -s_w e^{i\phi_2} & 0 & 0 & c_w \end{bmatrix} \begin{bmatrix} & & & 0 \\ & V_3 & & 0 \\ & & & 0 \\ 0 & 0 & 0 & 1 \end{bmatrix}, \tag{4.1}$$

where V_3 is the 3x3 matrix given in Eq. (2.1). The reason we put the additional phase ϕ_2, ϕ_3 where there are in Eq. (4.1) is to anticipate that the widening of the generation gap will keep its pace. Of course physical consequences are independent of ways of parameterization. The striking new phenomena are the proliferation of the number of $X_{CP's}$ they are

$$(X_{CP})_{ij}^{\alpha\beta} \equiv ImT_{ij}^{\alpha\beta} = (V_{\alpha j}^{*} V_{\alpha i} V_{\beta i}^{*} V_{\beta j}), \tag{4.2}$$

where α, β, i, j run from 1 to (N-1) with $\alpha < \beta, i < j$. So there are $[(N-1)(N-2)/2]^2$ number of $(X_{CP})_{ij}^{\alpha\beta}$, however totally there are only $(N-1)^2$ real parameters. Thus the X_{CP}'s saturate the whole parameter space of the quark mixing matrix! We see that the three-generation case is a rather singular case. Therefore the estimates of partial decay rates can be very different in the presence of higher than three generations of quarks. One important effect is that the partial decay rate differences in charm particle decays can be large too. Of special interest to note is that some of these large effects involving only tree graphs which are independent of the masses of the 4th or higher generations of quarks, and are dependent solely on the existence of higher generations of quarks. Partial decay rate difference between

$$F^{+} \to \overline{K}^{0} \pi^{+}, \; F^{-} \to K^{0} \pi^{-} \tag{4.3}$$

is such an example.

With this scenario of the possible existence of higher than three generations of quarks it is very worthwhile to be on the lookout for CP noninvariant decays in experiments where charm particles are or will be copiously produced.

CONCLUDING REMARKS:

We see that the Quark-Diagram Scheme has been very useful in our studies on weak decays and CP noninvariance. It provides a framework to advance our knowledge on nonleptonic decays: from experimental measurements to predict new experimental results; e.g. to test and give guidance to various theoretical models: e.g. provide experimental value for various amplitudes and final state interaction phases which any theoretical calculation must agree to.

For the urgently needed measurements of $|V_{ub}|$, $|V_{ub}/V_{cb}|$, the Quark Diagram Scheme has provided a useful model-independent framework to select out nonleptonic decays, and decay ratios, as given in Eqs. (2.5) and (2.8), which are directly proportional to $|V_{ub}|$ and $|V_{ub}/V_{cb}|$ respectively. Dedicated experimental measurements of these decays are very important for our understanding of the quark mixing phenomena and whether there are more than three generations of quarks. Charmless decays are useful for search of $|V_{ub}|$ only if the W-loop-diagram contributions with dominant mixing-matrix coefficients are negligible. To directly measure the size of the W-loop-diagram contribution, it is useful to study the decays which are purely from the W-loop diagrams as given in Eq. (2.6b).

In our search for CP noninvariance, the Quark Diagram Scheme has been very useful, providing guidance on what decay channels in heavy quark decays can possibly have nonzero partial decay rate differences. Using the scheme we see easily the qualitative reason why the beauty particle decays provide favorable reactions for the CP noninvariance search. Here a few more nice channels for experimental studies are pointed out, Eqs. (3.12), (3.13).

In the presence of higher then three generations of quarks, the number of CP nonvariance parameters increases from a unique X_{CP} to saturate the whole parameter space: e.g. nine X_{CP}'s for four generations of quarks, etc. This gives the possibility that the CP noinvariance of partial decay rate differences can also be appreciable in charm particle decays. It is therefore very worthwhile to be on the lookout for CP noninvariance decays in experiments where charm particles are or will be copiously produced.

TABLE I. Charm meson decays into a vector boson and a pseudoscalar meson.

	Experimental branching ratio (%)	Amplitudes with SU(3) breaking and final-state interactions[H]
	$\underline{D^+ \text{ decays}}$	
$\overline{K}^{*0}\pi^+$	$3.0 \pm 1.9 \pm 1.7$[A]	$(c_1)^2(a'+b')exp(i\delta_{3/2}^{\overline{K}^*\pi})$
$\rho^+\overline{K}^0$	$12.2 \pm 2.28 \pm 1.9$[A]	$(c_1)^2(a+b)exp(i\delta_{3/2}^{\rho K})$
- - - - -	- - - - -	- - - - -
$\phi\pi^+$	$0.93 \pm 0.26 \pm 0.17$[A]	$(s_1c_1)(b')exp(i\delta^{\phi\pi})$
$\overline{K}^{*0}K^+$	$0.53 \pm 0.24 \pm 0.14$[A]	$(s_1c_1)(a'-\overline{d}+\delta e)exp(i\delta_1^{\overline{K}^*K})$
	$\underline{D^0 \text{ decays}}$	
$\phi\overline{K}^0$	1.4 ± 0.5[B]	$(c_1)^2(\overline{c}\,')exp(i\delta^{\phi\overline{K}})$
$\omega\overline{K}^0$	$3.8 \pm 1.5 \pm 1.0$[A]	$(1/\sqrt{2})(c_1)^2(b+c)exp(i\delta^{\omega\overline{K}})$
$K^{*-}\pi^+$	$7.8 \pm 1.2 \pm 0.9$[A] $7.1 \pm 1.6 \pm 1.3$	$(c_1)^2\{(a'+c')-\frac{1}{3}(a'+b')[1-exp(i\Delta_{\overline{K}^*\pi})]\}exp(i\delta_{1/2}^{\overline{K}^*\pi})$
$\overline{K}^{*0}\pi^0$	$2.1 \pm 0.9 \pm 0.6$[A]	$(1/\sqrt{2})(c_1)^2\{(b'-c')-\frac{2}{3}(a'+b')[1-exp(i\Delta_{\overline{K}^*\pi})]\}exp(i\delta_{1/2}^{\overline{K}^*\pi})$
ρ^+K^-	$13.7 \pm 1.3 \pm 1.5$[A]	$(c_1)^2\{(a+c)-\frac{1}{3}(a+b)[1-exp(i\Delta_{\rho\overline{K}})]\}exp(i\delta_{1/2}^{\rho\overline{K}\pi})$
$\rho^0\overline{K}^0$	$1.3 \pm 0.4 \pm 0.3$[A]	$(1/\sqrt{2})(c_1)^2\{(b-c)-\frac{2}{3}(a+b)[1-exp(i\Delta_{\rho\overline{K}})]\}exp(i\delta_{1/2}^{\rho\overline{K}\pi})$
$\overline{K}^{*0}K^0$		$(c_1)^2\{(c-c')\}$
$K^{*0}\overline{K}^0$		$-(c_1)^2\{(c-c')\}$
$\overline{K}^{*0}\eta_8$		$\frac{1}{\sqrt{6}}(c_1)^2\{b'+c'-2\underline{c}\}$
$\overline{K}^{*0}\eta_0$		$\frac{1}{\sqrt{3}}(c_1)^2\{b'+c'+\underline{c}\}$
- - - - -	- - - - -	- - - - -
$\phi\pi^0$		$\frac{1}{\sqrt{2}}(s_1c_1)\{b'\}e^{i\delta\phi\pi}$ prediction: $Br(D^0 \to \phi\pi^0) \simeq 0.21\%$
	$\underline{F^+ \text{ decays}}$	
$\phi\pi^+$	$3.3 \pm 1.1^c \pm 4.4^d$ $13.0 \pm 3.0 \pm 4.0^e$	$(c_1)^2(a')exp(i\delta^{\phi\pi})$
$\overline{K}^{*0}K^+$	$(1.37 \pm 0.37) \times Br(F^+ \to \phi\pi^+)$ $(1.00 \pm 0.25 \pm 0.06) \times Br(F^+ \to \phi\pi^+)$[F]	$c_1^2\{b'+d\}e^{i\delta_1^{\overline{K}^*K}}$
- - - - -	- - - - -	- - - - -
$\rho^+\pi^0$		$(s_1c_1)^2\{d-d'\}e^{i\delta_1^{\rho\pi}}$
$\omega\pi^+$		$(s_1c_1)^2\{d+d'\}e^{i\delta_1^{\omega\pi}}$

[A]Reference 22.
[B]Reference 24; see also Refs. 22 and 23.
[C]Reference 25.
[D]Reference 23.
[E]Reference 26.
[F]Reference 35.
[H]$V_{us}V_{cs}^* \simeq -V_{ud}V_{cd}^* \simeq s_1c_1$ used;
[G]$\delta e \equiv \overline{e}-e$; $\delta f \equiv \overline{f}-f$; $\delta c \equiv \overline{c}-c$; the amplitudes with bars have $s\overline{s}$.

TABLE II. Charm meson decays into two pseudoscalars. The same notations a to f are used for amplitudes, but in general they have no relation to those in the PV decays in Table I.

	Experimental branching ratio (%) (Ref. 1)	Amplitudes with SU(3) breaking and final-state interactions[B]
	D^+ decays	
$\overline{K}^0\pi^+$	$3.5 \pm 0.5 \pm 0.4$	$(c_1)^2(a+b)exp(i\delta_{3/2}^{\overline{K}\pi})$
$\overline{K}^0K^+$	$1.11 \pm 0.34 \pm 0.21$	$(s_1c_1)(a-d+\delta e)exp(i\delta_1^{\overline{K}K})$
$\pi^0\pi^+$	≤ 0.53	$(1/\sqrt{2})(s_1c_1)(a+b)exp(i\delta_2^{\pi\pi})$
	D^0 decays	
$K^-\pi^+$	$4.9 \pm 0.4 \pm 0.4$	$(c_1)^2\{(a+c)-(a+b)\frac{1}{3}[1-exp(i\Delta_{\overline{K}\pi})]\}exp(i\delta_{1/2}^{\overline{K}\pi})$
$\overline{K}^0\pi^0$	$2.2 \pm 0.4 \pm 0.2$	$(1/\sqrt{2})(c_1)^2\{(b-c)-(a+\overline{b})\frac{2}{3}[1-exp(i\Delta_{\overline{K}\pi})]\}exp(i\delta_{1/2}^{\overline{K}\pi})$
$\overline{K}^0\eta$	$1.8 \pm 0.8 \pm 0.3$	$\cos\theta A(D^0 \to \overline{K}^0\eta_8) + \sin\theta A(D^0 \to \overline{K}^0\eta_0)$
$K^0\overline{K}^0$	≤ 0.62	$(s_1c_1)\{(-\delta e-2\delta f)+(a+\overline{c}-\delta e)\frac{1}{2}[1-exp(\Delta_{\overline{K}K})]\}exp(i\delta_0^{K\overline{K}})$
K^-K^+	$0.60 \pm 0.10 \pm 0.08$	$(s_1c_1)\{(a+c)+(\delta e+2\delta f)-(a+\overline{c}-\delta e)\frac{1}{2}[1-exp(i\Delta_{\overline{K}K})]\}exp(i\delta_0^{\overline{K}K})$
$\pi^+\pi^-$	$0.16 \pm 0.09 \pm 0.03$	$-(s_1c_1)\{(a+c)+(\delta e+2\delta f)-(a+b)\frac{1}{3}[1-exp(i\Delta_{\pi\pi})]\}exp(i\delta_0^{\pi\pi})$
$\pi^0\pi^0$		$\frac{1}{2}\sqrt{2}(s_1c_1)\{(b-c)+(\delta e+2\delta f)-(a+b)\frac{2}{3}[1-exp(i\Delta_{\pi\pi})]\}exp(i\delta_0^{\pi\pi})$
	F^+ decays	
$\overline{K}^0K^+$		$(c_1)^2\{b+d\}e^{i\delta_1^{\overline{K}K}}$
$\pi^0\pi^+$	0	
$\eta_8\pi^+$		$-\frac{\sqrt{2}}{\sqrt{3}}(c_1)^2\{a-d\}e^{i\delta^{\pi\eta_8}}$
$\eta_0\pi^+$		$\frac{1}{\sqrt{3}}(c_1)^2\{a+2d\}e^{i\delta^{\pi\eta_0}}$

[A] $V_{us}V_{cs}^* \simeq -V_{ud}V_{cd}^* \simeq s_1c_1$ used.

[B] $\delta e \equiv \overline{e} - e$; $\delta f \equiv \overline{f} - f$; $\delta c \equiv \overline{c} - c$; the amplitudes with tildes have $s\overline{s}$

ACKNOWLEDGEMENTS

I would also like to thank F. Botella, H.-Y. Cheng, and W.-Y. Keung for many collaborations on the subject, and especially H.-Y. Cheng for many collaborations in carrying out the program using the Quark Diagram Scheme

I would like to thank the organizers of the conference for their nice conference and for their warm hospitality.

REFERENCES

1. For reviews see L.-L. Chau, Phys. Rept. 95, 1 (1983); in "A Passion for Physics", Essays in Honor of Geoffrey Chew, World Scientific publisher, 1985, Ed. C. De Tar, Finkelstein and C.-I. Tan, p. 148; and in Proceedings of Kyoto Int. Sym., The Jubilee of the Meson Theory, Kyoto, Aug. 15-17, 1985; Prog. Theor. Phys. Sup. 85, 147 (1985).
2. Chau, L.-L. in *Proceedings of the 1980 Guangzhou Conference on Theoretical Particle Physics* (Van Nostrand Reinhold, New York, 1981, and Science Press of the People's Republic of China); and Phys. Rep. 95, 1 (1983).
3. Chau, L.-L., Cheng, H.-Y., Phys. Rev. Lett. 56, 1655 (1986)
4. Kobayashi, M. and Maskawa, T., Prog. Theor. Phys. 49, 652 (1973).
5. Behrends, S. et. al. Phys. Rev. Lett. 50, 881 (1983). For the b lifetime, Fernandez, E., Phys. Rev. Lett. 51, 1022 (1984); Lockyer, N.S. et. al., Phys. Rev. Lett. 51, 1316 (1983); For $\Gamma(b \to u)/\Gamma(b \to c)$, Klopfenstein, C. et al., Phys. Lett. 103B, 444 (1983); Chen, A. et al., Phys. Rev. Lett. 111B, 1084 (1984).
6. Ref. (1), and Chau, L.-L., (Wang) and Wilczek, F., Phys. Rev. Lett. 43, 816 (1979).
7. Christenson, J.H., Cronin, J.W., Fitch, V.L. and Turlay, R., Phys. Rev. Lett. 13, 138 (1964); and the classical theoretical discussions on CP violation: Lee, T.D., Oehme, R. and Yang, C.N., Phys. Rev. 106, 340 (1957); Wu, T.T., and Yang, C.N., Phys. Lett. 13, 380 (1964).
8. Sakharov, A., Zh. Eksp. Teor. Fiz. Pisma Red. 5, 32 (1967). For a review, see Goldhaber, M., Langacker, P., Slansky, R., Science, 210, 851 (1980).
9. Wolfenstein, L., Phys. Rev. Lett. 13, 562 (1964).
10. Gilman, F.J. and Wise, M.B., Phys. Lett. 83B, 83 (1979).
11. Black, J.K. et al., Phys. Rev. Lett. 54, 1628 (1985); Bernstein, R.H. et al. ibid, 54, 1631 (1985).
12. Current experiments measuring ϵ'/ϵ: CERN NA31; Fermilab 631; and the CERN LEAR experiment p. 82.
13. Chau, L.-L., Cheng, H.-Y., Phys. Rev. Lett. 54, 1768 (1985). For recent experimental results on $K \to \gamma\gamma$ see M. Holder's talk on the CERN NA31 experiment at the XXIII Int. Conf. on High Energy Physics, July 16-23, 1986, Berkeley.
14. For general discussions in the KM Scheme using the Quark Diagram Scheme to map out decays which have partial decay rate differences, L.-L. Chau Wang, AIP Conf. Proc. No. 72, Particle and Fields, Subseries No. 23, Virginia Polytechnic Inst. 1980, eds. G.B. Collins, Chang, L.N., and Ficenec, J.R.; For other developments see Ref. (15)-(17).
15. For earlier discussions see Sehgal, L.M. and Wolfenstein, L., Phys. Rev. 162, 1362 (1976), Overseth, O.E. and Pakvasa, S., Phys. Rev. 189, 1663 (1969). Pais, A. and Treiman, S.B., Phys. Rev. D12, 2744 (1975); Okun, L.B., Zakharov, V.I. and Pontecorvo, B.M. Lett. Nuovo Cim. 13, 218 (1975).
16. For time-like Penguin diagram contribution of Partial decay rate difference in the KM Scheme, see Barshay, and J. Geris, Phys. Lett. 84B, 319 (1979);

Bander, M., Silverman, D. and Soni, A., Phys. Rev. Lett. 43, 242 (1979); For tree graph contributions, see Barnabeu, J. and Jarlskog, C., Z. Phys. C8, 233 (1981).

17. For CP noninvariance in neutral B decays into CP self-conjugate state, see Bigi, I.I. and Sanda, A.I., Nucl. Phys. B193, 81 (1981).
18. For general estimates of CP noninvariance in charge $B^{\pm}$ decays, Chau, L.-L. and Cheng, H.-Y., Phys. Rev. Lett. 53, 1037 (1984)
19. Estimates of CP noninvariance in B^0 decays, including CP non-self-conjugate states, like $B_d^0 \to K^+\pi^-$, Chau, L.-L. and Cheng, H.-Y., Phys. Lett. 165B, 429 (1985).
20. Du, D.S., Dunietz, I. and Wu, D.D., Enrico Fermi Institute Report No. EFI 86-9 (1986); Dunitez, I. and Rosner, J.L., Phys. Rev. D34, 1404 (1986).
21. The detail calculations will be publications by Chau, L.-L. and Cheng, H.-Y.
22. Baltrusaitis, R. et al. (Mark III Collaboration), Phys. Rev. Lett. 55, 150 (1985); see also R.H. Schindler, Stanford Linear Accelerator Center Report No. SLAC-PUB-3799. 1985 (unpublished); D. Coward. Stanford Linear Accelerator Center Report No. SLAC-PUB-3818, 1985 (unpublished).
23. Chen, A. et al. (CLEO Collaboration), Phys. Rev. Lett. 51, 634 (1983); see also S. Stone, in *Proceedings of the Eleventh International Symposium on Lepton and Photon Interactions at High Energies*, edited by D.G. Cassel and D.L. Kreinick (F.R. Newman Laboratory of Nuclear Studies, Ithaca, 1983); E.H. Thorndike, in Proceedings of the International Lepton and Photon Conference, Kyoto, Japan, August 1985 (to be published).
24. Darden, C. et al. (Argus Collaboration), DESY Report No. DESY 84-04-3 (to be published); Albrecht, H. et al., Phys. Lett. 158B, 525 (1985).
25. Derrick, M. et al. (HRS Collaboration), Phys. Rev. Lett. 2568 (1985).
26. Althoff, M. et al. (TASSO Collaboration), Phys. Lett. 136B, 130 (1984).
27. Chau, L.-L. and Keung, W.-Y., Phys. Rev. Lett. 53, 1802 (1984).
28. Wolfenstein, L., Phys. Rev. Lett. 51, 1945 (1984).
29. For a recent thorough review on determination on mixing matrix, see Stone, S., "Experimental Determination of the KM Matrix", Cornell preprint CLNS-86/753, Oct. 1986.
30. Maiani, L., Phys. Lett. 62B, 183 (1976); Mignami, R., Lett. Al Nuovo Cimento, 28, 529 (1980).
31. Botella, F.J. and Chau, L.-L., Phys. Lett. 168B, 97 (1986).
32. See also He, X.-G. and Pakvasa, S. Phys. Lett. 156B, 236 (1985); Anselm, A.A. et al., Phys. Lett. 156B, 102 (1985); Gronau, M. and Schechter, J., Phys. Rev. D31, 1668 (1985); Harari, H. and Leurer, M., Weizmann Inst. of Sci. preprint, W15-86/24/May-Ph; Fritzsch, H. and Plankl, J., U. of Munchen and Max-Planck-Inst preprint MPI-PAE/PTH 20/86.
33. Chau, L.-L., Cheng, H.-Y. and Keung, W.-Y., "CP Violation in the Kaon System", Brookhaven, Indiana U., U. of Illinois preprint, 1986.
34. Chau, L.-L., Proc. of the VIth European Symposium on $N\overline{N}$, $Q\overline{Q}$ Interactions, Santiago de Compostela, Spain, Aug. 20-Sept. 3, 1982; and Proc. of the Eighth Int'l. Workshop on Weak Interactions and Neutrinos, Javea, Spain, Sept. 5-11, 1983; Proceedings of the 1984 Vanderbilt Conference, AIP Conference Proceedings No. 121, Ed. R.S. Panvini and G.B. Word.
35. Fermilab Experiment E691.

Results on Charmed Meson Decays from Mark III*

Steven R. Wasserbaech
Representing the Mark III Collaboration†

Abstract

We report recent results on charmed meson decays, obtained using the Mark III detector at SPEAR. The first topic discussed is the observation of $e^+e^- \to D_s D_s^*$ at $\sqrt{s} = 4.14\,\text{GeV}$. The D_s^* is detected as a peak in the mass distribution recoiling from $D_s^\pm \to \phi\pi^\pm$. The mass of the D_s^* is found to be $(2109.3 \pm 2.1 \pm 3.1)\,\text{MeV}/c^2$, yielding a D_s^*-D_s mass difference of $(137.9 \pm 2.1 \pm 4.3)\,\text{MeV}/c^2$. The production cross section times branching ratio is also measured. Next, a search for the decay $D^+ \to \mu^+\nu_\mu$ is described. A preliminary upper limit (90% CL) on $B(D^+ \to \mu^+\nu_\mu)$ of 8.4×10^{-4} is obtained, corresponding to an upper limit on the decay constant f_D of $340\,\text{MeV}/c^2$. Finally we present results of a search for the lepton family number violating decay $D^0 \to \mu e$. We find $B(D^0 \to \mu e) < 1.5 \times 10^{-4}$ at 90% CL.

⋆ Work supported in part by the Department of Energy, under contracts DE-AC03-76SF00515, DE-AC02-76ER01195, DE-AC03-81ER40050, DE-AM03-76SF0034, and by the National Science Foundation.

† Members of the Mark III Collaboration are D. Coffman, G. Dubois, G. Eigen, J. Hauser, D. Hitlin, C. Matthews, J.D. Richman, J.J. Russell, W.J. Wisniewski, Y. Zhu, *California Institute of Technology;* D.E. Dorfan, F. Grancagnolo, R.P. Hamilton, C.A. Heusch, L. Köpke, W.S. Lockman, R. Partridge, J. Perrier, H.F. Sadrozinski, M. Scarlatella, T.L. Schalk, A. Seiden, A. Weinstein, R. Xu, *University of California at Santa Cruz;* J.J. Becker, G. Blaylock, J.S. Brown, B. Eisenstein, T. Freese, G. Gladding, J. Izen, S.A. Plaetzer, C. Simopoulos, A.L. Spadafora, I.E. Stockdale, J.J. Thaler, B. Tripsas, A. Wattenberg, *University of Illinois, Champaign-Urbana;* T. Bolton, K.O. Bunnell, R. Cassell, D.H. Coward, K.F. Einsweiler, D. Favart, C. Grab, U. Mallik, R.F. Mozley, A. Odian, J. Parker, D. Pitman, R.H. Schindler, W. Stockhausen, W. Toki, F. Villa, S. Wasserbaech, N. Wermes, D.E. Wisinski, G. Wolf, *Stanford Linear Accelerator Center;* T.H. Burnett, V. Cook, A.L. Duncan, A.D. Li, R. Mir, P. Mockett, B. Nemati, L. Parrish, H. Willutzki, *University of Washington, Seattle.*

1. $e^+e^- \rightarrow D_s D_s^*$

In the quark model, the lowest-lying $c\bar{s}$ pseudoscalar meson, the D_s, has a higher mass vector meson partner, the D_s^*. The D_s^* mass is predicted[1] to lie 80 to 150 MeV/c^2 above that of the D_s. Evidence has been presented for a narrow state decaying into a D_s meson and a photon.[2] [3] Exclusive production of this state in association with the D_s in e^+e^- annihilation would provide new evidence that it is indeed the D_s^*. We report the first evidence of the exclusive reaction[4] $e^+e^- \rightarrow D_s^+ D_s^{*-}$, where the $D_s^\pm$ is observed in the decay

$$D_s^+ \rightarrow \phi\pi^+ \tag{1}$$

or in the cascade

$$D_s^{*-} \rightarrow \gamma D_s^-, \quad D_s^- \rightarrow \phi\pi^-. \tag{2}$$

A precise measurement of the D_s^* mass is also reported.

The data sample used in this analysis represents an integrated luminosity of $(6.30 \pm 0.46)\,\mathrm{pb}^{-1}$ at $\sqrt{s} = 4.14$ GeV. The analysis proceeds with the isolation of events containing one or more ϕ's. Charged kaons are identified using time-of-flight (TOF), and K^+K^- combinations with mass within 10 MeV/c^2 of the ϕ mass are selected. Candidate $\phi\pi^+$ decays are selected by combining a ϕ with each of the remaining charged tracks, assumed to be pions. A scatter plot of the $\phi\pi^+$ mass versus the recoil mass is shown in Fig. 1. Evidence for $D_s D_s^*$ production appears as a cluster of events near $M(\phi\pi^+) = 1.97$ GeV/c^2 and $M(\text{recoil}) = 2.10$ GeV/c^2. Another cluster near $M(\phi\pi^+) = 1.87$ GeV/c^2 and $M(\text{recoil}) = 2.01$ GeV/c^2 is evidence for production of D^+D^{*-}, with $D^+ \rightarrow \phi\pi^+$.[5] No significant evidence for $e^+e^- \rightarrow D_s^+ D_s^-$, $D_s^+ \rightarrow \phi\pi^+$ is observed.

The decay $D_s \rightarrow \phi\pi^+$ is isolated by requiring the recoil mass to lie between 2.04 and 2.18 GeV/c^2 (Fig. 2). A fit to this distribution yields 26.7 ± 5.2 (*stat.*) signal events above 5.6 background events. The fitted D_s mass is $(1972.4 \pm 3.7 \pm 3.7)$ MeV/c^2. The background shape is determined from the $\phi\pi^+$ mass distribution obtained by combining ϕ candidates with pions from different events.

The D_s^* signal in the recoil mass distribution is measured after restricting the $\phi\pi^+$ mass to the D_s region, 1.92 to 2.02 GeV/c^2. To improve the D_s^* mass resolution, a D_s mass[6] of 1971.4 MeV/c^2 is imposed as a constraint in the calculation of the recoil mass.[7] The resulting recoil mass distribution [Fig. 3] shows a narrow peak at 2.11 GeV/c^2 from reaction (1), on a broad structure between 2.07 and 2.15 GeV/c^2 from reaction (2). A fit to this distribution yields

$$M(D_s^*) = (2109.3 \pm 2.1 \pm 3.1)\,\mathrm{MeV}/c^2.$$

The shape of the signal distribution and the resolution (5.0 MeV/c^2) are determined from a Monte Carlo simulation which includes radiative corrections.[8] The result implies[9]

$$M(D_s^*) - M(D_s) = (137.9 \pm 2.1 \pm 4.3)\,\mathrm{MeV}/c^2.$$

A maximum likelihood calculation using the constrained recoil mass yields $\Gamma(D_s^*) < 22\,\mathrm{MeV}/c^2$ at 90% CL.

The production cross section times branching fraction is determined assuming $B(D_s^* \to \gamma D_s) = 100\%$. Using the number of observed $D_s^+ \to \phi\pi^+$ decays, and a $D_s^+ \to \phi\pi^+$ detection efficiency of 0.071, the result is

$$\sigma(e^+e^- \to D_s^+ D_s^{*-} + c.c.) \times B(D_s^+ \to \phi\pi^+) = (30 \pm 6 \pm 11)\,\mathrm{pb}.$$

We have also presented the following preliminary results:

$$\sigma(e^+e^- \to D_s^+ D_s^{*-} + c.c.) \times B(D_s^+ \to \bar{K}^{*0} K^+) = (31 \pm 6 \pm 11)\,\mathrm{pb},$$
$$\sigma(e^+e^- \to D_s^+ D_s^{*-} + c.c.) \times B(D_s^+ \to \bar{K}^{0} K^+) = (32 \pm 6 \pm 10)\,\mathrm{pb}.$$

In the analysis of these channels, candidate events are fit to the hypothesis $e^+e^- \to D_s^* D_s$, where the D_s^* is not reconstructed, and $D_s \to$ *mode*. These final states could be produced via the internal W emission and the annihilation diagrams.

The measured D_s^*-D_s mass difference can be compared with other vector-pseudoscalar splittings. For mesons containing at least one light quark, the mass-squared difference $\Delta_{M^2} = M^2(1^-) - M^2(0^-)$ is approximately constant.[10] This effect has motivated calculations of the mass-squared difference within models which assume a simple confining potential.[11] An approximately constant mass-squared difference is predicted for specific choices of α_s and the form of the potential. Our measurement of the D_s^*-D_s mass difference results in $\Delta_{M^2} = (0.563 \pm 0.020)(\mathrm{GeV}/c^2)^2$, which is consistent with this empirical rule.

2. $D^+ \to \mu^+\nu_\mu$ AND THE DECAY CONSTANT f_D

The D meson decay constant f_D is a direct measure of the overlap of the wavefunctions of the heavy and light quarks in the D meson,[12] and sets the scale for processes such as weak flavor annihilation and Pauli interference invoked to account for the differences in D^+ and D^0 lifetimes.[13] The decay constant also is essential in evaluating the magnitude of operators leading to $D^0\bar{D}^0$ and $B^0\bar{B}^0$ mixing.[14] The constant f_D may be unambiguously determined from the pure leptonic decay of the D^+:

$$\Gamma(D^+ \to \mu^+\nu_\mu) = \frac{G_F^2}{8\pi} f_D^2\, M_D\, M_\mu^2\, |V_{cd}|^2 \left(1 - \frac{M_\mu^2}{M_D^2}\right)^2.$$

The data employed (9.3 pb^{-1}) were obtained at $\sqrt{s} = 3.768$ GeV, where charmed D mesons are produced only in pairs. The search is carried out by isolating a sample of events in which a D^+ candidate is found, and then examining the recoil system for evidence of the $\mu^-\bar{\nu}_\mu$ decay. Seven hadronic D^+ tag modes are used, resulting in $2490 \pm 42 \pm 40$ D^+ candidates.

Candidate events are isolated by requiring the recoil system from a tag to have one track with charge opposite to that of the tag whose momentum lies between 0.775 and 1.125 GeV/c. The event is also required to have missing mass squared $M^2_{\rm miss}$ between -0.265 and $0.175\,({\rm GeV}/c^2)^2$, rejecting about 5% of the expected signal. Muon candidates are selected using the muon system, or (for tracks outside of the muon system acceptance) a combination of TOF, dE/dx, and calorimeter. In the latter case the event is required to contain no extra isolated photons.

Important sources of background to this process are $D^+ \to K^0_L\pi^+$, $D^+ \to K^0_L K^+$, $D^+ \to K^0_L\mu^+\nu$, $D^+ \to \pi^0\pi^+$, and $D^+ \to K^0_S\pi^+$. Decays with K^0_L's are rejected by the $M^2_{\rm miss}$ cut. The selection process reduces the total expected background to $1.16 \pm 0.16 \pm 0.20$ events.

When these cuts are applied to the data, no events are found to survive. The nearest event to the $M^2_{\rm miss}$ cut lies at $0.196\,({\rm GeV}/c^2)^2$. The observation of no events of the type $D^+ \to \mu^+\nu_\mu$ together with the background prediction yields a 90% CL upper limit of 1.35 signal events.[15] The acceptance for this decay mode is $0.72 \pm 0.01 \pm 0.05$. Dividing by the acceptance and the total number of D^+ tags[16] gives the preliminary limit $B(D^+ \to \mu^+\nu_\mu) < 8.4 \times 10^{-4}$ at 90% CL. Using a D^+ lifetime[17] of $(10.1^{+0.7}_{-0.6}) \times 10^{-13}$ s, and[18] $|V_{cd}|^2 = 0.0506 \pm 0.0065$, the branching ratio limit implies $f_D < 310\,{\rm MeV}/c^2$. When the errors on τ_{D^+} and $|V_{cd}|^2$ are included, we obtain $f_D < 340\,{\rm MeV}/c^2$ (90% CL).

This result does not probe the small values of f_D suggested by bag model[19] or QCD sum rule[20] calculations (150 to 280 MeV/c^2), but it restricts the range of values predicted by recent potential model[12] calculations (208 to 450 MeV/c^2). It also excludes the very high values of f_D which have been suggested[21] as an explanation for the large observed value of $\tau(D^+)/\tau(D^0)$.

3. SEARCH FOR $D^0 \to \mu e$

There has recently been great interest in searching for lepton family number violating processes such as $\mu \to e\gamma$, $\mu \to eee$, $K^0_L \to \mu e$, $D^0 \to \mu e$, and ν oscillations.[22] The decay $D^0 \to \mu e$ could be induced by massive leptoquarks whose existence is predicted in various extensions[23] [24] of the Standard Model. Within certain models,[23] scalar leptoquarks are expected to couple the "up-type" quarks (u, c, t) to charged leptons, and the "down-type" quarks (d, s, b) to neutral leptons. Thus, the decay $D^0 \to \mu e$ can be enhanced with respect to the experimentally more accessible $K^0_L \to \mu e$ decay.[25]

The data sample at $\sqrt{s} = 3.768\,\mathrm{GeV}$ corresponds to $41400\,^{+3100}_{-2700} \pm 2700$ produced D^0's.[26] The data are first searched for events containing at least two lepton candidates: one muon and one electron. The kinematics of the two-body D^0 decay require that both leptons have momentum $\mathrm{p} > 0.75\,\mathrm{GeV}/c$ in the laboratory frame. Leptons are selected on the basis of the momentum, energy deposited in the calorimeter, TOF, and range in the muon system.

The two principal sources of background to the decay $D^0 \rightarrow \mu e$ are hadronic charged two-body D^0 decays, and $\tau^+\tau^-$ pairs. Contamination from $D^0 \rightarrow K^-\pi^+$ decays is reduced by requiring candidate pairs to have invariant mass within $50\,\mathrm{MeV}/c^2$ of the D^0 mass. This cut does not reject the Cabibbo-suppressed decay $D^0 \rightarrow \pi^+\pi^-$. Since the τ background consists mainly of two-prong events accompanied by undetected neutrinos, requiring the missing energy[27] E_{miss} to be less than 1 GeV *in two-prong events only* eliminates this contamination. The expected background in the data sample is estimated to be $0.18 \pm 0.06 \pm 0.05$ $\pi^+\pi^-$ events.[26]

After all particle identification and kinematic cuts have been applied, the beam-constrained mass M_{bc} is calculated for each surviving candidate pair by constraining its energy to the beam energy. Two events with $M_{\mathrm{bc}} > 1.82\,\mathrm{GeV}/c^2$ are found. A study of the M_{bc} distribution of $D^0 \rightarrow K^-\pi^+$ in the same data sample shows that 90% of those two-body decays lie within $\pm 0.0055\,\mathrm{GeV}/c^2$ of the D^0 mass (Fig. 4). No μe candidate falls within this range. The efficiency for $D^0 \rightarrow \mu e$ is found to be $0.433 \pm 0.004 \pm 0.029$, while that for the background channel $D^0 \rightarrow \pi^+\pi^-$ is $0.0024 \pm 0.0004 \pm 0.0002$. After all analysis cuts, neither $D^0 \rightarrow K^-\pi^+$ decays nor $\tau^+\tau^-$ pair production contribute significantly.

The observation of no events of the type $D^0 \rightarrow \mu e$ yields a 90% CL upper limit ($N^{sb}_{0.9}$) of 2.30 on the total number of signal and background events. The upper limit on the branching fraction $\mathrm{B}(D^0 \rightarrow \mu e)$ is obtained by dividing by the efficiency and the total number of produced D^0's. When all systematic errors are included, we obtain $B(D^0 \rightarrow \mu e) < 1.5 \times 10^{-4}$ at 90% CL. This bound, which is model-independent, is approximately an order of magnitude lower than previous model-dependent measurements.[28]

Further details concerning these three topics are given elsewhere.[29] We gratefully acknowledge the efforts of the SPEAR staff. This work was supported by the Department of Energy, under contracts DE-AC03-76SF00515, DE-AC02-76ER01195, DE-AC03-81ER40050, DE-AM03-76SF00034 and by the National Science Foundation.

References

1. For a recent discussion see S. Godfrey and N. Isgur, Phys. Rev. **D32**, 189 (1985), and references therein.
2. The first evidence for the D_s^{*+} was reported by R. Brandelik *et al.*, Phys. Lett. **70B**, 132 (1977), and R. Brandelik *et al.*, Phys. Lett. **80B**, 412

(1979). These measurements were not confirmed by R. Partridge *et al.*, Phys. Rev. Lett. **47**, 760 (1981), and R.P. Horisberger, Ph.D. Thesis, SLAC Report No. 266 (1984) (unpublished).

3. H. Aihara *et al.*, Phys. Rev. Lett. **53**, 2465 (1984); H. Albrecht *et al.*, Phys. Lett. **146B**, 111 (1984).
4. Throughout this paper we adopt the convention that reference to a state also implies reference to its charge conjugate.
5. The Cabibbo-suppressed decay $D^+ \to \phi\pi^+$ has been observed by R. Bailey *et al.*, Phys. Lett. **139B**, 320 (1984) and R. M. Baltrusaitis *et al.*, Phys. Rev. Lett. **55**, 150 (1985).
6. A weighted average of $M(D_s) = (1971.4 \pm 1.7)\,\mathrm{MeV}/c^2$ is obtained using the measurements tabulated in M. Aguilar-Benitez *et al.* (Particle Data Group), Phys. Lett. **170B**, 1 (1986), replacing the previous ACCMOR value with their later measurement, $M(D_s) = (1972.7 \pm 1.5 \pm 1.0)\,\mathrm{MeV}/c^2$, H. Becker *et al.*, CERN-EP/86-172 (1986) (preprint, to be published in Phys. Lett.).
7. $M(\mathrm{recoil}) = \{ [\sqrt{s} - \sqrt{M^2(D_s) + p^2(D_s)}\,]^2 - p^2(D_s)\}^{\frac{1}{2}}$, where $M(D_s)$ is fixed.
8. In the Monte Carlo simulation, the D_s and D_s^* are assumed to be pseudoscalar and vector, respectively, with $B(D_s^* \to \gamma D_s) = 100\%$. Only initial state radiation is considered using a method from F. Behrends and R. Kleiss, Nucl. Phys. **B178**, 141 (1981).
9. For $e^+e^- \to D_s^* D_s$ at $\sqrt{s} = 4.14$ GeV, $\delta M(D_s^*) \cong -\delta M(D_s)$.
10. $\Delta_{M^2} = 0.575\,(\mathrm{GeV}/c^2)^2$ for ρ^0-π^0, $0.556\,(\mathrm{GeV}/c^2)^2$ for K^{*0}-K^0, and $0.551\,(\mathrm{GeV}/c^2)^2$ for D^{*0}-D^0. For a discussion of this empirical relation see A. Martin, Comm. Nucl. Part. Phys. **16**, 249 (1986).
11. K. Igi and S. Ono, Phys. Rev. **D32**, 232 (1985), and M. Frank and P. O'Donnell, CERN-TH-4367/86 (1986) (preprint).
12. S. N. Sinha, Alberta THY-3-86 (1986); P. J. O'Donnell, CERN-TH-4419/86 (1986); L. Maiani, *Proceedings of XXI International Conference on High Energy Physics* (Editions de Physique, Le Ulis, France, 1982) pp. 631-657; H. Krasemann, Phys. Lett. **96B**, 397 (1980).
13. R.M. Baltrusaitis *et al.*, Phys. Rev. Lett. **54**, 1976 (1985).
14. I.I. Bigi, G. Köpp, and P.M. Zerwas, Phys. Lett. **166B**, 238 (1986); J.F. Donoghue *et al.*, Phys. Rev. **D33**, 197 (1986).
15. The CL is defined as the probability that a given hypothesis [here, the sum of the signal (n_s) and background (n_b)] will give an observed number of events that is greater than the number actually seen by the experiment. See M. Aguilar-Benitez *et al.*, Rev. Mod. Phys. **56**, S46 (1984), A. G. Frodesen *et al.*, *Probability and Statistics in Particle Physics* (Universitettsforlaget, Bergen, 1979), pp. 167-168, 378-379. The limit on n_s at the 90% CL is

derived from the limit on $(n_s + n_b)$ by subtraction of n_b from 2.30 events. Including the errors on n_b gives a 90% CL upper limit on n_s of 1.35 events.

16. The errors on the number of tags and the acceptance are subtracted in order to give a more conservative limit. We thus use 0.662 for the acceptance, and 2408 as the number of tags.
17. V. Lüth, *Proceedings of International Symposium on Production and Decay of Heavy Flavors*, Heidelberg, 20-23 May 1986 (to be published), SLAC-PUB-4052 (1986).
18. M. Aguilar-Benitez *et al.*, Rev. Mod. Phys. **56**, S43 (1984).
19. E. Golowich, Phys. Lett. **91B**, 271 (1980); M. Claudson, HUTP-81/A016 (1982).
20. V.S. Mathur and M.T. Yamawaki, Phys. Rev. **D29**, 2057 (1984); V.A. Novikov *et al.*, Phys. Rev. Lett. **38**, 626 (1977).
21. M. Bander, D. Silverman, A. Soni, Phys. Rev. Lett. **44**, 7 (1980).
22. H.K. Walter, *Fifth Workshop on Grand Unification*, K. Kang, H. Fried, P. Frampton, eds., (World Scientific Press, Singapore,1984), p. 134; F. Boehm, *ibid.*, p. 315.
23. W. Buchmüller and D. Wyler, Phys. Lett. **B177**, 377 (1986); W. Buchmüller, CERN-TH-4499/86 (July 1986).
24. B.A. Campbell, J. Ellis, K. Enqvist, M.L. Gaillard, and D.V. Nanopoulos, in preparation.
25. M. Aguilar-Benitez, *et al.*, Phys. Lett. **B170**, 1 (1986).
26. R.M. Baltrusaitis *et al.*, Phys. Rev. Lett. **56**, 2140 (1986). The number of D^0's has been scaled to reflect the different integrated luminosity used in this analysis. The branching fractions employed: $B(D^0 \to K^-\pi^+) = 0.056 \pm 0.004 \pm 0.003$; $B(D^0 \to \pi^+\pi^-) = 0.0018 \pm 0.0006 \pm 0.0004$.
27. We define $E_{\text{miss}} = \sqrt{s} - (E_e + E_\mu + \sum_i E^i_{\text{neutral}})$. All neutral tracks are used in this calculation, which benefits from the full solid angle coverage of the calorimeter. No photon selection cuts are applied.
28. H. Palka *et al.*, CERN-EP/87-10 (1987); a 90% CL upper limit of 1.0×10^{-3} is reported; K. Riles *et al.*, SLAC-PUB-4156 (1986) give 2.1×10^{-3}. These limits depend on the production mechanisms, fragmentation functions, and D^0 branching ratios.
29. G.T. Blaylock *et al.*, SLAC-PUB-4158 (1987) (submitted to Phys. Rev. Lett.); W. Toki, *Proceedings of 14th SLAC Summer Institute on Particle Physics*, Stanford, 28 July-8 August 1986 (to be published), SLAC-PUB-4153 (1986); R.H. Schindler, *Proceedings of XXIII International Conference on High Energy Physics*, Berkeley, 16-23 July 1986 (to be published), SLAC-PUB-4055 (1986); J.J. Becker *et al.*, SLAC-PUB-4194 (1987) (submitted to Phys. Rev. Lett.).

Figures

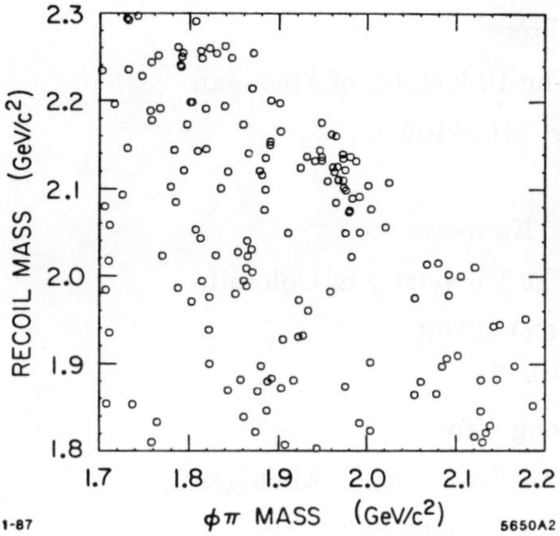

1. Scatter plot of $M(\phi\pi^+)$ versus M(recoil).

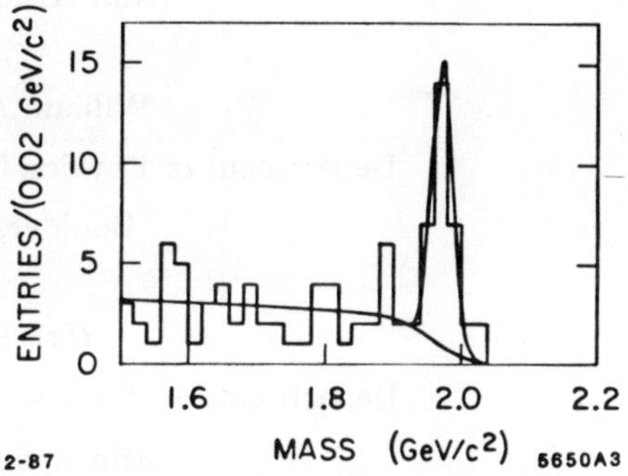

2. The projection of $M(\phi\pi^+)$ for $2.04 < M(\text{recoil}) < 2.18\,\text{GeV}/c^2$.

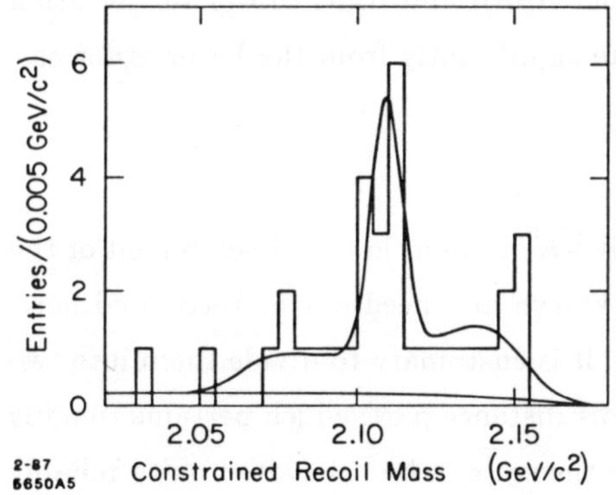

3. The recoil mass distribution with the D_s mass constrained at $1971.4\,\text{MeV}/c^2$.

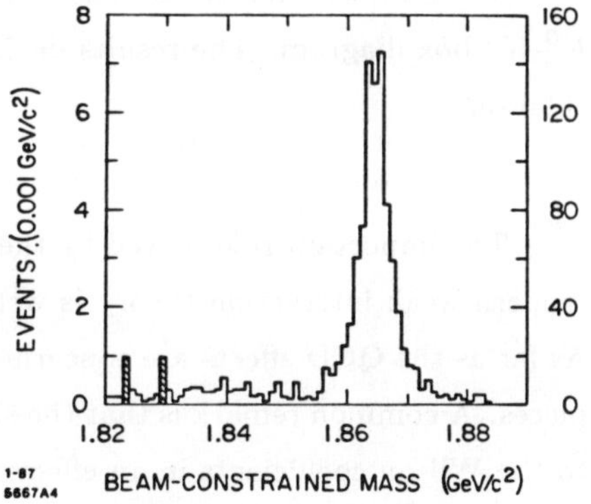

4. M_{bc} distribution for $K^-\pi^+$ events in the data (used to determine the $\pm 5.5\,\text{MeV}/c^2$ cut on M_{bc}). Superimposed are the two closest $D^0 \to \mu e$ candidates which pass all other cuts.

SUB-LEADING QCD CORRECTIONS TO THE K^0- $\overline{K^0}$ SYSTEM* †

J.-M. Frère*
Department of Physics, The University of Michigan
Ann Arbor, MI 48109

William A. Kaufman
Department of Physics, The University of Colorado
Boulder, CO 80309

York-Peng Yao
Department of Physics, The University of Michigan
Ann Arbor, MI 48109

ABSTRACT

We compute the corrections of order α_{QCD} and $\alpha^2_{QCD} ln M^2_W$ to the usual K^0-$\overline{K^0}$ box diagram. The results deviate significantly from the lower order calculation.

The important role played by the K^0-$\overline{K^0}$ system in the development of the modern weak interaction theory is well known and need not be recounted here. As far as the QCD effects are concerned, it is customary to divide them into two pieces. A common remark is that the short distance part, which pertains roughly to the Wilson coefficients in an effective theory, can be calculated quite reliably via renormalization group improvement, whereas an estimate of the long distance part, which has to do with operator matrix elements, is handicapped due to our lack of non-perturbative tools. Such a division is of course oversimplified; for

---

⋆ Work partially supported by the U.S. Department of Energy and NATO under contract No. 656/84

† Presented by W. A. Kaufman

* Permanent address: Service de Physique Theorique, U.L.B. Campus Plaine CP225, B-1050 Bruxelles, Belgium

example, there can be overlapping diagrams.

In this note we extend the short distance analysis. We shall find that the subleading contributions are not small.

Let us review briefly what has been accomplished previously in the short distance analysis. In the works of Gilman and Wise,[1] and Vysotskii,[2] the mass of the W bosons is taken to be the largest scale first. A leading logarithmic sum is performed with respect to $\alpha_s \ln M_W^2$, where α_s is the strong QCD coupling, evaluated at some appropriate scale. Then $\alpha_s \ln m_t^2$, $\alpha_s \ln m_b^2$, and $\alpha_s \ln m_c^2$ are successively summed until a three quark (u, d, s) effective theory results.

The $\Delta s = 2$ Lagrangian has the form

$$\begin{aligned} L_{\Delta s=2} = (G_F^2 M_W^2/4\pi^2)[\xi_c^2 \eta_1 m_c^2/M_W^2 + \xi_t^2 \eta_2 m_c^2/M_W^2 \\ + 2\xi_c \xi_t \eta_3 (m_c^2/M_W^2) \ln m_t^2/m_c^2] \bar{d}\gamma_\mu L s \bar{d}\gamma^\mu L s, \end{aligned} \quad (1)$$

where ξ_i are well known functions of the relevant K-M angles. η_i summarize the strong interaction effects. They have unit values in the absence of QCD.

In our present study, we shall sum over $(\alpha_s \ln M_W^2)^n$ and $\alpha_s(\alpha_s \ln M_W^2)^n$. The motivation to undertake such a study is partly due to our concern over the stability of these sums. As one may recall, if one performs only a one loop calculation without renormalization group improvement, the η_i's come out to have a sign opposite to the free quark results. Thus, even though renormalization group results restore the sign, higher order effects remain an issue and need to be investigated.

We have not attempted to sum heavy quark effects stepwise in the fashion of Gilman and Wise. This is partly because the task is quite involved. More importantly, because our calculation is a subleading one, we have some handle on what the scales Λ^2 should be in $\ln M_W^2/\Lambda^2$ when we approach new thresholds. Apparently, they are just the threshold factors. We can rearrange the series to decouple the heavy quarks smoothly as one steps down in momentum integration. We have compared our results with others and found agreement at a stage where such a comparison is meaningful.

It should be remarked that we have left out the double penguin related graphs, which may have some contributions.[3)] Let us now turn to describe some technical aspects of our work: an important observation to make in the analysis of the box diagram (Fig. 1) of the K^0-$\overline{K^0}$ system is that because of the GIM mechanism, the internal integration is $\sim dp^2/(p^2)^2$ for large momentum even after shrinking the W boson propagators to points. This good asymptotic behavior suggests that to include QCD effects, one should first make a short distance expansion in $1/M_W^2$ and $\ln M_W^2$ with respect to each W boson exchange. The true situation is somewhat more involved because of the existence of overlap diagrams, e.g., Fig. 2, for which the partition into $\Delta s = 1$ short distance and $\Delta s = 2$ long distance parts requires more care. This can in fact be easily handled[4] by Zimmermann oversubtraction identities. This elegant but rather painstaking technique to isolate effective vertices and to construct effective Lagrangians has been carried out. Let us not repeat it here, but save to say that we have analyzed all diagrams with two gluon corrections to the basic box diagram. We have used Feynman gauge for the gluon and W propagators, and neglected Higgs exchanges. Our conclusion is that contributions to order $(1/M_W^2)^2$ and leading and/or subleading in $\ln M_W^2$ come from Figs. 3 and 4. As checked by explicit calculation, the leading logarithms are readily taken into account by dressing both $\Delta s = 1$ vertices; on the other hand, special care must be paid to the overlap diagram (Fig. 5) to extract its subleading effects.

Fig. 4a gives the one and two loop QCD effects on $\Delta s = 1$ four quark operators. They have been to a large extent treated by Altarelli et al.[5)] However, we must take into account mass corrections and the existence of operators other than $(V-A)\otimes(V-A)$, which we need in our work. We call these 'finite parts', which can be obtained from an exact QCD one loop calculation.

In this analysis, we use dimensional regularization and modified minimal subtraction. We have taken the two loop results from Altarelli et al. in the same scheme. Furthermore, in order to have a smooth transition and decoupling when we cross quark thresholds, we shall rearrange the $\Delta s = 1$ Wilson coefficients so that the ratios $g^2(p^2+m_i^2)/g^2(M_W^2)$ appear ($i = u, c, t$).The difference will be absorbed into the 'finite parts'.

When we assemble two $\Delta s = 1$ vertices into a box diagram, some of the momenta flowing into the legs are vanishing (for external s and d) and others are equal to p. There may therefore be some doubt as to which value of g^2 should be used in computing the finite parts. We have chosen to use $g^2(p^2)$ consistently.

We now list our results. Let

$$O^{\pm} = \frac{1}{2}(\overline{q}_1\gamma_\mu Ls\overline{d}\gamma^\mu Lq_2 \pm \overline{q}_1\gamma_\mu Lq_2\overline{d}\gamma^\mu Ls), \tag{2}$$

where L is the left handed projection operator. We need the exact one loop results to $O(1/M_W^2)$ for each of the square vertices of Fig. 3a. For the left vertex we need to calculate the graphs of Fig. 6. They give

$$\begin{aligned} Fig.\ 6a = \frac{i}{16\pi^2}\frac{g_{EW}^2}{M_W^2}\Bigg\{ & 4\left[\frac{1}{3}O^+, -\frac{2}{3}O^-\right]\ln\frac{p^2+m_i^2}{M_W^2} \\ & -\left[\left(\frac{2}{9}+\frac{1}{3}T\right),\left(-\frac{2}{9}+\frac{2}{3}T\right)\right]\frac{1}{m_i}\gamma_\mu\gamma\cdot p\gamma_\nu L\otimes\gamma^\mu\gamma^\nu R\ B_i\Bigg\}, \end{aligned} \tag{3}$$

$$\begin{aligned} Fig.\ 6b = \frac{i}{16\pi^2}\frac{g_{EW}^2}{M_W^2}\Bigg\{ & \frac{4}{3}[O^+, O^-]\ C_i \\ & +\left[\left(\frac{8}{9}-\frac{1}{6}T\right),\left(\frac{4}{9}+\frac{1}{6}T\right)\right]\frac{1}{m_i}L\otimes\gamma\cdot pL\ B_i \\ & +\left[\left(\frac{8}{9}-\frac{1}{6}T\right),\left(\frac{4}{9}+\frac{1}{6}T\right)\right]\frac{1}{p^2}\gamma\cdot pL\otimes\gamma\cdot pL\ A_i\Bigg\}, \end{aligned} \tag{4}$$

$$\begin{aligned} Fig.\ 6c = \frac{i}{16\pi^2}\frac{g_{EW}^2}{M_W^2}\left[\frac{1}{3}O^+, -\frac{2}{3}O^-\right]\Bigg\{ & \frac{1}{2}\ln\frac{M_W^2}{p^2+m_i^2} \\ & +\frac{1}{2}\ln\frac{M_W^2}{p^2+m_j^2}+\frac{1}{2}\ln\frac{p^2+m_i^2}{m_s^2}+\frac{1}{2}\ln\frac{p^2+m_j^2}{m_s^2}\Bigg\}, \end{aligned} \tag{5}$$

where

$$T = t^a \otimes t_a,$$

$$A_i = \frac{1}{2} - \frac{m_i^2}{p^2} + \left(\frac{m_i^2}{p^2}\right)^2 \ln\frac{p^2+m_i^2}{m_i^2},$$

$$B_i = \frac{m_i^2}{p^2} - \left(\frac{m_i^2}{p^2}\right)^2 \ln\frac{p^2+m_i^2}{m_i^2}, \tag{6}$$

$$C_i = -1 + \frac{m_i^2}{p^2} - \left(\frac{m_i^2}{p^2}\right)^2 \ln\frac{p^2+m_i^2}{m_i^2} - \ln\frac{p^2+m_i^2}{\mu^2}.$$

There are two color factors given in square brackets. They go respectively with O^+ and O^- as basic vertices. We need these separate results in order to properly take into account the exponentiation of the one loop logarithms. We write down only the leading part of Fig. 6d:

$$Fig.\ 6d\ (leading) = \frac{i}{16\pi^2}\frac{g_{EW}^2}{M_W^2}\left[\frac{1}{3}O^+, -\frac{2}{3}O^-\right]\left\{\frac{1}{2}\ln\frac{M_W^2}{p^2+m_i^2} + \frac{1}{2}\ln\frac{M_W^2}{p^2+m_j^2}\right\}. \tag{7}$$

The 'finite part' will be treated separately in Fig. 5. Also, we need the self energy correction, which gives

$$\begin{aligned} Fig.\ 3b = \frac{-i}{16\pi^2}\Big\{&\gamma\cdot p\Big[1 - \frac{m_i^2}{p^2} + \left(\frac{m_i^2}{p^2}\right)^2 \ln\frac{p^2+m_i^2}{m_i^2} - \ln\frac{p^2+m_i^2}{\mu^2}\Big] \\ &+ 4m_i\Big[\frac{3}{2} - \frac{m_i^2}{p^2}\ln\frac{p^2+m_i^2}{m_i^2} - \ln\frac{p^2+m_i^2}{\mu^2}\Big]\Big\}. \end{aligned} \tag{8}$$

Now we can form the $\Delta s = 2$ process by contracting with the right vertex. We exponentiate the leading term and also make use of the results of Altarelli et al. This leads to a contribution

$$L^{(1)}_{\Delta s=2} = \frac{g_{EW}^4}{16\pi^2}\left(\frac{1}{M_W^2}\right)^2 \sum_{i,j}\int_0^\infty dp^2 \frac{(p^2)^2}{(p^2+m_i^2)(p^2+m_j^2)}\Bigg\{\frac{3}{2}(\chi_i\chi_j)^{-2/b_0}$$

$$\cdot\Bigg[1 + \frac{g^2(p^2)}{16\pi^2}\Big[4A_i + \frac{16}{3}B_i + \frac{8}{3}(C_i + D_i) + \frac{1}{3}\ln\frac{p^2+m_i^2}{m_s^2}$$

$$+ (i \to j)\Big] + \frac{g^2(p^2) - g^2(M_W^2)}{16\pi^2}\frac{2}{b_0}(\Gamma^+ + 2b_1)\Big]$$

$$- (\chi_i\chi_j)^{1/b_0}\Big[1 + \frac{g^2(p^2)}{16\pi^2}\Big[-\frac{8}{3}A_i - 3B_i + \frac{8}{3}(C_i + D_i) - \frac{1}{6}\ln\frac{p^2 + m_i^2}{m_s^2}$$

$$+ (i \to j)\Big] + \frac{g^2(p^2) - g^2(M_W^2)}{16\pi^2}\frac{1}{b_0}(\Gamma^+ + \Gamma^- - 2b_1)\Big] \tag{9}$$

$$+ \frac{1}{2}(\chi_i\chi_j)^{4/b_0}\Big[1 + \frac{g^2(p^2)}{16\pi^2}\Big[-\frac{4}{3}A_i - 4B_i + \frac{8}{3}(C_i + D_i) - \frac{2}{3}\ln\frac{p^2 + m_i^2}{m_s^2}$$

$$+ (i \to j)\Big] + \frac{g^2(p^2) - g^2(M_W^2)}{16\pi^2}\frac{2}{b_0}(\Gamma^- - 4b_1)\Big]\Big\}$$

$$\cdot\, \xi_i\xi_j\ \bar{d}\gamma_\mu L s \bar{d}\gamma^\mu L s,$$

in which the self energy effects are contained in

$$D_i = \frac{1}{2}\frac{1}{p^2 + m_i^2}\Big\{(-p^2 + m_i^2)\Big(1 - \frac{m_i^2}{p^2} + \big(\frac{m_i^2}{p^2}\big)^2\ln\frac{p^2 + m_i^2}{m_i^2} - \ln\frac{p^2 + m_i^2}{\mu^2}\Big)$$
$$- 8m_i^2\Big(\frac{3}{2} - \frac{m_i^2}{p^2}\ln\frac{p^2 + m_i^2}{m_i^2} - \ln\frac{p^2 + m_i^2}{\mu^2}\Big)\Big\}. \tag{10}$$

Here, the short distance physics is given by

$$\chi_i = \frac{g^2(p^2 + m_i^2)}{g^2(M_W^2)},$$
$$g^2(p^2) = g_0^2(p^2)\Big[1 - \frac{g_0^2(p^2)}{16\pi^2}\frac{b_1}{b_0}\ln\big(\ln\frac{p^2}{\Lambda^2}\big)\Big], \tag{11}$$

and the one loop running strong coupling constant is parameterized as

$$g_0^2(p^2) = \frac{16\pi^2}{b_0\ln(p^2/\Lambda^2)}, \tag{12}$$

with

$$b_0 = 11 - \frac{2}{3}f, \qquad b_1 = 102 - \frac{38}{3}f$$
$$\Gamma^+ = -\frac{221}{6} + \frac{20}{9}f, \qquad \Gamma^- = \frac{263}{3} - \frac{40}{9}f. \tag{13}$$

We take the number of flavors to be $f = 6$.

The final state interaction graphs give coefficients

$$Fig.\ 3d = \frac{g_{EW}^4}{16\pi^2}\Big(\frac{1}{M_W^2}\Big)^2 \sum_{i,j} \int_0^\infty dp^2 \frac{p^2}{p^2+m_j^2} \frac{g^2(p^2)}{16\pi^2}(6\ I + 2\ II - 2\ III)$$
$$\cdot \Big[\frac{3}{4}(\chi_i\chi_j)^{-2/b_0} + \frac{1}{6}(\chi_i\chi_j)^{1/b_0} + \frac{5}{12}(\chi_i\chi_j)^{4/b_0}\Big]\ \xi_i\xi_j, \tag{14}$$

$$Fig.\ 3f = \frac{g_{EW}^4}{16\pi^2}\Big(\frac{1}{M_W^2}\Big)^2 \sum_{i,j} \int_0^\infty dp^2 \frac{p^2}{p^2+m_j^2} \frac{g^2(p^2)}{16\pi^2}(12\ I - 8\ II' - 4\ III)$$
$$\cdot \Big[\frac{1}{2}(\chi_i\chi_j)^{-2/b_0} - \frac{1}{3}(\chi_i\chi_j)^{1/b_0} + \frac{1}{6}(\chi_i\chi_j)^{4/b_0}\Big]\ \xi_i\xi_j, \tag{15}$$

where

$$I = \frac{1}{2}\Big[\frac{1}{2} - \frac{m_i^2}{p^2} + \Big(\frac{m_i^2}{p^2}\Big)^2 \ln\frac{p^2+m_i^2}{m_i^2}\Big],$$

$$II = -\frac{1}{2}\Big[1 - \frac{3}{2}\frac{p^2}{p^2+m_i^2} + \frac{p^2}{p^2+m_i^2}\ln\frac{m_s^2}{m_i^2} - \frac{(p^2)^2+(m_i^2)^2}{p^2(p^2+m_i^2)}\ln\frac{p^2+m_i^2}{m_i^2}\Big],$$

$$II' = -\frac{1}{2}\frac{p^2}{p^2+m_i^2} + II, \tag{16}$$

$$III = \frac{1}{2}\Big[\frac{1}{2} - 2\frac{m_i^2}{p^2} + 3\Big(\frac{m_i^2}{p^2}\Big)^2 \ln\frac{p^2+m_i^2}{m_i^2} - \frac{(m_i^2)^2}{p^2(p^2+m_i^2)}\Big].$$

Actually Figs. 3d-g also yield other operators, such as $\bar{d}\gamma_\mu\gamma_\nu Rs\bar{d}\gamma^\mu\gamma^\nu Rs$. These cancel with potential mass singularities m_s^2, m_d^2 after integration over the gluon momentum. We have computed the coefficients to these operators and found them to be of order 10^{-2}, if no external momenta flow into the K^0-$\overline{K^0}$ system. In view of the potentially large values for the matrix elements in a vacuum saturation computation, these terms appear to be non-negligible. We would note, however, that the coefficients are very sensitive to a change in the external momenta which cut off the infrared. If we assume the K^0 mass as such a cutoff, for instance, these coefficients would be strongly suppressed. It is our opinion that these new operators should be dropped.

Finally, the overlap diagram contributes a finite part:

$$Fig.\ 5 = \frac{g_{EW}^4}{16\pi^2}\Big(\frac{1}{M_W^2}\Big)^2 \sum_{i,j} \int_0^\infty dp^2 \frac{M_W^2 p^2}{(p^2+m_i^2)(p^2+m_j^2)(p^2+M_W^2)} \frac{g^2(p^2)}{16\pi^2}$$

$$\cdot \Big\{ \frac{p^2}{M_W^2}(p^2+M_W^2)\Big(\ln\frac{p^2+m_i^2}{M_W^2} + \ln\frac{p^2+m_j^2}{M_W^2}\Big) + \frac{5}{3}M_W^2$$

$$- \frac{M_W^2}{p^2}\Big(p^2 + \frac{5}{3}M_W^2 - m_i^2 - m_j^2\Big)\ln\frac{p^2+M_W^2}{M_W^2}$$

$$+ \frac{p^2}{m_i^2-m_j^2}\Big[\Big(m_j^2 + 2\frac{(m_j^2)^2}{M_W^2} + \frac{5}{3}\frac{m_j^2 p^2}{M_W^2}\Big)\ln\frac{p^2+M_W^2}{p^2+m_j^2} - (j\to i)\Big]$$

$$+ \frac{p^2}{m_i^2-m_j^2}\Big(\frac{5}{3}p^2 + m_i^2 + m_j^2 + \frac{m_i^2 m_j^2}{M_W^2}\Big)\ln\frac{p^2+m_i^2}{p^2+m_j^2}$$

$$+ \frac{1}{m_i^2-m_j^2}\Big[\frac{5}{3}(m_j^2)^2 + \Big(-(m_j^2)^2 - \frac{5}{3}\frac{(m_j^2)^3}{p^2} + (m_i^2+m_j^2)\frac{(m_j^2)^2}{p^2}\Big)\ln\frac{p^2+m_j^2}{m_j^2}$$

$$- (j\leftrightarrow i)\Big] + \frac{4m_i^2 m_j^2}{m_i^2-m_j^2}\Big(\ln\frac{p^2+m_i^2}{p^2+m_j^2} + \frac{m_i^2}{p^2}\ln\frac{p^2+m_i^2}{m_i^2} - \frac{m_j^2}{p^2}\ln\frac{p^2+m_j^2}{m_j^2}\Big)\Big\}$$

$$\cdot \Big[\frac{7}{4}(\chi_i\chi_j)^{-2/b_0} - \frac{11}{6}(\chi_i\chi_j)^{1/b_0} + \frac{5}{12}(\chi_i\chi_j)^{4/b_0}\Big]\ \xi_i\xi_j. \tag{17}$$

The integration over p^2 is performed numerically, with a range from $\tilde{\Lambda}^2$ to $100M_W^2$.

We should perform two physical renormalizations. For mass renormalizations we make the replacement

$$D_i \to D_i + \frac{4m_i^2}{p^2+m_i^2}\Big(\frac{3}{2} - \frac{m_i^2}{\tilde{\Lambda}^2}\ln\frac{\tilde{\Lambda}^2+m_i^2}{m_i^2} - \ln\frac{\tilde{\Lambda}^2+m_i^2}{\mu^2}\Big), \tag{18}$$

where $\tilde{\Lambda}$ is an infrared boundary scale for integration which we vary in our numerical work (see later). In our final result the leading pieces should be multiplied by external wave function renormalizations

$$\left[1-\frac{g^2(\tilde{\Lambda}^2)}{16\pi^2}\frac{4}{3}\left(1+\frac{m_s^2}{\tilde{\Lambda}^2}+\left(\frac{m_s^2}{\tilde{\Lambda}^2}\right)^2\ln\frac{\tilde{\Lambda}^2+m_s^2}{m_s^2}-\ln\frac{\tilde{\Lambda}^2+m_s^2}{\mu^2}\right)\right], \qquad (19)$$

and a similar term with the d quark replacing the s quark.

We have tried to give a smooth transition across thresholds in χ by rearrangement of terms. We should do something likewise for $g^2(p^2)$, or vary the number of flavors taking part in the integration range. This proves to be cumbersome, in view of the various masses involved in the propagators. We have thus used six flavors throughout this calculation. (The numerical results are in fact quite insensitive to some change of effective flavors.)

Our results for η_i are listed in Table 1. In each case, we give we give the values of Λ_{QCD} in the $\overline{MS}$ scheme, the position of the strong infrared scale $\tilde{\Lambda}$ (at which the masses and external wave functions are normalized), and the mass of the top quark used. These may be compared with the values $\eta_1 \cong 0.69$, $\eta_2 \cong 0.59$, $\eta_3 \cong 0.41$ given by Gilman and Wise.

We note that the most striking difference between the present results and the leading log sum is in η_1, which is substantially reduced. As is well known,[6] the term proportional to η_1 in Eqn. (1) dominates the real part of the K^0- $\overline{K^0}$ mass matrix ($Re(M_{12}^{box})$), while $Im(M_{12}^{box})$ recieves contributions mainly from η_2 and η_3. As a result, we expect

$$\bar{\epsilon}_{box} = \frac{1}{2\sqrt{2}}\frac{Im(M_{12}^{box})}{Re(M_{12}^{box})} \qquad (20)$$

to be increased in the ratio $R = \eta_1^{leading\ log}/\eta_1$. Whereupon if the mass difference $\Delta M(K^0\text{-}\overline{K^0})$ were dominated by the box diagram, this would have resulted in a reduction by R^{-1} of the previously predicted ϵ'/ϵ. Of course, long distance effects should be kept in mind, and conventional wisdom is that they effect mainly $(Re\Delta M)$[7] in this case. It seems therefore more prudent to follow Ref. 6 and define ϵ_{box} with $Re(M_{12}^{exp})$ replacing $Re(M_{12}^{box})$ in the denominator.

Finally, a detailed study of the effects of using the η_i presented here on ϵ'/ϵ involves a complete discussion of the K-M matrix, which is beyond the scope of this paper.

REFERENCES

1. Gilman, F.J. and Wise, M. B., Phys. Rev. D27,1128 (1983).
2. Vysotskii, M. I., Yad. Fiz. 31, 1535 (1980) [Sov. J. Nucl. Phys. 31, 797 (1980)].
3. Eeg, J. O. and Picek, I., Phys. Lett. 160B,154 (1985); 177B, 432 (1986).
4. Kazama, Y. and Yao, Y.-P., Phys. Rev. D25, 1605 (1982).
5. Altarelli, G., Curci, G., Martinelli, G. and Petrarca, S., Nucl. Phys. B187, 461 (1981).
6. See, e.g., Donaghue, J. F., Golowich, E. and Holstein, B., Phys. Rep. 131, 319 (1986).
7. Frère, J.-M., Hagelin, J. and Sanda, A. I., Phys. Lett. 155B, 161 (1985). [Note that an unusual choice of the $\eta\eta'$ mixing angle here led to a different choice of relative sign for η_0 and η', and a negative value for ρ. Within that convention, a negative ρ is in agreement with the spectator quark model.]

Table 1: Various values of η_i. We use $m_u = 4$ Mev, $m_d = 8$ Mev, $m_c = 1.5$ Gev, $m_s = 150$ Mev and $M_W = 80$ Gev. $\tilde{\Lambda}$ is either fixed at 1 Gev or solved for by $\alpha_s(\tilde{\Lambda}) = 1$. Thus $\tilde{\Lambda} = 0.210$ Gev for $\Lambda_{QCD} = 0.1$ Gev and $\tilde{\Lambda} = 0.525$ Gev for $\Lambda_{QCD} = 0.25$ Gev.

m_t(Gev)	Λ_{QCD}(Gev)	$\tilde{\Lambda}$(Gev)	η_1	η_2	η_3
30	0.1	.210	.468	.700	.478
"	0.1	1	.510	.824	.562
"	0.25	.525	.319	.695	.378
"	0.25	1	.368	.782	.472
60	0.1	.210	.468	.729	.442
"	0.1	1	.510	.860	.523
"	0.25	.525	.319	.724	.349
"	0.25	1	.368	.817	.433

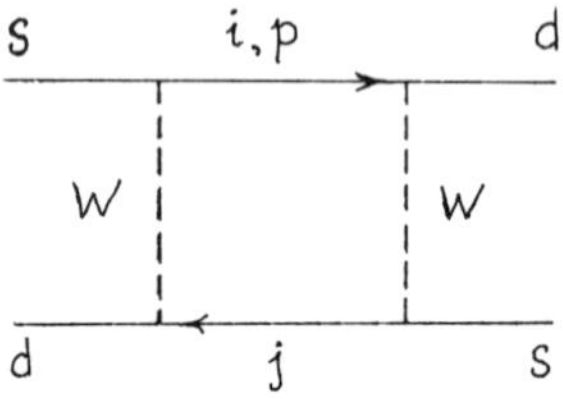

Fig. 1. The basic K^0-$\overline{K^0}$ diagram.

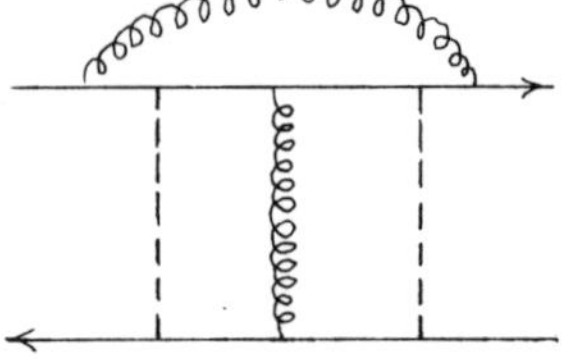

Fig.2. A diagram with overlapping QCD corrections.

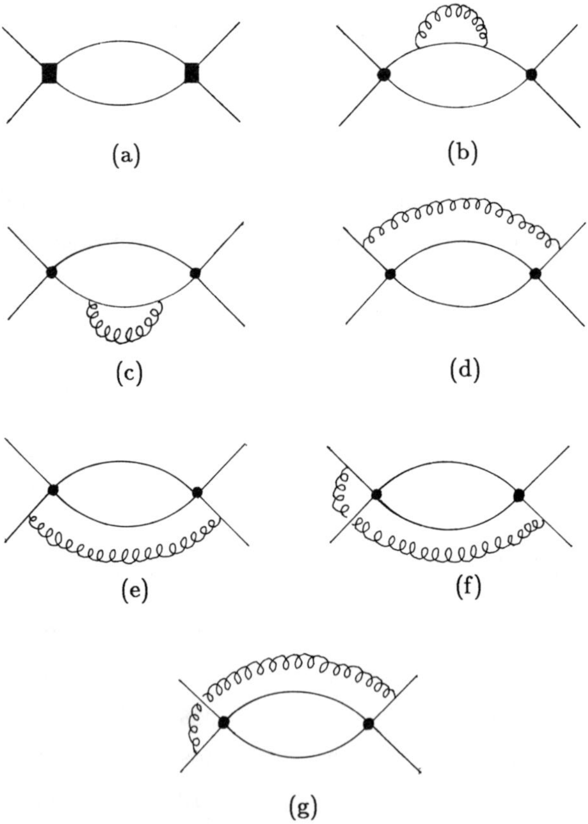

Fig. 3. QCD corrections that contribute to K^0- $\overline{K^0}$ in leading and subleading log.

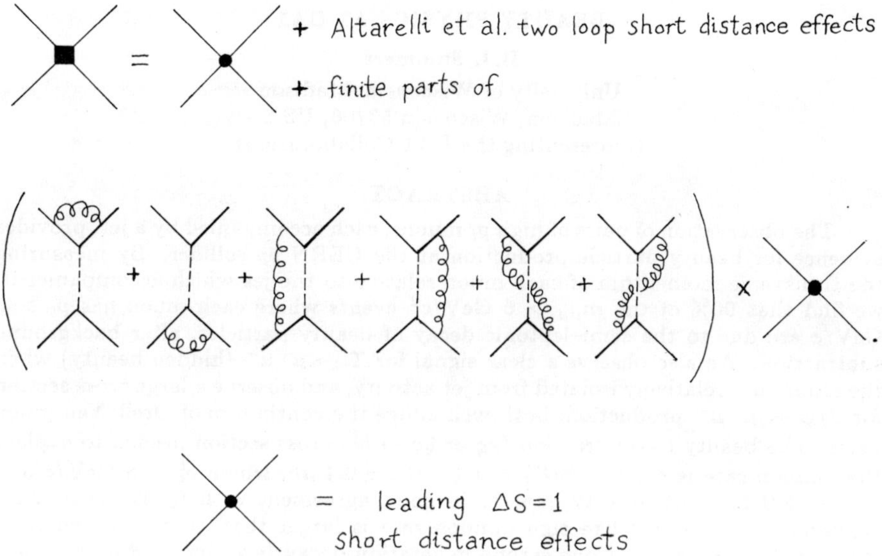

Fig.4. $\Delta s = 1$ diagrams that contribute to leading and subleading log.

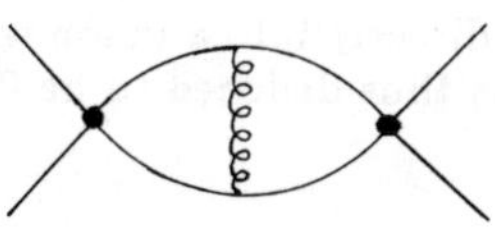

Fig. 5. Diagrams with an overlap in long and short distance physics.

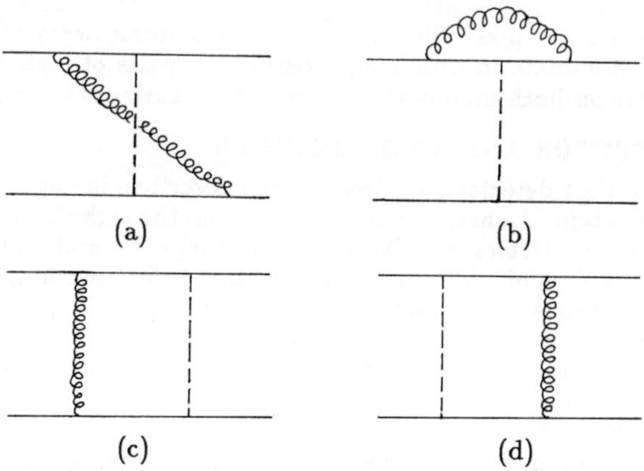

Fig. 6. Diagams we need to calculate to $O(M_W^{-2})$ for $\Delta s = 1$ vertices.

BEAUTY PHYSICS AT UA1

D.J. Summers
University of Wisconsin–Madison
Madison, Wisconsin 53706, USA
(representing the UA1 Collaboration)

ABSTRACT

The observation of pairs of high p_t muons, each accompanied by a jet, provides evidence for beauty particle production at the CERN $\bar{p}p$ collider. By measuring the transverse momentum of each muon relative to the jet which accompanies it, we find that 90% of the $m_{\mu\mu} > 6$ GeV/c^2 events where each muon has $p_t > 3$ GeV/c are due to the semi–leptonic decay of beauty particles, after background subtraction. We also observe a clear signal for $\Upsilon \rightarrow \mu^+\mu^-$ (hidden beauty) when the muons are relatively isolated from jet activity, and observe a large cross section for $J/\psi \rightarrow \mu^+\mu^-$ production; both well above the continuum of Drell–Yan muon pairs. The beauty flavor creation (gg or $\bar{q}q \rightarrow b\bar{b}$) cross section needed to explain the dimuon rate is $\sigma(\bar{p}p \rightarrow b\bar{b}X) = 1.1 \pm 0.1 \pm 0.4$ μb, where $p_t^b > 5$ GeV/c and $|\eta| < 2.0$ for both beauty quarks; in good agreement with QCD predictions. However, the observed like–sign dimuon rate is larger than that expected from events with one first and one second generation decay (e.g. $\bar{p}p \rightarrow b\bar{b}X$; $b \rightarrow \mu^-$, $\bar{b} \rightarrow \bar{c} \rightarrow \mu^-$). The predicted ratio of like–sign to unlike–sign dimuons is 0.26 ± 0.03. Experimentally, we measure $0.42 \pm 0.07 \pm 0.03$. A significant number of $B^0 \Longleftrightarrow \overline{B}^0$ transitions would explain the observed excess of like–sign events. The fraction of all beauty particles that decay directly into a muon with the opposite charge from that expected with no mixing is thus deduced to be 0.121 ± 0.047.

INTRODUCTION

The CERN $\bar{p}p$ collider is expected to to be a copious source of beauty particles with a predicted cross section of about 10 μb [1]. The b and c quark production cross sections are also expected to be roughly equal once the quark p_t is large enough to render b threshold effects unimportant. But because the fragmentation of b quarks is harder than c quarks [2] and because the b to c quark mass difference is greater than the c to s mass difference, the semi–leptonic decay of beauty quarks are expected to dominate in producing events with pairs of high p_t muons. Our $p_t > 3$ GeV/c cut on both muons also limits the background from $K, \pi \rightarrow \mu\nu$.

THE UA1 DETECTOR AND MUON TRIGGER

Because the UA1 detector has already been described in detail [3], let us focus on muon identification. A charged particle must pass through the large central drift chamber, the Central Detector (CD), the electromagnetic and hadronic calorimeters, and at least 60 cm of additional iron shielding before reaching the muon drift tubes. Its momentum is measured to an accuracy of about $\delta p/p = 0.005p$ (GeV/c) for 1 m long tracks, in the 0.7 T dipole magnetic field. It must pass through at least 9 nuclear interaction lengths of material which absorbs hadrons, and also results in a minimum dE/dx loss of $2\frac{1}{2}$ GeV or more. The muon trigger requires that muon tracks point back to an average vertex position within a ± 150 mr cone. Eight planes of muon drift tubes cover 75% of the solid angle and measure both the position and angle of tracks. Each drift tube module consists of two layers separated by 60 cm. Each layer has two drift tube planes per projection, staggered to resolve left–right ambiguities and to minimize overlapping dead spaces.

EVENT SELECTION

Our data was taken during collider runs in 1983 (108 nb^{-1} at $\sqrt{s} = 546$ GeV) and 1984-1985 (584 nb^{-1} at $\sqrt{s} = 630$ Gev) using one and two muon triggers. The total luminosity of 692 nb^{-1} produced 5 million muon triggers. Results from this muon data have recently been presented at conferences and in journals [4]. Dimuon events were selected requiring both muons to have $p_t > 3$ GeV/c, where p_t is the component of the momentum transverse to the beam direction. Both tracks also had to pass CD track length and quality cuts as well as CD to muon chamber matching cuts. This yielded 1175 dimuon candidates, which were scanned by physicists using a Megatek interactive graphics display. Events rejected by scanning were identified as: 9 cosmic rays, 3 double interactions, 56 cases of leakage through cracks, 63 K or π decays with a visibly kinked track or an abrupt momentum change, and 164 cases of mis-association of the muon and CD tracks (only 1/6 of the mis-associated events have $m_{\mu\mu} > 6$ GeV/c^2). These were removed leaving 880 scan–validated events. The mass distribution, uncorrected for acceptance, of 235 isolated (isolation defined later) unlike-sign dimuon events is shown in Fig. 1, where J/ψ, Υ, and Z^0 peaks are visible. The track quality and matching cuts are relaxed for the second muon in some of the 17 Z^0 events.

Finally, a $m_{\mu\mu} > 6$ GeV/c^2 cut was applied to eliminate dimuon events coming from either a single $b \to c$ cascade or a J/ψ or low mass Drell–Yan pairs. The $Z^0 \to \mu^+\mu^-$ events were also removed, leaving 512 high mass dimuons in the final sample. A Megatek picture of a typical event appears in Fig. 2.

BACKGROUND SOURCES

Because the UA1 detector has at least 9 interaction lengths of hadronic absorption, the major background to the prompt muon signal comes from $K^\pm$ and $\pi^\pm$ decay in flight. The background from shower leakage through the iron [5] and non-interacting hadrons is estimated at 2 events in the $m_{\mu\mu} > 6$ GeV/c^2 sample. Shower leakage particles must not only reach the muon chambers, but must also pass CD/muon chamber matching cuts in position *and* angle. Scanning effectively eliminates backgrounds from cosmic rays and shower leakage through cracks.

It is possible to mis-associate a muon track with an unrelated charged track which has enough momentum to reach the muon chambers if it only suffers ionization losses. We have counted 25 events in the sample where a nearby track in the CD could possibly form a second valid match to a muon track. Of these, 10 like-sign and 10 unlike-sign events, would either no longer be selected or would change sign if the other CD track were chosen. This gives an estimated mis–association background of 5 ± 5 like–sign and 5 ± 5 unlike–sign events.

The $\pi^\pm$ and $K^\pm$ decay background to the dimuon sample has been determined by simulating decays in real events [6]. The probability of decay in the CD barrel region is $0.11/p_t$ and $0.023/p_t$ for kaons and pions, respectively. The event sample used for the simulated decays contained single $p_t > 3$ GeV/c muon candidates, about 80% of which are background, mainly due to pion and kaon decays. Tracks of $p_t > 2$ GeV/c were chosen and their decay to a muon then simulated, including an allowance for decay inside calorimeters prior to shower commencement. Each such generated event was passed through the detector reconstruction and contributed to the background if it passed the same selection criteria as applied to the data, including scanning. The change in track direction at the point of a K or π decay was taken into account in the simulation and allowed to cause a higher or lower momentum as well as a kink which might be visible via scanning. The simulation assumed that the charged particle composition [7] is 58% $\pi^{\pm}$ (f_{π}), 21% $K^{\pm}$ (f_K), and 21% $p, \bar{p}$. The pion fraction was determined from parameterizations of the inclusive charged particle and inclusive pion p_t distributions of UA2, evaluated at 3 GeV/c. And the kaon fraction was determined assuming that the K/π ratio at fixed $m_T \equiv \sqrt{m^2 + p_t{}^2}$ is constant, and extrapolating the low p_t UA2 data to $p_t = 3$ GeV/c. The errors of the extrapolation are large, $\pm 17\%$ for K/π and $\pm 50\%$ for p/π. We estimate a background of 116 $K, \pi \to \mu\nu$ decays in the $m_{\mu\mu} >$ 6 GeV/c^2 sample, with a systematic error of 25% due to the uncertainty in the detector simulation, in the trigger and scanning efficiencies, and especially in f_{π} and f_K. The total background to the dimuon sample from all the above sources is estimated to be 127$\pm$32 events.

The exercise was repeated for inclusive muons ($p_t > 3$ GeV/c and $\mid \eta \mid < 1.5$), using the charged particle p_t distribution measured in minimum bias data as the parent sample. The result is shown in Fig. 3, where we find that the estimated background is close to the observed number of events in the region $3 < p_t < 4$ GeV/c, thus allowing little room for more background. At higher p_t, the background falls from 65% at $p_t \simeq 6$ GeV/c to less than 30% at $p_t \simeq 15$ GeV/c.

Finally, we have performed a second background calculation based on the observation that the CD/muon chamber matching is better for prompt muons than for background muons [8]. The shape of the matching χ^2 distribution for prompt muons was determined from the $J/\psi \to \mu^+\mu^-$ sample which has almost no background. The shape of the distribution for background muons was obtained from an inclusive $p_t > 3$ GeV/c muon sample which is estimated to contain about 80% background. A least squares fit was performed to determine the background fraction in the dimuon sample by fitting a weighted sum of the prompt muon and background muon matching χ^2 distributions to the observed dimuon distribution. The result of this fit is that the dimuon data contain 137$\pm$30 background events, where at least one one muon is not prompt.

Because the two background calculations are independent, their combination gives the expectation of 132$\pm$21 background events in the 512 event dimuon sample.

DRELL-YAN and Υ CROSS SECTIONS

To distinguish Drell-Yan dimuons and $\Upsilon \to \mu^+\mu^-$ from open heavy flavor, we measure the extent to which a muon is embedded in a jet (i.e. its isolation). Drell–Yan and Υ events are expected to mostly yield isolated muons. In Fig. 4 the scalar sum of the transverse energy found in the calorimeter cells in a cone of $\Delta R \equiv ((\Delta\eta)^2 + (\Delta\phi)^2)^{1/2} < 0.7$ around the muon, excluding the energy deposited by the muon itself, is plotted separately for like-sign and unlike-sign events. Here η is the pseudorapidity and ϕ is the azimuthal angle in radians. A clear enhancement appears in the unlike sign plot at $S < 9$ GeV^2 due to Drell–Yan and Υ decays, which is absent in the corresponding like-sign plot. This motivates defining the region $(\sum E_T(\mu_1))^2 + (\sum E_T(\mu_2))^2 < 9$ GeV^2 as isolated and the adjoint region as non-isolated. With this definition the 355 unlike-sign event sample contains 98 isolated and 257 non-isolated events. And the 157 event like-sign sample (78 ++ and 79 $--$ events) contains 15 isolated and 142 non-isolated events. This definition was tested by adding random pairs of $W \to \mu\nu$ events together; 82% of these *double events* pass the above isolation cut. The background sample contains 116 non-isolated and 16 isolated events, equally divided between like and unlike–sign.

Fig. 5 shows the mass distribution of the 98 isolated unlike-sign events. After subtracting the background, a three parameter fit has been made to determine the level of the Drell-Yan, $\Upsilon, \Upsilon', \Upsilon'' \to \mu^+\mu^-$ (1 : 0.3 : 0.15 production ratios assumed), and residual heavy flavor decays, using their individual mass distributions and allowing for the CD resolution [9]. After correcting for acceptance, we find

$$\sigma \cdot B\{\bar{p}p \to \Upsilon, \Upsilon', \Upsilon'' \to \mu^+\mu^-\} = 0.98 \pm 0.21 \pm 0.19 \text{ nb}$$

$$\sigma\{\text{Drell-Yan}; m_{\mu\mu} > 11\text{GeV/c}^2\} = 0.25 \pm 0.08 \pm 0.05 \text{ nb}$$

in good agreement with a gluon fusion model extrapolation from lower energy [10].

OPEN BEAUTY PRODUCTION

We have two pieces of direct evidence for open $b\bar{b}$ production in non–isolated dimuon events. First, the existence of a significant number of like-sign events $(142 - 58 = 84)$ which cannot come from the direct decays of $c\bar{c}$ events (given no D^0 mixing [11]) implies the presence of b quarks. Second, we use the transverse momentum of a muon relative to a beauty or charm jet to separate the two cases on a statistical basis [12]. Fig. 6 shows 175 non-isolated events in which the axis of a CD jet is within $\Delta R < 1.0$ of a muon. At least one $p_t > 1$ GeV/c track is also required in the jet in addition to the muon. To favor first generation decays, only the higher p_t muon is used for the like-sign events. By fitting decay Monte Carlo ${p_t}^{rel}$ distributions for the charm and beauty jets to the data and subtracting background, we find that the fraction of events due to $c\bar{c}$ is $(17 \pm 11)\%$ for the unlike-sign data and $(0 \pm 11)\%$ for the like-sign data. Combining both like- and unlike-sign samples leads to a global charm fraction of $c\bar{c}/(c\bar{c} + b\bar{b}) \sim 10\%$, which is in good agreement with QCD predictions [13].

The inclusive muon differential cross section, $d\sigma/dp_t$, for non-isolated events is plotted in Fig. 7 and compared to QCD predictions [13,14]. Also shown is $d\sigma/dp_t$ for the single muon events. Both distributions are corrected for acceptance and are background subtracted. Good agreement with the data is observed.

For the single muon prediction, higher order processes, flavor excitation and gluon splitting ($g \to b\bar{b}, c\bar{c}$), are important. In calculating gluon splitting, a 6 GeV/c^2 cutoff on the mass of the gluon is used, which is between the $b\bar{b}$ and $c\bar{c}$ thresholds. However, we have calculated that the underestimation of $\sigma(\bar{p}p \to g + X;\ g \to c\bar{c};\ c \to \mu\ ({p_t}^{\mu} > 6GeV/c))$ is less than 20%. In the dimuon events the higher order contributions are suppressed by the dimuon mass cut, but nonetheless account for about 25% of the predicted dimuon yield. Fig. 8 shows that higher order processes are needed to explain the observed pairs of muons with small azimuthal opening angles.

Normalizing the predicted number of dimuon events to the data yields a cross section for beauty flavor creation (gg or $\bar{q}q \to b\bar{b}$) of

$$\sigma(\bar{p}p \to b\bar{b}X;\ p_t^b > 5\ GeV/c, |\ \eta\ | < 2.0 \text{ for both beauty quarks}) = 1.1 \pm 0.1 \pm 0.4\ \mu\text{b}$$

The systematic error of 0.4 μb includes uncertainties in Monte Carlo parameters, the luminosity, and detector acceptance. ISAJET predicts 1.7 μb.

EVIDENCE FOR $B^0 \Longleftrightarrow \overline{B}^0$ OSCILLATIONS

If the observed ratio of like–sign to unlike–sign dimuons is larger than that expected from the fraction of events with one first and one second generation decay (e.g. $\bar{p}p \to b\bar{b}X$; $b \to c \to \mu^+$, $\bar{b} \to \mu^+$), then $B^0 \Longleftrightarrow \overline{B}^0$ may be required. We calculate this ratio by subtracting the expected number of background events from the observed number of non–isolated like and unlike–sign dimuons.

$$R = \frac{N(\pm\pm)}{N(+-)} = \frac{142 - 58}{257 - 58} = 0.42 \pm 0.07 \pm 0.03$$

To doublecheck, one may use the complete dimuon sample and subtract Drell–Yan (40 events) and Υ (45 events) contributions as well as the background.

$$R = \frac{N(\pm\pm)}{N(+-)} = \frac{157 - 64}{355 - 64 - 40 - 45} = 0.45 \pm 0.07 \pm 0.05$$

The systematic error is second and reflects the background subtraction uncertainty.

Now calculate a prediction for R without flavor mixing using ISAJET [13].

$$R = \frac{N(\pm\pm)}{N(+-)} = \frac{N_s}{N_f + N_c} = \frac{N_s/N_f}{1 + (N_c/N_f)} = \frac{0.30}{1 + 0.15} = 0.26 \pm 0.03,$$

where N_f is the predicted number of events passing our cuts from two first generation $b\bar{b}$ decays, N_s is the number from one first plus one second generation decay from $b\bar{b}$, and N_c is the number due to $c\bar{c}$. The error has been found using errors on the beauty (12% assumed) and charm muonic branching ratios and by varying N_c by 50%. The contributions from two second generation beauty decays and from double $b\bar{b}$ or $c\bar{c}$ production are very small. Mostly, the prediction of R depends on the muon momentum spectrum of B and D meson decays, which is experimentally known [2]. Others predict R values of 0.24±0.05 [14] and 0.25 [15,16].

A maximum likelihood fit [17] using ISAJET to find the fraction (χ) of all beauty particles that decay directly into a muon with the opposite charge from that expected with no mixing gives:

$$\chi = \frac{BR(b \to \overline{B}^0 \to B^0 \to \mu^+ + X)}{BR(b \to B \to \mu^\pm + X)} = 0.121 \pm 0.047,$$

which is related to R by : $$R = \frac{2\chi(1-\chi)N_f + [(1-\chi)^2 + \chi^2]N_s}{[(1-\chi)^2 + \chi^2]N_f + 2\chi(1-\chi)N_s + N_c}.$$

All of this has been used in the fit along with the observed and expected p_t spectra from first and second generation beauty decays. Second generation muons tend to have less momentum. The likelihood curve in Fig. 9 shows that no mixing (χ=0) is disfavored relative to the best fit for χ with a likelihood ratio of 1:73, or 2.9 standard deviations. Without the p_t spectra constraint, χ=0.10±0.05.

Experimentally, we do not distinguish between $B_d^0 \Longleftrightarrow \bar{B}_d^0$ and $B_s^0 \Longleftrightarrow \bar{B}_s^0$ contributions to χ. However, as an example, assume 20% B_s^0 production, that it has maximal mixing (χ_s=0.5), and that the B_d^0 does not mix (χ_d=0). Then we find χ=0.10, which is lower than but consistent with our observed value of 0.121±0.047.

We are also consistent with 90% confidence level limits on B^0 mixing from MARK II [18] ($\chi < 0.12$) and CLEO [19] ($\chi_d < 0.19$, $r_d < 0.24$; $\chi_d \equiv r_d/(1+r_d)$) as well as a recent observation of B_d^0 mixing by ARGUS [20] (χ_d=0.171±0.048±0.022, $r_d = 0.207 \pm 0.058 \pm 0.027$, assuming 55% $\Upsilon(4S) \rightarrow B^+B^-$ and 45% $\Upsilon(4S) \rightarrow B_d^0\bar{B}_d^0$).

REFERENCES

1. B. L. Combridge, Nuc. Phys. B151 (1979) 429.
2. A. Chen et al., Phys. Rev. Lett. 52 (1984) 1084; J. Green et al., Phys. Rev. Lett. 51 (1983) 347; S. Csorna et al., Phys. Rev. Lett. 54 (1985) 1894; W. Bacino et al., Phys. Rev. Lett. 43 (1979) 1073; R. Schindler et al., Phys. Rev. D 24 (1981) 78; R. Baltrusaitis et al., Phys. Rev. Lett. 54 (1985) 1976; S. Bethke, DESY preprint 85-067 (1985), and references therein.
3. A. Astbury et al.,UA1 Proposal, CERN/SPSC 78-06; M. Calvetti et al., Nuc. Inst. Meth. 174 (1980) 225; Nuc. Inst. Meth. 176 (1980) 175,217,223; M. Corden et al., Nuc. Inst. Meth. A238 (1985) 273; C. Cochet et al., Nuc. Inst. Meth. A243 (1986) 45; G. Bauer et al., Nuc. Inst. Meth. A253 (1987) 179.
4. C. Albajar et al., Beauty Production at the CERN Proton–Antiproton Collider, Phys. Lett. B186 (1987) 237; C. Albajar et al., Search for $B^0 - \bar{B}^0$ Oscillations at the CERN Proton–Antiproton Collider, Phys. Lett. B186 (1987) 247; T. Markiewicz, 23rd Int. Conf. on High–Energy Physics, Berkeley (July 1986) CERN–EP/86–175; M. Della Negra, 6th Topical Workshop on $p\bar{p}$ Collider Physics, Aachen (July 1986) CERN–EP/87–07.
5. R. Leuchs, Messung des Hadronischen Untergrundes bei Identifizierung von Myonen, Diplomarbeit in Physik, III Physikalisches Institut, Aachen (1982).
6. N. Ellis and M. Jimack, Estimation of the Background to the Dimuon Sample, UA1 TN 86/124 (1986); N. Ellis and M. Jimack, Estimation of the Background to the 1984 Inclusive $p_t > 6$ GeV/c Muon Sample, UA1 TN 86-79 (1986).
7. M. Banner et al., Phys. Lett. 122B (1983) 322, G.J. Alner et al., Nuc. Phys. B258 (1985) 505.
8. Hans–Günther Moser, Background Determination in Dimuons using Matching Distributions, UA1 Technical Note 87–04 (1987).
9. Hans–Günther Moser, Cross Sections for Drell Yan, Upsilon and $b\bar{b}$ Production, UA1 Technical Note 87–03(1987).
10. V. Barger and A. D. Martin, Phys. Rev. D 31 (1985) 1051.
11. W. Louis et al., Phys. Rev. Lett. 56 (1986) 1027.
12. H. Moser, Separation of $c\bar{c}$ and $b\bar{b}$ in in the Dimuons Using p_t of the Muon with respect to the Jet Axis, UA1 TN 87–02; M. Jimack, The Use of the p_t Relative Distribution to Separate $b\bar{b}$ from $c\bar{c}$ in Muon–Jet Data, UA1 TN 87–16.
13. F. Paige and S.D. Protopopescu (ISAJET Monte Carlo), BNL 38034(1986).
14. A. Ali, B. Van Eijk, and I. ten Have, Heavy Flavor Production in Large Transverse Momentum Proton–Antiproton Collisions, CERN–TH.4523/86 (1986); Bob Van Eijk, Heavy Flavour Production and Heavy Flavour Mixing at the CERN Proton–Antiproton Collider, Ph.D Thesis, NIKHEF (1987).
15. V. Barger and R. J. N. Phillips, preprint MAD/PH/155, 239, and 266.
16. F. Halzen and A. Martin, preprint DTP/84/14.
17. K. Wacker, Likelihood Analysis of the Mixing Parameter in Dimuon Events, UA1 Technical Note 87–27 (1987).
18. T. Schaad, W. B. Schmidke et al., Phys. Lett. 160B (1985) 188.
19. A. Bean et al., Phys. Rev. Lett. 58 (1987) 183.
20. H. Schröder (ARGUS), DESY Seminar, 24 February 1987.

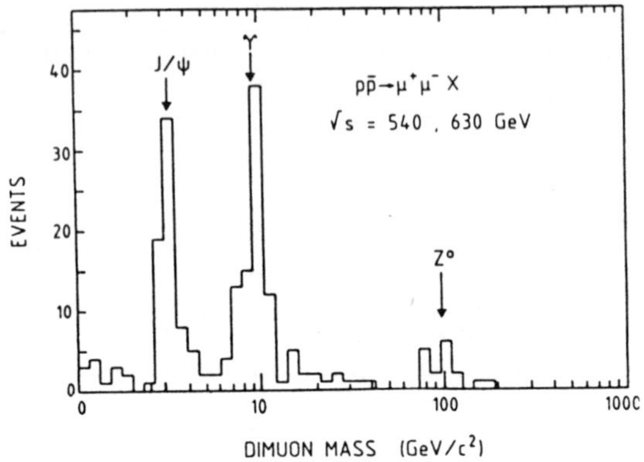

Fig. 1. Dimuon mass distribution for 235 isolated unlike–sign dimuon events with both $p_t^\mu > 3$ GeV/c. For $m_{\mu\mu} < 6$ GeV/c^2 events, the isolation criterion is $\Sigma E_t < 3$ GeV in a single cone of $\Delta R < 0.7$ around the *dimuon* direction. The mass scale is logarithmic.

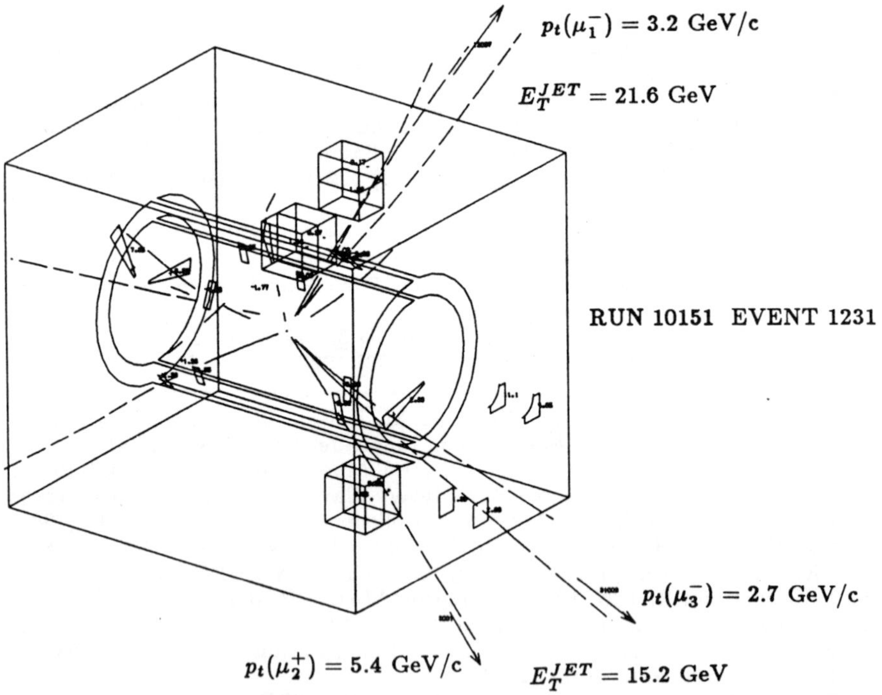

Fig. 2. Typical non–isolated dimuon event. $m(\mu_1^-, \mu_2^+) = 8.8 \pm 0.2$ GeV/c^2. This event also has a 3rd muon with $p_t < 3$ GeV/c. $m(\mu_2^+, \mu_3^-) = 2.84 \pm 0.09$ GeV/c^2.

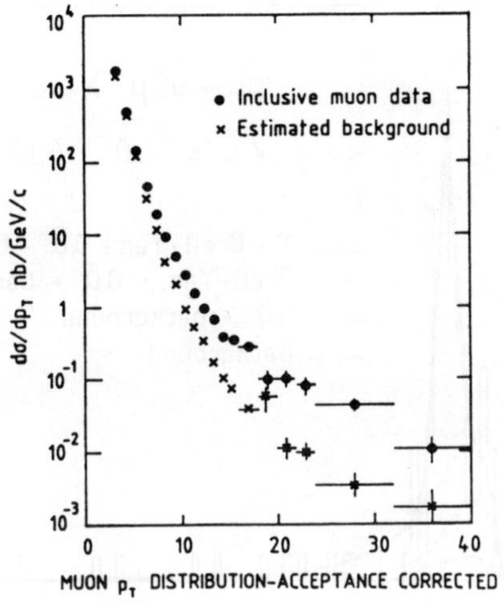

Fig. 3. Estimated background from $K, \pi \to \mu\nu$ decay compared to the inclusive muon data sample.

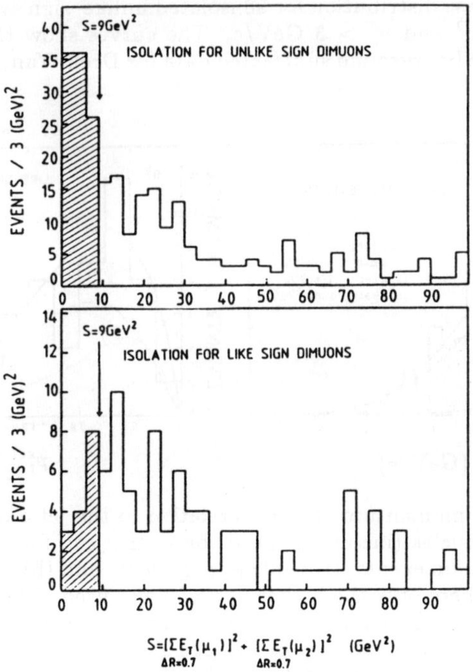

Fig. 4. Distribution of the isolation variable for 355 unlike–sign events and 157 like–sign events. The shaded regions, $S < 9$ GeV^2, define the isolated event samples.

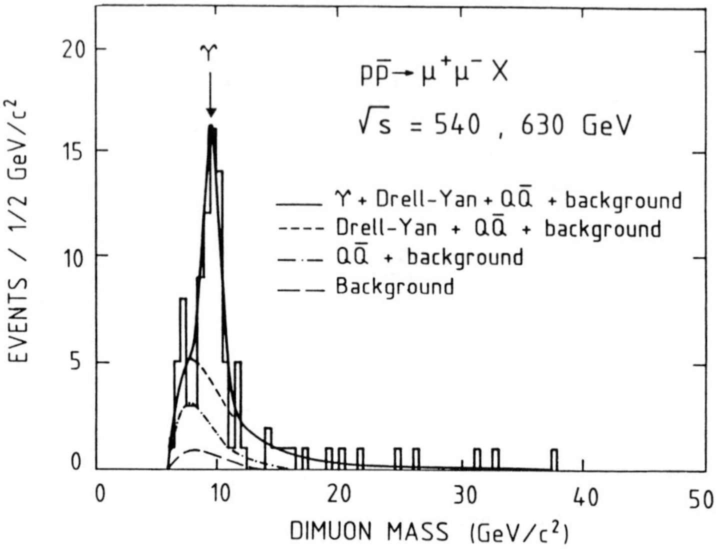

Fig. 5. Dimuon mass distribution for 98 isolated unlike–sign events with 50 GeV/c $> m_{\mu\mu} >$ 6 GeV/c^2 and $p_t^\mu >$ 3 GeV/c. The curves show the result of a three parameter fit to the background subtracted data for Drell–Yan, ϒ, and heavy flavor contributions.

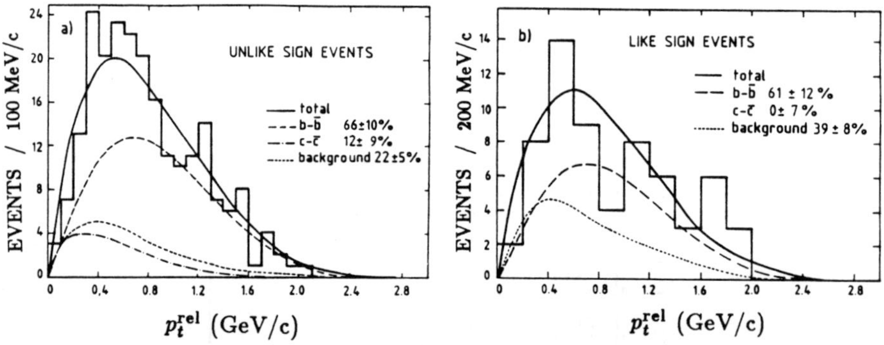

Fig. 6. Transverse momentum of muons relative to the jet axis, which is defined only by charged particles (including the muon in the jet), for (a) 226 muons from 113 unlike–sign dimuon events (two entries per event) and (b) 63 muons from like–sign muon events where only the higher p_t muon was used. The curves are a result of a two parameter fit to $c\bar{c}$ and $b\bar{b}$ contributions, where the background fraction has been fixed at estimated values of (a) 22±5% for unlike–sign events and (b) 39±8% for the like–sign events. After background subtraction, we find that the fraction of events due to $c\bar{c}$ is 15±11% for the unlike–sign data and < 18% for the like sign data at 90% C.L.

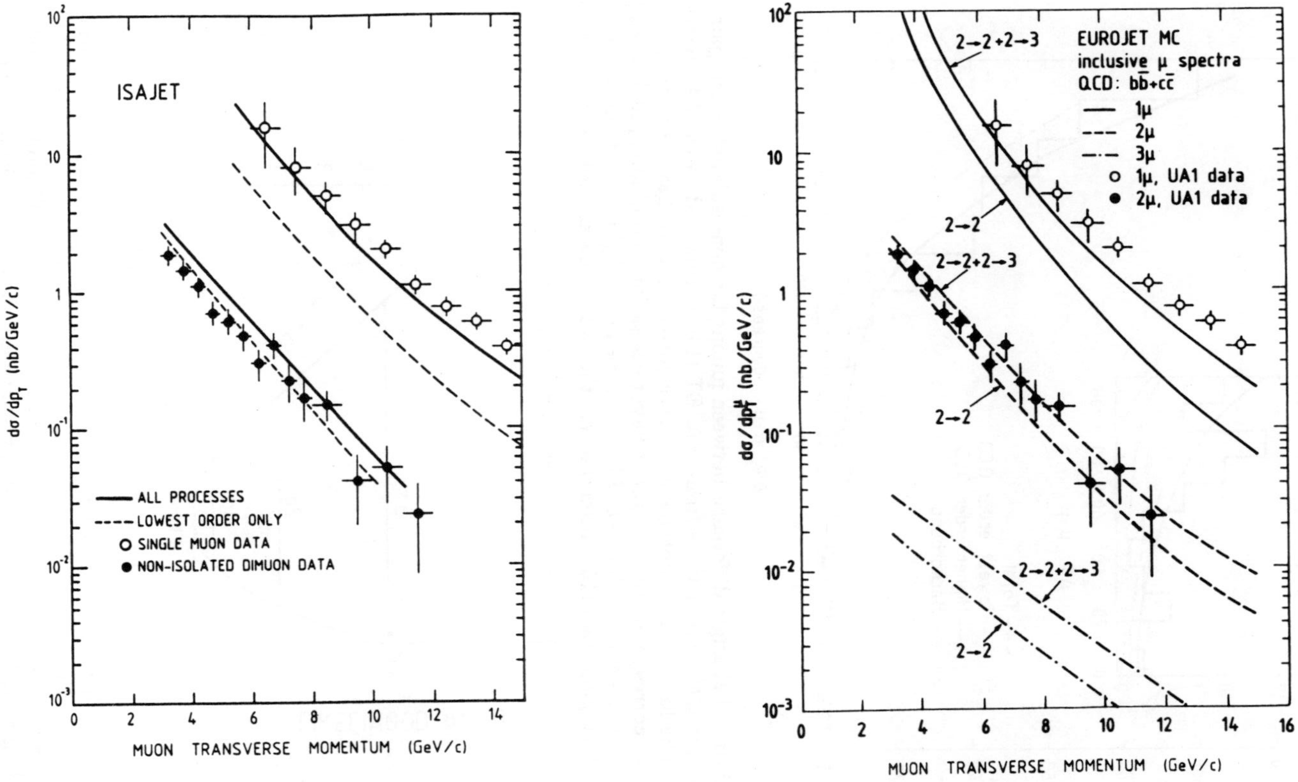

Fig. 7. Inclusive muon p_t distributions for single muon muon and non-isolated dimuon events ($m_{\mu\mu} > 6$ GeV/c^2). The curves are absolutely normalized QCD predictions from the ISAJET [13] and EUROJET [14] Monte Carlos.

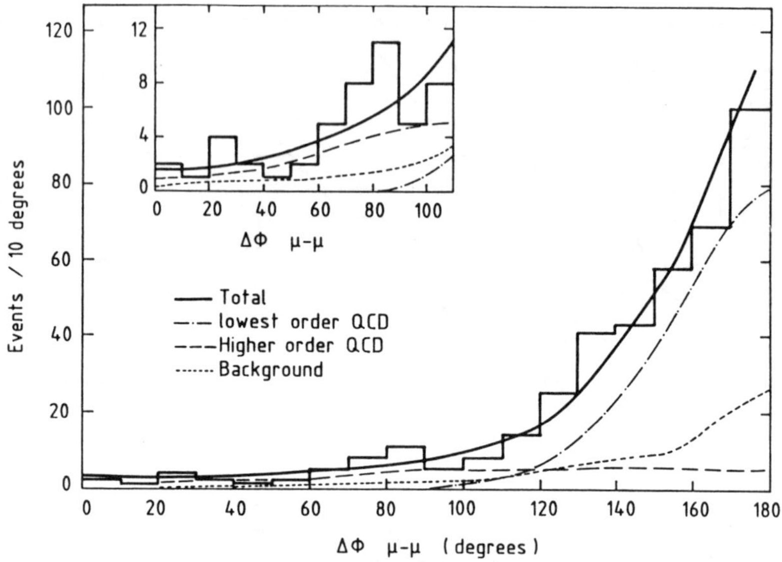

Fig. 8. Azimuthal angle difference between muons for non-isolated muon pairs. The curves are QCD predictions from ISAJET [13], normalized to the 399 events observed. The lowest order dot-dashed curve corresponds to flavor creation, the dashed curve corresponds to the sum of flavor excitation and gluon splitting, and the solid line is the sum of all three. Flavor excitation and gluon splitting are need to explain the region $\Delta\phi<110^0$ (insert) where the two muons are not back-to-back.

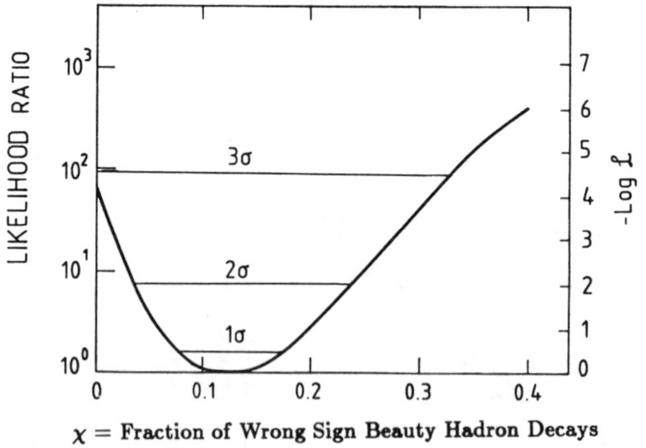

Fig. 9. Likelihood ratio as a fucntion of χ, resulting from a fit to the p_t distribution of first and second generation $b\bar{b}$ decays, $c\bar{c}$ decays, and dimuon background. The righthand scale shows the logarithm of the likelihood ratio. Likelihood ratio values corresponding to 1, 2, and 3 standard deviations are shown as horizontal lines.

CHARMLESS RARE B DECAYS, HEAVY QUARKS, AND A FOURTH GENERATION†

George Wei-Shu Hou

Department of Physics and Astronomy
University of Pittsburgh, Pittsburgh, PA 15260

A. Soni

Department of Physics
University of California, Los Angeles, CA 90024

ABSTRACT

Loop-induced rare B decays, viz. $b \to s\ell^+\ell^-$, $s\nu\bar{\nu}$, $s\gamma$, and QCD induced $b \to s$ transitions ("penguins"), are discussed. Because of the existence of the Z (and box) diagram contributions, $b \to s\ell^+\ell^-$, $s\nu\bar{\nu}$ decay rates grow rapidly with heavy internal quark masses m_Q. The on-shell process $b \to s\gamma$ has large QCD corrections, but depends mildly on m_Q. The QCD induced transition is dominated by a timelike gluon ($q^2 > 0$), which splits up into two quark or gluon partons, and has a very mild m_Q dependence. For 3 generations, with 25 GeV $\lesssim m_t \lesssim$ 200 GeV, the branching ratios of the inclusive processes range roughly as (1.5×10^{-6} - 1.5×10^{-5}), (8×10^{-6} - 1×10^{-4}), ~ (7×10^{-5} - 2×10^{-4}) and ~ (1% - 2%), respectively. If a 4th generation does exist, these rates could grow roughly by one order of magnitude, while the $s\ell^+\ell^-$, $s\nu\bar{\nu}$ modes may gain even more. A pragmatic parametrization of the 4×4 mixing matrix suitable for B physics is pointed out. Experimental search for these modes is strongly urged.

1. INTRODUCTION

With branching ratios (BRs) reaching down to the 10^{-11} - 10^{-12} level within a couple of years, at first sight it may seem that the kaon system is more interesting than the B system in providing tests of loop effects. Observe, however, that (i) $|V_{cb}|^2 \ll |V_{us}|^2$, so relatively the B lifetime is more prolonged; (ii) $m_B^2 \gg m_K^2$, making many more channels accessible for the B system, thus allowing for many new interesting possibilities; (iii) not only is the B lifetime more prolonged, but because $|V_{ts}V_{tb}/V_{cb}|^2/|V_{td}V_{ts}/V_{us}|^2 \sim 10^5$, loop induced rare B decays may be enhanced over the corresponding K decay modes by such a large factor, hence making B mesons a better probe of virtual

† Invited talk presented at the DPF meeting in Utah, Jan. 1987. Based on work done in colaboration with H. Steger and R. S. Willey.

effects. Given that the B meson has only been available for several years, it may well provide new clues to the unanswered questions of the Standard Model (SM).

The questions we would like to address are: What's the top quark mass? Is there a 4th generation? Is there a real deficit of charm in B decays?[1] We study loop induced $b \to s$ transitions, viz. the processes $b \to s\ell^+\ell^-$, $s\nu\bar{\nu}$,[2] $s\gamma$[3] and s + "g".[4] None of these processes (except $b \to s\nu\bar{\nu}$) are entirely new, but we have many new observations and insights that reveal the importance of these processes. Furthermore, we explore the impact that the existence of a 4th generation may have on these modes. As a consequence of such a study, we propose a "new" parametrization of the 4×4 mixing matrix that seems to be most suitable for the study of the B system. We will not study the more difficult and uncertain exclusive processes. Also left out are the discussions of B-$\bar{B}$ mixing and the question of CP violation in inclusive processes.[5]

To calculate the matrix elements for these processes, we need the effective vertices:

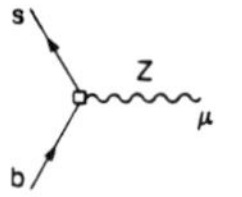

$$\frac{g^3/c_W}{16\pi^2}\, v_i \Gamma_i^{Z}(\xi)\ \bar{s}\gamma_\mu L b$$

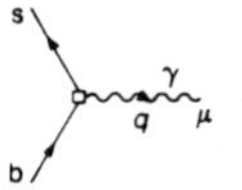

$$\frac{g^3 s_W}{16\pi^2(2M_W^2)}\, v_i\ \bar{s}\Big(F_1^{i}(\xi)(q^2\gamma_\mu - q_\mu \not{q})L + F_2^{i}\ i\sigma_{\mu\nu}q_\nu m_b R\Big)b$$

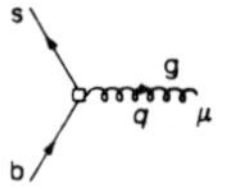

$$g_s\, \frac{g^2}{16\pi^2(2M_W^2)}\, v_i\ \bar{s}\, \frac{\lambda^a}{2}\Big(F_1^{i}(q^2\gamma_\mu - q_\mu \not{q})L + F_2^{i}\ i\sigma_{\mu\nu}q_\nu m_b R\Big)b$$

and the effective box graph

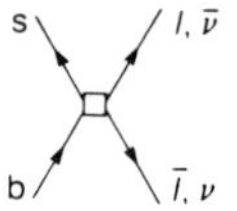

$$\pm\, \frac{g^4}{16\pi^2(2M_W^2)}\, v_i \begin{Bmatrix} \bar{C}_i^{\,box}(\xi) \\ \bar{D}_i^{\,box}(\xi) \end{Bmatrix} (\bar{s}\gamma_\mu L b) \begin{Bmatrix} \bar{\ell}\gamma_\mu L \ell \\ \bar{\nu}\gamma_\mu L \nu \end{Bmatrix}$$

where the form factors $F_1{}^{i}$, $F_2{}^{i}$, $\Gamma_i{}^{Z}$, $\bar{C}_i{}^{box}$, $\bar{D}_i{}^{box}$ etc. can be taken or adapted from Inami and Lim,[6] and are complicated functions of internal quark masses m_i. In the above, $v_i \equiv V_{is}^{*}V_{ib}$, i is summed from 2 to n (# of generations), ξ is the gauge parameter, and note that we have been a bit cavalier in using the same F_1, F_2 for both the $bs\gamma$ and bsg form factors; they are different. Note that F_1 for bsg is gauge independent while that for $bs\gamma$ is not.

We shall assume that the inclusive rate for each process is equal to the free quark result, e.g. $\Gamma(B \to K\ell^+\ell^- + X) \equiv \Gamma(b \to s\ell^+\ell^-)$. Also, to reduce uncertainties in estimating BRs, we shall always use the usual ansatz of taking ratios, e.g.

$$BR(B \to K\ell^+\ell^- + X) = \frac{\Gamma(b \to s\ell^+\ell^-)}{\Gamma(b \to (u,c)\ell\bar{\nu})} * BR^{expt.}(B \to \ell\bar{\nu} + X) \qquad (1)$$

2. PROCESSES $b \to s\ell^+\ell^-$, $s\nu\bar\nu$

Previous works[7] on $b \to s\ell^+\ell^-$ have not included the important Z^0 contribution. We point out that although the photonic F_1 term contains the leading-log that is important for low m_t, the Z^0 contribution can not be neglected when there are very heavy internal quarks, i.e. for $m_Q \gtrsim M_W$.

Using the bsγ, bsZ vertices and the box diagram contibution, we find the rate

$$\Gamma^{b\to s\ell\ell} = \chi^2\Big(|v_iA_i|^2 + |v_iB_i|^2 + 4s_W^2\, v_i(A_i+B_i)(v_jF_2{}^j) + 16s_W^4\,(\log\frac{m_b}{2m_\ell} - \frac{2}{3})|v_iF_2{}^i|^2\Big)\Gamma_o \qquad (2)$$

where $\chi \equiv g^2/16\pi^2$, $A_i \equiv \bar{C}_i - s_W^2(F_1{}^i + 2\bar{C}_i{}^Z)$, $B_i \equiv -s_W^2(F_1{}^i + 2\bar{C}_i{}^Z)$, $\bar{C}_i \equiv \bar{C}_i{}^Z + \bar{C}_i{}^{box}$ (these form factor combinations are now gauge invariant), $\Gamma_o = G_F{}^2m_b{}^5/(192\pi^3)$ and for each neutrino species, we get

$$\Gamma^{b\to s\nu\nu} = \Gamma^{b\to s\ell\ell}(s_W \to 0,\ \bar{C}_i \to \bar{D}_i) \qquad (3)$$

For 3 generations, because V_{ub} is small, it turns out that the BRs are not sensitive to the mild uncertainties in the 3×3 KM matrix. The results are plotted in Fig. 1, where for the $s\nu\bar\nu$ mode we have summed over three neutrino species. The qualitative nature can be understood as follows. For the $s\ell^+\ell^-$ mode, when m_t is not large, $x_t \equiv (m_t/M_W)^2 < 1$, the process is dominated by charm through the photon contribution since $F_1{}^i \sim 2Q\log x_i/3(x_i-1)$, while the other contributions are suppressed by powers of x_i. But as m_t grows, the Z contribution grows, growing ever more rapidly for $m_t > M_W$, asymptotically approaching a $m_t{}^4$ behavior in rates as $\bar{C}_i{}^Z \sim x_i/4$ for $x_i \gg 1$. Thus, for large m_t, the Z contributon takes over. The $s\nu\bar\nu$ mode does not have photonic contributions, and has an even more prominent m_t dependence. Of course, the curves are drawn up to $m_t = 500$ GeV just for the purpose of illustration, since from various grounds we already have a bound on m_t roughly as $m_t \lesssim 200$ GeV.[8]

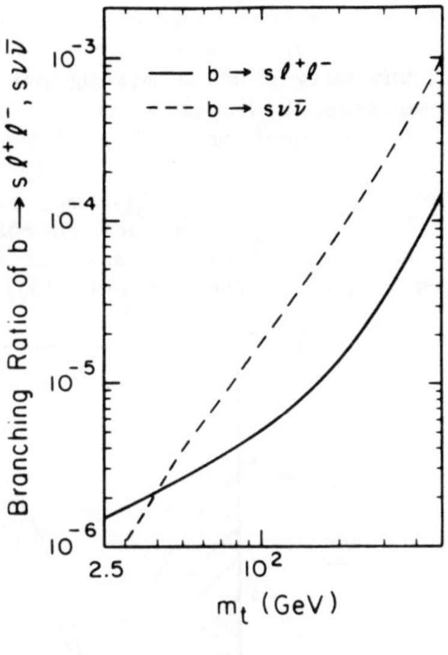

Fig. 1

Why is there such a strong heavy internal quark mass dependence? This apparent violation of the intuitive decoupling theorem[9] comes about because in a spontaneously broken gauge theory with fermion masses generated through Yukawa couplings with the scalar field developing a vacuum expectation value, the fermion mass is a coupling constant in disguise. Here it enters into the numerator of the amplitude through the longitudinal component of the W boson in the virtual loop. Thus, as m_t grows, not only we don't have a power $1/m_t$ suppression (for $m_t > M_W$)

from propagator effects, we actually have an amplitude growing as m_t^2. This is analogous to the box diagram contribution to the K_L-K_S mass difference, except now we have a rate $\propto$ (amplitude)$^2 \propto m_t^4$. In comparison, although similar effects do occur in the process $K \to \pi\nu\bar{\nu}$, but in that case the KM angles are constrained such that, to give the biggest rate it actually favors a lower m_t.

Thus, these decay modes are extremely interesting as they test the weak point of the underlying theory (SM), viz. the symmetry breaking and the fermion mass generation mechanism. If new heavy quarks get their masses from a different mechanism, their effects would be different. In this respect, it is natural to study the impact of a 4th generation of quarks (if they exist at all) on these decay modes.

But with a 4th generation, not only the new charge 2/3 quark mass $m_{t'}$ is unknown, we also have 3 more mixing angles and 2 more phases. Fortunately, the situation is not hopeless, since the following observations greatly reduce the number of essential independent parameters relevant to the problem:

(1) $|v_u| \equiv |V_{us}^* V_{ub}| \ll |v_c| \equiv |V_{cs}^* V_{cb}| \lesssim 0.05$ which is itself rather small. Thus, from unitarity,

$$v_u + v_c + v_t + v_{t'} = 0 \qquad (4)$$

and by ignoring v_u we can express v_t as

$$v_t = -(v_{t'} + v_c) \qquad (5)$$

and the only crucial mixing parameter is $v_{t'}$. We shall take $v_c \approx 0.05$ in subsequent discussions.

(2) inspection of Fig. 1 shows that the t contribution is relatively unimportant for say $m_t \lesssim M_W$. If there exist a 4th generation, it is plausible that $m_{t'} > M_W \gtrsim m_t$. Thus, $m_{t'}$ is the single important mass. We shall take $m_t \simeq 50$ GeV in most of the subsequent discussions.

We now plot BR vs. $v_{t'}$ for $|v_{t'}| \lesssim 0.2$ for $m_{t'} = 150$-500 GeV. In Figs. 2 and 3. The current CLEO limit[10] for the exclusive process B →

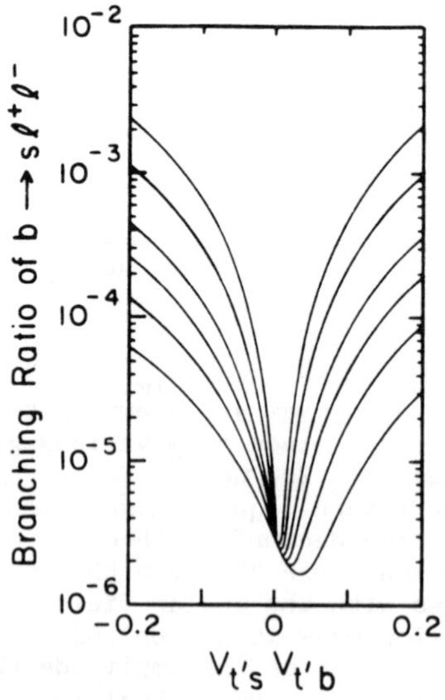

Fig. 2

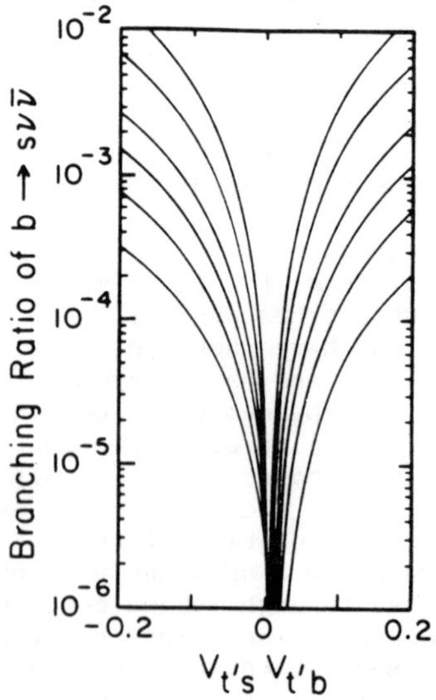

Fig. 3

Ke^+e^- is 2×10^{-4}. Even if one takes $\Gamma(B \to Ke^+e^- + X) \sim 2\,\Gamma(B \to Ke^+e^-)$, it is clear that we are not cutting into the unexplored parameter space of $v_{t'}$, $m_{t'}$. But, say this limit gets improved by more than an order of magnitude, we see that it then cuts deeply into the parameter space (if not yielding a positive signal already). Note that for 3 generations, the BR is less than 1.5×10^{-5}. Thus any observation beyond this level would indicate that a 4th generation may well exist.

As to the neutrino mode, it appears as if the experimental observtion would be quite hopeless at present. But recall that kaons have been around for decades, and only now can we explore the $K \to \pi\nu\bar{\nu}$ mode meaningfully. The search for the $b \to s\nu\bar{\nu}$ mode would thus be a challenge to our experimental colleagues. Note that, because of 3 ~ 4 species of neutrinos and the fact that the amplitude for each neutrino species is somewhat larger than the $s\ell\ell$ mode, the total $b \to s\nu\nu$ BR is more than 6 times as large as $B \to K\ell\ell$. It is clear that for these kind of modes, a BR that reaches easily above the 10^{-3} level cannot be called "rare" anymore. It would be a pitty if we can't study it.

3. PROCESS $b \to s\gamma$

Clearly, if there is a $b \to s\ell^+\ell^-$ mode there must also be the on-shell photon emission $b \to s\gamma$. This is a magnetic transition and only the F_2 form factor enters. The rate is

$$\Gamma^{b\to s\gamma} = \frac{3\alpha}{2\pi}\, |v_i F_2{}^i|^2\, \Gamma_0 \qquad (6)$$

The 3 generation case has been discussed before in the literature.[11] More recently, it has been pointed out that because the SM weak effect (the F_2 form factor) is suppressed by powers of x_i for small x_i, QCD corrections can bring in new $\log x_i$ terms at the α_s order which overwhelms the lowest order (in QCD) results.[12] The overall effect gives rise to a much milder m_t dependence for the process $b \to s\gamma$. The BR is roughly 10^{-4}. However,the result is now less precisely known. The current CLEO limit for the exclusive process $B \to K^*\gamma$ is 2×10^{-3}, but significant improvement is expected in several years, and could reach down to the 10^{-5} level. Thus, except for the usual problems of relating the exclusive rates to the inclusive, this mode should be seen soon, and thereby may provide a first test of loop effects (and the large QCD corrections) in the B system.

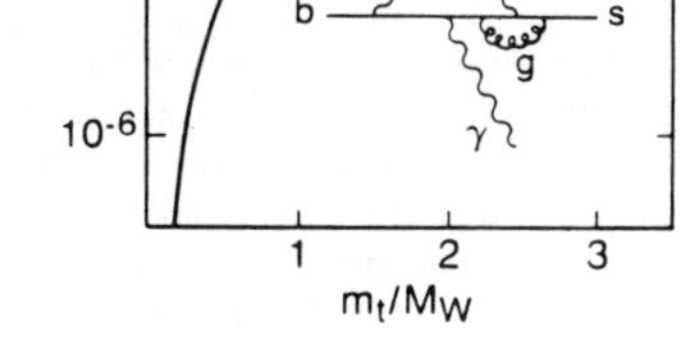

Fig. 4

The large QCD corrections to this process have a significant impact on the 4 generation result. From Fig. 4 it is seen that QCD effects are more prominent for low m_t, therefore, for 4 generations, the top contribution gets more enhancement compared to the t' contribution. But from eq. (5) it is seen that there is a tendency for the t and t' contributions to cancel. Thus, the fact that the t contribution gets more enhanced means that there is a much stronger cancellation between the t and t' contributions compared to the case without QCD corrections. This is illustrat-

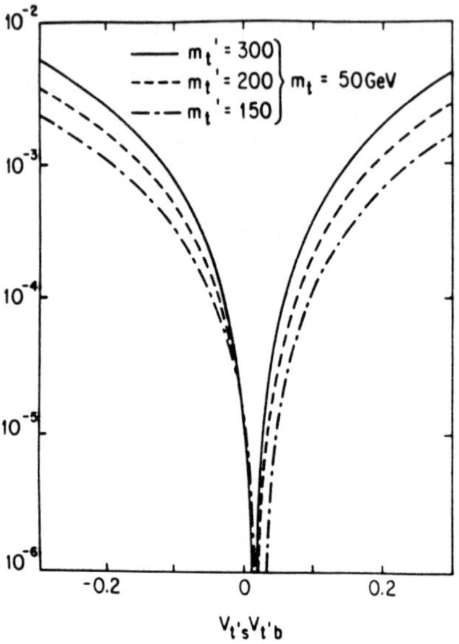

Fig. 5

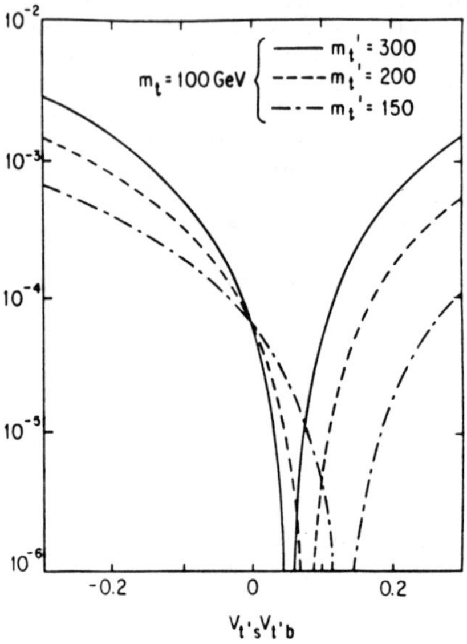

Fig. 6

ed in Figs. 5 and 6. Fig. 5 shows the result for $m_t = 50$ GeV without QCD corrections. Because of the funnel shape, this would have been a very powerful probe of the parameter $v_{t'} \equiv V_{t's}V_{t'b}$, in that if $b \to s\gamma$ is not observed say below 10^{-4} level, we could put very stringent bounds on $v_{t'}$, regardless of $m_{t'}$. Unfortunately, there are very large QCD corrections for $m_Q \lesssim 100$ GeV, therefore, for $m_t \lesssim 100$ GeV, the result is similar to that depicted in Fig. 6. There is now a much stronger $m_{t'}$ dependence, and although some qualitative features remain, unless one has a precise knowledge of the QCD corrections (and also knowledge of $m_{t'}$), this mode becomes less potent in probing $v_{t'}$. Still, note the general feature that the BR may gain roughly by an order of magnitude over the 3 generation result, but also could get very suppressed. Of course, if $v_{t'}$ vanishes, we recover the 3 generation result.

4. PROCESS b → s + "g"

This process is very relevant to the question of charm deficit[1] in B decays. It demands a much more detailed discussion than can be made here.[4] In the literature, QCD induced $b \to s$ transitions have been discussed before[13] in many occasions. For the specific problem of an overall inclusive BR, Eilam[14] has discussed the two processes

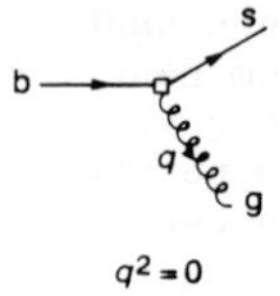

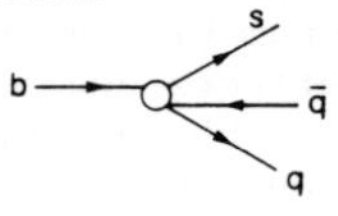

However, the first process has been slightly overestimated. For the second process, which was taken from the leading-log, all orders (LLAO) QCD renormalization group analysis of Guberina,[15] only the 4 quark

operators were taken into account, with questionable applicability for $m_t > M_W$. If one wants to explore the full range of heavy quark mass dependences, it is not clear whether the LLAO analysis would be more trustworthy than the "exact" 1-loop weak correction (but lowest order in QCD) calculation. Of greater importance, perhaps, is the fact that the <u>2 quark - 2 gluon operator</u> has never been taken into account before. We therefore study the parton level processes

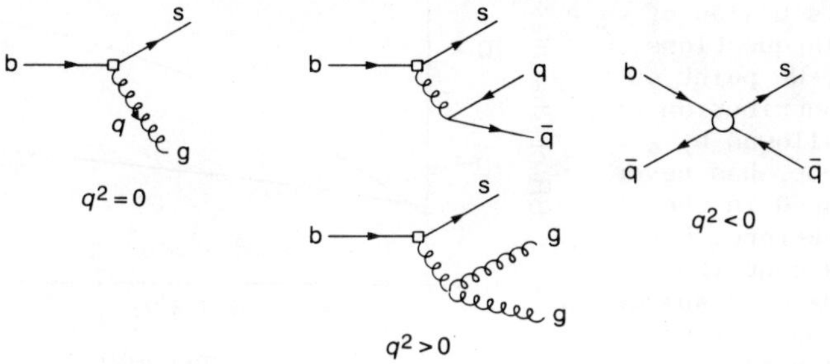

where for the $q^2 < 0$ process we use the Guberina result for purpose of comparison; it turns out to give negligible contribution. Note that the three momentum domains ($q^2 = 0$, > 0, < 0) are distinct at the parton level. Moreover, for $q^2 = 0$ ("on-shell" single gluon emission), only the magnetic form factor F_2 contributes, thus it is nontrivial to include the $q^2 > 0$ gluon split-up contribution, which does contain the leading F_1 form factor. We thus get the decay rate

$$\begin{aligned}\Gamma^{b\to s"g"} &= \frac{N^2-1}{N}\frac{3\alpha_s}{4\pi}\,|v_i\ F_2{}^i|^2\,\Gamma_o & (q^2 = 0)\\ &+ (n_f + N)\frac{N^2-1}{N}\frac{\alpha_s^2}{32\pi^2}\,|v_iF_1{}^i|^2\,\Gamma_o & (q^2 > 0)\\ &+ \left(\frac{f_B}{m_B}\right)^2\frac{(N^2-1)^2}{N^3}\,6\pi^2|v_ic_5{}^i|^2\,\Gamma_o & (q^2 < 0) \qquad (7)\end{aligned}$$

f_B is the meson decay constant, since the $q^2 < 0$ penguin is a two body annihilation process. The α_s dependence for the spacelike term is absorbed into the definition of c_5, which can be found in ref. 15. The form factors $F_1{}^i$ and $F_2{}^i$ can be derived from the corresponding photonic form factors by dropping the Q (internal quark electric charge) independent piece and then set Q = 1; both of them are now gauge invariant. We have maintained some generality in showing the result for SU(N) color group, and n_f noncharm light fermion flavors; in reality, N = 3 and n_f = 3 (u, d, s).

In arriving at eq. (7), many approximations have been made. First of all, $o(\alpha_s{}^2)$ QCD corrections for the $q^2 = 0$ case has been ignored at present. When included, it should exhibit similar behavior as in the photonic case,[12] except that the calculation would become even more complicated. However, as shown below, this term is not the dominant piece, therefore we are not making a serious error for not including these corrections. For the $q^2 > 0$ contribution, we have dropped the F_2 term in the amplitude.[16] Finally, we have tacitly assumed that the full hadronic rate is the same as the parton level decay rate, thereby assuming the partons fragment into hadrons with probability unity. But at the same time, by taking only an incoherent sum of the various processes, interference effects have been ignored. Note that no fragmentation model at present can take interference effects into account. For a more detailed discussion of these subtle questions, see ref. 4. We point out that the contribution of $b \to sg^*$, followed by $g^* \to gg$ split-up, has never been discussed in the literature before. For SU(3), this contribution is as strong as 3 species of light final state quarks ($b \to sq\bar{q}$).

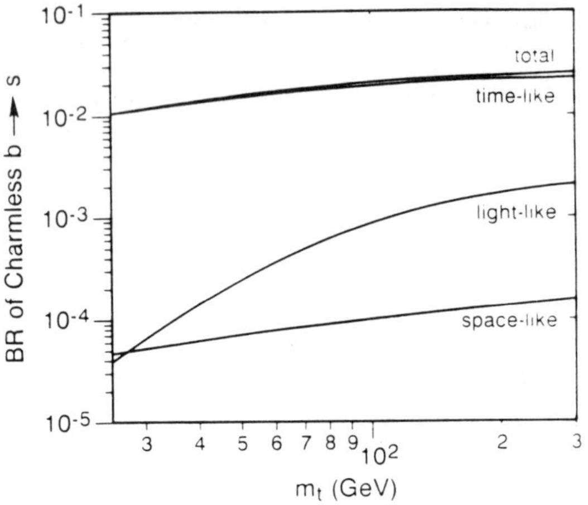

Fig. 7

With these provisions we plot the 3 generation result in Fig. 7. It is clear that the time-like gluon split-up ($q^2 > 0$) process dominates, even if the $q^2 = 0$ contribution gets slightly enhanced over the one plotted. For the $q^2 < 0$ process, although it has two body phase space,the $(f_B/m_B)^2$ factor leads to a very strong suppression and can be ignored. Note also that, comparing this LLAO curve to that of the "exact" (but w/o QCD correction) $q^2 > 0$ curve, the LLAO seems to have too hard an m_t dependence, thus casting some doubt on its validity for $m_t > M_W$.

Why does the $o(\alpha_s{}^2)$ process ($q^2 > 0$) dominate over the $o(\alpha_s)$ process ($q^2 = 0$)? As indicated earlier, this is because as q^2 vanishes, the F_1 contribution does not enter anymore. But in the $q^2 > 0$ process when q^2 is large, the F_1 term (which contains the leading-log) contributes strongly, and overcompensates not only for the extra power of α_s, but also for two body vs. three body phase space. This is a nontrivial result.

Thus, for the allowed m_t range, BR ranges from ~ 1 - 2%. This is not a "rare" decay anymore, but we need a way to measure inclusive hadronic rates (B to K without charm) directly. For the experimentally more accessible exclusive processes, not only one has smaller rates and the usual difficulties of evaluating hadronic matrix elements that involve two hadronic currents, but also it is difficult to incorporate the important $g^* \to gg$ split-up process into the calculation (which has not been taken into account in previous discussions).

Note that for large m_t (> 100 GeV) the dependence on heavy internal quark masses are very mild. This is because, unlike the photonic case, there is no WWg coupling, so $F_1{}^i$ approaches a constant for large m_Q. This has implications on the 4 generation result, which is plotted in Figs. 8 and 9 for m_t = 50 and 80 GeV, respectively.

For 50 GeV top, if KM angles are favorable, one may get a sizeable enhancement over the 3 generation result, and may well reach the 10% level. For heavier m_t, because of the flatness of the m_Q dependence of $F_1{}^Q$, there tends to be a stronger cancellation, and the BR range gets reduced. But still, a 5-10% BR for $b \to s$ w/o charm would constitute

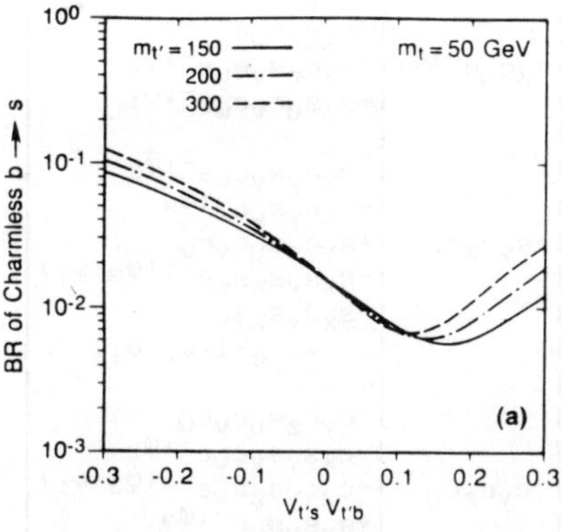

Fig. 8

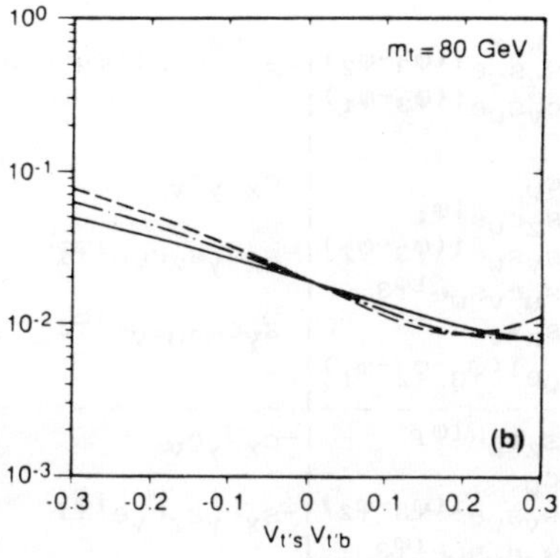

Fig. 9

very strong evidence for the existence of a 4th generation. The question is can one realistically search for this decay inclusively? Progress in experimental techniques is needed; we need to veto against charm.

It should be pointed out that, because of the extreme flatness of F_1^Q for large m_Q, as exemplified in Fig. 7, the 4th generation result will be very sensitive to the terms that were ignored in the discussion above. Although this is not essential for the 3 genenartion case, it may considerably enhance the 4 generation result.[16]

5. 4×4 MIXING MATRIX

It is clear that the B system will turn out a lot of new information. In particular, we have shown that these various modes are sensitive to the presence of a 4th generation; they may in fact provide the first indication of new physics.

Within the context of a 4 generation SM, what is the most pragmatic parametrization of the 4×4 KM matrix? Clearly, this is not a physics question, since physics does not depend on how one parametrizes the KM matrix. Rather, it is a matter of convenience, and we must adapt to the most readily available data in the next few years. Obviously, since V_{ub} and V_{cb} would still be of interest for a while to come, it would be useful to keep it as simple as possible. The element V_{tb} may be studied if t is produced, or through a detailed analysis of $W \to t\bar{b}$. There is a possibility that the 4th generation down quark b' may be lighter than the top. But barring this possibility, the elements $V_{ub'}$, $V_{cb'}$, $V_{tb'}$ may not be accessible for experimental study for a long time. But the B system is readily available, and we have demonstrated that the loop induced $b \to s$ transitions may give us information about the new mixing angles (and $m_{t'}$). Thus, there is some reason to keep the elements V_{ub}, V_{cb}, V_{tb}, $V_{t'd}$, $V_{t's}$, $V_{t'b}$, i.e. the 3rd column and 4th row of the 4×4 mixing matrix as simple as possible. This can be easily achieved for a unitary matrix, by rearrangimg the sequence of multiplication of the various rotation matrices which we take from Botella and Chau.[17] Thus

$$V_4 = \begin{bmatrix} & V_3 & & 0 \\ & & & 0 \\ & & & 0 \\ 0 & 0 & 0 & 1 \end{bmatrix} \begin{bmatrix} 1 & 0 & 0 & 0 \\ 0 & 1 & 0 & 0 \\ 0 & 0 & c_u & s_u \\ 0 & 0 & -s_u & c_u \end{bmatrix} \begin{bmatrix} 1 & 0 & 0 & 0 \\ 0 & c_v & 0 & s_v e^{-i\varphi_2} \\ 0 & 0 & 1 & 0 \\ 0 & -s_v e^{i\varphi_2} & 0 & c_v \end{bmatrix} \begin{bmatrix} c_w & 0 & 0 & s_w e^{-i\varphi_3} \\ 0 & 1 & 0 & 0 \\ 0 & 0 & 1 & 0 \\ -s_w e^{i\varphi_3} & 0 & 0 & c_w \end{bmatrix} =$$

$$\begin{bmatrix}
\begin{array}{l} c_xc_zc_w \\ -s_xc_zs_vs_we^{i(\varphi_3-\varphi_2)} \\ -s_zs_uc_vs_we^{i(\varphi_3-\varphi_1)} \end{array} &
\begin{array}{l} s_xc_zc_v \\ -s_zs_us_ve^{i(\varphi_2-\varphi_1)} \end{array} &
s_zc_ue^{-i\varphi_1} &
\begin{array}{l} c_xc_zs_we^{-i\varphi_3} \\ +s_xc_zs_vc_we^{-i\varphi_2} \\ +s_zs_uc_vc_we^{-i\varphi_1} \end{array} \\
\begin{array}{l} -s_xc_yc_w \\ -c_xs_ys_zc_we^{i\varphi_1} \\ -c_xc_ys_vs_we^{i(\varphi_3-\varphi_2)} \\ -s_yc_zs_uc_vs_we^{i\varphi_3} \\ +s_xs_ys_zs_v \\ \quad \ast s_we^{i(\varphi_3-\varphi_2+\varphi_1)} \end{array} &
\begin{array}{l} c_xc_yc_v \\ -s_xs_ys_zc_ve^{i\varphi_1} \\ -s_yc_zs_us_ve^{i\varphi_2} \end{array} &
s_yc_zc_u &
\begin{array}{l} c_xc_ys_vc_we^{-i\varphi_2} \\ -s_xc_ys_we^{-i\varphi_3} \\ +s_yc_zs_uc_vc_w \\ -c_xs_ys_zs_we^{-i(\varphi_3-\varphi_1)} \\ -s_xs_ys_zs_v \\ \quad \ast c_we^{-i(\varphi_2-\varphi_1)} \end{array} \\
\begin{array}{l} -c_xc_ys_zc_we^{i\varphi_1} \\ +s_xs_yc_w \\ +c_xs_ys_vs_we^{i(\varphi_3-\varphi_2)} \\ -c_yc_zs_uc_vs_we^{i\varphi_3} \\ +s_xc_ys_zs_v \\ \quad \ast s_we^{i(\varphi_3-\varphi_2+\varphi_1)} \end{array} &
\begin{array}{l} -c_xs_yc_v \\ -s_xc_ys_zc_ve^{i\varphi_1} \\ -c_yc_zs_us_ve^{i\varphi_2} \end{array} &
c_yc_zc_u &
\begin{array}{l} c_yc_zs_uc_vc_w \\ -c_xs_ys_vc_we^{-i\varphi_2} \\ -c_xc_ys_zs_we^{-i(\varphi_3-\varphi_1)} \\ +s_xs_ys_we^{-i\varphi_3} \\ +s_xc_ys_zs_v \\ \quad \ast c_we^{-i(\varphi_2-\varphi_1)} \end{array} \\
-c_uc_vs_we^{i\varphi_3} & -c_us_ve^{i\varphi_2} & -s_u & c_uc_vc_w
\end{bmatrix}$$

where

$$V_3 = \begin{bmatrix} c_xc_z & s_xc_z & s_ze^{-i\varphi_1} \\ -s_xc_y-c_xs_ys_ze^{i\varphi_1} & c_xc_y-s_xs_ys_ze^{i\varphi_1} & s_yc_z \\ s_xs_y-c_xc_ys_ze^{i\varphi_1} & -c_xs_y-s_xc_ys_ze^{i\varphi_1} & c_yc_z \end{bmatrix}$$

is the limiting 3 generation Maiani-Chau-Keung matrix[18]

Of course, no parametrization serves every purpose, and here we pay the price in the first three elements of the 1st and 4th columns. In contrast, the parametrization of Botella and Chau[17] keeps the 1st row and 4th column simple. Thus, although useful for studying b' decays (in the future), in their parametrization V_{cb} has two terms, V_{tb} and $V_{t'b}$ each are composed of 3 terms, while $V_{t's}$ and $V_{t'd}$ each have 5 separate terms. Therefore they are not particularly suitable for studying B physics (especially rare B physics), in the context of 4 generations. It is clear that, just like the case of B production, if b' is produced, b' decays should provide the best way for studying the elements $V_{ub'}$, $V_{cb'}$ and $V_{tb'}$. But before b' is produced, we would have to resort to utilizing loop induced rare B decays to provide us with information regarding the new mixing parameters. This is much like the situation before the advent of B physics, that we had to rely on constraints coming from the kaon system. Let us emphasize once again, though, that for the B system loop effects have larger BRs.

Our parametrization would not be useful for studies of kaon physics, but this is perhaps inevitable if there is a 4th generation, and perhaps one will have to make a choice between having the convenience in the K or the B system. Clearly, if all the new off-diagonal mixing elements are very suppressed (i.e., the 4th generation is effectively decoupled from the low energy world), all the parametrizations would become similar, reducing to the 3 generation result. Only through using and testing could one determine whether our parametrization is useful or not (if a 4th generation indeed exists).

6. SUMMARY OF RESULTS

Let us collect our results in the following table. QCD corrections for $b \to s\gamma$ has been taken into account. The experimental limits quoted are for the exclusive modes $B \to K\ell^+\ell^-$ or $B \to K^*\gamma$ (see ref. 10), and the reported deficit of charm (ref. 1).

	3 family m_t = 25-200GeV	4 family $\lvert v_{t'}\rvert<0.3, m_{t'}<300$GeV	Current limit
$b \to s\ell^+\ell^-$	$1.5\times10^{-6} - 1.5\times10^{-5}$	$2\times10^{-6} - 4\times10^{-4}$	$\lesssim 2\times10^{-4}$
$b \to s\nu\bar{\nu}$	$8\times10^{-6} - 1\times10^{-4}$	$< 3\times10^{-3}$	none
$b \to s\gamma^\dagger$	$\sim7\times10^{-5} - 2\times10^{-4}$	$< 3\times10^{-3}$	$\lesssim 2\times10^{-3}$
$b \to s$ "g"	1% - 2%	$\lesssim$ 10%	$\lesssim$ 50% (?)

7. CONCLUSIONS

- $b \to s\ell^+\ell^-$ can be used as a new way to set bounds on m_t.
- The m_t^4 dependence of the $b \to s\ell^+\ell^-$ rate probes the symmetry breaking and fermion mass generation mechanism. I.e., it can be turned around and used as evidence for or against Yukawa couplings.
- $b \to s\nu\bar{\nu}$ could be very interesting with the turning on of new facilities and detectors.
- $b \to s\gamma$ may be detected soon at CLEO and ARGUS. This mode may (?) be the first to show indications of a possible 4th generation, or lead to limits on $\lvert V_{t's}V_{t'b}\rvert$.
- QCD induced $b \to s$ transition without charm may not be rare at all. Once the current puzzle of the charm deficit is understood, one should try to measure this rate.
- With the potential activity in studying B physics in mind, we propose a pragmatic parametrization of the 4×4 mixing matrix.

If current experimental limits improve by an order of magnitude, we can probe heavy quark physics, set limits on m_t, and if a 4th generation exists, set limits on $\lvert V_{t's}V_{t'b}\rvert$, $\lvert V_{t'd}V_{t'b}\rvert$ and even $m_{t'}$, by studying RARE B DECAYS. QCD induced charmless B decays may not be rare at all.

We wish to thank H. Steger and R. S. Willey for collaboration and discussion. This work was supported by the NSF.

References

1. S. Stone, plenary talk, this proceedings. Also, talk by E.H. Thorndike, this proceedings.
2. W.S. Hou, R.S. Willey and A. Soni, PITT-86-08 and UCLA/86/TEP/43.
3. W.S. Hou, A. Soni and H. Steger, PITT-87-01 and UCLA/87/TEP/1.
4. W.S. Hou, A. Soni and H. Steger, PITT-87-02 and UCLA/87/TEP/3.
5. W.S. Hou and A. Soni, in progress.
6. T. Inami and C.S. Lim, Prog.Theo.Phys. 65, 297 (1981); Errata, ibid. 65, 1772 (1981).
7. N.G. Deshpande, G. Eilam, A. Soni and G.L. Kane, Phys.Rev. Lett. 57, 1106 (1986); P.J. O'Donnell, Phys. Lett. 175B, 369 (1986).
8. Aside from ρ parameter bounds, there is the bound coming from a recent detailed analysis of M_W, M_Z AND $\sin^2\theta_W$; see W. Marciano, this proceedings.
9. T. Appelquist and H.D. Politzer, Phys.Rev.Lett. 34, 43 (1975).
10. P. Avery et al., CLNS 86/737.
11. B.A. Campbell and P.J. O'Donnell, Phys.Rev. D25, 1989 (1982).
12. S. Bertolini, F. Borzumati and A. Masiero, CMU-HEP 86-19; N.G. Deshpande, P. Lo, J. Trampetic, G. Eilam and P. Singer, OITS 346.
13. M. Bander, D. Silverman and A. Soni, Phys.Rev.Lett. 43, 242 (1979); B. Guberina, R.D. Peccei and R. Rückl, Phys.Lett. 90B, 169 (1980); B. Guberina, Fizika 16, 49 (1984); G. Eilam, Phys. Rev.Lett. 49, 1478 (1982).
14. Last paper of ref. 12.
15. Third paper of ref. 12.
16. A more detailed analysis is given in ref. 4, which discusses the importance of interference effects.
17. F.J. Botella and L.-L. Chau, Phys.Lett. 168B, 97 (1986).
18. L. Maiani, in Proceedings of the 1977 International Symposium on Lepton and Photon Interactions at High Energies (DESY, Hamburg, 1977), p867.; L.-L. Chau and W.-Y. Keung, Phys.Rev. Lett. 53, 1802, (1984).

LIFETIMES OF HEAVY LEPTONS AND HADRONS*

VERA LÜTH
Stanford Linear Accelerator Center
Stanford University, Stanford, California, 94305

ABSTRACT

Recent measurements of the lifetimes of the τ lepton and charm and beauty particles are reviewed, with emphasis on the experimental techniques used for vertex detection.

1. Introduction

In the past few years, decays of heavy leptons and hadrons carrying heavy flavour quantum numbers have become a fashionable topic, for theorists and experimentalists alike. In the framework of the standard electro-weak theory, the flavour-changing transitions among quarks and leptons are described by their coupling to the charged weak boson, $W^{\pm}$. The parton diagrams for heavy leptons or quarks are identical to $\mu^{\pm}$ decay, and thus apart from differences in the couplings of the $W^{\pm}$ to various partons, the decay rates of the heavy quarks Q and leptons ℓ are calculable and closely related to the muon decay time τ_μ, namely

$$\tau_\ell = \frac{192\pi^3}{G_F^2 m_\ell^5} BR(\ell^- \to e^-\bar{\nu}_e\nu_\ell) = \tau_\mu \left(\frac{m_\mu}{m_\ell}\right)^5 BR(\ell^- \to e^-\bar{\nu}_e\nu_\ell)$$

$$\tau_Q = \tau_\mu \left(\frac{m_\mu}{m_Q}\right)^5 BR(Q \to e^-\bar{\nu}_e q) \Big/ \sum_q |V_{qQ}|^2 ,$$

where G_F is the Fermi constant, m represent the parton masses, and BR stands for the leptonic or semi-leptonic branching ratios. This simple relation is, however, only valid for free quarks and in the absence of flavour mixing and final state corrections due to the light quarks and gluons participating in the decay of a hadron. The differences in the couplings of the quarks to the $W^{\pm}$ are given by the Cabibbo-Kobayashi-Maskawa matrix elements V_{qQ}.[1] Precise measurements of the lifetimes of heavy leptons, mesons and baryons, combined with their branching ratios, can serve as a test of these simple assumptions, help to determine the elements of the quark mixing matrix V_{qQ}, and help in the formulation of theoretical models of weak decay.

* Work supported by the Department of Energy, contract DE-AC03-76SF00515.

From the point of view of an experimentalist the lifetime measurements are attractive because they represent a major challenge to the design and operation of detectors and to the data analysis. The principle difficulty experimentalists face in the detection of heavy lepton and hadron decays are their small production rates in hadron and photon interactions, and their short decay lengths. Measurements of the branching ratios and lifetimes of the τ lepton have remained an unchallenged domain of e^+e^- experiments, which benefit from the copious production of $\tau^+\tau^-$ pairs resulting in a very clean event topology. Signatures for heavy flavour particles can be derived from their relatively high mass and the weak nature of their decay. Masses of 2 GeV/c^2 and above give rise to large transverse momenta of the decay secondaries and small branching ratios for any particular decay mode. The weak coupling causes the emission of leptons ($e^\pm, \mu^\pm, \tau^\pm$, and neutrinos) and strange particles due to Cabibbo enhancement. Thus an experiment with good sensitivity requires a large-acceptance spectrometer with excellent momentum resolution and good particle identification, preferentially both for hadrons and leptons, and a vertex detector with superb resolution and granularity. Furthermore, measurements of beauty particle lifetimes in hadron beams will not be possible without selective and efficient triggers, or at least the possibility of a fast off-line filter of events.

The standard method to determine particle decay times is to measure the particle momentum and decay path, and thus requires an accurate determination of the production and decay vertices. If the decay products are not fully detected the momentum is often estimated from an unconstrained kinematic fit or from the measured effective mass and the momentum sum of the measured decay tracks. The accuracy of the estimate is tested by Monte Carlo simulation assuming a specific shape of the production spectrum. A more model independent estimate is based on the study of the decay length measured in the plane transverse to the beam in fixed target experiments. This method uses the fact that the transverse momentum distributions are well known. Another method relies on the measurement of the so-called impact parameter ρ[2] which is defined as the distance of closest approach of a track to the production vertex. ρ is proportional to the product of the decay length and the decay angle, and in the relativistic limit becomes insensitive to the momentum of the decaying particle. The clear advantage of this estimator is that it does not require a fully reconstructed decay or estimate of momentum, and can use individual tracks from hadronic or semileptonic decays, thus avoiding unacceptable losses due to small branching ratios and limited detector acceptance. Monte Carlo simulation is needed to relate the impact parameter to the decay time; this can be done to an accuracy of about 10%.

In the following, recent measurements of the τ, and the charm and beauty lifetimes will be reviewed, more information about the experiments and earlier results can be found in more detailed review articles.[3,4]

2. The Lifetime of the τ Lepton

Though we now have evidence for the decay of the Z^o and $W^{\pm}$ into $\tau^{\pm}$ leptons from the UA1 experiment at the SPS collider,[5] measurements of the τ branching ratios and lifetimes have been exclusively performed at the e^+e^- storage rings PEP, PETRA, DORIS and CESR, where large $\tau^+\tau^-$ pair production produces a clean event topology. All these experiments have improved the accuracy of the charge particle tracking by the installation of high precision drift chambers mounted on the outside of a thin-walled beam pipe. At present, the major limitation on the lifetime measurements is due to the fact that the production vertex is not observed, but is located somewhere inside the beam-beam interaction region. Tracks have to be extrapolated over a distance of several cm, depending on the radius of the beam pipe.

The MAC collaboration[6] has extracted the τ lifetime from the impact parameter distribution of all well measured charged particle tracks from τ decay. The production point is found using, with appropriate weights, the interception of the tracks and the beam ellipse. The event sample is selected by simple topology cuts, and contains less than 4% background. The resulting distribution for 6533 tracks is shown in Figure 1(a). The large statistics gain considerable precision; a clear shift and excess of tracks on the positive side are visible.The trimmed average is $44.5 \pm 2.4\ \mu$m, corresponding to a lifetime of

$$\tau(\tau^{\pm}) = 2.86 \pm 0.17 \pm 0.13 \cdot 10^{-13}\ \text{s.}^{\star}$$

The advantage of the impact parameter measurement is that systematic effects cancel to a large degree, and the trimmed average does not depend on the error on the impact parameter measurement, as long as all errors are distributed symmetrically around zero. The remaining systematic error is dominated by the uncertainty in the factor α which relates the impact parameter and lifetime; α is determined by Monte Carlo simulation.

The Mark II[7] and HRS[8] groups have derived the τ lifetime from the measured decay length distribution which is taken as the distance between the centre of the e^+e^- collision region and the vertex formed by the three well measured charged particles tracks from the decay $\tau^- \rightarrow \pi^-\pi^+\pi^-\nu_\tau$. The data from the Mark II group are shown in Figure 1(b). A maximum likelihood method fit is performed which takes into account the individual error of each decay length. The average decay length of $635 \pm 36\mu$m translates to an average lifetime of

$$\tau(\tau^{\pm}) = 2.86 \pm 0.16 \pm 0.25 \cdot 10^{-13}\ \text{s.}$$

The CLEO[9] and ARGUS[10] experiments have presented preliminary results based on very similar analyses for substantially larger samples of τ decays. Their data are

⋆ If there are two errors are quoted, the first represents the statistical, the second the systematic uncertainty.

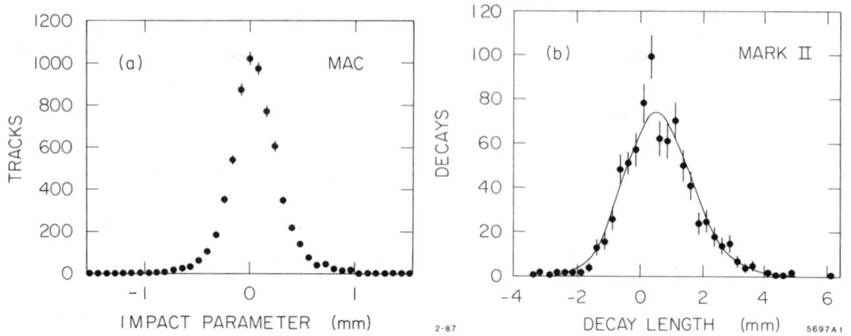

Fig. 1. Measurements of the $\tau^{\pm}$ lifetime: Distributions of (a) the impact parameter for 6533 tracks from the MAC, and (b) the decay lengths for 807 3-prong decays from the Mark II experiment.

recorded at lower energies, resulting in increased $\tau^+\tau^-$ production cross sections, reduced decay lengths and in larger background from hadronic final states.

A compilation of recent measurements of the $\tau^{\pm}$ lifetime is given in Table I. None of the results listed has been published yet, and most of these measurements can be viewed as a calibration for the high precision drift-chamber and the technique applied to measure small decay distances or impact parameters.

In the framework of the standard model the τ lifetime can be predicted, given the measured leptonic branching, $BR(\tau^- \to e^-\bar{\nu}_e\nu_\tau) = (17.93 \pm 0.36)\%$,[11] and the mass, $m_\tau = 1784.2 \pm 3.2$ MeV/c^2,[12] namely

$$\tau(\tau^{\pm})_{th} = 2.86 \pm 0.06 \cdot 10^{-13}\ \text{s}.$$

All experiments listed in Table I are in good agreement with each other, and the average differs by one standard deviation from the prediction. In fact, the measurements can be used to set limits on possible deviations from the basic assumptions of the standard model. The measurements confirm the equality of the coupling of the τ and μ leptons to the charged weak current to a level of 2%. The measurements also constrain the mass of the ν_τ to be less than 195 MeV/c^2 at 90% confidence level; a limit which is not competitive with recent direct measurements.

3. Lifetimes of Charm Particles

In the following, an overview over experiments measuring lifetimes of the charm mesons, D^0, D^+, and D_s^+, (commonly referred to as F^+)† and charm baryons, Λ_c^+,

† Throughout this report reference to particles like D^0, D^+, D_s^+ implies also the charge conjugate states $\overline{D}^0, D^-, D_s^-$, unless explicitly stated.

Table I: Measurements of the τ Lepton Lifetime

Experiment	Reference	Lifetime (10^{-13} s)
MAC	6	2.86±0.17±0.13
Mark II	7	2.86±0.16±0.25
HRS	8	2.85±0.15±0.11
CLEO	9	3.33±0.14±0.20
ARGUS	10	3.02±0.19±0.12
Average		2.97±0.10

Ξ_c^+, and Ω_c^o, will be given. The experiments are grouped as to the apparatus used for vertex detection.

3.1 *Emulsion Experiments*

In recent years, nuclear emulsions have been revived as active targets for lifetime experiments because of their superb spatial resolution (better than 1 μm) and granularity. With the addition of high resolution tracking external to the emulsion stacks, computer-aided scanning has substantially enhanced the analysing power of this technique. Two experiments have recently reported results based on data recorded many years ago.

The E-531[13] experiment employed an emulsion in the ν beam at FNAL. The group recently published 58 D^0 decays and 47 decays of charged charm particles, among them 6 unique D_s^+, 13 unique Λ_c, and 28 decays that are consistent with D^+, but also compatible with the kinematics of D_s^+, and/or Λ_c^+ decay. This Λ_c^+ and D_s^+ contamination, which is estimated from a fit to the decay time distribution to be 4.8 ± 5.0 events, is responsible for the larger error in the D^+ lifetime. A likelihood fit to the observed decay time distributions results in

$$\tau(D^0) = 4.3 \pm^{0.7}_{0.5} \pm^{0.1}_{0.2} \cdot 10^{-13}\ \text{s} \qquad \text{and} \qquad \tau(D^+) = 11.1 \pm^{4.4}_{2.9} \cdot 10^{-13}\ \text{s}.$$

The lifetime ratio is calculated to be $2.6\pm^{1.1}_{0.8}$. The average lifetimes of the fitted D_s^+ and Λ_c^+ decays are $2.6 \pm^{1.6}_{1.1} \cdot 10^{-13}$ s and $2.0 \pm^{0.7}_{0.6} \cdot 10^{-13}$ s, respectively.

The WA-58[14] group exposed a thin emulsion target to a photon beam at the CERN SPS. Their results, listed in Tables III and IV, are based on 45 events containing 27 D^+, 44 D^0, and 11 Λ_c^+ decays. While the observation of two secondary vertices in most of the events greatly enhances the purity of the charm selection, the identification of individual decay modes remains difficult, partially due to the limited mass resolution and particle identification of the Omega spectrometer. Only

8 D^+, 8 D^0, and 2 Λ_c^+ decays are identified by unambiguous fits to decay modes involving two or more charged and no neutral secondaries.

3.2 Bubble Chamber Experiments

The use of bubble chambers as active targets has the advantage that within a small fiducial volume, tracks can be accurately measured and clearly associated with the production or decay vertices. In the small bubble chamber LEBC, the single track resolution has been pushed to a few μm, by the introduction of laser optics, by improved HPD measuring machines, and by optimum operating conditions resulting in small bubble diameters and high bubble density. At Fermilab, holography is being used in the 15 feet bubble chamber.

The NA-27 group used the hydrogen bubble chamber, LEBC, in the European Hybrid Spectrometer and has recently completed the analysis of data recorded in a 360 GeV π^- beam.[18] The experimenters rely on the excellent vertex resolution and picture quality to select a sample of clean charm decays. They apply several different techniques to derive lifetime estimates for charm particle decays that cannot be constrained by kinematics. In particular, they have reduced the model dependence by combining the distribution of decay lengths[19] measured in the plane transverse to the beam with the measured exponential form of the transverse momentum distribution. The transverse decay length distributions for the neutral and charged charm sample are shown in Figure 2.

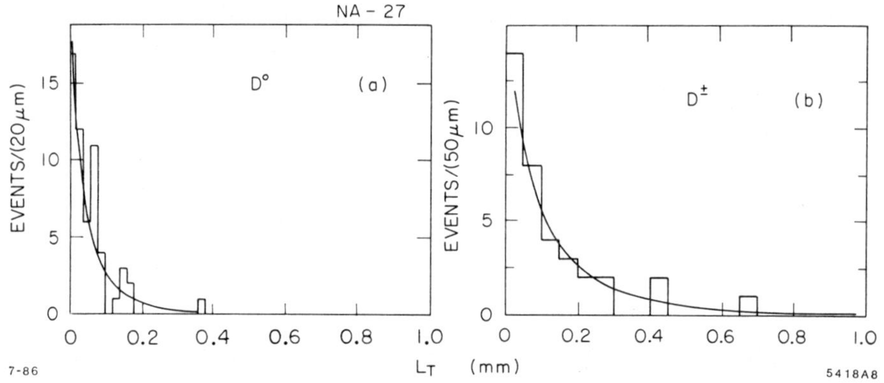

Fig. 2. NA-27: Transverse decay length distributions for selected D meson decays.

The length l_T has been corrected, event by event, for the minimum detectable length. A maximum likelihood fit gives

$$\tau(D^0) = 4.1 \pm^{0.7}_{0.6} \cdot 10^{-13}\ \text{s} \qquad \text{and} \qquad \tau(D^+) = 10.7 \pm^{2.8}_{1.8} \cdot 10^{-13}\ \text{s}.$$

In the D^+ sample, there is no evidence for a short-lived component due to D_s^+ or Λ_c^+ contamination. The ratio of D^+ to D^0 lifetimes is $2.6 \pm^{0.8}_{0.6}$.

The NA-27 has recently reported a result[20] on the Λ_c^+ lifetime, based on nine three-prong vertices that have been unambiguously fitted to Cabibbo-favoured decay modes, $\tau(\Lambda_c^+) = 1.2 \pm^{0.5}_{0.3} \cdot 10^{-13}$ s.

3.3 Experiments with Silicon Vertex Detectors

The use of Silicon as an active target and as a high resolution tracking device was pioneered by two groups at CERN. The NA-1 group[21,22] uses a target made of 40 silicon wafers, 300 μm thick and spaced by 100 μm, to detect multiple vertices in an event with tracks reconstructed in the downstream spectrometer. Since there are two charm decays per event, the association of the decay length and the charged decay secondaries often remains ambiguous. This problem is overcome by selecting exclusive $D^0\overline{D}^0$ production and $D^{*+} \to D^0\pi^+$ decays.

The NA-11 group was the first to use silicon microstrips to reconstruct secondary vertices and to demonstrate $>$ 99% efficiency and single track resolution of 5 μm per plane.[23] A system of on-line processors was used to trigger on prompt electrons or more than one kaon, detected by a calorimeter and Cerenkov counters downstream. Recently, the group has presented a sample of 69 semi-leptonic decays $D^+ \to \overline{K}^{*0}(890)\ e^+\nu_e$.[25] The decay times are estimated using the observed invariant mass and momentum, and corrected for the minimum detectable decay time compatible with the vertex cuts. The resulting lifetime of $\tau(D^+) = 11.2 \pm^{1.6}_{1.3} \pm 0.8 \cdot 10^{-13}$ s compares well with the earlier result based on 28 $D^+ \to K^-\pi^+\pi^+$ decays.

In an attempt to detect secondary vertices at the trigger level, the same group, under the label NA-32, installed an active target of 14 finely segmented silicon counters.[26] While the on-line charm selection did not produce satisfactory results, a total of $38 \cdot 10^6$ were recorded with an interaction trigger in a 200 GeV hadron beam. At present 11 million π^- and 5 million K^- interactions have been analysed resulting in 98 fully reconstructed D decays. Two or more decay tracks are required to form a vertex that is separated from the interaction point by a distance of at least 3 mm. The D momentum must point back to the production point to within a few μm. Preliminary data, shown in Figure 3, demonstrate the cleanliness of the samples. There are 46 D^+ and 52 D^0 decays, above a background of 4 and 10 events, respectively. To correct for detection losses at short decay distances the measured decay times are corrected for the smallest detectable decay time, $t_{corr} = t - t_{min}$, compatible with the selection criteria. Maximum likelihood fits to the background subtracted time distributions yield

$$\tau(D^0) = 3.9 \pm^{0.6}_{0.5} \cdot 10^{-13} \text{ s} \quad \text{and} \quad \tau(D^+) = 9.8 \pm^{1.9}_{1.5} \cdot 10^{-13} \text{ s.}$$

The systematic errors are estimated to be substantially smaller than the statistical errors quoted. The measurements translate into a lifetime ratio of $\tau(D^+)/\tau(D^0) = 2.5 \pm 0.6$.

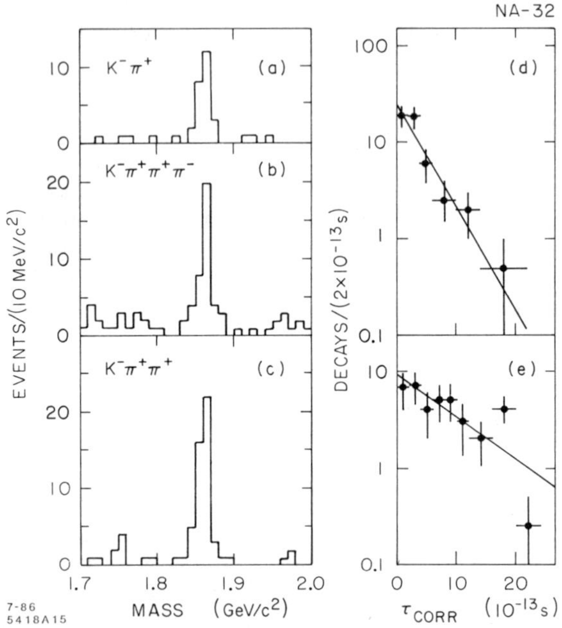

Fig. 3. NA-32: Effective mass and corrected decay time distributions for the selected D mesons.

The NA-32 group[27] has applied the same analysis to search for the decay $D_s^+ \to K^-K^+\pi^+$ and finds 12 events above a background of less than one. The mass is 1972.8±2.1 MeV/c^2. Due to ambiguities in the kaon and proton identification a few events are compatible with the decay modes $D^+ \to K^-\pi^+\pi^+$ or $\Lambda_c^+ \to K^-p\pi^+$. They are excluded from the sample. A fit to the corrected decay time distribution gives $\tau(D_s^+) = 3.4 \pm^{1.3}_{0.9} \cdot 10^{-13}$ s. If we combine these 12 decays with the 9 decays observed by NA-11 in the same detector the weighted average is

$$\tau(D_s^+) = 3.3 \pm^{1.0}_{0.6} \cdot 10^{-13} \text{ s}.$$

In 1985 the NA-32 collaboration[28] improved the resolution and granularity of the vertex detector by the addition of two CCDs at a distance of 10 mm and 20 mm from a 2.5 mm thick Cu target, that was exposed to a 230 GeV negative hadron beam. The trigger required at least two particles without a signal in the threshold Cerenkov counters, thus increasing the D_s^+ and Λ_c signals by about a factor of 12. This is the first use of Charged Coupled Devices as tracking detectors for minimum ionizing particles. At present roughly 20% of the 16 million triggers have been processed and the results are presented in Figure 4. Figure 4(a) shows a decay $\Lambda_c^+ \to pK^-\pi^+$ projected onto a plane parallel to the CCDs. With an incident intensity of 10^6 particles per 2.5 s beam pulse there are on the average 2 hits/mm^2, compared to a cell density of 2100/mm^2 in the active area of 8.8×2.4 mm^2. The primary vertex is determined with a precision of 2 μm in the plane transverse and 60 μm along the direction of the incoming beam. A total of 14 Λ_c decays have been observed (including one background event), their masses are peaked at 2285.6 ± 1.1 MeV/c^2. Three decays with ambiguous particle identification are compatible with D^+ or D_s^+ decay to $KK\pi$ and they are excluded from the lifetime determination. A fit gives

$$\tau(\Lambda_c^+) = 1.4 \pm^{0.5}_{0.3} \pm 0.3 \cdot 10^{-13} \text{ s},$$

where the systematic error reflects the sensitivity of the fit to changes in the event selection and assumed position error.

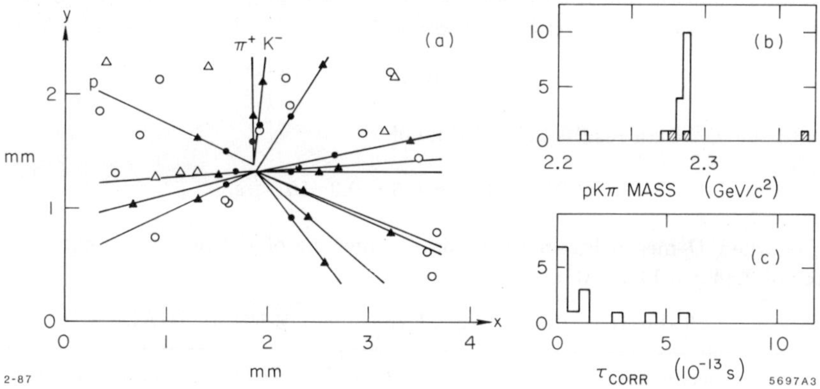

Fig. 4. NA-32: Measurement of the Λ_c lifetime, (a) display of a single event detected in the CCDs in the projection transverse to the beam, (b) effective mass and (c) decay time distribution. Hits in the first and second CCD are marked as triangles and circles, respectively. The open symbols mark hits that are not associated with this event. They are due to additional beam tracks passing during the active time of the CCDs.

Over the past six months, the E-691 group from Fermilab has reported lifetime measurements with high statistical and systematic accuracy.[29] The present analysis is based on up to 45% of a total 10^8 interactions recorded by the Tagged Photon Spectrometer. Three triplets of Silicon strip detectors with 50 μm spacing and digital read-out were installed to improve the charged particle tracking close to the target. The event selection is remarkably straightforward and designed to minimize systematic errors in the determination of the lifetimes. (1) Tracks from the decay of a charm particle are required to form a good secondary vertex, all other tracks are used to form the primary vertex. (2) The impact parameter of the reconstructed charm candidate relative to the primary vertex is not to exceed 80 μm. (3) The decay path is required to be larger than l_{min}. This distance is chosen to be typically 6-10 times the resolution σ_z, in order to reduce the background for each decay mode to an acceptable level. The proper time is calculated using the measured momentum and the distance from l_{min} to the decay vertex. (4) The particle masses had to be consistent with the Cerenkov counter pulse heights. The number of signal and background events, and the l_{min} cut are given in Table II. There are three independent samples for the D^0 and one for the D^+, their statistics and the cleanliness are remarkable. The decay time distributions are fit to a sum of signal and background. The fit takes into account the resolution, acceptance, detection efficiency and the measured background distributions. The three D^0 samples are statistically independent, they have different corrections and background. The fact that all three samples agree provides a check on the systematic uncertainties. The background subtraction for the two D^* modes is negligible; for the decay $D^0 \to K^-\pi^+$ it causes a shift by 0.05 ps. A global fit to all three D^0 subsamples gives

$$\tau(D^0) = 4.35 \pm 0.15 \pm 0.10 \cdot 10^{-13} \text{ s.}$$

A fit to the D^+ data results in a lifetime of

$$\tau(D^+) = 10.6 \pm 0.5 \pm 0.3 \cdot 10^{-13} \text{ s.}$$

The charged D meson lifetime exceeds the lifetime of the neutral D meson by a factor of $2.44 \pm 0.14 \pm 0.08$.

Table II: Results from Experiment E-691 at FNAL

Decay Mode	Vertex Cut ℓ_{min}/σ_z	Decays	Background	Lifetime (10^{-13} s)
$D^o \to K^-\pi^+$	8	746	450 ± 10	4.3 ± 0.2
$D^{*+} \to D^o\pi^+ \to K^-\pi^+\pi^+$	5	385	16 ± 2	4.6 ± 0.3
$D^{*+} \to D^o\pi^+ \to K^-\pi^+\pi^+\pi^-\pi^+$	7	228	25 ± 3	4.0 ± 0.3
$D^+ \to K^-\pi^+\pi^+$	10	969	383 ± 9	10.6 ± 0.5
$D_s^+ \to \phi\pi^+$	6	61	21 ± 3	$4.9 \pm^{0.8}_{0.7}$
$D_s^+ \to \overline{K}^{*o}K^+$	10	38	15 ± 3	$4.6 \pm^{1.0}_{0.8}$
$\Lambda_c^+ \to pK^-\pi^+$	6	65	100 ± 10	$2.0 \pm^{0.5}_{0.4}$

The same data sample was used to study the D_s^+ lifetime.[30] The D_s^+ mesons were identified by two different decay modes, $D_s^+ \to \phi\pi^+ \to K^-K^+\pi^+$ and $D_s^+ \to$

$\overline{K}^{*o} K^+ \to K^- K^+ \pi^+$. The mass spectra and corrected decay time distributions for the two sample are shown in Figure 5. There are two well separated mass peaks, one from the Cabibbo-suppressed decay of the D^+, the other from the decay of the D_s^+. The decay time distributions include only decays within the mass region 1.953 – 1.985 GeV/c^2. The number of signal and background events are given in Table II. A maximum likelihood fit to the total sample of 99 D_s^+ decays gives a mean lifetime of

$$\tau(D_s^+) = 4.8 \pm^{0.6}_{0.5} \pm 0.2 \cdot 10^{-13} \text{ s}.$$

The main contributions to the systematic error arise from the corrections for efficiency and resolution, –0.050±0.017 ps, and from the background subtraction, ±0.12 ps. Decays $D^+ \to K^-\pi^+\pi^+$ and $\Lambda_c \to pK^-\pi^+$ are estimated to contribute less than one background event to the total D_s^+ sample.

At this conference, the E-691 group[31] presented the first, still preliminary results on the lifetime of the Λ_c^+ (cdu) baryon, based on 50% of the total data sample. The analysis follows the same procedure outlined above for charm mesons. The results, based on the decay mode $\Lambda_c^+ \to pK^-\pi^+$, are given in Table II and in Figure 6. The background is substantially larger than for the charm meson decays, because of the smaller production cross section and the shorter lifetime. The systematic error, quoted as $\pm 0.3 \cdot 10^{-13}$ s, is still under study.

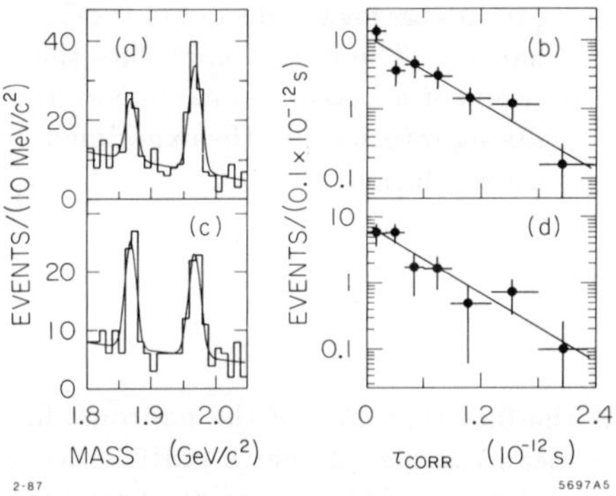

Fig. 5. E-691: Effective mass and decay time distribution for the decays $D_s^+ \to \phi\pi^+$ and $D_s^+ \to \overline{K}^{*o} K^+$.

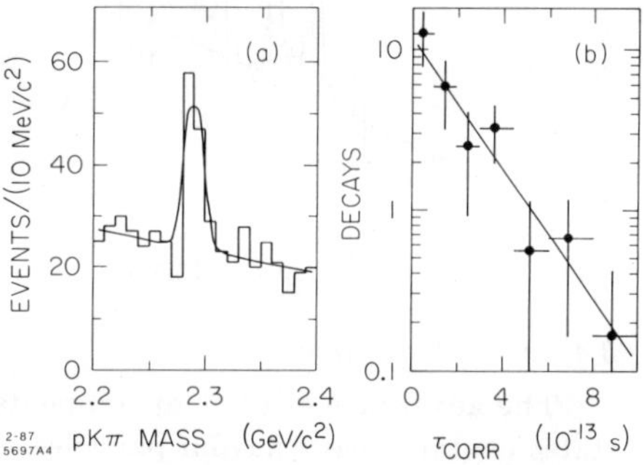

Fig. 6. E-691: Effective mass and decay time distribution for selected Λ_c^+ decays.

The E-400 group at Fermilab presented a first result on the production and lifetime of the Ξ_c^+ (csu) baryon at this conference.[32] The data were recorded in a high energy neutron beam by a large spectrometer with two magnets and three Cerenkov counters, placed downstream of a segmented target and a vertex detector of Silicon wafers and 9 planes of MWPC with 0.25 mm wire spacing. The resolution of the impact parameter was typically 60 μm, the average error on the longitudinal position of the interaction point was 1 mm. Figure 7 shows the $\Lambda^o K^-\pi^+\pi^-$ effective mass for candidates for the decay of the Ξ_c^+ baryon. For these candidates, the $K\pi\pi$ vertex is required to be separated by 3σ from the primary vertex and the Λ^o momentum has to exceed the momentum of both pions and point back to the secondary vertex. The mass distribution shows two narrow peaks, separated by 72 ± 8 MeV/c^2. There are 32 ± 8 and 27 ± 11 events in two peaks. They are interpreted as two Cabibbo-favoured decays, $\Xi_c^+ \to \Lambda^o K^-\pi^+\pi^-$ and $\Xi_c^+ \to \Sigma^o K^-\pi^+\pi^-$. The separation of the two peaks corresponds to the energy of the missing photon from the decay $\Sigma^o \to \Lambda^o\gamma$. A maximum likelihood for a sum of a polynomial background and two Gaussian peaks yields a Ξ_c^+ mass of 2448 ± 5 MeV/c^2. The uncertainty of the absolute mass scale is estimated to be 30 MeV/c^2. The average lifetime for all decays was determined from a comparison of signal and background events with Monte Carlo simulations, the result,

$$\tau(\Xi_c^+) = 6.2 \pm^{1.8}_{1.6} \cdot 10^{-13} \text{ s},$$

compares well with an earlier measurement from the CERN hyperon experiment.[33]

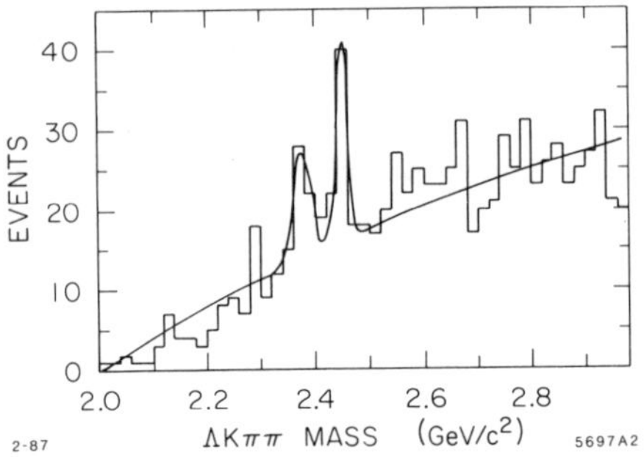

Fig. 7. E-400: Effective mass of candidates for the decays of the Ξ_c^+ baryon. The curve represents the result of a maximum likelihood fit taking into account the experimental resolution.

3.4 e^+e^- Experiments

The advantage of e^+e^- experiments is the fact that 45% of the hadronic final states contain heavy flavour particles, and clean samples of charm particle decays can be obtained on the basis of kinematics alone, avoiding losses at short decay distances.

The DELCO group has published[34] a measurement of the D^0 lifetime based on the measurement of the impact parameter of the K^- and π^+ relative to the beam centre that is monitored by a set of four electrodes placed inside the vacuum chamber. A multi-cell Cerenkov counter and a very loose cut on the $D^{*+} - D^0$ mass difference are employed to select the decay modes $D^0 \to K^-\pi^+ +$ neutrals. The average impact parameter of 151.7 ± 42.5 μm translates to a lifetime $\tau(D^0) = 4.6 \pm 1.5 \pm^{0.7}_{0.6} \cdot 10^{-13}$ s. Systematic studies show that the measurement is largely bias free and insensitive to small errors in alignment or resolution.

The Mark II collaboration[35] recently published a measurement of the D^0 and D^+ lifetimes, the HRS,[36–38] TASSO,[39,40] CLEO,[41] and ARGUS[42] groups presented preliminary results on D^0, D^+, and D_s^+ lifetimes at recent conferences. Since space is limited and the analyses are similar for these experiments, only the CLEO analysis will be described in detail. The charged particle tracking in the CLEO detector relies on two cylindrical drift chambers, a 10 layer vertex chamber with a 90 μm resolution and a larger volume chamber with 17 layers and 140 μm resolution. The chambers are operated in a 10 kGauss magnetic field and are instrumented to measure drift time and specific ionization. The extrapolation error for a single high momentum track is approximately 100 μm. Candidates for the following charm meson decay modes are selected (a) $D^{*+} \to D^0\pi^+$, $D^0 \to K^-\pi^+$, (b) $D^+ \to K^-\pi^+\pi^+$, and (c) $D_s^+ \to \Phi\pi^+$, $\Phi \to K^-K^+$. The kaons are identified by time-of-flight measurements or by dE/dx in the drift chamber gas. Appropriate cuts on angles, momenta, and masses are applied to select the D^{*+} and Φ decays. The decay point of a charm meson candidate is determined as the fitted intersection of all charged secondaries. In Figure 8 the distributions in effective mass and decay distance are shown for the three decay modes measured by the CLEO collaboration. There are 247 D^0, 317 D^+ and 87 D_s^+ decays above backgrounds of 28, 279, and 54, respectively. Although the measurement of the decay distances is limited by the detector resolution and the beam size, the displacement of the nearly Gaussian distributions to positive values is apparent. The observed distributions are fitted to a sum of two contributions, the charm particle decay distribution and the background distribution. The background contribution is measured separately from candidate decays that pass all the selection criteria except for having an invariant mass outside the signal region. Their decay time distributions are centred on zero (within errors). The results are still preliminary, $\tau(D^0) = 5.0 \pm 0.7 \pm 0.4 \cdot 10^{-13}$ s, $\tau(D^+) = 11.4 \pm 1.6 \pm 1.0 \cdot 10^{-13}$ s, and $\tau(D_s^+) = 4.6 \pm 2.1 \pm 0.5 \cdot 10^{-13}$ s. The systematic errors include the uncertainties in the detector resolution, the background subtraction, and the beam position. The lifetime ratios are

$$\tau(D^+)/\tau(D^0) = 2.3 \pm 0.5 \qquad \text{and} \qquad \tau(D_s^+)/\tau(D^0) = 0.9 \pm 0.5.$$

2.5 *Summary on Charm Particle Lifetimes*

A compilation of lifetime measurements of charm mesons is given in Table III. There are now 16 experiments contributing 2414 D^0, 1743 D^+, and 320 D_s^+ decays, substantially more than a year ago. The improvement is mainly due to the high statistics, high resolution results from the E-691 experiment at Fermilab. In attempting to combine the available information, averages and combined errors have been calculated. All individual results, whether final or preliminary, have been weighted by the inverse square of the fractional error, a recipe that in the limit of perfect resolution and negligible acceptance corrections corresponds to the number of events in the sample, which is the correct weight for an exponential distribution.

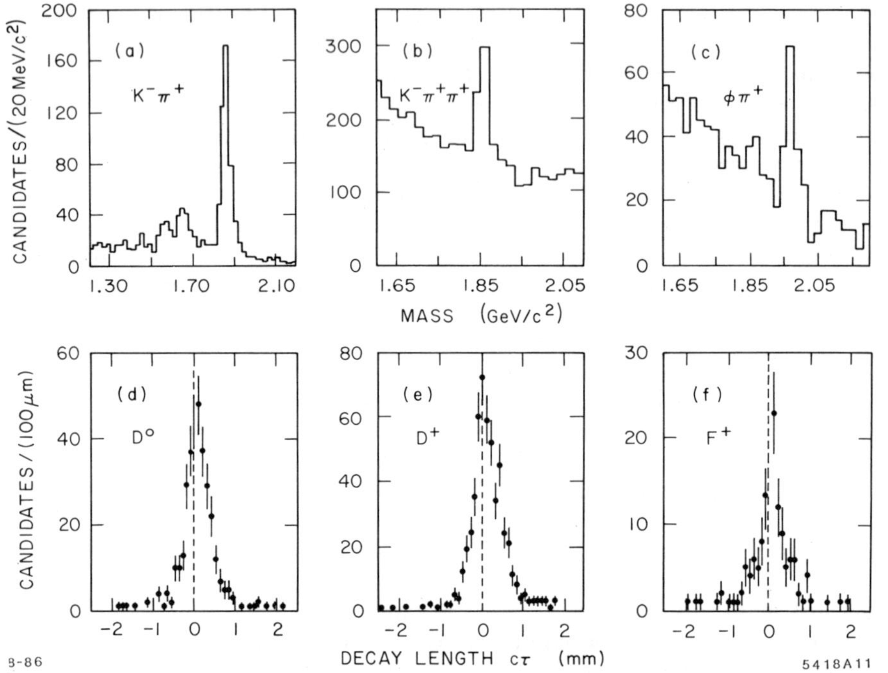

Fig. 8. CLEO: Effective mass and decay length distributions for D^0, D^+ and D_s^+ decays.

(For large Gaussian errors like in e^+e^- experiments, however, the inverse square of the total error is the more appropriate weight.) Using the combined statistical and systematic errors quoted, the best estimate for the lifetime of the charmed mesons are, in units of 10^{-13} s,

$$\tau(D^0) = 4.42\pm^{0.15}_{0.14}\ , \qquad \tau(D^+) = 10.3 \pm 0.4\ , \qquad \tau(D_s^+) = 4.4 \pm^{0.5}_{0.4}\ .$$

The lifetime for the charged and neutral D mesons are clearly different, the average of the ratio measured by individual experiments is

$$\tau(D^+)/\tau(D^0) = 2.27 \pm 0.13\ .$$

The D^+/D^0 lifetime ratio can also be inferred from the semi-leptonic branching ratios. The Mark III group[43] has reported $BR(D^+ \to e^+X) = 0.170 \pm 0.019 \pm 0.007$, and $BR(D^0 \to e^+X) = 0.075 \pm 0.011 \pm 0.004$, and the ratio $BR(D^+ \to e^+X)/BR(D^0 \to e^+X) = 2.3 \pm^{0.5}_{0.4} \pm 0.1$, in agreement with the ratio from the direct measurements. Since the purely leptonic widths of the charm mesons are negligible and the semi-leptonic partial widths of the D^0 and D^+ should be nearly equal (unless Cabibbo-suppressed processes play a major role in D^+ decay), a difference in the lifetime and semi-leptonic branching ratios implies a difference in the non-leptonic width of the two states. Numerous theoretical explanations have been proposed to either enhance the D^0 or suppress the D^+ hadronic width and they were discussed at length at this conference. A resolution of this and other questions will require detailed study of exclusive decay modes.

Table III: Measurements of Lifetimes of Charmed Mesons

		D^+		D^o		D_s^+	
Experiment	Ref.	Decays	$\tau(10^{-13}$ s)	Decays	$\tau(10^{-13}$ s)	Decays	$\tau(10^{-13}$ s)
E-531	13	23	$11.1\pm^{4.4}_{2.9}$	58	$4.3\pm^{0.7}_{0.5}\pm^{0.1}_{0.2}$	6	$2.6\pm^{1.6}_{1.1}$
WA-58	14	27	$5.0\pm^{1.5}_{1.0}\pm1.9$	44	$3.6\pm^{1.2}_{0.8}\pm0.7$		
SHF	15	48	$8.6\pm1.3\pm^{0.7}_{0.3}$	50	$6.1\pm0.9\pm0.3$		
NA-16	16	15	$8.4\pm^{3.5}_{2.2}$	16	$4.1\pm^{1.3}_{1.0}$		
NA-18	17	7	$6.3\pm^{4.9}_{2.3}\pm1.5$	9	$4.1\pm^{2.6}_{1.3}\pm0.5$		
NA-27	18	40	$10.7\pm^{2.8}_{1.8}$	60	$4.1\pm^{0.7}_{0.6}$		
NA-1	21,22	98	$9.5\pm^{3.1}_{1.9}$	51	$4.3\pm^{1.4}_{0.9}$		
NA-11	23,24	28	$10.6\pm^{3.6}_{2.4}\pm1.6$	26	$3.7\pm^{1.0}_{0.7}\pm0.5$	9	$3.2\pm^{1.5}_{1.0}$
	25	69	$11.2\pm^{1.6}_{1.3}\pm0.8$				
NA-32	26,27	42	$9.8\pm^{1.9}_{1.5}$	42	$3.9\pm^{0.6}_{0.5}$	12	$3.4\pm^{1.3}_{0.9}$
E-691	29,30	969	$10.6\pm0.5\pm0.3$	1360	$4.35\pm0.15\pm0.10$	99	$4.8\pm^{0.6}_{0.5}\pm0.2$
DELCO	34				$4.6\pm1.5\pm^{0.7}_{0.6}$		
MKII	35	16	$8.9\pm^{3.8}_{2.7}\pm1.3$	66	$4.7\pm^{0.9}_{0.8}\pm0.5$		
HRS	36,37,38	114	$8.1\pm1.2\pm1.6$	53	$4.2\pm0.9\pm0.6$	13	$3.5\pm^{2.4}_{1.8}\pm0.9$
TASSO	39,40			13	$4.3\pm^{2.0}_{1.4}\pm0.8$	7	$3.4\pm^{2.9}_{1.6}\pm0.7$
CLEO	41	247	$11.4\pm1.6\pm0.7$	317	$5.0\pm0.7\pm0.4$	87	$4.6\pm2.2\pm0.5$
ARGUS	42			249	$4.89\pm0.65\pm0.60$	86	$4.4\pm1.4\pm1.0$
Total		1743	$10.29\pm^{0.44}_{0.40}$	2414	$4.42\pm^{0.15}_{0.14}$	319	$4.37\pm^{0.48}_{0.38}$

Measurements of the D_s^+ lifetime have in the past suffered from extremely low statistics and the contamination from D^+ and Λ_c^+ decays. New data from the fixed target experiments E-691 and NA-32 show that the D_s^+ and D^0 lifetimes are comparable.

Among the ten $J^P = 1/2^+$ charmed baryon states predicted by SU_4 only the lowest mass state, Λ_c^+ (cud),* is well established at a mass of 2285.6±1.8 MeV/c^2.[44] Five experiments have contributed to the measurement of the Λ_c^+ lifetime, in particular NA-32 with high resolution data recorded with a set of CCDs. The CERN hy-

* The quark contents of the hyperon states is indicated to explain the nomenclature.

peron experiment WA-62[45] was the first to present evidence for the strange charm baryons Ξ_c^+ (csu) at a mass of 2460 ± 15 MeV/c^2 and Ω_c^0 (css) at 2740 ± 10 MeV/c^2. The experimenters observe a clear shift to positive values in the background subtracted decay length distribution for the decay $\Xi_c^+ \rightarrow \Lambda^0 K^- \pi^+ \pi^+$. A fit yields $\tau(\Xi_c^+) = 4.8 \pm^{2.1}_{1.0} \pm^{2.0}_{1.0} \cdot 10^{-13}$ s. The three reconstructed decays $\Omega_c^0 \rightarrow \Xi^- K^- \pi^+ \pi^+$ have an average decay time of $7.9 \pm 2.8 \pm 2.0 \cdot 10^{-13}$ s. Theoretical estimates for the ratio of the Ξ_c^+ to Λ_c^+ lifetimes vary between two and four.[46]

Table IV: Measurements of Lifetimes of Charmed Baryons

		Λ_c^+		Ξ_c^+		Ω_c^0	
Experiment	Ref.	Decays	$\tau(10^{-13}$ s)	Decays	$\tau(10^{-13}$ s)	Decays	$\tau(10^{-13}$ s)
E-531	13	13	$2.0\pm^{0.7}_{0.6}$				
WA-58	14	11	$2.3\pm^{0.9}_{0.6}\pm0.4$				
NA-27	22	9	$1.2\pm^{0.5}_{0.3}$				
NA-32	28	14	$1.4\pm^{0.5}_{0.3} \pm 0.3$				
E-691	31	65	$2.0\pm^{0.5}_{0.4} \pm 0.3$				
E-400	32			59	$6.2\pm^{1.8}_{1.6}$		
WA-62	33,44			82	$4.8\pm^{2.1}_{1.0}\pm^{2.0}_{1.0}$	3	$7.9 \pm 2.8 \pm 2.0$
Total		112	$1.8\pm^{0.3}_{0.2}$	141	$5.7\pm^{1.6}_{1.1}$	3	7.9 ± 3.4

4. Lifetimes of Beauty Particles

It has been four years since the MAC[47] and Mark II[48] collaborations first reported lifetimes of beauty particles in the range of 10^{-12} sec, substantially longer than anticipated. These first measurements were confirmed by other experiments at PEP and PETRA, and now updates of earlier results with additional data, improved detectors and refinements in the analysis are available. Recent measurements from the four PEP experiments are shown in Figure 9.

The MAC collaboration[49] has completed an analysis based on the total data sample collected at PEP, 30% of which was recorded with a high resolution vertex detector. This device was installed on the outside of a vacuum pipe of 3.5 cm radius, and consists of 6 layers of thin-walled tubes counters and is operated at a pressure of 4 atmospheres. Each tube provides a position measurement with a resolution of 50 μm. The error on the impact parameter was improved from 350 μm for data recorded without the vertex chamber to 90 μm for the data with the vertex chamber. Multiple scattering contributes 360 μm/p(GeV) and 65 μm/p(GeV), respectively, to these errors. Hadronic events containing beauty particles were tagged by a muon or electron with a large transverse momentum, with respect to the thrust axis. The sample consists of 462 events, 152 with the vertex chamber in operation, and is expected to contain 70% $b\bar{b}$ and 16% $c\bar{c}$ events. The impact parameters

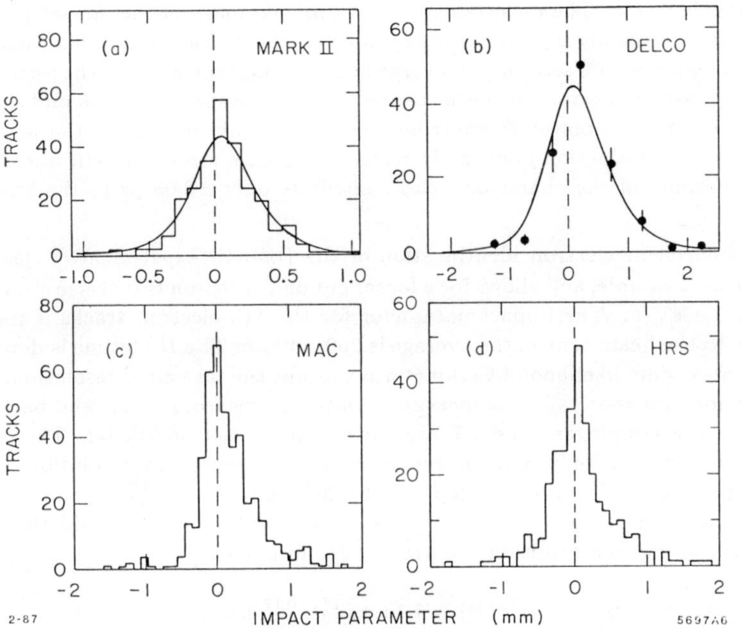

Fig. 9. Measurements of the beauty lifetime: Impact parameter distributions of the four PEP experiments.

of all well-measured tracks with momentum above 0.5 GeV/c were measured in the plane transverse to the beam. There are 1558 and 441 tracks in the two data samples with impact parameters of less than 4 mm and 3 mm, respectively. The B production point was determined from remaining tracks in the event. This reduces the uncertainty in the impact parameter by about a factor of three compared to the measurement relative to the beam centre. The more precise vertex chamber data show not only a positive displacement, but also a clear tail on the positive side. To provide a robust and precise measure of this distribution the means were determined after 10% of the tracks were removed symmetrically from the tails. The trimmed means are 154 ± 20 μm for the early data and 129 ± 14 μm for the vertex chamber data. B lifetimes were obtained by adjusting its value in the Monte Carlo simulation to reproduce the measured trimmed means. The results are $\tau(B) = 1.24 \pm 0.29$ ps and $\tau(B) = 1.35 \pm 0.30$ ps, for the two subsamples, and for the total sample the lifetime is

$$\tau(B) = \{1.29 \pm 0.20(stat) \pm 0.07(syst)\} \cdot (1.00 \pm 0.15) \text{ ps.}$$

The complete electron sample yields 0.92±0.35 ps, while the muon data yield 1.30±0.25 ps. The systematic error has been separated into an additive term and an overall scale factor. The uncertainty in the trimmed mean of the impact parameter distribution is mainly due to the uncertainty in the B fragmentation. This is estimated to cause a 10% error in the overall scale. The uncertainty in the purity of the sample arises from errors in the measured leptonic branching ratios and detection efficiencies, it is estimated to contribute $\pm 7\%$. The uncertainty in the determination of the B production point adds $\pm 7\%$ to the scale error. Present uncertainties in the lifetimes of the charm particles contribute only ± 0.05 ps to the systematic error.

The excellent electron identification of the DELCO experiment[50] leads to a very clean $b\bar{b}$ sample, and allows for a looser cut on the lepton transverse momentum, namely 1 GeV/c. The impact parameter for the 113 electron tracks is measured relative to the beam centre, the average is $259 \pm 49\ \mu$m The B lifetime is determined from a maximum likelihood fit taking into account the measured resolution, including the non-Gaussian tails, the measured contributions from charm and background events in the sample, and the ± 3 mm cut on the impact parameter. The principle systematic errors arise from the uncertainty in the experimental resolution, $\pm^{0.07}_{0.04}$ ps, the fragmentation function and leptonic branching ratios, $\pm^{0.07}_{0.12}$ ps, and the Monte Carlo modelling of the jet axis, $\pm^{0.03}_{0.00}$ ps. The authors choose to add these errors linearly, leading to a result on the average B lifetime of

$$\tau(B) = 1.17 \pm^{0.27}_{0.22} \pm^{0.17}_{0.16} \text{ ps.}$$

The Mark II group[51] has tripled the data sample since its first publication. The analysis is very similar to that of the DELCO group. A final publication with substantially improved resolution is in preparation.

The HRS group[52] has just completed the B lifetime analysis. They measure an average impact parameter of $80 \pm 27\ \mu$m for 301 electrons with high transverse momentum.

The JADE experiment,[53] combines dE/dx information from the drift-chamber with the lead glass signals associated with a track to obtain good electron identification. In addition, the group eliminates three-jet events to reduce the contamination of the electrons by high p_t hadrons and to assure a correct determination of the thrust axis. The lifetime is derived from the average impact parameter by comparison with Monte Carlo simulated events.

The TASSO collaboration[54] has recently presented an update on their B lifetime measurement based on twice the number of events previously published.[55] The $b\bar{b}$ events are selected by a cut on the product of the sphericities of the two jets in a frame that approximates the rest frame of the produced B mesons. This technique gives a higher efficiency but lower purity than the selection of leptons with high transverse momentum. The impact parameter distribution of all tracks in the B enriched sample shows a marked excess at positive values compared to the B depleted sample. The average B lifetime is determined by comparing the average impact parameter of 91 ± 17 μm with the Monte Carlo predictions for different $\tau(B)$. The systematic error is dominated by the uncertainty in the purity of the sample.

In addition the TASSO group explored two other methods to determine the decay distance in B decay. For both of these methods no particular cuts were applied to select $b\bar{b}$ events, but the complete hadron sample was included. Consequently, these measurements rely on a Monte Carlo simulation to reproduce the B decay multiplicities and fragmentation as well as the detector resolution and details of the track fitting. The results are still preliminary and should be considered as a check and confirmation of the impact parameter measurement.

The first method selects the best 3-prong vertex in each jet and calculates the decay length using the sphericity axis as the approximate B direction. There are 3106 vertices, the mean decay length is 141 ± 16 μm. A Monte Carlo prediction with the B lifetime and fragmentation function as free parameters is adjusted to fit the measured distribution. The fit yields $\tau(B) = 1.50 \pm^{0.37}_{0.29} \pm 0.29$ ps.

In the second method the decay vertex is defined as the weighted mean intersection of all tracks in each jet with the sphericity axis. In addition to the track error and angle, its rapidity is used as a weight to enhance to the contribution from high momentum tracks. A dipole moment is defined as the weighted distance between the vertices of the two jets. On the average 8 tracks are used per event, with an impact parameter resolution of typically 200 μm. The measured dipole moments of 4874 events have a mean of 328 ± 28 μm. The lifetime is estimated to be $\tau(B) = 1.62 \pm^{0.33}_{0.29} \pm 0.25$ ps.

A compilation of the six recent measurements of the average lifetime of B hadrons produced in e^+e^- annihilation is given in Table V. The experiments agree very well, though the systematic errors remain substantial because of uncertainties in the sample purity, resolution, fragmentation and decay of B hadrons. Many features of the modelling of the hadronic final states are common among the experiments, as are some aspects of detector design and analysis, leading to errors that are not totally independent. The weighted average of the measurements is

$$\tau(B) = 1.17 \pm 0.14 \cdot 10^{-12} \text{ s.}$$

The only information on the lifetimes of individual B mesons has been obtained by the CLEO collaboration.[56] The number of di-lepton events from B-decays translates to a limit $0.48 < \tau(B^0)/\tau(B^+) < 1.9$.

The B lifetime can be related to the Cabibbo-Kobayashi-Maskawa matrix elements V_{cb} and V_{ub} which represent the coupling of the b quark to charged weak current, in particular,

$$1/\tau_b = [A \cdot |V_{cb}|^2 + B \cdot |V_{ub}|^2] \; / \; BR(b \to Xe\nu) \;\cdot\; 10^{14} \text{ sec}^{-1},$$

where the coefficients depend on the quark masses m_u, m_c, m_b and QCD corrections. They have been calculated and reproduce the lepton spectra in semileptonic decays

Table V: Measurements of the Average Lifetimes of Beauty Particles

	MARK II	MAC	DELCO	HRS	JADE	TASSO
Luminosity pb^{-1}	200	220	214	200	63	25
Vertex Chamber Inner Radius (cm)	10.1	4.8	12.1	9.0	40.0	8.1
ID Lepton e	Pb/LA	Pb/Gas	Cerenkov	Pb/Scint.	Pb/Glass	Pb/LA
μ	Fe	Magn. Fe			Fe	Fe
Cuts p (GeV/c)	2.0	2.0	1.0	2.0	2.0	
p_t (GeV/c)	1.5	1.5	1.0	1.5	1.8	
# Leptons	284	561	113	301	113	
Signal Fraction	0.64	0.70	0.79	0.53	0.75	0.30
Resolution (μm)	90	50	150	100	450	140
Impact Parameter (μm)	80±17	129±19	249±49	80±27	282±66	92±17
Lifetime (10^{-12}s)	0.85 ± 0.17	1.29 ± 0.20	$1.17 \pm^{0.27}_{0.22}$	$1.02 \pm^{0.41}_{0.37}$	$1.80 \pm^{0.51}_{0.30}$	1.57 ± 0.32
Syst. Error	±0.21	±0.17	$\pm^{0.17}_{0.16}$		±0.40	$\pm^{0.37}_{0.34}$

of charm and beauty mesons, $A = 0.58$ and $B = 1.18$.[57] Using the average of the semi-leptonic branching ratios of B mesons of $(11.8 \pm 0.3 \pm 0.6)\%$[58] one obtains

$$\tau_b = [4.9 \cdot |V_{cb}|^2 + 10.0 \cdot |V_{ub}|^2]^{-1} \cdot 10^{-14} \text{ sec.}$$

In the limit of no $b \to u$ transitions one obtains

$$|V_{cb}| = 0.044 \pm 0.003 \pm 0.005,$$

where the first error quoted represents the experimental, the second the theoretical uncertainties. Based on the conservative limit on the $b \to u$ transitions presented here by the CLEO group, we obtain

$$|V_{ub}| < 0.012 \qquad (90\% \text{ C.L.}).$$

With this additional input and the unitarity condition, the absolute values of all elements of the Cabibbo-Kobayashi-Maskawamatrix can be determined or severely constrained. In fact, the matrix becomes almost diagonal, and thus there is very little mixing between the second and the third generations of quarks.[59] The small value of $|V_{cb}|$ imposes constraints on the top mass, the ratio ϵ'/ϵ in K^0 decay and CP violation in beauty meson decay.

4. Conclusion

The present status of lifetime measurements of heavy flavour particles can be summarized as follows:

- Lifetimes of different charm particles are different,

$$\tau(D^+) > \tau(D^0) \geq \tau(\Lambda_c).$$

This observation indicates either problems with the naive parton description including short distance QCD effects or the need for W exchange and annihilation diagrams, or both. There is new information on many exclusive decays available from the Mark III and ARGUS experiments. In particular, the observation of $D^0 \to K_s\Phi$ supports contributions from W exchange, while the relatively large branching ratios for $D^0 \to \overline{K}^0\pi^0$ and $D^+ \to \Phi\pi^+$ suggest the absence of colour suppression. The large rate of $D^+ \to \overline{K}^0 K^+$ relative to $\overline{K}^0\pi^+$ could be explained by interference in D^+ decay and may be contributing to its reduced hadronic width.

- Six measurements at e^+e^- storage rings agree on an average lifetime of the B hadrons produced of $(1.17 \pm^0.14)$ ps. There is still only one directly observed, hadro-produced $B^-\overline{B}^o$ event.[60]

In summary, measurements of charm particle lifetimes have substantially improved over the last year, but orders of magnitude more data are needed to study lifetime of individual charm baryons and beauty mesons. There are many experiments presently under way, NA-14 and NA-32 at CERN, E-653, E-687, E-690 and E-769 at Fermilab, and ARGUS at DESY and CLEO at Cornell. At SLC and LEP substantial production rates and high precision vertex detectors will become available. Thus, before the end of this decade, measurements of heavy quark lifetimes and couplings, possibly including the top, can be expected. It is an enormous effort, but it has been and will continue to be fun!

Acknowledgements

I should like to thank many of my friends and colleagues who contributed the results of many years of hard work to this review.

References

1. M. Kobayashi, T. Maskawa, Prog. Theor. Phys. 652 (1973) 49.
2. S. Petrera and G. Romano, Nucl. Instrum. Methods 174 (1980) 61.
3. R. Sidwell, N.W. Reay and N.R. Stanton, Ann. Rev. Nucl. Part. Sci. 33 (1983) 539.
4. V. Lüth, *Proceeding of the Int. Conference on Physics in Collision V,* Autun, France (1985).
5. C. Albajar *et al.*, CERN-EP 86-81, submitted to Phys. Lett. (1986).
6. D.M. Ritson, *Proceedings of the 23rd Int. Conference on High Energy Physics,* Berkeley, CA (1986).

7. J. Jaros, *Proceedings of the Int. Conference on Physics in Collision*, Santa Cruz, CA (1984).
8. D. Blockus, private communication.
9. S.E. Csorna et al., Contribution to the Int. Conference on High Energy Physics, Berkeley, July 16-23, 1986.
10. P. Padley, contribution to this Conference.
11. P. Burchat, in *Proceedings of the Int. Conference on High Energy Physics*, Berkeley, CA (1986).
12. Particle Data Group, M. Aguilar-Benitez *et al.*, Phys. Lett. 170B (1986) 1.
13. N. Ushida *et al.*, Phys. Rev. Lett. 56 (1986) 1767 and ibid. 56 (1986) 1771.
14. M. Adamovich *et al.*, CERN/EP 86-77, and V. Castillo Gimenez, Thesis Doctoral, Universitad de Valencia (1986).
15. K. Abe *et al.*, Phys. Rev. D33 (1985) 1.
16. M. Aguilar-Benitez *et al.*, Phys. Lett. 122B (1983) 312.
17. A. Badertscher *et al.*, Phys. Lett. 123B (1982) 471.
18. M. Aguilar-Benitez *et al.*, CERN/EP 86-167, submitted to Z. Phys. (1986).
19. K. Roberts, Ph.D. Thesis, University of Liverpool (1986).
20. M. Aguilar-Benitez *et al.*, CERN EP/87-17 (1987).
21. E. Albini *et al.*, Phys. Lett. 110B (1982) 339.
22. S.R. Amendolia *et al.*, Paper submitted to the Int. Conference on High Energy Physics, Berkeley, July 17-23, 1986.
23. R. Bailey *et al.*, Zeitsch. Physik C28 (1985) 357.
24. E. Belau *et al.*, Contribution to Int. Conference on Hadron Spectroscopy, College Park, MD (1985).
25. H. Palka *et al.*, Contribution to Int. Conference on High Energy Physics, Berkeley, July 17-23, 1986.
26. H. Becker *et al.*, Contribution to Int. Conference on High Energy Physics, Berkeley, July 17-23, 1986.
27. H. Becker *et al.*, Phys. Lett. 184B (1987) 277.
28. S. Barlag *et al.*, Phys. Lett. 184B (1987) 283.
29. J.C. dos Anjos *et al.*, Phys. Rev. Lett. 58 (1987) 311; and M. Witherell, presentation at this Conference.
30. J.C. dos Anjos *et al.*, FERMILAB-PUB-87/29E (1987).
31. L. Cremaldi, presentation at this Conference.
32. P. Coteus, presentation at this Conference, COLO-HEP-140 (1986).
33. S. Biagi *et al.*, Phys. Lett. 122B (1983) 455.

34. H. Yamamoto *et al.*, Phys. Rev. D22 (1985) 2901.

35. L. Gladney *et al.*, Phys. Rev. D34 (1986) 2601.

36. S. Abachi *et al.*, ANL-HEP-CP-8673, Contribution to the Int. Conference on High Energy Physics, Berkeley, July 17-23, 1986.

37. S. Abachi *et al.*, ANL-HEP-CP-8662, Contribution to the Int. Conference on High Energy Physics, Berkeley, July 17-23, 1986.

38. S. Abachi *et al.*, ANL-HEP-CP-8660, Contribution to the Int. Conference on High Energy Physics, Berkeley, July 17-23, 1986.

39. M. Althoff *et al.*, Zeitsch. Physik C32 (1986) 1986.

40. W. Braunschweig *et al.*, RAL 86-109 (1986).

41. S.E. Csorna *et al.*, Cornell University CLNS 86/747, Contribution to the Int. Conference on High Energy Physics, Berkeley, July 16-23,1986.

42. P. Padley, presentation at this Conference.

43. R.M. Baltrusaitis *et al.*, Phys. Rev. Lett. 54 (1985) 1976.

44. Particle Data Group, Rev. Mod. Phys. 56 (1984) 1.

45. S.F. Biagi *et al.*, Phys. Lett. 122B (1983) 455, and 150B (1985) 230.

46. J.G. Körner *et al.*, Zeitsch. Physik C28 (1985) 175.

47. E. Fernandez *et al.*, Phys. Rev. Lett. 51 (1983) 1022.

48. N.S. Lockyer *et al.*, Phys. Rev. Lett. 51 (1983) 1316.

49. W.W. Ash *et al.*, SLAC-PUB 4123, submitted to Phys. Rev. Lett. (1986).

50. D.E. Klem *et al.*, SLAC PUB-4025 (1986), submitted to Phys. Rev. Lett.

51. J. Jaros, Int. Conference on Physics in Collision IV, Santa Cruz, CA (1984).

52. D. Blockus *et al.*, ANL-HEP-PR 86-144 (1987).

53. W. Bartel *et al.*, DESY 86-001 (1986).

54. A. Caldwell, Paper presented at the Int. Conference on High Energy Physics, Berkeley, July 17-23, 1986.

55. M. Althoff *et al.*, Phys. Lett. 149B (1984) 524.

56. D. Kreinick, *Proceedings of the Int. Symposium on production and Decay of Heavy Hadrons,* Heidelberg, Germany (1986).

57. B. Grinstein, M.B. Wiser and N. Isgur, Phys. Rev. Lett. 56 (1986) 298.

58. S. Weseler, *Proceedings of the Int. Symposium on Production and Decay of Heavy Hadrons*, Heidelberg, Germany (1986).

59. K. Kleinknecht, *Proceedings of the Int. Symposium on Production and Decay of Heavy Hadrons,* Heidelberg, Germany (1986).

60. J.P. Albanese *et al.*, Phys. Lett. 158B (1985) 186.

THE MYSTERY OF THE MISSING D'S

Edward H. Thorndike
Department of Physics and Astronomy
University of Rochester
Rochester, New York 14627, U.S.A.

1. THE MYSTERY

From the lepton momentum spectrum in semileptonic B decay we have learned that $\Gamma(b\to uW^-)/\Gamma(b\to cW^-)$ is small, probably less than 0.04. Consequently there should be lots of charmed particles, in particular D's, in B decay. A substantial fraction appears to be missing.

1.1 Charm from $b\to cW^-$

The quark diagrams for B decay are shown in Fig. 1. In the absence of $q\bar{q}$ popping or color mixing, the c quark at the lower vertex will form a D^0 or D^+ meson (diagram a). If $u\bar{u}$ or $d\bar{d}$ pairs are popped between the c quark and the spectator antiquark, then D^0 or D^+ mesons will still be formed, along with extra π's. Should an $s\bar{s}$ pair be popped (diag b), then a D_s meson will appear, along with a kaon. Baryons are known to occur among the B decay products. The most plausible diagram for this (diag c) leads to charmed baryon production. CLEO results support the viewpoint that whenever a baryon-antibaryon pair is produced, one of them is a charmed baryon. The W^- in b decay is expected, from phase space considerations, to materialize as a $\bar{c}s$ pair 15% of the time. This $\bar{c}s$ pair will often form a D_s or D_s* (diag d), though on occasions, with the popping of a $q\bar{q}$ pair (diag e) it will form $\bar{D}$ K. The $\bar{c}$ from the W^- can also combine with the c from the lower vertex to form a ψ or other $c\bar{c}$ bound state (diag f). Finally, the decay can proceed through the penguin diagram (diag g), leading to a final state with neither open nor hidden charm.

Table 1 gives measured and inferred yields of charmed particles in B decay. From it an accounting of the fate of the c quark in $b\to cW^-$ can be attempted. Both CLEO[1] and ARGUS[2] have measured the yields of D^0 and D^+ in B decay. These measurements require a subtraction for $\bar{D}$ produced from $W^-\to\bar{c}s$, which I estimate at 4±2%. CLEO has interpreted its baryon signal in terms of a charmed baryon yield,[3] at 8±3%. Both CLEO[4] and ARGUS[5] measure a ψ yield near 1%. The contribution from unseen $c\bar{c}$ bound states is comparable,[6] leading to a contribution of 2%. Production of D_s at the lower vertex

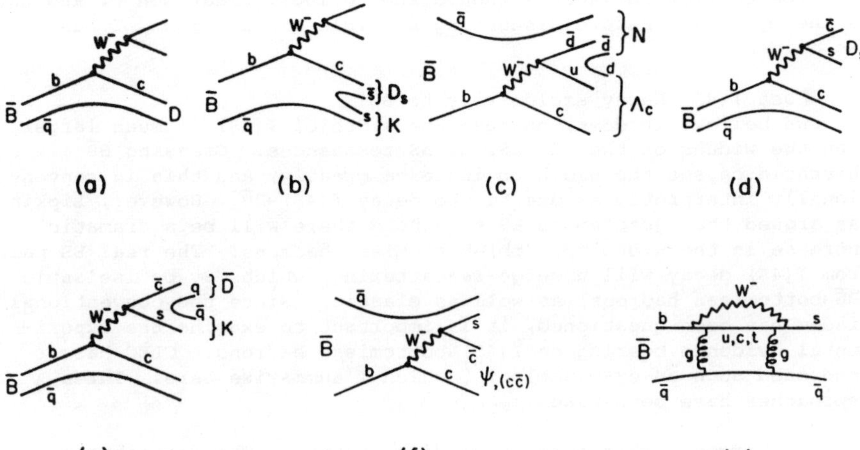

Fig. 1. Quark diagrams for B decay.

Table 1. Measured and inferred yields of charmed particles in B decay. The first three rows are measurements, the next four are estimates (see text). The $B \to D^0X$ and $B \to D^+X$ yields use Mark III branching ratios for D^0 and D^+ decay (Ref. 11). The second error given reflects Mark III's quoted errors.

	CLEO		ARGUS
$B \to D^0X$	39 ± 5 ± 4%		50 ± 11 ± 5%
$B \to D^+X$	17 ± 4 ± 2%		23 ± 9 ± 2%
B → charmed baryon	8 ± 3%		
$B \to c \to D_s$		1 ± 1%	
$B \to (c\bar{c})X$		2 ± 1/2%	
Penguins		1 ± 1%	
less, $b \to cW^-$, $W^- \to \bar{c}s \to \bar{D}$		−4 ± 2%	
Sum	64 ± 8 ± 6%		81 ± 15 ± 7%

(diag b) requires popping of an $s\bar{s}$ pair, and should be suppressed relative to production of Dπ at the lower vertex, itself believed to be small. I estimate 1±1% for this fate for the c quark. Finally, penguins are estimated[7] at 1±1%.

The entries in Table 1 should sum to 100%. They don't, and this is the mystery. Before discussing possible explanations, I must take two asides.

1.2 Does $\Upsilon(4S)$ Decay Exclusively to $B\bar{B}$?

We believe it does, because the width of $\Upsilon(4S)$ is much larger than the widths of the 1S, 2S, or 3S resonances. Crossing $B\bar{B}$ threshold causes the width to increase greatly, and this is conventionally interpreted as due to the decay $\Upsilon(4S)\to B\bar{B}$. However, Lipkin[8] has argued that just above $B\bar{B}$ threshold there will be a dramatic increase in the width for $(b\bar{b})\to$bottomless hadrons. The real $B\bar{B}$ pair from $\Upsilon(4S)$ decay will undergo rescattering, which may be inelastic ($B\bar{B}\to$bottomless hadrons) as well as elastic. Since the conventional wisdom has been questioned, it is important to examine the experimental evidence bearing on $\Upsilon(4S)\to$bottomless hadrons. CLEO has conducted such an examination,[9] which I summarize here. Three approaches have been taken.

1.2.1 Charged particle momentum distribution. The distribution in $x=p/E_{beam}$ from the reaction $\Upsilon(4S)\to B\bar{B}$ cannot extend above $x=0.5$, while the distribution from $\Upsilon(4S)\to$bottomless hadrons can. CLEO's x distribution, shown in Fig. 2, cuts off sharply at $x=0.5$, giving no positive evidence for bottomless decays. To convert this result into a quantitative statement (an upper limit), one requires a model for the bottomless final state. If that final state resembles continuum events, then the data in Fig. 2 yields an upper limit on bottomless $\Upsilon(4S)$ decays of 4%. If the bottomless final state resembles ggg events, then the upper limit deteriorates to 13%. In the unlikely case that the bottomless state has an even softer spectrum, then the limit deteriorates even further.

1.2.2 Comparing dilepton and single-lepton rates. The second method of looking for the decay $\Upsilon(4S)\to$bottomless hadrons involves the yields of leptons and dileptons in $\Upsilon(4S)$ decay. If all leptons come from $B\bar{B}$ events, then the lepton yield N_l may be written

$$N_l = 2\,\bar{b}_B \varepsilon_B\,(1-f)\,N_{4S}, \tag{1}$$

while the yield of dileptons from parallel decays (one lepton from each B) N_{ll} may be written

$$N_{ll} = \overline{b_B^2}\,\varepsilon_B^2\,(1-f)\,N_{4S}, \tag{2}$$

where $\bar{b}_B$ ($\overline{b_B^2}$) is the mean (mean squared) B semileptonic decay branching ratio, ε_B is the efficiency for detecting leptons from B decay, N_{4S} is the number of detected $\Upsilon(4S)$ events and f is the fraction of $\Upsilon(4S)$ decays that lead to states without b-flavored hadrons. Defining

$$\alpha = 4\,N_{4S}N_{ll}/N_l^2, \tag{3}$$

one readily shows that

$$f = 1-(\overline{b_B^2}/(\bar{b}_B)^2)/\alpha. \tag{4}$$

The ratio of the mean square to the square of the mean B semileptonic branching fraction $\overline{b}_B^{\,2}/(\overline{b}_B)^2$ is necessarily greater than 1.0, so an upper limit on α provides a upper limit on f.

$$f > 1-1/\alpha. \qquad (5)$$

If there are leptons from the bottomless decays of $\Upsilon(4S)$, then the above equations require minor modifications. For plausible models of bottomless $\Upsilon(4S)$ decay, these modifications have a very small effect.

CLEO data, combining the 1982 and 1985 $\Upsilon(4S)$ data samples, give an upper limit on α of 1.18, which translates into an upper limit on bottomless $\Upsilon(4S)$ decays of 16%. If one allows for the presence of leptons in the bottomless final state, this limit changes only slightly, to 17%.

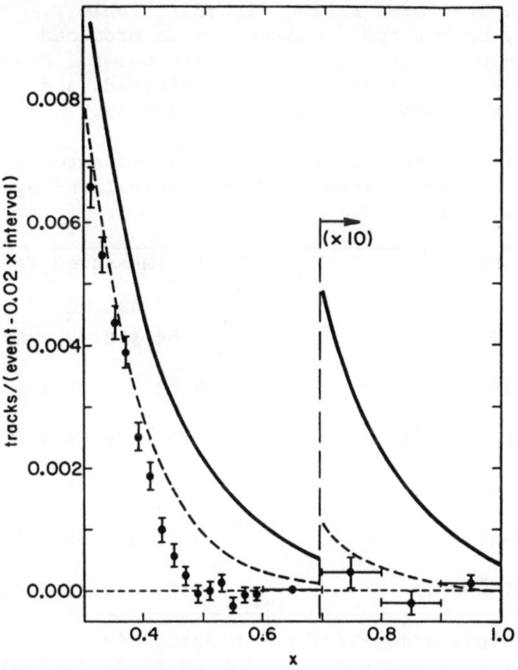

Fig 2. Distribution in $x=p/E_{beam}$ for charged particles from $\Upsilon(4S)\to B\overline{B}$, from CLEO.

1.2.3 <u>Search for $\Upsilon(4S)\to\Upsilon(1S)$ + hadrons</u>. If $\Upsilon(4S)$ decays to bound $b\overline{b}$ states, then $\Upsilon(1S)$ will appear, either as a direct decay product or as a secondary product from some heavier bound $b\overline{b}$ state. CLEO has conducted two searches. In the first, they looked for $\Upsilon(4S)\to\pi^+\pi^-\Upsilon(1S)$; $\Upsilon(1S)\to\mu^+\mu^-$ or e^+e^-. They found no evidence for a signal, and obtained an upper limit of $\Gamma[\Upsilon(4S)\to\pi^+\pi^-\Upsilon(1S)]/\Gamma[\Upsilon(4S)\to\text{all})] < 0.25\%$. In the second search, they looked for $\Upsilon(4S)\to\Upsilon(1S)\pi^+\pi^-X$; $\Upsilon(1S)\to\mu^+\mu^-$. Again they found no evidence of a signal, and obtained an upper limit of 1% on the branching ratio.

1.2.4 Conclusions. There is no evidence for $\Upsilon(4S)$ decays to bottomless states. A model-independent upper limit of 17% has been obtained. If continuum events are a reasonable model for the bottomless states, then an upper limit of 4% is obtained.

1.3 Branching Ratios for $B \to$Kaons.

Heretofore, measured inclusive kaon branching ratios have been sums over K^+ or K^-, and therefore not capable of distinguishing between s and $\bar{s}$ quarks. CLEO has recently developed a technique for measuring the separate branching ratios.[10] They use $B\bar{B}$ events containing identified lepton-kaon pairs. The lepton tags the flavor of one of the B's. Any kaon from that B recoils against the lepton. Kaons from the other B are distributed uniformly in angle with respect to the lepton. By studying lepton-kaon angular correlations, CLEO obtained the branching ratios listed in Table 2.

In b decay both s and $\bar{s}$ quarks can be produced. The excess of s over $\bar{s}$ is due entirely to the decay chain $b \to c \to s$. From the $K^- - K^+$ differences, the $s-\bar{s}$ differences can be inferred, and $b \to c$ determined. CLEO finds $\Gamma(b \to cW^-)/\Gamma(b \to \text{all}) = 0.98 \pm 0.20$.

Table 2. Branching ratios for $B \to$ kaons, as measured by CLEO (Ref. 10). The errors shown are statistical and systematic, respectively.

Branching Ratio	Measured Value
$\Gamma(\bar{B} \to l^- K^- X)/\Gamma(\bar{B} \to l^- X)$	$0.54 \pm 0.07 \pm 0.06$
$\Gamma(\bar{B} \to l^- K^+ X)/\Gamma(\bar{B} \to l^- X)$	$0.10 \pm 0.05 \pm 0.02$
$\Gamma(\bar{B} \to l^- K^0/\bar{K}^0 X)/\Gamma(\bar{B} \to l^- X)$	$0.39 \pm 0.06 \pm 0.04$
$\Gamma(\bar{B} \to K^- X)/\Gamma(\bar{B} \to \text{all})$	$0.66 \pm 0.05 \pm 0.07$
$\Gamma(\bar{B} \to K^+ X)/\Gamma(\bar{B} \to \text{all})$	$0.19 \pm 0.05 \pm 0.02$
$\Gamma(\bar{B} \to K^0/\bar{K}^0 X)/\Gamma(\bar{B} \to \text{all})$	$0.63 \pm 0.06 \pm 0.06$

1.4 Possible Explanations for the Missing D's.

We are now in a position to list possible explanations as to why the entries in Table 1 sum to only 64%, rather than 100%.

1. There could be a huge $b \to u$ contribution. It would have to exceed 20% to reduce the discrepancy in Table 1 to a 1.5 σ effect. A 20% $b \to u$ contribution is not inconsistent with the measurement from kaon yields given in Section 1.3 above, but it is totally inconsistent with the measurement from the lepton spectrum, to be discussed in Section 2.

2. There could be a large contribution from $\Upsilon(4S) \to$bottomless hadrons. Again, it would have to exceed 20% to reduce the discrepancy in Table 1 to a 1.5 σ effect. This figure is not seriously inconsistent with the model-independent limit obtained in Section 1.2 above, but it is inconsistent with the model-dependent limit.

3. CLEO could have made an error in their $B \to D^0 X$ and $B \to D^+ X$ inclusive branching ratios. The error would have to exceed a factor 1.3 to reduce the discrepancy in Table 1 to a 1.5 σ effect.

4. In measuring the inclusive D branching ratios, CLEO uses the $D^0 \to K^- \pi^+$ and $D^+ \to K^- \pi^+ \pi^+$ decay modes. To obtain their numbers, CLEO uses the branching ratios for these modes measured by Mark III.[11] Mark III could have made an error in their measurements. A factor exceeding 1.3 is required, to reduce the Table 1 discrepancy to less than 1.5 σ.

5. Some combination of the preceeding four possibilities might be operating, perhaps with added contributions from: (a) the $b \to c \to D_s$ contribution being larger than estimated (2% would not be out of the question); (b) penguins being larger than estimated (again, 2% is a possibility); and (c) the $W^- \to \bar{c}s\bar{D}$ contribution being smaller than estimated (2% is not unreasonable).

As an example of explanation 5, if the Mark III measurements were high by a factor 1.15, the $b \to u$ contribution were 5%, the $\Upsilon(4S) \to$ bottomless hadrons contribution were 5%, and $b \to c \to D_s$, penguins, and $W^- \to \bar{c}s \to \bar{D}$ together contributed an extra 3%, then the discrepancy of Table 1 would be reduced to a 1.3 σ effect. None of the individual items mentioned would be particularly surprising.

Hadronic decays of B admit too many partial explanations. Were this the only example of "missing D's," we might conclude that there was no mystery. However, semileptonic decays of B show a similar shortage of D's. While the statistical accuracy is not as good, the process is much easier to interpret.

1.5 Charm from $b \to c l \nu$.

CLEO has measured the D^0 yield in $\Upsilon(4S)$ events containing leptons,[1] and determined that (35±9±3.5)% of the semileptonic B decays contain a D^0. This measurement was done with leptons above 1.0 GeV/c; indeed, the preponderance of leptons was above 1.4 GeV/c.

Unfortunately, there are no measurements of the D^+ branching ratio in semileptonic B decay. I therefore use two indirect arguments to arrive at an upper limit for that number.

1. One expects the dominant semileptonic decays to be $\bar{B}^0 \to D^+ l^- \nu$, $\bar{B}^0 \to D^{*+} l^- \nu$, $B^- \to D^0 l^- \nu$ and $B^- \to D^{*0} l^- \nu$. To the extent that this is the case, B^- contributes only D^0, while $\bar{B}^0$ contributes D^0 and D^+. One expects the production of $B^0\bar{B}^0$ events from $\Upsilon(4S)$ to be, if anything, lower than the production of B^+B^- events, since $\bar{B}^0$ is heavier than B^-. Further, one expects the semileptonic decay of the neutral B's to be, if anything, smaller than that of charged B's, since an annihilation diagram can contribute to the hadronic decay of neutral B's. To obtain a conservative upper limit on the number of D^+ from semileptonic B decay, I assume equal production rates and equal semileptonic decay branching ratios for charged and neutral B's. Finally, one expects $B \to D^* l \nu$ to be more prevalent than $B \to D l \nu$ (by a factor 3 from spin considerations; by a factor 3.8 from a quark model computation).[12] Again being conservative, I assume equal frequency of D and D*. With these assumptions, one obtains 0.6 for an upper limit on the D^+/D^0 ratio in semileptonic B decay. A more likely value is 0.45.

2. The kaon branching ratios given in Table 2 show that there are more K^- than $K^0 + \bar{K}^0$ in semileptonic $\bar{B}$ decay, in particular

$\Gamma(\overline{B}\to l^- K^0/\overline{K}^0 X)/\Gamma(\overline{B}\to l^- K^- X) = 0.72\pm0.16$. Since D^0 makes fewer $K^0+\overline{K}^0$ than K^- (the ratio is 0.71), while D^+ makes more $K^0+\overline{K}^0$ than K^- (the ratio is 2.1), this demonstrates an excess of D^0 over D^+ in semileptonic $\overline{B}$ decay. A quantitative treatment yields $D^0/(D^0+D^+) = 0.98\pm0.20$, $D^+/D^0 < 0.52$.

Remaining conservative, I take the least restrictive of the two upper limits for D^+/D^0 in semileptonic B decay, i.e. 0.6. Using that number in conjunction with the CLEO measurement for $B\to D^0 X l\nu$, one obtains $\Gamma(B\to DXl\nu)/\Gamma(B\to Xl\nu) \le (56\pm14\pm6)\%$. To this we must add damn little. Charmed baryons can be ignored, because leptons from $\overline{B}\to\Lambda_c \overline{P} X l\nu$ will be low momentum, due to the high mass of $\Lambda_c \overline{p}$. $B\to KD_s l\nu$ will be small because it involves popping an $s\overline{s}$ pair between c quark and spectator antiquark (unlikely) and will give rise to somewhat low momentum leptons (KD_s is heavier than D). A figure of 1±1% seems reasonable for the $B\to K^- D_s l\nu$ contribution. Because we are dealing with a semileptonic decay, all the processes due to $W^-\to\overline{c}s$ are absent, as are penguins and $\Upsilon(4S)\to$bottomless hadrons.

We find a sum of $<(57\pm14\pm6)\%$, where we expect 100%. There are only three possible explanations.

1. There could be an error in the CLEO $B\to D^0 X l\nu$ branching ratio. An error of a factor 1.3 is required to boost the 57% to within 1.5 σ of 100%.

2. There could be an error in the Mark III $D^0\to K^-\pi^+$ branching ratio, used by CLEO in measuring $B\to D^0 X l\nu$. Again, an error of a factor 1.3 is required to get within 1.5 σ of 100%.

3. There could be a huge $b\to u$ contribution; 20% is required to shrink the 100% figure to within 1.5 σ of 57%.

While explanation 1 is a perfectly reasonable possibility, I have nothing useful to say about it (other than that the measurement should be carefully checked). The other two explanations are examined further in the next two sections.

2. DETERMINING $(b\to u)/(b\to c)$ FROM $B\to Xl\nu$.

Since the u quark is much lighter than the c quark, the decay $B\to X_u l\nu$ is expected to have a stiffer lepton momentum spectrum and a

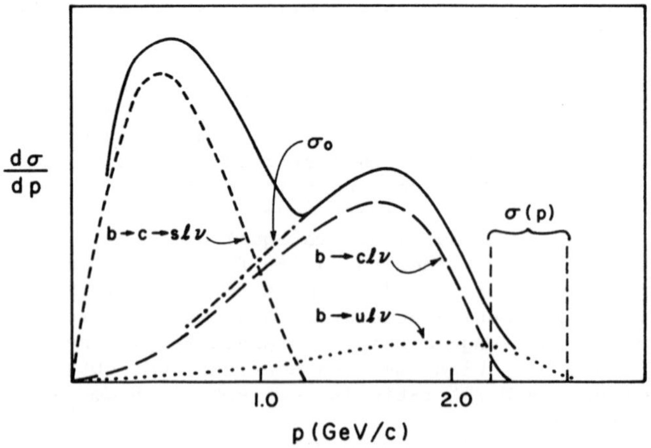

Fig. 3. Components of the lepton spectrum from B decay.

higher end point than the decay $B \to X_c l\nu$. This principle has been exploited by CLEO[13] and CUSB[14] to obtain upper limits on $(b\to u)/(b\to c)$. The method used in the past was to fit the measured lepton spectrum over a wide momentum range to a mix of $b\to ul\nu$, $b\to cl\nu$, and $b\to c\to sl\nu$. This method depends, in a complicated way, on the theoretical models chosen for $B\to X_c l\nu$ and $B\to X_u l\nu$.

2.1 CLEO's New Method and Result

Recently, CLEO has developed a different method[15] of analyzing the lepton spectrum, which has the advantages of less sensitivity to the model for $B\to X_c l\nu$, and simple dependence on the model for $B\to X_u l\nu$. This method is illustrated by Fig. 3, which shows the components of the lepton spectrum due to $b\to ul\nu$, $b\to cl\nu$, and $b\to c\to sl\nu$. The primary spectrum (the sum of $b\to ul\nu$ and $b\to cl\nu$) and the total spectrum (the sum of all three pieces) are also shown. Consider the lepton yield in a momentum interval near the end point of the spectrum (e.g., 2.3 to 2.6 GeV/c). This yield, $\sigma(p)$, may be written

$$\sigma(p) = \sigma_o(b\to u)f_u + \sigma_o(b\to c)f_c. \qquad (6)$$

In this expression, σ_o is the integrated primary lepton yield, $(b\to u)$ is the fraction of the primary yield that comes from $b\to ul\nu$, and f_u is the fraction of the $b\to ul\nu$ spectrum that lies in the chosen momentum interval. $(b\to c)$ and f_c have the corresponding definitions in terms of $b\to cl\nu$. Equation (6) is readily solved to yield

$$(b\to u)/(b\to all) = (\sigma(p)/\sigma_o - f_c)/(f_u - f_c) \cdot \qquad (7)$$

$\sigma(p)$ is obtained directly from data. While σ_o uses a model of the primary lepton spectrum to separate primary and secondary spectra, and to extrapolate to zero momentum, there is negligible model dependence in this procedure.

Three ingredients determine f_c: the shape of the lepton spectrum for $B\to X_c l\nu$ in the B rest frame, the Doppler smearing as the lepton is boosted from the B rest frame to the lab frame, and the smearing due to the momentum resolultion of the detector. The last two items are well understood; the first only partially. The measured lepton momentum spectrum constrains the shape of the lepton spectrum in the B rest frame, and so all models for $B\to X_c l\nu$ which adequately fit the spectrum below 2.2 GeV/c give similar values for f_c for the interval 2.3-2.6 GeV/c. The range of values for f_c obtained by considering many models is taken as the error on f_c, and gives the dependence of this method of analysis on the model for $B\to X_c l\nu$. This model dependence, while significant, is not large.

Using the method just described, and using the momentum interval 2.3 to 2.6 GeV/c, CLEO obtains[15]

$$(b\to u)/(b\to c) = (0.011 \pm 0.014)[0.083/(f_u - 0.002)]$$

The quoted error includes the $B\to X_c l\nu$ model dependence. This value translates into an upper limit of

$$(b\to u)/(b\to c) < 0.031\ [0.083/(f_u - 0.002)]$$

2.2 Determining f_u.

While f_c is largely determined from the measured lepton spectrum below the end point region, f_u must be obtained strictly from theory. The only constraint one has is that the theory for $B \to X_u l\nu$ give sensible results when applied to $B \to X_c l\nu$, where one does have data against which to compare.

I will examine two theoretical models. The first is the spectator quark model, with gluon corrections as calculated by Altarelli et al.[16] This model gives a sum over hadronic final states, and should be valid where those final states are closely spaced. It contains two parameters, m_b, the b quark mass, and P_F, the Fermi momentum factor. The second model is a nonrelativistic quark model (Grinstein, Wise, and Isgur[12]), using wave functions to describe the behavior of quarks within hadrons. This model calculates transitions to specific hadronic final states, and should be valid for widely spaced low lying states. The mistaken impression has been given that these two models give substantially different results. As we shall see, this is not the case.

For $B \to X_c l\nu$, one expects X_c to be dominated by D and D*. The non-relativistic quark model should be valid, but perhaps the spectator quark model should not. The calculated spectra are given in Fig. 4a. For very reasonable parameters (m_b = 5.0 GeV, P_F = 225 MeV/c), spectator quark model gives results indistinguishable from the nonrelativistic quark model. (The spectrum from a stationary free b quark is also shown in Fig. 4a. It differs from the other two.)

For $B \to X_u l\nu$, one does not expect X_u to be dominated by a few resonances. π and ρ will contribute, but so will heavier mesons, and perhaps also multipion nonresonant states. The spectator quark model might not be a good description for hadronic masses below 1.0 GeV,

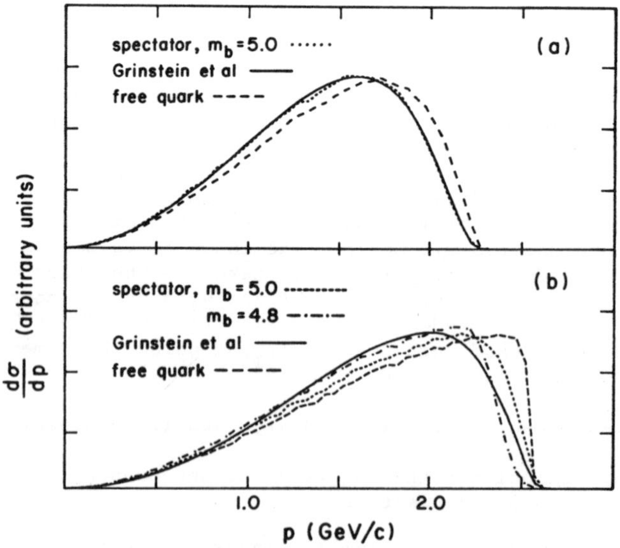

Fig. 4. Calculated spectra for (a) $B \to X_c l\nu$ and (b) $B \to X_u l\nu$.

where π and ρ dominate, but should do fine above 1.0 GeV, where there are many closely spaced states. The parameters that fit $B \to X_c l\nu$ should be appropriate. To be conservative, I will also use a lower b quark mass, 4.8 GeV. The nonrelativistic quark model should do fine for π and ρ, but might work less well for heavy hadronic masses. Grinstein et al. have included in their calculation all mesonic states below 1.6 GeV. It appears that hadronic states above 1.6 GeV cannot be neglected, and might conceivably account for as much as half of the contribution to the lepton spectrum. Figure 4b shows the spectra predicted by the spectator quark model with P_F = 225 MeV/c, m_b = 5.0 GeV, and with P_F = 225 MeV/c, m_b = 4.8 GeV, and the spectrum predicted by the nonrelativistic quark model including all hadronic states up to 1.6 GeV. (For reference, the spectrum from a stationary b quark is also shown. It differs markedly from the other three.)

Values of f_u obtained from the various models are given in Table 3. For the nonrelativistic quark model I consider two extreme cases: There is no contribution from hadronic states with masses above 1.6 GeV; half the contribution comes from hadronic states with masses above 1.6 GeV. For the latter case, the spectator quark model with m_b = 4.8 GeV is used for the hadronic mass region above 1.6 GeV.

I believe the true value of f_u is bracketed by those given in Table 3. In that case, (b→u)/(b→c) is less than 0.06, and cannot explain the missing D's.

Table 3. Values of f_u

Model	f_u
Spectator quark	
with m_b = 5.0 GeV	0.117
with m_b = 4.8 GeV	0.073
Nonrelativistic quark	
assuming no contribution from M_{had} > 1.6 GeV	0.085
assuming 50% contribution from M_{had} > 1.6 GeV	0.050

3. BRANCHING RATIOS FOR $D^0 \to K^-\pi^+$ AND $D^+ \to K^-\pi^+\pi^+$

3.1 Mark II, Lead Glass Wall, and Mark III Measurements.

All three of the above groups obtain their D's by running at ψ"(3770). Mark II[17] and LGW[18] compare the number of times $D^0 \to K^-\pi^+$ and $D^+ \to K^-\pi^+\pi^+$ are reconstructed with cross section for producting ψ", making the assumption that ψ" decays to $D\bar{D}$ 100% of the time. Mark III[11] compares the number of fully reconstructed $D\bar{D}$ events (double tags) with the number of times a single D is reconstructed (single tags), and obtains their results independent of the ψ" production cross section and of any assumptions about ψ" decay.

Measurements by the three groups are compared in Table 4. One notes excellent agreement in the product of production cross section times branching ratio, but disagreement in the cross section, and consequently disagreement in the branching ratio. This disagreement could be explained if ψ" did not decay to $D\bar{D}$ 100% of the time.

Table 4. Comparison of Lead Glass Wall (Ref. 18), Mark II (Ref. 17), and Mark III (Ref. 11) D branching ratio determinations.

	LGW	M_K II	M_K III
$\sigma(D^0)$ (nb)	11.5 ± 2.5	8.0 ± 1.6	4.5 ± 0.5
$\sigma \cdot B(K^-\pi^+)$	0.25 ± 0.05	0.24 ± 0.02	0.25 ± 0.02
$B(K^-\pi^+)$%	2.2 ± 0.6	3.0 ± 0.6	5.6 ± 0.5
$\sigma(D^+)$ (nb)	9.0 ± 2.0	6.0 ± 1.2	4.5 ± 0.5
$\sigma \cdot B(K^-\pi^+\pi^+)$	0.36 ± 0.06	0.38 ± 0.05	0.39 ± 0.03
$B(K^-\pi^+\pi^+)$ (%)	3.9 ± 1.0	6.3 ± 1.5	11.6 ± 2.5

3.2 Does ψ" Decay Exclusively to $D\bar{D}$?

We believe it does, for the same reasons we believe ϒ(4S) decays exclusively to $B\bar{B}$. But Lipkin's arguments[8] are just as applicable here, and in fact were motivated by the discrepancy between Mark III and the earlier D branching ratio measurements. I therefore present two pieces of experimental evidence on the subject.

3.2.1 Comparing dilepton and single lepton rates. The formulas derived for ϒ(4S) decay to bottomless states apply also to the decay of ψ" to charmless states. The relevant equation is

$$f = 1 - [\overline{b_D^{\,2}}/(\bar{b}_D)^2/\alpha \tag{8}$$

The ratio of the mean square to the square of the mean D semileptonic branching ratio can be evaluated, using the measured lifetimes[19] of the charged and neutral D's, and using the mix of $D^0\bar{D}^0$ and D^+D^- from the ψ" decay as predicted by Eichten et al.[20] One obtains $\overline{b_D^{\,2}}/(\bar{b}_D)^2 = 1.13 \pm 0.04$. The only published results on dileptons from ψ" are by the DELCO collaboration.[21] DELCO found the ratio of charged to neutral D semileptonic branching fractions to be greater than 4.3 at 95% confidence level. Present data[19] give a value near 2. This discrepancy could be explained if there were charmless decay of ψ". From Fig. 1 of Ref. 21, one sees that the DELCO data prefer that one of the two D semileptonic branching ratios vanish, which suggests*

*Figure 1 of Ref. 21 plots the semileptonic decay branching ratios for charged and neutral D's against each other, and shows the bands in this space allowed by their single lepton and dilepton measurements. The centers of the bands cross at slightly negative values of one of the two branching ratios, and a computation of $\overline{b^2}/(\bar{b})^2$ at the intersection points yields values that range between 2.1 and 2.7, and average to 2.5. The apparent value for $\overline{b^2}/(\bar{b})^2$, arrived at under the assumption that all ψ" decays are to $D\bar{D}$, is α, as can be seen by setting f=0 in Eq. (8). To be conservative, I take α = 2.0.

that $\alpha > 2$. The error on α is dominated by the 28% error in the number of dileptons. Using $\alpha = 2.0\pm0.6$ in Eq. (8) one obtains $f=0.4\pm0.2$.

3.2.2 Comparison of D semileptonic decay branching ratio, as measured at ψ'' and in continuum. Some measurements of the D semileptonic decay branching ratio using ψ'' as the source of D mesons have assumed that ψ'' always decays to $D\overline{D}$. Allowing for the possibility of charmless ψ'' decays, one has

$$b_{\psi''} = (1-f)\ \overline{b}_D^{\ \psi''}\ , \tag{9}$$

where $b_{\psi''}$ is the apparent D semileptonic decay branching ratio measured at ψ'' under the assumption of no bottomless decays, and $\overline{b}_D^{\ \psi''}$ is the D mean semileptonic decay branching ratio averaged over the $D\overline{D}$ decays of ψ''. The mean semileptonic decay branching ratio for continuum-produced charmed hadrons, $\overline{b}_c^{\ cont}$, will not equal $\overline{b}_D^{\ \psi''}$. Charged D's have a substantially larger semileptonic decay branching ratio than do D^0, D_s and Λ_c, and the mix of these particles differs between the continuum and ψ''. I write $\overline{b}_c^{\ cont} = \kappa\overline{b}_D^{\ \psi''}$, and calculate† κ to be 0.85 ± 0.03. Combining the above expression with Eq. (9), one obtains

$$f = 1 - (b_{\psi''}/\overline{b}_c^{\ cont})\,\kappa\ . \tag{10}$$

The mean D semileptonic decay branching ratio has been measured from the inclusive lepton yield at ψ'' by LGW[22] (7.2 ± 2.8)%, and DELCO[23] (8.0 ± 1.5)%. The weighted mean of these measurements is $b_{\psi''} = (7.8 \pm 1.3)$%. The mean semileptonic decay branching ratio for continuum-produced c-flavored hadrons, $\overline{b}_c^{\ cont}$, has been measured by many groups.[24] The weighted mean of those measurements gives $\overline{b}_c^{\ cont} = (8.9 \pm 0.6)$%. Because the $D^\pm$ semileptonic decay branching ratio is twice as large as those of D^0, D_s, and Λ_c, and because ψ'' contains a larger fraction of $D^\pm$ than does the continuum, one expects a larger semileptonic branching ratio fromn ψ'' than from the continuum, as reflected in κ being significantly less than 1. The measurements show the opposite. Inserting them into Eq. (10), one finds $f = 0.26 \pm 0.14$.

3.2.3 Conclusions. Both methods of looking for charmless decays suggest a positive effect, though neither supplies compelling evidence. Where needed, I will use their weighted average $f=0.3\pm0.1$.

---

†Necessary ingredients are (i) The ratio of the semileptonic decay branching ratios of D^+ and D^0, which I take from the lifetime measurements (Ref. 19) as 2.07 ± 0.23; (ii) the ratios of the semileptonic decay branching ratios of F^+ and Λ_c to D^0, which I take from lifetime measurements (Ref. 19) as ≤ 1.0; (iii) the mix of $D^0\overline{D}^0$ and D^+D^- on the ψ'', which I take from Eichten et al.(Ref. 20) to be $D^0\overline{D}^0{:}D^+D^- = 1.36{:}1$; (iv) the mix of D^+, D^0, F^+, and charmed baryons on the continuum. I take (80 ± 5)% primary production of D/D^*, split 1/4-3/4, and split equally between charged and neutral mesons. The remaining (20 ± 5)% is F and Λ_c, the precise mix being unimportant since their semileptonic decay branching ratios are approximately equal. With this input, I obtain $\kappa_c = 0.85\pm0.003$.

Because charmless decays of ψ'' appear to be present, the Mark II and LGW measurements of branching ratios for $D^0 \to K^- \pi^+$ and $D^+ \to K^- \pi^+ \pi^+$ require corrections.

3.3 Branching Ratios from D* Cross Sections.

The D* production cross sections in e^+e^- collisions have been measured at center of mass energies between 3.9 and 4.3 GeV, by two groups. In an experiment done many years ago at SPEAR, the Mark II collaboration measured[25] the D* production cross section by reconstructing $D^0 \to K^- \pi^+$ and $D^+ \to K^- \pi^+ \pi^+$, and then untangling the final states $D\bar{D}$, $D^*\bar{D}$, and $D^*\bar{D}^*$ by examining the spectrum of the effective mass recoiling against the reconstructed D. In a recently completed analysis, also of data taken at SPEAR, the Crystal Ball collaboration has measured[26] the D* production cross section by examining the π^0 kinetic energy distribution. There is a strong peak below 40 MeV that arises from the decays $D^{*0} \to D^0 \pi^0$ and $D^{*+} \to D^+ \pi^0$. This peak tags D*'s. The Mark II measurements depend linearly on the $D^0 \to K^- \pi^+$ and $D^+ \to K^- \pi^+ \pi^+$ branching ratios, while the recent Crystal Ball measurements are independent of these branching ratios. A comparison of the cross sections as measured by the two groups yields, in some average way, the $D^0 \to K^- \pi^+$ and $D^+ \to K^- \pi^+ \pi^+$ decay branching ratios.

The cross sections obtained by the two groups are compared in Fig. 5. Agreement is good. The ratio of Mark II cross section to Crystal Ball cross section is 0.85 ± 0.25. If one had used the Mark III $D^0 \to K^- \pi^+$ and $D^+ \to K^- \pi^+ \pi^+$ branching ratios[11] in computing the Mark II cross sections, then the Mark II points would be lower by a factor of 2, in clear disagreement with Crystal Ball. Since Mark II used branching ratios of 3.0% for $D^0 \to K^- \pi^+$ and 6.3% for $D^+ \to K^- \pi^+ \pi^+$, the Mark II-Crystal Ball comparison leads to branching ratios of 3.0% × (.85 ± .25) = 2.6 ± 0.8% for $D^0 \to K^- \pi^+$ and 6.3×(.85 ± 2.5)=5.4±1.6% for $D^+ \to K^- \pi^+ \pi^+$.

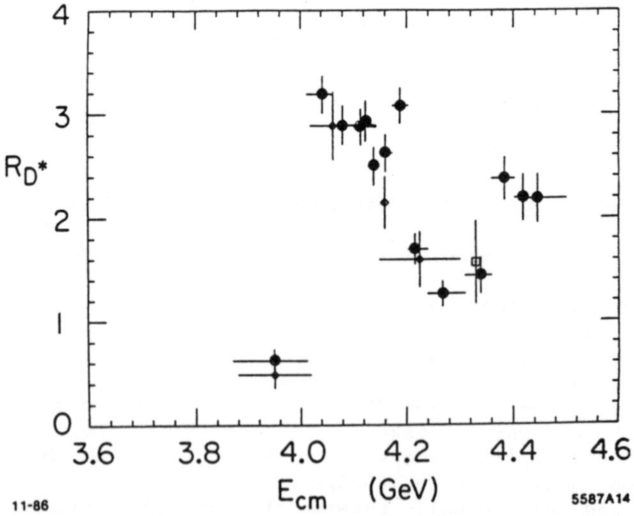

Fig 5. D* production cross sections obtained by Crystal Ball (Ref. 26) solid points, and Mark II (Ref. 25) diamonds.

3.4 Comparison of Branching Ratio Measurements.

The branching ratios as measured by Mark II,[17)] Lead Glass Wall,[18)] Mark II-Crystal Ball D* comparison,[25,26)] and Mark III[11)] are given in Table 5. For Mark II and LGW, the numbers are given as reported and as corrected for a 30% ψ"→charmless hadrons possibility. Weighted averages are also computed, both for the assumption that ψ" decays 100% to $D\overline{D}$ and for the assumption that it decays to $D\overline{D}$ 70% of the time. Two remarks are in order: (1) The Mark III experiment is technically the strongest of the four. (2) The Mark III experiment disagrees with the mean of the other three.

Table 5. Branching ratios for $D^0 \to K^-\pi^+$ and $D^+ \to K^-\pi^+\pi^+$

	$D^0 \to K^-\pi^+$	$D^+ \to K^-\pi^+\pi^+$
Mark II (Ref. 17)		
as reported	3.0 ± 0.6%	6.3 ± 1.5%
30% charmless	4.3 ± 0.9%	9.0 ± 2.1%
LGW (Ref. 18)		
as reported	2.2 ± 0.6%	3.9 ± 1.0%
30% charmless	3.1 ± 0.9%	5.6 ± 1.4%
Mark II-Crystal Ball D*comparison (Refs.25,26)	2.6 ± 0.8%	5.4 ± 1.6%
Mark III (Ref. 11)	5.6 ± 0.5%	11.6 ± 1.6%
Weighted average, assuming ψ"→$D\overline{D}$ 100%	(3.7 ± 0.3)%	(6.0 ± 0.7)%
Weighted average, assuming ψ"→$D\overline{D}$ 70%	(4.4 ± 0.4)%	(7.6 ± 0.8)%

4. A VERDICT?

Having reviewed all of the evidence, can we come to a verdict?

1. One possibility is that the Mark III branching ratios for $D^0 \to K^-\pi^+$ and $D^+ \to K^-\pi^+\pi^+$ are wrong, by a factor between 1.5 and 2.0. If this were the case, it would explain the missing D's both in semileptonic and in hadronic B decay.

2. If the Mark III branching ratios are correct, then:
   - a) There must be a mistake in at least one of the experiments measuring the D* production cross section;
   - AND b) Either there is a mistake in both the Mark II and LGW measurement of $D^0 \to K^-\pi^+$ and $D^+ \to K^-\pi^+\pi^+$ branching ratios, or ψ" decays to charmless states at least 30% of the time.
   - AND c) Either the CLEO $B \to D^0 l\nu$ branching ratio is wrong or something absolutely nuts is going on in the decay $B \to X_u l\nu$, causing X_u to be very massive.

Of the two possibilities listed, parsimony favors the first. However, I can find no flaw in the Mark III experiment.

ACKNOWLEDGEMENT

I would like to thank my CLEO collaborators, especially Ronald Poling and Sheldon Stone, for many useful discussions. My research has been supported by the National Science Foundation.

REFERENCES

1) Bortoletto, D. et al., (CLEO), Phys. Rev. D35, 19 (1987); Word, G., contribution to this conference.
2) Schubert, K. R., Proc. of XXIII International Conference on HEP, Berkeley, CA (to be published).
3) Blucher, E., contribution to this conference.
4) Haas, P. et al., (CLEO), Phys. Rev. Lett. 55, 1248 (1985); Alam, M.S. et al., (CLEO), Phys. Rev. D34, 3279 (1986).
5) Albrecht, H. et al., (ARGUS), Phys. Lett. 162B, 395 (1985).
6) Kuhn, J.H. et al., Z.Phys.C5, 117 (1980).
7) Eilam, G., Phys. Rev. Lett. 49, 1478 (1982).
8) Lipkin, H., Argonne Report No. ANL-HEP-PR-76-43, 1986.
9) Behrends, S. et al., (CLEO), contribution to the 23rd International Conference on HEP, Berkeley, CA, 1986 (unpublished).
10) Alam, M.S. et al., (CLEO), submitted to Phys. Rev. Lett.; Tipton, P .L., Ph.D. Thesis, Univ. of Rochester, 1987.
11) Baltrusaitis, R.M. et al., (Mark III), Phys. Rev. Lett. 56, 2140 (1986).
12) Grinstein, B., Wise, M.B., and Isgur, N., Phys. Rev. Lett. 56, 298 (1986); and Calif. Inst. Technology-Univ. Toronto Report No. CALT-68-1311, UTPT-85-37 (1986).
13) Chen, A. et al., (CLEO), Phys. Rev. Lett. 52, 1084 (1984).
14) Klopfenstein, C. et al., (CUSB), Phys. Lett. 130B, 444 (1983).
15) Guida, J.A., Ph.D. Thesis, Univ. of Rochester, 1987.
16) Altarelli, G. et al., Nucl. Phys. B208, 365 (1982).
17) Schindler, R. H. et al., (Mark II), Phys. Rev. D24, 78 (1981).
18) Peruzzi, I. et al., (LGW), Phys. Rev. Lett. 39, 1301 (1977); Scharre, D. L. et al., Phys. Rev. Lett. 40, 74 (1978).
19) E. H. Thorndike in Proc. of 1985 International Symposium on Lepton and Photon Interactions at High Energies, ed. M. Konuma and K. Takahashi (Kyoto), 1986, p.406.
20) E. Eichten et al., Phys. Rev. D21, 203 (1980).
21) W. Bacino et al., (DELCO), Phys. Rev. Lett. 45, 329 (1980).
22) Feller, J.M. et al., (LGW), Phys. Rev. Lett. 40, 274 (1978).
23) Bacino, W. et al., (DELCO), Phys. Rev. Lett. 43, 1073, (1979).
24) Aihara, H. et al., (TPC), Phys. Rev. D31, 2719 (1985); Althoff, M. et al., (TASSO), Z.Phys.C22, 219 (1984); Phys. Lett. 146B, 443 (1984); Behrend, H.J. et al., (CELLO), Z.Phys.C19, 291 (1983); Adeva, B. et al., (MARK J), Phys. Rev. Lett. 51, 443 (1983); Hanson, G.G. (Mark II), Proc. Int. Europhysics conf. High-Energy Physics, Brighton, p.147-51, Didcot: Rutherford Appleton Lab. (1983); Koop, D. E. et al., (DELCO), Phys. Rev. Lett. 52, 970 (1984); Isen, J.M., DESY report DESY 84-104 (October 1984); Tipton, P.L., (CLEO) Univ. Rochester Qualifying Exam Brief, December 1983 (unpublished).
25) Coles, M.W. et al., (Mark II), Phys. Rev. D26, 2190 (1982).
26) Osterheld, A. et al., (Crystal Ball), SLAC-pub 4160 (1986), and submitted to Phys. Rev. D.

II. Electroweak Interactions 2

Conveners — M. Strovink and G. Kane

LOW THRUST HADRONIC EVENTS with ISOLATED μ from PETRA

Min Chen

Department of Physics
M.I.T., Cambridge, Mass02139

Abstract

A systematic search of new partcles at PETRA is performed, particularly on the excess of low thrust inclusive muon events first reported by MARKJ at the highest PETRA energy. The probability for this excess to be statistical fluctuation is less than 1/100000.

I. INTRODUCTION.

PETRA scanned with center of mass energy from 12 GeV to 46.78 GeV and found no new threshold due to pair production of either new heavy leptons or top quarks. Table I shows a summary(1) of the lower mass limits at 95% confidence level of many types of new particles postulated by models listed at the left most column:

Table I

LOWER MASS LIMITS for NEW PARTICLES

Motivation	**Particles**	**Mass (GeV)** 95% lower limit
Standard model	Top quark pair mass	>46.67 (from R)
	1S toponium	>46.67 (from scan)
Extended theory	$H^{\pm}$ or technipion	18
	pair of B' ($Q = 1/3$)	45.5
	new sequential lepton	22.7 (41 from UA1)
	$\tilde{e}$ from pair production	22
	$\tilde{e}$ from single γ	> 40 for light $\tilde{\gamma}$
	$\tilde{\mu}$	21
	$\tilde{\tau}$	18
	$\tilde{q}(\rightarrow q + \tilde{\gamma})$	21
SUSY	$\tilde{\gamma}(\rightarrow \tilde{G} + \gamma$	22
	$\tilde{H}^{\pm}$	22.5
	$\tilde{Z}^{0}$	$> 35(m_{\tilde{e}} < 40GeV)$
	$\tilde{W}^{\pm}$ (pair production)	22
	$\tilde{W}^{\pm}$	40 for light sneutrino
Long-lived	$'\mu'$ like	20
QED	e^{*}	> 70 if $\lambda = 1$
violating	μ^{*}	> 28 if $\lambda = 1$
excited quarks	q^{*}	22
or leptons	Leptoquark	21

As seen, the limit for top quark pair production is at 46.6 GeV. That for ground state toponium, assuming that the product of its partial width to e+ e- and its hadronic branching ratio is about 3.5 keV as extrapolated from charmonia and bottonia(2) decays, is also at 46.6 GeV. The corresponding limit for a fourth family charge 1/3 quark B' is at 45.5 GeV.

The four experiments at PETRA (MARKJ, JADE, CELLO and TASSO) have used the reaction e+ e- ----> μ + hadrons, to search for new particles. The low thrust inclusive μ rate, among the many experimental signatures(1) investigated, is a particularly sensitive tool to look for new physics since experimentally muons can be clearly identified and the detection efficiency and resolution of inclusive muon events have been well calibrated by comparing data at lower energies with Monte Carlo predictions. As one result, the production and fragmentation parameters concerning B and C quarks have been determined(3) at 35 and 44 GeV. Since the masses of all the five known quarks are small in comparison with the beam energy, the thrust distribution of inclusive muon events tends to peak at high values. Any significant increase at higher energies in the low thrust region thus could indicate the production of some new heavy particles.

In the Spring of 84, MARKJ reported(4) the observation of an excess of low thrust inclusive muon events at the highest PETRA energy, $\sqrt{s} > 46.3$ GeV. This observation was later confirmed by the JADE collaboration(5) but not by the other two experiments(6,7). JADE(5), CELLO(6) and TASSO(7) also found no excess in the inclusive electron data. In this report we investigate the significance of the results and the method used to obtain each result. The compiled results from all four experiments are presented and discussed.

II. STATUS of the MARKJ and JADE LOW THRUST INCLUSIVE μ EVENTS.

Both MARKJ and JADE have observed a significant excess in the low thrust inclusive muon data at $\sqrt{s} > 46.3$ GeV. The measured cross sections for inclusive muon events, with $T < 0.8$ and $\cos\delta < 0.7$, where δ is the angle between the thrust axis and the muon direction, are shown in Fig. 1 as functions of the c.m. energy from 36.9 GeV to 46.78 GeV. At the highest energy the cross section is significantly higher than what one would expect using the lower energy data or Monte Carlo calculation based on five quark production.

In Fig. 2 we show the differential cross section of inclusive muon events from MARKJ as a function of thrust for the low energy data with 36.9 GeV $\leq \sqrt{s} \leq$ 46.3 GeV which agrees with the Monte Carlo simulations with 5 flavors but not with that of the 6 flavors. The total integrated luminosity is 30 pb-1. Fig. 3 shows the MARKJ highest energy data, 46.3 GeV $\leq \sqrt{s} \leq$ 46.78 GeV, with a luminosity of 2.8 pb-1. The histogram in this figure is the data from the lower energy interval. In this higher energy interval, the cross section for thrust values below 0.8 is larger than that in the lower energy region 8 events observed whereas 1.76± 0.3 are expected using the rate at lower energy. Similarly for JADE, six are observed with 2.2 expected corresponding to 1.8 pb-1. In addition we find that most of the μ in these events are isolated from the hadronic and electromagnetic energy flow. Fig. 4 shows the energy flow of each event as well as the energy and position of the muon. The muon almost never overlaps with the most energetic jet or energy cluster. Each of the events A, B and C from MARKJ and events A, D and E from JADE further contains an isolated electromagnetic shower indicating the presence of an isolated high energy (≈ 10 GeV) electron or photon. The detail information about the event shape and the μ momenta for the MARKJ events are listed in Table II.

Table II. Parameters for the MARKJ events. Thrust, major and minor are the % of energy projected along the 3-principal axes of the energy flow of an event; pt is the μ momentum projected perpendicular to the thrust axis; pout is the μ momentum out of the event plane; Ec is the energy including the μ ionization energy inside a 30 degree cone of the μ direction.

Event	A	B	C	D	E	F	G
Thrust	0.79	0.74	0.72	0.69	0.73	0.73	0.77
Major	0.32	0.56	0.43	0.57	0.53	0.51	0.42
Minor	0.19	0.17	0.30	0.25	0.24	0.19	0.17
O	0.13	0.39	0.13	0.32	0.29	0.32	0.25
O_B	0.25	0.49	0.26	0.36	0.38	0.33	0.41
p_μ [GeV]	2.87	2.40	3.04	4.25	2.18 4.46	16.1	2.67
$p_\perp$ [GeV]	2.64	1.93	2.51	4.20	1.78 4.20	15.7	2.40
p_{out} [GeV]	1.10	0.39	2.25	3.57	1.02 1.70	7.69	0.89
T_μ	0.46	0.63	0.48	0.35	0.48 0.51	0.48	0.48
E_c	0.06	0.11	0.04	0.01	0.07 0.03	0.02	0.11
$\|\cos\delta\|$	0.38	0.60	0.56	0.16	0.58 0.34	0.22	0.44

Some of the interesting features of these events are the following: they are planar, 3-jet like with isolated muons; three of the MARKJ events (A, B and C) contain high energy isolated electrons or photons. So are three JADE events (A, D and E).

The data are then displayed against thrust and the angle δ of a muon track with respect to the thrust axis as shown in Fig. 5 and 6 for the events in the low and high energy samples. We see that seven of the MARKJ events and five of the JADE events observed at high energies are with thrust less than 0.8 and $\cos\delta < 0.7$ while 0.76 ± 0.17 and 0.56 ± 0.18 events are respectively expected by extrapolating from the lower energy data of the two groups.

MarkJ has examined data in this and in neighboring energy regions to see if the events are due to imperfections in the detector. Each events contains a clean track in each of the outside and middle muon chambers as well as in the vertex detector. The timing of the muon counters, after correction for time of flight, are within 1 ns of the beam crossing. Both measurements show that the events are not background from one of the beams, nor from cosmic rays. Other background such as hadron decays and punch through was calculated by a detailed Monte Carlo program and checked against lower energy data. They do not rise significantly from 43 GeV to 46.78 GeV. Furthermore punch through muons are typically found in the midst of hadron jets, whereas in these 12 events from JADE and MARKJ, the 14 muons are isolated from the main clusters of energy. MARKJ also investigated the effect of the beam associated noise on the data, using records of the detector contents at random beam crossing times, which are uncorrelated with the usual triggers and excluded that as a source for these events.

MARKJ has also studied the production via two photon processes, such as eecc, $ee\tau\tau$ and $ee\mu\mu$ etc., as possible sources of this excess. A detailed Monte Carlo studies show that the expected numbers of events due to these processes are negligible.

III. STATUS of the LOW THRUST INCLUSIVE MUON EVENTS from PETRA.

Recently CELLO and TASSO have performed a similar analysis and report that no excess is observed. In Table III, we show for each group the accumulated lumunosity, the overall μ detection efficiency, the expected and observed low thrust isolated muon events. As a measure of the relative sensitivity of the experiments, we define the effective luminosity as the product of the integrated luminosity and the overall acceptance. The rightmost column shows the result from all experiments at PETRA. We find that with an effective luminosity of 5.4 pb-1 and a total of 96 inclusive muon events at $\sqrt{s} > 46.3$ GeV, PETRA observed 14 low thrust events with isolated muons while 2.4 ± 0.3 is expected. The probabilty of the high energy region to be explained by the same physics as established using the data $39.79 < \sqrt{s} < 46.3$ GeV is less than one part in one million. Since originally the MARKJ excess was based on the selection of low thrust only, in the last row of Table III we also show the result with thrust < 0.8

for MARKJ but with both thrust and δ cuts for the others. The probabilty of 15 events observed with 3.5 ± 0.47 expected is less than ten parts in a million. We conclude that the PETRA inclusive μ data as a whole shows a significant evidence of new physics present at the energy region $\sqrt{s}$ > 46.3 GeV.

TABLE III

Low THRUST INCLUSIVE MUONS EVENT from PETRA at $\sqrt{s}$ > 46.3 GeV to show the lumunosity, the overal μ acceptance, the expected and observed low thrust isolated muon events from each group from PETRA. The relative sensitivity of the experiments is proportional to the effective luminosity which is the product of the integrated luminosity and the overall μ acceptance. The rightmost column shows the combined results from all four experiments at PETRA.

Groups	MARKJ	JADE	CELLO	TASSO	PETRA
Lumi(pb-1)	2.8	1.8	2.1	1.1	7.8
μ acceptance	70%	62%	72%	29%	
Lumi* μ accp	2	1.1	1.5	.3	4.9 pb-1
expected events for t<0.8 and |cosδ|<0.7	0.76±0.17	0.56±0.2	0.6-0.7	0.3	2.4±0.3
observed	7 (8)#	5	1	1	14 / 15

Prob. of observing 14 with 2.4±0.3 expected : < one in a million
15# 3.5±0.47 : < ten in a million

* CELLO endcap was not used for the result discussed here.
\# if for MARKJ data we use the selection thrust < 0.8 only.

IV. THREE PUZZLES.

We have shown in the previous section that the low thrust inclusive μ events from MARKJ and JADE, also from the compiled data from all four experiments at PETRA, show a statistically significant excess at $\sqrt{s}$ > 46.3 GeV. There are three frequently raised questions: 1. Why are no events seen in the inclusive electron channel? 2. Why doesn't TASSO see such an excess? 3. Why neither CELLO?

The inclusive electron events requires very different experimental detection technique from inclusive muons. Consequently, it has very different acceptance and we can not draw any conclusions at this moment before a careful comparison is made.

To answer the second question, we use Table III to compute the number of events JADE and TASSO would have observed if their effective luminosity were the same as MARKJ's luminosity, i.e. 2 pb-1. We find that the effective number of observed events per 2 pb-1 for TASSO is 6.7 ± 6.7 (scaled up from the one event observed by the ratio of luminosities) which is in good agreement with with the 7 events observed by MARKJ and the 9 events, (also scaled up from the 5 observed events) of JADE. The TASSO data point at this high energy is also shown in Fig. 1.

In order to find out how significant is CELLO's non-observation of such an excess, we compute the relative acceptances of the experiments by feeding all 14 low thrust inclusive μ events from JADE and MARKJ through the analysis chain of each group with the exception of TASSO who is yet to complete this analysis. We find that the relative acceptances are CELLO : JADE : MARKJ = 52% : 65% : 59%. Using the measured low thrust inclusive μ rate at PETRA, we can now compute the expected number of events for CELLO. We find that CELLO would expected to find 4 events. The probability for one out of four experiments to find less than two events when four are expected is about 25% thus quite probable. Therefore there is no significant contradiction among the four experiments and it should be valid to use the compiled data from PETRA as a whole, which shows a significant excess.

V. WHAT the EXCESS IS NOT LIKELY to BE?

Due to limited machine beam time, PETRA could not determine the origin of the excess of low thrust inclusive μ's. Instead, we briefly listed here what we believe they could not be.

1. They are not fake muons, such as decays, punch through or beam background.
2. They are not simply statistical fluctuation since the probability is less than a few parts in a million when all experiments at PETRA are included.
3. They are not just from two photon processes such as, e e τ τ, e e c c or e e μ μ etc. since they all have negligiably small cross sections and very small acceptance.
4. They are not from the decays of top quarks or a new kind of high mass leptons. However, a new charge 1/3 quark with mass ≈ 23 GeV reproduces almost all of the one dimentional distributions such as thrust, oblateness although they predict too many events in the region with $T < 0.8$ and $\cos\delta > 0.7$, as shown in Table IV.
5. μ-e universality is not proven to be violated yet.
6. High order QED processes and QCD such as
   e + e ---> q q + gluon
   with gluon--> c + c --> μ + X

have to be investigated, but none is expected to show such a dramatic rise between 44 and 46.3 GeV.

Number of muons				
Data High	Data Low	MC b' direct	MC b' cascade	Combined Low+b'
All inclusive muon events				
31	24.4 ± 1.0	6.6 ± 0.4	2.7 ± 0.2	33.7
$T < 0.8$				
9	2.4 ± 0.4	5.1 ± 0.3	2.0 ± 0.2	9.5
$T < 0.8$ and $\|\cos\delta\| < 0.7$				
8	0.8 ± 0.2	3.0 ± 0.3	1.0 ± 0.1	4.8
$T < 0.8$ and $\|\cos\delta\| > 0.7$				
1	1.6 ± 0.3	2.2 ± 0.2	0.9 ± 0.1	4.7
$O > 0.2$				
7	2.6 ± 0.4	2.8 ± 0.3	1.2 ± 0.1	6.6
$O_B > 0.3$				
6	2.4 ± 0.3	1.9 ± 0.2	0.8 ± 0.1	5.1
$O_B < 0.3$ and $T < 0.8$				
3	0.9 ± 0.2	3.1 ± 0.3	1.1 ± 0.1	5.2
$p_\mu > 5$ GeV				
9	6.7 ± 0.5	4.7 ± 0.3	1.6 ± 0.1	13.0
$p_\perp > 3$ GeV				
4	2.0 ± 0.3	4.6 ± 0.3	1.6 ± 0.1	8.2
$T_\mu < 0.7$				
9	2.6 ± 0.4	5.5 ± 0.4	2.1 ± 0.2	10.3
$E_c < 0.08$				
8	3.2 ± 0.4	4.9 ± 0.3	1.7 ± 0.1	9.7
$p_{out} > 1.5$				
4	1.9 ± 0.3	3.6 ± 0.3	1.2 ± 0.1	6.8
$p_\perp > 3.0$ GeV and $T < 0.8$				
3	0.5 ± 0.1	3.7 ± 0.3	1.2 ± 0.1	5.4

Table IV

The number of muons within different cuts measured and expected for L=2.8 pb^{-1} . High means $\sqrt{s} \geq 46.30$ GeV and Low $36.90 \leq \sqrt{s} < 46.30$ GeV. The prediction from Low energy data does not include the correction for an increase in the fraction of inclusive muon events

This region, $\sqrt{s} > 46.3$ GeV, clearly must be carefully studied again. It is too bad PETRA failed to re-examine it in the one and half years following the observation of the excess. Now it is up to the new colliders KEK, SLC and LEP to find out the real origin of these events. From the experience of PETRA, one cannot over emphasize here the importance of having a clean μ detector with large acceptance. On the other hand, UA1 at the collider could also be sensitive to this type of events with isolated μ and jets.

REFERENCES.

1. B. Adeva et. al., Phys. Rev. Lett. 53(1984)1806;
   Phys. Lett. 152B(1985)439; Phys. Rev. 34(1986)681;
   H.J. Behrend et. al., DESY 86-050(1986);
   W. Bartel et. al., Phys. Lett. 152B(1985)385;
   G. Bartha et. al., Phys. Rev. Lett. 56(1986)685;
   E. Fernanez et al., Phys. Rev. Lett. 54(1985)1118;
   B. Naroska, DESY 86-113(1986).
   G. Alexander et. al., Phys. Lett. B, 181(1986) 179.
   B. Zhou, M.I.T. thesis(1987).
2. D.R. Yennie, Phys. Rev. Lett. 34(1975)239;
   M. Krammer, H. Krasemann and S. Ono, DESY 80-25(1980).
3. W. Bartel et. al., DESY 86-129(1986).
4. A. Bohm, Proceeding of the XX Recontre de Moriond,1985.
   B. Adeva et. al. Phys. Rev. 34(1986)681.
5. M. Kuhlen, Proceeding of the XXI Recontre de Moriond,1986.
6. D. Mellor, private communication.
7. CELLO Collaboration, contribution to the Berkeley Conference,1986.

Fig. 1 The measured cross sections from MARKJ and JADE for inclusive muon events, with $T < 0.8$ and $|\cos\delta| < 0.7$, as a function of the c.m. energy from 36.9 GeV to 46.78 GeV. The TASSO data at the highest energy is also shown.

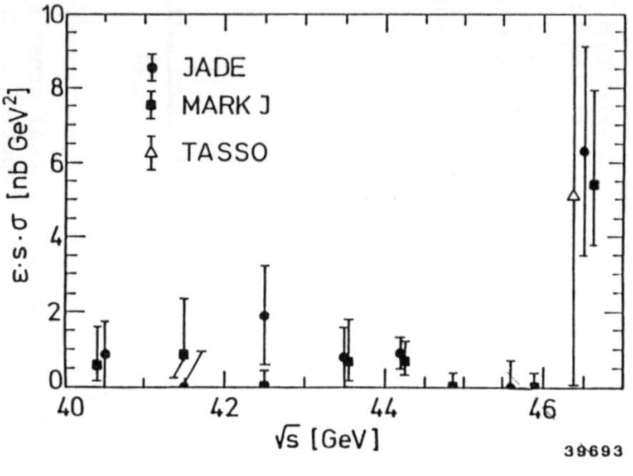

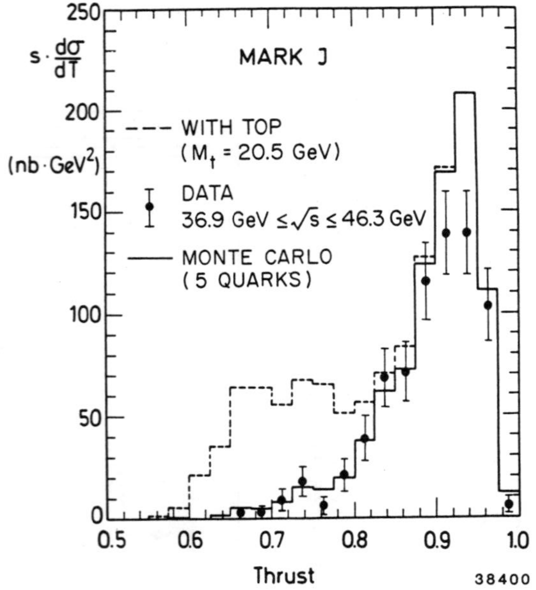

Fig. 2 The differential cross section of inclusive muon events as a function of thrust for the energy interval of 36.9 GeV ≤ $\sqrt{s}$ ≤ 46.3 GeV for MARKJ. Monte Carlo simulations with 5 flavors and 6 flavors are shown in solid and dashed histograms respectively

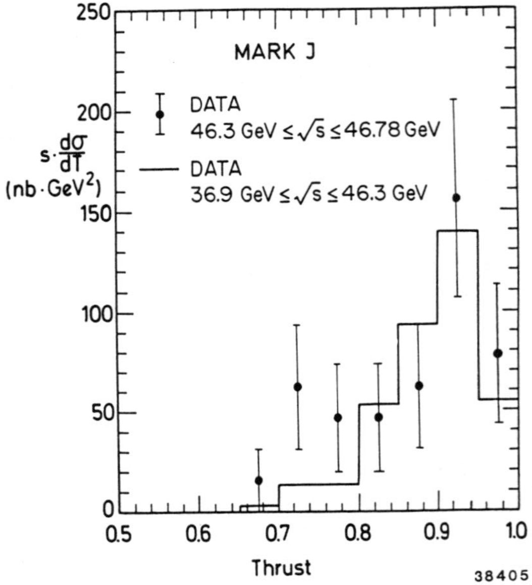

Fig. 3 The thrust distribution of the highest energy data, 46.3 GeV ≤ $\sqrt{s}$ ≤ 46.78 GeV, with a luminosity of 2.8 pb-1 for MARKJ to be compared with the histogram, which is the data from the lower energy interval.

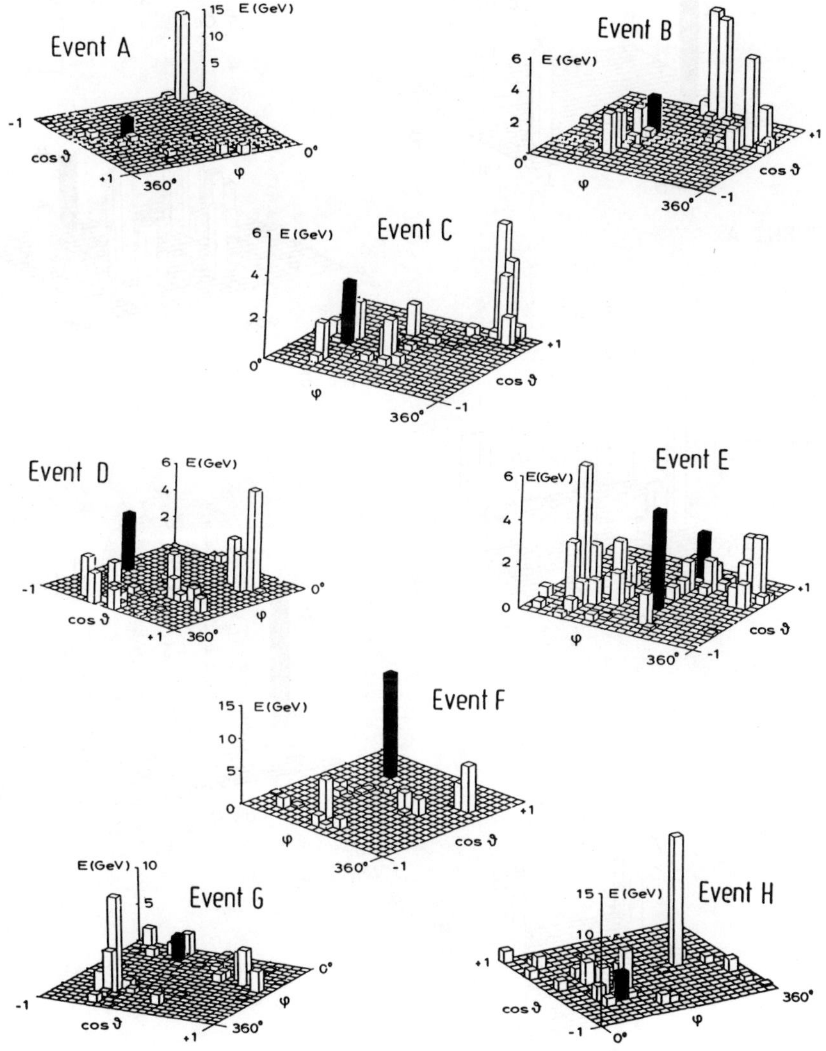

Fig. 4a The energy flow of MARKJ low thrust inclusive μ events. The energy and position of the muons are shown in solid.

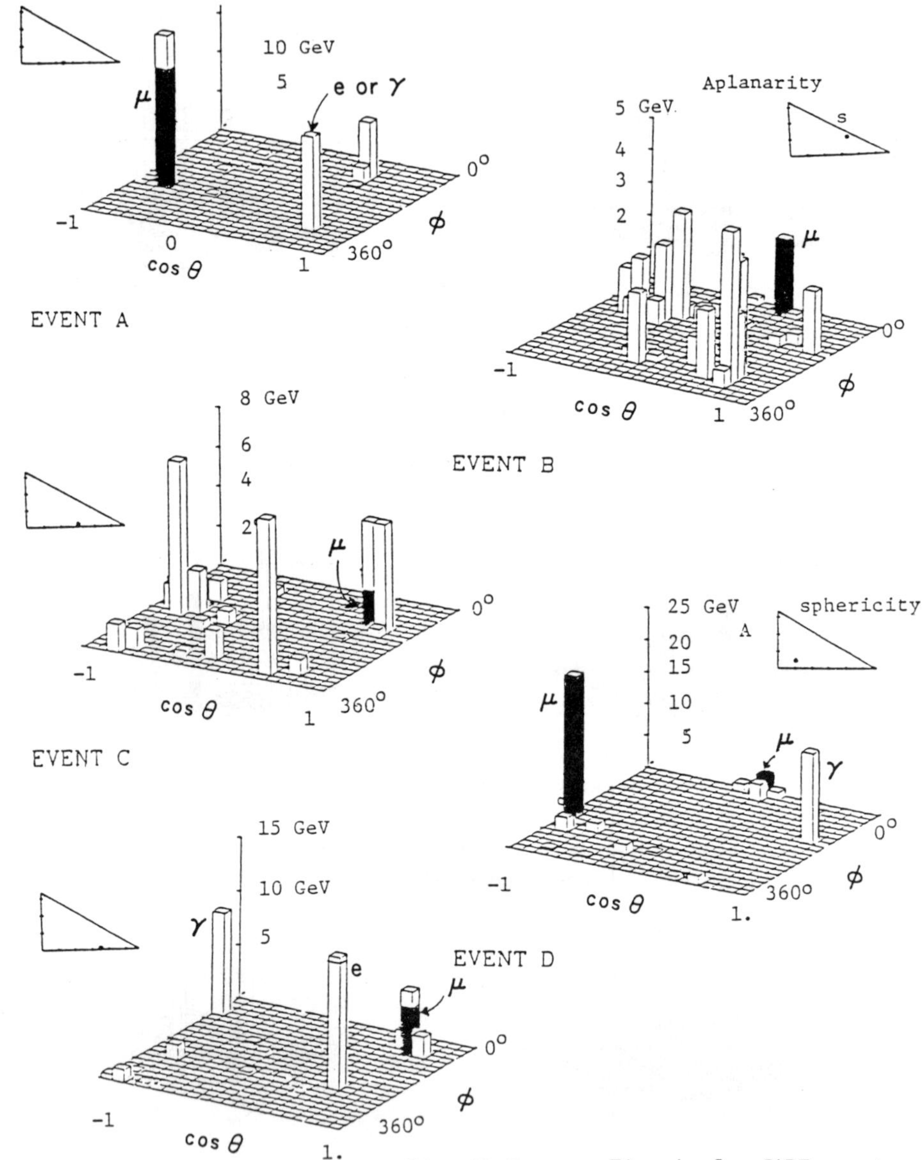

Fig. 4b Same as Fig. 4a for JADE events.

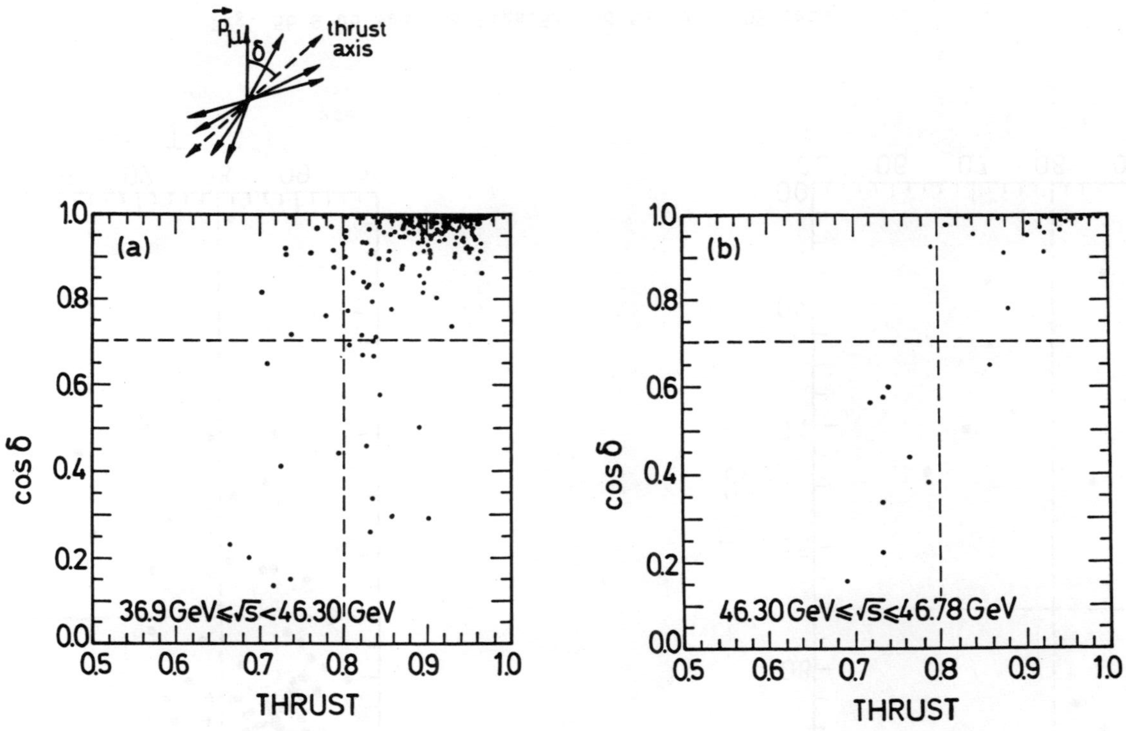

Figs. 5a & 6a The two dimensional distributions versus thrust and cos δ for the MARKJ data in the low and high energy samples respectively.

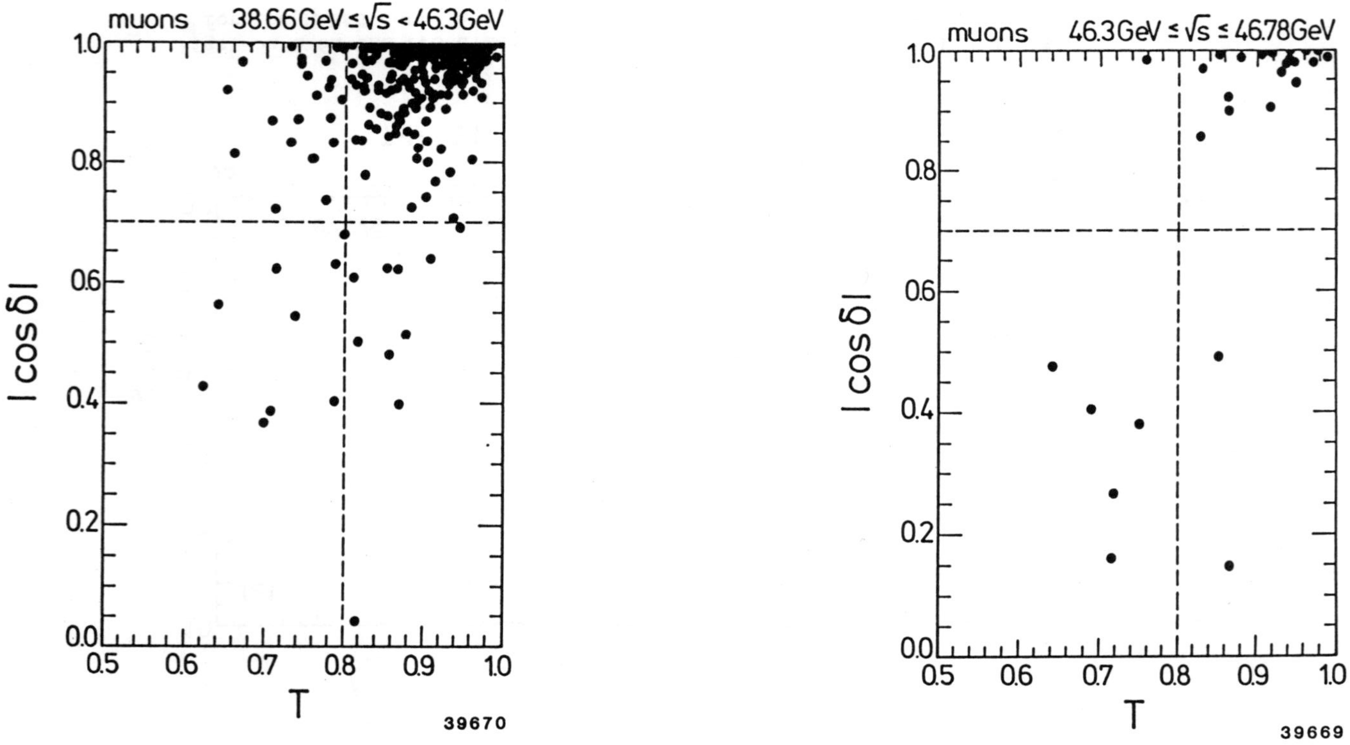

Fig. 5b & 6b Same as Figs.5a and 6a for JADE data.

Missing Transverse Energy Events and a Search for Additional Neutrino Families and a 4th Generation Lepton*

David B. Cline

UNIVERSITY OF CALIFORNIA
Los Angeles
California

and

UNIVERSITY OF WISCONSIN
Madison
Wisconsin

Abstract

We report on the most recent analysis of the missing transverse energy events in the UA1 experiment and discuss the search for a 4th generation heavy lepton from these data. We also report the latest results on the number of neutrino species from Z° decay using the ratio of widths technique. A summary of the worlds data on δN_{ν} is given.

---

* For the UA1 Group: An invited talk at the Salt Lake City, DPF Meeting, January 1987.

1. *Missing Energy Events in UA1*

Table 1 summarizes the cuts and the final sample of events obtained in the UA1 experiments from 1983-1985. An example of a missing energy event that is due to a $W \rightarrow \tau + \nu$ is shown in Figure 1. A non-τ-like event (monojet) is shown in Figure 2. The objective of the missing transverse energy event selection is to define an event sample for which the background due to fluctuations in the detector response is small, thereby isolating events containing one or more energetic neutrinos (or other non-interacting particles). The missing transverse energy events were selected using the following criteria:[1,2]

1. *missing transverse energy*, $E_T^{miss} > 15$ GeV and $N_\sigma > 4$, where N_σ is defined as $E_T^{miss}/(0.7\sqrt{\Sigma E_T})$. ΣE_T is the scalar sum of transverse energy observed in all calorimeter cells where transverse energies are measured in GeV. N_σ is thus a measure of the experimental significance of the missing transverse energy.

2. *one or more jets* observed in the calorimeters with $E_T > 12$ GeV and $|\eta| < 2.5$.

3. *one or more tracks* observed in the drift chambers with $p_T > 1$ GeV/c within a cone of size $\Delta R = 0.4$ centered on the calorimeter jet axis ($\Delta R^2 \equiv \Delta\phi^2 + \Delta\eta^2$, where ϕ is the azimuthal angle and η is the pseudorapidity).

4. *isolation of the missing transverse energy*, by rejecting events with a jet observed in the calorimeter with $E_T > 8$ GeV and $|\eta| < 2.5$ (or a jet observed in the central detector drift chambers with $p_T > 5$ GeV/c) within ± 30 degrees in ϕ of the missing transverse energy direction.

5. *veto of coplanar jet activity*, by rejecting events with a calorimeter jet of $E_T > 8$ GeV and $|\eta| < 2.5$ (or a central detector jet with $p_T > 5$ GeV/c) within 150–210 degrees in ϕ of the direction of the highest E_T jet.

6. *veto of events with an electron or muon candidate* in order to remove contributions from $W \rightarrow e\nu$ or $\mu\nu$ or heavy flavour semi-leptonic decays. For this purpose we have developed an algorithm for removing electrons or muons having tracks with $p_T > 10$ GeV/c. In addition, events were

scanned and removed if they had muon candidates with $p_T > 3$ GeV/c. The cuts used to define electrons and muons for veto purposes in this analysis are less stringent, especially on isolation requirements, than those used to select low background samples of W events.

For additional information see references 1 and 2. A total of 56 events satisfied the cuts described above and in table 1. We may now make use of measured properties of semi-hadronic τ-decays, namely the low multiplicity, low mass jet, to separate out the production and decay of a $W^{\pm}$ into tau-lepton plus neutrino, through the process:

$$p\bar{p} \to W + X; W \to \tau v; \tau \to \text{hadrons} + v.$$

The signature is a high transverse momentum narrow hadronic jet with associated missing transverse energy. For this purpose, we define three variables particularly suitable for separating high p_T τ-jets from ordinary jets:

1. *fraction of jet energy* (F), as measured in the calorimetry, contained within a cone of size $\Delta R = 0.4$ relative to that contained in the jet definition cone of $\Delta R = 1.0$,

2. *angular separation* ($r = \Delta R$) between the leading track of the jet, as measured in the central detector drift chambers, and the jet axis as measured in the calorimeters, and

3. *charged particle multiplicity* (n) of tracks with $p_T > 1$ GeV/c within a cone of $\Delta R = 0.4$ around the calorimeter jet axis.

Using the τ Monte Carlo described later, we have produced the distributions of F, r and n for the jet coming from the τ decay. We calculate a τ-log-likelihood, $L_\tau = \ln(f_F f_r f_n)$, for the highest E_T jet in an event, where f_F, f_r and f_n are the values of the τ Monte Carlo distributions for the measured values of F, r and n for this jet. The distributions of L_τ for the data, τ Monte Carlo, and jet data are shown in Figure 2; the range of L_τ is -16 to +4. The probability that a jet comes from a τ decay is greater with increasing L_τ. The cut $L_\tau > 0$ keeps 78% of the $\tau \to$ hadrons + v decays and rejects 89% of the ordinary jet events. Therefore, we define our final τ sample as those events with $L_\tau > 0$ (32 events). The remaining events are used in the search for[1,2]:

$$W \to L + \nu_L$$

and

$$Z \rightarrow \nu\bar{\nu}$$

2. *Technique to Search for the 4th Generation Lepton in W Decay*

The search for a 4th generation of leptons and quarks is of great importance for the standard model. In the past the lepton of each generation was detected first (electron, before the neutron, μ before strangeness and charm, the τ before the b or t quarks). It is possible this will happen again due to the unique signature of the heavy lepton process. Previously, it was pointed out that the $W^{\pm}$ particle decays provide a unique mass range in which one can search for new leptons.[1,2,3] Furthermore, we expect copious production of $W^{\pm}$ particles at the CERN and Fermilab $\bar{p}p$ Colliders in the next few years. We follow the initial ideas of early papers and carry out a study of the experimental method needed to separate the signal from the production and decay of the new lepton into one or two jet final states from the backgrounds.[3,4,5] We have applied this analysis to the UA1 detector at CERN, however the technique is general enough to be applied to the other detectors at CERN and Fermilab.

We consider the decays of the $W^{\pm}$ particles

$$W^{+} \rightarrow L^{+} + \nu_{L}$$

$$W^{-} \rightarrow L^{-} + \bar{\nu}_{L}$$

where L is a new sequential lepton with mass in the range 23 to 70 GeV/c^2 and ν_L, $\bar{\nu}_L$ are a fourth generation of neutrinos and anti-neutrinos which are assumed to be massless. The hadronic decays of the $L^{\pm}$ are expected to be[3,4,5]

$$L^{+} \rightarrow u + \bar{d} + \bar{\nu}_{L}$$

$$\rightarrow c + \bar{s} + \bar{\nu}_{L}$$

$$L^{-} \rightarrow \bar{u} + d + \nu_{L}$$

Two techniques have been proposed to detect the L in the W decay debris.

1. Observation of an excess of single lepton events with missing energy in a region where $W \rightarrow \ell + \bar{\nu}_{\ell}$ is suppressed and reduction of the corresponding background from $W \rightarrow \tau + \bar{\nu}_{\tau} \rightarrow \ell + \bar{\nu}_{\ell} + \nu_{\tau} + \bar{\nu}_{\tau}$ with a microvertex detector.

2. Observation of one or two jet events with missing energy from the L decay. So far only technique 2 has been applied in the UA1 experiment. The experimental technique proposed here makes use of a background suppression by detecting two jets from the heavy lepton production and decay that are on the same side of the $\bar{p}p$ line of interaction (i.e. within 180° of each other). As we will show this selection and certain requirements on the jets from the decay serve to reduce the expected backgrounds. Fig. 3 illustrates the technique proposed here.

In this analysis we make use of a modified ISAJET Monte Carlo. The full matrix element of the decay including the polarization of the L has been included with both right and wrong helicity states generated. In Reference 5 it was shown that the W decay gives a kinematic window up to ~(65–70) GeV/c^2 to search for the new lepton. In this paper we will report on calculations up to a lepton mass of 55 GeV/c^2. Higher mass calculations will be reported in a subsequent report.

3. *Limits on M_L*

For this analysis only events with $\ln \tau < 0$ are used.[2] A careful calculation of the background and a subtraction from the data yields the curves shown in Fig. 4 for $M_L = 35$ GeV/c^2 and 55 GeV/c^2. The subsequent limit on the L mass is

$$M_L > 41 \text{ GeV } 90\% \text{ CL}$$

including systematic errors.[2]

4. *Techniques to Count Neutrino Species*

Several techniques have been invented to search for additional neutrino families including

$$e^+e^- \rightarrow \nu\bar{\nu}\gamma$$

$$\bar{p}p \rightarrow Z_{\hookrightarrow \nu\bar{\nu}} + \text{gluon}$$

$$(\Gamma_Z/\Gamma_W) \text{ technique to measure the Z width}$$

In Table 2 we report the sensitivity of these techniques to additional neutrino families. These range from a 6% effect to a ~25% effect per unit family increase.

5. *Limits on δN_ν*

The ratio of widths technique yields the best limits for present $\bar{p}p$ collider data.[6] Fig. 5 shows the most recent fit to the UA1 and UA2 data values of R where

$$R = N(W \to \ell_\nu)/N(Z \to e^+e^-)$$

when fully corrected for acceptance and efficiency.[7] This results in a $\delta N_\nu = (N_\nu - 3)$ value that depends on the t quark mass but reaches a value of

$$\delta N_\nu < 5$$

for $M_t > 50$ GeV. When combined with the results from e^+e^- collisions this points to the overall value of less than or equal to 4 neutrino families. This is in good agreement with cosmological limits.[9] (See Fig. 6)

6. *Future Search for L and δN_ν*

We may anticipate the different mass ranges to be searched for the 4th generation lepton L as shown in Figure 7. The present UA1 limit appears to limit the search region for e^+e^- colliders, whereas W decay can extend the range to 65–70 GeV. Beyond this, LEP II is needed. Finally, if $M_L > 100$ GeV it will be necessary to discover this lepton indirectly by the observation of

$$\delta N_\nu = 1$$

or wait until a new e^+e^- high energy collider is constructed. As shown in Figure 7 all techniques reach some intrinsic limit.[10] I wish to thank my colleagues in UA1 for helpful discussion.

References

1. Events with Large Missing Transverse Energy at the CERN Collider: $W \to \tau\nu$ Decay and Test of τ-μ-e Universality at $Q^2 = M_w^2$, C. Abajan et al, CERN-EP/86–81 Submitted to Physics Letters

2. Events with Large Missing Transverse Energy at the CERN Collider: Search for the Decays of $W^\pm$ into Heavy Leptons and of the Z° into Non-Interacting Particles; C. Abajan et al CERN EP/86–82, submitted to Physics Letters.

3. D. Cline and C. Rubbia, Phys. Lett. 127B, 277 (1983). D. Cline, Recontre de Moriond, 1983 (published in the 1983 Proceedings).

4. V. Barger et al Phys. Lett. 133B, 449 (1983)

5. The Search for the 4th Generation Leptons from $W^\pm$ Decay: A Method to Extract the Signal, D. Cline and M. Mohammadi, Wisconsin preprint WISC EX-86-269

6. D. Cline and J. Rohlf, UA1 TN 83–40 (1983); K. Hikasa, P.R. D29, 1939 (1984); N. Deshpande et al. P.P.L. 54 1757, (1985); F. Halzen, K. Mursada, Phys. Rev. Lett. 51, 857, (1983).

7. Steve Geer - Private Communication

8. Limits on Neutrino, Selectrons and Winos from Single-Photon Searches, T.L. Lavine, UWISC Preprint MAC-748, 1986.

9. G. Steigman, D.N. Schramm and J. Gunn, 1977, Phys. Lett. 866, 502. J. Yang, M.S. Turner, G. Steigman, D.N. Schramm and K.A. Olive, Ap. J. 281, 493 (1984); A.M. Boesgaard and G. Steigman, Ann. Rev. Astron. Astrophysics 23, 319 (1985).

10. See D. Cline in Proceedings of the Aachen Antiproton-Proton Collider Conference 1986 (to be published)

TABLE 1

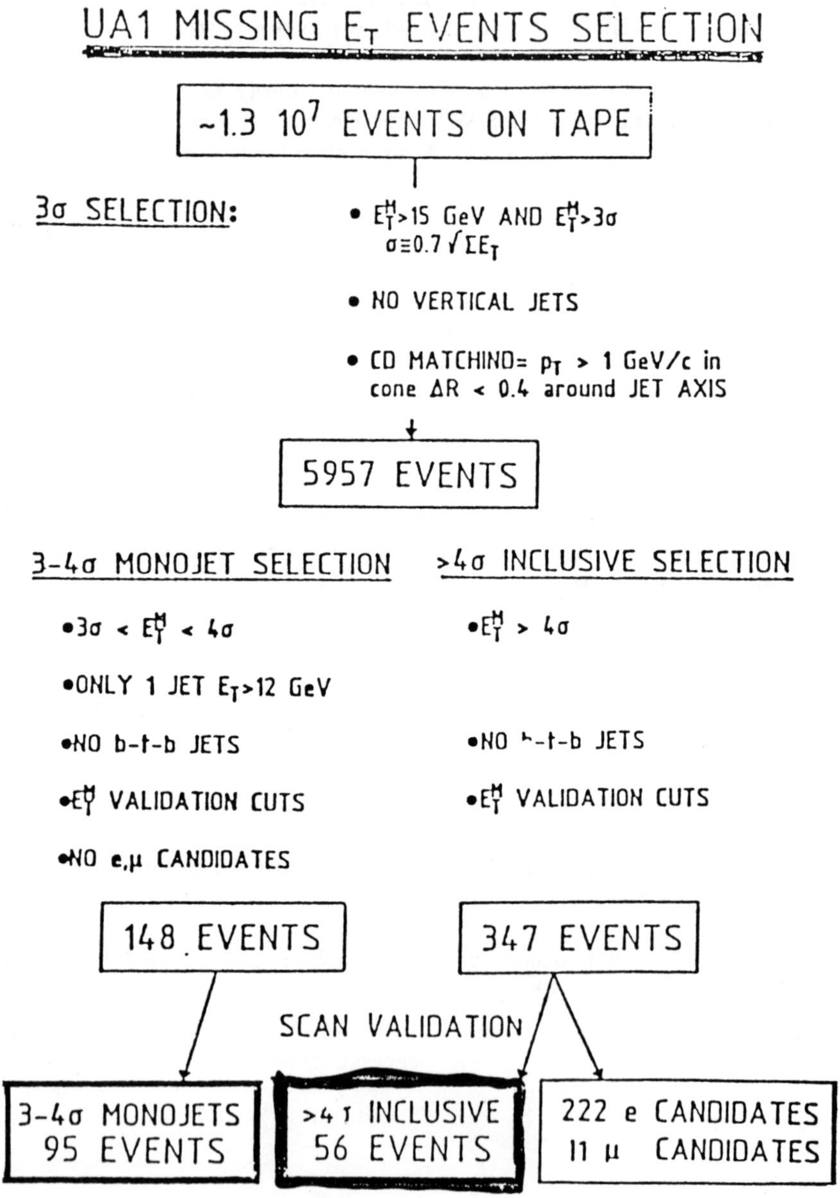
UA1 MISSING E_T EVENTS SELECTION
~1.3 10^7 EVENTS ON TAPE
3σ SELECTION:
• E_T^M>15 GeV AND E_T^M>3σ
σ≡0.7 $\sqrt{\Sigma E_T}$
• NO VERTICAL JETS
• CD MATCHINO= p_T > 1 GeV/c in cone ΔR < 0.4 around JET AXIS
5957 EVENTS
3-4σ MONOJET SELECTION
•3σ < E_T^M < 4σ
•ONLY 1 JET E_T>12 GeV
•NO b-t-b JETS
•E_T^M VALIDATION CUTS
•NO e,μ CANDIDATES
>4σ INCLUSIVE SELECTION
•E_T^M > 4σ
•NO b-t-b JETS
•E_T^M VALIDATION CUTS
148 EVENTS
347 EVENTS
SCAN VALIDATION
3-4σ MONOJETS
95 EVENTS
>4σ INCLUSIVE
56 EVENTS
222 e CANDIDATES
11 μ CANDIDATES

TABLE 2

Techniques to Search for Additional Neutrino Families

Process	Dependence on δN_ν	r
$e^+e^- \to \nu\nu\gamma$	$\sigma = \alpha G F^2 s(1.75 + \frac{\delta N_\nu}{4})$	0.13
Γ_Z/Γ_W	$R = 2.7\,x\,\frac{3.3\pm0.2}{[QCD]}\left[\frac{2.83+0.18\,\delta N_\nu}{2.82}\right]$	≤ 0.06
$\bar{p}p \to Z + g \ldots$	$\sigma \sim (A + B\delta N_\nu)^*$	
$\hookrightarrow \nu\bar{\nu}$	$A \sim 10$	0.1
	$B \sim 1$	to
	[UA1 Data]	~ 0.2
Γ_Z Measurement (Direct)	$\Gamma_Z = 2.83 + 0.18\,\delta N_\nu$	0.06

*Improved calorimetry can change the ratio of B/A to ~ 1

$$r = \frac{[d\sigma/_{d[\delta N_\nu]}]\; N_\nu = 3}{[\sigma]\; N_\nu = 3}$$

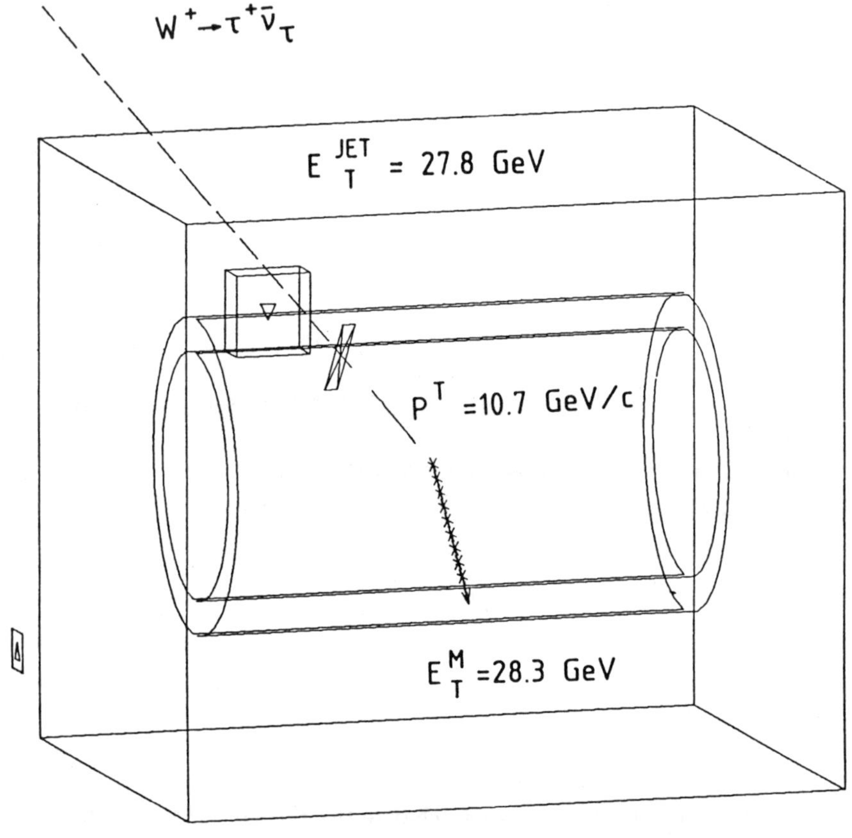

FIGURE 1

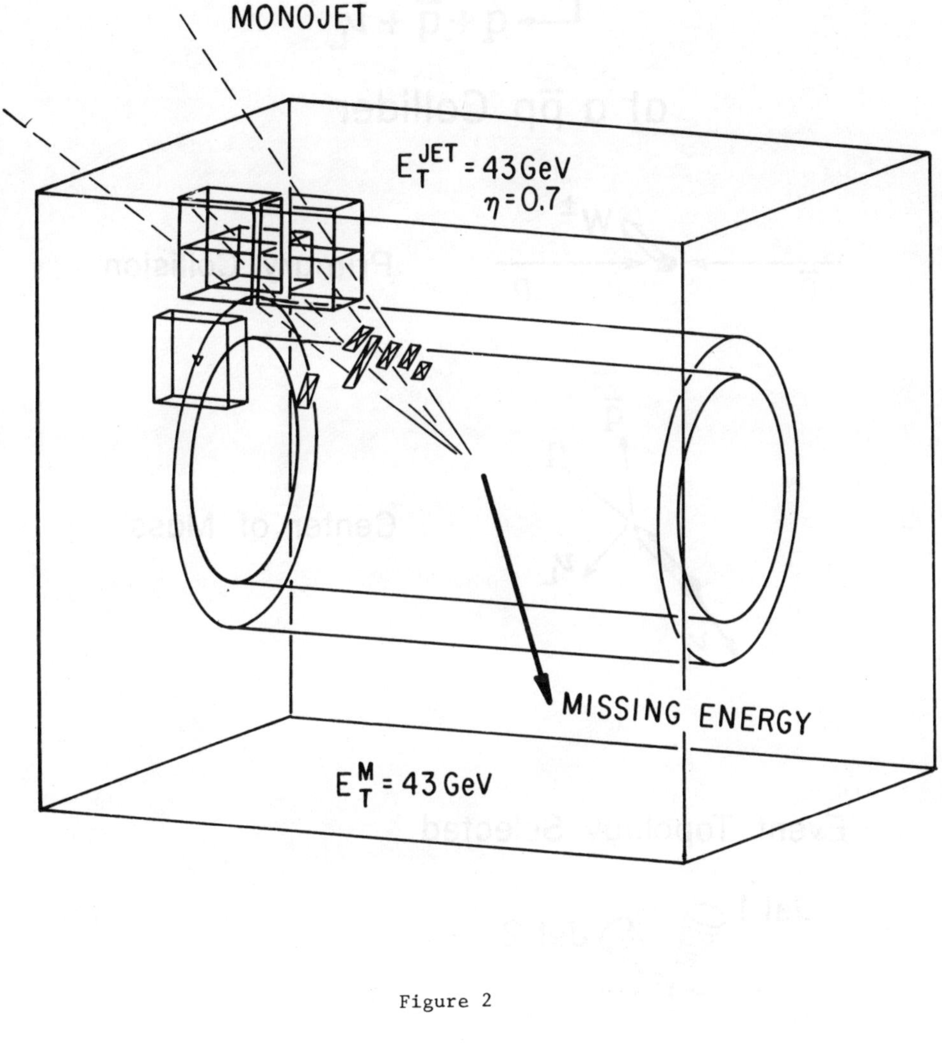

Figure 2

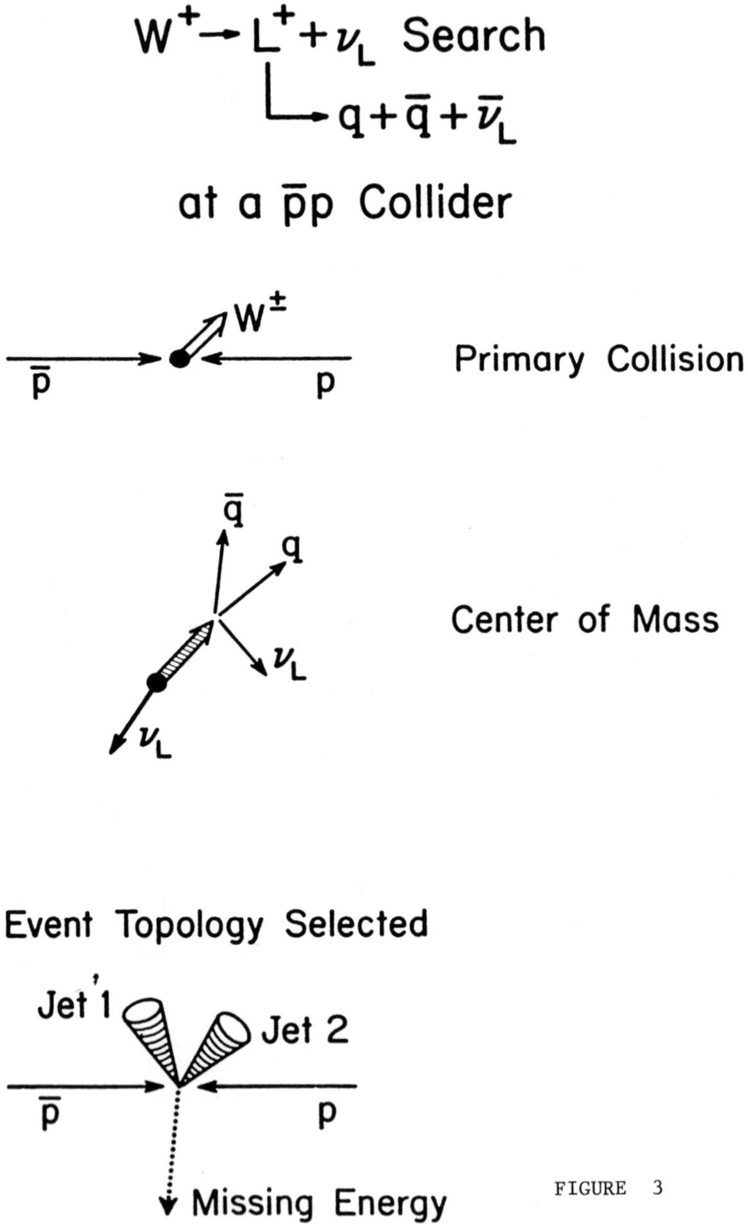
$W^+ \to L^+ + \nu_L$ Search
$\to q + \bar{q} + \bar{\nu}_L$
at a $\bar{p}p$ Collider
$W^\pm$
$\bar{p}$
p
Primary Collision
$\bar{q}$
q
ν_L
ν_L
Center of Mass
Event Topology Selected
Jet 1
Jet 2
$\bar{p}$
p
Missing Energy

FIGURE 3

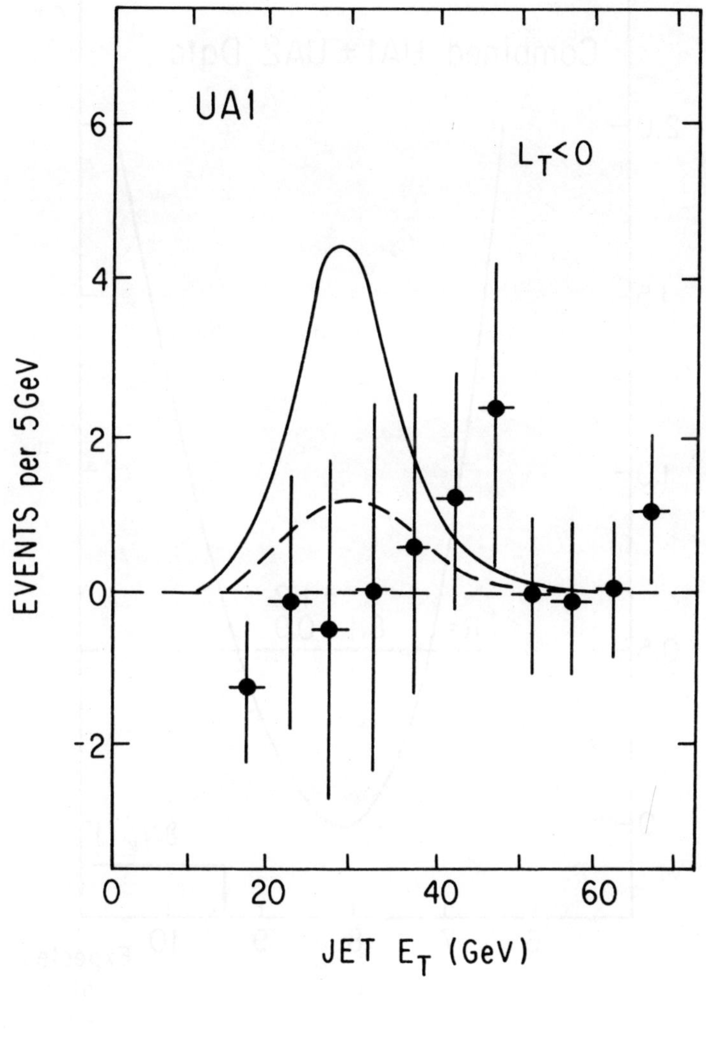

Figure 4

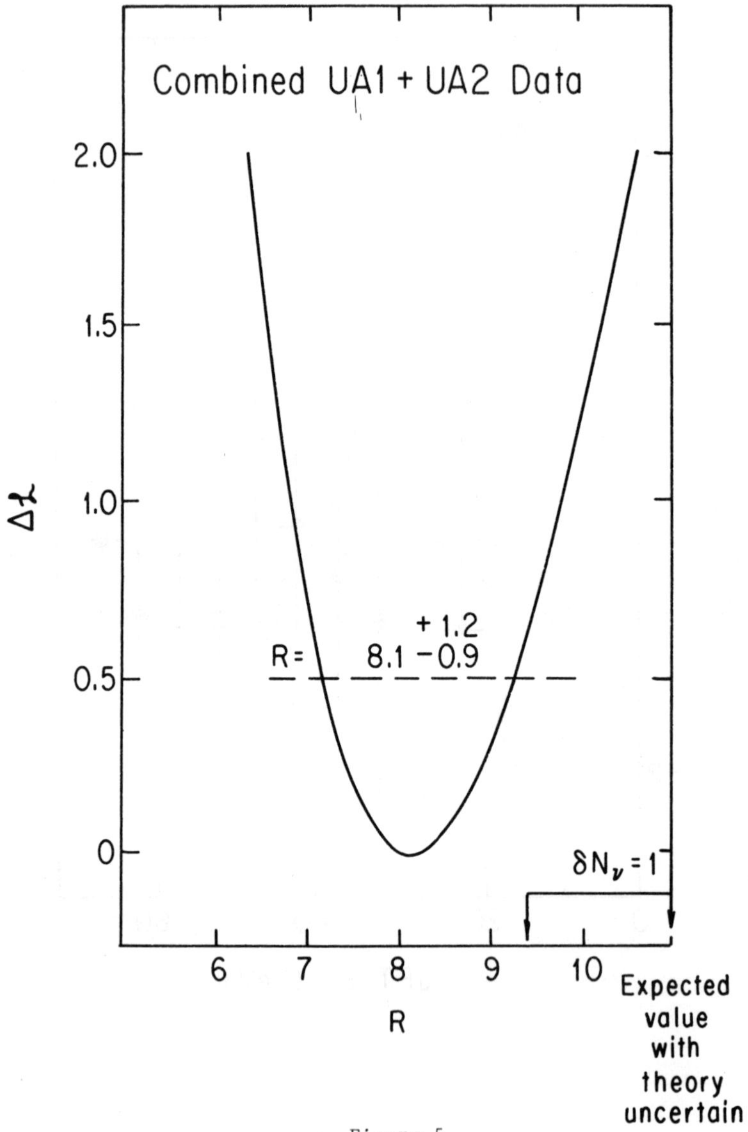
Combined UA1 + UA2 Data
2.0
1.5
1.0
0.5
0
Δℒ
R = 8.1 +1.2 −0.9
δN_ν = 1
6
7
8
9
10
R
Expected value with theory uncertain

Figure 5

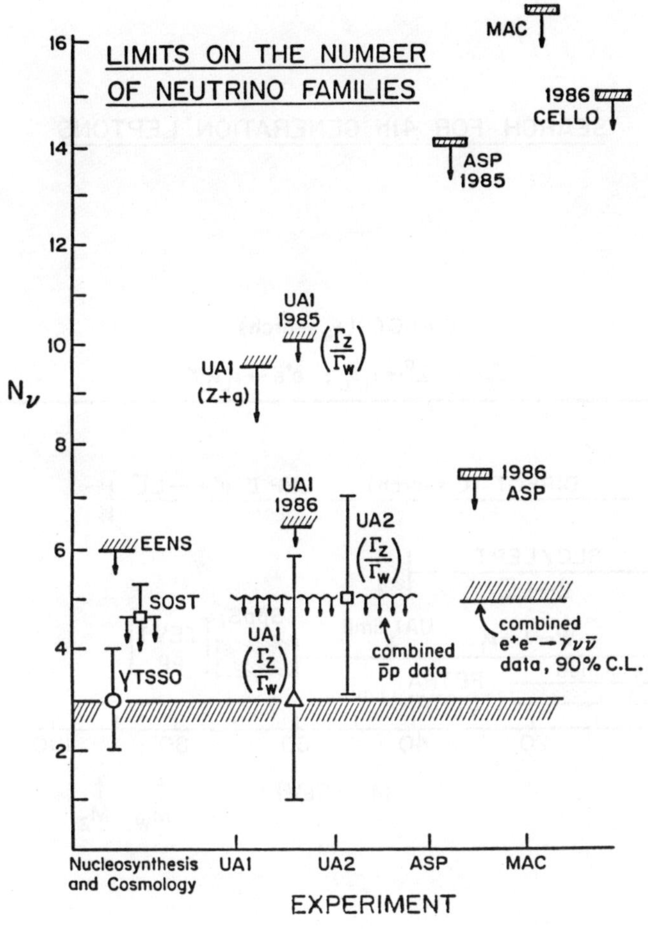
LIMITS ON THE NUMBER
OF NEUTRINO FAMILIES
16
14
12
10
8
6
4
2
N_ν
MAC
1986
CELLO
ASP
1985
UA1
1985
$\left(\frac{\Gamma_Z}{\Gamma_W}\right)$
UA1
(Z+g)
UA1
1986
UA2
$\left(\frac{\Gamma_Z}{\Gamma_W}\right)$
1986
ASP
EENS
SOST
UA1
$\left(\frac{\Gamma_Z}{\Gamma_W}\right)$
combined
$\bar{p}p$ data
combined
$e^+e^- \to \gamma\nu\bar{\nu}$
data, 90% C.L.
YTSSO
Nucleosynthesis
and Cosmology
UA1
UA2
ASP
MAC
EXPERIMENT

FIGURE 6

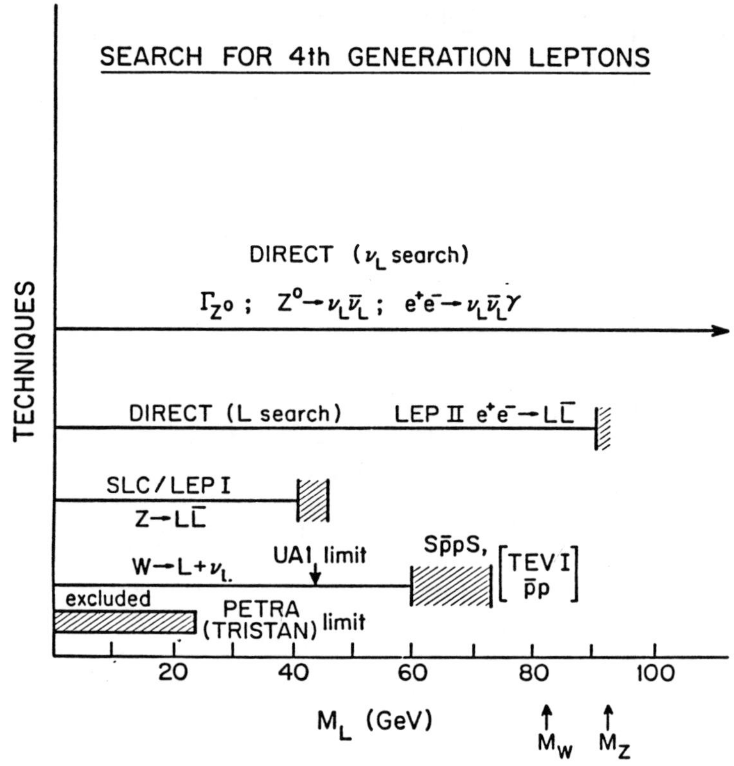
SEARCH FOR 4th GENERATION LEPTONS
TECHNIQUES
DIRECT (ν_L search)
Γ_{Z^0} ; $Z^0 \to \nu_L \bar{\nu}_L$; $e^+e^- \to \nu_L \bar{\nu}_L \gamma$
DIRECT (L search)
LEP II $e^+e^- \to L\bar{L}$
SLC/LEP I
$Z \to L\bar{L}$
$W \to L + \nu_L$
UA1 limit
$S\bar{p}pS$,
TEV I $\bar{p}p$
excluded
PETRA (TRISTAN) limit
20
40
60
80
100
M_L (GeV)
M_W
M_Z

FIGURE 7

W and Z Properties in UA1

UA1 Collaboration

Presented by
G. Bauer
Harvard University
Cambridge, MA

I. Introduction

Since the first observation of the Intermediate Vector Bosons [1] of the electroweak theory [2], the CERN $\bar{p}p$ collider has continued to provide a growing sample of $W^{\pm}$ and Z^0 bosons. The current results on $W^{\pm}$ and Z^0 properties measured in the UA1 detector are presented here. The focus will be upon the electroweak results; the production properties can be found in another presentation at this conference [3].

We only sketch an outline of the UA1 detector here, for it has been discussed in detail elsewhere [4]. The core is a large drift chamber immersed in a uniform dipole magnetic field which provides a momentum measurement of charged particles. Around the drift chamber is an electromagnetic calorimeter, followed by a hadronic calorimeter. The hermetic calorimeter coverage permits the reliable measurement of the apparent E_T imbalance due to high E_T neutrinos. Behind the hadron calorimeter lies three planes (per projection) of limited streamer tubes interleaved in an iron wall, followed by four planes of chambers for the tracking and triggering of muons.

Data has been accumulated at $\sqrt{s}$ of 546 and 630 GeV. The analysis extends over three data sets corresponding to electrons, muons and taus. We briefly outline these data sets below; more detail can be found in previous publications [3,5-7].

The electron data required a cluster of electromagnetic (e.m.) calorimeter cells to have an E_T greater than 15 GeV, and having a matching $p_T > 7$ GeV/c track. The cluster is required to pass standard isolation criteria, and have a shower shape consistent with that expected for an isolated e.m. cluster. The resulting 266 events with a missing E_T above 15 GeV form our $W^{\pm} \rightarrow e^{\pm}\nu$ sample. Events with a second isolated e.m. cluster with $E_T > 15$ GeV, and two-cluster mass above 70 GeV/c^2 (to eliminate Drell-Yan pairs), compose the $Z^0 \rightarrow e^+e^-$ sample (32 events).

For the muon sample, the muon chamber track is required to match a track in the central chamber with p_T > 15 GeV/c and which passes isolation criteria. Application of a missing energy cut of 15 GeV produces a sample of 57 $W \rightarrow \mu \nu$ events. The signature for the Z^0 is so distinctive that the missing E_T requirement in the W selection need only be replaced by the presence of a second track with p_T > 15 GeV/c in the central chamber to produce a Z^0 sample of 21 events.

The $W \rightarrow \tau \nu$ sample is obtained from the electron, jet, or the missing E_T triggers. We required E_T^{miss} > 15 GeV and 4σ from zero, with at least one jet with E_T > 12 GeV (at least one p_T > 1 GeV/c track). The background from jet-jet fluctuations is reduced by vetoing events with a jet within $\pm 30°$ azimuthally of the missing E_T, or opposite the highest E_T jet. To produce a sample of $W \rightarrow \tau \nu$ events a "τ-likelihood" [7] cut is applied, which reflects the distinctive characteristics of the hadronic decay modes of the τ (low charged multiplicity and narrowness of the jet). A $L_\tau > 0$ cut (retaining 78% of the $W \rightarrow \tau \nu \rightarrow$ hadrons) resulted in 32 events.

II. W and Z^0 Masses and Standard Model Parameters

The Z^0 has been observed in both the e and μ channels, and its mass has been determined separately for each. A subsample of 14* well measured electron events was fit, resulting in a mass value of 93.0 ±1.4 (±3.0) GeV/c^2 (first error is statistical, second is systematic). The fit for the 17 well measured muon events yields a mass of 94.1 $^{+8.4}/_{-6.6}$ (±2.8) GeV/c^2. If we assume that these events have no missing energy, we can impose overall energy balance in the event, which means that the total transverse energy of the calorimeters is equal and opposite to the Z^0 p_T. This calorimeteric measure of the Z^0 p_T is often more precise than that determined by the charged tracks. This additional constraint can be applied to the mass fit resulting in a value of 90.7 $^{+5.2}/_{-4.8}$ (±3.2) GeV/c^2. These determinations of m_Z are consistent with each other and with the result 91.5 ±1.2 (±1.7) GeV/c^2 obtained by UA2 [8]. The observed width of the Z^0 is dominated by the detector resolution, but from the electron fit, an upper limit of 8.3 GeV (90% C.L.) is derived.

The determination of m_W is slightly complicated by the emission of a ν. Although the E_T is determined from the missing transverse energy, the longitudinal component of the ν is unknown and the two-body mass can not be constructed. Instead we use the transverse mass:

---

* The mass fits presented in this talk do not yet include the 1985 electron data.

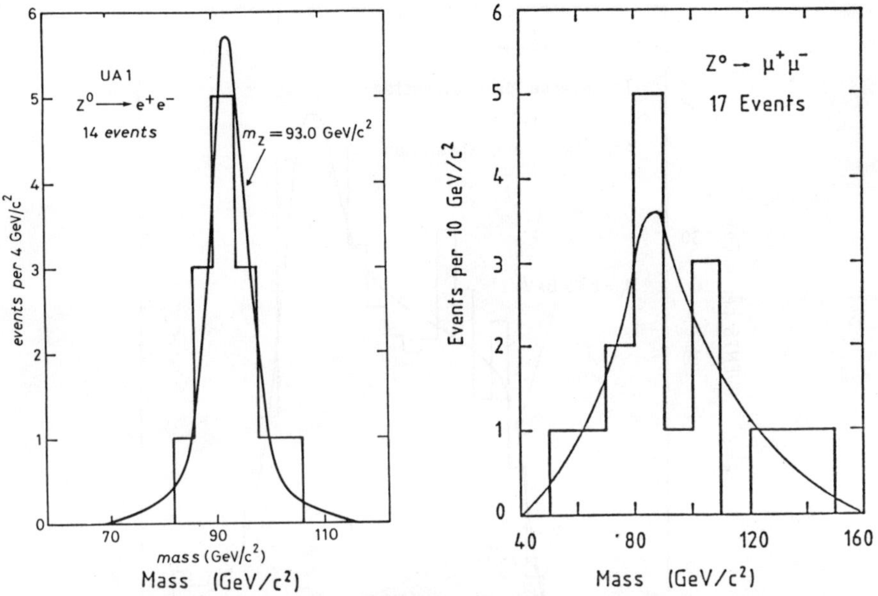

Fig. 1 Dilepton mass distribution in the electron and muon channels for the Z^0 with the fit results superimposed.

$$m_T = \sqrt{2E_T^e E_T^\nu (1 - \cos\phi_{e\nu})}$$

where the E_T's for the electron and neutrino are used, and $\phi_{e\nu}$ is the transverse opening angle of the pair. The m_T distribution for the $W \rightarrow e\nu$ decays is shown in Fig. 2. The background, about 10% of the signal, arises principally from jet-jet fluctuations and $W \rightarrow \tau\nu \rightarrow e\nu\nu$. Since these events are associated with low E_T electrons and neutrinos, we apply an E_T cut of 30 GeV to both particles to form the enhanced transverse mass distribution, shown in Fig. 3a. The fit for this distribution results in $m_W = 83.5\ ^{+1.1}/_{-1.0}$ ($\pm$2.7) GeV/c^2, and $\Gamma_W \leq 6.5$ GeV (90% C.L.).

The m_T distributions for $W \rightarrow \mu\nu$ and $\tau\nu$ are shown in Fig 3b-c. The curve for the muon data is the result of a fit yielding $m_W = 81.8\ ^{+6.0}/_{-5.3}$ ($\pm$2.6) GeV/c^2. The curve superimposed on the tau data corresponds to a W mass of 83.5 GeV/c^2. The data can be seen to prefer a higher mass, and a likelihood fit results in a value of 89 $\pm$3 ($\pm$6) GeV/c^2. Again, the measurements for each channel are in good agreement with each other, and also with the UA2 value, 80.2 $\pm$0.6 ($\pm$0.5 $\pm$1.3) GeV/c^2 [8].

The measurements of the $W^\pm$ and Z^0 masses provides an important means to test the Standard Model. The weak mixing angle is defined by the relation:

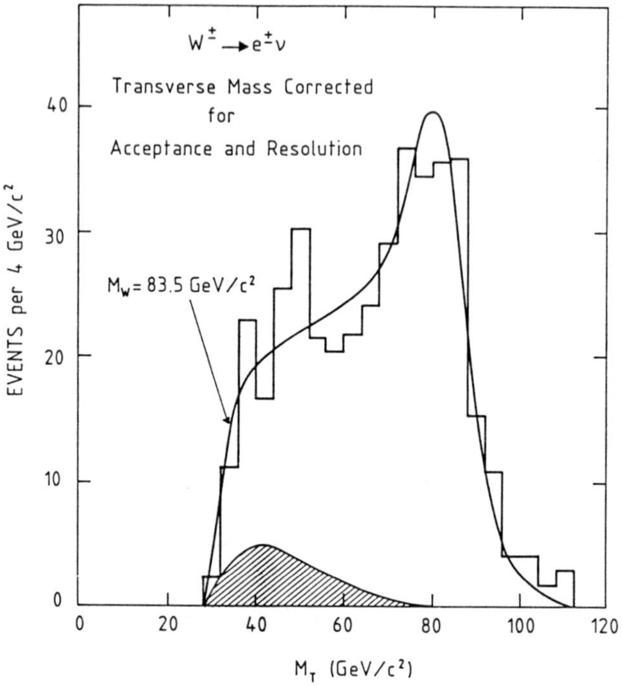

Fig. 2 Transverse mass distributions for $W \rightarrow e\nu$ events.

$$\sin^2\hat{\Theta}_W \equiv \frac{(38.65\ \mathrm{GeV})^2}{M_W^2}$$

or, in an alternative definition is related to the boson masses by:

$$\sin^2\Theta_W \equiv 1 - \frac{M_W^2}{M_Z^2}$$

Table 1 shows a compilation of these parameters, and good agreement is seen with the value of $\sin^2\Theta_W$ obtained by low-energy ν-scattering experiments:

0.225 ± 0.005 (±0.003 theor.)	CDHS [9]
0.236 ± 0.005 (±0.003 theor.)	CHARM [10]

Finally, we can define the parameter ρ,

$$\rho \equiv \frac{M_W^2}{M_Z^2} \sec^2\hat{\Theta}_W$$

The minimal Standard Model predicts $\rho = 1$, and deviations from this value are

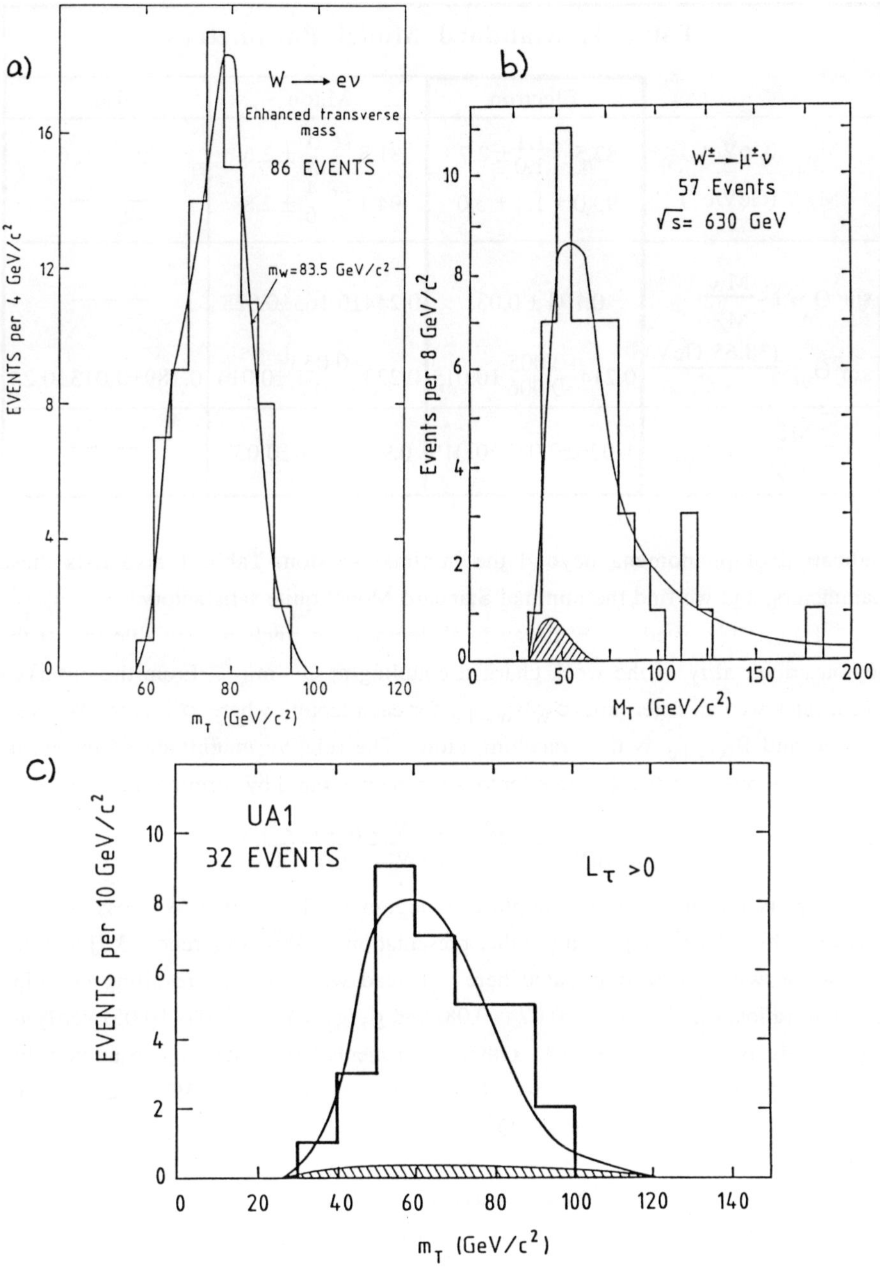

Fig. 3 Dilepton transverse mass distributions for $W\rightarrow l\nu$: a) electron data (enhanced distribution), b) muon data, c) tau's from missing E_T selection.

Table 1: Standard Model Parameters

	Electron	Muon	Tau
M_W (GeV/c^2)	$83.5\ ^{+1.1}_{-1.0} \pm 2.7$	$81.8\ ^{+6.0}_{-5.3} \pm 2.6$	$89 \pm 3 \pm 6$
M_Z (GeV/c^2)	$93.0 \pm 1.4 \pm 3.0$	$94.1\ ^{+8.4}_{-6.6} \pm 2.8$	———
$\sin^2\theta_W \equiv 1 - \frac{M_W^2}{M_Z^2}$	0.194 ± 0.031	$0.244 \pm 0.163 \pm 0.023$	———
$\sin^2\hat{\theta}_W \equiv \frac{(38.65\ \text{GeV})^2}{M_W^2}$	$0.214\ ^{+0.005}_{-0.006} \pm 0.015$	$0.223\ ^{+0.033}_{-0.029} \pm 0.014$	$0.189 \pm 0.013 \pm 0.25$
$\rho \equiv \frac{M_W^2}{M_Z^2} \sec^2\hat{\theta}_W$	$1.026 \pm 0.037 \pm 0.019$	$0.97 \pm 0.19 \pm 0.03$	———

indicative of phenomena beyond the minimal version. Table 1 also lists these parameters, and we find the minimal Standard Model quite satisfactory.

Having observed the W decay in all leptonic channels we are able to test the lepton universality of the weak charged coupling at $Q^2 = m_W^2$. From the observed event rates we can determine $\sigma_W B_{W\to l\nu}$ for each lepton, where σ_W is the W cross section and $B_{W\to l\nu}$ is the branching ratio. The relative magnitude of the weak charged coupling for the different leptons can be measured by forming the ratio:

$$R_{e\mu} = \frac{\sigma B(W\to \mu\nu)}{\sigma B(W\to e\nu)} = \frac{\Gamma(W\to \mu\nu)}{\Gamma(W\to e\nu)} = \left(\frac{g_\mu}{g_e}\right)^2$$

where g_l is the electroweak coupling to lepton l. The actual W cross sections measured by UA1 are reported in other presentations at this conference [3,7] and that discussion will not be duplicated here. Instead we quote the resulting coupling constant ratios: $g_\mu/g_e = 1.05 \pm 0.07\ (\pm 0.08)$ and $g_\tau/g_e = 1.01 \pm 0.09\ (\pm 0.05)$ verifying lepton universality to about 10%. Other experiments have tested lepton universality to greater precision, ~0.6% for e-μ (from π decays [11]) and ~5% for e-τ (from τ decays [12]), but at much lower Q^2.

III. Decay Angular Distributions

One of the most distinctive features of the leptons observed in W decay is the angular asymmetry that arises due to the parity violating nature of the weak interaction. W production arises via quark-antiquark annihilation, since the W couples only to left- (right-)handed (anti-)fermions the W^- (W^+) is polarized along the direction of the p ($\bar{p}$). With the W so polarized, the lepton, being left-handed, will preferentially be emitted in the direction of the outgoing proton, i.e. e^- in the out going p-direction, and e^+ in that of the $\bar{p}$. The angular distribution is predicted to be of the form $(1 + \cos\Theta^*)^2$, where Θ^* is the angle of the e^- (e^+) with respect to the outgoing p ($\bar{p}$) in the W rest frame. The measured angular distribution for the electron events with well determined charges is shown in Fig. 4, and it agrees well with the theory. The asymmetry, $A=(N_+-N_-)/(N_++N_-)$ (where N_+ is the number of events with $\cos\Theta^* > 0$, and N_- for those less than zero) is found to be 0.77 ± 0.04, compared to the predicted value of 0.75. From a model independent perspective it has been shown [13] that a particle of arbitrary spin, J, produced by a system with global helicity μ, and decaying into a system with helicity λ, will have

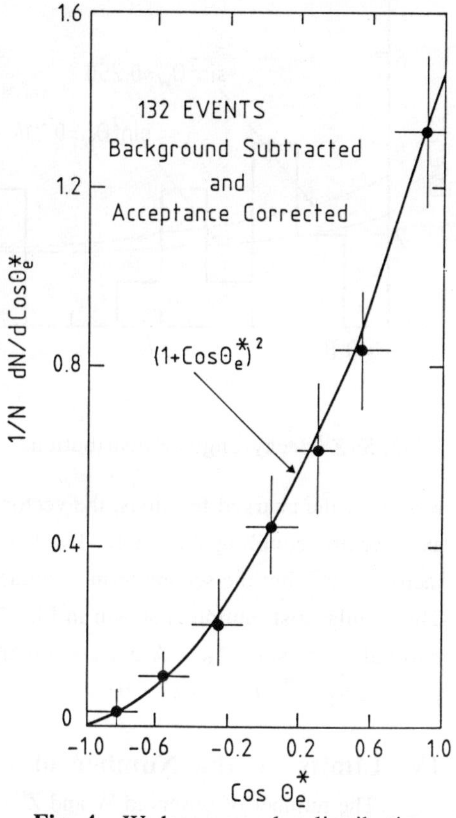

Fig. 4 W decay angular distribution.

$$\langle\cos\Theta^*\rangle = \frac{\langle\lambda\rangle\langle\mu\rangle}{J(J+1)}$$

In the Standard Model $\langle\mu\rangle = \langle\lambda\rangle = -1$ and J = 1, resulting in $\langle\cos\Theta^*\rangle = 0.5$, whereas $\langle\cos\Theta^*\rangle = 0$ for J = 0, and for $J \geq 2$, $\langle\cos\Theta_1^*\rangle \leq 1/6$. The data yields $\langle\cos\Theta^*\rangle = 0.43 \pm 0.07$, which supports the spin-1 assignment for the W and maximal helicity of the initial and final states.

Unlike the W, the Z^0 is expected to have a small asymmetry since it is not restricted to left handed couplings. Specifically, the angular distribution is given by

$$\frac{d\sigma}{d\cos\Theta_1^*} = (V_i^2 + A_i^2)(V_j^2 + A_j^2)(1 + \cos^2\Theta_1^*) + 8A_i^2 V_i^2 A_j^2 V_j^2 \cos\Theta_1^*$$

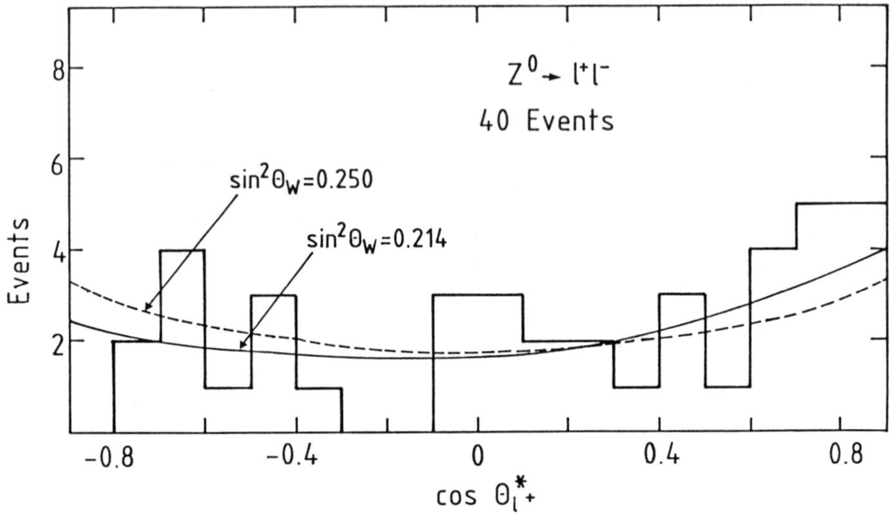

Fig. 5 Z^0 decay angular distribution.

where for the charged fermions, the vector term $V_f = \pm (1 - 4\,|Q_f|\,\sin^2\Theta_W)$, and the axial vector coupling $A_f = \pm\, 1$. The first term in the angular distribution is symmetric in Θ^*, but the second term introduces a small asymmetry dependent on Θ_W. The angular distribution is shown in Fig. 5 with theoretical curves superimposed for two values of $\sin^2\Theta_W$. A net asymmetry is observed, $A_Z = 0.30 \pm 0.15$, corresponding to $\sin^2\Theta_W = 0.18 \pm 0.04$, in agreement with earlier determinations.

IV. Limits on the Number of Light Neutrinos

The number of observed W and Z^0 events is, aside from detection efficiencies, dependent upon the cross sections and leptonic branching ratios, and the branching ratios are in turn dependent upon all the allowed decay modes. If we form the ratio:

$$R = \frac{\sigma_W B(W\to l\,\nu)}{\sigma_Z B(Z\to l^+l^-)} = \frac{\sigma_W}{\sigma_Z}\;\frac{\Gamma\,(W\to l\nu\,)}{\Gamma\,(Z\to l^+l^-)}\;\frac{\Gamma_Z^{Tot}}{\Gamma_W^{Tot}}$$

we see that R is dependent upon the product of three ratios. The first two, the ratio of production cross sections and the leptonic decay widths, are calculable within the Standard Model. While the third term, the ratio of the total widths, is dependent upon all allowed decay modes. The value of R, obtained from the cross sections, constrains new decay modes of the W and Z^0; of particular interest is $Z^0\to \nu\,\nu$ for

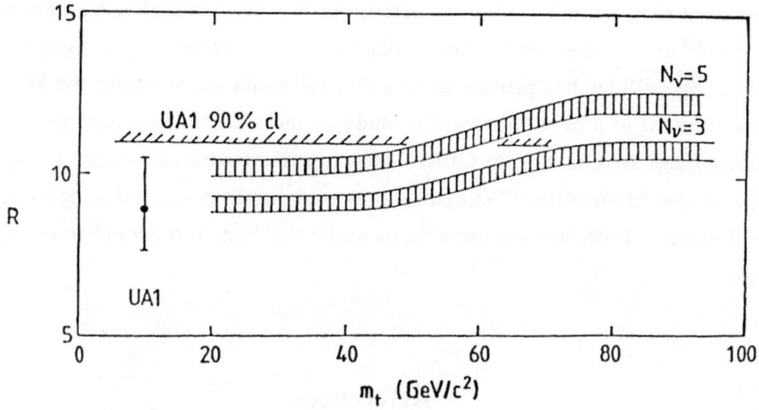

Fig. 6 The measured value of R, its 90% C.L., and the theoretical value as a function of m_t for 3 and 5 neutrinos. Shaded bands indicate the range of theoretical uncertainty (from Ref. 15).

new light neutrinos. Assuming that there no other unexpected paricles entering in either the W or Z^0 decays, such as a new sequential lepton (current mass limit is 41 GeV/c^2 [7]), the ratio of the total widths is only dependent upon the number of neutrinos and the top mass (assuming $Z \rightarrow t\bar{t}$ or $W \rightarrow t\bar{b}$ is kinematically allowed). Combining the electron and muon data [3] in a log likelihood fit, we arrive at a ratio of $8.9\ ^{+1.6}/_{-1.3}$ (or $R \leq 11.1$ at 90% C.L.).

The principal "theoretical" limitation in this method is due to the error in the production cross sections that arises from the uncertainty in the structure functions. Several authors [14,15] have conducted studies of this method and its limitations. The results of the most recent analysis [15] is shown in Fig. 6, where we see that the errors are large enough that m_t is not quite constrained, but there can be no more than 7 neutrinos (90% C.L.) for a light top.

V. Conclusions

We have observed the W in all three leptonic decay channels, and the Z^0 in the electron and muon channels. The mass values and the resultant Standard Model parameters are well described by the minimal version of SU(2)xU(1). A comparison of the three leptonic decay modes of the W was used to test the universality of the charged weak coupling among the three leptons, and was found to be the same as the electron to 10% at $Q^2=m_W{}^2$. The decay angular distribution of the W is observed to have the large asymmetry expected from (V-A), and the W is explicitly found to be spin-1. The Z^0 angular distribution is also found to be in agreement with the theory, and was used as a second method to determine $\sin^2 \Theta_W$. Finally, a comparison of the W and Z^0 event rates can be used to place limits on the number of light neutrinos.

The $\bar{p}p$ collider has proven to be a fruitful arena for studying the W and Z^0. We look forward to a new phase in the study of the weak bosons next year with the operation of the ACOL ring at CERN, which will provide an order of magnitude increase in the luminosity. With this increased luminosity, and a new uranium-TMP calorimeter [16], we will be able to probe the Standard Model to ever greater precision.

References

[1] G. Arnison *et al.* (UA1 Collaboration), Phys. Lett. **122B** (1983) 103; M. Banner *et al.* (UA2 Collaboration), Phys. Lett. **122B** (1983) 476; G. Arnison *et al.* (UA1 Collaboration), Phys. Lett. **126B** (1983) 398; P. Bagnaia *et al.* (UA2 Collaboration), Phys. Lett. **129B** (1983) 130.

[2] S. Weinberg, Phys. Rev. Lett. **19** (1967) 1264; A. Salam, *Proc. 8th Nobel Symp.*, Apenasgarden, 1968 (Almqvist and Wiksell, Stockholm, 1968), p. 367.

[3] D. Smith (UA1 Collaboration), these proceedings.

[4] M. Barranco Luque *et al.*, Nucl. Instr. and Meth., **176** (1980) 175; C. Cochet *et al.*, Nucl. Instr. and Meth., **A243** (1986) 45; M. Corden *et al.*, Nucl. Instr. and Meth., **A238** (1985) 273; K. Eggert *et al.*, Nucl. Instr. and Meth., **176**, (1980) 217; G. Bauer *et al.*, CERN-EP/86-161 (Nucl. Instr. and Meth., in press).

[5] G. Arnison *et al.* (UA1 Collaboration), Europhys. Lett. **1** (1986) 327.

[6] G. Arnison *et al.* (UA1 Collaboration), Phys. Lett. **134B** (1984) 469; G. Arnison *et al.* (UA1 Collaboration), Phys. Lett. **147B** (1984) 241.

[7] D. Cline (UA1 Collaboration), these proceedings; C. Albajar *et al.* (UA1 Collaboration), CERN-EP/86-81 (Phys. Lett., in press).

[8] K. Einsweiler (UA2 collaboration), these proceedings.

[9] H. Abramowicz *et al.* (CDHS Collaboration) Phys. Rev. Lett. **57** (1986) 298.

[10] K. Winter (CHARM Collaboration), *Proc. 13th Int. Conf. on High-Energy Physics*, Berkeley, 1986, in press.

[11] D.A. Bryman *et al.*, Phys. Rev. Lett. **50** (1983) 7.

[12] B. Barish *et al.*, CALT 68-1321 (1985).

[13] M. Jacob, Nuovo Cimento **9** (1985) 826.

[14] F. Halzen and K. Mursula, Phys. Rev. Lett. **51** (1983) 857; K. Hikasa, Phys. Rev. **D29** (1984) 1939; N.G. Deshpande *et al.*, Phys. Rev. Lett. **54** (1985) 1757; F. Halzen, Phys. Lett. **182B** (1986) 388.

[15] A.D. Martin, R.G. Roberts, and W.J. Stirling, RAL-87-002 (also DTH/87/2) 1987.

[16] J.D. Dowell (UA1 Collaboration), *Proc. 6th Topical Workshop on Proton-Antiproton Collider Physics*, Aachen, 1986 (CERN-EP/86-193).

ELECTROWEAK RADIATIVE CORRECTIONS
and Neutral Current Phenomenology

William J. Marciano
Brookhaven National Laboratory
Upton, New York 11973

Precision measurements of $\sin^2\theta_W$ including the effects of radiative corrections are surveyed. A world average $\sin^2\theta_W = 0.231\pm 0.004$ is obtained. Comparison of deep-inelastic $\nu_\mu N$ scattering and m_W or m_Z is shown to test the standard model at the quantum loop level and constrain new physics. Implications for extra Z' bosons and grand unified theories are briefly discussed.

Introduction

Neutral current measurements of $\sin^2\theta_W$ and direct determinations of m_W and m_Z have now become precise enough to start testing the standard $SU(2)_L \times U(1)$ model at the level of its $O(\alpha)$ radiative corrections. Those tests confirm the standard model[1] and constrain new physics that might be appended to it. Furthermore, they now provide an accurate value for $\sin^2\theta_W$ which can be used to rule out or at least constrain some grand unified theories. In this talk, I will survey the present status of the most precise $\sin^2\theta_W$ experiments, focusing on the effect of radiative corrections and discuss some constraints on new physics implied by those measurements. The results quoted are part of an ongoing analysis by the collaboration in Ref. 1. Final results along with a detailed discussion of their implications will be published shortly.

Listed in Table 1 are the values of $\sin^2\theta_W^{\text{unc.}}$ obtained from various types of experiments before taking electroweak radiative corrections into account. The numerical shift due to $O(\alpha)$ electroweak corrections and the final $\sin^2\theta_W \equiv 1 - m_W^2/m_Z^2$ values are also given. Those radiative corrections were evaluated in the standard model assuming $m_t \simeq 45\text{GeV}$ and $m_H \simeq 100\text{GeV}$. The effect of relaxing the assumption $m_t = 45\text{GeV}$ will be subsequently described. (Throughout this

Table 1: Values of $\sin^2\theta_W \equiv 1 - m_W^2/m_Z^2$ before and after electroweak radiative corrections are included for a variety of experiments. The values $m_t \simeq 45$GeV and $m_H \simeq 100$GeV were employed in the radiative corrections.

Experiment	$\sin^2\theta_W^{\text{unc.}}$	Rad. Cor.	$\sin^2\theta_W$
Cs Atomic P.V. (Paris[2]- Boulder[3])	0.221 ± 0.027	+0.008	0.229 ± 0.027
eD Asymmetry (Yale - SLAC[4])	0.224 ± 0.020	-0.005	0.219 ± 0.020
$\overset{(-)}{\nu}_\mu e$ (BNL[5]- CHARM[6])	0.212 ± 0.023	$\lesssim$ 0.002	0.214 ± 0.023
$\overset{(-)}{\nu}_\mu P$ (BNL[7])	0.220 ± 0.016 $^{+0.023}_{-0.031}$	+0.002	0.222 ± 0.016 $^{+0.023}_{-0.031}$
$\nu_\mu N$ deep-inel. (CDHS,[8] CHARM,[9]) (CCFRR,[10] FMM[11])	0.241 ± 0.006	-0.009	0.232 ± 0.006
$m_W = 80.9 \pm 1.4$GeV (UA1[12]- UA2[13])	0.212 ± 0.008	+0.017	0.229 ± 0.008
$m_Z = 91.9 \pm 1.8$GeV (UA1[12]- UA2[13])	0.208 ± 0.011	+0.022	0.230 ± 0.011
World Average*			0.231 ± 0.004

*(Including all other experiments)

paper, I assume that ordinary QED corrections are separately accounted for by experimentalists.)

Let me begin by briefly commenting on each of the $\sin^2\theta_W$ measurements illustrated in Table 1.

Atomic Parity Violation: Cesium is the simplest heavy atomic system in which parity violation has been observed. Experiments in Paris and Boulder find for the so-called weak charge[2,3]

$$Q_W(Cs) = -71.7 \pm 5.8 \quad \text{(Exp. Ave.)} \tag{1a}$$

where the error comes from experiment and atomic theory. That is to be compared with the standard model prediction (for $\sin^2\theta_W \equiv 1 - m_W^2/m_Z^2$)

$$Q_W\,(Cs) = -23 - 220\sin^2\theta_W^{\text{unc.}} \quad \text{(tree level)} \tag{1b}$$

$$Q_W\,(Cs) = -22.5 - 215\sin^2\theta_W \quad (\text{inc. R.C.}^{14}) \tag{1c}$$

Comparison of these quantities leads to the values for $\sin^2\theta_W^{\text{unc.}}$ and $\sin^2\theta_W$ in Table 1. The good agreement between $\sin^2\theta_W$ extracted in this way and the world average value provides one of the best constraints for models with extra neutral gauge bosons (see discussion below). Further measurements and continued scrutiny of the atomic theory may reduce the error in $Q_W\,(Cs)$ by a factor of 3 or 4.

eD Asymmetry: The asymmetry,

$$A \equiv (d\sigma_R - d\sigma_L)\,/\,(d\sigma_R + d\sigma_L)\,,$$

in the scattering of polarized electrons on an unpolarized deuterium target was measured a number of years ago at SLAC. Without taking electroweak radiative corrections into account they found[4]

$$\sin^2\theta_W^{\text{unc.}} = 0.224 \pm 0.020 \tag{2a}$$

Radiative corrections[15,16] reduce the predicted asymmetry and lead to

$$\sin^2\theta_W = 0.219 \pm 0.020. \tag{2b}$$

$\overset{(-)}{\nu}_\mu e$ **Scattering:** Experiments at BNL[5] and CERN[6] have measured the quantity

$$R \equiv \frac{\sigma\,(\nu_\mu e \to \nu_\mu e)}{\sigma\,(\bar{\nu}_\mu e \to \bar{\nu}_\mu e)} = \frac{3\left(1 - 4s^2 + 16/3s^4\right)}{\left(1 - 4s^2 + 16s^4\right)} \tag{3a}$$

$$s^2 \equiv \sin^2\theta_W$$

to be

$$R = 1.38^{+0.40}_{-0.31} \pm 0.17 \quad \text{(BNL Exp.)} \tag{3b}$$

$$R = 1.26^{+0.72}_{-0.45} \quad \text{(CHARM Collab.)} \tag{3c}$$

Those measurements lead to the value of $\sin^2\theta_W$ in Table 1. Radiative corrections are small[17] for $m_t \simeq 45$GeV. A CHARM II experiment at CERN hopes to collect several thousand $\overset{(-)}{\nu}_\mu e$ events and thereby lower the statistical error in $\sin^2\theta_W$ obtained from *Eq.* (3*a*) to ± 0.005.

$\overset{(-)}{\nu}_\mu P$ **Elastic Scattering**: Recent analysis of the ratio $\sigma^{el}(\nu_\mu P)/\sigma^{el}(\bar{\nu}_\mu P)$ by a BNL[7] experimental collaboration yields the value of $\sin^2\theta_W$ in Table 1. A complete analysis of the radiative corrections has not been carried out, but many corrections cancel in such a ratio. Continuing data analysis should lower the statistical error in $\sin^2\theta_W$ considerably.

Deep-Inelastic $\nu_\mu N$ Scattering: New high statistics results for

$$R_\nu \equiv \sigma(\nu_\mu N \to \nu_\mu X)/\sigma(\nu_\mu N \to \mu X)$$

have been reported.[8,9,10] That quantity must be corrected for sea quark effects, QCD corrections, charm thresholds, etc.[1] After all such corrections are made, the experimental value for R_ν can be compared with the standard model valence quark prediction[18,19]

$$R_\nu = \rho_{\nu N}^2\left[\frac{1}{2} - \kappa\sin^2\theta_W + \frac{20}{27}\kappa^2\sin^4\theta_W\right] \tag{4a}$$

$$\rho_{\nu N}^2 = 1 + O(\alpha) \simeq 0.98 \tag{4b}$$

$$\kappa = 1 + O(\alpha) \simeq 0.99 \tag{4c}$$

and $\sin^2\theta_W$ can be extracted. Setting $\rho_{\nu N} = \kappa = 1$ in Eq. (4a) leads to $\sin^2\theta_W^{\text{unc.}}$ given in Table 1. (It is dominated by the recent results from CDHS[8] and CHARM.[9]) Electroweak radiative corrections[16,18,19] tend to reduce $\sin^2\theta_W$ by about 4% relative to $\sin^2\theta_W^{\text{unc.}}$. In fact, the shift $\Delta s^2 \equiv \sin^2\theta_W^{\text{unc.}}(\nu N) - \sin^2\theta_W$ can be approximated by

$$\frac{\Delta s^2}{s^2} \simeq -4\Delta\rho + \Delta\kappa \tag{5}$$

where $\Delta\rho = \rho_{\nu N} - 1, \quad \Delta\kappa = \kappa - 1$.

$W^\pm$ and Z Masses: The natural relationship (with $\Delta r = 0$)

$$m_W^2 = m_Z^2\cos^2\theta_W = \frac{\pi\alpha}{\sqrt{2}G_\mu\sin^2\theta_W(1-\Delta r)} \tag{6}$$

is modified by $O(\alpha)$ electroweak radiative corrections embodied in Δr. An updated evaluation of those corrections gives[18,20,21,22]

$$\Delta r = 0.072 \pm 0.002 \tag{7}$$

Setting $\Delta r = 0$ in Eq. (6) and using $\alpha = 1/137.036$, $G_\mu = 1.16636 \times 10^{-5}\text{GeV}^{-2}$ leads to the lowest order relations

$$\sin^2\theta_W^{\text{unc.}} = \left(\frac{37.281\text{GeV}}{m_W}\right)^2 \tag{8a}$$

$$\sin^2 2\theta_W^{\text{unc.}} = \left(\frac{74.562\text{GeV}}{m_Z}\right)^2 \tag{8b}$$

Including the Δr value in Eq. (7) modifies them to

$$\sin^2\theta_W = \left(\frac{38.70\text{GeV}}{m_W}\right)^2 \tag{9a}$$

$$\sin^2 2\theta_W = \left(\frac{77.40\text{GeV}}{m_Z}\right)^2 \tag{9b}$$

Employing the average m_W and m_Z masses in Table 1 then leads to the values of $\sin^2\theta_W^{\text{unc.}}$ and $\sin^2\theta_W$ given there.

Glancing at the $\sin^2\theta_W^{\text{unc.}}$ and $\sin^2\theta_W$ columns in Table 1, it is clear that the uncorrected $\sin^2\theta_W^{\text{unc.}}$ value obtained from deep-inelastic $\nu_\mu N$ scattering differs from the m_W and m_Z values, but the $\sin^2\theta_W$ results are all in good agreement after radiative corrections are accounted for. The standard model is, therefore, being tested at the level of its $O(\alpha)$ electroweak radiative corrections.

To make the above comparison more quantitative, consider the quantities

$$\Delta r + \frac{\Delta s^2}{s^2} \simeq 1 - \left(\frac{37.281\text{GeV}}{m_W \sin\theta_W^{\text{unc.}}(\nu N)}\right)^2 \tag{10a}$$

$$\Delta r + \frac{\Delta \sin^2 2\theta_W}{\sin^2 2\theta_W} \simeq 1 - \left(\frac{74.562\text{GeV}}{m_Z \sin 2\theta_W^{\text{unc.}}(\nu N)}\right)^2 \tag{10b}$$

where $\sin^2\theta_W^{\text{unc.}}(\nu N)$ specifies the uncorrected value extracted from deep-inelastic data. Theory predicts (for $m_t \simeq 45\text{GeV}$) that these radiative corrections are 0.11 and 0.10 respectively. Inserting the values of $\sin^2\theta_W^{\text{unc.}}(\nu N)$, m_W and m_Z from Table 1 into these expressions gives 0.12 ± 0.04 and 0.10 ± 0.04. The agreement between theory and experiment is very good. It illustrates quite nicely the need for radiative corrections.

New physics appendages to the standard model or changing the input m_t value would modify the theoretical predictions in Eq. (10). We can, therefore, use the present good agreement with experiment as a constraint. For example, if m_t were very large, the prediction for Δr would decrease whereas Δs^2 hardly changes. For $m_t^2/m_W^2 \gg 1$, one finds the leading term in the Δr shift is given by[21]

$$-\frac{3\alpha}{16\pi}\frac{\cos^2\theta_W}{\sin^4\theta_W}\frac{m_t^2}{m_W^2} \tag{11}$$

The exact dependence on m_t can be found in Ref. 21. From the present experimental value of Δr, one finds the constraint

$$m_t \lesssim 180\text{GeV} \tag{12}$$

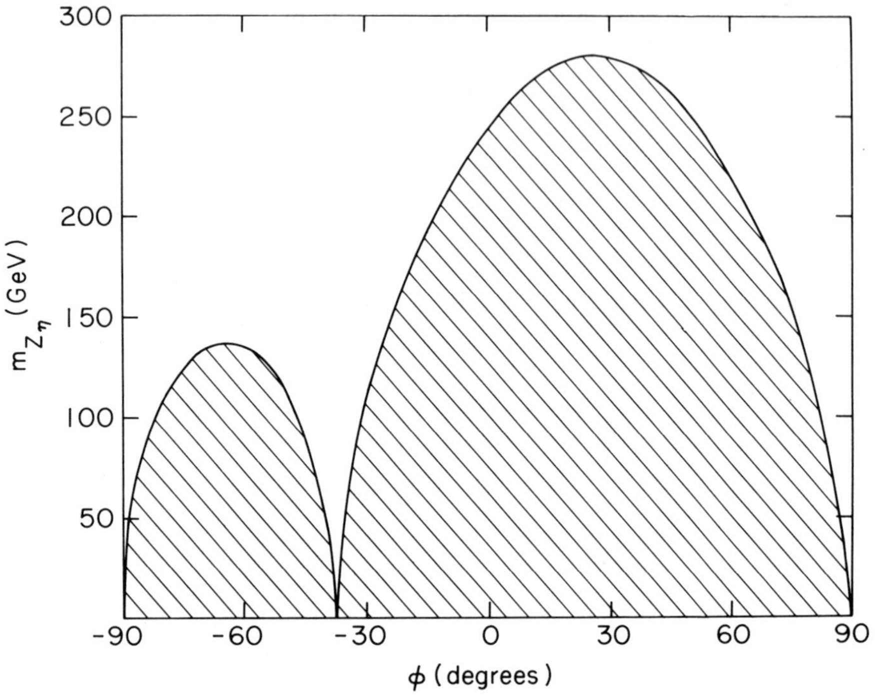

Figure 1: 90% C.L. constraints on m_{Z_η} from Cs atomic parity violation experiments.

That bound also applies to a 4th generation mass difference $m_{t'} - m_{b'}$. In the case of a heavy 4th generation charged lepton L with a massless neutrino partner, it becomes

$$m_L \lesssim 300\text{GeV} \tag{13}$$

Those constraints can be further improved by examining the effect of changing m_t on all experimental determinations of $\sin^2 \theta_W$. Results from such a global analysis will be reported in Ref. 1.

The good agreement between the various experiments in Table I can also be used to provide bounds on extra Z' gauge bosons.[23] Here, I would like to illustrate the important role low energy experiments such as atomic parity violation play in those constraints. Consider for example the E_6 model breaking $E_6 \to SU(3)_C \times SU(2)_L \times U(1) \times U(1)_\chi \times U(1)_\psi$ with two extra neutral gauge bosons

$$Z_\eta = Z_\chi \cos\phi + Z_\psi \sin\phi \tag{14a}$$

$$Z_{\eta'} = -Z_{\chi} \sin\phi + Z_{\psi} \cos\phi \tag{14b}$$

In such a scenario, the predicted weak charge in Eq. (1c) changes by (assuming no mixing of Z_η or $Z_{\eta'}$ with the ordinary Z boson)

$$\Delta Q_W(Cs) = +\frac{844}{3} tan^2\theta_W \left\{ \frac{m_W^2}{m_{Z_\eta}^2} \left(\cos^2\phi + \sqrt{\frac{5}{3}} \sin\phi \cos\phi \right) + \frac{m_W^2}{m_{Z_{\eta'}}^2} \left(\sin^2\phi - \sqrt{\frac{5}{3}} \sin\phi \cos\phi \right) \right\} \tag{15}$$

If one further assumes that $m_{Z_{\eta'}}^2 \gg m_{Z_\eta}^2$, then the good agreement between experiment and standard model theory leads to the 90% confidence level constraints on m_{Z_η} in Fig. 1. Note that the mass bounds are already quite good for a range of ϕ values ($\phi = 0$ is the SO(10) model). A factor of four reduction in the experimental error on $Q_W(Cs)$ combined with further atomic theory scrutiny would push up the bounds by a factor of 2. So, atomic parity violation experiments have the potential of becoming one of our best probes of extra Z' bosons.

As a final comment, I would like to call attention to the fact that the world average

$$\sin^2\theta_W = 0.231 \pm 0.004 \tag{16}$$

is now rather well determined and somewhat higher than in the past. The value in Eq. (16) has important implications for grand unified theories (GUTS). Indeed, it contradicts the minimal SU(5) model's prediction[24]

$$\sin^2\theta_W = 0.216^{+0.003}_{-0.005} \quad \text{(Minimal SU (5))} \tag{17}$$

Of course, the minimal SU(5) model has already been ruled out by proton decay experiments.[25] In the case of non-minimal GUTS, Eq. (17) can be used to constrain their additional mass scales. For example, it is in good agreement with supersymmetric GUTS with $m_{SUSY} \gtrsim 1$TeV. Another way of illustrating the disagreement between minimal SU(5) (or any GUT with a "great desert" scenario) and low energy phenominology is to evolve the standard model's gauge couplings to high energies and see if they meet at some unification point. Enploying the low energy $\overline{MS}$ coupling values

$$\alpha_1(m_W) = \frac{5\alpha(m_W)}{3\cos^2\theta_W(m_W)} = 0.0169 \pm 0.0001 \tag{18a}$$

$$\alpha_2(m_W) = \frac{\alpha(m_W)}{\sin^2\theta_W(m_W)} = 0.0342 \pm 0.0006 \tag{18b}$$

$$\alpha_3(m_W) = 0.107^{+0.013}_{-0.008} \tag{18c}$$

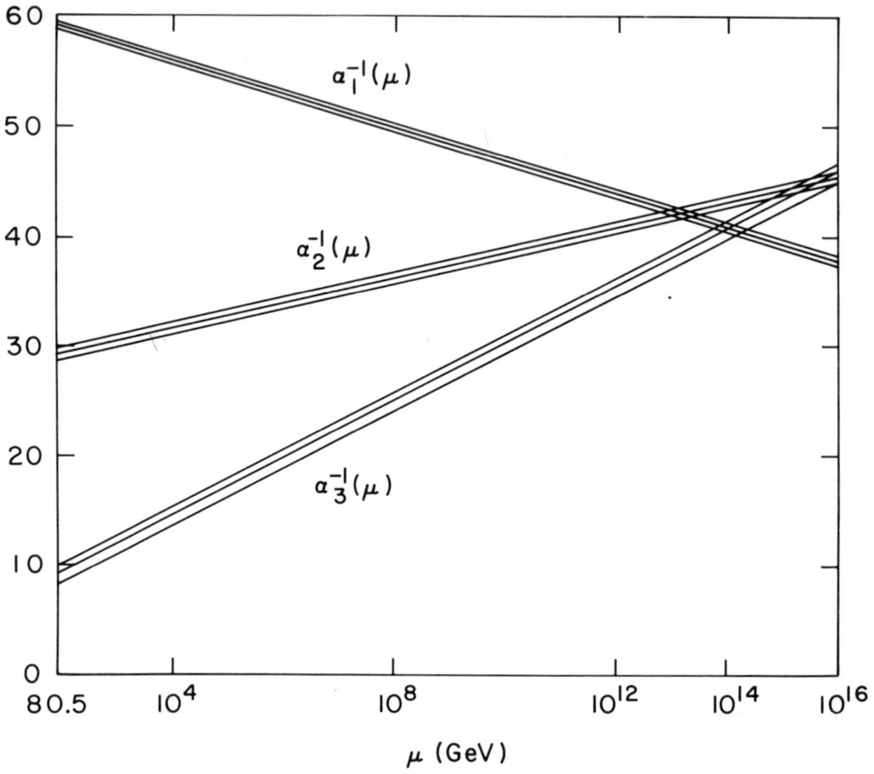

Figure 2: Coupling evolution of gauge couplings under the assumption of no new physics beyond the standard model.

found using $\alpha^{-1}(m_W) = 127.76 \pm 0.30$, $\sin^2\theta_W(m_W) = 0.229 \pm 0.004$ and $\Lambda^{(4)}_{\overline{MS}} = 150^{+150}_{-75}$MeV, and evolving them to higher energies via

$$\mu\frac{\partial}{\partial\mu}\alpha_i = b_i\alpha_i^2 + O\left(\alpha_i^2\alpha_j\right), \quad i,j = 1,2,3 \tag{19}$$

$$\begin{pmatrix} b_1 \\ b_2 \\ b_3 \end{pmatrix} = -\frac{1}{2\pi}\begin{pmatrix} -41/10 \\ 19/6 \\ 7 \end{pmatrix}$$

one finds the curves in Fig. 2. Note that the couplings fail to meet at a common point for the given uncertainties. They are, however, in the neighborhood of one another near 10^{15}GeV. I interpret that proximity as a strong indication of grand unification but with the requirement of some new physics between m_W and

10^{15}GeV that modifies the evolution equation in Eq.(19) and leads to unification. At issue is: What new physics enters and at what energy is it manifested? Hopefully, the coming generation of high energy colliders and precision measurements will help provide an answer.

Acknowledgment

Work supported by the Department of Energy under Contract DE-AC02-76CH00016.

References

1. U. Amaldi, A. Böhm, S. Durkin, P. Langacker, A. Mann, W. Marciano, A. Sirlin, and H. Williams, Review Article (in progress).
2. C. Bouchiat and C.A. Piketty, LPTENS preprint (1986); M.A. Bouchiat, J. Guéna and L. Pottier, J. Physique 46, 1897 (1985).
3. S. Gilbert, M. Noecker, R. Watts and C. Wieman, Phys. Rev. Lett. 55, 2680 (1985).
4. C.Y. Prescott et al., Phys. Lett. 77B, 347 (1978); 84B, 524 (1979).
5. L.A. Ahrens et al., Phys. Rev. Lett. 54, 18 (1985).
6. F. Bergsma et al., Phys. Lett. B147, 481 (1984).
7. K. Abe et al., Phys. Rev. Lett. 56, 1107 (1986).
8. H. Abramowicz et al., Phys. Rev. Lett. 57, 298 (1986).
9. J. Allaby et al., CERN preprint (1986); K. Winter in Proceedings of 1986 Conf. on High Energy Physics, Berkeley..
10. P. Reutens et al., Phys. Lett. 152B, 404 (1985); F. Merritt in Proceedings of 1986 Conf. on High Energy Physics, Berkeley.
11. D. Bogert et al., Phys. Rev. Lett. 55, 1969 (1985).
12. S. Geer, in Proceedings of 1986 Conf. on High Energy Physics, Berkeley.
13. A. Roussarie, in Proceedings of 1986 Conf. on High Energy Physics, Berkeley.
14. W.J. Marciano and A. Sirlin, Phys. Rev. D27, 552 (1983); D29, 75 (1984).
15. W.J. Marciano and A. Sirlin, Proc. *Cornell Z^0 Theory Workshop* (1981).
16. J. Wheater and C.H. Llewellyn Smith, Nucl. Phys. B209, 27 (1982).
17. S. Sarantakos, A. Sirlin and W. Marciano, Nucl. Phys. B217, 84 (1983).

18. W.J. Marciano and A. Sirlin, Phys. Rev. D22, 2695 (1980).

19. A. Sirlin and W.J. Marciano, Nucl. Phys. B189, 442 (1981).

20. A. Sirlin, Phys. Rev. D22, 971 (1980).

21. W.J. Marciano and A. Sirlin, Phys. Rev. D29, 945 (1984).

22. F. Jegerlehner, BERN preprint (1985).

23. L.S. Durkin and P. Langacker, Phys. Lett. 166B, 436 (1986).

24. W.J. Marciano and A. Sirlin, Phys. Rev. Lett. 46, 163 (1981).

25. M. Goldhaber, in Proceedings of 1986 Conf. on High Energy Physics, Berkeley.

ELASTIC SCATTERING OF MUON NEUTRINOS AT B.N.L.

K. Abe, L. A. Ahrens, K. Amako, S. H. Aronson, E. W. Beier, J. L. Callas, D. Cutts, M. Diwan, L. S. Durkin[a], B. G. Gibbard, S. M. Heagy[b], D. Hedin, J. S. Hoftun, M. Hurley[c], S. Kabe, Y. Kurihara, R. E. Lanou, A. K. Mann, M. D. Marx, M. J. Murtagh, Y. Nagashima, F. M. Newcomer, T. Shinkawa, E. Stern, Y. Suzuki, S. Terada, D. H. White[d], H. H. Williams, Y. Yamaguchi, and T. York

Physics Department, Brookhaven National Laboratory, Upton, NY 11973
Dept. of Physics, Brown University, Providence, RI 02912
Dept. of Physics, Hiroshima University, Hiroshima 730, Japan
National Lab. for High Energy Physics (KEK), Ibaraki-Ken 305, Japan
Physics Department, Osaka University, Toyanaka, Osaka 560, Japan
Dept. of Physics, University of Pennsylvania, Philadelphia, PA 19104
Dept. of Physics, State University of New York, Stony Brook, NY 11794

Presented by L. S. Durkin[a]

In this paper measurement of the purely leptonic reactions: $\nu_\mu e \to \nu_\mu e$ and $\bar{\nu}_\mu e \to \bar{\nu}_\mu e$, and the semileptonic reactions: $\nu_\mu p \to \nu_\mu p$ and $\bar{\nu}_\mu p \to \bar{\nu}_\mu p$ are presented and discussed in terms of the standard model. The data also places limits of interest on the magnitudes of a possible neutrino charge radius and a neutrino magnetic moment.

The data reported here were taken at the Brookhaven National Laboratory Alternate Gradient Synchrotron (A.G.S.) in a 170 ton, fine-grained target detector. The neutrino beam and the apparatus and its performance have been described in detail in earlier publications.[1,2]

---

[a]Now at Ohio State University, Columbus, Ohio 43210
[b]Now at Lockheed Aerospace, Sunnyvale, CA 94304
[c]Now at M.I.T. Lincholn Laboratory, Lexington, MA 02173
[d]Now at Los Alamos National Lab., MS H846, Los Alamos, NM 87544

In the standard electroweak model, the differential cross sections for neutrino electron elastic scattering are given by

$$\frac{d\sigma(\overset{(-)}{\nu}_\mu e)}{dY} = \frac{G_F^2 m_e E_\nu}{2\pi}\left[(g_V \pm g_A)^2 + (g_V \mp g_A)^2 (1-Y)^2\right]$$

where $Y = 1-E_e\theta_e^2/2m_e$ and ; the vector and axial vector coupling constants are related to $\sin^2\theta w$ by: $g_V = \frac{1}{2} - 2\sin^2\theta w$ and $g_A = -\frac{1}{2}$. Within this model a precision measurement of $\sin^2\theta w$ comes from a measurement of the ratio of the cross sections, $R = \sigma(\nu_\mu e)/\sigma(\bar{\nu}_\mu e)$.

The $\nu_\mu e$ and $\bar{\nu}_\mu e$ data were collected from 2.26×10^{19} protons on the neutrino producing target (P.O.T.) and 3.4×10^{19} P.O.T. respectively. The neutrino electron elastic scattering data analysis, procedures, and the treatment of backgrounds have been described previously.[3,4]

Shown in Figure 1 and 2 are the final $\nu_\mu e(\bar{\nu}_\mu e)$ data as a function of θ_e^2 in the interval $\theta_e^2<0.03$ rad with $210\text{MeV}<E_e<2100\text{MeV}$. The neutrino (antineutrino) data has been divided into "electron" and "gamma" like showers based on ionization information near the front of the showers. The presence of a signal from $\nu_\mu e\to\nu_\mu e$ and $\bar{\nu}_\mu e\to\bar{\nu}_\mu e$ is evident since kinematics require that $\theta_e^2\leq 2m_e/E_e$, and thus the signal is confined almost completely within the interval $0<\theta_e^2<.005$ rad^2. The "gamma" and "electron" distributions show a background slowly rising with decreasing θ_e^2 which consists mainly of: 1) photons from π^0 mesons produced in neutral current reactions, and 2) electrons from $\nu_e n\to e^- p$ ($\bar{\nu}_e p\to e^+ n$) induced by a small contamination of $\nu_e(\bar{\nu}_e)$ in the neutrino (antineutrino) beam. After background subtraction a signal of 107 ± 15 $\nu_\mu e$ and 59 ± 10 $\bar{\nu}_\mu e$ events remain.

To determine the cross sections $\sigma(\nu e)$ and $\sigma(\bar{\nu}_\mu e)$ and their ratio, we have indirectly measured the integrated incident neutrino fluxes using samples of quasielastic events $\bar{\nu}_\mu p\to\mu^+ n$ and $\nu_\mu n\to\mu^- p$ with small momentum transfer, $Q^2<0.4$ $(\text{GeV/c})^2$. Samples of 3.38×10^4 $\mu^+ n$ and 8.87×10^4 $\mu^- p$ (after background subtraction) were obtained. Correcting the signal and normalization samples for acceptances and efficiencies yields the total cross sections:

$$\sigma(\bar{\nu}_\mu e)/E_{\bar{\nu}_\mu} = [1.16 \pm .20(\text{stat}) \pm .14(\text{syst})] \times 10^{-42}\ \text{cm}^2/\text{GeV}$$

$$\sigma(\nu_\mu e)/E_{\nu_\mu} = [1.85 \pm .25(\text{stat}) \pm .27(\text{syst})] \times 10^{-42}\ \text{cm}^2/\text{GeV}$$

The fully corrected ratios of the normalized cross sections for $\nu_\mu e$ and $\bar{\nu}_\mu e$ scattering yield directly the ratio, R:

$$R = \sigma(\nu_\mu e)/\sigma(\bar{\nu}_\mu e) = 1.60 \begin{matrix} + .41 \\ - .31 \end{matrix} \pm .17$$
$$\qquad\qquad (\text{stat})\ (\text{syst})$$

The systematic error on this ratio is smaller than the systematic error on the individual cross sections due to cancellation of systematic uncertainties (e.g. acceptances, energy scales, ...) in the ratio. Comparing this ratio[5] to the standard model cross section we obtain

$$\sin^2\theta w = .192 \begin{matrix} + .027 \\ - .022 \end{matrix} (\text{stat}) \pm .013\ (\text{syst})$$

This value is dominated by systematic errors.

The excellent experimental resolution in θ_e (13 mrad/$\sqrt{E_e}$) combined with the large $\nu_\mu e$ statistics led us to consider the θe^2 dependence on the two Y terms in $d\sigma(\nu_\mu e)/dY$; namely the flat Y term and the term in $(1-Y)^2$.[6] Monte Carlo shapes corresponding to the two terms are shown in addition to the data points in Figures 3a and 3b.

Fitting the relative contribution to the θ_e^2 shapes to the data points in the θ_e^2 distribution determines $(g_V+g_A)^2$ and $(g_V-g_A)^2$, and yields the four 67% C.L. regions shown in Figure 4. Only one region is seen to be consistent with $\nu_\mu e \to \nu_\mu e$ in the standard model with $\sin^2\theta w$=.231, the world average value.[7] The region also contains the more precise values: g_V=−.079±.060 and g_A=−.483±.042 determined previously from the $\nu_\mu e$ and $\bar{\nu}_\mu e$ cross sections.[8] This calculation is the first use of a Y-distribution in a νe elastic scattering measurement.

The $\overset{(-)}{\nu}_\mu e$ elastic scattering also proves to be a sensitive probe of neutrino structure.[6] Any gauge theory of weak and electromagnetic interactions allows a non-zero charge radius, $\langle r^2 \rangle$. The additional contribution to the $\nu_\mu e$ cross section is:

$$\frac{2}{3\pi} m_e E_\nu \left[\frac{\pi}{\sqrt{2}} \alpha_{em} G_F \langle r^2\rangle \left(\frac{8}{3}\sin^2\theta w-1\right) + \frac{\pi^2}{2}\frac{4}{9}\alpha_{em}^2 \langle r^2\rangle\right]$$

where for reasonable values of $\langle r^2\rangle$ the interference term dominates. Fixing $\sin^2\theta w=.232\pm.008$ from measurements of the W and Z masses[9] and assuming $\langle r^2\rangle$ positive a 90% C.L. limit is found fitting our data: $\langle r^2\rangle \leq 0.81\times10^{-32}$ cm^2. This value is superior to the limit found in inelastic μ-N and ν_μ-N scattering.

In the standard model with massless, left-handed neutrinos, the magnetic dipole moment, defined as $fe/2m_e$ is identically zero. The contribution to $d\sigma(\nu_\mu e)/dY$, of a possible magnetic dipole moment is: $\alpha_{em} f^2((1-Y)^2/Y]$. Fixing $\sin^2\theta w$ as in the charge radius limit, we obtain a 90% C.L. limit on f of: $f<.95\times10^{-9}$ Bohr Magnatons. This limit approaches in sensitivity the bound, $f<10^{-10}$, obtained from astrophysical arguments.

The νp and $\bar{\nu}p$ elastic scattering data were obtained from separate exposures of $.55\times10^{19}$ P.O.T. and 2.5×10^{19} P.O.T. respectively. The $\nu_\mu p$ and $\bar{\nu}_\mu p$ differential cross sections depend on the quantities: G_F, M_V and M_A, and μ_p and μ_n, in addition to $\sin^2\theta w$. Details of these differential cross sections are given in reference 10. All these quantities are known to a precision better than that required by the measurement with the exception of M_A which we chose to constrain to the world average value.

The neutrino proton elastic scattering data analysis, procedures, and the treatment of backgrounds have been described previously[11,12]. Elastic proton scattering candidates were single tracks ($Q^2 \geq .35(\text{GeV/c})^2$) fully contained in our detector. To separate protons from pions (muons), separate confidence levels (C.L.) for each hypothesis were constructed for each candidate event from the observed energy deposits in the scintillators and PDT's. The results are shown in Figure 5. Clear separation between protons and pions (muons) is observed.

After data selection the resulting samples of 1686 $\nu_\mu p$ and 1821 $\bar{\nu}_\mu p$ candidates contain backgrounds from: a) $\nu_\mu n \to \nu_\mu n$, b) neutral current single pion production, and c) a small contribution from multipion reactions. In addition corrections were necessary for elastic scatters produced by the measured contamination of "wrong-helicity" neutrinos in the beam. After background subtraction final samples of 951 $\nu_\mu p \to \nu_\mu p$ and 776 $\bar{\nu}_\mu p \to \bar{\nu}_\mu p$ remain.

To determine the differential cross section the integrated incident neutrino flux was calculated as in the $\nu_\mu e$ analysis using quasielastic events. Shown in Figure 6 are the extracted absolute differential cross sections $d\sigma(\nu_\mu p)/dQ^2$ and $d\sigma(\bar{\nu}_\mu p)/dQ^2$. The error bars are statistical in nature with an additional shape systematic error of 18% and 8% added to the points at Q^2=.45 and .55(GeV/c)^2 respectively. An additional 11% scale uncertainty has been assigned to each of the cross section measurements. Constraining M_A to the present world average value and fitting the differential cross sections to the standard model we obtain:

$$\sin^2\theta w = .220 \pm .016(\text{stat}) \begin{array}{l} +\ .023 \\ -\ .031 \end{array} (\text{syst})$$

The solid curves in Figure 5 represent this best fit to the data. Unlike the $\nu_\mu e$ analysis the result is dominated by systematic uncertainties.

If $\sin^2\theta w$ is fixed to the value .220 and M_A is constrained to the world average value, a search may be made for additional terms in the axial vector form factor, $G_A(Q^2)$, due to neutral current contributions from heavy quark currents or a "non-standard" isoscalar axial vector current[10]. Writing $G_A(Q^2)=G_A^3(Q^2)(1+\eta)$, one finds $\eta=.12\pm.07$, or equivalently $0.0<\eta<.25$ at 95% C.L. . The value extracted for $\sin^2\theta w$ is insensitive to the choice of any η in this range.

There is seen to be good agreement between $\sin^2\theta w$ obtained from $\overset{(-)}{\nu}_\mu e$ and $\overset{(-)}{\nu}_\mu p$ elastic scattering. Furthermore these values of $\sin^2\theta w$ are in good agreement with values of $\sin^2\theta w$ determined from masses of the W- and Z-bosons[9], and inelastic electron-deuteron scattering, and from deep inelastic scattering[7]. Within present experimental errors of about seven percent, the weak neutral current parameter is a universal constant.

We are grateful to P. Langacker and W. J. Marciano for many informative discussions. This work was supported in part by the U. S. Department of Energy and the Japanese Ministry of Science and Culture through the Japan-U.S.A. Cooperative Research Project on High Energy Physics.

REFERENCES

1. L. A. Ahrens et al., Phys. Rev. D34, 74 (1986).
2. L. A. Ahrens et al., preprint E-734-86-2 (Accepted for publication in N.I.M.).
3. L. A. Ahrens et al., Phys. Rev. Lett. 54, 18 (1985).
4. L. A. Ahrens et al., Phys. Rev. Lett. 51, 1514 (1983).
5. There is only one previous measurement of R which yields: $\sin^2\theta_W = .215 \pm .032 \pm .013$. Details of this measurement can be found in F. Bergsma et al., Phys. Lett. 147B, 481 (1984).
6. K. Abe et al., Phys. Rev. Lett. 58, 636 (1987).
7. "Status of the Weak Neutral Current," U. Amaldi et al., (to be published).
8. L. A. Ahrens et al., "Perspectives in Electroweak Interactions," XX Recontre de Moriond, ed. J. Tran Thanh Van, p. 45 (1985).
9. C. Rubbia, "Proceedings of the 1985 International Symposium on Lepton and Photon Interactions," Kyoto, Japan (to be published) L. DiLella, ibid.
10. J. E. Kim et al., Rev. Mod. Phys. 53, 211 (1981).
11. K. Abe et al., Phys. Rev. Lett. 56, 1107 (1986).
12. L. A. Ahrens et al., "Measurement of Neutrino-Proton and Antineutrino-Proton Elastic Scattering", (accepted for publication in Phys. Rev.)

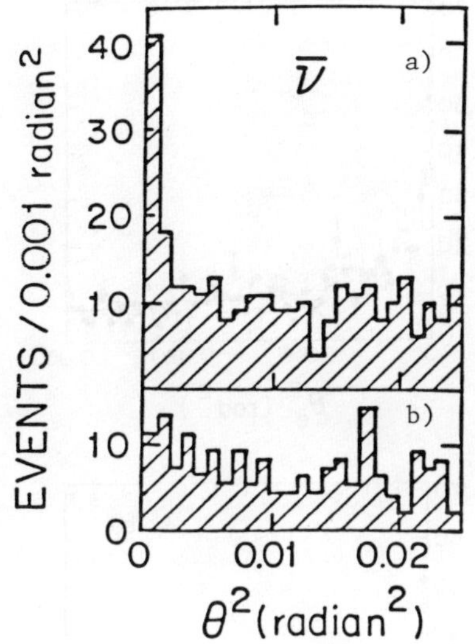

Fig. 1 The $\bar{\nu}_\mu e$ distribution in θ_e^2 for a) predominantly electron events, and b) predominantly photon events.

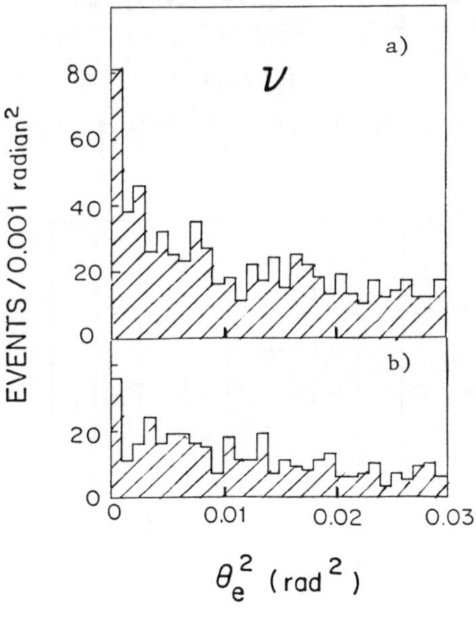

Fig. 2 The $\nu_\mu e$ distribution in θ_e for a) predominantly electron events, and b) predominantly photon events.

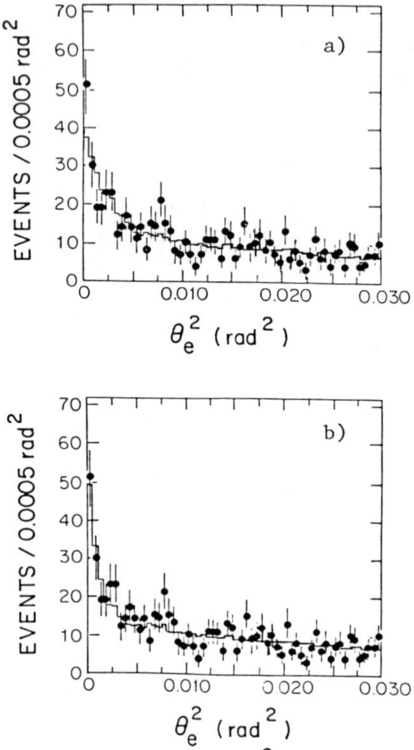

Fig. 3 The observed distribution in θ_e^2(points) and the Monte Carlo distribution (histogram) for: a) a $(1-Y)^2$ term and b) a flat term plus background.

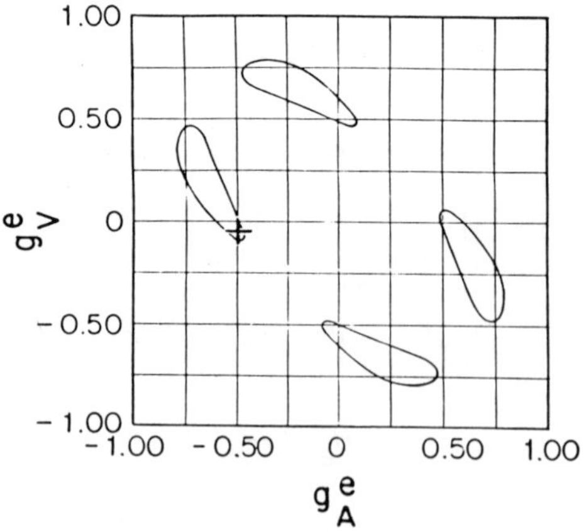

Fig. 4 The four areas in g_V vs g_A space (67% C.L.) allowed by the measurement of $d\sigma(\nu_\mu e)/dY$.

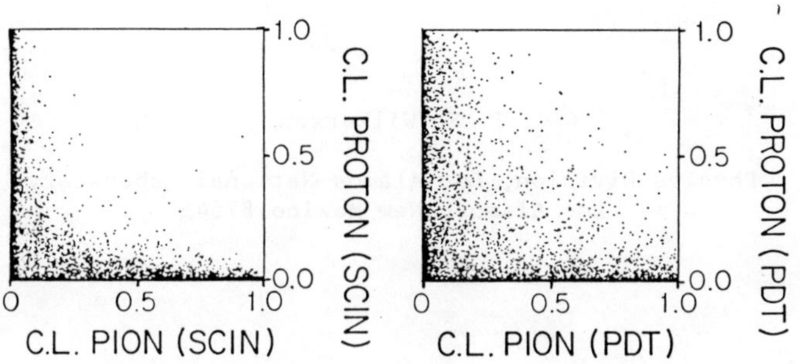

Fig. 5 C.L. for proton hypothesis vs C.L. for pion hypothesis by measurements in the scintillators (left) and PDT's (right).

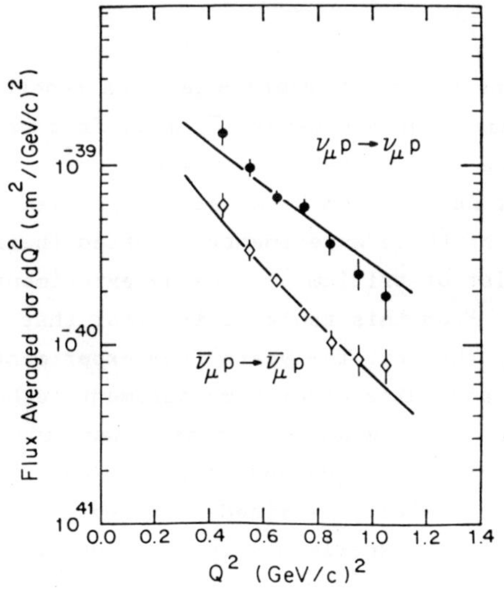

Fig. 6 The measured (data points) flux averaged differential cross sections for $\nu_\mu P$ and $\bar{\nu}_\mu P$ elastic scattering.

LIMITS ON THE ELECTRON-ANTINEUTRINO MASS

J. F. Wilkerson

Physics Division, Los Alamos National Laboratory
Los Alamos, New Mexico 87545

ABSTRACT

New measurements near the endpoint of the tritium beta-decay spectrum are producing limits on the electron-antineutrino mass which are below the central mass value of 30 eV reported by ITEP[1]. The factors that influence the neutrino mass sensitivity of tritium beta decay measurements will be discussed followed by a review of the current experimental results.

1. INTRODUCTION

It has been nearly seven years since the report by the ITEP group in Moscow of evidence for a non-zero $\overline{\nu}_e$ mass from measurements near the endpoint of the tritium beta-decay spectrum[2]. Motivated by this result a number of new tritium beta decay experiments were begun and during 1986 a few of these experiments reported their first results. Table 1 gives a list of tritium beta decay experiments that are currently active. From this table it is clear that we may expect many new results during the next few years from experiments using radically different sources and quite diverse measurement techniques.

The important consequences a non-zero neutrino mass would have in our understanding of physics demand that each experiment, independent of its results, be carefully examined. However, to perform a careful evaluation of any these experiments requires understanding the factors that influence the neutrino mass sensitivity. Thus, measurement considerations will be discussed before reviewing the experiments that have reported results.

Table 1.

Investigator	Location	Spectrometer	Source	Status*
Lyubimov (ITEP)	Moscow	Toroidal Mag	Valine-T	R
Bowles, Robertson, JFW	Los Alamos	Toroidal Mag	T_2, T gas	R
Kundig	Zurich	Toroidal Mag	C-T	R
Stoeffl	Livermore	Toroidal Mag	T_2, T gas	U
Ohshima (INS)	Tokyo	$\pi\sqrt{2}$ Mag	Arachidic Acid-T	R
Sun (IAE)	Beijing	$\pi\sqrt{2}$ Mag	PAD organic	P
Daniel	Munich	$\pi\sqrt{13/2}$ mag	Hf-T	P
Clark, Frisch	Yorktown (IBM)	Retarding E-S	?-T	T
Fackler	Livermore	Retarding E-S	Al_2O_3-T, T_2 solid	T
Lobashev	Moscow	Retarding E-S	T_2, T gas	U
Bonn, Otten	Mainz	Retarding E-S	T_2, T gas	U
Boyd	Columbus	Retarding E-S	T_2 solid	U
Jelly	Oxford	Cyl Mirror E-S	Cd Palmitate-T	T
Wellenstein	Brandeis	Cyl Mirror E-S	T_2 gas	U
Simpson	Guelph	Silicon	Si-T	C
Derbin-Popeka	Leningrad	Silicon	Si-T	C
Shang	Beijing	Silicon	Si-T	P

* R = Result P = Preliminary Result T = Testing U = Under construction C = Complete

2. TRITIUM BETA DECAY MEASUREMENT CONSIDERATIONS

Tritium beta decay experiments are sensitive to $\bar{\nu}_e$ mass in the energy region of the beta spectrum from a few m_ν below the endpoint energy to the endpoint energy, E_o, of the decay. The actual experimental procedure is to measure the energy region about the endpoint and also far below the endpoint. Then, the curve derived from extrapolating to the endpoint from the measurements far below the endpoint can be compared with the measurements near the endpoint

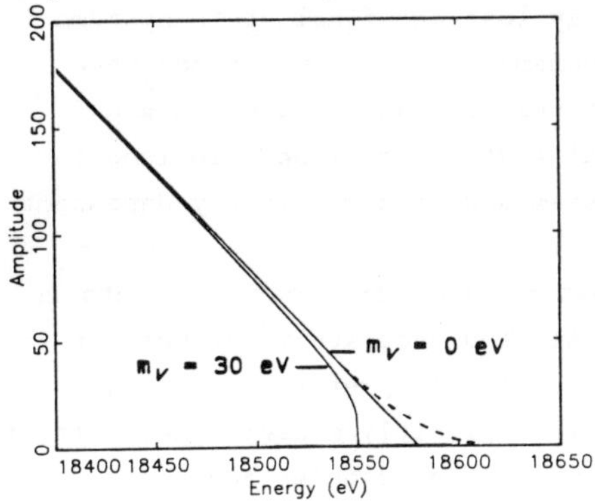

Fig. 1. Kurie plot showing the effect of neutrino mass on the beta spectrum of tritium. The dashed line represents the influence of resolution smearing and atomic final state effects.

yielding a sensitive determination of the neutrino mass. This effect is illustrated in Fig. 1 where Kurie plots of tritium decay spectra assuming m_ν = 0 eV and m_ν = 30 eV are shown for the energy region around E_o. This limited region of $\overline{\nu}_e$ mass sensitivity imposes two principal difficulties on any attempt to make a measurement. First, the decay rate in the energy region near the endpoint is a very small fraction of the total decay rate. For example, the decay rate in the last 100 eV of the spectrum below E_o is about 2×10^{-7} the total decay rate. Thus, acquiring sufficient statistics with good signal-to-background (S:B) is one of the primary considerations in making a tritium beta decay measurement.

The second consideration for a measurement is that all possible systematic effects which modify the beta spectrum must be accurately accounted for. Notice in Fig. 1 that in the Kurie plot representation, (a linearization of the beta decay probability function) one expects that for the ideal case the zero neutrino mass decay spectrum is a straight line while the finite neutrino mass decay spectrum curves downwards. However, in actual measurements, resolution broadening effects and decays to the different atomic final states of the source introduce an upward curvature to the measured decay spectrum (the dashed curve in Fig. 1). Thus, one must accurately account for all possible systematic effects, since an underestimation of these effects will result in an underestimation of the neutrino mass and conversely an overestimation of the effects results in an overestimation of the neutrino mass.

2.1 Systematic Effects

Systematic effects are either source related or of instrumental origin. All sources introduce energy loss and final state systematic effects into a beta spectrum measurement. Solid sources may have additional non-negligible contributions from backscattering and surface contamination. Instrumental systematic effects include the finite energy resolution of the system and possibly energy-dependent extraction efficiencies.

Because the elimination of most of these systematic effects is impossible, one would like to minimize and accurately account for them in a model independent manner. It is clearly preferable to design into the experiment the capability to explicitly measure these effects since calculations often introduce model dependencies and the corresponding uncertainties in the determination of m_ν.

2.1.1 Final state effects. When a tritium atom decays to a $^3He^+$ ion there is a probability of populating any of the energetically allowed atomic final states of the daughter $^3He^+$ ion. The observed beta spectrum is actually the sum of all of the individual branches to all possible final states. Extracting a reliable value for m_ν requires the precise knowledge of the branching ratios and the energies for all possible final decay states. This is because the reported neutrino mass is of the same order as the binding energy of electrons in ^3He.

The only sources for which the decay probabilities can be accurately determined are for atomic tritium[3] and molecular tritium[4,5] where the uncertainties are at the level of ~1 eV. For complex sources such as the tritiated valine molecule used by the ITEP group[1] or tritium implanted in carbon used in the Zurich experiment[6] the final state effects are difficult to calculate and exact calculations are impossible. Hence, model dependent uncertainties are introduced into the m_ν value or limit for all sources except ionic, atomic or molecular tritium.

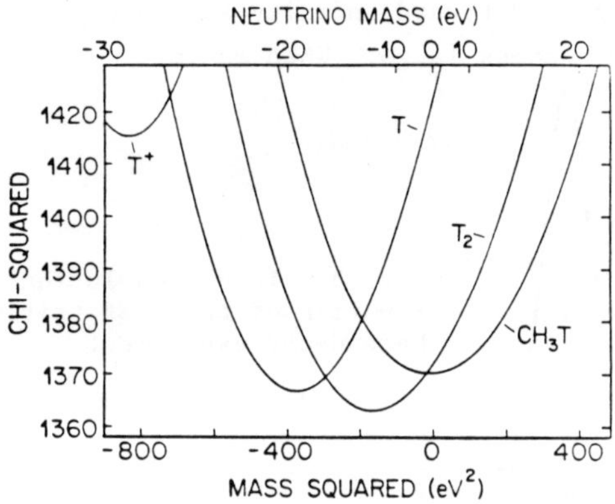

Fig. 2. χ^2 vs m_ν^2 from analysis of Fritschi et al.[6]. Note for different final state assumptions the large change in best fit m_ν, but small change in χ^2.

The absolutely critical role that final state effects play in determining an experiment's best value for m_ν can be seen in Fig. 2. This figure of χ^2 vs m_ν^2 from Fritschi et al.[6] displays various χ^2 curves assuming different final states configurations. Although there are almost no differences in the best χ^2 values for the different configurations there are significant differences in the best m_ν^2 values. For example assuming molecular final states instead of atomic final states results in a shift of the best value of m_ν by 15 eV.

2.1.2 Total resolution function. The total resolution function (TRF) is defined as the convolution of the instrumental resolution and the total source energy loss. In an analogous fashion to final state effects, an accurate knowledge of the TRF is absolutely imperative.

Most experiments have devised methods to directly measure instrumental resolution. But, ascertaining the energy loss contributions has proven more difficult. With solid sources the total energy loss is a combination of energy loss in the target, energy loss of electrons that backscatter from the source backing, and energy loss in any surface contamination.

A few comments on the parametrization of the TRF are warranted. The full width half maximum (FWHM) of the TRF is a very poor description of the distribution. For example the ITEP TRF shown in Fig. 3 has a FWHM of 22 eV but the second moment, σ^2, of the distribution is 2250 eV^2. (The effect of this size TRF on the spectrum if expressed in neutrino mass is ~67 eV.) However, one must also use caution in using the σ^2 parametrization and explicitly define the energy region used for its calculation. This is because σ^2 of TRFs that include backscattering tend toward infinity as the energy region used for the calculation becomes large. When making

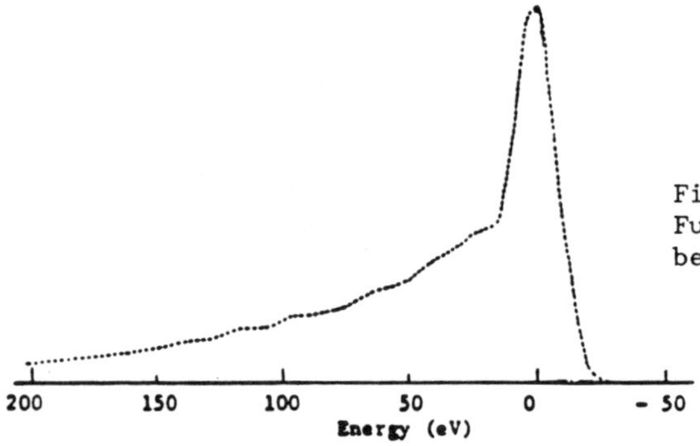

Fig. 3. Total Resolution Function of the ITEP tritium beta decay experiment.

comparisons between different experiments the best method is to examine the actual TRF and not just the FWHM or the second moment parameters. Finally, often when tritium beta decay experiments are being discussed there is emphasis on the instrumental resolution while source energy loss is ignored. Clearly the TRF of an experiment is the relevant distribution to examine.

2.2 Summary Of Measurement Considerations

A statistically significant result that either implies a finite neutrino mass or rules out a neutrino mass is meaningless if all systematic effects have not been accurately and completely accounted for. Hence, the elimination or minimization and total understanding of all systematic effects is crucial in obtaining a believable result.

If experiments push the m_ν value towards zero the limitations imposed by systematic effects will dominate and must not be ignored.

3. ANALYSIS CONSIDERATIONS

Before reviewing the experimental results a few comments with regards to the analysis of the data are in order. Most of the experiments analyze the data by simultaneously fitting several parameters of the Fermi decay probability function

$$N(E) = C\, F(Z,R,E)\, p_e\, E\, \Sigma_i w_i (E_o - E_i - E)[(E_o - E_i - E)^2 - m_\nu{}^2 c^4]^{1/2}$$
$$\times\, [1 + \alpha_1 (E_o - E) + \alpha_2 (E_o - E)^2] \quad ; \quad E \le E_o - E_i - m_\nu c^2$$

by using a maximum likelihood method. In their analysis, most of the experiments fit to the neutrino mass m_ν, the endpoint energy E_o, and amplitude C. They may additionally fit to a collection of the following parameters; the background *BG*, a linear correction coefficient α_1 and a quadratic correction coefficient α_2. The use of these coefficients α_1 and α_2 must have some physical motivation.

Finally, a few general comments on the analysis:

- The maximum likelihood estimator must employee the proper statistics. For example, in the Los Alamos analysis[7], it was discovered from Monte Carlo simulation studies that using the Gaussian statistics based χ^2 estimator introduced a non-negligible systematic shift to the best fit m_ν value. The use of Poisson based maximum likelihood estimator eliminated the shift error.
- Regardless of the parameters fit, the results from the analysis should not depend (within statistics) on the portion of the spectrum analyzed.
- Determining the error of a result requires careful analyses of all systematic errors using the maximum likelihood analysis code. Attempting to make approximations and simplifications for analytical analyses of errors is not only dangerous but usually wrong.

4. CURRENT RESULTS

In addition to a new ITEP m_ν value, three experimental groups have reported limits on m_ν from tritium beta decay measurements in 1986. The unique aspects and salient features of these four experiments will be discussed below. Table 2 offers a summary of some of the pertinent parameters from these experiments. Because of space limitations, experiments reporting preliminary results and the less sensitive Si detector experiments will not be examined. (Reference 8 contains reports from many of these experiments.)

Table 2.

Expt.	Last 100 eV S:B	Total Counts	TRF σ^{2*} eV^2	FWHM eV	Source Thickness $\mu g/cm^2$	No Loss %	Limit or Value eV
LANL	4:1	.6k	540	36	0.1	83	< 27 95% CL
ITEP	16:1	130k	2250	22	~4	37	30 ± 2
Zurich	?	~1000k	550	27	~4	60	< 18
INS	1:1	5k	1197	14	1.5	60 - 65	< 32 95% CL

* σ^2 was calculated over the TRF energy range of +50 to -150 eV.

It is interesting to note (Table 1) that the groups that have results are using some type of magnetic spectrometers. Although the electrostatic spectrometer experiments have excellent instrumental resolution, thus far all of these experiments have encountered difficulties and delays, usually arising from background problems. It is encouraging however, that three of these experiments have started taking preliminary data with tritium sources and may produce results in 1987.

4.1 Los Alamos Experiment

Of the experiments with results or taking data the Los Alamos experiment is the only one using a gaseous molecular tritium source and hence the only experiment with model independent results. In addition to the clear advantage of using a source with simple, well understood final states, the use of a gaseous source eliminates backscattering and surface contamination that add uncertainties to solid source based measurements. Furthermore, the use of a pure tritium source yields the highest specific activity, thus minimizing source energy loss.

The Los Alamos apparatus, which has been described in detail elsewhere[7)], consists of an extended source and a toroidal spectrometer. The instrumental resolution function is directly measured using a short-lived (1.8 hours) gaseous ^{83m}Kr isomer 17.835(20) keV K-conversion line. Calculations to determine the energy loss contribution to the TRF are based directly on doubly differential cross-section for electrons scattering from H_2.

The Los Alamos data consist of four data sets, each of 3-4 days duration (Fig. 4). In the analysis of the data m_ν, E_o, C, *BG*, and α_2 were fit. The inclusion of the α_2 term was based on the energy dependent extraction efficiency of the apparatus. The uncertainty in the final result is predominantly statistical. An upper limit on the mass of the electron antineutrino is found to be 26.8 eV at the 95% confidence level (CL) or 23.3 eV at the 90% CL Improvements now in progress to the apparatus are expected to result in a sensitivity to neutrino mass in the vicinity of 10 eV.

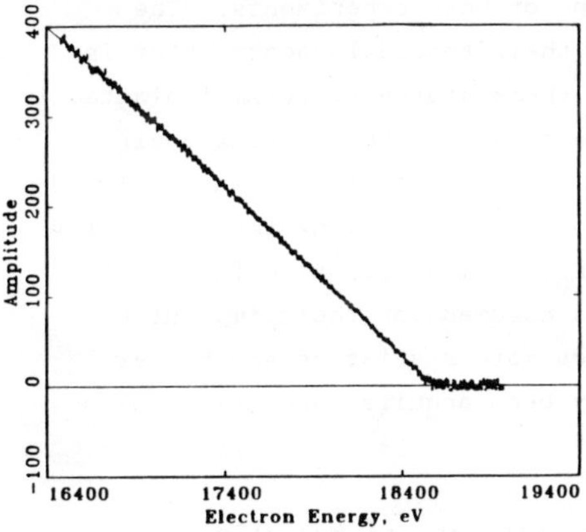

Fig. 4. Kurie plot of one of the LANL data sets.

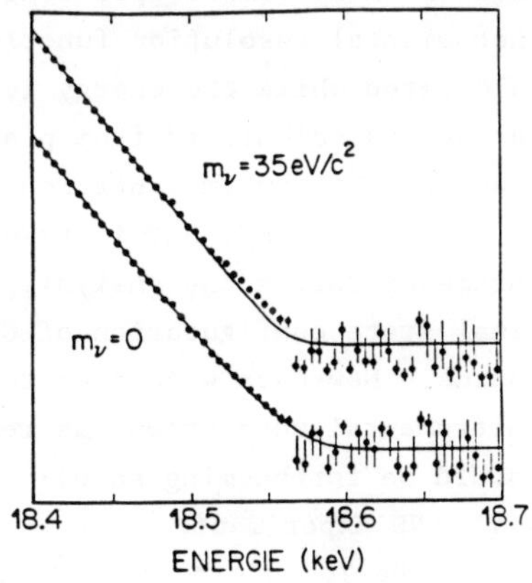

Fig. 5. Zurich data with fits for m_ν = 0 and m_ν = 35 eV.

4.2 ITEP Experiment

The ITEP experiment using a complex amino acid tritiated valine source and a toroidal spectrometer has implemented a number of improvements since their initial measurement. They have substantially improved their S:B ratio and implemented methods to directly measure their instrumental resolution function and source energy loss contribution. The fitting parameters m_ν, E_o, C, α_2, and (in the 1986 analysis) α_1 were used in the analysis. The model dependent result of 30 ± 2 eV is statistically significant, but systematic uncertainties remain, especially from the use of the complex valine source.

The ITEP experiment also claims a model independent limit on the neutrino mass of $17 < m_\nu < 40$ eV. This limit is obtained by analyzing the data while compressing or expanding the energy scale of the valine final state spectrum and observing at what points the fit E_o is incompatible with E_o determined from ion cyclotron resonance measurements of the T-^{3}He mass difference.

4.3 Zurich Experiment

Like the ITEP experiment, the results from the Zurich measurement are not statistics limited (Fig. 5), but their upper limit on m_ν of 18 eV (no confidence level quoted) is in direct disagreement with the ITEP result. Since the statistical evidence to support both claims is very strong, the difference between the two results must arise from systematic problems in one or both experiments. The instrumental resolution function of their toroidal spectrometer is calculated while the energy loss of their source (tritium implanted in carbon) is calculated from plasmon excitation theory using their measured tritium implantation depth. Backscattering is not included in the Zurich TRF, but is taken into account by using the α_1 fitting parameter during the analysis (m_ν, E_o, C, & *BG* are also fit). The final state configuration of CH_3T is assumed for their implanted source. New data with over ten times more statistics and better instrumental resolution has recently been acquired and new results should be forthcoming soon.

4.4 INS Experiment

The Tokyo experiment[9)] uses an air core $\pi\sqrt{2}$ magnetic spectrometer and a novel mono-layer Cd salt of arachidic acid containing tritium. By replacing the natural Cd with ^{109}Cd, measurements with the Ag KLL Auger lines allow determination of instrumental resolution and of source energy losses. Although calculations of FS effects are in progress for the actual source, the present results used the identical valine final state calculations used by the ITEP group. The parameters m_ν, E_o, C, & α_2 were fit in the data analysis which yields a model dependent upper limit on m_ν of 32 eV at the 95% CL. The INS group plans to acquire increased statistics which should improve their current limit.

A few final observations on the present measurements:

- Exchanging the ITEP and Zurich TRFs would essentially produce an exchange of their results.
- Notice in Table 2 that the values for source thickness and percent of electrons emerging from the source with no energy loss seem to be

inconsistent for the solid source based ITEP, Zurich, and INS experiments.

5. CONCLUSIONS

Based on current results the $\bar{\nu}_e$ mass question is still open. In 1987 a more definitive answer on the $\bar{\nu}_e$ mass should be reached. However, one must beware of the problem of "intellectual phase locking" where results agree with the expected value and are not rigorously examined. All of the experiments must be critically examined regardless of their results. Sufficient statistics are necessary for a reliable measurement, but as the limit on m_ν is further reduced the role of systematic effects will become increasingly important. Elimination or the direct measurements of systematic effects is the crucial requirement in making a reliable determination of the $\bar{\nu}_e$ mass.

REFERENCES

1. V. Lyubimov, in *Nuclear Beta Decays and Neutrino*, eds. T. Kotani, H. Ejiri, & E. Takasugi, p. 273 (World Scientific, 1986).
2. V. A. Lyubimov et al., Zh. Eksp. Teor. Fiz. 81, 1158 (1981).
3. R. D. Williams and S. E. Koonin, Phys. Rev. C27, 1815 (1983).
4. R. L. Martin and J. S. Cohen, Phys. Lett. 110A, 95 (1985).
5. W. Kolos et al., Phys. Rev. A31, 551 (1985); O. Fackler et al., Phys. Rev. Lett. 55, 1388 (1985); K. Szalewicz et al., submitted to Phys. Rev. A. (1986).
6. M. Fritschi et al., Phys. Lett 173B, 485 (1986).
7. T. J. Bowles et al., in *Nuclear Beta Decays and Neutrino*, eds. T. Kotani, H. Ejiri, & E. Takasugi, p. 304 (World Scientific, 1986); D. A. Knapp, Ph.D. Dissertation LA-10877-T, Princeton University, 1986 (unpublished).
8. *Nuclear Beta Decays and Neutrino*, eds. T. Kotani, H. Ejiri, & E. Takasugi (World Scientific, 1986).
9. H. Kawakami et al., in *Nuclear Beta Decays and Neutrino*, eds. T. Kotani, H. Ejiri, & E. Takasugi, p. 294 (World Scientific, 1986).

III. Beyond the Standard Model: Phenomenology and Experiment

Conveners — E. Kolb and D. Sinclair

UNDERGROUND SEARCH FOR MUONS CORRELATED WITH CYGNUS X-3

R.M. Bionta[11], G. Blewitt[4], C.B. Bratton[5], D. Casper[2,13], A. Ciocio[13], R. Claus[13], M. Crouch[9], S.T. Dye[6], S. Errede[10], G.W. Foster[14], W. Gajewski[1], K.S. Ganezer[1], M. Goldhaber[3], T.J. Haines[1], T.W. Jones[7], D. Kielczewska[8], W.R. Kropp[1], J.G. Learned[6], J.M. LoSecco[12], J. Matthews[2], H.S. Park[11], L.R. Price[1], F. Reines[1], J. Schultz[1], S. Seidel[2,13], E. Shumard[15], D. Sinclair[2], H.W. Sobel[1], J.L. Stone[13], L. Sulak[13], R. Svoboda[1], G. Thornton[2], J.C. van der Velde[2], and C. Wuest[11]

1. The University of California, Irvine; 2. The University of Michigan; 3. Brookhaven National Laboratory; 4. California Institute of Technology; 5. Cleveland State University; 6. The University of Hawaii; 7. University College, London; 8. Warsaw University; 9. Case Western Reserve; 10. The University of Illinois; 11. Lawrence Livermore National Laboratory; 12. Notre Dame University; 13. Boston University; 14. Fermi National Accelerator Laboratory; 15. A.T.T. Bell Laboratory

ABSTRACT

We have examined muon data collected with the IMB (Irvine-Michigan-Brookhaven) proton decay detector in search of a muon flux associated with the x-ray binary Cygnus X-3. A sample of 3.6×10^4 muons from the direction of Cygnus X-3 during the period September 1982 until April 1984 shows no evidence of its characteristic 4.8 hour periodicity. In the phase interval 0.65 to 0.90 where a signal has been reported by other workers at a similar depth we obtain a 90% confidence level limit to the flux of 6.9×10^{-11} cm^{-2}s^{-1} for a depth of 1800 meters water equivalent. Three subsets of the data taken during Cygnus X-3 radio bursts have been separately analyzed. We find some evidence for an excess in the phase interval 0.50-0.60 at the time of the October 1983 outburst. We examine this excess in detail.

INTRODUCTION

Two experiments[1–4] have recently reported underground muons from the direction of Cygnus X-3 (right ascension 20.5 hr, declination 40.9°) that exhibit its characteristic 4.8 hour x-ray period.[5] Although Cygnus X-3 has been observed above ground in airshowers at TeV and PeV energies,[6] the reported underground muon flux is not consistent with the flux and assumed gamma ray origin of these showers. If confirmed, the underground results would indicate new physics, such as a new neutral particle or unexpected interactions of a known neutral. The variability of Cygnus X-3 observed by the surface experiments apparently also extends to the underground signal.[2] Thus, although the Kamioka[7] and Fréjus[8] experiments have reported upper limits that are below the fluxes found by the Soudan and NUSEX nucleon stability experiments these limits apply to a later time and so do not preclude, for example, a variable source that may have been mostly in a low state since late 1983. The on-time of the IMB experiment has a better temporal overlap with both of the experiments that have reported positive results. We have data at depths similar to the Soudan experiment. We examine these data for evidence of a signal at the same phase.

The Soudan group has also reported[9] a burst of muons occuring at the time of a large radio outburst of Cygnus X-3 in October 1985. Although we have no data from the period, we use data taken at the times of three previous radio outbursts to search for muon bursts at those times.

DATA ANALYSIS

The IMB proton decay detector[10] is located in the Morton-Thiokol salt mine near Fairport Harbor, Ohio (41.7°N, 81.3°W) at a depth of 1570 m water equivalent (mwe). The detector is a rectangular tank ($22.5 \times 17 \times 18$ m^3) filled with 8000 metric tons of purified water. At the time these data were taken 2048 five-inch photomultipliers (PMs) were arrayed on the 6 sides of the tank on a ~ 1 m grid. These PMs provide both pulse height and timing information.

After cuts on the number of triggering PMs, the directions of through going muons were determined by identifying the cluster of early triggering PMs as the muon entry point and the cluster of brightest PMs as the exit point. The fit is further constrained by the PM times in both clusters. A fiducial cut is made requiring the point of closest approach to the center of the detector to be at least 2 m inside the detector from the PM planes. This fitting procedure has an 80% efficiency for saving through-going single muons which satisfy the fiducial requirement. The final sample is predominantly single muons of energy greater than 2 GeV at the detector. The angular resolution of the fitter was determined by first applying the fitter to Monte Carlo generated data. This result was then calibrated by a comparison of the difference between the muon directions obtained by pairs of different fitters when applied to both Monte Carlo and real data. A 7° angular cut is used in the analysis presented here. This value gives the optimal signal to noise for a point source of muons and has a 70% acceptance efficiency for such a point source.

During normal data taking we use on-line cuts designed to reject the majority of the cosmic ray muons which trigger the detector at $\sim$ 2.7 Hz. In this mode muons may be saved for several reasons. Two samples of such muons are included in this analysis. They were saved because: 1) one event in 64 is saved as a random sample (referred to as the random-save sample) and 2) events occuring within 10 ms of each other are saved (the 10ms sample). This second sample has not been analyzed in full. In addition, data is taken without cuts for a nominal one day per week. This "uncut" data provides our largest data sample for the present analysis.

The on-source muons, i.e. those with directions within 7° of Cygnus X-3, are binned according to their phase in the 4.8 hour Cygnus X-3 orbital period using the ephemeris of van der Klis and Bonnet-Bidaud.[5] To obtain the expected phase distribution in the absence of a source, the muon rate as a function of local coordinates was tabulated using all the muons in the relevant data sample. This produces a table of the muon rates that takes into account the detector angular response, the cosmic ray muon flux and any variations in the overburden. Then, at fixed intervals during the on-time, the position of Cygnus X-3 was determined and the number of muons expected from that direction was obtained from the rate table and added to the background phasogram at the appropriate phase. The effective off-source to on-source exposure ratio is $\sim 10 : 1$ for the uncut sample for which the analyzed data sets were preferentially chosen to have Cygnus X-3 at high altitudes, and $\sim 20:1$ for the other samples.

RESULTS AT 1570 mwe

The IMB detector, at a depth of 1570 mwe, is slightly shallower than the 1800 mwe of the Soudan experiment. For the period September 1981 through November 1983 Soudan,[2] using an on-source sample of 1183 single muons, found an excess flux within 3° of Cygnus X-3 in the phase interval 0.65 to 0.90, corresponding to an average flux over the whole 4.8 hour period of 7.3×10^{-11} cm^{-2}s^{-1} (+50 − 25%) for the source at a constant effective depth of 1800 mwe. The IMB uncut sample contains 36046 on-source muons, obtained in 655 hours livetime from September 1982 through April 1984 with an average on-source detector area of 265 m^2. The mean depth penetrated by these muons is 1800 mwe. The phase distribution of these muons and the calculated background is shown in Fig. 1. There is a slight net excess of 1.8σ. We find no significant excess in the phase interval 0.65-0.90 (8749 observed, 8704 expected) and obtain a corresponding upper limit at the 90% confidence level (C.L.) of 5.6×10^{-11} cm^{-2}s^{-1} for depths greater than 1570 mwe. To convert this to a flux at a fixed depth requires a knowledge of the zenith angle (θ_z) distribution of the Cygnus X-3 muons. The Soudan flux was obtained assuming a $\cos^3\theta_z$ dependence,[2] this being approximately the distribution of the cosmic ray muon flux out to $\sim 60°$. Under the same assumption our flux limit is 6.9×10^{-11} cm^{-2}s^{-1} at 1800 mwe.

The flux limits derived from the uncut data strictly apply only to a non-variable source because of the poor total time coverage of the sample. The random-save muon sample is much smaller than the uncut sample but is more continuous ($\sim$ 360 live days from September 1982 through June 1984). The background subtracted phasogram for the 3902 on-source muons in this sample is shown in Fig. 2. The 90% C.L. flux limit for phases 0.65-0.90 is 1.8×10^{-10} cm^{-2}s^{-1}.

CORRELATIONS WITH RADIO OUTBURTST

The Soudan group interprets the arrival time distribution of their on-source, in-phase muons in terms of a long term variability of the source which is active for only $\sim$ 7% of the time.[2] During these active times the flux would be

$\sim 10^{-9}$ cm^{-2}s^{-1}. That group has also recently reported observing a burst of muons from the direction of Cygnus X-3 near a phase of 0.7, coincident with a major radio outburst of the system in October 1985.[9] The muon burst is spread over $\sim$ 4 weeks. This suggests that other muon burst periods may also coincide with radio outbursts. We have examined data from around the times of three previous radio outbursts[11,12] for which we have data. Because we do not know the exact nature of any correlation between the radio outbursts and muon bursts we use data from two weeks before and after the radio outburst times. Although the Soudan muon burst in October 1985 occured in a small range near phase 0.7, the other bursts might not be at the same phase. The radio outbursts for which we have data occur during the period of the earlier Soudan observations, and so we again look for an effect in the phase interval 0.65 to 0.90.

The phase distributions for the random save sample at these times are shown in Fig. 3. The flux limits for the phase interval 0.65 to 0.90 are 1.1×10^{-9} cm^{-2}s^{-1} for September/October 1982, 2.1×10^{-9} cm^{-2}s^{-1} for February 1983 and 7.2×10^{-10} for September/October 1983.

The September/October 1983 random-save data (Fig. 3c) does show a large excess (4.2σ) in the phase interval 0.50 -0.65 and we have examined this more closely.

Because the Cygnus X-3 period is indistinguishable from one fifth of a day over such short periods of time, we investigate the possibility of a diurnal origin of the excess by dividing the data into two 12 hour periods each day. To maximize the statistics we use data from 1 to 12 October and 10 phase bins. These phase distributions (Fig. 4) show that both halves of the day contribute to the excess. This is not consistent with a diurnal origin which would given an excess in only one of two halves.

As a further check we analyzed events that were saved because their trigger occurred within 10 ms of a previous trigger, for September and October 1983. This provides a second independent sample of about the same size as the random-save sample. The phase distribution for these events from 1 to 12 October (Fig. 5a) shows only a slight excess (2.2σ) at 0.50 to 0.60. The phase distribution for the sum of the two samples is shown in Fig. 5b. From Fig. 6 it appears that the excess occurs in two distinct episodes, the first from about 2-5 October and the second from 9-11 October. It should be noted, however, that the sample does not include data from 28 September to 1 October, or from 6 October.

The excess between phases 0.50 and 0.60 in Figures 5b is 4.3σ. The statistical significance is reduced by the a posteriori choice of the final data subset and of the phase bin. This excess would correspond to an average flux over the twelve days of 1.0×10^{-9} $cm^{-2}s^{-1}$ at 1800 mwe.

SUMMARY

We find no evidence of a muon flux associated with Cygnus X-3 at the phase at which the Soudan experiment found an excess. The upper limit from our total data sample is similar to the flux measured by Soudan. We have also searched for a signal at the times of three Cygnus X-3 radio bursts. Once again we find no evidence for an excess in the phase interval 0.65-0.90. These results are summarized in table 1. We have found an interesting correlation between a 4σ excess phase 0.50-0.60 and the October 1983 radio outburst. We know of no other reported underground effects which might be associated with this observation. The Soudan data for the same period have been presented at this meeting[13] show no excess at phase 0.50-0.60.

We are grateful to our host, Morton Thiokol Inc. This work was supported in part by the U.S. Department of Energy.

REFERENCES

1. M.L. Marshak et al., *Phys. Rev. Lett.* 54, 2079 (1985).
2. M.L. Marshak et al., *Phys Rev. Lett.* 55, 1965 (1985).
3. G. Battistoni et al., *Phys. Lett.* 155B, 465 (1985).
4. G. Battistoni et al., Proc. 19th International Cosmic Ray Conference, La Jolla, U.S.A., 1, 62 (1985).
5. M. van der Klis and J.M. Bonnet-Bidaud, *Astron. Astrophysics* 95, L5 (1981).
6. A.A. Watson, Proc. 19th International Cosmic Ray Conference, La Jolla, U.S.A., 9, 111 (1985).
   K.E. Turver, ibid, 9, 399.
   A.M. Hillas, ibid, 9, 407.
7. Y. Oyama et al., *Phys. Rev. Lett.* 56, 991 (1986).
8. Ch. Berger et al., *Phys. Lett.* 174B, 118 (1986)
9. M.L. Marshak et al., 23rd ICHEP, Berkeley (1986).
10. R. Bionta et al., *Phys. Rev. Lett.* 51, 27 (1983).
11. B.J. Geldzahler et al., *Astrophys. J.* 273, L65 (1983).
12. K.J. Johnston et al., *Astrophys. J.* 309, 707 (1986).
13. H. Courant et al., this meeting.

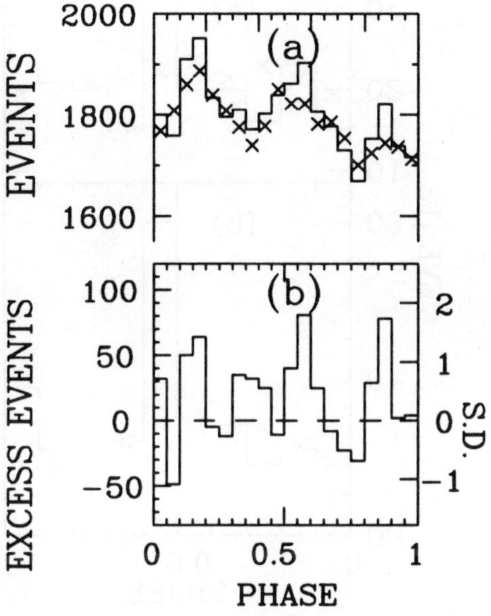

Fig. 1. a) The phase distribution of muons with directions within 7 deg. of Cyg X-3 from the uncut sample (solid histogram), and the expected distribution in the absence of a source (crosses); b) the background subtracted phase distribution. The right hand scale is in approximate standard deviations (S.D.)

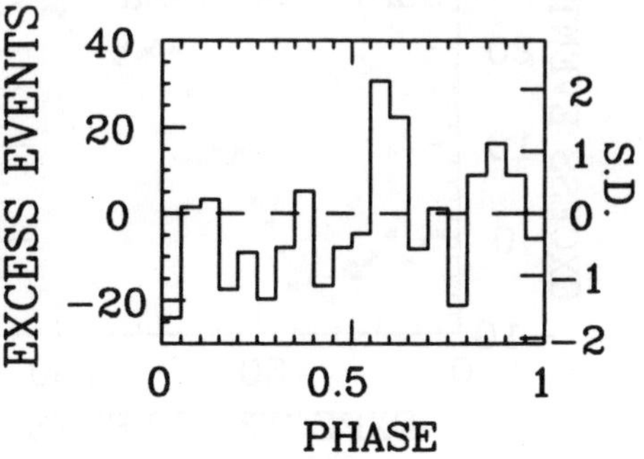

Fig. 2. The on-source background subtracted phase distribution for the random save muons.

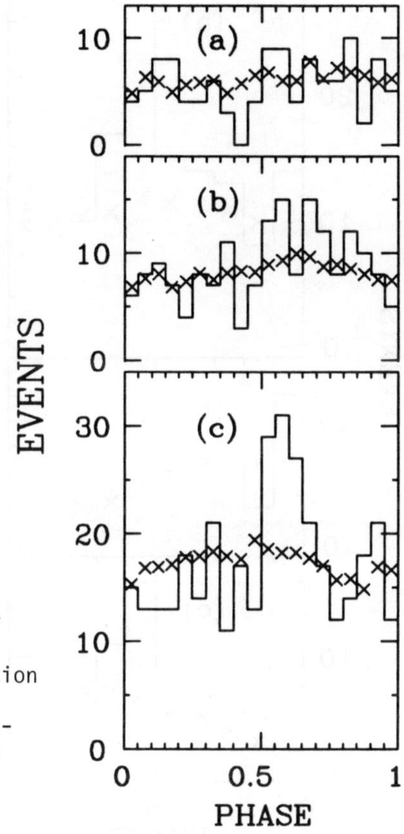

Fig. 3. The phase distributions of the random save muons at Cyg X-3 radio flare times.
a) 280 hours livetime from 14 September to 8 October 1982;
b) 403 hours livetime from 2 February to 3 March 1983;
c) 690 hours livetime from 17 September to 31 October 1983.

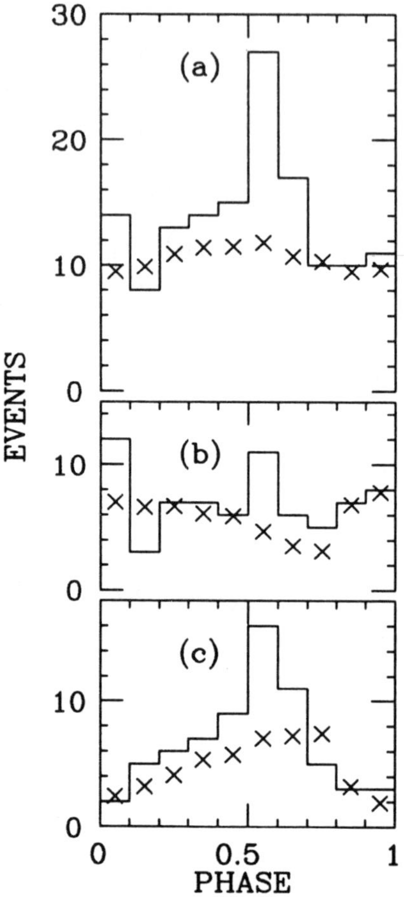

Fig. 4. The phase distribution for the random save muons for 1-12 October 1983. The solid histogram is the observed distribution, the crosses the expected.
a) No restriction on the time of day (203 hours on-time);
b) For data between 0030 and 1230 UT (101 hours);
c) For data between 1230 and 0030 UT (102 hours).

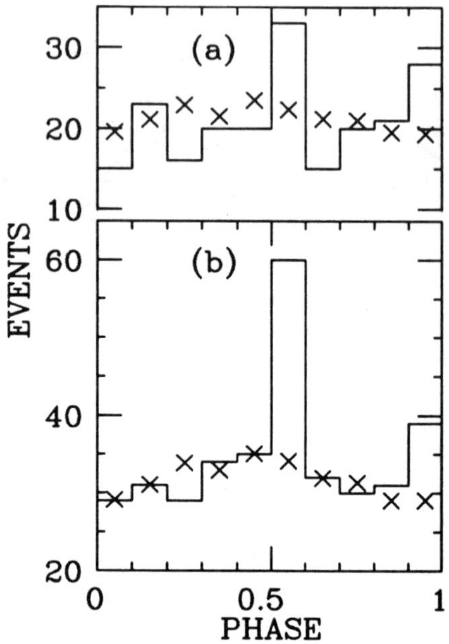

Fig. 5. a) The phase distribution for the 10 ms sample between 1 and 12 October 1983 (198 hours on-time);
b) The sum of the data in Figs 4a and 5a.

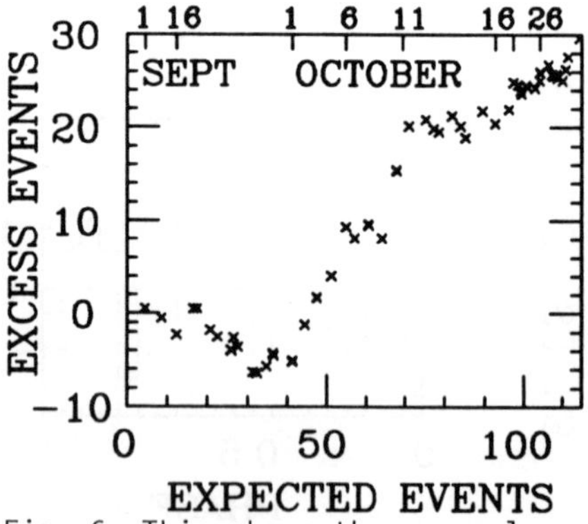

Fig. 6. This shows the accumulation of the excess between phase 0.50 and 0.60 from the beginning of September and the end of October 1983. Each cross represents one data tape.

EXPERIMENTAL EVIDENCE SUGGESTIVE OF NEUTRINOLESS ββ-DECAY WITH A NEAR MASSLESS PARTICLE

F.T. Avignone III[†], R.L. Brodzinski[††], H.S. Miley[†], and J.H. Reeves[††]

[†]Department of Physics and Astronomy, University of South Carolina, Columbia, SC 29208 USA

[††]Pacific Northwest Laboratory, Richland, Washington 99352 USA

The energy spectrum of the PNL-USC low-background Ge detector shows evidence of a shoulder in the vicinity of the ^{76}Ge ββ-decay Q-value. This energy region is above all observable γ ray lines. If attributed to 0ν ββ-decay of ^{76}Ge with Majoron emission, the net count rate corresponds to $T_{1/2}^{0\nu,B} = (6\pm1) \times 10^{20}$ yr and coupling constant $g_{ee} = (8\pm1)\times10^{-4}$.

Neutrinoless (0ν) ββ-decay can be engendered by a number of mechanisms including Majorana ν-mass, right-handed ν-currents, Higgs exchange, and the emission of a massless Goldstone boson. Recently, Chikashige, Mohapatra and Peccei[1] considered the consequences of lepton number being a spontaneously broken global symmetry in the context of simple gauge models. They introduced the name Majoron for the resulting massless Goldstone boson. Gelmini and Roncadelli[2], and independently but subsequently, Georgi, Glashow and Nussinov[3] proposed a similar model which contains a complex Higgs triplet in addition to the usual doublet. In this model, a nonzero vacuum expectation value of the triplet field spontaneously breaks B-L global symmetry, giving mass to light Majorana neutrinos and resulting in a real massless Goldstone boson, the Majoron. They point out that this particle could well be coupled weakly enough to other particles that it has thus far escaped detection.

Neutrinoless ββ-decay with the emission of a Majoron was first discussed by Georgi, Glashow, and Nussinov[3] and subsequently by Vergados[4], Haxton and Stephenson[5], and Doi, Kotani and Takasugi[6]. Using the results given in reference (3), the ratio of the rate of 0ν ββ-decay with the emission of a Majoron, Γ(B), to that driven by Majorana ν-mass, $\Gamma(m_\nu)$, can be expressed as follows:

$$\Gamma(B)/\Gamma(m_\nu) = 1.21 \times 10^{-3}\ g_{ee}^2 \left(\frac{m_e}{m_\nu}\right)^2 \varepsilon^2\ R(\varepsilon), \tag{1}$$

where,

$$R(\varepsilon) = \left\{ \frac{\varepsilon^4+14\varepsilon^3+84\varepsilon^2+210\varepsilon+210}{\varepsilon^4+10\varepsilon^3+40\varepsilon^2+60\varepsilon+30} \right\} . \tag{2}$$

In Equation (1), g_{ee} is the coupling constant of Majorons to electron-neutrinos, and ε is the ββ-decay energy in units of electron rest mass energy. It should be pointed out that the numerical coefficient in equation (1) differs from those derived from the results given references (4) and (6) by factors of 8 and 4 respectively. The definition of g_{ee} and the results given in reference (3) have been used in all of the analyses discussed here.

Limits have been placed on g_{ee} by Klimenko, Pomanski and Smolnikov[7] ($<1.5 \times 10^{-3}$) using the decay of ^{150}Nd, and by Elliot, Hahn and Moe[8] ($<5\times10^{-4}$) using the decay of ^{82}Se. The most stringent limits from the decay of ^{76}Ge have been published recently by Caldwell et al.[9] ($T_{1/2}^{0\nu,B} \gtrsim 6\times10^{20}$ yr and $g_{ee} < 1.1\times10^{-3}$). In the case of ^{82}Se[8], there are candidate events which can be used to calculate a definite half-life and its error. An analysis similar to that presented here would be more difficult when applied to the spectrum presented in reference (9), because there are γ ray peaks above the ββ-decay Q-value at 2041 keV.

Corrections to the spectrum for contributions from radioimpurities can only be made if the location and distribution of each isotope were precisely known. The data from the PNL-USC ^{76}Ge ββ-decay experiment[10] are unique in this regard. There are no observable γ ray lines above that from the 1461 keV γ ray following the decay of ^{40}K (see Figure 1). The main reasons for publishing the present results is to encourage efforts to eliminate interfering background from other experiments and to stimulate independent searches for similar effects in the spectra from the ββ-decay of other isotopes as well as other ^{76}Ge experiments.

The description of our experimental apparatus has recently been published and will not be repeated here[10, 11]. It is, however, important to point out that the data used in these analyses were taken in two parts. The first data collection period was equivalent to 281 days during an elapsed time of 294 days, during which the inner shield was a copper collar more than 7 cm thick[10]. This was followed by a period of 129 days during which only the very low energy portion of the spectrum was investigated. The final data of interest here were collected in the next 121 day period. During the final period, the copper inner shield had been replaced by bricks cast from 448 year old lead. Replacing the copper with the ancient lead reduced the ^{60}Co, produced by the cosmogenic reaction ^{63}Cu(n,α)^{60}Co, by a factor of 5.4. Some copper still remains in component parts of the detector.

The raw data were combined into 100 keV bins as shown in Figure 2. The data up to 1440 keV are taken from the final 121 day counting period during which the ^{60}Co was much reduced. The data above this energy are from the full 402 days to gain statistical significance. A good fit to the data above the ββ-decay end point energy at 2041 keV is obtained with a straight line. The maximum of the theoretical energy spectrum for the two electrons emitted in 0ν ββ-decay with a Majoron is located at approximately 1600 keV. A curve of the shape given in reference (6), plus the straight line, was fit to the data between 1600 keV and 3950 keV. A three parameter likelihood function was calculated, and the variances obtained were added in quadrature to the Poisson statistical fluctuation of the most probable area under the theoretical curve. The resulting number of events which could be attributed to such an effect is 196±28 counts in 402 days. This corresponds to 588±84 events in the entire three body decay spectrum or 1.46±0.21 events per day. An analysis of the first 281 days of data, taken with the copper inner shield, yields a rate of 1.7±0.3 day^{-1} under the spectral bump in question. The rate during the final 121 days, during which the inner shield was 448 year old lead, was 1.4±0.4 day^{-1}. These results eliminate cosmogenic radioactivity in the copper shield as a probable cause of the observed effect, while they also place constraints on the half-life of radioactive candidates which might mimick an exotic decay. The number of ^{76}Ge atoms in the fiducial volume is 4.3×10^{23}, and if these events are in fact due to this exotic decay process, the half-life is $(6\pm1)\times10^{20}$ yr. Combining equation (1) with the calculations of Haxton and Stephenson[5], the following relation is obtained:

$$g_{ee}^2 = 3.5\times10^{14}\ \mathrm{yr}/T_{\frac{1}{2}}^{0\nu,B}\ \mathrm{yr}. \qquad (3)$$

Accordingly, $g_{ee} = (8\pm1)\times10^{-4}$ which is only slightly smaller than the estimated limit given in reference (3). If this effect is indeed due to 0ν ββ-decay with the emission of a Majoron, most of the conclusions of reference (3) remain unchanged, and the present data are not in conflict with the other observations discussed there.

A serious caution is in order. In actual fact, the only discovery which can be unambiguously claimed here is the existence of a sudden sharp shoulder on the background spectrum resembling the expected shape of a most interesting exotic decay. Possible background effects which might also explain these results will now be examined.

The only background in the spectrum above the 1461 keV γ ray peak is the broad 5.3 MeV α-peak and its degraded continuum as shown in Figure 1. Documentation that this background is from the α-decay of ^{210}Po, resulting from the decay of ^{210}Pb in ordinary soft solder in proximity to the detector, is given in refs. 10 and 11 where it was shown that the ^{210}Po was brought to the surface when the solder was melted, and it decays with $T_{\frac{1}{2}}$ = 138 days. If the effect is due to some process involving α-particles, it would have changed over the 3.9 half-lives of ^{210}Po during which the data were collected. No change in rate, nor in shape, was observed. The α-particles do not appear to provide an explanation.

Next, there are eight known γ ray peaks in the spectrum due to the decay of ^{40}K, ^{60}Co, ^{58}Co, ^{56}Co, ^{65}Zn and ^{54}Mn in the detector components. In addition there is a γ ray at 662±1 keV which was erroneously identified as that from the decay of ^{137}Cs[10]. It was later noted that this γ ray decays in several weeks following installation and is from some unidentified source perhaps associated with radon daughters. The ^{40}K is probably associated with a resistor in the first stage of amplification, but the decay energy is insufficient to interfere in the region of interest. The isotope ^{65}Zn is cosmologically produced in the Ge crystal itself via the reaction $^{70}Ge(n,\alpha 2n)^{65}Zn$ prior to the time the Ge was installed in the mine. The total energy available in this decay is 1349 keV, again too small to explain any part of the observed effect.

The remaining isotopes were cosmogenically produced primarily in the 7.3 cm thick cylindrical copper inner shield which was installed for the first 281 day data collection period. The isotopes $^{56,58,60}Co$ and ^{54}Mn are primarily produced by cosmic ray neutrons via the following reactions: $^{63}Cu(n,\alpha 4n)^{56}Co$, $^{63}Cu(n,\alpha 2n)\,^{58}Co$, $^{63}Cu(n,\alpha)^{60}Co$, and $^{63}Cu(n,2\alpha 2n)^{54}Mn$. All of these isotopes, with the exception of ^{54}Mn, have decay energies sufficently high to cause background in the region of interest. Two of the remaining three candidates, ^{56}Co and ^{58}Co, can be eliminated because of their short half-lives, 77 and 71 days, respectively. The effect in question is just as pronounced in the data taken more than 200 days after the copper was installed in the mine, but these isotopes did decay by the predicted factor. The remaining candidate, ^{60}Co, has a well known γ ray cascade with total energy high enough to provide background in the region of interest. However, after the copper collar was removed the ^{60}Co rate was reduced by a factor of 5.4, but the effect did not change accordingly. The change in inner shield configuration also eliminates candidates in the lead shielding since no observable change was noticed when the copper was replaced by lead.

The possibility that the observed effect is due to radioactive isotopes internal to the crystal, formed either by cosmogenic isotopes or by contamination of the Ge itself, was carefully considered. A thorough search was made of all isotopes with $Q>1700$ keV and $T_{\frac{1}{2}}>200$ days. This half-life would correspond to a decrease in the effect by a factor of 3.3 between our first data collection period with the copper shield and the later one with the old lead inner shield.

However, the comparison of the event rates during these data collection periods precludes such a reduction with a level of confidence of 97.30%. There are a total of twenty four candidates excluding α-emmitters because of the absence of α-lines in the spectrum. These are the 16 isotopes listed in Table I and ^{50}V, ^{92}Nb, ^{96}Zr, ^{150}Eu, ^{152}Eu, ^{172}Hf, ^{194}Hg and ^{207}Bi. No decay of ^{96}Zr has been observed, while the rest were easily eliminated on the grounds of strong electron-capture branches which would have resulted in easily observable γ-ray lines. Each of the remaining background candidates was hypothesized as the cause of the observed spectral bump. A detailed Monte Carlo calculation was made for each candidate, which included all significant decay branches and assumed a uniform distribution of the isotope in the Ge. The resulting spectral response functions were fit to the data and subjected to a standard goodness-of-fit test based on the χ^2 distribution. The quantity $Q(\chi^2,n)$ is the probability that values of χ^2 equal to or greater than that resulting from the fit could be observed due to statistical fluctuations alone. Two of the candidates, ^{22}Na and ^{26}Al, were eliminated based on the absence of sharp γ ray lines at the predicted levels of the Monte Carlo calculations. The results are presented in Table I.

It is very interesting to hypothesize that the observed excess of counts from about 2041 keV, the $\beta\beta$-decay energy of ^{76}Ge, down to approximately 1600 keV is due to ordinary 2ν $\beta\beta$-decay. This portion of the spectrum yields 196±28 counts above the continuum. This energy interval corresponds to about 1.3% of the total theoretical 2ν $\beta\beta$-decay spectrum. If these counts are attributed to this decay mode, the corresponding half-life is 2×10^{19} years. This and other experiments[9, 10] have been used to set a lower limit of $\sim10^{20}$ years for 2ν $\beta\beta$-decay of ^{76}Ge using a much larger fraction of the spectrum. In fact, a predicted spectrum corresponding to a half-life of 2×10^{19} yr. falls well above the entire continuum down to about 500 keV. These data were also subjected to a χ^2 goodness-of-fit test, and the result is also given in Table I. This test very strongly mitigates against this hypothesis. The theoretical prediction for this decay half-life is 4.15×10^{20} yr.[5]. The experimental limit on the 2ν

$\beta\beta$-decay half-life of ^{82}Se[8, 12] is approximately a factor of 6 above the theoretical calculation of reference (5). While a comparison between different isotopes is not conclusive, it is still highly unlikely that the ^{76}Ge 2ν $\beta\beta$-decay rate would be a factor of 20 greater than shell model predictions while that of ^{82}Se is a factor of 6 lower.

Finally, there could well be a similar effect observed in another experiment. The UCSB-LBL ^{76}Ge $\beta\beta$-decay experiment also has an excess of counts in the same energy region[13]. It has approximately the same number of events per kg of Ge per year as the bump in the PNL-USC spectrum, but does not exhibit a dip below 1600 keV, which may be due to background sources not present in our detector.

The recent data from the UC Irvine ^{82}Se $\beta\beta$-decay experiment[12], has also been subjected to a maximum likelihood analysis, assuming all events above 2 Mev are due to 0ν $\beta\beta$-decay with Majoron emission. No substraction of expected 2ν $\beta\beta$-decay was attempted because its contribution here is small. The result of our reanalysis of the Irvine data is 11.0±3.3 counts in that portion of the spectrum above the maximum in the theoretical distribution. This corresponds to 33±10 0ν $\beta\beta$-decays with the emission of Majorons in 5426 live hours of counting. Using the parameters given in reference (8), $T_{\frac{1}{2}}^{0\nu,B}(^{82}Se) = (1.9\pm0.6) \times 10^{20}$ yr. which corresponds to $g_{ee} = (9^{+2}_{-1}) \times 10^{-4}$, in agreement with the value obtained from our data. However, the UC Irvine group has made no claim that the events are in fact due to 0ν $\beta\beta$-decay.

We cannot at this point claim to have unambiguously observed 0ν $\beta\beta$-decay resulting in two electrons and a Majoron. Unfortunately, because of the broad distribution in total electron energy inherent in this process, conclusive evidence will require verification by a series of experiments involving a variety of $\beta\beta$-decay unstable iosotopes. It will require that anomalous "shoulders" in the spectra appear with the proper shape and at the correct energies and that they yield half-lives which correspond to values of g_{ee} which are consistent from experiment to experiment. While not conclusive, the evidence presented here is very suggestive and is certainly strong enough to justify intense efforts to produce searches of improved sensitivity. We have recently eliminated the α-decay background[11] and will count for several months with the new version of the prototype. If this phase confirms our background reduction goals[10, 11], the volume of Ge will be increased which will result in a significantly more sensitive test of the shape of the continous two-electron energy spectrum and more sensitive searches for all modes of ββ-decay of ^{76}Ge.

The authors would like to acknowledge very helpful conversations with Andrzej Drukier, Richard Fernholz, Graciella Gelmini, Tom Girard, Wick Haxton, Mike Moe and Shmuel Nussinov. This work was supported by the U.S. Department of Energy under Contract No. DE-AC06-76RLO 1830 and the National Science Foundation under Grant No. PHY-8405654.

REFERENCES

[1] Y. Chikashige, R.N. Mohapatra and R.D. Peccei, Phys. Lett. 98B 264 (1981).

[2] G.B. Gelmini and M. Roncadelli, Phys. Lett. 99B 411 (1981).

[3] Howard M. Georgi, Sheldon Lee Glashow, and Shmuel Nussinov, Nucl. Phys. B193 297 (1981).

[4] J.D. Vergados, Phys. Lett. 109B 96 (1982).

[5] W.C. Haxton and G.J. Stephenson, Jr., Prog. Part. Nucl. Phys. 12 409 (1984).

[6] Masaru Doi, Tsuneyuki Kotani, and Eiichi Takasugi, Prog. Theor. Phys. Supplement 83 1 (1985).

[7] A.A. Klimenko, A.A. Pomansky and A.A. Smolnikov, in Proc. of the Eleventh Int. Conf. on Neutrino Physics and Astrophysics, Dortmund, W. Germany, 1984, ed. K. Kleinecht and E.A. Paschos, World Scientific, Singapore 161 (1985).

[8] S.R. Elliot, A.A. Hahn, and M.K. Moe, Phys. Rev. Lett. 56 2582 (1986).

[9] D.O. Caldwell, R.M. Eisberg, D.M. Grumm, D.L. Hale, M.S. Witherell, F.S. Goulding, D.A. Landis, N.W. Madden, D.F. Malone, R.H. Pehl, and A.R. Smith, Phys. Rev. D33 2737 (1986).

[10] F.T. Avignone, III, R.L. Brodzinski, J.C. Evans, Jr., W.K. Hensley, H.S. Miley, and J.H. Reeves, Phys. Rev. C34 666 (1986).

[11] R.L. Brodzinski, J.H. Reeves, F.T. Avignone, III, and H.S. Miley, Pacific Northwest Preprint PNL-SA-14236, to be published in Nucl. Instrum. and Meth.

[12] M.K. Moe, Proc. Intersections Between Particle and Nuclear Physics, Lake Louise Canada, 1986, AIP Conference Proceedings 150, 1012, ed. Donald Geesaman (1986).

[13] M. Witherell (Private Communication).

Table I. Results of Chi-Square hypothesis tests of the Monte-Carlo response functions as the observed spectral shape.

Isotope	Exclusion CL	Isotope	Exclusion CL
^{22}Na	99.990%	^{106}Rh(Ru)	99.986%
^{26}Al	>99.999	^{110m}Ag	99.530
^{42}Ar	>99.999	^{126}Sn	99.77
^{44}Ti	99.40	^{134}Cs	99.82
^{60}Co	99.996	^{154}Eu	97.98
^{60}Co(external)	99.390	^{166}Ho	99.58
^{68}Ge	98.43	^{182}Ta	99.998
^{90}Sr	97.0	^{194}Os	97.040
^{94}Nb	98.40	2ν $\beta\beta$-decay of ^{76}Ge	>99.999

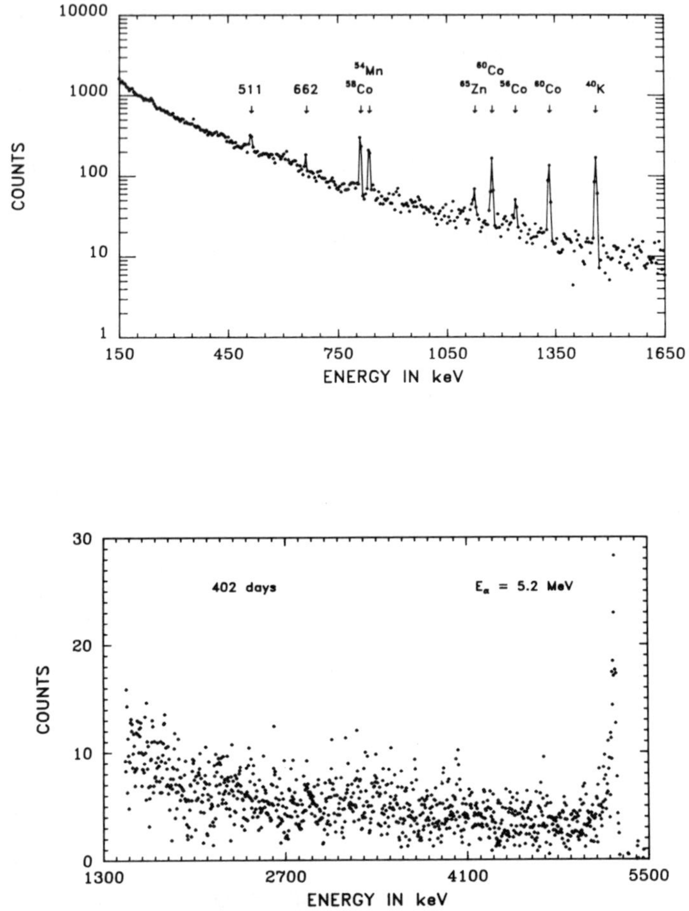

Figure 1. Background in the PNL-USC 135 cm^3 prototype Ge detector after 402 days of counting. Clearly visable are the α peak, its continuum, and γ ray peaks from ^{40}K and from cosmogenically produced isotopes.

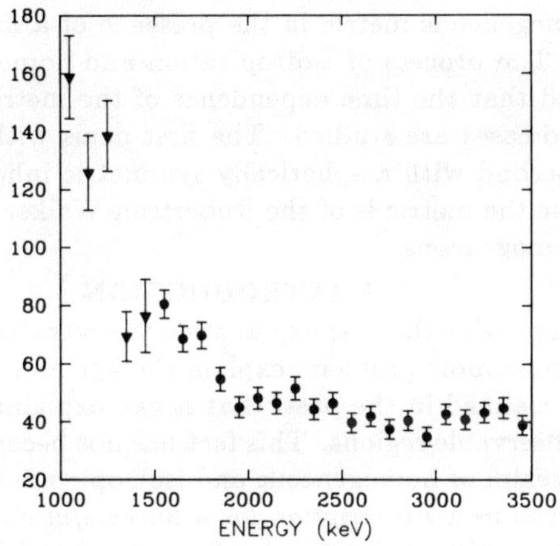

Figure 2. The triangles represent data taken during the final 121 day collection period which have been corrected for γ ray lines and the experimentally determined ^{40}K Compton continuum. The data from 1550 keV and higher were collected during the entire 402 day collection period and are normalized 121 days.

INFLATION IN TOLMAN-BONDI COSMOLOGICAL MODELS

Jaime A. Stein-Schabes

Theoretical Astrophysics Group
Fermi National Accelerator Laboratory
P.O.Box 500,Batavia, Illinois 60510

ABSTRACT

Exact analytical solutions of Einstein's equations are found for a spherically symmetric inhomogeneous metric in the presence of a massless scalar field with a flat potential. The process of isotropization and homogenization is studied in detail. It is found that the time dependence of the metric becomes de Sitter for large times. Two cases are studied. The first deals with a homogeneous scalar field, while the second with a spherically symmetric inhomogeneous scalar field. In the former case the metric is of the Robertson-Walker form, while the latter is intrinsically inhomogeneous.

I. INTRODUCTION

Inflation can provide the best explanation to several cosmic conundrums[1)–5)]. It can solve the monopole problem, explain the age and flatness of the Universe. And it has been claimed in the past, that it can explained the homogeneity and isotropy of the observable regions. This fact has not become well established until recently. In the realm of homogeneous and isotropic model P. Steinhardt and M. Turner[6)] have given us a *Prescription for a Successful New Inflation* which, when fallow step by step, will produce and inflationary model. This prescription has recently been extended to encompass a large class of both inhomogeneous and anisotropic models. The extension comes in the form of a theorem, the so called No-Hair theorem. In a loose manner it states that any inhomogeneous space time that is not to positively curved, in the presence of a cosmological constant, with an energy momentum tensor that satisfies the strong and dominant energy conditions, will approach the de Sitter solution asymptotically. The theorem has been recently proven by Lars Jensen and myself[7)], and it generalizes an earlier version that holds for homogeneous space-times proven by R. Wald[8)] Even though these results indicate that dynamically the universe goes from an inhomogeneous and anisotropic phase to one which is isotropic and homogeneous on scales of the observable Universe, it gives no details on the evolution of the scalar field or any other content of the Universe. In order to learn more about the evolutionary details of the isotropization process we have to study specific examples. Most inflationary models have been constructed assuming the background space-time metric is homogeneous, either one of the Bianchi models or a Kantowski-Sacks model[9),10)]. For these models it has been shown that once inflation starts the process of isotropization is remarkably efficient. Furthermore, once inflation has successfully ended the universe remains isotropic for a very long time. Very little analytical work has been done to solve the problem in the case of inhomogeneous space-times, but some numerical results have been obtained[11)] for the case of an inhomogeneous scalar field, however the background metric is still homogeneous and isotropic.

In this talk I would like to report on some work I have done[12] in which Einstein's equations are solved explicitly when gravity is minimally coupled to a scalar field that has a flat potential and the metric is that of an inhomogeneous spherically symmetric space-time, the so called Tolman-Bondi[13].

II. THE MODEL

The Lagrangian will be that of gravity minimally coupled to a scalar field $\phi(r,t)$ with a potential $V(\phi)$,

$$S = \int \sqrt{-g}\left(R - \frac{1}{2}g^{\mu\nu}\partial_\mu\phi\partial_\nu\phi - V(\phi)\right)d^4x \tag{2.1}$$

Where $g = det(g_{\mu\nu})$ and R is the Ricci scalar. Units are taken so that such that $16\pi G = c = \hbar = 1$. From the action we get,

$$R_{\mu\nu} - \frac{1}{2}g_{\mu\nu}R = T_{\mu\nu} \tag{2.2}$$

with

$$T_{\mu\nu} = \frac{1}{2}\partial_\mu\phi\partial_\nu\phi - \left(\frac{1}{4}(\partial_\alpha\phi\partial^\alpha\phi) + \frac{1}{2}V(\phi)\right)g_{\mu\nu} \tag{2.3}$$

$$\frac{1}{\sqrt{-g}}\partial_\mu\left(\sqrt{-g}\partial^\mu\phi\right) = -\frac{dV(\phi)}{d\phi} \tag{2.4}$$

The metric is given by the Tolman-Bondi line element,

$$ds^2 = -dt^2 + X^2(r,t)dr^2 + Y^2(r,t)(d\theta^2 + \sin^2\theta d\varphi^2) \tag{2.5}$$

with r, θ, and φ the normal spherical coordinates (in the comoving frame). The field equations can be written as (ϕ will be rescale to 2ϕ),

$$2\frac{\dot{X}}{X}\frac{\dot{Y}}{Y} + \left(\frac{\dot{Y}}{Y}\right)^2 + \frac{1}{Y^2} - \frac{1}{X^2}\left(2\frac{Y''}{Y} + \left(\frac{Y'}{Y}\right)^2 - 2\frac{X'}{X}\frac{Y'}{Y}\right) = (\dot{\phi})^2 + \frac{(\phi')^2}{X^2} + \frac{1}{2}V(\phi) \tag{2.6}$$

$$2\frac{\ddot{Y}}{Y} + \left(\frac{\dot{Y}}{Y}\right)^2 + \frac{1}{Y^2} - \frac{1}{X^2}\left(\frac{Y'}{Y}\right)^2 = -(\dot{\phi})^2 - \frac{(\phi')^2}{X^2} + \frac{1}{2}V(\phi) \tag{2.7}$$

$$\frac{\ddot{X}}{X} + \frac{\ddot{Y}}{Y} + \frac{\dot{X}}{X}\frac{\dot{Y}}{Y} - \frac{1}{X^2}\left(\frac{Y''}{Y} - \frac{X'}{X}\frac{Y'}{Y}\right) = -(\dot{\phi})^2 + \frac{(\phi')^2}{X^2} + \frac{1}{2}V(\phi) \tag{2.8}$$

$$\frac{\dot{Y}'}{Y} - \frac{\dot{X}}{X}\frac{Y'}{Y} = -\dot{\phi}\phi' \tag{2.9}$$

$$\ddot{\phi} + \left(\frac{\dot{X}}{X} + 2\frac{\dot{Y}}{Y}\right)\dot{\phi} - \frac{1}{X^2}\left(\phi'' + \left(3\frac{X'}{X} + 2\frac{Y'}{Y}\right)\phi'\right) = -\frac{dV(\phi)}{d\phi} \tag{2.10}$$

where ($\cdot$, $'$) will denote time and space derivatives. I will solve these equation for two cases, when $\phi = \phi(t)$, and when $\phi = \phi(r,t)$. We shall also assume that for the epoch of interest, the potential can be well approximated by a constant value $V(\phi) \simeq 2\Lambda$. This can be justified by noticing that in general the quantity multiplying the $\dot{\phi}$ term in eq.(2.10) acts as a friction term and it has been shown elsewhere[7),8)], that this term is always larger that in the homogeneous and isotropic model. What this means is that the field will always move slower in an anisotropic and inhomogeneous model, so spending more time than usual on the flat part of the potential.

III. THE HOMOGENEOUS CASE: $\phi = \phi(t)$

In this case it is easy to see that eqs. (2.9) and (2.10) immediately integrate to give respectively

$$X(r,t) = f(r)Y'(r,t) \tag{3.1}$$

$$\dot{\phi}(t) = \frac{l(r)}{XY^2} = \frac{l(r)}{f(r)}\frac{1}{Y^2Y'} \tag{3.2}$$

with $l(r)$ and $f(r)$ arbitrary integration functions. Eq.(3.2) is a very interesting equation as it forces the spatial dependence of $Y(r,t)$ to be such that $\dot{\phi}$ is only a function of t. Taking now the following combination of equations (2.6) – (2.7) – 2 × (2.8) we get

$$\frac{\ddot{X}}{X} + 2\frac{\ddot{Y}}{Y} = -2(\dot{\phi})^2 + \Lambda \tag{3.3}$$

using eqs.(3.1) and (3.3) we can decouple the equations and get one for $Y(r,t)$,

$$\frac{\ddot{Y}'}{Y'} + 2\frac{\ddot{Y}}{Y} = \frac{l^2(r)}{f^2(r)}\frac{1}{Y^4(Y')^2} + \Lambda \tag{3.4}$$

To solve this partial differential equation we use separation of variables. We look for solutions of the form $Y(r,t) = Y_r(r)Y_t(t)$, this allows a separation into two systems of ordinary differential equations.

$$3\frac{\ddot{Y}_t}{Y_t} - \frac{m_1}{Y_t^6} = \Lambda \tag{3.5}$$

$$Y_r'Y_r^2 = \sqrt{\frac{-2f^2(r)}{m_1 l^2(r)}} \tag{3.6}$$

eq.(3.6) requires $m_1 < 0$. Eq.(3.5) has a general first integral

$$\left(\frac{\dot{Y}_t}{Y_t}\right)^2 = \frac{\Lambda}{3} - \frac{k}{Y_t^2} - \frac{m_1}{6Y_t^6} \tag{3.7}$$

with k an integration constant. The general solution to this equation involve elliptical functions. Since this is a well studied equation and its asymptotic behaviour well known we shall not say more about its general solution. These have a stable asymptotic limit of the form $Y_t = Y_0 e^{\sqrt{\frac{\Lambda}{3}}t}$ which is the same time evolution as in the de Sitter model. Then by eqs. (3.1) the time evolution for $X(r,t)$ is the same as that for Y_t, namely $X(r,t) = f(r)Y_r'Y_t \rightarrow X_r(r)e^{\sqrt{\frac{\Lambda}{3}}t}$ where $f(r)$ is an arbitrary integration function. From these and eq.(3.2) we get for $\dot{\phi}$

$$\dot{\phi}(t) = \sqrt{\frac{-m_1}{2}}\frac{1}{Y_t^3} \tag{3.8}$$

which is perfectly consistent with the functional dependence of $\dot{\phi}$. For the spatial part of Y_r eq.(3.6) cannot be integrated until we specify the form of the arbitrary function. The fact that the universe is expanding exponentially fast, makes the scalar field ϕ go rapidly to a constant value ϕ_0. The ansatz used to find the solution for $X(r,t)$ and $Y(r,t)$ together with the assumed homogeneity of the scalar field determined the dynamics of the model to be identical to the homogeneous and isotropic model for any potential $V(\phi)$. From all the above considerations we find that $X_t = Y_t$ and since the scalar field evolution is determined by (2.10) (without the spatial derivatives), it becomes the same equation as for the homogeneous scalar field. Furthermore, if we run the solution through the constraint equations we immediately find that $f^2(r) = (1 - kY_r^2)^{-1}$. This is a direct consequence of the separability of the metric components into space and time parts, which immediately forces the Universe to expand isotropically with an expansion rate that approaches the de Sitter one. These equations can be integrated without assuming the functions are separable into their space and time parts, but the result is exactly the same[12)]. The physical reason is that both the scalar field and the cosmological constant produce "isotropic forces" so there is nothing to keep the model from becoming isotropic as it evolves and it would seem that any initial anisotropy and inhomogeneity present in the model must be put in by hand. It is then impossible to have an anisotropic and inhomogeneous universe if the scalar field is to be homogeneous. We can now write the full solution for the metric,

$$ds^2 = -dt^2 + X_t^2 X_r^2 dr^2 + Y_t^2 Y_r^2 (d\theta^2 + \sin^2\theta d\varphi^2) \tag{3.9}$$

with the help of eq.(3.1) and using the freedom to redefine the coordinate system, we introduce a new radial variable $\rho = Y_r$ then eq.(3.9) becomes in the large time limit

$$ds^2 = -dt^2 + e^{\sqrt{\frac{\Lambda}{3}}t}\left[\frac{d\rho^2}{1 - k\rho^2} + \rho^2(d\theta^2 + \sin^2\theta d\varphi^2)\right] \tag{3.10}$$

which is exactly the de Sitter solution in its open, flat or closed version. What we can conclude in this case is that the model was never inhomogeneous but rather looked like one due to a bad choice of coordinates, so even if we assumed that the metric was more general than the standard model, the equations only accepted the homogeneous and isotropic solution. The flatness of the model is then explained as in the standard FRW model. By calculating the three-curvature scalar we get that asymptotically it goes to zero like ${}^{(3)}R \simeq e^{-2\sqrt{\frac{\Lambda}{3}}t}$, however the individual components of the the three-Ricci tensor do not vanish in the comoving frame (unless, of course, $k = 0$) However, when a measurement is done in the observers frame then all components of the curvature tensor vanish asymptotically fast in the same way that the curvature scalar.

IV. THE INHOMOGENEOUS CASE: $\phi = \phi(r,t)$

In this case we look for solutions of the form,

$$X(r,t) = X_r(r)X_t(t) \tag{4.1}$$

$$Y(r,t) = Y_r(r)Y_t(t) \tag{4.2}$$

This decomposition of the scale factors forces the scalar field to be of the following form (this can be immediately seen from eqs. (3.3), (4.1) and (4.2)), $\phi(r,t) = \phi_r(r) + \phi_t(t)$. Eq.(2.9) is again immediately separable and integrable,

$$\frac{Y_r'}{Y_r} = -\phi_r' \tag{4.3}$$

$$\frac{\dot{Y}_t}{Y_t} - \frac{\dot{X}_t}{X_t} = \dot{\phi}_t \tag{4.4}$$

we now use eq.(2.10), which again splits naturally into the following equations,

$$\ddot{\phi}_t + \left(\frac{\dot{X}_t}{X_t} + 2\frac{\dot{Y}_t}{Y_t}\right)\dot{\phi}_t = \frac{m_2}{X_t^2} \tag{4.5}$$

$$\phi_r'' + \left(3\frac{X_r'}{X_r} + 2\frac{Y_r'}{Y_r}\right)\phi_r' = m_2 X_r^2 \tag{4.6}$$

with m_2 the separation constant. To complete the system of equations we use (3.3) and the space part of (2.8) (of course the time-dependent part of this equation becomes one more of the constraint equations),

$$\frac{\ddot{X}_t}{X_t} + 2\frac{\ddot{Y}_t}{Y_t} = -2(\dot{\phi}_t)^2 + \Lambda \tag{4.7}$$

$$\frac{Y_r''}{Y_r} - \frac{X_r'}{X_r}\frac{Y_r'}{Y_r} + (\phi_r')^2 = m_3 X_r^2 \tag{4.8}$$

So, the problem has been reduced to solve eqs.(4.3), (4.6) and (4.8) for the spatial part and eqs.(4.4), (4.6) and (4.9) for the time part. If we define $a \equiv \frac{1}{4}(m_2+m_3)$, we can formally solve the spatial part to give,

$$X_r'(\ln(Y_r))' = -aX_r^3 \tag{4.9a}$$

$$(X_r')^2 = \left[C_1 X_r^{\frac{2m_3}{a}} + \frac{2a^2}{m_3}\right] X_r^4 \tag{4.9b}$$

Both Y_r and ϕ_r are given for the case $a \neq 0$ by

$$Y_r = Y_0 e^{-a\int \frac{X_r^3}{X_r'} dr} \tag{4.10a}$$

$$\phi_r = \phi_0 + a \int \frac{X_r^3}{X_r'} dr \tag{4.10b}$$

The general solution to (4.9b) is given in terms of Gauss hypergeometric functions, however, to construct the general solution would only obscure the features we want to highlight. Instead, we shall explore the consequences when special values are chosen for these constants.

i) a = 0 : In these case we can see from eq.(4.9a) that $X_r' = 0$, then

$$Y_r = \begin{cases} Y_0 \cos^{\frac{1}{2}}\left(\sqrt{2m_2} r\right), & m_2 > 0 \\ Y_0 r^{\frac{1}{2}}, & m_2 = 0 \\ Y_0 \sinh^{\frac{1}{2}}\left(\sqrt{-2m_2} r\right), & m_2 < 0 \end{cases} \tag{4.11}$$

this in turn gives for ϕ_r

$$\phi_r = \begin{cases} \phi_0 - \frac{1}{2}\ln\left(\cos\left(\sqrt{2m_2} r\right)\right), & m_2 > 0 \\ \phi_0 - \frac{1}{2}\ln(r), & m_2 = 0 \\ \phi_0 - \frac{1}{2}\ln\left(\sinh\left(\sqrt{-2m_2} r\right)\right), & m_2 < 0 \end{cases} \tag{4.12}$$

ii) $a = \pm\frac{2m_3}{n}$: For this case the solution is a finite series that depends on the value of n. For simplicity let $n = \pm 2$, then the solution is given by,

$$X_r^2 = \begin{cases} \frac{b}{1 - C_1 b^2 r^2} & n = 2, \quad b > 0 \\ \frac{1}{b} sec\left(\frac{\sqrt{C_1}}{b} r\right) & n = -2, \quad b > 0 \end{cases} \tag{4.13}$$

Y_r and ϕ_r can now be obtained from (4.10a) and (4.10b). For the time part we get,

$$\phi_t = \phi_0 + ln\left(\frac{Y_t}{X_t}\right) \tag{4.14}$$

where X_t is given by

$$X_t = \begin{cases} \sqrt{\frac{5m_2}{2\Lambda}} \cosh\left(\sqrt{\frac{2\Lambda}{5}} t\right), & m_2 > 0 \\ e^{\sqrt{\frac{2\Lambda}{5}} t}, & m_2 = 0 \\ \sqrt{\frac{-5m_2}{2\Lambda}} \sinh\left(\sqrt{\frac{2\Lambda}{5}} t\right), & m_2 < 0 \end{cases} \tag{4.15}$$

Since all the solutions tend asymptotically towards the case $m_2 = 0$ I shall only calculate Y_t and ϕ_t in this limit

$$Y_t = Y_0 e^{\sqrt{\frac{9\Lambda}{40}}t} \sinh^{\frac{1}{2}}\left(\sqrt{\frac{9\Lambda}{10}}t\right) \tag{4.16}$$

$$\phi_t = \phi_0 + \sqrt{\frac{9\Lambda}{40}}t - \frac{1}{2}\ln\sinh\left(\sqrt{\frac{9\Lambda}{10}}t\right) \tag{4.17}$$

For large times $X_t \simeq Y_t \simeq e^{\sqrt{\frac{2\Lambda}{5}}t}$ and $\phi_t = \phi_0$. Now we can construct the full solution, which in the large time limit takes the following form,

$$ds^2 = -dt^2 + e^{2\sqrt{\frac{2\Lambda}{5}}t}\left[X_r^2 dr^2 + Y_r^2(d\theta^2 + \sin^2\theta d\varphi^2)\right] \tag{4.18}$$

By introducing a new variable $Y_r = \rho$ we get,

$$ds^2 = -dt^2 + e^{2\sqrt{\frac{2\Lambda}{5}}t}\left[Z_\rho^2 d\rho^2 + \rho^2(d\theta^2 + \sin^2\theta d\varphi^2)\right] \tag{4.19}$$

$$Z_\rho^2 = \frac{Z_0^2}{\rho^2}(X_\rho^{\frac{2m_3}{a}} + \frac{2a^2}{m_3}) \tag{4.20}$$

with X_ρ given by eq.(4.9b). The curvature scalar goes to zero asymptotically, just as in the homogeneous case. In order to understand the results obtained we must remember that the coordinates used are comoving coordinates, so even though inhomogeneities are not disappearing inside a comoving volume they certainly are for an observer measuring physical volume. Inhomogeneities get freeze out and then just pushed outside the observable horizon.

V. CONCLUSIONS

By explicitly solving Einstein's equations for an inhomogeneous spherically symmetric metric of the Tolman-Bondi type coupled to a real scalar field with a potential that has a flat region, I studied in detailed the isotropization and homogenization of the universe on scales of the order of the horizon today. Two cases were studied, the first with a homogeneous scalar field, the second with a general one. In the first case the homogeneity severely restricts the functional dependence of the metric coefficients with time. In fact, it forces the model to expand isotropically for all time, the rate at which it does it approaches the de Sitter rate asymptotically. The second case is more interesting as truly inhomogeneous solutions can be found. The time evolution was very similar to the homogeneous case. The solutions found have one less arbitrary function than the general solution should. It would nevertheless, be extremely useful to know if these solutions are stable against small perturbations that do not decouple the functional dependence of the metric components into their space and time parts. I have also shown that in all cases studied, the universe tends to a state of isotropic expansion which approaches the de Sitter rate for large times. In this sense the

Universe is "isotropizing". However, the homogeneity and flatness are a bit more subtle. The important point to notice is that the time evolution decouples from the spatial variation, the former becomes identical to the de Sitter case. The inhomogeneities are frozen out and then are pushed out of the observers horizon by the rapid expansion of the Universe. So we could conclude that globally the models do not become homogeneous. However, Inflation can still do the trick. The three-space is not maximally symmetric as we would expected if the No-Hair Theorem was absolutely correct. Nevertheless, since the three space is only expanding in a conformal way, then an observer living in this three space would see the universe around it becoming homogeneous, isotropic and flat.

Acknowledgments : I would like to thank Rocky Kolb for the invitation to the DPF meeting in Salt Lake City, Utah January 1987. This work was supported in part by DOE and NASA.

REFERENCES

1) Guth, A.H.,Phys. Rev. **D23**, 347 (1981).
2) Linde, A.D., Phys. Lett. **108B**, 389 (1982).
3) Albrecht, A. and Steinhardt, P.J., Phys. Rev. Lett. **48**, 1220 (1982).
4) Turner, M.S., *The Inflationary Paradigm*, Proceedings of the Cargèse School on Fundamental Physics and Cosmology, eds. J. Audouze and J. Tran Thanh Van, Editions Frontieres, Gif-Sur-Yvette (1985).
5) Blau, S.K. and Guth, A.H., *Inflationary Cosmology*, To be published in **300 Years of Gravitation** eds. S. Hawking and W.Israel, CUP 1987.
6) Steinhardt, P. and Turner, M.S. Phys.Rev. **D29**, 2162 (1984).
7) Jensen, L.G. and Stein-Schabes, J.A., Phys. Rev. **D35**,4 (Feb 1987).
- Jensen, L.G. and Stein-Schabes, J.A., Proceedings of the first course of the International School of Particle Astrophysics, Erice,Italy 1986.
8) Wald, R.W., Phys. Rev. **D28**, 2118 (1983).
9) Jensen,L.G. and Stein-Schabes, J.A., Phys. Rev. **D34**,931 (1986).
10) Turner,M.S. and Widrow,L., Phys. Rev. Lett. **57**, 2237 (1986).
11) Albrecht, A., Brandenberger, R.H. and Matzner, R.A., Phys.Rev. **D32**,6 (1985).
12) Stein-Schabes, J.A., Phys. Rev. **D** (in press) 1987.
13) Tolman,R.C., Proc. Nat. Acad. Sci. **16**, 511 (1930).
- Bondi,H., Mon. Not. R. Astr. Soc. **107**, 410 (1947).

PAIR PRODUCTION OF LONGITUDINAL W AND Z BOSONS AND THE GOLDSTONE-BOSON EQUIVALENCE THEOREM

Scott S. D. Willenbrock

Physics Department, University of Wisconsin
Madison, Wisconsin 53706

ABSTRACT

W and Z boson pairs are often the signal for new particle production, *e.g.* heavy Higgs bosons, new Z' bosons, heavy quark and lepton pairs, and heavy wino pairs. We argue that these W and Z bosons are predominantly longitudinally polarized, while those produced from background processes are primarily transversely polarized. This may be understood by implementing the equivalence of longitudinally-polarized vector bosons and Goldstone bosons.

1. INTRODUCTION

The signature for the production of many types of new particles at future colliders will be W and Z boson pairs. There will also be W and Z boson pairs produced by conventional background processes. Since we would like to separate the signal from the background, we are led to search for some qualitative differences between the gauge boson pairs produced by these two sources.

In this talk I would like to espouse one such difference. As we will see, the W and Z bosons which are produced in the decay of heavy particles are predominantly longitudinally polarized, while those produced via ordinary processes involving light particles are primarily transversely polarized.[1] Since longitudinally- and transversely-polarized gauge bosons have different decay distributions, we can use this information to help distinguish the signal from the background. This method was suggested by Duncan, Kane, and Repko in their work on Higgs boson decays to W boson pairs.[2]

The intuition for this result comes from the Goldstone-boson equivalence theorem, which relates the properties of longitudinally-polarized gauge bosons to the corresponding Goldstone bosons which are eaten by the Higgs mechanism to provide the longitudinal degrees of freedom. This theorem was suggested a long time ago,[3] and was extended and proven rigorously in the recent work of Chanowitz and Gaillard.[4] We will use this theorem to understand why the signal and background W and Z boson pairs have such different polarization properties.

The remainder of this talk is organized as follows. In Section II we demonstrate that W and Z boson pairs will be the signal of the production of many types of new particles. In Section III we state the Goldstone-boson equivalence theorem and show its relevance to W and Z boson pair production. In Sections IV and V we discuss Z and W boson pair production from quark-antiquark annihilation. In Section VI we discuss a less-well-known background process—W and Z boson pairs from gluon fusion via a virtual quark loop. We make a concluding remark in Section VII.

We will restrict our discussion to a high-energy hadron collider since it is in this type of machine that backgrounds are particularly severe. We focus on the proposed Superconducting Super Collider (SSC), a $\sqrt{s} = 40$ TeV pp collider, although our remarks are valid for any hadron collider capable of pair-producing W and Z bosons. A more detailed discussion of the topics dealt with in this talk may be found in Ref. 5.

2. NEW PHYSICS SIGNATURES

The classic example of W and Z boson pairs signaling new physics is the decay of a heavy ($m_H > 2M_W$) Higgs boson.[2] The W and Z boson pair decay modes dominate the total width. If there is no Higgs boson, then the weak interaction will become strong at about 1 TeV and this will manifest itself as an enhanced cross section for WW and ZZ scattering.[4,6] In either case, the sector of the standard model responsible for symmetry breaking is best studied in W and Z boson pair production.

Many grand unified models require the existence of additional U(1) gauge groups at the weak scale. These groups are broken, resulting in additional Z' bosons.[7] These neutral bosons can be copiously produced at hadron colliders, and their decays into W boson pairs studied.[8] This decay mode is important because it depends on the amount of mixing between the new Z' bosons and the usual Z boson.

New heavy fermions will decay into W bosons and possibly even Z bosons. Since new fermions are generally pair-produced, the signal will again be W and Z boson pairs. Some possibilities for new fermions include a heavy top quark ($m_t + m_b > M_W$), fourth generation quarks and leptons, exotic fermions associated with grand unified theories, and fermionic partners of W and Z bosons (winos and zinos) which occur in supersymmetric models.

It is not surprising that so many new particles will decay into W (and possibly Z) bosons. After all, the weak interaction is responsible for the decay of all known unstable fundamental particles. The only new twist here is that heavy particles will decay into *real* W and Z bosons rather than virtual ones.

In the next section we will show that the W and Z bosons that result from the decay of heavy particles are predominantly longitudinally polarized. This may be understood in terms of the Goldstone-boson equivalence theorem, which we will now state.

3. GOLDSTONE BOSON EQUIVALENCE THEOREM

The equivalence theorem states that an amplitude involving external longitudinally-polarized gauge bosons is equivalent to the same amplitude with the external gauge bosons replaced by the corresponding Goldstone bosons, plus corrections of order M/E, where M and E denote the gauge boson mass and energy, respectively.[3,4] Thus, at high gauge boson energies, longitudinally-polarized gauge bosons behave like Goldstone bosons. They remember that they were Goldstone bosons before they were transmogrified by the Higgs mechanism.

Goldstone bosons couple to matter in a manner similar to the Higgs boson, since originally they all came from an SU(2) doublet Higgs field. Since the Higgs boson couples preferentially to heavy particles, the same is true of the Goldstone bosons. For example, the coupling of the Higgs boson to Goldstone-boson pairs is enhanced by a factor of $(m_H/M_W)^2$. Similarly, the coupling of a new Z' boson to Goldstone boson pairs is enhanced by $(M_{Z'}/M_W)^2$. The coupling of a heavy fermion to a Goldstone boson is enhanced by (m_f/M_W). Hence, by the equivalence theorem, a heavy Higgs boson, a new heavy Z' boson, and heavy fermions decay predominantly to longitudinally-polarized W and Z bosons.

On the other hand, the coupling of longitudinally-polarized W and Z bosons to ordinary light particles is suppressed. Since the equivalence theorem is only accurate to order M/E, we expect the suppression factor to be M/E. We will now demonstrate this in two background processes, the pair production of Z and W bosons from quark-antiquark annihilation.

4. Z BOSON PAIR PRODUCTION

The principal Z boson pair production process is quark-antiquark annihilation. We will consider the production of (a) two transversely polarized Z bosons, (b) one transverse and one longitudinal, and (c) two longitudinal Z bosons. Using the equivalence theorem, these processes may be represented by the Feynman diagrams in Figs. 1(a)–1(c), where the dashed lines represent Goldstone bosons.

Recall that the coupling of Goldstone bosons to light particles is suppressed by M/E, where M and E are the (equivalent) gauge boson mass and energy, respectively. Hence, for each additional longitudinally-polarized Z boson in the final state, we expect the amplitude to be suppressed by an additional power of M_Z/E_Z. We therefore expect $\sigma_{TT} > \sigma_{LT} > \sigma_{LL}$ (T = transverse, L = longitudinal), with the ratio of cross sections increasing with increasing Z boson energy.

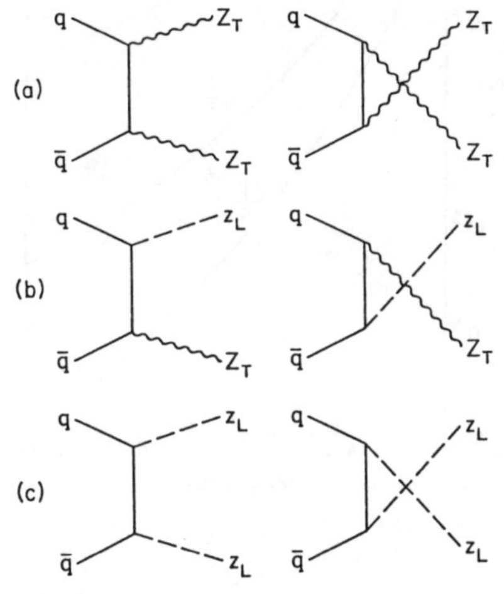

Fig. 1. Feynman diagrams in the equivalent Goldstone boson theory for Z boson pair production: (a) two transversely-polarized Z bosons, (b) one transverse and one longitudinal, and (c) two longitudinal Z bosons.

Figure 2 shows the invariant mass distribution of Z boson pairs produced by quark-antiquark annihilation at the SSC. The relative magnitudes of the three polarization states behave exactly as we have anticipated. Note that at an invariant mass of 1 TeV, the cross section for two longitudinally polarized Z bosons is about three orders of magnitude less than the cross section for two transversely polarized Z bosons. Thus the polarization state is an extremely powerful tool for eliminating backgrounds, as I had advertised in the introduction.

We now go on to study W boson pair production. Due to the presence of a non-abelian vertex, it is more subtle than Z boson pair production.

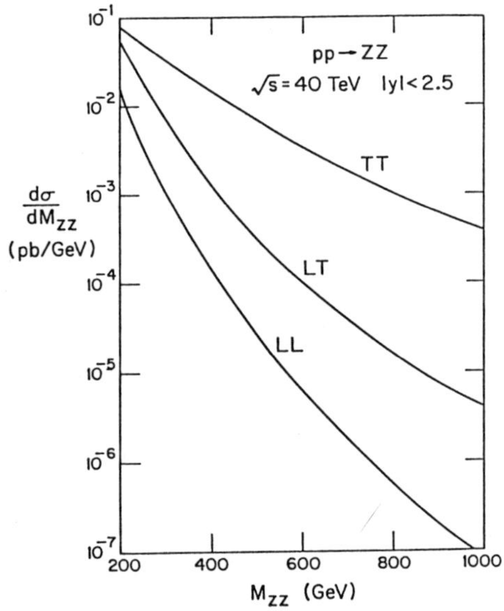

Fig. 2. Invariant mass distribution for Z boson pair production at the SSC. L (longitudinal) and T (transverse) refer to the polarization of the Z bosons.

5. *W* BOSON PAIR PRODUCTION

The relevant Feynman diagrams in the equivalent Goldstone boson theory are given in Fig. 3 for (a) two transversely polarized W bosons, (b) one transverse and one longitudinal, and (c) both longitudinal. We will treat each process separately in the following.

If both W bosons are transversely polarized, they may have the same or opposite helicities. At high W boson energies, it is only the amplitude for opposite helicities which is unsuppressed.[9] Thus the final state total angular momentum is at least $J = 2$. Therefore, the s channel diagram does not contribute at all in the high energy limit, since it is entirely $J = 1$. The cross section for two transversely-polarized W bosons is therefore dominated by the t-channel diagram.

Now consider case (b), where one W boson is transversely polarized and the other is longitudinally polarized. The Goldstone boson coupling to the quark in the t channel diagram is suppressed by M_W/E_W. The Goldstone boson coupling to the gauge bosons in the s channel diagrams is proportional to M_W; however, since a tri-linear gauge-boson coupling is linear in the four-momenta of the gauge bosons, this coupling is also suppressed by M_W/E_W. We therefore expect this polarization state to be suppressed with respect to the doubly-transverse mode, especially at large W boson pair invariant mass.

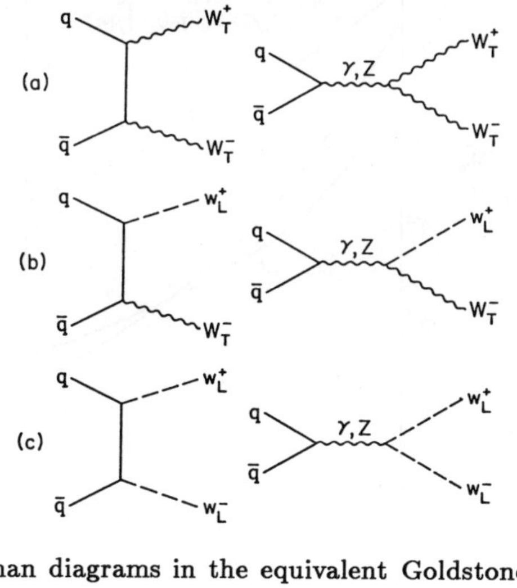

Fig. 3. Feynman diagrams in the equivalent Goldstone boson theory for W boson pair production: (a) Two transversely-polarized W bosons, (b) one transverse and one longitudinal, and (c) two longitudinal W bosons.

Finally, consider the production of two longitudinally polarized W bosons, case (c).[10] The t channel diagram is doubly suppressed, so we can ignore it at high W boson energies. The s channel diagram is unsuppressed, however. The photon and Z boson couple to the Goldstone bosons with electroweak strength, just as they do to any charged particle. Hence this polarization state is dominated by the s-channel diagrams.

We conclude that both the doubly-transverse and doubly-longitudinal states will be unsuppressed at high W boson energies. However, since the doubly-longitudinal amplitude is mediated by neutral currents, we expect it to be numerically smaller than the doubly-transverse amplitude, which involves only charged currents.

In Fig. 4 we show the invariant mass distributions for the three different polarization states at the SSC.[2] The mixed longitudinal-transverse state is suppressed at high invariant mass as anticipated. Furthermore, the doubly-longitudinal mode is indeed smaller than the doubly transverse mode by about an order of magnitude. The doubly-transverse mode is further enhanced at high invariant masses by the t-channel propagator, which peaks in the forward direction. The result is that the doubly-longitudinal mode is suppressed, which is exactly what we want phenomenologically.

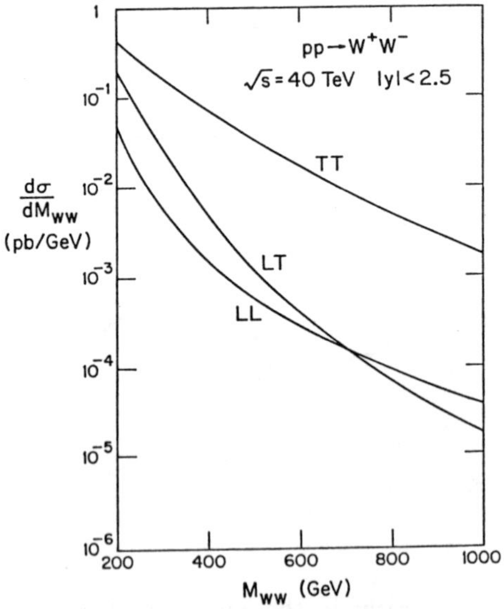

Fig. 4. Invariant mass distribution for W boson pair production at the SSC. L (longitudinal) and T (transverse) refer to the polarization of the W bosons.

6. GLUON FUSION

The last W and Z boson pair production mechanism I will discuss is not usually considered as a background process. W and Z bosons may be pair produced from gluon fusion via an intermediate quark loop, as shown in Fig. 5(a). At first sight this process appears negligible. In fact, if you cut the box in half vertically, you see that the right-hand-side is just the quark-antiquark annihilation process we discussed in the previous two sections. We've now added two more strong couplings and a loop, so the cross section will be suppressed by a factor of roughly $\alpha_s^2/4\pi^2$ relative to quark-antiquark annihilation. However, the gluon-gluon luminosity is almost 100 times as large as the quark-antiquark luminosity.[11] Furthermore, the loop implies a coherent sum of all flavors of quarks, which is a factor of n_f^2 in the cross section. Using $\alpha_s = 0.1$ and $n_f = 6$, we see that gluon fusion is naively as large as quark-antiquark annihilation. It may therefore be an important background process.

The box diagram is difficult to calculate and a definitive study on it does not yet exist, although there are some interesting partial results.[12] Fortunately, if we are only concerned with longitudinally-polarized W and Z bosons, we can use the equivalence theorem to conclude that the box is a negligible source of gauge boson pairs at high gauge-boson energies. The equivalent Goldstone boson diagram is shown in Fig. 5(b). Since it contains two suppressed couplings, it is of order $(M/E)^2$. Therefore the cross section will be suppressed by $(M/E)^4$. Similarly, the cross section for one transversely polarized and one longitudinally polarized gauge boson will be suppressed by $(M/E)^2$. Thus we do not need to be concerned that longitudinally-polarized gauge bosons produced from gluon fusion will be a large background to new particle production.

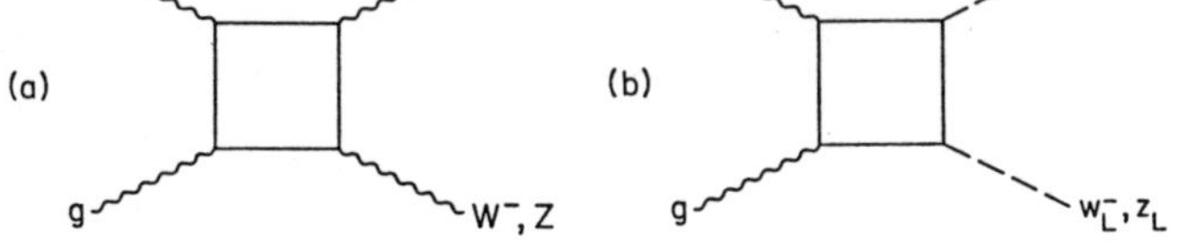

Fig. 5. W and Z boson pair production from gluon fusion via a quark loop: (a) a typical Feynman diagram, (b) the same diagram in the equivalent Goldstone boson theory.

7. CONCLUSION

The major role of today's colliders—the CERN $Sp\bar{p}S$, the Fermilab Tevatron, and the Stanford Linear Collider—is to produce weak gauge bosons and to look for new particles in their decay products. In future colliders, the situation will be reversed—weak gauge bosons will be the decay products of new, heavy particles. In this talk I have shown that these gauge bosons will act like Goldstone bosons, and that we can use this fact to separate them from ordinary background gauge bosons.

REFERENCES

1. Koval'chuk, V. A. and Rekalo, M. P., Sov. J. Nucl. Phys. **32**, 870 (1980); Bilchak, C. L., Brown, R. W., and Stroughair, J. D., Phys. Rev. **D29**, 375 (1984).
2. Duncan, M. J., Kane, G. L., and Repko, W. W., Nucl. Phys. **B272**, 517 (1986). For Z boson pairs, see Duncan, M. J., Phys. Lett. **179B**, 393 (1986); Abbasabadi, A. and Repko, W. W., Michigan State University report MSU/TH-Dec 86.
3. Cornwall, J. M., Levin, D. N., and Tiktopoulos, G., Phys. Rev. **D10**, 1145 (1974); Vayonakis, C. E., Lett. Nuovo Cim. **17**, 383 (1976); Lee, B. W., Quigg, C., and Thacker, H., Phys. Rev. **D16**, 1519 (1977); Chanowitz, M. S., Furman, M., and Hinchliffe, I., Phys. Lett. **78B**, 285 (1978); Nucl. Phys. **B153**, 402 1979.
4. Chanowitz, M. S. and Gaillard, M. K., Nucl. Phys. **B261**, 379 (1985). See also Gounaris, G. J., Kögerler, R., and Neufeld, H., Phys. Rev. **D34**, 3257 (1986).
5. Willenbrock, S.S.D., University of Wisconsin report MAD/PH/333 (1987).
6. Chanowitz, M. S., these proceedings.
7. For a review of grand unified theories, see Langacker, P., Phys. Rep. **72**, 185 (1981).
8. Nandi, S., Oklahoma State University report 184 (1986); Adeva, B., Del Aguila, F., Nanopoulos, D. V., and Zwirner, F., CERN report TH.4535/86; Dicus, D. A., Wilson, S. L., and Vega, R., University of Texas report DOE-ER40200-093.
9. Hagiwara, K., Peccei, R. D., Zeppenfeld, D., and Hikasa, K., University of Wisconsin report MAD/PH/279 (1986).
10. Chanowitz, M. S. and Gaillard, M. K., Phys. Lett. **142B**, 85 (1984).
11. Eichten, E., Hinchliffe, I., Lane, K., and Quigg, C., Rev. Mod. Phys. **56**, 579 (1984).
12. Amettler, Ll., Gava, E., Paver, N., and Treleani, D., Phys. Rev. **D32**, 1699 (1985); Pumplin, J. C., Repko, W. W., and Kane, G. L., Proc. 1986 Snowmass Summer Study on the Physics of the SSC.

PROSPECTS FOR OBSERVATION OF SUPERSYMMETRY AT THE TEVATRON

R. Arnowitt* and Pran Nath**

*Center for Theoretical Physics
Department of Physics, Texas A&M University
College Station, TX 77843

**Lyman Laboratory of Physics
Harvard University, Cambridge, MA 02138+ and
Department of Physics, Northeastern University
Boston, MA 02115

ABSTRACT

Wino and Zino production at the Tevatron give rise to supersymmetry trilepton, dilepton and jet events with missing energy. If the Tevatron operates at peak luminosity of $10^{31}\text{cm}^{-2}\text{sec}^{-1}$, the trilepton signal is sensitive to Winos with mass up to 70-80 GeV, the dilepton signal up to 50-60 GeV, and the monojet signal up to 45-50 GeV.

1. INTRODUCTION

The search for supersymmetric particles has been the subject of much experimental consideration at high energy colliders. Pair production of squarks ($\tilde{q}$) and gluinos ($\tilde{g}$) followed by the decays

$$\tilde{q} \to q + \tilde{\gamma} \quad ; \quad \tilde{g} \to q + \bar{q} + \tilde{\gamma} \tag{1}$$

gives rise to monojet and <u>copious</u> dijet production. The UA1 analysis of their monojet and dijet data at the CERN $Sp\bar{p}S$ had led to the following lower bound on the squark and gluino masses[1)]

$$m_{\tilde{q}} \geq 70 \text{ GeV} ; \; m_{\tilde{g}} \gtrsim 60 \text{ GeV} \tag{2}$$

Within the framework of N=1 Supergravity[2)], there exists a general result that models with light photinos possess one Wino ($\tilde{W}$) lying below the W boson and one Zino ($\tilde{Z}$) below the Z boson[3)]:

$$m_{\tilde{W}} \leq M_W \quad ; \quad m_{\tilde{Z}} \leq M_Z \tag{3}$$

Thus $\tilde{W}\tilde{\gamma}$, $\tilde{W}\tilde{W}$ and $\tilde{W}\tilde{Z}$ production is possible at the $Sp\bar{p}S$. Hadronic decays then lead to monojets and <u>few</u> dijets.

2. SUSY PRODUCTION CROSS SECTIONS AT THE TEVATRON

The existance of lowlying Wino and Zinos obeying Eq.(3) is one of the strongest predictions of N=1 supergravity models. While the Sp$\bar{p}$S, SLC and LEPI machines will be able to detect Winos if $m_{\tilde{W}} \lesssim 50$ GeV, the Tevatron offers the extra possibility of detecting Winos up to the theoretical maximum mass, i.e. $m_{\tilde{W}} \approx 70$-80 GeV, provided it operated at peak luminosity of $10^{31}\,\mathrm{cm}^{-2}\mathrm{sec}^{-1}$ and with full lepton detection capabilities. In the following analysis we consider the minimal N=1 supergravity model[2)] with two Higgs doublets whose neutral components have approximately equal VEVs, (as is the case for the tree breaking, dimensional transmutation and No Scale models). Some of the conclusions also apply to models where one VEV is much smaller than the other (i.e. renormalization group models).

The basic processes involving Winos and Zinos at the Tevatron are

$$p + \bar{p} \rightarrow \tilde{W} + \tilde{\gamma} \tag{7a}$$

$$p + \bar{p} \rightarrow \tilde{W} + \tilde{Z} \tag{7b}$$

$$p + \bar{p} \rightarrow \tilde{W} + \tilde{W} \tag{7c}$$

which proceed through vector boson s-channel poles and squark t and u channel poles. These processed have been considered in Refs. (6,7). In general the vector boson graph strongly dominates and so the cross sections are insensitive to the squark mass. They do, however, depend sensitively on $m_{\tilde{W}}$, and also on the sign of $m_{\tilde{\gamma}}$. (We chose the sign leading to the smallest production cross sections. The alternate choice would increase cross sections presented below and hence the SUSY signal by ≈50% or more.) There is considerable theoretical error in the normalization of the W and Z

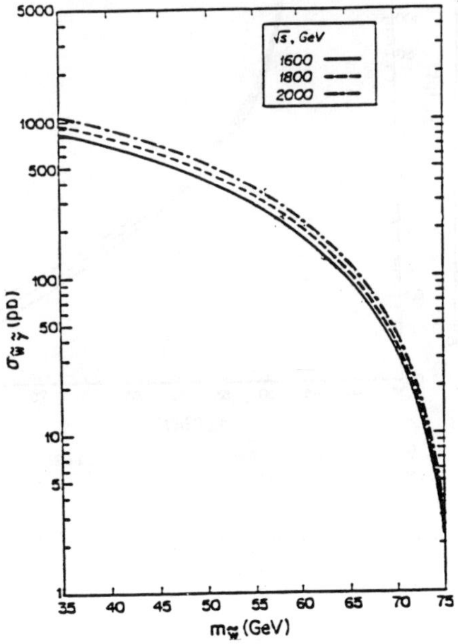

Fig. 1. $p\bar{p} \rightarrow \tilde{W}\tilde{\gamma}$ cross section a function of $\tilde{W}$ for $m_{\tilde{\gamma}} = 7$ GeV.

production cross section σ_W and σ_Z[8]. In order to minimize this error, we have normalized the production cross sections so that they agree with the values of σ_W and σ_Z measured by UA1 at $\sqrt{s}$ = 630, and then theoretically calculated the increase to the Tevatron energy. We also model experimental cuts on the UA1 apparatus. Once CDF is fully operational at the Tevatron, one can avoid this by using the experimental values of σ_W and σ_Z at 2 TeV and the precise CDF cuts. Thus the results being presented here are meant to be indicative of what one might expect at the Tevatron, rather than a specific prediction.

The production cross sections for Eqs. (7) are given Figs. 1,2, 3. There is a rapid fall off above the W and Z poles in Figs. 2 and 3, though a considerable tail survives at high Wino mass.

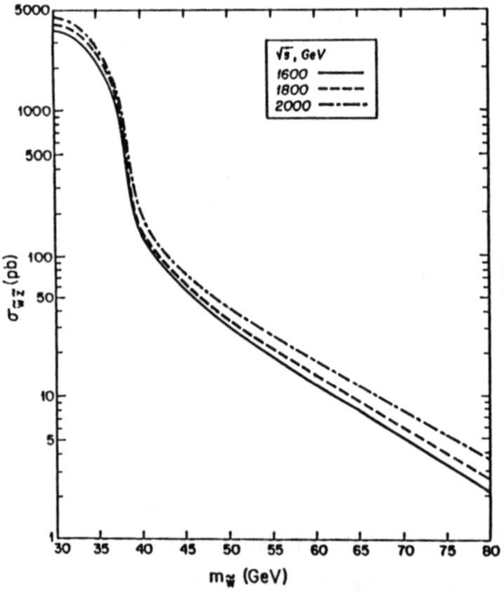

Fig. 2. $p\bar{p} \to \tilde{W}\tilde{Z}$ cross section for $m_{\tilde{\gamma}}$ = 7 GeV.

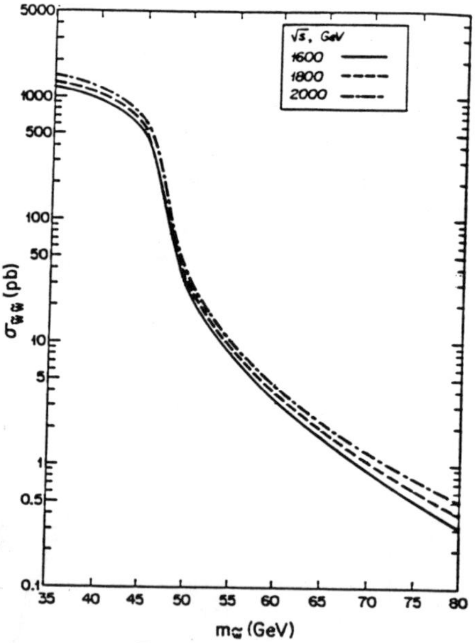

Fig. 3 $p\bar{p} \to \tilde{W}\tilde{W}$ cross section for $m_{\tilde{\gamma}}$ = 7 GeV.

3. TRILEPTON EVENTS

The most promising SUSY signal is probably the trilepton events. These arise from Eq. (7b) followed by the leptonic decays

$$\tilde{W} \to \ell_1 + \nu_1 + \tilde{\gamma} \tag{8a}$$

$$\tilde{Z} \to \ell_2 + \bar{\ell}_2 + \tilde{\gamma} \tag{8b}$$

For light photinos, the Wino branching ratios are approximately mass independent, while the Zino branching ratios depend on both the Wino and squark mass. One may estimate the trilepton cross section $\sigma_{3\ell}$ as follows:

$$\sigma_{3\ell} = \sigma_{\tilde{W}\tilde{Z}} \sum B(\tilde{W} \to \ell_i \nu_i \tilde{\gamma}) B(\tilde{Z} \to \ell_j \bar{\ell}_j \tilde{\gamma}) P_{ij}(\tilde{W},\tilde{Z}) \tag{9}$$

where B are the branching ratios, i and j sum over e and μ leptons, and P_{ij} is the efficiency for detecting leptons ℓ_i, ℓ_j and $\bar{\ell}_j$. Modeling P_{ij} on the UA1 apparatus we find $P_{ij} \simeq 0.2$ for $m_{\tilde{W}} = 35$ GeV and $P_{ij} \sim 0.4$ for $m_{\tilde{W}} = 80$ GeV. The results are given in Table I. The cross section decreases with increasing squark and Wino masses.

Table I. Trilepton cross section at $\sqrt{s} = 2$ TeV for $m_{\tilde{q}} = 150,250$ GeV ($m_{\tilde{\gamma}} = 7$ GeV).

$m_{\tilde{W}}$ (GeV)	$\sigma_{3\ell}$(pb) 150	250
40	1.18	0.22
50	0.41	0.12
60	0.25	0.10
70	0.15	0.08
80	0.09	0.06

At peak luminosity, a run of 10^7 sec. would accumulate an intergrated luminosity of $L = 10^2 pb^{-1}$. Thus there are SUSY trilepton events even for the highest values of $m_{\tilde{W}}$ provided $m_{\tilde{q}} \lesssim 350$ GeV. These events are expected to be hadronically quiet

Standard Model backgrounds include WZ production (followed by leptonic decays of the W and Z) and heavy flavor prodution (c, b, t). Using[10] $\sigma_{WZ} = 2.0$ pb at $\sqrt{s} = 2$ TeV, we find $\sigma_{3\ell} \simeq 0.009$ which would correspond to less than 1 event for $L = 10^2$ pb^{-1}. We roughly estimate other backgrounds should also be $\lesssim 1$ event.

4. DILEPTON EVENTS

Dilepton events arise from Eq. (7c) followed by the leptonic decays Eq (8a) of the Wino, and from Eq. (7b) where the Wino decays leptonically while the Zino decays according to

$$\tilde{Z} \to q_1 + \bar{q}_2 + \tilde{W} \to q_1 + \vec{q}_2 + (\ell_2 + \nu_2 + \tilde{\gamma}) \tag{10}$$

In the latter mode, the quarks come off at low energy and usually do not pass the UA1 jet criteria, leaving an effective dilepton event. The most striking signal occurs when the leptons are of unlike type i.e. $\ell_1 = e(\mu)$ and $\ell_2 = \mu(e)$. Then one has

$$p + \bar{p} \to e^{\pm} + \mu^{\pm} + \not{p}_T \tag{11}$$

The decay branching ratios for Eq. (10) depend on both the Wino and squark mass. The e-μ dilepton cross section is given in Table II. It is a decreasing function of the Wino mass, but an increasing function of the squark mass, since the decay rate of $\tilde{Z} \to q_1 \bar{q}_2 \tilde{W}$ increases as[9] $(m_{\tilde{4}})^4$. Thus for $L = 10^2\ pb^{-1}$ there are SUSY dilepton events for $m_{\tilde{W}} \lesssim 65 - 70$ GeV. The dilepton Standard Model backgrounds, however, are larger than corresponding trilepton backgrounds.

Table II. Dilepton cross section at $\sqrt{s}$ = 2 TeV for $m_{\tilde{q}}$ = 150,350 GeV ($m_{\tilde{q}}$ = 7 GeV).

$m_{\tilde{W}}$ (GeV)	$\sigma_{e\mu}$(pb) 150	350
45	6.52	6.68
50	0.42	0.53
55	0.13	0.21
60	0.06	0.11
65	0.03	0.07

These are: (i) $p + \bar{p} \to W + W$ followed by leptonic decay of the W bosons; (ii) Possible τ' (fourth sequential lepton) production via $p + \bar{p} \to Z \to \tau' + \bar{\tau}'$; (iii) Possible $t\bar{t}$ production with either $t,\bar{t}$ leptonic decay or $t \to W + b$ decay if $m_t > M_W + m_b$; (iv) $c\bar{c}$ and $b\bar{b}$ production. Using σ_{WW}= 7.5 pb at $\sqrt{s}$ = 2 TeV[10], one finds a dilepton rate for (i) of $\sigma_{e\mu}(WW)$ = 0.09pb. Thus the SUSY dilepton rate will be greater than a 3σ effect over the WW dileptons for $m_{\tilde{W}} \lesssim 60$ GeV for heavy squarks ($m_{\tilde{q}} \gtrsim 350$ GeV) and for $m_{\tilde{W}} \lesssim 55$ GeV for lighter squarks. Experimentally[1], $m_{\tau'} \gtrsim 41$ GeV and we find background (ii) negligible for $m_{\tau'} \gtrsim 47$ GeV. We roughly estimate backgrounds (iii) and (iv) to be $\approx(1-5)\ \sigma_{e\mu}(WW)$. However, the SUSY events (11) should be hadronically quiet, and it may be possible to eliminate such backgrounds by a cut on total E_T[a1]. If heavy flavor production events cannot be separated from the SUSY events, then the above limits on $m_{\tilde{W}}$ would be reduced by ~ 5 GeV.

5. MONOJETS AND DIJETS

Monojets and dijets arise from all three production processes of Eqs. (7) followed by hadronic decays of the Winos and Zinos. The monojet and dijet cross sections for $\not{p}_T > 34$ GeV at $\sqrt{2}$ = 2 TeV are

a1. We should like to thank D. Dibitonto for pointing this possibility out to us.

Table III. SUSY Monojet (σ_{mj}) and dijet (σ_{Dj}) cross section at $\sqrt{s}$ = 2 TeV for $\not{p}_T > 34$ GeV

$m_{\tilde{W}}$	σ_{Mj}(pb)	σ_{Dj}(pb)
40	86.7	23.8
45	51.7	13.7
50	16.7	1.0
55	10.9	0.8
60	6.9	0.5

given in Table III, assuming a UA1 type jet algorithm. (The dijet cross section is probably an over estimate, as UA1 isolation cuts were not included incalculating σ_{Dj}.) Extrapolating the UA1 estimates[5] for Standard Model monojets to the Tevatron energy we find we find σ_{Mj}(S.M.) $\doteq$ (100 ± 15) pb (where the error is mainly from uncertainty in energy scale). Thus, it would be difficult to detect a Wino monojet signal unless $m_{\tilde{W}} \lesssim 50$ GeV. In addition, monojets and dijets could arise from squark or gluino production. These monojets would be characterized by being hard ($\not{p}_T \gtrsim 50$ GeV) and by having a large dijet to monojet ratio.[11] Such missing energy events would be observable at the Tevatron provided $m_{\tilde{q}}$, $m_{\tilde{g}} \lesssim 150 - 175$ GeV.

5. CONCLUSIONS

One of the strongest predictions of N=1 supergravity models with light photinos is the presence of a Wino with mass $m_{\tilde{W}} < M_W$. The testing of such a prediction is thus of importance. Such a particle produces signals that could be detected at the Tevatron running at peak luminosity $10^{31}\,cm^{-2}sec^{-1}$ and with full muon detection capability. For SUSY signals to be 3σ or larger above the Standard Model backgrounds, we estimate the following Wino mass ranges that can be detected at the Tevatron:

(1) Trilepton events: $m_{\tilde{W}} \lesssim$ 70-80 GeV for $m_{\tilde{q}} \lesssim 350$ GeV

(2) Dilepton events: $m_{\tilde{W}} \lesssim$ 55-60 GeV for $m_{\tilde{q}} \gtrsim 350$ GeV
$m_{\tilde{W}} \lesssim$ 50-55 GeV for $m_{\tilde{q}} \lesssim 350$ GeV

(3) Monojet Events: $m_{\tilde{W}} \lesssim$ 45-50 GeV

The trilepton and dilepton events should be hadronically quiet. The monojet events should have $\not{p}_T \lesssim 50$ GeV and be accompanied by very few dijets. More detailed studies of background phenomena could possibly raise the above Wino mass limits.

REFERENCES

† Visiting Scholar

1. Honma, A., XXIII International Conference on High Energy Physics, Berkeley (1986); CERN EP/86-153.

2. For a review of N=1 Supergravity models, see Nath, P., Arnowitt, R., and Chameseddine, A., "Applied N=1 Supergravity" (World Scientific, Singapore, 1984); Nilles, H. P., Phys. Rep. C110, 1 (1984).

3. Weinberg, S., Phys. Rev. Lett. 50, 387 (1983); Arnowitt, R., Chamseddine, A. H., and Nath, P., Phys. Rev. Lett. 50, 232 (1983).

4. Arnowitt, R. and Nath, P., to be pub. Phys. Rev. D35 (1987), Northeastern NUB#2699 (1986); Arnowitt, R. and Nath, P. XXIII International Conference on High Energy Physics, Berkeley (1986), Northeastern NUB#2704 (1986).

5. J. Rohlf, Proc. of Oregon Meeting of Division of Particles and Fields, Eugene OR 1985 (World Scientific, Singapore, 1986). This analysis has the advantage of not depending on parton distributions or fragmentation models as the cut is on $\not{p}_T$ (rather than L_τ).

6. Nath, P. and Arnowitt, R., CTP-TAMU-21/86; NUB#2716; HUTP-87/A006 (1986).

7. Baer, H., Hagiwara, K., and Tata, X., MAD/PH/298; ANL-HEP-PR-86-97; KEK-TH136 (1986); Gluck, M, Godbole, R. M. and Reya E., Dortmund Do-Th 86/25 (1986).

8. Alterelli, G., Ellis, R., Greco, M., and Marinelli G., Nucl. Phys. B246 , 12(1984). See also Baer, H. and Berger, E.L., Phys. Rev. D34, 1361 (1986).

9. Chamseddine, A. H., Nath, P. and Arnowitt, R., Phys. Lett. 129B 445 (1983); Nath, P., Arnowitt, R., and Chamseddine, A. H., HUPT-83/A077; NUB#2588 (1983) unpublished.

10. Eichten, E. Fermilab - Pub 85/178 - T (Enratum), unpublished.

11. Baer.et al[8]; Reya, E. and Roy, D.P., DO-Th86/06; TIFR/TH/86 - 16 (1986).

IV. Superunification Theory

Conveners — P. Nelson and P. Ramond

A New Topological Term in 2d Field Theory

Philip Nelson
Lyman Laboratory of Physics
Harvard University
Cambridge MA 02138

This talk describes some work done jointly with L. Alvarez-Gaumé, J.-B. Bost, G. Moore, and C. Vafa [1][2], building on [3] and [4].

We were concerned with establishing bosonization results on two-dimensional surfaces with complicated topology. Far from being a mere curiosity, bosonization is of great interest in string theory. For example, bosonization has been used in light-cone gauge to prove the equivalence of the Green-Schwarz and NSR superstring [5][6]. Bosonization also plays a key role in understanding the gauge- and super-symmetry of the heterotic string [7] and in formulating the covariant fermion emission vertex [8][9].

The papers [1], [2] generalize existing results on Fermi-Bose equivalence for Fermi fields of any spin on the sphere [10]–[13]. In this talk I will only discuss a subproblem, that of bosonizing spin 1/2 on the torus. It turns out that this problem is only slightly more difficult than the sphere case. One needs a way to "tell" the bosonic theory which of the various spin structures it is to mimic; this is accomplished by adding to the bosonic action a new global term. The new term is already familiar to mathematicians as the parity of a spin structure; it has an immediate generalization to any genus surface.

1. Bosonic action

We will let Σ denote any compact Riemann surface, a 2d "spacetime" in the euclidean formalism. We begin by reviewing the situation when Σ is the sphere.

We will consider only free fermions, that is, fermions interacting with background gauge and gravitational fields but without self-interactions such as mass or quartic terms. This is the case of interest for the NSR superstring in flat spacetime.

The prototypical fermionic system one might wish to bonsonize has action

$$S_f = \int_\Sigma \bar{\psi}\, i\partial\!\!\!/\psi \quad .$$

In euclidean path integrals ψ and $\bar{\psi}$ are independent, and it has become traditional to rename the fields, with $\psi_1 \mapsto c \quad \psi_2 \mapsto \bar{b} \quad \bar{\psi}_1 \mapsto b$ and $\bar{\psi}_2 \mapsto \bar{c}$. Using complex notation and rescaling fields we then have

$$S_f = \frac{i}{2\pi} \int_\Sigma (b\bar{\partial}c + \bar{b}\partial\bar{c}) \quad , \tag{1.1}$$

where b and c are sections of a spin bundle L, and $\bar{b}$ and $\bar{c}$ are sections of $\bar{L}$. $\bar{\partial}$ is the Cauchy-Riemann operator. The integrand is a (1,1)-form and so can be integrated over Σ without the use of any metric. That is, (1.1) is classically conformally invariant. It is also invariant under the global chiral transformation $b \mapsto e^{i\alpha}b,\ c \mapsto e^{-i\alpha}c$.

Similarly the prototypical scalar action for a single real field φ has the form

$$\begin{aligned} S_1 &= \int_\Sigma 2\pi\vec{\nabla}\varphi\cdot\vec{\nabla}\varphi\; d(\mathrm{vol}) \\ &= \int_\Sigma 4\pi i\,\partial\varphi\wedge\bar{\partial}\varphi \quad . \end{aligned} \tag{1.2}$$

The second form makes it clear that S_1 is also classically conformally invariant. It also has an invariance under shifts $\varphi \mapsto \varphi +$ constant.

The operator analysis of bosonization on $2d$ Minkowski spaces teaches us two important physical lessons. (See for example [14]–[18],[10].) First of all, the correspondence should be between the fermionic bilinears and *exponentials* of φ, properly normal-ordered. Secondly, the bosonic field φ is properly to be regarded as a *periodic*, or circle-valued, field. This fact is compatible with the first if the normalization of φ is chosen such that its ambiguity does not affect its exponential. We can see the periodicity of φ either in the periodic sine-Gordon potential of [14] or in the case where spacetime is the Minkowski cylinder [10]. In either case the crucial physical basis of bosonization is that fermions in (1.1) correspond to *solitons*, or states where φ is multiple-valued, in (1.2). In particular, when spacetime has noncontractible loops the partition function of the bosonic system gets important contributions from soliton sectors.

On the sphere, of course, the second observation above is immaterial, since there are no noncontractible loops on Σ and hence no solitons. To make the first observation concrete in our present notation we will assume that we have

$$\begin{aligned} b\bar{b} &\propto e^{q\varphi} \\ c\bar{c} &\propto e^{-q\varphi} \end{aligned} \tag{1.3}$$

and find q, starting with the case of spin-1/2. First note that (1.2) gives a two-point function with singularity

$$\langle \varphi(z)\varphi(w)\rangle \sim -\frac{1}{16\pi^2}\log|z-w|^2 \ .$$

In proving this we have used the fact that

$$\bar{\partial}_P \partial_P \log|z_P - z_Q|^2 = 2\pi i\, \delta_Q(P) \tag{1.4}$$

where $\delta_Q(P)$, the delta function, is a (1,1)-form at P. (1.4) is easily shown by integrating on a small disk and using deRham's theorem.

The classical stress tensor of (1.2) is

$$T = -8\pi^2 \partial\varphi\partial\varphi \quad , \tag{1.5}$$

a $(2,0)$-tensor. Its quantum version looks the same but with normal ordering to remove self-contractions. T is defined so that in operator products [12]

$$T(z)\psi(w) \sim \frac{h}{(z-w)^2}\,\psi + \text{less singular terms} \quad .$$

where h is the spin of ψ. One can check the normalization of (1.5) by showing that $\partial\varphi$ has spin one. Then

$$T(z)e^{\pm q\varphi(w)} \sim \left(\frac{1}{z-w}\right)^2 \frac{q^2}{16\pi^2}\cdot\left(-\frac{1}{2}\right) + \cdots \quad .$$

Thus choosing $q = 4\pi i$ gives $b\,,c$ of spin one half.

It is now simple to show the equivalence of (1.1) and (1.2), still in spin 1/2. First, the zero-point functions agree up to an overall multiplicative constant:

$$Z_b[g] = \text{const.} \cdot Z_f[g] \ . \tag{1.6}$$

Each side of (1.6) is a functional of a metric chosen on Σ to regularize the theories. (1.6) holds because both bosons and nonchiral fermions are free of gravitational anomalies, and both have the *same* anomalous variation under Weyl transformations. Since on the sphere every metric g is related to any reference g_0 by coordinate and Weyl transformations [19], we see that (1.6) really does hold up to a constant. As for the higher correlation functions, they follow from the fact that

$$\begin{aligned} \langle e^{4\pi i\varphi(z)} e^{-4\pi i\varphi(w)} \rangle &= \exp[16\pi^2 \langle \varphi(z)\varphi(w) \rangle] \sim \frac{1}{|z-w|^2} \\ &\sim \langle b\bar{b}(z) c\bar{c}(w) \rangle \ . \end{aligned} \tag{1.7}$$

For spins other than one half, the bosonic action (1.2) must be supplemented by a term S_2 explicitly breaking invariance under shifts of φ, since the corresponding transformation of b, c is anomalous. We cannot discuss this term here; see [1],[2].

2. The Torus

When we move up in complexity from the sphere to the torus we at once encounter two novel features. First, it is no longer true that every metric is related to every other by coordinate and Weyl symmetries: a residual "moduli space" of inequivalent metrics remains [19]. Secondly, there is now a variety of inequivalent bundles over Σ in which b and c could take values. Thus we need some way to tell the bosons what spin structure they are to mimic. Fortunately on the torus we still have a canonical formalism, which we can use to address the problem.

A spin bundle L is a bundle whose square is the cotangent to Σ. On the torus the cotangent is trivial, so that we can parametrize the four possible choices for L by measuring the difference between L and one particular spin bundle, the trivial one. If we take the torus defined by the unit square in $\mathbb{C}$, we then have

$$\begin{aligned} b(1) &= e^{2\pi i\theta}\, b(0) \\ b(i) &= e^{-2\pi i\phi}\, b(0) \end{aligned} \ , \tag{2.1}$$

and similarly for c. In this talk I will restrict to untwisted spin bundles, i.e. θ, $\phi = 0$ or $\frac{1}{2}$. The field c then lives in the bundle $K \otimes L^{-1} \cong L$. Also we will not consider $\theta = \phi = 0$ since with this choice $Z_f = 0$ due to the fermion zero mode; that is, we consider only the three "even" spin structures.

Certainly (1.6) cannot hold as it stands on the torus, since one side depends on θ, ϕ while the other does not. Instead one expects that the bosonic theory with action S_1 should give the *sum* over all spin structures of the corresponding fermionic theories. Detailed calculation affirms this expectation [3]. To bosonize just one spin structure one must add to S_1 a new term S_3 depending on θ, ϕ. We will find S_3 by canonically quantizing and applying the physical lesson that fermions correspond to solitons of the field φ.

Since moduli will not play an important role in this talk we will continue to take the torus to be the unit square in C, with identifications and (2.1). We will quantize with euclidean time running up the imaginary axis. Then the partition functions $Z_f(\theta, \phi)$ are traces over the Ramond and Neveu-Schwarz Hilbert spaces, for $\theta = 0$, $\frac{1}{2}$ respectively. ϕ on the other hand denotes the boundary conditions in time. We then have

$$Z_f(\tfrac{1}{2}, \tfrac{1}{2}) = Tr_{NS} e^{-H}$$
$$Z_f(\tfrac{1}{2}, 0) = Tr_{NS} (-)^F e^{-H}$$

Hence $Z_f(\frac{1}{2}, \frac{1}{2}) \pm Z_f(\frac{1}{2}, 0)$ is a trace over the even (resp. odd) fermion-number space. It must therefore in the bosonic language receive contributions only from states of even (resp. odd) soliton number.

A functional integral over φ includes a sum over *all* soliton sectors. Thus our modification to the action S_3 must have the effect of weighting the various winding sectors in such a way as to cancel the odd-soliton contributions to $Z_b(\frac{1}{2}, 0) + Z_b(\frac{1}{2}, \frac{1}{2})$, and so on. A possible set of weighting factors is shown in Fig. 1. In the left column we have shown the spin structures for the fermionic system. On the right the boxes represent the contributions to the bosonic path integral from the winding sectors with $(n, m) = (\frac{1}{2}\cdot\text{even}\ ,\ \frac{1}{2}\cdot\text{even}\ )$, $(\frac{1}{2}\cdot\text{even}, \frac{1}{2}\cdot\text{odd})$, and so on. Each box on the right thus represents a sum $Z_{b,i}^{\text{partial}}$, $i = 1, \ldots 4$ over

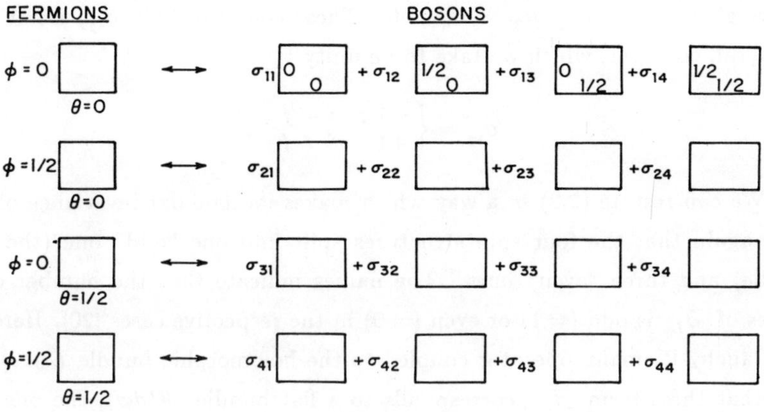

Fig. 1: Weighting the soliton sectors.

an infinite subclass of field configurations. The effects of S_3 are in the phases σ_{ij}: we have $Z_b(\frac{1}{2},0) = \sum_{i=1}^{4} \sigma_{3i} Z_{b,i}^{\text{partial}}$, etc.

The conditions that $Z_b(\frac{1}{2},0) \pm Z_b(\frac{1}{2},\frac{1}{2})$ have no odd (resp. even) soliton contribution now imply that in these combinations $Z_{b,3}^{\text{partial}}$ and $Z_{b,4}^{\text{partial}}$ must cancel (resp. $Z_{b,1}^{\text{partial}}$ and $Z_{b,2}^{\text{partial}}$), i.e.

$$\sigma_{33} = -\sigma_{43}\,, \qquad \sigma_{34} = -\sigma_{44}$$

$$\sigma_{31} = \sigma_{41}\,, \qquad \sigma_{32} = \sigma_{42}$$

We can get further conditions by letting the torus degenerate with very long time. Then only the ground state contributes to the fermionic partition function. With zeta-function regularization the split ground state of the Ramond sector has nonzero energy, so that $Z_f(0,\phi) \to 0$. Take $\phi = \frac{1}{2}$. On the bosonic side, only the zero-soliton sectors contribute, but they do so independently of the time winding $2m$, in the limit. Thus the two contributing partial sums $Z_{b,1}^{\text{partial}}$ and $Z_{b,2}^{\text{partial}}$ must cancel from $Z_b(0,\frac{1}{2})$, so that $\sigma_{21} = -\sigma_{22}$.

We must also impose the condition of modular invariance on the σ_{ij}. Requiring for example that our prescription be unchanged when we exchange the roles of space and time gives relations like $\sigma_{21} = \sigma_{31}$, $\sigma_{22} = \sigma_{33}$, etc. Requiring that the torus with corners 0, 1, $i+2$, $i+1$ give the same answers as the unit square gives $\sigma_{31} = \sigma_{41}$, $\sigma_{33} = \sigma_{44}$, etc. These conditions fix σ_{ij}, $i \neq 1$ up to an overall constant, which we take to be unity:[1]

$$\sigma_{ij} = \begin{cases} -1\,, & i = j \\ +1\,, & i \neq j \end{cases} \quad . \tag{2.2}$$

We can restate (2.2) in a way which makes its modular invariance obvious. Note again that the four spin structures split into one "odd" one (the trivial bundle) and three "even" ones. The names indicate that the number of zero modes of $\bar{\partial}_L$ is odd ($= 1$) or even ($= 0$) in the respective cases [20]. Here $\bar{\partial}_L$ is the Cauchy-Riemann operator coupled to the holomorphic bundle L [19]. Note also that the 1-form $d\varphi$ corresponds to a flat bundle $F(d\varphi)$, the one whose holonomy around a loop γ is given by $d\varphi$. The prescription (2.2) simply says that we must add to S_1 the topological term

$$S_3 = i\pi\sigma(L \otimes F(d\varphi)) \quad , \tag{2.3}$$

where $\sigma(L')$ is 0 or 1, depending on whether L' is even or odd. Note that $L \otimes F(d\varphi)$ really is a spin bundle, since with our normalization $d\varphi$ defines a half-integral cohomology class.

The proof that $S_1 + S_3$ is the correct bosonic action, as well as the generalization to arbitrary twists θ, ϕ, comes after we generalize everything to genus $g \geq 1$ [1],[2]. We emphasize, however that (1.2) and (2.3) are by now a very plausible prescription on the torus, and that in higher genus essentially no new physics will be needed.

3. Higher Genus

Various problems beset us when we try to generalize this discussion to world-sheets with more than one handle. For one thing, the anomaly term S_2 alluded to above becomes problematical since it involves not $d\varphi$ but φ itself, and so is ill-defined for multiple-valued field configurations φ. Secondly, when the cotangent to Σ is no longer trivial, then the trivial bundle is not a spin bundle and we

[1] We can use the same reasoning to fix the σ_{1j}. However, to fix the relative sign of σ_{1j} relative to the others we must interpolate between the spin structures, as we do in [1],[2].

cannot refer all spin bundles to it as we did in (2.1). Instead we need to choose a reference spin bundle — but any such choice automatically breaks coordinate invariance, since a diffeomorphism of Σ in general transforms a chosen spin structure into another, different one. Remarkably, these two problems cancel each other. The extra term in S_2 needed to make it well-defined gives it a coordinate transformation which cancels that of S_3 when a suitable reference spin bundle is chosen.

Also on higher genus surfaces there is no obvious operator formalism. Questions about bosonization then turn into path-integral questions about functional determinants. To prove that bosonization works we had to have recourse to methods of algebraic geometry. These are described in an introductory way in [19], and the full details of their application are in [2].

References

[1] L. Alvarez-Gaumé, J. B. Bost, G. Moore, P. Nelson, and C. Vafa, "Bosonization in arbitrary genus," Phys. Lett. **B178** (1986) 41.

[2] L. Alvarez-Gaumé, J. B. Bost, G. Moore, P. Nelson, and C. Vafa, "Bosonization on higher genus Riemann surfaces," Harvard Preprint HUTP-86/A062.

[3] L. Alvarez-Gaumé, G. Moore, and C. Vafa, "Theta functions, modular invariance, and strings," Comm. Math. Phys. **106**(1986)1.

[4] J. Bost and P. Nelson, "Spin-1/2 bosonization on compact surfaces," Phys. Rev. Lett. **57** (1986) 795.

[5] R. Shankar, Phys. Lett. **92B**, 333 (1980); E. Witten, in *Fourth workshop on grand unification, ed. H. Weldon, P. Langacker, and P. Steinhardt*, (Birkhauser, 1983).

[6] R. Nepomechie, "Nonabelian bosonization, triality, and superstring theory," Phys. Lett. **178B** (1986) 207; **180B** (1986) 423; L. Brown and R. Nepomechie, Non-abelian bosonization: current correlation functions," Washington preprint.

[7] D. Gross, J. Harvey, E. Martinec, and R. Rohm, Phys. Rev. Lett. **54**(1985)502; Nucl. Phys. **B256**(1985)253; **B267**(1986)75.

[8] D. Friedan, E. Martinec, and S. Shenker, Phys. Lett.**B160**(1985)55.

[9] V. Knizhnik, "Covariant superstring amplitudes from the sum over fermionic surfaces," Phys. Lett. **178B** (1986) 21.

[10] For a history, see D. Wolf and J. Zittartz, Zeit. Phys. **B51** (1983) 65.

[11] R. Marnelius, Nucl. Phys. **B211** (1983) 14.

[12] D. Friedan, E. Martinec, and S. Shenker, "Conformal invariance, supersymmetry, and string theory," Nucl. Phys. **B271** (1986)93.

[13] W. Siegel and B. Zwiebach, Nucl. Phys. **B263** (1986) 105.

[14] S. Coleman, Phys. Rev. **D11** (1975) 2088.

[15] S. Mandelstam, Phys. Rev. **D11** (1975) 3026.

[16] T. Banks *et al.*, Nucl. Phys. **B108** (1976) 119.

[17] I. Frenkel, J. Funct. Anal. **44** (1981) 259.

[18] P. Goddard and D. Olive, in *Vertex operators in mathematics and physics*, ed. J. Lepowsky, S. Mandelstam, and I. M. Singer (Springer, 1985).

[19] P. Nelson, "Lectures on strings and moduli space" HUTP-86/A047, to appear in Phys. Reports.

[20] M. Atiyah, "Riemann surfaces and spin structures," Ann. Sci. Ec. Norm. Sup. 4e série **4** (1971) 47.

CONSTRUCTION OF FOUR DIMENSIONAL FERMIONIC STRING MODELS

H. Kawai, D. C. Lewellen, J. Schwartz, D. Senechal and S.-H. H. Tye

Newman Laboratory of Nuclear Studies
Cornell University
Ithaca, NY 14853
USA

ABSTRACT

The "spin structure" construction of four-dimensional fermionic string models is briefly summarized and discussed.

Superstring theories offer the possibility of describing all known forces in a single mathematically consistent framework.[1] In particular, fermionic closed string models automatically include Einstein gravity as an effective low energy theory and appear to be completely finite and anomaly free.[2] Using the spin-structure construction described in Ref. 3] it has proved easy to construct many heterotic type string models (that is with world-sheet supersymmetric right moving excitations and bosonic left movers) in four space-time dimensions which are phenomenologically viable, e.g., are non-supersymmetric or possess N=1 space-time supersymmetry and contain chiral fermions and potentially realistic gauge groups. The near uniqueness of heterotic string models in ten dimensions[4] is entirely absent in four dimensions. The chief source of the abundance of candidate models is the large number of left moving degrees of freedom (typically a rank 22 gauge group).

Let us begin with a summary of the spin structure construction formalism. We direct the reader to Ref. 3] for the pertinent derivations, notations, and further references. In order to insure a physical theory which is Lorentz invariant, satisfies positivity, and is finite, we construct a two dimensional theory on the string world sheet which is reparametrization invariant, world-sheet supersymmetric, super-conformal invariant and modular invariant. For simplicity, we employ the Ramond-Neveu-Schwarz (RNS) formalism[5] for the space-time fermions and represent all of the string's internal degrees of freedom with world sheet fermions; thus in four dimensions, in the light cone gauge which we employ, there are, besides the two transverse bosonic string coordinates, two right moving RNS fermions representing space-time degrees of freedom and an additional eighteen right moving and forty-four left moving Majorana fermions on the string world sheet. This number of fermionic variables is required for (super)conformal invariance.

In order to fully determine a heterotic type string model we must specify the boundary conditions (spin structures) of all of the two-dimensional fermions around noncontractible loops on the string world sheet. Modular invariance in fact requires that we take a linear combination of contributions with different sets of boundary conditions since each piece individually is not invariant under modular transformations. In particular the fermionic contribution to the one loop vacuum amplitude is a sum of partition functions each determined by a set of spin-structures (s.s.)

$$Z_{\text{fermion}} = \sum_{(s.s.)} C(s.s.)Z(s.s.) \ . \tag{1}$$

The coefficients C(s.s.) should be chosen consistent with one loop modular invariance and also so that Z_{fermion} can be interpreted as a sensible sum of physical states in four dimensional space-time, i.e., the sum over spin-structures for one direction on the torus should naturally form a projection onto a subspace of world-sheet states with correct space-time statistics. In the simplest example this is just the GSO projection[6]. Explicit one-loop modular invariance and a physically sensible projection are sufficient to guarantee multiloop modular invariance as well[3]. Subject to these requirements the general solution to (1) can be written down and its content reduced to a set of rules giving the spectrum of states included in the projection formed in (1). The fermion boundary conditions are conveniently represented by 64-dimensional vectors whose ℓ^{th} component takes the value 0 or 1/2 depending on whether the ℓth fermion has periodic or antiperiodic boundary conditions. (We restrict ourselves to these possibilities here for simplicity, though in general we may choose any boundary conditions for which the supercurrent S(z) is periodic or antiperiodic on the string world-sheet[3].) The appropriate sum over sets of boundary conditions runs over all boundary condition vectors spanned by some set of basis vectors (W_i) which includes, in particular, the vector W_0 with all components equal to 1/2. The form of S(z) and W_i, together with a matrix of constants k_{ij} which determines in part the projection formed in (1), specifies a given string model. To guarantee modular invariance and physically sensible projections these quantities must satisfy the constraints,

$$k_{ij} + k_{ji} = W_i \cdot W_j \pmod 1 \tag{2}$$

$$2k_{ij} = 0 \pmod 1 \tag{3}$$

$$k_{ii} + k_{io} + s_i - \frac{1}{2} W_i \cdot W_i = 0 \pmod 1 \ . \tag{4}$$

The quantity s_i appears to insure that states contribute to the partition function (1) with the correct space-time statistics. It takes the value 0 or 1/2 depending on whether the first components of W_i (which pertain to right moving space-time degrees of freedom) agree or not, i.e., $s_i = W_i^1$ (mod 1). The inner product between W vectors in (2) and (4) is defined with Lorentzian signature (left components minus right components) and, conforming to the notation of Ref. 3], implicitly includes an overall factor of 1/2 since we define our W_i to describe real world sheet fermions rather than complex ones, i.e. $W_i \cdot W_j = \frac{1}{2} [\sum_{l=1}^{20} W_i^l W_j^l - \sum_{l=21}^{64} W_i^l W_j^l]$.

In addition to (2), (3), and (4), we must choose the set (W_i) to be consistent with the quartic constraints[3] and world-sheet supersymmetry. Specifically the world-sheet supercurrent for right movers must be either periodic or antiperiodic around the world-sheet torus. The expression for the supercurrent S(z) in terms of the world-sheet fermions is given by[7,8]

$$S(z) = \frac{1}{3} i \sum_{3 \leq j,k,l \leq 20} f^{jkl} \psi^j \psi^k \psi^l + \sum_{i=1,2} \psi^i \partial_z \chi^i . \qquad (5)$$

Super-Virasoro algebra requires (up to some overall normalization) that the f^{ijk} be the structure constants of a semi-simple Lie group.[7,8] The only semi-simple Lie algebras of dimension eighteen are $[SU(2)]^6$, $SU(3) \times SO(5)$ and $SU(4) \times SU(2)$. Note that in D space-time dimensions the f^{ijk} are structure constants of a dimension 3(10-D) semi-simple Lie group so that if $D \geq 6$ the only possible group structure for both the left and right moving supercurrents is $[SU(2)]^{10-D}$

Eq. (5) represents a strong constraint on the allowed choices of spin-structures. Assume that S(z) is periodic (antiperiodic) around one of the noncontractible loops of the world-sheet torus. Then from (1) ψ^1 and ψ^2 must have periodic (antiperiodic) boundary conditions and so the first two components of W are 0 (1/2). Group the remaining world-sheet fermions into two sets, those with periodic (antiperiodic) boundary conditions, labeled by a,b,c, and those with antiperiodic (periodic) boundary conditions, labeled by i,j,k. S(z) will be periodic (antiperiodic) if the structure constants of the form f^{abi} and f^{ijk} are all zero. This is just the statement that f^{abc} are the structure constants of a symmetric subgroup, H, of the original group, G, with all of the f's as structure constants (i.e. G/H is a symmetric space).

In addition, we must insure in any sector that the fermion number operators commute. In fact, a careful analysis shows that modular invariance and a physically sensible projection require the commutation of such operators even between different sectors. This condition is equivalent to choosing basis vectors such that given any four, the number of simultaneously nonzero real fermion components is even,

$$\sum_{\{\ell:\text{real}\}} 8W_i^\ell W_j^\ell W_k^\ell W_n^\ell \in Z \quad \text{for any } i, j, k, n .$$

This is the quartic constraint which must be satisfied. Now we are ready to consider model construction.

To get massless fermions in four dimensional heterotic type string models, we must choose $[SU(2)]^6$ case. Together, they have an $[SU(2)]^6 \times S_6$ (permutation) symmetry. Writing the six triplets as three triplets of complex fermions and the 44 right moving fermions as 22 complex fermions, the general form of the W_i is[3]

$$W_i = (S_i(a_1b_1c_1)(a_2b_2c_2)(a_3b_3c_3)|d_1 \ldots d_{22}),$$
$$a + b + c = S \pmod 1). \tag{10}$$

Here $S_i = 0$ or $1/2$ is the only space-time component and the constraint on the triplet structure for the right movers is a consequence of the local world-sheet supersymmetry of the covariant two-dimensional theory.

As we always have the vector $W_0 = ((\frac{1}{2})^{10}|(\frac{1}{2})^{22})$. The vacuum energy level of the (right, left) movers is $(-\frac{1}{2}, -1)$. Taken by itself this gives a tachyonic model with gauge group $[SO(3)]^6 \times SO(44)$ and no massless fermions. The SO(44) symmetry arises naturally from the left movers; the $[SO(3)]^6$ gauge bosons are obtained by the left movers (to get a space-time vector) and exciting the right-moving internal degrees of freedom. Adding $W_1 = (0(0\frac{1}{2}\frac{1}{2})^3|(\frac{1}{2})^{22})$ removes the tachyons and adds a sector of massless fermions, in fact giving an $N = 4$ supersymmetric SO(44) model. All that remains of the $[SO(3)]^6$ symmetry are the six vector bosons in the $N = 4$ supergravity multiplet.

By adding new vectors to the set W_0, W_1, we can construct $N = 2$ supersymmetric, $N = 1$ supersymmetric (both chiral and non-chiral) as well as non-supersymmetric models. If the rank of the gauge group is 22, the group is always simply-laced. In terms of RNS fermions, the rank of the gauge group can be reduced. In fact, the rank of the gauge group from the left movers can be reduced from 22 all the way down to zero. For rank less than 22, the gauge group can be non-simply laced.

We have written a computer program which can generate consistent models efficiently. The list of $N = 1$ supersymmetric chiral models is very long (of the order of ten thousand of them). The number of $N = 4$ supersymmetric models originally proposed by Narain[9] is around twelve hundred.

String theories with both left and right moving sectors possessing world-sheet supersymmetry (e.g. the type II strings in ten dimensions)[1] offer an attractive alternative to the multitude of heterotic string models since they have fewer degrees of freedom constrained by more world-sheet symmetries. It is interesting to ask whether such models in four dimensions retain enough flexibility to include something as rich and complex as the SU(3) × SU(2) × U(1) standard model. It turns out that the most interesting of these models share a significant property of heterotic strings which distinguishes them from Type II models in ten dimensions and their compactifications: the spin-structures and world sheet supercurrents are asymmetric between the left and right movers. To underscore this important distinction we refer to these as Type III string models.

To explicitly construct a class of finite and anomaly free Type III models, those which can be represented with free world sheet fermions, we employ the spin-structure construction technique described earlier. Models with N = 8, 4, 2, and 1 supersymmetry as well as non-supersymmetric models can easily be constructed; however, chiral models have so far eluded our attempts described in Ref. 8].

One intriguing possibility is the construction of orientable open string models. In ten dimensions, the only supersymmetric Type II string models have N = 2. Since an open string has only N = 1 supersymmetry, we cannot construct an orientable string as originally discussed in Ref. 1]. However, let us compactify the Type II string to 4 dimensions. Then the closed string has N = 8 supersymmetry while any open string has only N = 4 supersymmetry. However, in 4 dimensions, we can introduce Z_2 twists to the closed string so that the supersymmetry can be reduced to N = 4. This opens the possibility that an orientable open string with N = 4 supersymmetry can be consistent, which further suggests the existence of an orientable open string with N = 1 supersymmetry. This possibility is presently under investigation.

This research is supported by the National Science Foundation.

REFERENCES

1. See, e.g., Schwarz, J.H., Phys. Rep. 89, 223 (1982); Green, M.B., Surveys in High Energy Phys. 3, 127 (1983); Green, M.B., Schwarz, J.H. and Witten, E., "Superstring Theory", Cambridge University Press (1986).

2. Green, M.B. and Schwarz, J.H., Nucl. Phys. B255, 92 (1985).

3. Kawai, H., Lewellen, D.C., and Tye, S.-H.H. Tye, Phys. Rev. D34, 3794 (1986); Phys. Rev. Lett. 57, 1832 (1986); Cornell preprint CLNS 86/751 (1986), to be published in Nucl. Phys. See also Seiberg, N. and Witten, E., Nucl. Phys. B276, 272 (1986); Parkes, A., "Two-Loop Modular Invariance and Spin-Statistics", CERN-TH 4525/86 (1986); Mueller, M. and Witten, E., Princeton preprint (1986); Lerche, W., Lust, D. and Schellekens, A.N., CERN-TH 4590/86 (1986); Antoniadis, I., Bachas, C.P. and Kounnas, C., LBL-22709 (1986).

4. Gross, D.J., Harvey, J.A., Martinec, E. and Rohm, R., Phys. Rev. Lett. 54, 502 (1985); Nucl. Phys. B256 (1985) and B267, 75 (1986).

5. Ramond, P., Phys. Rev. D3, 2415 (1971); Neveu, A. and Schwarz, J.H., Nucl. Phys. B31, 86 (1971).

6. Gliozzi, F., Scherk, J. and Olive, D., Nucl. Phys. B122, 253 (1977).

7. Goddard, P., Nahm, W. and Olive, D., Phys. Lett. 160B, 111 (1985); Goddard, P., Kent, A. and Olive, D., Comm. Math. Phys. 103, 105 (1986); Antoniadis, I., Bachas, C., Kounnas, C. and Windey, P., Phys. Lett. 171B, 51 (1986); Blum, R., Dolan, L. and Goddard, P., unpublished.

8. Kawai, H., Lewellen, D.C. and Tye, S.-H.H. Tye, CLNS 87/760 (1987).

9. Narain, K.S., Phys. Lett. 169B, 41 (1986); Narain, K.S., Sarmadi, M.H. and Witten, E., Nucl. Phys. B279, 369 (1987).

The Equivalence of Polyakov and Light-Cone Bosonic String Theory*

Steven B. Giddings†
Joseph Henry Laboratories
Princeton University, Princeton NJ 08544

Historically there have been two functional integral approaches developed to treat the loop expansion in string theory. The interacting string picture,[1] which was developed first, is closely tied to the physical picture of strings propagating in spacetime and undergoing occasional interactions. In contrast, the Polyakov approach[2] involves a sum over geometrical surfaces and hence the physical picture of string propagation is more obscured. A long-standing question has been whether these two formalisms are in fact equivalent; here I will summarize some recent work establishing the equality of amplitudes calculated by either prescription.[3,4]

There are both physical and mathematical motivations for wanting to know that these formalisms are equivalent. On the physical side, this result is one way to establish the unitarity of the Polyakov Ansatz. Since the interacting string picture originates from a treatment of strings in a physical gauge, it gives manifestly unitary amplitudes; hence the equivalence with the Polyakov amplitudes establishes unitarity of those amplitudes. Furthermore this construction involves a treatment of the light-cone world sheets from the Polyakov point of view; this can in turn be used to gain insight[5] into possible formulations of string field theory involving the light-cone world sheets.[6] On the mathematical side, equivalence of these two formalisms could hardly be expected if the light-cone world sheets didn't provide a single cover of moduli space. However, this is a non-trivial statement. Moduli space is topologically quite complicated (its Euler character is $\chi = \zeta(1-2g)$ for genus

---

★ Invited talk presented at the 1987 Annual Meeting of the APS Division of Particles and Fields, Salt Lake City, Utah, and also at the Conference on Superstrings, ITP Santa Barbara.

† Address after July 1, 1987: Lyman Laboratory of Physics, Harvard University, Cambridge MA 02138.

g), and so in general it is hard to find simple parametrizations of it. Until it was shown that physicists had held in their hands for several years the explicit light-cone parametrization of moduli space, mathematicians only knew of essentially one explicit parametrization[7] (and still none for the moduli space of a surface without marked points). The light-cone parametrization in fact provides a cell decomposition of moduli space, and thus may be useful for extracting topological information about moduli space. Finally, the Polyakov integration measure over moduli space has been shown to be the square of a unique holomorphic form on moduli space, and its equivalence with the light-cone measure gives a simple expression for this form which involves only the determinant of the scalar Laplacian.

There are two main steps necessary in establishing the equivalence of the interacting string and Polyakov amplitudes. Both approaches involve the integral of some integration measure over some finite dimensional space. Hence equality is established if we a) demonstrate the equivalence of the integration regions and b) demonstrate the equivalence of the integration measures. A Polyakov amplitude for a process with n external states and g loops is written abstractly as a functional integral over the space of all metrics; this is gauge fixed by picking a slice to which any metric can be brought by a combined Weyl rescaling and diffeomorphism. The functional integral is then[8] (in D=26)

$$\begin{aligned} A_P(1,\cdots,n) = \int_{\mathcal{M}_{g,n}} [dm] \frac{det\langle \mu_\alpha, \phi_\beta\rangle}{det(\phi_\alpha,\phi_\beta)^{\frac{1}{2}}} (det P_1^+ P_1)^{\frac{1}{2}} \\ \left(\frac{2\pi}{\int_M d^2\xi\sqrt{g}} det'\triangle\right)^{-13} \langle \mathcal{V}_1\cdots\mathcal{V}_n\rangle \end{aligned} \tag{1}$$

Here $\mathcal{M}_{g,n}$ denotes the moduli space, which is represented by the gauge-fixing slice and has coordinates m. The Beltrami differentials μ_α are the tangent vectors to the slice corresponding to unit deformations in the coordinates m, and the quadratic differentials ϕ_α are the basis covectors for moduli space. Thus the expression involving μ and ϕ represents the 'volume element' on moduli space in the coordinates m. The term $det(P_1^+ P_1)$ is a gauge-fixing determinant. The determinant of the Laplacian comes from the functional integral of the maps of the world-sheet into spacetime, as does the correlation function involving the external states (here represented by wave functionals $\mathcal{V}$). The latter functional integral is computed using the Polyakov action $S_P = \frac{1}{2}\int_M d^2\xi\sqrt{g}g^{mn}\partial_m X^\mu \partial_n X_\mu$. In the interacting string picture the analogous amplitude is

$$A_I(1,\cdots,n) = \int [ds]\left(\frac{2\pi}{\int_M d^2\xi\sqrt{g}}det'\Delta\right)^{-12}\langle \mathcal{W}_1\cdots\mathcal{W}_n\rangle \tag{2}$$

Here the integration is over the parameters s parametrizing the light-cone diagrams (to be described shortly); from the functional integral over the 24 physical degrees of freedom we find the factor involving the determinant of the Laplacian and also the correlation function from the external state wave functionals $\mathcal{W}$. This functional integral is computed using the light-cone action $S_I = \frac{1}{2}\int d\tau d\sigma\left[(\partial_\tau X^i)^2 + (\partial_\sigma X^i)^2\right]$. The external state correlation functions are equal in the two approaches (this can be checked explicitly for tachyonic external states, and follows in general by factorization). Therefore to show equivalence I will first argue that the light-cone diagrams cover moduli space exactly once;[3] then compution of the Polyakov measure in the light-cone coordinates for moduli space will show the equality of the two integration measures.[4]

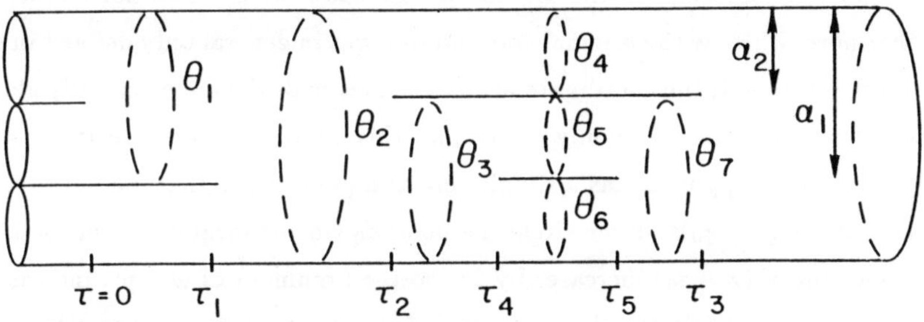

Fig. 1. A light-cone diagram with three incoming and one outgoing strings, and two loops.

A typical light-cone diagram is shown in Fig. 1. The parameters labelling such a diagram (with g loops and n external states) are the radii of the external strings α_i, $i = 1,\cdots n$, the radii of the internal strings, α_I, $I = 1\cdots g$, the interaction times (one sets the first interaction time to zero) τ_a, $a = 1,\cdots,2g+n-3$, and the twist angles of the internal strings, θ_α, $\alpha = 1,\cdots 3g+n-3$. For a given scattering process the α_i, which are taken proportional to the P^+ momenta of the external strings, are held fixed. The other parameters are integrated over; since there are $6g-6+2n$ of them, which is precisely the real dimension of moduli space, it is at least conceivable that as these parameters range freely we could cover moduli space. The basic idea behind the proof that we do in fact cover moduli space once is that for a given Riemann surface with some specified points, *i.e.* for a given point in moduli space, and for a given choice of α_i (*i.e.* external P^+) there is a unique *canonical* Abelian differential* ω. Furthermore, such an Abelian differential is in one-to-one correspondence with a light-cone diagram; as we will see, ω induces the metric on the diagram. This will show that light-cone diagrams are in one-to-one correspondence with points in moduli space, and so we have a single cover.

To see how this works, suppose we're given any Abelian differential $\omega = \omega_z dz$ on some Riemann surface; here z is a local complex parameter. Define a new complex coordinate $w = \int^z \omega$, so $dw = \omega$, and take the *flat* metric in the coordinate w:

$$ds^2 = dw d\bar{w} = |\omega_z|^2 dz d\bar{z} \ .$$

This metric is conformal to the original metric $dzd\bar{z}$, with conformal factor $|\omega_z|^2$. Furthermore, whereas the original coordiante z was in general only defined in some small neighborhood, the coordinate w and the metric $dwd\bar{w}$ are essentially globally defined because ω was. The only places that this all breaks down are at the zeroes and poles of ω. Suppose ω has a simple zero at a point z_0, so that $dw \sim (z-z_0)dz$, or $(w-w_0) \sim (z-z_0)^2$. If we circle the point z_0 (in the locally flat metric $dzd\bar{z}$), the argument of $(z-z_0)$ increases by 2π, so the argument of $w-w_0$ increases by 4π. But this is precisely the change in angle found in the flat metric on a light-cone diagram if we circle one of the interaction points. So simple zeroes of an Abelian differential correspond to light-cone interaction points. Further, if ω has a simple pole, so that $dw \sim \frac{dz}{z-z_0}$, then $w - w_0 \sim ln(z-z_0)$. The logarithm induces a conformal transformation taking us from a punctured neighborhood of the point z_0 to a cylinder shooting off to infinity, with z_0 being mapped to infinity. So we find that simple poles in an Abelian differential correspond to external states on a string diagram.

⋆ An Abelian differential is just another word for a meromorphic one form.

In general there are many Abelian differentials on a Riemann surface; clearly we must look for more restrictions. On a light-cone diagram let $w = \tau + i\sigma$ be the natural complex parameter; τ is the light-cone time, and σ labels position along the string at fixed time. Now consider a loop A circling one of the internal strings on the light-cone diagram; the integral $\int_A dw$ clearly gives i times the length of the internal string. Similarly for the integral around a loop C about an external string we just get i times its length. Now consider the integral around a loop B circling one of the internal loops of the string diagram (pictured as horizontal slits in Fig. 1). Since the net change in light-cone time is zero for a closed loop, $\int_B dw$ again gives a pure imaginary number related to the twists of the internal strings. Since $dw = \omega$, and since the set of all loops A, B and C form a homology basis for the Riemann surface, we see that we require an Abelian differential with simple poles of fixed real residues (*i.e.* external P^+), and *pure imaginary periods* around any homology cycle.

It's easy to show that such a differential always exists and is unique. In fact, the space of differentials with simple poles at some specified points P and Q is $g+1$ dimensional by the Riemann-Roch theorem. By adding such differentials we can always make a differential μ with simple poles at any specified points and with any given residues (as long as they sum to zero – P^+ conservation). The most general such differential is then $\omega = \mu + \sum_{I=1}^{g} E_I \omega_I$, where E_I are g arbitrary complex constants, and ω_I are the g linearly independent holomorphic differentials on the surface. The condition of pure imaginary periods gives $2g$ real conditions (g for the A curves and g for the B curves) that uniquely fix the constants E_I. All we have to do to solve for the E_I is invert a matrix; this turns out to be the imaginary part of the period matrix, which is positive definite for a non-singular Riemann surface. This gives us the unique canonical differential ω, and after resolving a couple of technical details, it is easily seen that this differential gives us a light-cone diagram. Conversely, a light-cone diagram (with almost everywhere flat metric) clearly determines a Riemann surface and a canonical Abelian differential $\omega = dw$, so we find that the light-cone diagrams do in fact give us a single cover of the moduli space for a Riemann surface with two or more marked points. Further, since the coordinates (τ, α, θ) are given by the periods of the canonical Abelian differential or by the its integral between its zeroes, these coordinates are nicely related to the natural complex coordinates on moduli space.

Having dispensed with the issue of the integration region, we now turn to the issue of the measure. I will show how to compute the Polyakov measure in the light-cone coordinates for moduli space, and show that it is equivalent to the light-cone measure in (2). The basic idea here is that the Polyakov measure involves the determinant of the vector Laplacian, $P_1^+ P_1$; however, on a flat surface there is little to distinguish vectors from scalars. The light-cone surface is flat except for isolated points, so one might expect the determinant of the vector Laplacian (to the one half power, since a vector has two components) to be equal to the determinant of the scalar Laplacian, up to a correction factor due to the curvature singularities. The latter factor will cancel the finite dimensional determinant in (1) involving μ and ϕ, and the determinant of the scalar Laplacian will cancel one power of the determinant in (1), leaving us with the determinant to the minus twelfth power as in (2).

To actually show that this works we must compute the various factors in (1). I will sketch this calculation; further details can be found in Ref. 4. In calculating the finite dimensional determinant, we will need a basis for the quadratic differentials (*i.e.* the cotangent vectors to moduli space). These are constructed using the Abelian differentials we already have in hand. A set of g linearly independent quadratic differentials is $\{\phi_{Izz}\} = \{\omega_z \omega_{Iz}\}$. The other $2g+n-3$ differentials are found by noting that the canonical differential has $2g+n-2$ simple zeroes (light-cone interaction points) $P_0, P_1, \cdots P_a$. Take the $2g+n-3$ Abelian differentials $\omega_{P_0 P_a}$ that have simple poles at points P_0 and P_a; then the set $\{\phi_{azz}\} = \{\omega_z \omega_{P_0 P_a z}\}$ consists of quadratic differentials holomorphic on the Riemann surface (minus the external state points) and linearly independent of each other and of the ϕ_I. With this form for the quadratic differentials the computation of the overlap matrix $(\phi_\alpha, \phi_\beta)$ becomes particularly simple; we have

$$(\phi_\alpha, \bar{\phi}_\beta) = \int d^2\xi \sqrt{g} (g^{z\bar{z}})^2 \phi_{\alpha zz} \bar{\phi}_{\beta \bar{z}\bar{z}} = 2i \int \omega_\alpha \wedge \bar{\omega}_\beta \ ,$$

where we use the fact that the metric on the light-cone diagram is $ds^2 = |\omega|^2$. Notice that the canonical Abelian differential has completely disappeared from this formula. From this simple expression these matrix elements can be computed using Riemann's bilinear relations (see *e.g.* Ref. 9), and we find

$$[det(\phi_\alpha,\phi_\beta)]^{\frac{1}{2}} = 2^g(4\pi)^{2g+n-3}det(Im\Omega)det[G_{P_0}(P_a,P_b)] \tag{3}$$

where Ω_{IJ} is the period matrix for the Riemann surface, and $G_{P_0}(P_a,P_b)$ is a Green's function for the scalar Laplacian, $\triangle_P G_{P_0}(P,Q) = -2\pi(\delta(P,Q)-\delta(P,P_0))$. This Green's function is constructed by taking the real part of the integral of the Abelian differential with simple poles at Q and P_0 and with pure imaginary periods, $G_{P_0}(P,Q) = Re\int_{P_0}^{P}\omega_{QP_0}$.

Next we compute the determinant $\langle\mu_\alpha,\phi_\beta\rangle$. To do this we must construct the Beltrami differentials μ_α corresponding to deformations of the light-cone diagram induced by unit changes of the coordinates τ,α and $\alpha\theta$. The Beltrami differentials corresponding to deformations of complex structure are induced by *discontinuous* deformation vector fields[5]. To visualize this think of the twist deformation on the torus; we cut the torus, twist one side while holding the other fixed, and then reglue. The corresponding vector field is discontinuous because of the cut. In our case we deform the light-cone diagram by deforming the Euclidean cylinders comprising the diagram. There are three fundamental deformations to consider: the *twist* deformations (twisting a cylinder); the *stretch* deformations (lengthening the cylinder); and the *shift* deformations (changing the radius of the cylinder). These change the coordinates θ,τ and α respectively. These deformation vector fields are easily written down, and their corresponding Beltrami differentials are given by $\mu_{\bar{z}}^{z} = \nabla_{\bar{z}}V^z$. The overlap of such a differential with a holomorphic quadratic differential is

$$\langle\mu,\phi\rangle = 2i\int dz\wedge d\bar{z}(\nabla_{\bar{z}}V^z)\phi_{zz} = -2i\oint_C dz(disc\ V^z)\phi_{zz}\ ;$$

here I have integrated by parts and used the fact that ϕ_{zz} is holomorphic. We are left with a boundary term, given by the integral over the contour C along which V^z is discontinuous. For unit deformations of τ,α and $\alpha\theta$ the discontinuity in V^z

will be unity in the w coordinates; using $V^z = V^w/\omega_z$ we find

$$\langle \mu, \phi_\alpha \rangle = 2i \oint_C \omega_\alpha \ .$$

Thus we reduce the evaluation of the $\langle \mu, \phi \rangle$ overlap to the problem of integrating the differentials $\omega_I, \omega_{P_0 P_a}$ over various homology cycles on the surface. The required integrals are simple to perform; they just give elements of the period matrix or residues at the poles of $\omega_{P_0 P_a}$. The determinant is then easily calculated, and the answer is

$$det\langle \mu_\alpha, \phi_\beta \rangle = (8\pi)^g (4\pi)^{4g+2n-6} det(Im\Omega) \ .$$

Combining this with (3) then gives the complete finite dimensional determinant,

$$\frac{det\langle \mu_\alpha, \phi_\beta \rangle}{det(\phi_\alpha, \phi_\beta)^{1/2}} = (4\pi)^{3g+n-3} det^{-1}\{G_{P_0}(P_a, P_b)\} \ . \qquad (4)$$

Next we must compute $det(P_1^+ P_1)^{1/2}$; we now know that we should find the determinant of the scalar Laplacian times a factor cancelling (4). The vector and scalar Laplacians clearly act on different spaces, so to compare the two operators we must supply a prescription to convert vectors into scalars. Given a vector V^z, the obvious way to find a pair of scalars is simply to take the components of the vector in the light-cone coordinates; thus for each vector we have one complex scalar given by

$$V^z = \psi/\omega_z \ , \ V^{\bar{z}} = \bar{\psi}/\omega_{\bar{z}} \ . \qquad (5)$$

The space of vectors on which the determinant of $P_1^+ P_1$ is taken is the space of regular diffeomorphisms of the surface. Since ω vanishes at the points P_0, P_a, we must take ψ to vanish there also so that V^z is regular. It is easy to show that on a vector field such as (5), $P_1^+ P_1$ just reduces to the scalar Laplacian acting on ψ, plus correction terms due to the curvature singularities at the interaction points. The correction terms involve delta functions at the points P_0, P_a however, so they drop out when multiplying a function that vanishes at these points. The result is that $det(P_1^+ P_1)^{1/2} = det^* \Delta$, where det^* is the determinant taken over the space of scalar functions vanishing at the points P_0, P_a. This determinant in turn can be expressed as a path integral

$$(det^*\triangle)^{-1/2} = \int \mathcal{D}\psi e^{-I_0[\psi]}\delta[\psi(P_0)]\delta[\psi(P_1)]\cdots\delta[\psi(P_{2g+n-3})]$$

with action $I_0 = \frac{1}{2}\int d^2\xi\sqrt{g}g^{mn}\partial_m\psi\partial_n\psi$. The delta functions can be represented by their Fourier transforms, $\delta[\psi(P_a)] = \frac{1}{2\pi}\int d\varpi e^{i\varpi\psi(P_a)}$. With these inserted in the path integral, we can complete the square (using the Green's function $G_{P_0}(P,Q)$); we then integrate over the variables ϖ to find

$$(detP_1^+P_1)^{1/2} = det^*\triangle = \left(\frac{2\pi}{\int_M d^2\xi\sqrt{g}}det'\triangle\right)det\{G_{P_0}(P_a,P_b)\}\ . \qquad (6)$$

Finally we combine (1), (4), and (6) to find the Polyakov amplitude in the light-cone coordinates on moduli space:

$$A_P(1,\cdots,n) = \int[d\tau][d\alpha][\alpha d\theta]\left(\frac{2\pi}{\int_M d^2\xi\sqrt{g}}det'\triangle\right)^{-12}\langle\mathcal{V}_1\cdots\mathcal{V}_n\rangle\ .$$

This shows us that the Polyakov measure for integration over moduli space in fact equals the measure deduced from the light-cone approach.

I have not quite shown that the amplitudes are equivalent; we must still compare the momentum and polarization dependent X correlation functions in the two approaches. As mentioned above, these correlation functions agree when all external states are tachyons.[1] The equivalence of amplitudes with general physical states then follows by factorization of amplitudes with purely tachyonic external states. This can be made even more explicit by use of the DDF construction[10] which essentially originates from factorization of amplitudes with external tachyons and transverse states at mass level zero. The DDF construction provides a means of comparing arbitrary light-cone states with physical states in the Polyakov formalism; this is elaborated upon in Ref. 4.

One other point that should be made is that the amplitude (1) involves the functional integral over a non-compact surface with wave-functions specified on the boundary; we would like to compare this to the corresponding Polyakov amplitude with vertex operators inserted on a compact Riemann surface. To do this one must conformally map the semi-infinite cylinders corresponding to external states to small boundaries on the Riemann surface; one then converts the wave-function into a vertex operator (or vice versa) by sewing in a small disk with the vertex operator inserted. To recover the standard result we must also show how the integration measure over the $3g-3+n$ dimensional moduli space of the punctured surface reduces to the integration measure over the $3g-3$ dimensional moduli space of the unpunctured surface, together with the local integration measure $\sqrt{g(\xi_i)}d^2\xi_i$ for the integration of each of the vertex operators over the Riemann surface. A fuller account of the resolution of both of these issues is given in Ref. 4.

Finally one should argue that unitarity is compatible with the integration over one copy of moduli space. There are two points to check. The first is that I have shown equivalence of the two expressions (1),(2) only up to a constant factor $b^{2g-2}c^n$. However, the factor b, since its power is the Euler character of the surface, can be absorbed by a redefintion of the string loop coupling constant. The factor c is then absorbed in defining the relative normalization of states in the light-cone picture and in the Polyakov picture. Since c is a constant universal to all amplitudes it can be fixed once and for all at tree level, and thus doesn't spoil unitarity. The second point is that if we integrate τ, α, and θ over their complete ranges (*i.e.* $\tau \in (-\infty, \infty)$, $\alpha \in (0, \alpha_{total})$, $\theta \in (0, 2\pi)$), we cover the moduli space for the corresponding surface several times. The key is to divide out by a discrete symmetry factor to ensure that we count each *geometrical* light-cone diagram only once. The same Feynman symmetry factor should also appear in (2), so unitary amplitudes are in fact obtained from the integration over a single copy of moduli space.

I have argued that the Polyakov and light-cone formalisms are equivalent, and in particular that the Polyakov amplitudes are unitary. There are several possible extensions of this work. It would seem possible to show the equivalence of the Polyakov and light-cone amplitudes for the open and unoriented bosonic strings. Such a proof could be made in the fashion described above, taking advantage of the close relationship between the world-sheets for the closed and open or unoriented strings. In fact, it is a corollary to the theorem of Ref. 3 that the light-cone diagrams for the oriented open string (including also the closed string and closed string interaction, together with the usual open string - closed string interactions and the four-point open string interaction) provide a single cover of moduli space for Riemann surfaces with an appropriate number of boundary components, and with at least two punctures. This follows from the observation that to each Riemann surface with boundary there is associated a unique closed Riemann double, and the Riemann double will have a unique light-cone diagram. One might hope to prove similar theorems for unoriented strings, in which case the equivalence of the two formalisms reduces to the comparison of the measures. A more important problem is to prove the unitarity of the superstring and of the heterotic string. This could perhaps also be done along the lines discussed in this talk, provided one had a sufficient understanding of the superstring measure (in both the light-cone and covariant formalisms). Now one would need a theorem regarding a single cover of supermoduli space using a superspace extension of the light-cone diagrams. Because the topology of a super Riemann surface comes from the body directions, it seems that it shouldn't be difficult to extend the cover of moduli space to a cover of supermoduli space. Last of all, it would be desireable (and likely plausible) to find a manifestly covariant proof of the unitarity of the Polyakov formalism.

Applications of this work could be possibly be made along several lines. Since we have an explicit parametrization of moduli space, together with a fairly simple expression for the Polyakov integration measure, it may prove possible to do explicit calculations for Polyakov scattering amplitudes using the light-cone coordinates for moduli space. Even though there is no known explicit parametrization for the moduli space of an unpunctured surface, it is conceivable that even *vacuum* amplitudes could be computed using the light-cone parametrization. To do this one would put wavefunctions corresponding to the SL_2 invariant vacuum on the punctures. Such states don't care about their position on the world sheet, so the integral over the light-cone parameters would reduce to an integral over the moduli for the unpunctured Riemann surface, times a simple volume factor for each external state. Finally, this work can be applied along the lines of Ref. 5 to infer the ghost insertions that should come from a gauge-fixed string field theory based on the light-cone world sheets. These ghost insertions should differ from those in Ref. 6 by pieces that decouple in the path integral, and could perhaps lend further insight into string field theory constructions.

ACKNOWLEDGEMENTS

I would like to thank my collaborators, E. D'Hoker and S. Wolpert, for many valuable conversations. Thanks also to S. Mandelstam and J. Polchinski for useful conversations, and to the organizers of the Conference on Superstrings and of the DPF meeting for stimulating conferences.

REFERENCES

1. S. Mandelstam, *Nucl. Phys* **B64** (1973), 205; *Phys. Rep.* **13C** (1974),259; in *Unified String Theories*, ed. M. Green and D. Gross (World Scientific, 1986); and in *Second Nobel Symposium on Elementary Particle Physics*, Marstrand, June 1986, and Berkeley preprint UCB-PTH-86/16.
2. A. Polyakov, *Phys. Lett.* **B103** (1981), 207.
3. S. Giddings and S. Wolpert, Princeton preprint PUPT-1025, to appear in *Comm. Math. Phys.* **109**.
4. E. D'Hoker and S. Giddings, Princeton preprint PUPT-1046, submitted to *Nucl. Phys.*
5. S. Giddings and E. Martinec, *Nucl. Phys.* **B278** (1986), 91.
6. H. Hata, K. Itoh, T. Kugo, H. Kunitomo, and K. Ogawa, Kyoto University preprints; M.A. Awada, *Phys. Lett.* **172B** (1986), 32, and Cambridge preprints; and A. Neveu and P. West, *Phys. Lett.* **168B** (1986), 192 and CERN preprints.
7. See *e.g.* J. Harer *The Cohomology of the Moduli Space of Curves*, C.I.M.E. Lecture notes.
8. D. Friedan in 1982 Les Houches Summer School *Recent Advances in Field Theory and Statistical Mechanics*, J.-B. Zuber and R. Stora (eds.), North Holland (1984); O. Alvarez, *Nucl. Phys.* **B216** (1983), 125; J. Polchinski, *Comm. Math. Phys.* **104** (1986), 37; G. Moore and P. Nelson, *Nucl.*

Phys. **B266** (1986), 58; E. D'Hoker and D.H. Phong, *Nucl. Phys.* **B269** (1986), 205.

9. H. Farkas and I. Kra, *Riemann Surfaces*, Springer Verlag (1980).
10. E. Del Giudice, P. Di Vecchia and S. Fubini, *Ann. Phys. (NY)* **70** (1972), 378.

THIRRING STRINGS

Jonathan Bagger

Lyman Laboratory of Physics
Harvard University
Cambridge, Massachusetts 02138

1. Introduction

In this talk I will discuss work done in collaboration with Dennis Nemeschansky, Nathan Seiberg and Shimon Yankielowicz [1]. We have shown that closed bosonic strings propagating in the presence of toroidal background fields have an alternative description in terms of twisted Thirring models [6] – [7] (see also [8]). We call these models Thirring strings [7], but the techniques we use date back to the days of current algebra. Our results can be easily extended to include superstrings [9] and heterotic strings [10] as well.

Recall that a closed bosonic string propagating on a torus is described by a two-dimensional nonlinear sigma model, with action

$$S = \frac{1}{2\pi}\int d\sigma d\tau \left[G_{\mu\nu}\,\partial_\alpha X^\mu \partial_\alpha X^\nu + B_{\mu\nu}\,\epsilon^{\alpha\beta}\partial_\alpha X^\mu \partial_\beta X^\nu\right] . \qquad (1.1)$$

The metric $G_{\mu\nu}$ and the torsion potential $B_{\mu\nu}$ are taken to be constants. We work at the classical string level, so the base space has the topology of a cylinder, $S^1 \times R$. The target space is the product of $(26-d)$-dimensional Minkowski space and a d-dimensional torus, $\mathcal{M}_{26-d} \times T^d$. Our notation is such that the d compact coordinates X^μ are identified under the lattice translations

$$X^\mu \rightarrow X^\mu + 2\pi n^\mu , \qquad (1.2)$$

where the n^μ are normalized to be integers.

Narain [11], and Narain, Sarmadi and Witten [12] have shown that these models are in one-to-one correspondence with the set of $2d$-dimensional even, self-dual Lorentzian lattices. This description in terms of Lorentzian lattices provides a second, alternative realization of the physics associated with strings propagating on background tori.

In this talk we will describe a third way to realize the same physics. We will show that a given bosonic model, specified by a metric $G_{\mu\nu}$ and torsion potential $B_{\mu\nu}$, is equivalent to a family of twisted Thirring models. These Thirring models are not *new* strings, but rather they are just fermionic realizations of a bosonic string propagating on a torus.

We will see that a given boson model has many different fermionic descriptions, parametrized by d^2 independent Thirring couplings. It is always possible to choose the couplings so that the fermionic twists are simple, corresponding to independent Ramond and Neveu-Schwarz sectors. With this choice, the fermions are interacting. Conversely, it is also possible to choose the couplings so that the fermions are free, but then the twists are very complicated. In fact, we will see that the different fermionic descriptions are in one-to-one correspondence with the even, self-dual Lorentzian lattices of dimension $2d$.

2. Bosonic string on S^1

To begin, let us consider the simplest torus – the circle S^1. We would like to derive the appropriate U(1) × U(1) current algebra. For convenience, we scale the field X so that it is periodic under shifts by $2\pi R$. This scaling eliminates the metric from the action (1.1), giving

$$S = \frac{1}{2\pi}\int d\tau d\sigma\ \partial_\alpha X \partial_\alpha X\ . \tag{2.1}$$

As before, the base space is a cylinder ($0 \leq \sigma < \pi$), and the target space is a circle of radius R. For simplicity, we do not discuss the 25 free scalar fields that span $\mathcal{M}_{25}$.

The scalar field X is free, so its mode expansion can be written in terms of left- and right-moving coordinates:

$$\begin{aligned} X &= X_L + X_R \\ X_L &= X_{0L} + P_{0L}\, u + \textit{oscillators} \\ X_R &= X_{0R} + P_{0R}\, v + \textit{oscillators}\ , \end{aligned} \tag{2.2}$$

where $u = \tau + \sigma$ and $v = \tau - \sigma$. The left- and right-moving zero modes can be decomposed as follows [3]:

$$\begin{aligned} X_{0L} &= (X_0 + \widetilde{X}_0)/2 \\ X_{0R} &= (X_0 - \widetilde{X}_0)/2\ , \\ P_{0L} &= (P_0 + \widetilde{P}_0)/2 \\ P_{0R} &= (P_0 - \widetilde{P}_0)/2\ . \end{aligned} \tag{2.3}$$

The zero mode X_0 corresponds to the center of mass of the string, and its conjugate momentum is P_0. The variable $\widetilde{X}_0$ has no obvious physical interpretation, but it is conjugate to the momentum $\widetilde{P}_0$. Therefore we postulate the commutators

$$[P_0, X_0] = -i, \qquad [\widetilde{P}_0, \widetilde{X}_0] = -i, \tag{2.4}$$

where all other commutators vanish. These are sufficient to show that

$$[P_{0L}, X_{0L}] = -\frac{i}{2}, \qquad [P_{0R}, X_{0R}] = -\frac{i}{2}. \tag{2.5}$$

The commutators (2.5) are exactly what one would find from a constrained Hamiltonian analysis, as was done with the left-movers in the original heterotic string.

Let us compare the above expansion for X with the usual expansion for a boson on a torus:

$$X = X_0 + M\tau/R + 2LR\sigma + oscillators. \tag{2.6}$$

Comparing with (2.2) and (2.3), we see that P_0 has eigenvalue M/R, and that $\widetilde{P}_0$ has eigenvalue $2LR$. The quantization of P_0 is consistent with the fact that X_0 lies on a circle of radius R. Similarly, the fact that $\widetilde{P}_0$ is quantized in units of $2LR$ implies that $\widetilde{X}_0$ lies on a circle of radius $1/2R$.

Let us now solve the model using current algebra [6]. We start by defining the left- and right-moving currents,

$$\begin{aligned} J_L(u) &= P_L(u) = \frac{1}{\pi}\,\partial_u X_L \\ J_R(v) &= -P_R(v) = -\frac{1}{\pi}\,\partial_v X_R. \end{aligned} \tag{2.7}$$

These currents generate translations on the target space, and are symmetries of the action (1.1). With the above commutators, it is easy to show that

$$\begin{aligned} [J_L(u), J_L(u')] &= \frac{i}{2\pi}\,\delta'(u-u') \\ [J_R(v), J_R(v')] &= \frac{i}{2\pi}\,\delta'(v-v'). \end{aligned} \tag{2.8}$$

All other commutators vanish. The currents $J_L(u)$ and $J_R(v)$ generate a U(1) $\times$ U(1) current algebra, or a U(1) $\times$ U(1) Kac-Moody symmetry. (For a recent review, see [5].)

The theory described above has a current-current interaction, so the energy-momentum tensor is of Sugawara form:

$$\begin{aligned} \Theta_L(u) &= 2\pi :J_L(u)^2: \\ \Theta_R(v) &= 2\pi :J_R(v)^2:. \end{aligned} \tag{2.9}$$

All normal ordering is done with respect to the creation and annihilation operators of the *free* currents. The energy-momentum commutators can be calculated with the help of (2.8) and (2.9):

$$\begin{aligned}[]
[\Theta_L(u), \Theta_L(u')] &= 2i\,\{\Theta_L(u) + \Theta_L(u')\}\,\delta'(u-u') - \frac{i}{6\pi}\,\delta'''(u-u') \\
[\Theta_R(v), \Theta_R(v')] &= 2i\,\{\Theta_R(v) + \Theta_R(v')\}\,\delta'(v-v') - \frac{i}{6\pi}\,\delta'''(v-v')\,.
\end{aligned} \tag{2.10}$$

These commutators generate a Virasoro algebra, and the δ''' term is the conformal anomaly.

Now that we have the full current algebra, we have all we need to solve the model in terms of currents. The Hamiltonian is given by

$$H = \frac{1}{2}J_{0L}^2 + \frac{1}{2}J_{0R}^2 + N_L + N_R\,, \tag{2.11}$$

where J_{nL} and J_{nR} are the Fourier modes of the currents, and

$$\begin{aligned}
N_L &= \sum_{n>0} J_{-nL}J_{nL} \\
N_R &= \sum_{n>0} J_{-nR}J_{nR}\,.
\end{aligned} \tag{2.12}$$

Given (2.2), (2.3) (2.6) and (2.7), we find J_{0L} and J_{0R} in terms of M and L:

$$\begin{aligned}
J_{0L} &= LR + M/2R \\
J_{0R} &= LR - M/2R
\end{aligned} \tag{2.13}$$

Substituting (2.13) into (2.11), we recover the usual spectrum for a bosonic string on a circle of radius R:

$$\frac{1}{4}(mass)^2 = H = N_L + N_R + (M/2R)^2 + (LR)^2\,. \tag{2.14}$$

3. Fermionic description of bosonic string on S^1

Let us now try to describe the same system in terms of fermions. We take our fermionic action to be as follows:

$$S = \frac{1}{\pi}\int d\tau d\sigma\,\left[i\psi_L^*\partial_v\psi_L + i\psi_R^*\partial_u\psi_R - h\,\psi_L^*\psi_L\,\psi_R^*\psi_R\right]\,. \tag{3.1}$$

Here h is the Thirring coupling, and ψ_L and ψ_R are the left- and right-handed fermions. The action (3.1) gives rise to the following equations of motion:

$$\begin{aligned} \partial_v \psi_L &= -\,2\pi i\, h\ :J_R(v)\psi_L: \\ \partial_u \psi_R &= -\,2\pi i\, h\ :J_L(u)\psi_R:\ . \end{aligned} \tag{3.2}$$

Here $J_L(u) = \psi_L^* \psi_L / 2\pi$ and $J_R(v) = \psi_R^* \psi_R / 2\pi$. These products are very singular. To solve the Thirring model, we never need the expressions for th. about the action (3.1). Since the fermions are not free, all normal ordering is done with respect to the *current* creation and annihilation operators [6].

Let us solve the Thirring model from first principles, later making contact with the bosonic model. As before, we use the powerful techniques of current algebra. The current-current commutators follow from the Kac-Moody symmetry

$$\begin{aligned} [J_L(u),\, J_L(u')] &= \frac{i}{2\pi}\,\delta'(u-u') \\ [J_R(v),\, J_R(v')] &= \frac{i}{2\pi}\,\delta'(v-v')\ . \end{aligned} \tag{3.3}$$

Now, however, we must also define current-fermion commutators. The obvious choice is given by

$$\begin{aligned} [J_L(u),\, \psi_L(u',\, v')] &= -\,\frac{a}{2}\,\psi_L(u',\, v')\ \delta(u-u') \\ [J_R(v),\, \psi_L(u',\, v')] &= -\,\frac{b}{2}\,\psi_L(u',\, v')\ \delta(v-v') \\ [J_R(v),\, \psi_R(u',\, v')] &= -\,\frac{c}{2}\,\psi_R(u',\, v')\ \delta(v-v') \\ [J_L(u),\, \psi_R(u',\, v')] &= -\,\frac{d}{2}\,\psi_R(u',\, v')\ \delta(u-u')\ . \end{aligned} \tag{3.4}$$

The energy-momentum tensor is as before, so we find

$$\begin{aligned} [H+P,\, \psi_L] &= -\,2i\,\partial_u\psi_L &&= -\,2\pi a\ :J_L(u)\psi_L: \\ [H-P,\, \psi_L] &= -\,2i\,\partial_v\psi_L &&= -\,2\pi b\ :J_R(v)\psi_L: \\ [H-P,\, \psi_R] &= -\,2i\,\partial_v\psi_R &&= -\,2\pi c\ :J_R(v)\psi_R: \\ [H+P,\, \psi_R] &= -\,2i\,\partial_u\psi_R &&= -\,2\pi d\ :J_L(u)\psi_R: \\ [D+M,\, \psi_L] &= -\,2iu\,\partial_u\psi_L &&- \frac{i}{4}\,a^2\,\psi_L \\ [D-M,\, \psi_L] &= -\,2iv\,\partial_v\psi_L &&- \frac{i}{4}\,b^2\,\psi_L \\ [D-M,\, \psi_R] &= -\,2iv\,\partial_v\psi_R &&+ \frac{i}{4}\,c^2\,\psi_R \\ [D+M,\, \psi_R] &= -\,2iu\,\partial_u\psi_R &&+ \frac{i}{4}\,d^2\,\psi_R\ . \end{aligned} \tag{3.5}$$

Demanding ψ_L and ψ_R satisfy the equations of motion (3.2) implies $b = d = 2h$, where h is the Thirring coupling. Demanding they have spin-$\frac{1}{2}$ implies $a^2 - b^2 = c^2 - d^2 = 4$. These conditions are satisfied if

$$\begin{aligned} a &= z + 1/z \\ b &= z - 1/z \\ c &= z + 1/z \\ d &= z - 1/z , \end{aligned} \tag{3.6}$$

where z is fixed by the Thirring coupling h.

We are now ready to solve the model. Because we have the same current algebra as in the bosonic case, the Hamiltonian is also the same:

$$H = \frac{1}{2} J_{0L}^2 + \frac{1}{2} J_{0R}^2 + N_L + N_R , \tag{3.7}$$

To find the spectrum, we must examine the zero modes J_{0L} and J_{0R}. These zero modes measure the "charges" of the left- and right-moving vacua.

Let us assume the left-moving vacuum has charge $-Q_L$, and the right-moving vacuum has charge $-Q_R$. States built off these vacua with the *current* creation operators have the same charges as the vacua, since $[J_{0L}, J_{-nL}] = [J_{0R}\, J_{-nR}] = 0$ for $n > 0$. States built off these vacua with *fermions* have different charges, since

$$\begin{aligned} [J_{0L}, \psi_L] &= -\frac{a}{2}\psi_L \\ [J_{0R}, \psi_L] &= -\frac{b}{2}\psi_L \\ [J_{0R}, \psi_R] &= -\frac{a}{2}\psi_R \\ [J_{0L}, \psi_R] &= -\frac{b}{2}\psi_R . \end{aligned} \tag{3.8}$$

A sector created by n applications of ψ_L and m applications of ψ_R has charges $J_{0L} = -Q_L - na/2 - mb/2$ and $J_{0R} = -Q_R - ma/2 - nb/2$. Each sector contains an entire Fock space built with the current creation operators.

With these ideas in mind, we see that the full Hamiltonian is given by

$$H = \frac{1}{2}\left(Q_L + n\frac{a}{2} + m\frac{b}{2}\right)^2 + \frac{1}{2}\left(Q_R + m\frac{a}{2} + n\frac{b}{2}\right)^2 + N_L + N_R . \tag{3.9}$$

It is this spectrum that must agree with the bosonic spectrum (2.14).

Before matching (3.9) to the bosonic spectrum, it is convenient to introduce the Mandelstam operators [2]. The Mandelstam operators are explicit representations of the fermions ψ_L and ψ_R in terms of the bosonic field X:

$$\begin{aligned} \psi_L &= :\exp[-\,iaX_L(u) + ibX_R(v)]: \\ \psi_R &= :\exp[+\,iaX_R(v) - ibX_L(u)]: . \end{aligned} \tag{3.10}$$

The normal ordering in (3.10) is with respect to the creation and annihilation pieces of X_L and X_R – but not the zero modes (see [1] for details). The Mandelstam operators obey the fermionic current algebra. They are normalized so that they obey the correct canonical commutation relations when $b = 0$.

Let us check to see whether the Mandelstam operators are well defined under shifts in X_0 and $\widetilde{X}_0$. Under shifts X_0 by $2\pi R$, we find

$$\begin{aligned} \psi_L &\rightarrow \psi_L \, e^{-i2\pi R/z} \\ \psi_R &\rightarrow \psi_R \, e^{+i2\pi R/z} \, , \end{aligned} \tag{3.11}$$

where we have used the solution for a and b in terms of z. Under shifts in $\widetilde{X}_0$ by π/R, we have

$$\begin{aligned} \psi_L &\rightarrow \psi_L \, e^{-i\pi z/R} \\ \psi_R &\rightarrow \psi_R \, e^{-i\pi z/R} \, . \end{aligned} \tag{3.12}$$

Thus we see that ψ_L and ψ_R are not appropriate operators for describing strings on a torus. The good quantum operators are products of ψ's, invariant under the above shifts. We must therefore project onto states invariant under (3.11) and (3.12).

The necessity for the projection can also be seen from another point of view. Let us look at the boundary conditions for ψ_L and ψ_R (at fixed τ), where ψ_L and ψ_R act on a state with well-defined L and M. The boundary conditions follow from the normal-ordered form of the operators. We find

$$\begin{aligned} \psi_L(\pi) &= - \psi_L(0) \, e^{-iM\pi z/R \, - \, 2iL\pi R/z} \\ \psi_R(\pi) &= - \psi_R(0) \, e^{-iM\pi z/R \, + \, 2iL\pi R/z} \, , \end{aligned} \tag{3.13}$$

For general z, the fermions ψ_L and ψ_R obey very complicated boundary conditions. Now, the original bosonic theory was modular invariant, so the fermionic theory ought to be modular invariant as well. Modular invariance relates boundary conditions in σ to boundary conditions in τ. Therefore a sum over boundary conditions in σ must be accompanied by a sum over boundary conditions in τ. The sum over the τ boundary conditions gives the generalized GSO projection discussed above. The GSO projection is necessary to ensure the modular invariance of the fermionic theory.

Let us see how this works for two extreme cases. For the first, we take $z = 1$. This implies that $b = d = 0$, so the fermions are *free*. The good quantum operators are products of the fermions, invariant under shifts in X_0 and $\widetilde{X}_0$. Under shifts in X_0, the fermions ψ_L and ψ_R transform as follows,

$$\begin{aligned} \psi_L &\rightarrow \psi_L \, e^{-2\pi i R} \\ \psi_R &\rightarrow \psi_R \, e^{+2\pi i R} \, . \end{aligned} \tag{3.14}$$

For R an integer, (3.14) does not constrain the spectrum. Under shifts in $\widetilde{X}_0$, the story is not quite so simple. Now ψ_L and ψ_R twist as follows,

$$\begin{aligned} \psi_L &\to \psi_L \, e^{-\pi i/R} \\ \psi_R &\to \psi_R \, e^{-\pi i/R} \, . \end{aligned} \tag{3.15}$$

The equations (3.14) and (3.15) determine the generalized GSO condition for arbitrary radius R. For R an integer, (3.15) shows that ψ_L and ψ_R twist by a $2R$ root of unity. Note that for irrational R, the GSO condition can only be fulfilled for infinite products of fermion operators!

Let us now consider the choice $z = R$. In this case the fermions are *interacting* (for $R \neq 1$). Under shifts in X_0, (3.11) tells us

$$\begin{aligned} \psi_L &\to \psi_L \\ \psi_R &\to \psi_R \, . \end{aligned} \tag{3.16}$$

Under shifts in $\widetilde{X}_0$, (3.12) implies

$$\begin{aligned} \psi_L &\to -\psi_L \\ \psi_R &\to -\psi_R \, . \end{aligned} \tag{3.17}$$

This holds true for any radius R. The GSO condition is simply $(-1)^{F_L+F_R} = 1$.

It is clear that this choice for z gives just two sectors – Neveu-Schwarz and Ramond. However, to solve the model, we must use the full power of current algebra, for now the fermions are interacting. The spectrum is found from the current-algebra Hamiltonian,

$$H = \frac{1}{2}\left(Q_L + n\frac{a}{2} + m\frac{b}{2}\right)^2 + \frac{1}{2}\left(Q_R + m\frac{a}{2} + n\frac{b}{2}\right)^2 + N_L + N_R \, , \tag{3.18}$$

where $a = R + 1/R$ and $b = R - 1/R$. The values of n and m are restricted by the GSO condition, so $n + m = 0$, mod 2. Modular invariance tells us that there are two sectors, with $Q_L = \eta/R$, $Q_R = -\eta/R$, for $\eta = 0, \frac{1}{2}$. Setting $n = L + M$, $m = L - M$, and substituting above, we find

$$\begin{aligned} H &= \frac{1}{2}\left(\frac{\eta}{R} + \frac{M}{R} + RL\right)^2 + \frac{1}{2}\left(\frac{\eta}{R} + \frac{M}{R} - RL\right)^2 + N_L + N_R \\ &= \frac{1}{R^2}(\eta + M)^2 + (RL)^2 + N_L + N_R \, . \end{aligned} \tag{3.19}$$

Counting both sectors ($\eta = 0, \frac{1}{2}$), we see that this theory matches the bosonic spectrum for any radius R.

4. String on an arbitrary torus T^d

The general procedure described above can be readily extended to an arbitrary torus T^d. There are now d left-moving currents $J_L^a(u)$ and d right-moving counterparts $J_R^a(v)$. These currents obey a $\mathrm{U}(1)^d \times \mathrm{U}(1)^d$ Kac-Moody symmetry. As above, we use the same current algebra to describe the bosonic and fermionic theories. In the fermionic theory, there are d left- and right-handed Weyl fermions ψ_L^i and ψ_R^i. These fermions have commutation relations with the currents that are described by *matrices* A^{ai}, B^{ai}, C^{ai}, and D^{ai}. Demanding ψ_L^i and ψ_R^i satisfy the equations of motion fixes B^{ai} and D^{ai} in terms of the Thirring coupling. Demanding they have spin-$\frac{1}{2}$ requires $\sum_a A^{ai}A^{ai} - \sum_a B^{ai}B^{ai} = \sum_a C^{ai}C^{ai} - \sum_a D^{ai}D^{ai} = 4$.

The current-current and current-fermion commutators are all we need to solve the model. As above, the Hamiltonian is of Sugawara form:

$$H \;=\; \frac{1}{2}\,J_{0L}^a J_{0L}^a \;+\; \frac{1}{2}\,J_{0R}^a J_{0R}^a \;+\; N_L \;+\; N_R\;. \tag{4.1}$$

In the bosonic theory, the zero modes J_{0L}^a and J_{0R}^a are given in terms of the momenta and winding numbers. In the fermionic theory, they measure the charges of the left- and right-moving vacua. A sector created by n^i applications of ψ_L^i and m^i applications of ψ_R^i has charges $J_{0L}^a = -Q_L^a - A^{ai}n^i/2 - D^{ai}m^i/2$ and $J_{0R}^a = -Q_R^a - C^{ai}m^i/2 - B^{ai}n^i/2$. The vacuum charges Q_L^a and Q_R^a, and the occupation numbers n^i and m^i, are chosen to reconstruct the bosonic spectrum.

To proceed further, we consider the Mandelstam operators. As in the one-flavor case, they are exponentials of the bosonic fields:

$$\begin{aligned}\psi_L^i &= :\exp[-\,iA^{ai}X_L^a(u) \;+\; iB^{ai}X_R^a(v)]:\\ \psi_R^i &= :\exp[+\,iC^{ai}X_R^a(v) \;-\; iD^{ai}X_L^a(u)]:\;.\end{aligned} \tag{4.2}$$

It is not hard to show that these Mandelstam operators are ill defined under shifts in X_0^a and $\widetilde{X}_0^a$. The good operators are products of the fermions, invariant under the shifts.

The general shifts are not very illuminating; they are given in Ref. [1]. It is possible, however, to choose simple cases in which the analysis is easier. One simple case corresponds to free fermions with complicated twists. Another gives interacting fermions with simple twists. All in all, there is a d^2 parameter family of choices, corresponding to the d^2 elements in an $\mathrm{SO}(d,d)$ matrix.

Thus far we have only discussed the fermionization of bosonic strings propagating on $\mathcal{M}_{26-d} \times T^d$. It is straightforward to generalize our results to include superstrings [9] and heterotic strings [10]. The main difference for superstrings is that the total space must have dimension 10, and not 26. Heterotic strings are a little more complicated because of the possibility of symmetry breaking by Wilson lines. The details are given in Ref. [1]

5. Interaction picture fermions

In the previous sections we gave a fermionic description of toroidal compactification. The fermions we constructed were fully interacting Heisenberg operators. In this section we will take a somewhat different point of view. We will now take our fermions in the interaction picture. This will give fresh insight into our previous results, and will illuminate the connection between our work and that of Narain, Sarmadi and Witten [12].

For simplicity, let us start by returning to the one-flavor case – a single boson on a circle of radius R. The action is as above,

$$S = \frac{1}{2\pi} \int d\tau d\sigma \; \partial_\alpha X \partial_\alpha X \; , \tag{5.1}$$

where X is identified with $X + 2\pi R$. Changing to the variables $\phi = rX/R$, we find

$$S = \frac{1}{2\pi} \left(\frac{R}{r}\right)^2 \int d\tau d\sigma \; \partial_\alpha \phi \partial_\alpha \phi \; , \tag{5.2}$$

where ϕ is identified with $\phi + 2\pi r$ – it is a boson on a circle of radius r. We can write S as the sum of a free part

$$S_0 = \frac{1}{2\pi} \int d\tau d\sigma \; \partial_\alpha \phi \partial_\alpha \phi \tag{5.3}$$

and an interaction part

$$S_I = \frac{1}{2\pi} \left[\left(\frac{R}{r}\right)^2 - 1 \right] \int d\tau d\sigma \; \partial_\alpha \phi \partial_\alpha \phi \tag{5.4}$$

These trivial field redefinitions do not change the fact that we are describing a boson on a circle of radius R.

Let us now describe this model in terms of fermions. We could proceed as in previous sections, and construct the Heisenberg operators

$$\begin{aligned} \psi_L &= : \exp[-\, 2iX_L(u)] : \\ \psi_R &= : \exp[+\, 2iX_R(v)] : , \end{aligned} \tag{5.5}$$

in terms of our original X. Instead, let us construct a new set of Mandelstam operators, in terms of the new boson ϕ:

$$\begin{aligned} \lambda_L &= : \exp[-\, 2i\phi_L(u)] : \\ \lambda_R &= : \exp[+\, 2i\phi_R(v)] : \end{aligned} \tag{5.6}$$

As with the operators (5.5), these operators are also fermions – they have spin one-half and they satisfy canonical anticommutation relations. In general, both sets of fermions are subject to complicated boundary conditions. They also suffer from phase ambiguities, so modular invariance demands that we perform a generalized GSO projection.

What is the relation between the operators (5.6) – which depend on the unphysical parameter r – and those of our original model, which describe a free boson on a circle of radius R? The answer is that the new operators can be thought of as *interaction picture* fermions describing the same physics [3].

To see this, let us write the action (5.2) in terms of the light cone coordinates u and v,

$$\begin{aligned} S &= S_0 + S_I \\ S_0 &= \frac{2}{\pi}\int d\sigma d\tau\ \partial_u\phi\partial_v\phi \\ S_I &= \frac{2}{\pi}\left[\left(\frac{R}{r}\right)^2 - 1\right]\int d\sigma d\tau\ \partial_u\phi\,\partial_v\phi\ . \end{aligned} \tag{5.7}$$

The free part of (5.7) fermionizes into a free action for the interaction picture fermions. The interaction part gives a Thirring coupling. Together, they give a full fermionic action of the following form:

$$S = \frac{1}{\pi}\int d\tau d\sigma\ \left[i\lambda_L^*\partial_v\lambda_L + i\lambda_R^*\partial_u\lambda_R - h\,\lambda_L^*\lambda_L\,\lambda_R^*\lambda_R\right]. \tag{5.8}$$

The Thirring coupling h can be found from the action S_I. Using the bosonization formulae

$$\begin{aligned} J_L(u) &= \frac{1}{\pi}\partial_u\phi_L = \frac{1}{2\pi} :\lambda_L^*\lambda_L: \\ J_R(v) &= -\frac{1}{\pi}\partial_v\phi_R = \frac{1}{2\pi} :\lambda_R^*\lambda_R: , \end{aligned} \tag{5.9}$$

we find

$$h = \frac{1}{2}\left[\left(\frac{R}{r}\right)^2 - 1\right]. \tag{5.10}$$

This value for the Thirring coupling in terms R and r differs from the value obtained in section 3. The difference stems from the fact that the currents are normalized differently. To relate the couplings, we must compare the equations of motion. From (5.8) we find

$$\begin{aligned} \partial_v\lambda_L &= -2\pi i\,h\ :J_R(v)\lambda_L: \\ \partial_u\lambda_R &= -2\pi i\,h\ :J_L(u)\lambda_R: . \end{aligned} \tag{5.11}$$

This is identical to (3.2), except for the fact that the currents in this analysis differ from those in section 3 by a factor of r/R. Taking this into account, we see that the interaction picture fermions indeed describe the physics of a boson of radius R.

So far our discussion has been a standard exercise in bosonization. The only new feature is the arbitrary parameter r. This parameter appears in two places: in the boundary conditions of the fermions and in the interaction term in the action. Two cases of special interest are $r = R$, when the interaction vanishes but the fermion is subject to complicated boundary conditions; and $r = \frac{1}{2}$ or $r = 1$, where the sum over the spin structures is simple but there is a non-trivial four-Fermi interaction.

Clearly, all physical results should be independent of the parameter r. Therefore we have an infinite set of fermionic descriptions of the same model, labeled by the value of r. This is reminiscent of the situation in the previous sections. The difference is that here we are using the interaction picture, and previously we were using the Heisenberg picture. To understand the connection between these two approaches, we shall describe the model in the language of Narain, Sarmadi and Witten.

Let us start with the boson ϕ, which takes its values on a circle of radius r, and split it into ϕ_L and ϕ_R. After the standard quantization, it is easy to show that (ϕ_L, ϕ_R) lies on a two-dimensional even self-dual Lorentzian lattice:

$$\begin{aligned}\begin{pmatrix}\phi_L\\ \phi_R\end{pmatrix} &\to \begin{pmatrix}\phi_L\\ \phi_R\end{pmatrix} + \pi n r\begin{pmatrix}1\\ 1\end{pmatrix} + \frac{\pi m}{2r}\begin{pmatrix}1\\ -1\end{pmatrix}\\ &= \begin{pmatrix}\phi_L\\ \phi_R\end{pmatrix} + \pi n\, e_1 + \pi m\, e_2\,.\end{aligned} \tag{5.12}$$

Here n and m are integers, and e_1 and e_2 are the two vectors that generate the lattice. Different models with different radii r_1 and r_2 have different identification rules (5.12). The different fields are related by the SO(1,1) Lorentzian rotation

$$\begin{pmatrix}\phi_L\\ \phi_R\end{pmatrix}_{r_2} = \begin{pmatrix}\cosh\alpha & \sinh\alpha\\ \sinh\alpha & \cosh\alpha\end{pmatrix}\begin{pmatrix}\phi_L\\ \phi_R\end{pmatrix}_{r_1}, \tag{5.13}$$

where the "rotation angle" is $\alpha = \log(r_2/r_1)$.

Let us return now to our fermionic problem. The interaction picture operators $\lambda_L =: \exp[+2i\phi_L]$: and $\lambda_R =: \exp[-2i\phi_R]$: describe a boson of radius R, provided they interact through a Thirring interaction with coupling constant (5.10). The corresponding Heisenberg picture operators must have the same boundary conditions and the same phase ambiguities as their interaction picture counterparts. The Heisenberg operators describe the model after the Thirring interaction has been taken into account, and should be written in terms of the original field X. The transformation from (ϕ_L, ϕ_R) to (X_L, X_R) is an SO(1,1) rotation of the form (5.13). Therefore the Heisenberg picture operators are given by

$$\begin{aligned}\psi_L &= :\exp[-\,2i\cosh\alpha X_L \,+\, 2i\sinh\alpha X_R] :\\ \psi_R &= :\exp[+\,2i\cosh\alpha X_R \,-\, 2i\sinh\alpha X_L] :,\end{aligned} \tag{5.14}$$

where $\alpha = \log(r/R)$.

The Lorentzian rotation that takes us from (5.6) to (5.14) does not commute with the Hamiltonian. The interaction picture operators evolve according to the free Hamiltonian. After the Lorentzian rotation, the Heisenberg operators evolve according to the full Hamiltonian, including the interaction terms. From this point of view, solving the Thirring model amounts to performing an SO(1,1) rotation on the interaction picture fields.

The generalization of the previous discussion to the multi-flavor case is straightforward. The general Lorentzian lattice has p positive metric and q negative metric directions, corresponding to p left-moving and q right-moving coordinates. We wish to describe the model associated with the $p+q$ -dimensional Lorentzian lattice Λ, spanned by the $p+q$ coordinates $(X_L^a, X_R^{b'})$.

Let us describe this model in terms of fields $(\phi_L^a, \phi_R^{b'})$, taking values on *another* $p+q$ -dimensional self-dual Lorentzian lattice λ (the analog of the boson on a circle of radius r). The interaction picture fermions are

$$\begin{aligned} \lambda_L^a &= :\exp[-\,2i\phi_L^a]: \\ \lambda_R^{b'} &= :\exp[+\,2i\phi_R^{b'}]: \;. \end{aligned} \tag{5.15}$$

As before, the operators λ_L^a and $\lambda_R^{b'}$ are subject to various boundary conditions, and they have phase ambiguities.

The Heisenberg picture operators are written in terms of the original variables $(X_L^a, X_R^{b'})$. The original variables are related to $(\phi_L^a, \phi_R^{b'})$ by an SO(p,q) rotation. Therefore the Heisenberg picture fermions are

$$\begin{aligned} \psi_L^i &= :\exp[-\,2iA^{ai}X_L^a \;+\; 2iB^{b'i}X_R^{b'}]: \\ \psi_R^{j'} &= :\exp[+\,2iC^{b'j'}X_R^{b'} \;-\; 2iD^{aj'}X_L^a]: \;. \end{aligned} \tag{5.16}$$

where a, $i = 1, \ldots, p$, and b', $j' = 1, \ldots, q$. The matrix

$$M = \begin{pmatrix} A & B \\ D & C \end{pmatrix} \tag{5.17}$$

is an SO(p,q) matrix, so it satisfies

$$M\, g\, M^T \;=\; g \;, \tag{5.18}$$

with $g = \mathrm{diag}(1, -1)$. The dimensions of the sub-matrices 1 and -1 are $p \times p$ and $q \times q$, respectively.

As in the one-flavor case, the interaction picture and Heisenberg picture fermions coincide if $\lambda = \Lambda$. Then $A = 1$, $B = 0$, $C = 1$ and $D = 0$, and the fermions are free. Alternatively, we can take λ to be such that the interaction picture fermions satisfy simple boundary conditions. This requires a Thirring interaction to reproduce the original bosonic theory. As before, the corresponding Heisenberg picture operators are interacting, with simple boundary conditions.

6. Conclusion

In this talk we have seen that the toroidal compactification of strings – bosonic, heterotic and supersymmetric – has an alternative description in terms of twisted Thirring models. We have found that a given bosonic background has an infinite number of fermionic descriptions. These formulations differ by $SO(p,q)$ rotations of the original Lorentzian lattice – that is to say, they correspond to the infinite number of frames we can use to construct interaction picture fermions. Different frames give rise to different boundary conditions on the fermions, both in the interaction and Heisenberg pictures.

With this understanding of the fermion operators, we can also describe a fermionic formulation of orbifolds [13]. Orbifolds are characterized by discrete symmetries that are used to divide a Lorentzian lattice. This division leads to more identification rules for the bosons $(X_L^a, X_R^{b'})$. These identification rules give rise to new boundary conditions for the fermions. They also lead to new phase ambiguities which must be removed by appropriate projectors. It is not hard to see that the formalism presented here can be used to fermionize any orbifold, including the left-right asymmetric orbifolds of reference [14]. It is a straightforward exercise to check that this construction also agrees with the fermionic construction of the orbifolds given in [15].

This work was supported by the National Science Foundation, grant number NSF-PHY82-15249.

References

[1] J. Bagger, D. Nemeschansky, N. Seiberg and S. Yankielowicz, Harvard preprint HUTP/86-A088, to be published in Nucl. Phys. B.

[2] S. Mandelstam, Phys. Rev. D7 (1973) 3763, 3777; D11 (1975) 3026; S. Coleman, Phys. Rev. D11 (1975) 2088; M. Halpern, Phys. Rev. D12 (1975) 1684.

[3] T. Banks, D. Horn and H. Neuberger, Nucl. Phys. B108 (1976) 199.

[4] I. Frenkel and V. Kac, Invent. Math. 62 (1980) 23; G. Segal, Comm. Math. Phys. 80 (1981) 301; P. Goddard and D. Olive, in *Vertex Operators in Mathematics and Physics*, eds. J. Lepowsky, S. Mandelstam and I.M. Singer, Springer-Verlag (1984).

[5] P. Goddard and D. Olive, Int. J. of Mod. Phys. A1 (1986) 303, and references therein.

[6] G.F. Dell'Antonio, Y. Frishman and D. Zwanziger, Phys. Rev. D6 (1972) 988.

[7] I. Andric and J.L. Gervais, Nucl. Phys. B42 (1972) 240; S.H.H. Tye, Phys. Rev. D9 (1974) 2821.

[8] H. Neuberger, A. Niemi and G. Semenoff, Phys. Lett. 181B (1986) 244; H. Kawai, D.C. Lewellen and S.H.H. Tye, Cornell preprint CLNS-86-751.

[9] J.H. Schwarz, Phys. Rep. 89 (1982) 223; M.B. Green, Surveys in H. E. Phys. 3 (1982) 127.

[10] D.J. Gross, J.A. Harvey, E. Martinec and R. Rohm, Phys. Rev. Lett. 54 (1985) 502; Nucl. Phys. B256 (1985) 253; B267 (1986) 75.

[11] K.S. Narain, Phys. Lett. B169 (1986) 41.

[12] K.S. Narain, M.H. Sarmadi and E. Witten, Nucl. Phys. B279 (1987) 369.

[13] L. Dixon, J.A. Harvey, C. Vafa, and E. Witten, Nucl. Phys. B261 (1985) 678; B274 (1986) 285.

[14] H. Kawai, D.C. Lewellen and S.H.H. Tye, Phys. Rev. Lett. 57 (1986) 1832, 58 (1987) 429; M. Mueller and E. Witten, Phys. Lett. 182B (1986) 28; K.S. Narain, M.H. Sarmadi and C. Vafa, Harvard preprint HUTP/86-A089.

[15] S. Elitzur, E. Gross, E. Rabinovici and N. Seiberg, Weizmann preprint WIS-86/32/June-PH.

Exotic Spheres as Topological Defects in String Theory

RYAN ROHM*

California Institute of Technology
Pasadena, California 91125†

and

Institute for Theoretical Physics
University of California
Santa Barbara, California 93106

ABSTRACT

I expand on Witten's suggestion for topological gravitational defects in higher-dimensional theories; they should consist of exotic differential structures on a manifold with the same topology as the vacuum. Classification of possible defects and their physical manifestations is discussed.

String theory seems to have the usual spectrum of gauge-field topological defects[1] such as monopoles, cosmic strings, and so forth, at least from our understanding of the low-energy limit. Similarity between the manifestations of Yang-Mills theories and general relativity in string theory—for instance, the parallel ways in which gauge field and gravitational curvatures enter the effective action—suggests the existence of gravitational topological defects, with structures similar to those found in the Yang-Mills sector.

Some reasons why we might be interested in investigating such topological defects:

* Weingart fellow
† Permanent address

-Phenomenology: In principal, topological defects of this sort might give rise to observable solitons. At the current stage of development it is difficult even to show the existence of stable solitons, so that the question of possible observations is premature.

-Consistency: The structure of topological defects is closely connected to that of anomalies; even in consistent theories they can shed light on the mechanisms of anomaly cancellation.

-Deeper understanding: Although in principle the study of the spectrum of topological defects might shed light on the underlying structure of the theory, at the present level of understanding we are unable to make use of this connection. Progress is more likely to occur in the opposite direction.

This is because the very notion of 'topological' defect requires some outline of the formal approach intended: to define what the allowed configurations are and the allowed transitions between them. (It does not necessarily involve a detailed specification of the underlying dynamics, so that the set of topological defects and that of stable solitons may not precisely overlap.) Lacking a comprehensive formal approach to string theory, we instead work with the low-energy approximations to the string theory given by effective field theories, and thus have nothing rigorous to say about the underlying string theory.

There are basically two formal approaches to the question of possible topological defects, which arise most naturally from the point of view of the Hamiltonian or path-integral approach to quantum gravity; each yields a slightly different picture of the topological defects.

In the Hamiltonian approach, we require that the global topology of space remains fixed and assume smooth evolution from our initial configuration (assuming that the underlying theory prevents the formation of singularities). As a consequence, any topological defects which can be pair-produced in such a theory must be free of 'handles'— the space-time manifold containing the pair of (soliton+antisoliton) must be diffeomorphic† to the vacuum configuration.[3] This ap-

† The requirement is actually h-cobordism[2], which is equivalent to diffeomorphism except in four dimensions.

proach is 'minimal'—because of locality and cluster-decomposition, such configurations are necessarily possible. In this case, the source of the defect lies in the structure of the tangent bundle—exotic differential structures.

The existence of different differential structures on the same topological manifold—essentially, different notions of which continuous functions are differentiable—is analogous to the existence of different spin structures for fermion fields. In that case one starts with a smooth tangent bundle, with structure group SO(n), and attempts to define the spinor bundle by lifting the action of SO(n) to its double cover Spin(n). Spin structures exist if a certain obstruction (the second Stiefel-Whitney class) vanishes, and have a certain discrete set of choices, depending on the 1-dimensional homology of the manifold. Similarly, one can start with a topological tangent bundle, in which the overlap functions are not linear transformations on a vector bundle but general homeomorphisms, and look for a topologically equivalent vector bundle structure; again, one finds certain topological obstructions to the existence of a smooth structure, and a certain set of discrete choices in the case that one exists.

In a path-integral approach to quantum gravity, one does not usually make restrictions on the space-time topology, except to require certain boundary conditions; for instance one might require that the initial and final spatial configurations have some particular topology. The consequences of this for the spectrum of topological defects were recently examined by Bais, Gomez, and Rubakov,[4] who found that the possible classes of defects are classified by cobordism: two 'soliton' spatial configurations could mix if they combine to form the boundary of a space-time manifold, or A$\sim$B if A+(-B)=boundary(M). In particular, a configuration is in the same class as the vacuum if it is a boundary already by itself; as a consequence, in this formalism all of the exotic sphere defects are topologically trivial. (As also pointed out in reference 4, if one restricts consideration to spin manifolds, there are additional stable configurations, some of which could be represented by exotic differential structures.)

In this paper we will work under the assumption that the correct approach to string theory does not allow the topology of the space-time to change. There are important reasons for skepticism about the correctness of this approach, or any other current formulation. Since no fundamental geometrical formulation of strings is known at present, or even known to exist, we do not have a rigorous framework for examining the stability of solitons or topological nontriviality. (This also applies, of course, to gauge-field defects.) What is worse in this case is that the natural size for topological defects is of order the Planck length, so that not only do we have to contend with unknown effects of string physics, but they are certainly important as well. In string theory there seems to be a tendency toward discrete mathematics, as evidenced in the classification of conformal field theory models and in the orbifold compactifications, so that one might worry that string configurations may not correspond to any smooth manifold, or only correspond in a different way. Finally, a more philosophical problem is that if we fix the space-time topology, we have no possibility of explaining how one Calabi-Yau manifold out of ten-thousand gets picked out.

Despite these difficulties I believe it is interesting to pursue this problem. For one thing, the possibility of defects in the smooth structure of the space-time manifold can be considered outside its application to string theory, and it would have subtle and interesting effects in classical gravity. There is also a somewhat speculative reason for thinking that some of the qualitative features of smooth-structure defects will apply to string theories. Since the Riemannian geometry on which the low-energy physics is based becomes ill-defined at short distances, one might attempt to replace it by some notion like those proposed for string field theory;[5] as in algebraic geometry, one replaces the points of a space by the set of functions vanishing at a particular point, from which the usual geometry can be recovered by purely algebraic ideas. In this framework, exotic differential structures might produce local perturbations in the algebra of smooth functions—thus, natural algebraically-defined defects in space-time.

As one might expect from analogous questions arising in gauge theories, there is a close relationship between topological defects and anomalies; for instance, the phenomena of fermion zero-modes on monopoles, strings, etc. can be understood through the same mathematical framework as anomalies. The motivation for the present work arises from the demonstration by Witten[3][6] of the cancellation of global gravitational anomalies in the point field theory approximations to string theories. These potential anomalies arise from topologically nontrivial diffeomorphisms of S^{10}, which can be studied by using them to construct exotic differential structures on S^{11}; one might expect them as well to be related to the existence of lower-dimensional topological defects, there being some nontrivial relations between exotic spheres of different dimensions.

We will now (finally) begin to examine the justification for exotic differential structures with a little mathematical background. Since in solving a problem it is important to start by asking the right question, we should make sure to ask our question in the right mathematical context; there is a tendency for physicists to blur the boundaries between different categories. As an admonitory example, consider one of the most well-known bits of necessary information, namely the fact that the nth homotopy group of n-sphere is Z, the group of integers under addition, generated by the identity map and classified by the winding number. This operation resides naturally in the topological category; if we require that the maps be differentiable, we find that the maps of unit winding number are not all equivalent, but can be further divided into classes according to the discrete components of the diffeomorphism group ($\pi_0(DIFF(S^n))$). These discrete components are then joined by paths corresponding to continuous maps which are not diffeomorphisms.

There are then two different categories one can deal with. Topological manifolds are defined by giving a topological space together with continuous maps from the manifold to R^n. Differential manifolds have the structure of topological manifolds, with the additional requirement that the composite map from R^n to R^n on adjoining coordinate patches be infinitely differentiable.

The Poincare Conjecture attempts to relate two other notions of identity for a manifold, postulating that if the homotopy group of a three- manifold is the same as that of a three-sphere, then there exists a continuous map to the three-sphere. It has so far resisted all efforts to prove or disprove it. Ironically, the generalization to higher (and lower) dimensions is known to be true. One can ask an analogous question in the category of differentiable manifolds, namely, does the existence of a continuous map from S^n to M^n imply the existence of a differentiable map? The answer, as perhaps one could guess from the warning about $\pi_n(S^n)$, is not necessarily, for n≥7; for example, in d=7,8,9,10,11 dimensions there are 28,2,8,6,992 distinct differential structures, respectively[7]. Therefore, topological manifolds and differentiable manifolds are not the same: in fact, some manifolds admit no smooth structure whatsoever. The situation in four dimensions is even stranger, and not well understood; Euclidean R^4 admits whole continuous families of distinct smooth structures, and compact 4-manifolds may have countably infinite sets of differential structures.[8]

To summarize what one expects to result from the mathematical analysis, one wants to obtain some way of classifying the possible differential structures, in order to know what to sum over. The basic procedure in the case of the n-sphere was described by Kervaire and Milnor,[7] in which case the set of differential structures forms an Abelian group. One wants as well to understand how to 'detect' exotic differential structures, by their effect on the wavefunctions of various fields, in order to understand the physical properties of the structures obtained in this way.

One cannot distinguish between different smooth structures by topological invariants, since the underlying manifold is the same by definition. One way of constructing invariants which describe the difference between different smooth structures is in terms of topological invariants of smooth manifolds which have the given smooth manifold as a boundary.[3] Another approach, and one more closely related to our physical goal, is to look for analytic invariants—effects of the smooth structure on the types of fields which can be smoothly defined on the given manifold. Some examples of this are the fermion zero-modes manifested by some of the 9- and 10-dimensional exotic spheres; there are also restrictions on the sorts of metrics which may be smoothly defined (an exotic sphere is never 'round').

In order to apply these ideas to the physics of compactified theories, we must understand how to construct exotic differential structures on manifolds other than the sphere. There is an action of the group of smooth structures of the n-sphere on the set of smooth structures of an n-manifold: cut out an ordinary disk from the manifold and from the sphere, and join them together at the boundary. This operation yields pointlike topological defects—instantons or 'lumps', depending on whether we start with the spacetime manifold or with some spatial section. As discussed by Witten[3] and Killingback,[9] there are five possible instantons and seven monopole-like defects which can be constructed in this way.

In general this action of the group of exotic spheres on manifolds of the same dimension does not exhaust all the possible smooth structures, and it may also be the case that some of the nontrivial smooth structures on S^n may be trivial on M^n. One generalization of this idea is to cut out some topologically nontrivial submanifold N^k, together with a neighborhood, and then to glue it back in with a twist: we replace the disk normal to N by one with a nontrivial smooth structure. In order to begin to understand what sort of defects are possible, we can properly represent the boundary conditions by looking at exotic smooth structures on $S^p \times S^q$; this construction was studied by Schultz[10] and by De Sapio[11] They found that all of the distinct differential structures could be described by combining an exotic q-sphere structure on S^q $(q \geq p)$ with an exotic (p+q)-sphere structure glued into the total space. One application of their work is that the exotic differential structures on S^7 are trivialized if we take the cross product with S^p, $p \geq 2$; we are then unlikely to find any nontrivial defects from the seven-spheres.* This work suggests that there may be three basic types of topological defects arising from exotic differential structures: those with total dimension 2, giving rise to stringlike, point or instanton defects (depending on the number of dimensions embedded internally), and arising from the 8-dimensional exotic sphere; those with total dimension 1, giving monopole or instanton defects, arising from one of the 7 9-spheres; or instantons (as discussed by Witten and Killingback) arising from one of the five 10-spheres. I will describe the classification of such defects more carefully in a paper currently in preparation.

Before ending this discussion I should emphasize that have not considered here whether one will find stable solitonlike solutions associated with these defects. In the context of the classical Einstein theory, one would not expect to find solutions of the field equations corresponding to a soliton at rest in empty space, as a consequence of the positive energy theorem and related considerations. It is interesting to consider the associated classical initial-value problem: if one starts out at t=0 with some metric on a background with nontrivial smooth structure, what is the subsequent evolution? Naively one expects that irregularities would radiate away, and at long times the configuration would approach a static black hole. In view of the higher-curvature terms in the string effective action, one might expect the soliton to stabilize at some finite size, of order the string scale.

* This result also suggests that compactification of d=11 supergravity on an exotic seven-sphere is indistinguishable from the ordinary seven-sphere compactification

Acknowledgements

I would like to thank my fellow participants at the ITP Program on Superstrings for many useful discussions. This research was supported in part by the U.S. Department of Energy under contract DEAC 03-81-ER40050, and by the NSF under Grant No. PHY82-17853, supplemented by funds from the National Aeronautics and Space Administration.

REFERENCES

1. Wen, X.-G. and E. Witten, *Nucl.Phys.* **B261** (1985), 651 ; Rohm, R. and E. Witten, "The antisymmetric tensor field in superstring theory", *Ann. Phys.* **170** (1986), 454.
2. Smale, S., "On the structure of manifolds", *Amer. J. Math.* **84** (1962), 387-399.
3. Witten, E., "Global Gravitational Anomalies", *Commun. Math. Phys.* **100** (1985), 197.
4. Bais, F.A., C. Gomez and V. A. Rubakov, "On the Global Stability of Gravitational Lumps", *Nucl. Phys.* **282** (1987), 531.
5. Witten, E., "Noncommutative geometry and string field theory", *Nucl. Phys.* **B268** (1986), 253.
6. Witten, E., "Topological Tools in Ten Dimensional Physics," in Unified String Theories, edited by M. Green and D. Gross, World Scientific, 1986.
7. Kervaire, M. and J. Milnor, "Groups of Homotopy Spheres I", *Ann. Math.* **77** (1964), 504-537
8. Donaldson, S.K., "An application of gauge theory to the topology of 4-manifolds", *J. Diff. Geo.* **18** (1983), 269-316; Freedman, M., "The topology of four-dimensional manifolds", *J. Diff. Geo.* **17** (1984), 357-454.
9. Killingback, T. P., "Topological Sectors in Supersymmetric String Theory," Princeton University preprint PUPT-86-1187, July, 1986.
10. Schultz, R. E., "Smooth structures on $S^p \times S^q$", *Ann. of Math.* **90** (1969), 187-198.

11. De Sapio, R., "Differential Structures on a product of spheres", *Comm. Helv. Math.* **44**(1969), 61; "Differential Structures on a Product of Spheres II", *Ann. of Math.* **89** (1969), 305-313.

V. Field Theory and Theory of the Strong Interactions

Conveners — P. MacKenzie and D. Soper

EVALUATION OF PERTURBATIVE QCD IN MONTE-CARLO EVENT GENERATORS

John C. Collins

Department of Physics, Illinois Institute of Technology
Chicago, IL 60616, U.S.A.*
and
Fermi National Accelerator Laboratory, Batavia, IL 60510, U.S.A.

ABSTRACT

I examine the extent to which Monte-Carlo simulations reproduce the predictions of perturbative QCD especially in the case of hadron-hadron scattering. Although the Monte-Carlos have great success in reproducing most of the qualitative features of the theory, they do not fully incorporate even the leading logarithmic approximation. Work is needed to give a systematic method for inclusion of both leading and non-leading effects. As the energy of accelerators gets higher, perturbative phenomena become increasingly independent of intrinsically non-perturbative phenomena; thus the extent to which Monte-Carlo simulations get the theory correct rather than merely parameterizing data increases in importance. Furthermore, at the higher energies there are increasingly important QCD background to new physics processes; thus it is essential to make quantitatively correct predictions.

1. INTRODUCTION

There are two schools of calculator in perturbative QCD: those who use the 'analytic'[1] methods and those who use the Monte-Carlo[2] methods. We may caricature these by saying that the analytic methods provide accurate calculations of (almost) nothing, while the Monte-Carlo calculations provide inaccurate calculations of everything. Particularly because they provide calculations of everything the Monte Carlos have become a favorite tool of experimentalists and many theorists for the comparison of QCD and the real world. But precisely because of their pervasiveness, it is also important to understand how well they actually reproduce the predictions of the theory, and to see how they can be improved.

This is what I will attempt to do in this paper. I will discuss only the Monte-

---

* Permanent address

Carlo event generators for hadron-hadron collisions. This is the area in which the Monte Carlos are least well developed. Moreover, because the highest energy accelerators are hadron-hadron colliders, it is there that the differences between the Monte Carlo calculations and the real QCD theory are likely to become most apparent. The review is aimed at both Monte-Carlo experts and physicists who are interested in the physics of QCD. I will refer to Webber's review[2] for the theory of the Monte Carlos and for references to the original literature.

Let us first remind ourselves of the main differences between the analytic and the Monte-Carlo methods.

1. The analytic calculations start from explicit factorization formulae for specific cross sections. Only certain kinds of cross section can be treated, mostly inclusive ones. The Monte Carlos start from a probabilistic algorithm that generates complete events.
2. There is a more-or-less complete proof of the analytic factorization formula for many processes. But at best the Monte-Carlo algorithms are proved at the leading logarithm level. In certain key aspects, the current versions of these algorithms do not even correctly include the leading logarithm approximation, as we will see later.
3. The perturbatively calculable parts of the analytic formulae, viz. the hard scattering cross sections and the evolution kernels, have systematic expansions in powers of the strong interaction coupling α_s. Thus analytic calculations can be improved in accuracy by using more terms in the expansions. However, there is no *systematic* method yet known for including higher order corrections in the Monte Carlos.
4. In many cases, large logarithms of ratios of kinematic variables mess up the convergence of perturbation expansions. Such a case is the Drell-Yan cross section at low transverse momentum. In the best cases, like Drell-Yan, the analytic formulae can be improved so that the large logarithms are effectively resummed; one really generates a new factorization formula in which the perturbative kernels no longer have large logarithms; the logarithms are generated by integrations over the kernels. By contrast, the Monte Carlos provide a leading logarithm resummation of such large logarithms. It is not always manifest that they do this correctly, and certainly one does *not* in general know how to make systematic improvements.

It is clear, certainly, that the Monte Carlos are very useful, especially when one needs a simulation of the typically complicated events that are produced in high energy collisions. Qualitatively they reflect important features of QCD. They are needed in the design of detectors[3,4], to see whether proposed designs will properly perform its required function when presented with actual events. A second use is to provide input to a simulated detector so that efficiency corrections, etc, can be estimated. Provided that the corrections are insensitive to the precise details of the events (e.g., multi-particle correlations), and that the Monte Carlo used is tuned to fit data, this is a legitimate use.

But when a Monte Carlo is more intimately involved in the physics analysis, as so often happens now, the Monte Carlo must accurately reflect the real QCD theory. The questions I will address are: (a) How valid are the Monte Carlos as calculations of QCD? (b) How much improvement is needed?

This paper is laid out as follows. First, in Sec. 2, I will explain the basis of the Monte-Carlo algorithms, and explain the particular difficulties with gauge theories. Then in Sec. 3, I will give a brief overview of the structure of a typical Monte-Carlo. In Sec. 4, I will show how the Monte Carlos fail to include the whole of the leading logarithm approximation. Sec. 5 will expose the areas in which improvements are needed. In the final section, I will summarize the outlook for the subject.

It should be noted that I will be taking the point of view of a devil's advocate. Lest my criticisms be considered harsh, it should be remembered that a good Monte Carlo is hard to write and maintain. Moreover considerable progress has been made in improving them and increasing the amount of real theory contained in them. In principle, they are not just models, but should be a valid method of calculation from the fundamental QCD hamiltonian.

2. WHY MONTE CARLOS WORK

First let us consider a model field theory (e.g., $(\phi^3)_6$, but not QCD) that has no gauge fields. Then it is relatively easy[5,6,7,8,9] to derive a factorization theorem for many interesting processes – Drell-Yan, jet production, etc. The structure of a typical factorization theorem is illustrated in Fig. 1. One parton out of each incoming hadron enters a hard scattering. Out of the hard scattering emerges some number of high p_T partons. These partons fragment into jets of hadrons. Let Q be a scale characteristic of the virtualities and transverse momenta of the hard scattering. Then the hard scattering can be calculated in a power series in the effective coupling $\alpha_s(Q)$. The distributions of partons in the incoming hadrons and of hadrons in the outgoing jets are non-perturbative. These distributions depend on the scale Q, and their Q-dependence is given by a renormalization group equation called the Altarelli-Parisi equation[10,11,12]. The kernel of this equation has a useful perturbative expansion in powers of $\alpha_s(Q)$. Corrections to the factorization formulae are higher twist, that is, suppressed by a power of Q.

The standard factorization theorems are for inclusive processes: one sums over many unobserved particles. However, one can derive further factorization theorems that apply to the jets. The result is summarized in the jet calculus[13]. One gets the picture summarized in Fig. 2. The partons coming into, or going out of, the hard scattering have lower virtualities than that of the hard scattering. One factors the associated jets into high and low virtuality pieces. Thus in Fig. 2, the lines have decreasing virtuality as one goes away from the hard scattering. The correct picture for the jets is of a cascade decay.

This picture leads immediately to a Monte-Carlo setup for the calculation[2], where one works with probabilities for decays. The jet calculus ensures that the kernels, the decay probabilities, etc can be expanded in powers of α_s at an

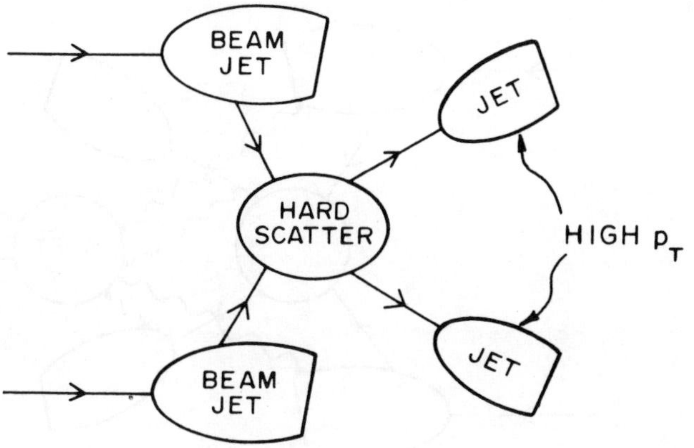

Fig. 1. Simple factorization theorem.

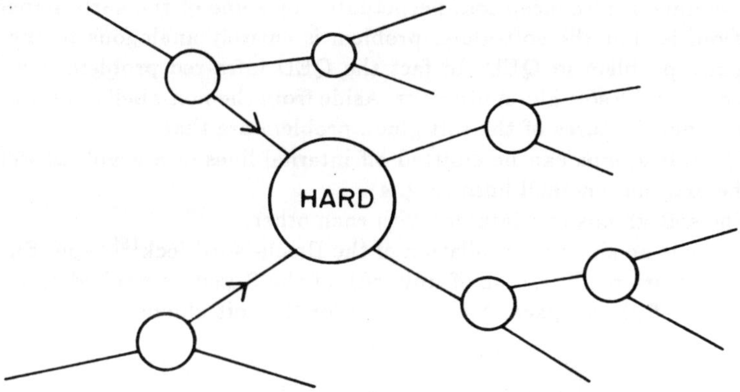

Fig. 2. Multiple factorization for jet calculus.

appropriate scale.

2.1 QCD and Soft Gluons

In QCD (or in any other field theory with elementary gauge fields), the basic hard scattering picture, Fig. 1 gets considerably modified.[8,14] At the leading twist level, soft gluons can attach to any of the jet factors in a graph (Fig. 3). What is meant by a soft gluon is one all of whose momentum components are much less that the total center-of-mass energy. Thus a 10 GeV gluon is soft when one is considering production of $\frac{1}{2}$ TeV jets at the Tevatron ($\sqrt{s} = 2\,\mathrm{TeV}$).

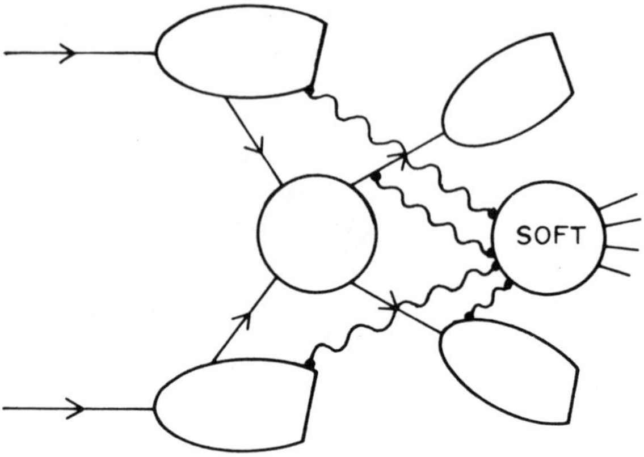

Fig. 3. Typical leading region for hard process in QCD.

A common misconception, perpetuated by some of the early papers on factorization, is that the soft gluon problem is entirely analogous to the infra-red divergence problem in QED. In fact the QED infra-red problem is a relatively *trivial* case of the soft gluon problem. Aside from the non-abelian nature of QCD, the main new features of the soft gluon problem are that

1. The soft gluons can be emitted off internal lines of a graph as well as from the external on-shell lines (Fig. 4).
2. The soft gluons can interact with each other.
3. There is no simple cancellation of the Bloch-Nordsieck[15] type. Such cancellations are really a case of unitarity of the S-matrix applied to a restricted region of phase space. More is need for the soft gluons.

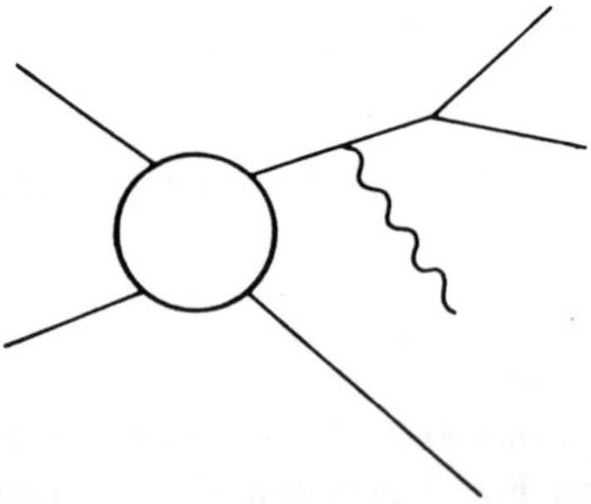

Fig. 4. Emission of soft gluons can be off internal lines.

The factorization theorems for hard processes can indeed be proved in QCD, but only after much labor. Most of the complete results are recent[16,17,18]. One needs to sum over attachments of soft gluons to jets. These sums involve the use of gauge invariance, as expressed in particular types of Ward identities. After these sums, one gets factors to which a unitarity sum can be applied to prove cancellation.

Fancier factorization theorems can be proved, e.g., for Drell-Yan at low transverse momentum[19], where the effects of the soft gluons do not cancel, but merely factorize.

At this stage, it should be clear that the iterated factorization such as is needed for the jet calculus and the Monte-Carlo algorithms do not hold as *simple* corollaries of the basic factorization theorem. If one uses the leading logarithm approximation and neglects azimuthal dependence in the showering, then a result is obtained[2] that fits in with a jet calculus structure, and can easily been turned into a Monte-Carlo algorithm.

The leading logarithm approximation is correct over the biggest regions of phase space for each graph.

3. STRUCTURE OF A MONTE-CARLO EVENT GENERATOR

The basic structure of a Monte-Carlo event generator can be summarized by Fig. 1, which represents a typical factorization formula for a *non-gauge* theory.

First one specifies a 'trigger' condition, that states what kind of hard scattering one is going to work with. This condition might be, for example, that one has a jet of greater than some specified transverse energy, E_{Tmin}. Or it might be that one has a Drell-Yan pair of invariant mass, Q, bigger than some value.

Then one generates the basic hard scattering with a probability distribution given by the hard scattering cross section convoluted with some parton distributions.

The partons associated with the beam jets and with the outgoing high p_T jets are generated by an probabilistic algorithm that implements the multiple-factorization that is the result of the jet calculus. This gives the showering associated with the initial- and final-state evolution.

The showering is stopped at some stage, and the partons in the final state are turned into hadrons. There are several methods used for the hadronization: independent fragmentation[20], string models[21], cluster algorithms[22].

Finally, some provision must be made for the fragmentation of the beam jets, ideally so that when no hard scattering occurs, a reasonable approximation to normal 'minimum bias' events is generated.

Monte Carlos do not always incorporate the state of the art in these areas. For example, independent fragmentation may be used, even though it is generally agreed that it is not a correct method in QCD[23]. Showering in the initial state has not been commonly included until recently[3,4]. Over a limited range of energy, it is often possible to tune up a Monte Carlo to agree with a set of data, even if the algorithm used is not correct.

4. WHERE THE MONTE CARLOS DON'T INCLUDE THE LLA

It can easily found out from, say, Webber's review[2] that there are two areas where none of the existing Monte Carlos do not even agree with the leading logarithm approximation. These both concern the azimuthal distributions of the partons produced in the showering. The first is concerned with the collinear fragmentation of partons, while the second concerns the physics of the soft gluons.

4.1 Collinear Decays

The problem of collinear decays can be explained by the example illustrated in Fig. 5 of a system of zero angular momentum that decays to two transversely polarized particles. Each of these then decays to two other partons. Let us work in the center-of-mass frame of the two gluons. Then the fact that the gluons have spin 1 restricts the azimuthal distribution of the final particles to be:

$$\frac{dN}{d\phi_a d\phi_b} \propto 1 + A\cos 2(\phi_a - \phi_b). \tag{1}$$

The coefficient A may have any value between $+1$ and -1. This happens even if the two decays are completely independent. This example embodies the physics of the 'paradox' of Einstein, Podolsky and Rosen. That A is in general non-zero is a purely quantum mechanical effect, and occurs even if the flight times are long so that the sequential decays can otherwise be thought of purely classically.

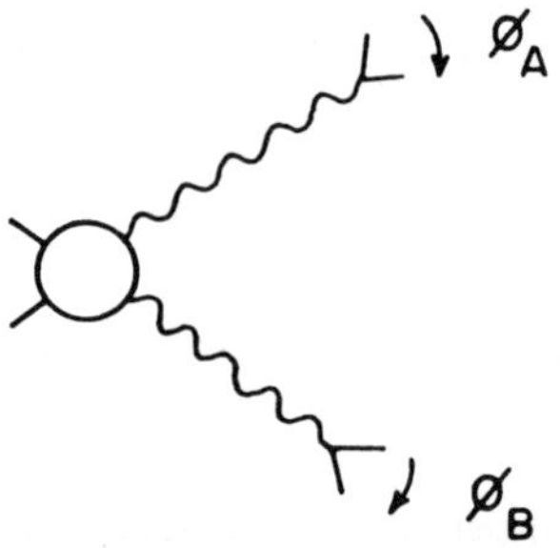

Fig. 5. Sequential decay that allows azimuthal correlations.

If one works in a field theory, then even in a kinematic region where the standard factorization theorem applies, A can be nonzero; indeed it can even take one of the limiting values ± 1. The effect is specific to spin-1 gluons: if the gluons were replaced by quarks, then the value of A would be of higher twist. Current Monte Carlos uniformly replace A by zero. This approximation can be dangerously bad whenever one uses the correlations between several particles to get information on the spin of a parent particle, as in the process:

$$\text{Higgs} \to W^+ + W^- \to 4 \text{ jets}. \tag{2}$$

(To detect this decay of the Higgs, one obviously must subtract out the background due to purely QCD processes that create four jets, and one would typically wish to discriminate against the background on the basis of spin information.)

4.2 Soft Gluon Emission

A second case where azimuthal correlations are neglected is in the emission of soft gluons. Suppose we have an initial parton P decaying into two collinear partons A and B, with an angle θ between them. Now consider a soft gluon emitted off the $A + B$ system; we let the angles between the soft gluon and the partons A and B be θ_A and θ_B. Now the soft gluon cannot be uniquely associated with A or B, so that a priori one cannot use the simple showering algorithm that would be correct in a nongauge theory. However,[2] the probability P for the process may written as the sum of two terms, P_A and P_B, which have the following property after azimuthal averaging: The average of P_A (P_B) is the probability for collinear emission of the gluon off A (B) with θ_A (θ_B) restricted to be less than θ. This result[2] enables the soft gluon emission, with the neglect of azimuthal correlations, to be incorporated in a standard Monte Carlo very easily.

5. BEYOND THE LLA

5.1 Higher Order Corrections

Consider the integral:

$$\int_0^\infty dx \, \frac{Q}{(x+m)\,(x+Q)} \; = \int_m^Q \frac{dx}{x} \; + \; \text{finite (as } Q \to \infty). \qquad (3)$$

This is a model of the momentum integral in a typical Feynman graph for a hard process, with x representing the virtuality of one of the lines. The first term on the right is the leading logarithm approximation. Since the finite remainder arises from the endpoints of the integration, one expects to be able to absorb the remainder into higher order corrections to the hard scattering and to the evolution kernels.

These corrections will evidently be particularly significant when one goes out on the tails of distributions and thereby samples the endpoints of the integrations. Consider for example Fig. 6[24], which gives the aplanarity distributions in e^+e^- annihilation at 34 GeV. Both the data and two calculations are plotted. One calculation is the full α_s^2 result for this distribution, while the other is obtained from a Monte Carlo calculation, using only the LLA. The LLA is evidently substantially inaccurate.

Since the large aplanarity tail is where one might look for new physics (e.g., a top quark), one must include the higher order corrections. Similarly, if one wishes to use a Monte Carlo for measuring Λ from data, one must include non-leading terms.

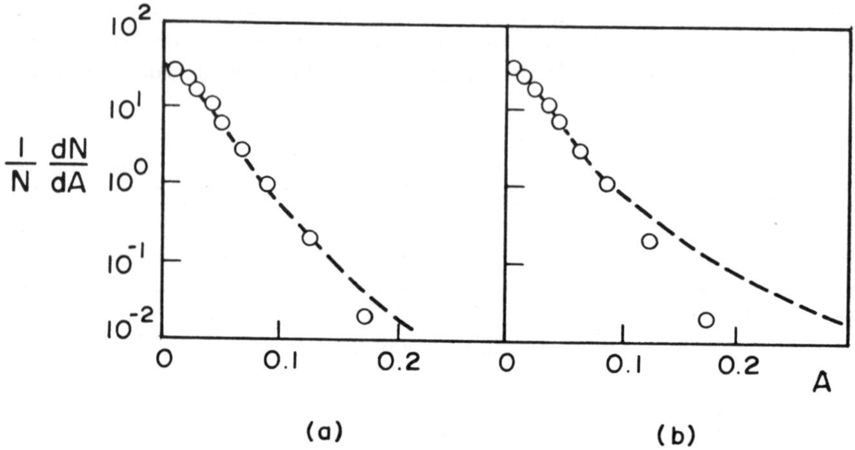

Fig. 6. Aplanarity distribution in e^+e^- annihilation: theory and experiment. The dots are the data at 34 GeV. The dashed line in graph (a) is the result of a full $O(\alpha_s^2)$ calculation; the dashed line in graph (b) is the result of a LLA shower calculation.

The problem in incorporating higher order corrections is that one must avoid double counting. In the analytic methods, given a high order graph, there is a well-defined procedure for subtracting off the contribution that the graph makes in the LLA. Before obtaining the final result there are many cancellations of soft gluon effects.

There is no known *systematic* method of doing this for the Monte-Carlo algorithms. One problem is that after subtraction the remainder is not always positive. A negative probability cannot be directly incorporated into a Monte Carlo.

5.2 Soft Gluons

Since soft gluons cannot really be uniquely assigned to jets, except by convention, one really needs a better treatment than the one usually given[2]. There are issues here that transcend the area of higher order corrections just discussed. In the analytic methods, even when they are applied to soft gluon emission, one works with an inclusive cross section that is not sensitive to the fate of the soft gluons, but at most to their effects on the hard partons (e.g., their transverse momentum distributions). One cannot afford this luxury in the Monte Carlos.

One particular problem is that after emission, the soft gluons can interact among themselves.

This and related issues were raised by Brodsky[25].

Since there is a substantial amount of technology already available for treating soft gluons, as in the Drell-Yan process, one should be able to make progress here. It is sometimes said that, since there are many interference effects associated with soft gluons, it is impossible to treat them with a Monte Carlo. But cross sections are always positive, so it is not a priori impossible to develop a sophisticated Monte Carlo that correctly includes interference.

5.3 Beam Jets and Hadronization

Inevitably one must choose a cutoff Q_0 below which one does not use perturbative methods, but has some parameterization of the nonperturbative physics. Ideally, the results of a calculation should be independent of this cutoff. It follows that models of the nonperturbative physics should be compatible with the perturbative physics. For example, in the analytic methods, the simplest factorization theorems force the nonperturbative physics to be in the form of parton distributions and fragmentation functions.

Certain models are thereby excluded, for example independent fragmentation[23]. The structure of perturbative QCD gives results that are much more like a string model.

5.4 Beam Jets

One must convert the remnants of the incoming hadrons into some part of the final state. This is mostly nonperturbative physics, but not entirely, as we will see. A substantial contribution to event structure is made by the beam jets. A simple way of making them is to generate a distribution of particles that is uniform in rapidity and is Gaussian in transverse momentum, with a fixed width. Such a parameterization (which is conveniently labelled as single Pomeron exchange) disagrees with experiment. To get the large fluctuations observed, one must allow a variable number of exchanged Pomerons.

One can get a perturbative model[26] for the multi-Pomeron exchange by permitting hard scattering into jets to be used down to low transverse momentum (around 1.5 GeV). There is a large enough probability of production of such minijets that one must allow for more than one hard scattering occurring simultaneously in the same hadron-hadron collision. The ladder graphs for hard scattering to minijets provide a perturbative model for the Pomeron that has many good properties[27]. As Sjöstrand[26] showed, the multiple hard scattering model is able to reproduce many features of the data – e.g., forward-backward correlations, KNO multiplicity distribution.

However, it appears that to fit the data, the probability of a double hard scattering must be rather higher than one's first guess; that is, the partons are confined in a rather smaller space in a hadron than the naïve radius. The current version of ISAJET[28] incorporates a phenomenologically similar fudge in its multiple Pomeron model to fit the activity associated with hard scattering.

There is in fact a reasonable rationalization of this effect. To get a hard scattering, the incoming hadrons must have a smaller impact parameter than normal. Thus the condition that there be a hard scattering biases the event towards more overlap of the hadrons and thus to a large number of exchanged

Pomerons.

Clearly, further work should have the aims both of improving the perturbative treatments and of meshing these with the nonperturbative models for the Pomeron. These perturbative and nonperturbative treatments will be compatible in the sense defined earlier.

6. OUTLOOK

The Monte Carlo event generators clearly incorporate much QCD physics that is correct qualitatively. Comparison with experiment has forced the authors of the Monte Carlos to include known physics that was initially missing from the programs.

At collider energies there are many quantitative details, e.g., jet profiles, that are governed by perturbative physics. But much of this detail is also soft physics. Thus at $\sqrt{s} = 2$ TeV, a gluon of E_T is both perturbative and soft.

Since at high energies there is much perturbative controllable physics, one should aim at better quantitative results in the Monte Carlos. For example, one needs:

1. Systematic methods to included higher order corrections (also known as non-leading logarithms).
2. In particular, one must do this for the soft gluon emission.
3. The fudge on the azimuthal distributions must be removed.
4. There must be a much more systematic derivation of the underlying algorithms.
5. The physics of the beam jets should be better tied to perturbative Reggeon physics.

ACKNOWLEDGMENTS

This work was supported in part by the Department of Energy and by the John Simon Guggenheim Memorial Foundation. I would like to thank many colleagues for extensive discussions on the subject matter treated, especially T.D. Gottschalk, A.H. Mueller, F.E. Paige, D.E. Soper, and B.R. Webber.

LITERATURE CITED

1] E.g., Mueller, A.H., Phys. Rep. 73, 237 (1981).

2] Webber, B., in Annual Reviews of Nuclear and Particle Science, (1986), and references therein.

3] Proceedings of 1984 DPF Summer Study of the Design and Utilization of the Superconducting Super Collider, Donaldson, R., Morfin, J., (eds.),

4] Proceedings of 1986 DPF Snowmass Summer Study, Donaldson, R., (ed.).

5] Efremov, A.V. and Radyushkin, A.V., Teor. Mat. Fiz. 44, 17 (1980) [Eng. transl.: Theor. Math. Phys. 44, 573 (1981)].
6] Efremov, A.V. and Radyushkin, A.V., Teor. Mat. Fiz. 44, 157 (1980) [Eng. transl: Theor. Math. Phys. 44, 664 (1981)].
7] Amati, D., Petronzio, R., and Veneziano, G., Nucl. Phys. B140, 54 (1978) and B146, 29 (1978).
8] Libby, S. and Sterman, G., Phys. Rev. D18, 3252, 4737 (1978).
9] Ellis, R.K., Georgi, H., Machacek, M., Politzer, H.D. and Ross, G.G., Nucl. Phys. B152, 285 (1979).
10] Gribov, V.N. and Lipatov, L.N., Sov. J. Nucl. Phys. 46, 438 (1972).
11] Altarelli, G. and Parisi, G., Nucl. Phys. B126, 298 (1977).
12] Johnson, P.W. and Tung, W.-T., Phys. Rev. D16, 1769 (1977).
13] Konishi, K., Ukawa, A. and Veneziano, G., Nucl. Phys. B157, 45 (1979).
14] Efremov, A.V. and Radyushkin, A.V., Teor. Mat. Fiz. 44, 327 (1980). [Eng. transl.: Theor. Math. Phys. 44, 774 (1981)].
15] Bloch, F. and Nordsieck, A., Phys. Rev. 52, 54 (1937); Yennie, D., Frautschi and S.C., Suura, H., Ann. of Phys. 13, 379 (1961); Kinoshita, T., J. Math. Phys. 3, 650 (1962); Lee, T.D. and Nauenberg, M., Phys. Rev. 133, 1549 (1964).
16] Collins, J.C. and Sterman, G., Nucl. Phys. B185, 172 (1981).
17] Bodwin, G., Phys. Rev. D31, 2616 (1985).
18] Collins, J.C., Soper D.E. and Sterman, G., Nucl. Phys. B250, 199 (1985).
19] Collins, J.C., Soper, D.E. and Sterman, G., Nucl. Phys. B261, 104 (1985).
20] Field, R.D. and Feynman, R.P., Nucl. Phys. B136, 1 (1978).
21] Andersson, B., Gustafson, G., Ingelman G. and Sjöstrand, T., Phys. Rep. 97, 33 (1983).
22] Webber, B.R., Nucl. Phys. B238, 492 (1984).
23] Cox, B., Proceedings of the Oregon Workshop on Super High Energy Physics, 120 (1985); Sjöstrand, T., Z. Phys. C26, 93 (1984); Phys. Lett. 142B, 420 (1984); Gottschalk, T.D., Proceedings of the Oregon Workshop on Super High Energy Physics 94 (1985); Corcoran, M.D., Phys. Rev. D32, 592 (1985).
24] Gottschalk, T.D., in Ref. 4.
25] Brodsky, S.J., in Ref. 3.
26] Sjöstrand, T., Fermilab preprint FERMILAB-PUB-85/119-T (1985).
27] Gribov, L.V.,Levin, E.M. and Ryskin, M.G., Phys. Rep. 100, 1 (1983).
28] Paige F.E. and Protopopescu, S.D., Proceedings of the UCLA Workshop on SSC Physics (1986).

PHOTONS IN HARD COLLISIONS : AN OPTIMIZED QCD APPROACH

Patrick Aurenche
LAPP, BP. 909, 74019 Annecy-le-Vieux Cedex
FRANCE

The work reported here has been done in collaboration with R. Baier, P. Chiappetta, A. Douiri, M. Fontannaz and D. Schiff.

1. INTRODUCTION

In the following we discuss the case of hard hadronic collisions where at least one photon is involved in the initial and/or the final state. The processes fall in three classes:

- prompt photon reactions: $h+h' \to \gamma+X$, $h+h' \to \gamma+\gamma+X$
- photoproduction reactions: $\gamma+h \to h'+X$, $\gamma+h \to \gamma+X$, $\gamma+h \to \gamma+h'+X$
- photon-photon reactions: $\gamma+\gamma \to h+X$.

The study of this processes is interesting for many reasons. In hard collisions, the photon couples to the quarks taking part in the hard subscattering so that an identification of the processes at the parton level is possible in a more direct way than in purely hadronic reactions. Furthermore, the theoretical predictions are more reliable since they involve less structure functions. It is also important to remark that these reactions are related to one another by crossing symmetry at the parton level so that the comparison between theory and experiment for all these processes will provide a very stringent test of QCD. Besides the above general properties several specific reasons make the study of these reactions particularly valuable. For example, when only one photon is involved, the cross-sections are proportional to the strong interaction coupling α_s and they are therefore very sensitive to the value of Λ. Also, some of them are linear in the gluon distribution in the hadrons $F_{G/h}(x,M)$ which is, consequently, directly related to the p_T dependence of the cross-sections. (We recall that Λ and $F_{G/h}(x,M)$ are usually determined from Deep

Inelastic Scattering or Drell-Yan type processes, where they enter as non leading effects and are therefore not well constrained there.) Furthermore photoproduction reactions can be used to constrain Λ and the gluon fragmentation function. The reactions with two photons are used to measure the quantity Σe_q^4, where e_q is the quark charge, which allows to distinguish the Integrally Charged Quark (ICQ) models from the Standard (QCD) model. The recent NA14 data[1] on $\gamma+N \to \gamma+X$ definitely rule out ICQ models.

Concerning theoretical predictions it is well known that there are ambiguities in the results of perturbation theory. Indeed when one calculates higher-order terms divergencies arise which can be removed at the cost of introducing arbitrary constants : the renormalization scale μ which appears in $\alpha_s(\mu/\Lambda)$ and the factorization scale M which enters the structure function $F_{i/h}(x,M)$. The calculated differential cross-section then becomes a function of these unphysical parameters: $d\sigma/dyd\vec{p}_T\ (s,\ p_T,\ y;\ \mu/p_T,\ M/p_T)$.

In the following, we shall briefly describe the calculation[2),3)] and discuss in some details the rôle of μ and M. Various ways to remove the ambiguities associated to μ and M will be presented: this is the object of the optimization procedure. The phenomenology of the reactions $h+h' \to \gamma+X$ and $\gamma+h \to h'+X$ will then be reviewed. In a final section the relationship between optimization and summation of higher order terms will be illustrated using Drell-Yan processes as an example.

2. THE PERTURBATIVE RESULTS

We discuss in the following the hadroproduction of a single photon at large p_T (the case of photoproduction is similar). At the lowest order, two classes of diagrams contribute: q-$\bar{q}$ annihilation (Fig. 1-a)

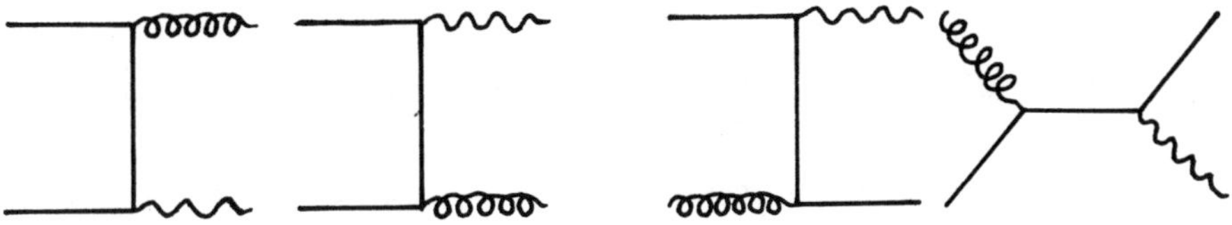

a) b)

Fig. 1

and "QCD Compton" (Fig. 1-b). The differential cross-section, in the parton model, is written as

$$d\sigma = \alpha \frac{\alpha_s}{\pi} \sum_{i,j} \int dx_1\, dx_2\, F_{i/h}(x_1)\, F_{j/h'}(x_2)\, d\hat{\sigma}^{ij} \tag{1}$$

where $F_{i/h}(x)$ is the density of parton i in hadron h and x is the fraction of momentum of h carried by i. It is interesting to note that for $\bar{p}p$ or π^-p the annihilation diagrams are dominant (for $x_T = 2p_T/\sqrt{s} > .2$) whereas they are negligible in the case of pp collisions (for the whole x_T range): the latter case can thus provide a good measurement of the gluon distribution in the proton.

At the second order in α_s, the diagrammatic structure becomes much more complicated. One can distinguish four classes of diagrams:

- Radiative corrections to "QCD Compton" scattering: the corresponding real and virtual diagrams are shown in Fig. 2-a,b. (In b) the diagrams with self energy loops on the external legs have been omitted.)
- Radiative corrections to q-$\bar{q}$ annihilation: the real diagrams are shown in Fig. 2-c, and the virtual ones are obtained from Fig. 2-b by crossing.
- Gluon-gluon fusion (Fig. 2-d).
- Bremstrahlung diagrams in q-q scattering (Fig. 2-e).

Note that the triple-gluon coupling appears and the results will then be a test of the non-abelian nature of QCD. The calculation is performed using the dimensional regularization method $(n = 4-2\varepsilon)$ with vanishing parton masses $(m_q = m_G = 0)$.

When evaluating loop diagrams such as those of Fig. 3-b, ultraviolet divergencies arise: they manifest themselves as term proportional to $1/\varepsilon$ multiplying the corresponding lowest order diagram. Keeping only relevant pieces the diagrams of Fig. 3 lead to the following results

$$d\hat{\sigma} \sim \alpha \frac{\alpha_s}{\pi} \left(d\hat{\sigma}^{q\bar{q}} + \frac{\alpha_s}{\pi} \left(\frac{(4\pi)^{\varepsilon}}{\varepsilon\Gamma(1-\varepsilon)} \left(1 + \varepsilon \ln \frac{\mu^2}{p_T^2}\right) \frac{b}{2}\, d\hat{\sigma}^{q\bar{q}} + f \right)\right) \tag{2}$$

with $b = (11N_c - 2N_f)/6$ the coefficient of the first term in the β-function expansion, μ an arbitrary scale introduced to give the

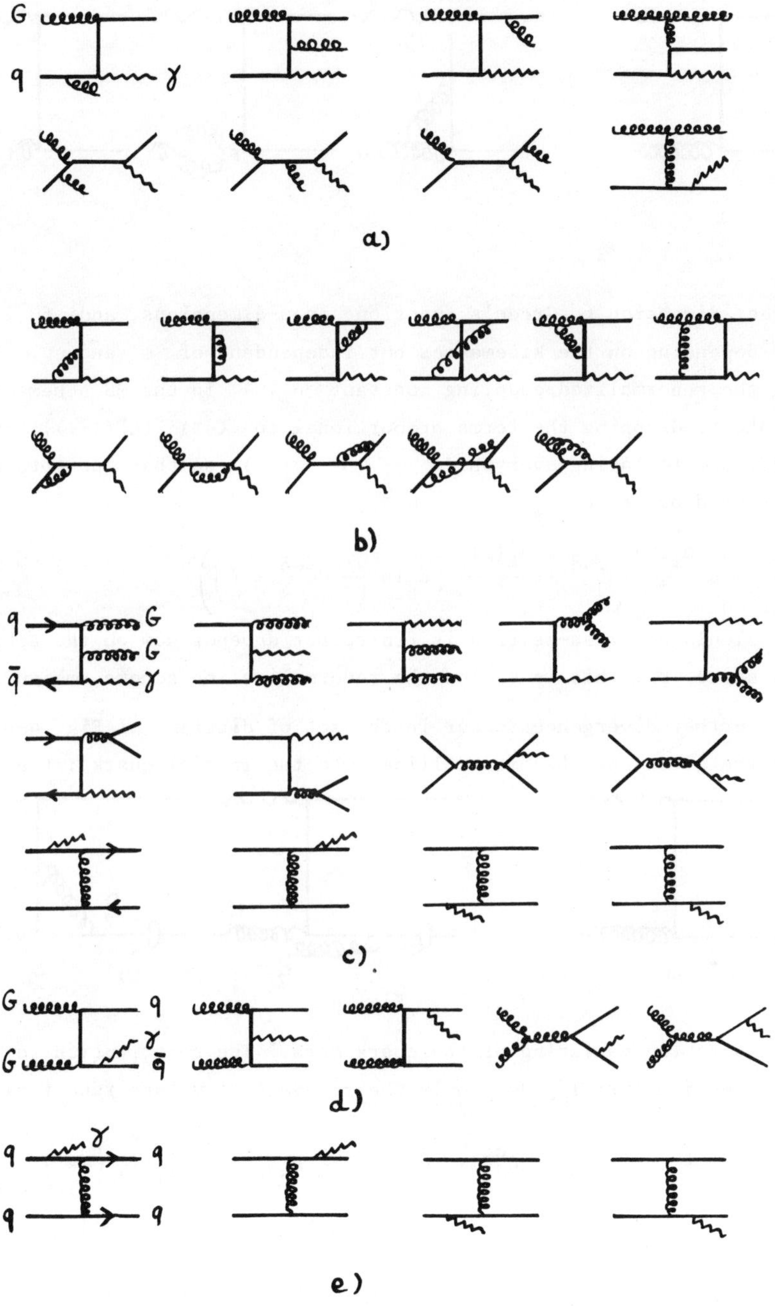

Fig. 2

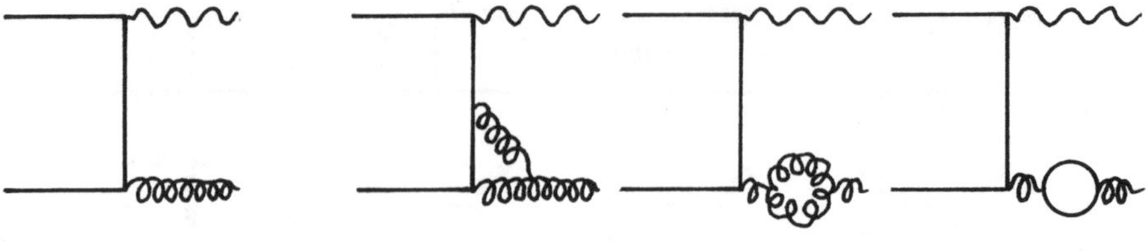

a) b)

Fig. 3

correct dimension to Green's functions in n-dimensions, and f is a term depending on the kinematics but independent of ε and μ. Defining the renormalized coupling constant $\alpha_s(\mu)$ in the $\overline{MS}$ scheme amounts to dropping the terms proportional to $(4\pi)^\varepsilon/(\varepsilon\Gamma(1-\varepsilon))$ in Eq.(2) and replacing everywhere α_s by $\alpha_s(\mu)$ so that one obtains at the second order in α_s

$$d\hat{\sigma} \sim \alpha \frac{\alpha_s(\mu)}{\pi}\left(d\hat{\sigma}^{q\bar{q}} + \frac{\alpha_s(\mu)}{\pi}\left(b \ln \frac{\mu}{p_T}\, d\hat{\sigma}^{q\bar{q}} + f\right)\right) \tag{3}$$

The calculated cross-section is finite but depends now on the arbitrary mass scale μ : this is called the renormalization scheme ambiguity[4-7].

Further divergences occur in the set of diagrams of Fig. 4-b, when the gluon p_2 becomes collinear to the initial quark : they are

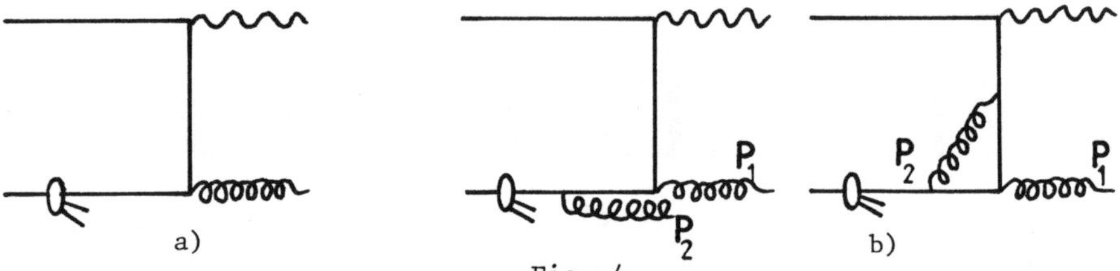

a) b)

Fig. 4

related to the vanishing of the quark mass. The contribution to the cross-section (writing down only the relevant structure function) is

$$d\sigma \sim \alpha \frac{\alpha_s(\mu)}{\pi} \int dx \left[F_{q/h}(x)\, d\hat{\sigma}^{q\bar{q}} + \frac{\alpha_s(\mu)}{\pi} \int \frac{dz}{z} F_{q/h}(z) \left(\frac{(4\pi)^\varepsilon}{\varepsilon\Gamma(1-\varepsilon)} \left(\frac{\mu^2}{p_T^2}\right)^\varepsilon \cdot \frac{1}{2} P_{qq}\left(\frac{x}{z}\right) d\hat{\sigma}^{q\bar{q}} + g\left(\frac{x}{z}\right)\right)\right] \tag{4}$$

where $P_{qq}(x)$ is the usual Gribov-Lipatov-Altarelli-Parisi splitting function. As usual one defines the "dressed" structure function

$F_{q/h}(x,M)$, with scaling violations set by the arbitrary scale M, by

$$F_{q/h}(x,M) = \int \frac{dz}{z} F_{q/h}(z) \left[\delta(1-\frac{x}{z}) - \frac{\alpha_s(\mu)}{\pi} \frac{(4\pi)^{\varepsilon}}{\varepsilon\Gamma(1-\varepsilon)} \left(\frac{\mu^2}{M^2}\right)^{\varepsilon} \frac{1}{2} P_{qq}(\frac{x}{z})\right] \quad (5)$$

Introducing $F_{q/h}(x,M)$ into eq.(4) one recovers a finite result:

$$d\sigma \sim \alpha \frac{\alpha_s(\mu)}{\pi} \int dx \frac{dz}{z} F_{q/h}(z,M) \left[\delta(1-\frac{x}{z}) + \frac{\alpha_s(\mu)}{\pi} \left(P_{qq}(\frac{x}{z}) \ln \frac{p_T}{M} d\hat{\sigma}^{q\bar{q}} + g(\frac{x}{z})\right)\right] \quad (6)$$

The arbitrariness in M reflects the factorization scheme ambiguity[8-9].

One is now ready to give the general form of the cross-section at $\mathcal{O}(\alpha^2)$

$$\frac{d^3\sigma}{dy d\vec{p}_T} := \sigma^{(2)} = a(\mu)F(M) \otimes F(M) \otimes \left[\hat{\sigma} + a(\mu)\left(b \ln \frac{\mu}{p_T} \hat{\sigma} + 2P_{qq} \ln \frac{p_T}{M} \otimes \hat{\sigma} + \mathbb{C}\right)\right] \quad (7)$$

For simplicity we have devoted the convolutions by the symbol $\otimes$, while the notation $F(M)$ stands for $F_{q/h}(x,M)$ and $a(\mu)$ for $\alpha_s(\mu)/\pi$. We have kept in eq.(7) only the contribution arising from valence q and $\bar{q}$ (non singlet component). Similar expressions can of course be written for general initial states. The quantity $\mathbb{C}$ is finite, independent of μ and M. The dependence of the higher order correction

$$r_1\left(\frac{\mu}{p_T}, \frac{M}{p_T}\right) = b \ln \frac{\mu}{p_T} \hat{\sigma} + 2P_{qq} \ln \frac{p_T}{M} \otimes \hat{\sigma} + \mathbb{C} \quad (8)$$

on the arbitrary parameters is explicitly written out. The dependence on μ and M of the full cross-section is then obtained once we recall the usual evolution equations[10] for

- the coupling: $da(\mu)/d\ln\mu = -ba^2(\mu)(1+ca(\mu))$, $c = \frac{153-19N_f}{2(33-2N_f)}$ (9)

and

- the structure function: $dF(M)/d\ln M = a(M)\, P_{qq} \otimes F(M)$ (10)

Turning back to eq.(7) and considering only the leading logarithm terms $\sigma^{(1)} = a(\mu)F(M) \otimes F(M)\hat{\sigma}$, one finds that it is monotonous in μ and M : it decreases when μ or M increase (one recalls that P_{qq} is effectively negative in the kinematical range of interest). However the higher order term $r_1(\mu,M)$ obviously increases under the same conditions so that the variation of $r_1(\mu,M)$ approximately compensates that of $\sigma^{(1)}$ and the cross-section calculated at $\mathcal{O}(a^2)$ becomes more stable. Although it can be shown from eqs (7)-(9) that the variation of $\sigma^{(2)}$ with respect to changes in μ and M is of order a^3, nevertheless in some regions of the parameter space the variation may be rather large. This is illustrated in the following figures. In Fig. 5-a, we show the behaviour of $d\sigma/(dyd\vec{p}_T)$ for $p+p \to \gamma+X$ at $\sqrt{s}$ = 63 GeV, p_T = 10 GeV/c and $y = 0$ in the range $.03 < a(\mu) < .11$ and $.1 \leq C_M = M^2/p_T^2 \leq 10$. At small M values the cross-section goes through a maximum shown by the dark "plaquette". When M is varied the line of maxima in $a(\mu)$ (the ridge line) goes through a minimum as shown by the darkened area which corresponds to a saddle point of the surface. Similar features occur for $d\sigma/(dyd\vec{p}_T)$ in $p+\bar{p} \to \gamma+X$ at $\sqrt{s}$ = 630 GeV and p_T = 45 GeV/c with the qualitative difference that the variations are somewhat less pronounced than in the previous case.

Perturbation theory does not specify which is the best prediction since μ and M are not determined beyond the vague requirement that they should be "of the order of the large mass scales involved in the process". To make a unique prediction requires going beyond perturbation theory as it is done in the optimization method.

3. OPTIMIZED PERTURBATIVE QCD

In the following we shall discuss several approaches starting with the method initiated by Stevenson[6] concerning the renormalization scheme ambiguity and extended by Politzer[8] to solve the factorization scheme ambiguity. The previous authors observe that the physical cross-section, as well as the theoretical one evaluated to all orders of perturbation theory, are independent of the scales μ and M. However the perturbative series truncated at the second order (or any finite order) does depend on μ and M. The "best" prediction, according to Stevenson and Politzer, is the prediction for which the calculated cross-section enjoys the same properties as the physical cross-section, namely independence on the unphysical parameters. This leads

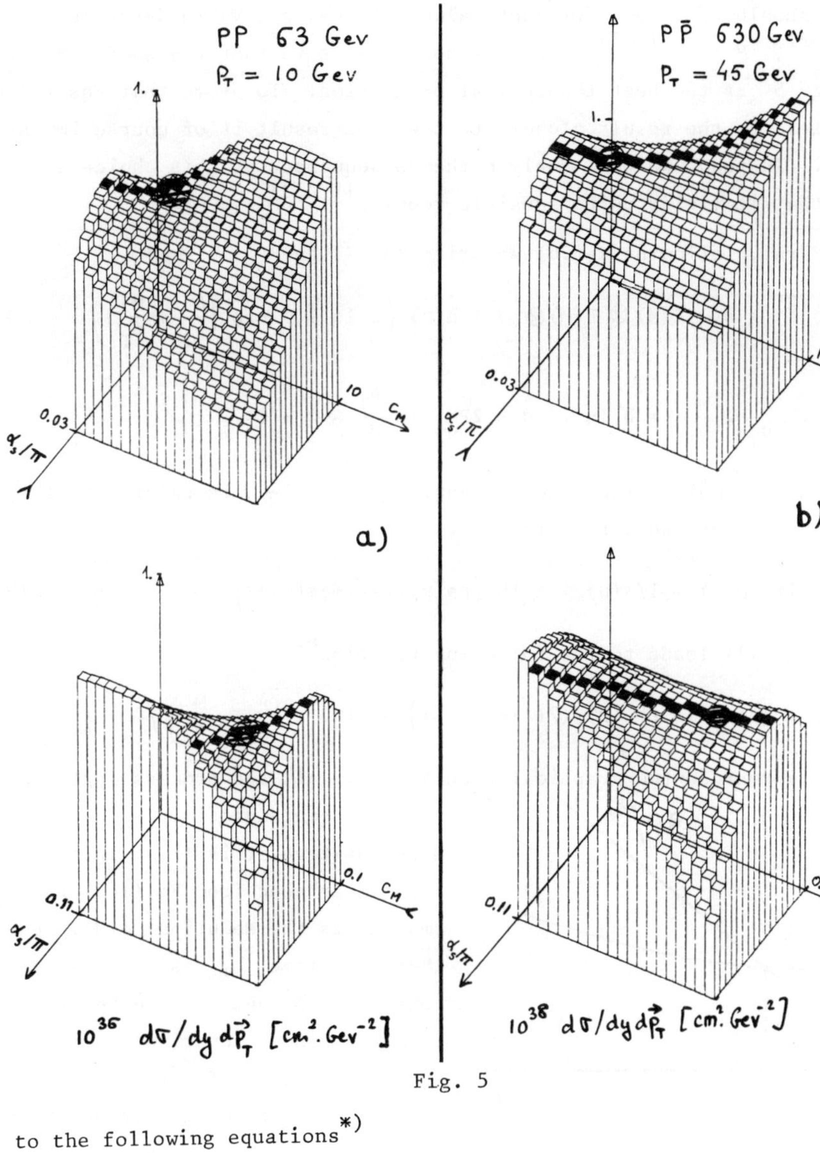

Fig. 5

to the following equations*)

$$d\sigma^{(2)}/d\ln\mu = 0 \tag{11}$$

$$d\sigma^{(2)}/d\ln M = 0 \tag{12}$$

---

*) One could also optimize with respect to the factorization prescription[8]. We use here the non-universal convention for the quarks and the universal convention for the gluon.

which should be solved for each values of $\sqrt{s}$, p_T, y to determine $\mu = \mu^{OPT}$, $M = M^{OPT}$ [11]. This method selects in fact the saddle point of Fig. 5 as the best theoretical prediction. To prove that Eqs (11)-(12) lead to the result closest to the true result is of course impossible. Instead, we will analyse the consequences of this choice and show that it yields quite sensible results[11,2,3].

First, following Ref.6) we write eq.(7) as

$$\sigma^{(2)} = a(\mu)F(M) \otimes F(M) \otimes \left[\hat{\sigma} + a(\mu)\left(b \ln \frac{\mu}{\Lambda}\hat{\sigma} - \rho_1\left(\frac{\Lambda}{p_T}, \frac{M}{p_T}\right)\right)\right] \quad (13)$$

with

$$\rho_1\left(\frac{\Lambda}{p_T}, \frac{M}{p_T}\right) = b \ln \frac{p_T}{\Lambda}\hat{\sigma} - 2P_{qq} \ln \frac{p_T}{M} \otimes \hat{\sigma} - \mathbb{C} \quad (14)$$

being a renormalization scheme invariant. $\Lambda \equiv \Lambda_{\overline{MS}}$ is related to the coupling by the usual formula

$$b \ln (\mu/\Lambda) = 1/a(\mu) + c \ln \left(ba(\mu)/(2(1+ca(\mu)))\right) \quad (15)$$

Equation (11) leads to the following relation[6]

$$1 + ca(\mu) \ln \left(ba(\mu)/(2(1+ca(\mu)))\right) = a(\mu)\, \rho_1\left(\frac{\Lambda}{p_T}, \frac{M}{p_T}\right) \Big/ \hat{\sigma} \quad (16)$$

which is represented graphically on Fig. 6. All the process dependence is contained in $\rho_1/\hat{\sigma}$. The more positive it is (i.e. large p_T, small M as seen from Eq.(14)) the smaller is the solution $\bar{a}(M)$. Furthermore since ρ_1 as well as the constants b and c are renormalization scheme invariant so is $\bar{a}(M)$: the value of $\bar{a}(M)$ does not depend on the renormalization scheme as it is expected of a physical quantity.

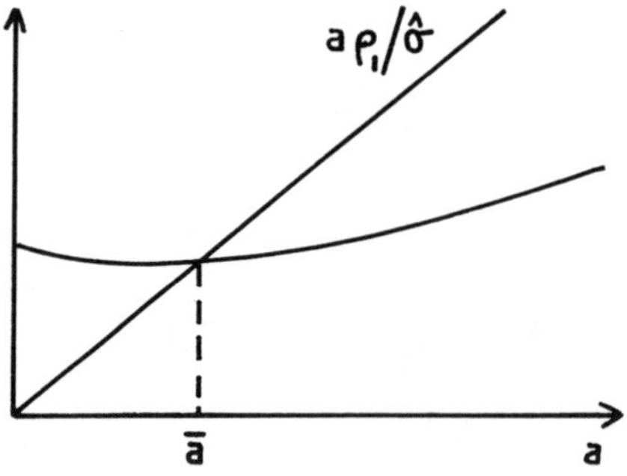

Fig. 6

Plugging eq.(16) into eq.(13) one obtains the result:

$$\sigma^{(2)} = \bar{a}(M)\; F(M) \otimes F(M) \otimes \hat{\sigma} \left(1 - \bar{a}(M) \frac{c/2}{1 + c\bar{a}(M)}\right) \quad (17)$$

The optimization with respect to M is then done numerically with

$M = M^{OPT}$ being determined by the minimum of the RHS of eq.(17). At the optimum point, the cross-section takes a very simple form being proportional to the Born cross-section (evaluated at the scales μ^{OPT} and M^{OPT} which contain all the process dependence). Furthermore the higher order corrections are small, at most 15%[11]. The saddle point, selected by the optimization, leads to a perturbative series with a good apparent convergence.

As a variation on the method with two scales described above one can, before applying the optimization procedure, eliminate the scale M in favour of μ using the renormalization group equations (i.e. set $M = \mu$ in eq.(7)) and then apply eq.(11). It can be shown that the resulting cross-section is always larger than in the two scale problem within the following inequality $(\sigma^{OPT}(1\text{ scale}) - \sigma^{OPT}(2\text{ scale}))/\sigma^{OPT} \leq 10\%$ for all real photon reactions considered[11].

An alternative to the Stevenson-Politzer method is the approach advocated by Grunberg[5]: keeping only the scale μ, one introduces a scheme invariant effective charge so that the cross-section takes the form

$$\sigma^G = a^{eff}(\mu_G)\; F(\mu_G) \otimes F(\mu_G) \otimes \hat{\sigma} \tag{18}$$

i.e. the scale is chosen so that the higher order correction vanishes (fastest apparent convergence). One can show that the results are very close to the previous ones: indeed one finds $|\sigma^G - \sigma^{OPT}(2\text{scales})|/\sigma^{OPT} \leq 10\%$ for all the reactions considered[11].

The above results are illustrated in Fig. 7 by the study of the non singlet contribution to prompt photon production at $\sqrt{s} = 63$ GeV, $p_T = 10$ GeV/c. The lines of equal cross-sections when $a(\mu)$ and M^2 are varied are shown in Fig. 7-a and the lines of equal corrections ($K-1 = a(\mu)\, r_1/\hat{\sigma}$) are shown in Fig. 7-b. The results of the various optimization procedures are indicated and although the corresponding scales may be rather different the stability of the predictions is obvious; they also correspond to small corrections. This is to be contrasted with the "standard" result obtained for $\mu = M = p_T$: it is in a region of instability where corrections are large. Very similar results are obtained concerning photoproduction results[2]. It is not

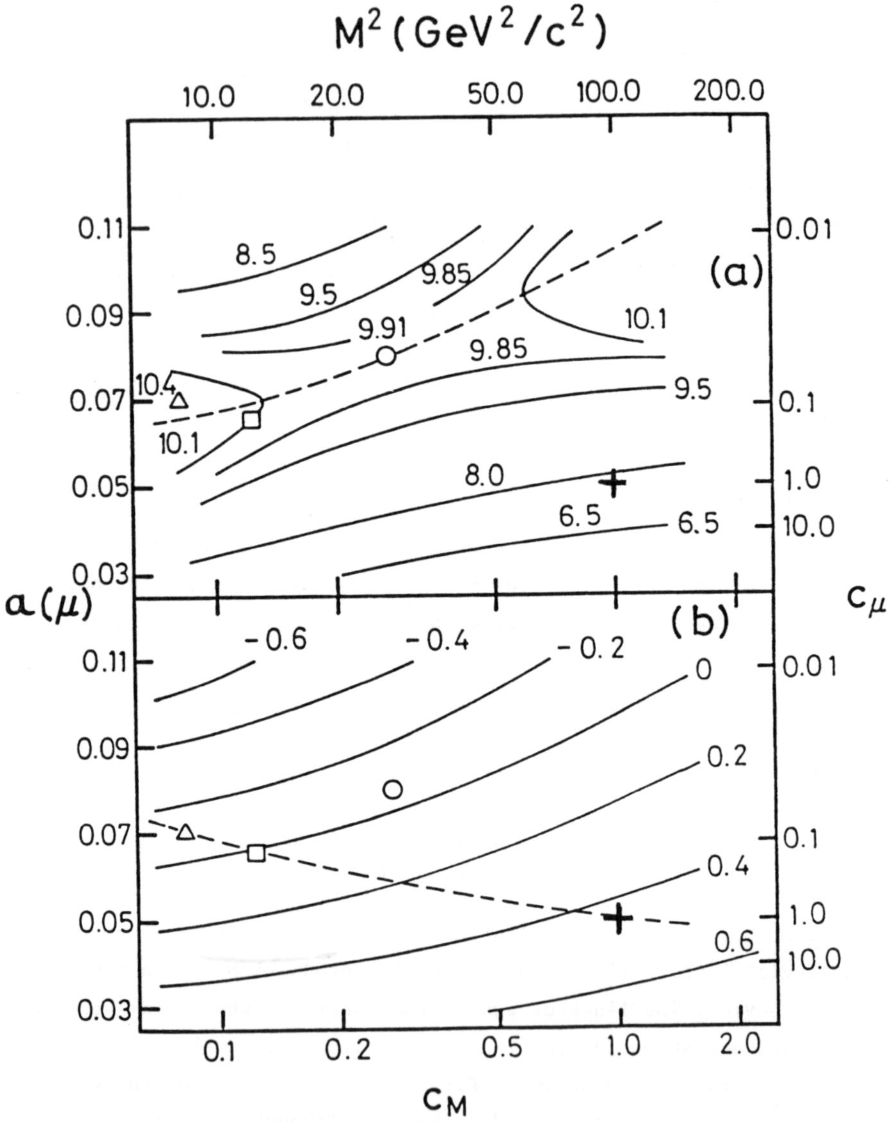

Fig. 7 a) The differential cross-section $d\sigma/dyd\vec{p}_T$ for $\bar{p}+p \to \gamma+X - p+p \to \gamma+X$ at $\sqrt{s} = 63$ GeV, $p_T = 10$ GeV/c and $y = 0$.
b) The correction factor $(K-1)$ under the same conditions.
The various symbols are : O, the two scale optimization result; Δ, the one scale optimization; □ , Grunberg result; +, the standard scale $(\mu = M = p_T)$ result.

surprising since the behaviour of the cross-section around the saddle point are determined by the coefficients b, c and the effective P_{qq} (or more generally P_{ij}) value[2].

To summarize one can say that the optimization allows to make unambiguous predictions which are independent of the details of the method used. The region of maximum stability under changes of μ and M is very close to the regions of vanishing corrections. The method also sets its own limits in the sense that there exists regions where eqs (11),(12) cannot be satisfied (generally at low x_T) : in such cases we conclude that perturbation theory cannot be trusted.

4. PHENOMENOLOGY

In the following table we give a list of experiments which have measured or will measure the single photon spectrum in hadronic collisions[12-22]. The data cover a wide range of energy and x_T with a

	$\sqrt{s}$ [GeV]	Projectile	+ Target	x_T range
E629	19.4	p	C	
NA3	19.4	$\pi^{\mp}$,p	C	
WA70	23.	$\pi^{\mp}$,p	H_2	$.25 \leq x_T \leq .6$
NA24	24.	$\pi^{\mp}$,p	H_2	
UA6	24.	p,p	H_2	
E705	24.	$\pi^{\mp}$,p,$\bar{p}$	H_2,D_2	$.25 \leq x_T \leq .5$
E706	$\lesssim$ 41	$\pi^{\mp}$,p	C	
R806, R808	53.,63.	p,$\bar{p}$	p	$.1 \leq x_T \leq .35$
ISR-110	63.	p	p	
UA1, UA2	540,630	$\bar{p}$	p	$.07 \leq x_T \leq .2$

number of beams $(p,\bar{p},\pi^{\pm})$.

We compare the optimized QCD predictions to data using two sets of parton distributions obtained (from fits to Deep Inelastic and Drell-Yan processes) by Duke and Owens[23]: set 1 corresponds to Λ = .2 GeV and a "soft" gluon while set 2 has Λ = .4 GeV and a "hard" gluon. Figures 8 show a good agreement of p-p data with set 1 whereas set 2

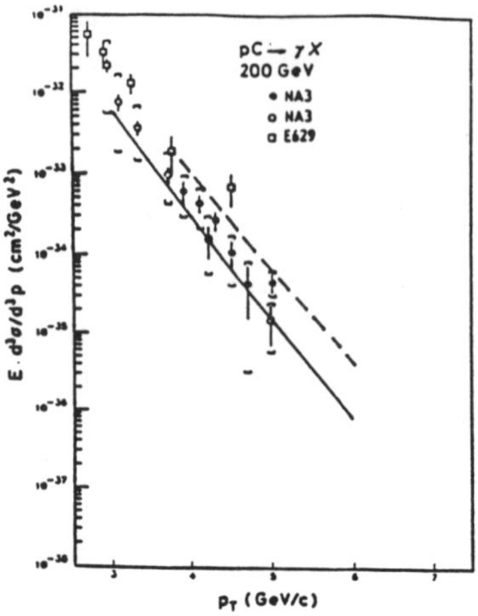

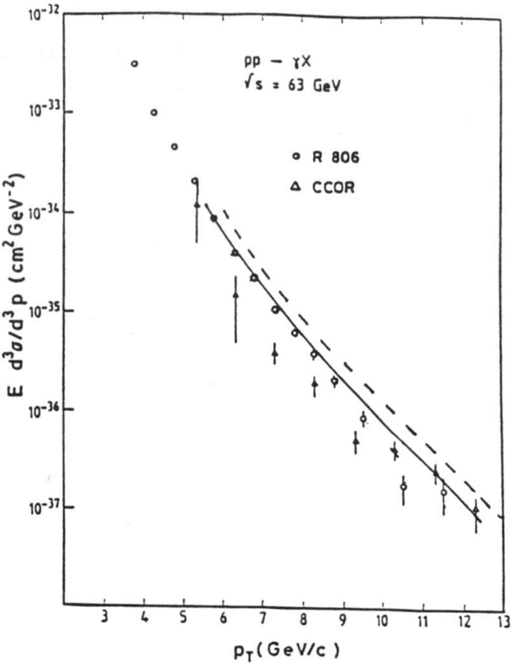

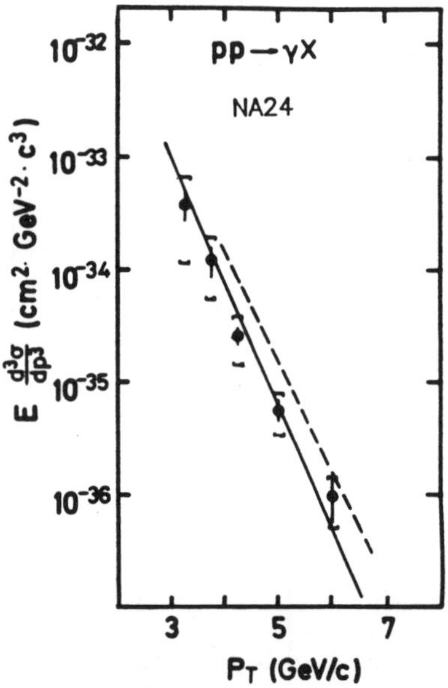

Fig. 8

Comparison of the two scale optimized predictions with the data. The solid line is obtained when using Duke-Owens, set 1 structure functions and the dashed line when using Duke-Owens, set 2.

yields higher cross-sections and can be excluded. At higher energy, the UA2 data cannot distinguish between set 1 (Fig. 9) and set 2 (not shown) as they give very similar results in the p_T range covered by the data. On Fig. 10 one shows the comparison with the preliminary π p data of NA24 and WA70. Again a good agreement is observed with set 1.

In the future, with the experimental results becoming more and more precise it should be possible to constrain the parameters entering

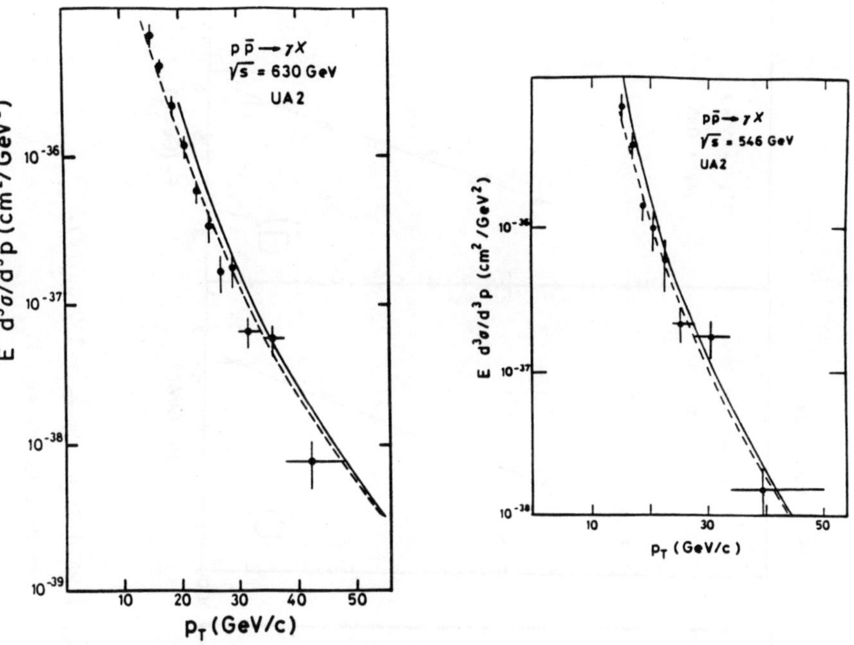

Fig. 9 : The cross-sections of single inclusive photon (solid line) and "isolated photon" (dashed line) compared to UA2 data. The theoretical predictions are based on Duke-Owens set 1 structure functions.

the QCD predictions : for example, the value of Λ can be obtained from p-$\bar{p}$ data at low energy. The p-p data will then constrain the gluon distribution in the proton. To illustrate the sensitivity of the predictions on Λ we have calculated the $p+p \to \gamma+X$ cross-section at $\sqrt{s}$ = 40 GeV using two different values of Λ in α_s, keeping the usual Duke-Owens 1 parametrization of the structure functions : with Λ = 300 MeV rather than Λ = 100 MeV the cross-section increases by more than 50% over the p_T range considered.

Similar phenomenology can be done with photoproduction reactions. Using NA14 data[24] on $\gamma+N \to \pi^o+X$ it was found that a good description of the data was obtained with $\Lambda = 120 \pm {}^{105}_{50}$ MeV and $zD_{\pi/G}(z,25) \sim$ $\sim(1-z)^{1\pm^{.65}_{.0}}$ [25].

Preliminary

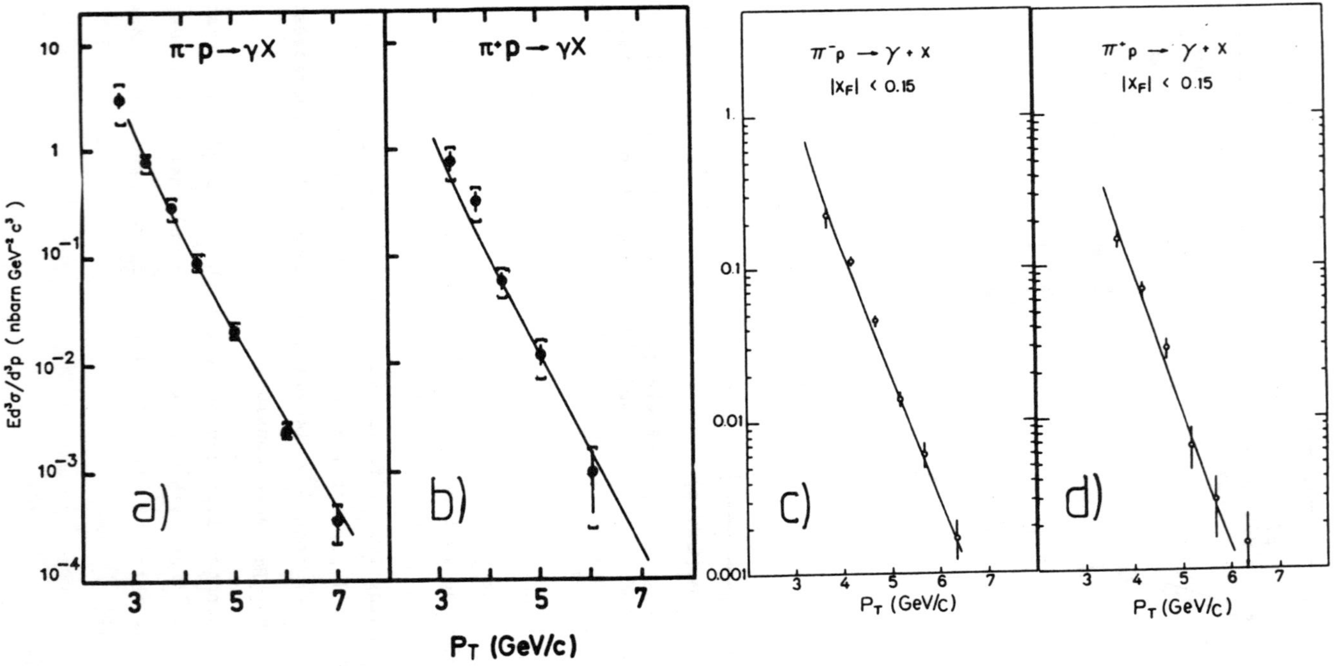

Fig. 10 : a,b) NA24 data (P_{beam} = 300 GeV/c); c,d) WA70 data (P_{beam} = 280 GeV/c). The curves are the theoretical results using Duke Owens set 1.

5. DRELL-YAN PRODUCTION AND OPTIMIZATION

The optimization procedure can also be applied to the Drell-Yan cross-section calculated to $\mathcal{O}(a)$[11,27]. Only the factorization scheme ambiguity is considered since no renormalization is required at this order. Using the lowest order evolution equations it is found that all optimization methods yield the same result i.e. at the stable point the $\mathcal{O}(a)$ terms vanish. The optimization amounts to a summation of higher order terms. In particular it can be shown that it correctly sums the $a\pi^2$ terms as well as the $a\ln^2 n$ terms (where n is the moment) calculated in Ref. 26. Finally it is worth noting that it improves agreement with data at low energies[27].

6. CONCLUSIONS

We have presented the higher order QCD calculation for processes involving real photons and emphasized the ambiguities which affect the results and weaken the predictive power of the perturbative calculation. Different ways to lift these ambiguities are analyzed and shown to give very similar results. When confronted to the data, the theoretical predictions favor a small value of Λ($\Lambda_{\overline{MS}}$ = 200 MeV) and a soft gluon structure function in the proton. When the results of the present and near future high statistics experiments are available it will be possible to determine Λ as well as the gluon densities in the hadrons and the gluon fragmentation function very precisely.

REFERENCES

1) NA14 Collaboration, Astbury P. et al., Phys. Lett. 152B, 419 (1985).

2) Aurenche P., Baier R., Douiri A., Fontannaz M. and Schiff D., LPTHE Orsay 86/24 (1986).

3) Aurenche P., Baier R., Douiri A., Fontannaz M. and Schiff D., Phys. Lett. 140B, 87 (1984); work in preparation.

4) Celmaster W. and Sivers D., Ann. of Phys. 143, 1 (1982).

5) Grunberg G., Phys.Lett. 95B, 70 (1980); Phys.Rev. D29, 2315 (1984).

6) Stevenson P.M., Phys.Rev. D23, 2916 (1981); Nucl.Phys. B203 472 (1982); B231, 65 (1984).

7) Pennington M.R., Phys.Rev. D26, 2048 (1982); Dhar A. and Gupta V., Phys.Rev. D29, 2822 (1984).

8) Politzer H.D., Nucl.Phys. B194, 493 (1982); Stevenson H.D. and Politzer H.D., Rice Univ. preprint DOE/ER/05096. 22, to appear in Nucl. Phys.

9) Duke D.W. and Kimel J.D., Phys.Rev. D25, 2960 (1982); Nakkagawa H. and Niegawa A., Phys.Lett. 119B, 415 (1982); idem with Yokota H., Phys.Rev. D34, 244 (1986).

10) For a collection of papers, see "Perturbative Quantum Chromodynamics", M. Jacob ed., North-Holland.

11) Aurenche P. Baier R., Fontannaz M. and Schiff D., LPTHE Orsay 86/14 (1986).

12) E629 Collaboration, Mc Laughlin M. et al., Phys.Rev.Lett. 51, 971 (1983).

13) NA3 Collaboration, Badier J. et al, Z.Phys. C31, 341 (1986).

14) NA24 Collaboration, de Marzo C. et al., MPI.PAE/Exp.El.168 (1986).

15) WA70 Collaboration, Bonesini M. et al., XXIII Int. Conf. on High Energy Physics, Berkeley, USA (1986).

16) UA6 Collaboration, Snow G., Talk at this Conference.

17) E706 Collaboration, Ferbel T. et al., XVI Int. Symp. on Multiparticle Dynamics, Kyriat Anavim, Israel (1985).

18) CCOR Collaboration, Angelis A.L.S. et al., Phys.Lett. 98B, 115 (1981).

19) R806 Collaboration, Anassontzis E. et al., Z.Phys. C13, 277 (1982).

20) AFS Collaboration, Åkesson T. et al., CERN-EP/86-195 (Nov. 1986).

21) CMOR Collaboration, Pope B.G., Talk at this Conference.

22) UA2 Collaboration, Appel J.A. et al., Phys.Lett. 176B, 239 (1986).

23) Duke D.W. and Owens J.F., Phys.Rev. D30, 49 (1984); Owens J.F., Phys.Rev. D30, 943 (1984).

24) NA14 Collaboration, Augé E. et al., Phys.Lett. 168B, 163 (1986).

25) Richard F., private communication; see Ref. 2).

26) Van Neerven W.L., Phys.Lett. 147B, 175 (1984); Sterman G., Stony Brook preprint, ITP-SB-86-52.

27) Aurenche P. and Chiappetta P., CERN-TH-4452/86.

UNORTHODOX LATTICE FERMION DERIVATIVES AND THEIR SHORTCOMINGS*

Geoffrey T. Bodwin
High Energy Physics Division—Argonne National Laboratory
Argonne, Illinois 60439

Eve V. Kovacs
Department of Physics—University of Illinois at Chicago
Chicago, Illinois 60680

ABSTRACT

We discuss some recent proposals for incorporating fermions into lattice field theory and analyze them in the context of weak coupling perturbation theory.

1. INTRODUCTION

In recent years, much effort has been devoted to finding lattice fermion formulations that solve the spectrum doubling problem without destroying chiral invariance. Three candidate formulations of current interest are the DWY (SLAC) derivative[1)], the Quinn-Weinstein variant of the DWY derivative[2)], and the recent proposal of Rebbi.[3)] While all of these schemes eliminate the spectrum doubling in the case of free fermions, it is unclear whether they represent satisfactory lattice transcriptions in the case of a fully-interacting gauge theory. In the present work, we attempt to resolve this issue by making use of perturbation theory—in some cases to all orders—to analyze the various proposals.

2. REVIEW OF THE DOUBLING PROBLEM

The basic problem in transcribing any continuum field theory onto the lattice is to replace derivatives by an appropriate finite difference operator. Three local operators that yield the derivative in the continuum limit are

$$\nabla^+_\mu f(x) = \frac{1}{a}[f(x+a_\mu) - f(x)] \tag{1a}$$

$$\nabla^-_\mu f(x) = \frac{1}{a}[f(x) - f(x-a_\mu)] \tag{1b}$$

$$\nabla^\pm_\mu f(x) = \frac{1}{2a}[f(x+a_\mu) - f(x-a_\mu)], \tag{1c}$$

where a_μ is a unit vector in the μ-direction whose length is equal to the lattice spacing a. The operators ∇^+ and ∇^- are the nearest neighbor forward and back-

*Talk presented by Eve V. Kovacs.

ward differences, respectively; $\nabla^{\pm}$ is the second-nearest neighbor difference. If we take bilinears in these operators to form a lattice Laplacian, the only Hermitian combinations are $\sum_\mu(\nabla^{\pm}_\mu)^2$ and $\sum_\mu \nabla^+_\mu\nabla^-_\mu$. The momentum space representations of these operators are as follows:

$$-\sum_\mu(\nabla^{\pm}_\mu)^2 \longrightarrow \frac{1}{a^2}\sum_\mu \sin^2(k_\mu a) \tag{2a}$$

$$-\sum_\mu \nabla^+_\mu\nabla^-_\mu \longrightarrow \frac{1}{a^2}\sum_\mu 4\sin^2(\tfrac{1}{2}k_\mu a). \tag{2b}$$

In Fig. 1 we have plotted the square root of these momentum space representations for one component of momentum k_μ. For both of these finite difference Laplacians, the region near $k_\mu = 0$ is linear and therefore yields the usual continuum dispersion relation. However, for $(\nabla^{\pm}_\mu)^2$ there is an additional zero-crossing at $k_\mu = \pm\pi/a$. In the neighborhood of this extra zero-crossing one also obtains the continuum dispersion relation. Thus, this region of momentum space could potentially give rise to extra particle species in the continuum limit. Of course, $(\nabla^+_\mu\nabla^-_\mu)^{1/2}$ does not suffer from the problem of exta zero crossings, and it provides a satisfactory lattice transcription for the Klein-Gordon operator.

In the case of the continuum Dirac operator, $\sum_\mu -i\gamma_\mu\partial_\mu + m$, one has an expression that is linear in the derivative. (Throughout this paper we use a Euclidean formulation of field theory; the γ's are the anti-Hermitian Dirac matrices satisfying $\{\gamma_\mu,\gamma_\nu\} = -2\delta_{\mu\nu}$.) The most obvious lattice transcription is achieved by substituting one of the lattice derivatives of Eq. (1). However, the only Hermitian choice is the second-nearest neighbor operator $\nabla^{\pm}_\mu$ of Eq. (1c), which yields

$$\sum_\mu -i\gamma_\mu\partial_\mu + m \longrightarrow \frac{1}{a}\sum_\mu \gamma_\mu \sin(k_\mu a) + m. \tag{3}$$

Since $\nabla^{\pm}_\mu(k)$ has zeros at $k_\mu = 0$ and $k_\mu = \pm\pi/a$, this so-called 'naive' fermion derivative leads to a doubled spectrum, with 2^d species occurring in d dimensions. For example, in two dimensions there are four species, which come from the momentum regions (k_0, k_1) near $(0,0)$, $(0,\pi/a)$, $(\pi/a, 0)$, and $(\pi/a,\pi/a)$.

One solution to the spectrum doubling problem was proposed by Wilson[4)] and involves adding to the naive derivative an extra momentum dependent term $2/a\sin^2(k_\mu a/2)$ as follows:

$$\sum_\mu -i\gamma_\mu\partial_\mu + m \longrightarrow \frac{1}{a}\sum_\mu[\gamma_\mu \sin(k_\mu a) + 2\sin^2(\tfrac{1}{2}k_\mu a)] + m. \tag{4}$$

This extra term vanishes in the continuum limit for any finite k, but gives the extra species from k_μ near $\pm\pi/a$ a mass proportional to $1/a$. Hence, the extra species decouple in the continuum limit. Unfortunately, the Wilson mass term explicitly breaks the chiral symmetry that is present in the massless continuum theory. Hence, the Wilson fermions are not protected from mass renormalization even in the limit of zero bare mass. Therefore, in order to maintain a vanishing renormalized mass, one must adjust the bare mass parameter as the limit $a \to 0$ is taken . It turns out that in the continuum limit the Wilson derivative yields the correct chiral anomaly.

The Kogut-Susskind[5] derivative is something of a compromise between the Wilson and naive derivatives. It partially eliminates the species doubling by distributing the components of the Dirac spinor across several lattice sites. Some doubling still occurs because certain of the spinor degrees of freedom are reinterpreted as extra flavors. The full chiral symmetry of the theory is now a *flavor* chiral symmetry. In the Kogut-Susskind scheme, only one flavor nonsinglet component of this symmetry is realized. The momentum space representation of the Dirac operator contains a Wilson-like mass term that has been multiplied by γ_5 and a flavor matrix τ_μ:

$$\sum_\mu -i\gamma_\mu\partial_\mu + m \longrightarrow \frac{1}{a}\sum_\mu[\gamma_\mu \sin(k_\mu a) - 2\tau_\mu\gamma_5 \sin^2(\tfrac{1}{2}k_\mu a)] + m. \tag{5}$$

Attempts to find a chirally symmetric lattice theory with an undoubled spectrum and the correct chiral anomaly have led to a number of 'no-go' theorems. These theorems fall into two general categories. The first category is typified by the theorems of Nielsen and Ninomiya[6,7] and Karsten and Smit.[8] Loosely speaking, they conclude that any local lattice derivative either breaks chiral invariance or exhibits spectrum doubling. This result can be understood intuitively by trying to construct a momentum space version of the lattice derivative that passes linearly through zero at $k_\mu = 0$. Since a derivative that is local in position space must have a continuous Fourier transform, the lattice periodicity implies that the derivative function in momentum space must have at least one additional zero crossing. The second category of 'no-go' theorem is typified by the result of Ninomiya and Tan.[8] They use topological arguments to prove that a lattice theory with a continuous chiral symmetry and the correct anomaly is impossible. The properties of the naive, Wilson, and Kogut-Susskind derivatives are all compatible with the 'no-go' theorems. The proposals for the fermion derivative that are the subject of this paper all circumvent the first category of 'no-go' theorem by virtue of their nonlocality. Even the applicability of the second category of 'no-go' theorem could be in doubt, owing to the singular nature of the continuum limit or the peculiar chiral structure of these lattice schemes.

3. RESULTS

3.1 The DWY (SLAC) Derivative

The momentum space representation of the DWY derivative, $D^{DWY}(k)$, is shown in Fig. 2. It duplicates the continuum derivative within the boundaries of the Brillouin zone, but, owing to the requirement of lattice periodicity, it is discontinuous at the zone boundaries. These discontinuities imply that the DWY derivative is non-local in position space. Hence, it evades the first category of 'no-go' theorem. The DWY derivative was originally proposed in the context of a Hamiltonian formulation. Its extension to the Lagrangian formulation of lattice field theory has been discussed by Rabin.[10] All of the results that we report here are derived from the Lagrangian formulation.

The discontinuous nature of the DWY derivative is a potential source of difficulty when one takes the limit $a \to 0$. Indeed, Karsten and Smit[11,12], found that the one loop contributions to the vacuum polarization and the anomaly in weak-coupling perturbation theory contain non-covariant and non-renormalizable ultraviolet divergences that arise from the discontinuities. The peculiar nature of the continuum limit has cast some doubt on the applicability of the second category of 'no-go' theorem.

Rabin[10] has proposed that the difficulties encountered by Karsten and Smit are not fundamental to the DWY Lagrangian theory. Rather, he suggests, they are artifacts of the perturbation expansion that can be removed through a partial resummation of that expansion. The resummation has the effect of softening the discontinuities in the DWY derivative, as indicated, for example, by the dashed curve in Fig. 3. We denote this 'smeared' DWY derivative by $\mathcal{D}^{DWY}(k)$. Because $\mathcal{D}^{DWY}(k)$ is a continuous function, it has an extra zero-crossing at $k_\mu = \pm\pi/a$, and thus could lead to spectrum doubling. However, in the limit $a \to 0$, the slope at $k_\mu = \pm\pi/a$ can, in general, tend to infinity, so the final outcome is not obvious.

Our approach[13] is to analyze the DWY derivative in the context of a model that is completely soluable in the continuum, namely the Schwinger model[14] (two-dimensional quantum electrodynamics). We find, as expected, that the lattice version of this model is also completely soluable in the continuum limit. Our analysis makes use of Rabin's resummation procedure as a technical device to control singularities that would otherwise appear in some of the amplitudes. In this study we examine the mass gap, the anomaly, and the chiral order parameter $\langle\overline{\psi}\psi\rangle$. The lattice diagrams that one needs to consider in order to evaluate these quantities are shown in Figs. 4-6, respectively. In Figs. 4(a) and 4(b) we present the nonseagull and seagull graphs, respectively, that contribute to the vacuum polarization and the anomaly in two dimensions. The anomaly graphs are obtained by inserting a factor γ_5 to the right of one of the vertices in the vacuum polarization graphs. This insertion is indicated by a 5 in the figures. In Fig. 5 we show the complete photon propagator, which, as usual, is obtained from a geometric series involving the vacuum polarization. This series gives rise to the Schwinger model mass gap. Fig. 6 contains the graphs corresponding to the four-point function $\langle\overline{\psi}(x)\psi(x)\overline{\psi}(0)\psi(0)\rangle$, from which one can determine $\langle\overline{\psi}\psi\rangle$. The blobs on the photon lines denote complete photon propagators. For the details of these calculations we refer the reader to Ref. 13. Here we merely present a few of the essential features.

In the Schwinger model, the 'smeared' DWY derivative $\mathcal{D}^{DWY}(k)$ takes on precisely the form shown in Fig. 3. The singularity at $k_\mu = \pm\pi/a$ is smoothed out over a region of width $e\ln^{1/2}(1/\mu a)$, where e is the coupling constant and μ is the mass gap. In the limit $a \to 0$, the width of the smeared region tends to infinity, but not as rapidly as the width of the Brillouin zone, which grows like $1/a$. Hence, the slope of $\mathcal{D}^{DWY}(k)$ at $k_\mu = \pm\pi/a$ tends to infinity in the continuum limit.

As explained in detail in Ref. 13, the only superficially divergent graphs in the continuum Schwinger model are the $O(e^2)$ fermion loops. In a lattice model these graphs are potentially logarithmically divergent in the limit $a \to 0$. However, the graphs of interest contain at least one vector vertex, and if they are evaluated at zero external momentum, these graphs vanish by virtue of gauge invariance. We can, therefore, subtract from the potentially divergent graphs their values at zero external momentum without changing the final result. In the subtracted graphs the potential for logarithmic divergences is removed. Then, for all the graphs in the theory, detailed power counting arguments show that, for any smooth lattice derivative D_μ, the leading contributions in the limit $a \to 0$ come from the linear regions in the neighborhoods of the zero-crossings of the D_μ. In particular, near the ith zero-crossing, located at $k_\mu = \overline{k}_{i\mu}$ we may approximate the fermion propagator as follows:

$$\frac{1}{\sum_\mu \gamma_\mu D_\mu(k) + m} = \frac{1}{\sum_\mu c_{i\mu}\gamma_\mu(k_\mu - \overline{k}_{i\mu}) + m}, \tag{6a}$$

where

$$c_{i\mu} = D'_\mu(k)|_{k=\bar{k}_i} \tag{6b}$$

is the slope of the lattice derivative function at the zero-crossing. In particular, for $D^{DWY}(k)$ in the limit $a \to 0$, $c_{i\mu} = 1$ for $\bar{k}_{i\mu} = 0$ and $c_{i\mu} \to -\infty$ for $\bar{k}_{i\mu} = \pm\pi/a$. One might think that this infinite coefficient in the denominator for k_μ near $\pm\pi/a$ would suppress the contributions from the extra species. However, the coupling to the gauge field can also become large near $k_\mu = \pm\pi/a$. To see this, note that, in general, gauge invariance requires that the $O(e)$ fermion-photon vertex be of the form

$$V_\mu^{(1)}(k,l) = -e\gamma_\mu \frac{D_\mu(k+l) - D_\mu(k)}{(2/a)\sin(\frac{1}{2}l_\mu a)}, \tag{7}$$

where, as shown in Fig. 4(a), k is the fermion momentum, l is the photon momentum, and μ is the index that carries the photon's polarization. For k_μ near $\bar{k}_{i\mu}$, this vertex becomes

$$V_\mu^{(1)}(k,l) \approx -e\gamma_\mu c_{i\mu}. \tag{8}$$

We see that in the case of the 'smeared' DWY derivative, for k_μ near $\pm\pi/a$, the large coefficients cancel between the propagators and the vertices. Thus, each of the species associated with (k_0, k_1) near $(0,0)$, $(0,\pi/a)$, $(\pi/a, 0)$, and $(\pi/a,\pi/a)$ could, in principle, contribute to the amplitudes.

In Table 1 we show the results of our Schwinger model analysis for the DWY derivative, along with some results obtained from similar analyses of the Wilson, naive, and Kogut-Susskind derivatives. The quantities given in the table are the constants of proportionality between the lattice expressions and the continuum ones. It turns out that the value of the vacuum polarization $\Pi_{\mu\nu}$ controls both the mass gap and $\langle\bar{\psi}\psi\rangle$, with the mass gap being proportional to the square root of the vacuum polarization and $\langle\bar{\psi}\psi\rangle$ vanishing unless the vacuum polarization takes on the continuum value.

Table 1. The values of $\Pi_{\mu\nu}$, the mass gap, the anomaly, and the chiral order parameter $\langle\bar{\psi}\psi\rangle$ obtained from the continuum limits of the various lattice versions of the Schwinger model.

Derivative	$\Pi_{\mu\nu}$	Mass Gap	Anomaly	$\langle\bar{\psi}\psi\rangle$
Wilson	1	1	1	1
Naive	4	2	0	0
Kogut-Susskind	2	$\sqrt{2}$	2	0
DWY	2	$\sqrt{2}$	0	0

As expected, the Wilson theory correctly reproduces the continuum results.

The naive lattice theory exhibits the anticipated spectrum doubling: each of the four species contribute to the vacuum polarization an amount equal to the continuum vacuum polarization, thus doubling the mass gap; the species from (k_0, k_1) near $(0,0)$ and $(\pi/a,\pi/a)$, which have positive chiral charge, each contribute to the anomaly graph an amount that is equal to the continuum contribution, while the species from (k_0, k_1) near $(0,\pi/a)$ and $(\pi/a, 0)$, which have negative chiral charge, each contribute an amount that is equal in magnitude and opposite in sign to the continuum contribution.

Owing to its two flavors, the Kogut-Susskind derivative also displays spectrum doubling, but with only half as many species as the naive derivative. Somewhat surprisingly, the flavor singlet anomaly is also doubled, but $\langle\overline{\psi}\psi\rangle$ vanishes. This is a consequence of the fact that the remnant of chiral symmetry in the Kogut-Susskind theory is a continuous flavor *nonsinglet* chiral symmetry. Thus, the flavor *nonsinglet* axial current, rather than the flavor *singlet* axial current, is conserved in the Kogut-Susskind theory. On the other hand, since $\langle\overline{\psi}\psi\rangle$ is rotated by transformations under a continuous symmetry (i.e. the flavor nonsinglet symmetry), it cannot develop spontaneously a vacuum expectation value in two dimensions.

The DWY derivative also exhibits spectrum doubling, but it turns out that only the species from (k_0, k_1) near $(0,0)$ and $(\pi/a, \pi/a)$ contribute to the vacuum polarization—each yielding a contribution equal to the continuum one. In the case of the anomaly graph, all four species contribute to give the following result:

$$A^{DWY}_{\mu\nu}(l) = \frac{2e^2}{\pi}(\epsilon_{\mu\nu} - \frac{\sum_\alpha \epsilon_{\mu\alpha} l_\alpha l_\nu}{l^2}) - \frac{2e^2}{\pi}(\epsilon_{\mu\nu} - \frac{\sum_\alpha \epsilon_{\mu\alpha} l_\alpha l_\nu}{l^2_\nu}). \tag{9}$$

Here μ is the index associated with the axial vector current and ν is the index associated with the vector current. The first term in parentheses comes from the species with (k_0, k_1) near $(0,0)$ and $(\pi/a, \pi/a)$. Each gives a contribution that is equal to the continuum one. The second term in parentheses is noncovariant, even in the continuum limit, and arises from the species with (k_0, k_1) near $(0, \pi/a)$ and $(\pi/a, 0)$. The noncovariance is a reflection of the fact that these species correspond to particles for which the velocity of light is either infinite or infinitessimal in the limit $a \to 0$. Note, however, that

$$\sum_\mu k_\mu A^{DWY}_{\mu\nu} = 0, \tag{10}$$

with the contributions of the noncovariant species cancelling the contributions of the covariant ones. Thus the axial vector current has no anomaly. This cancellation of the anomaly by infinite velocity species was anticipated by Ninomiya and Tan.[8] Our results confirm that their 'no-go' theorem holds even in the case of the DWY derivative with its tricky continuum limit.

3.2 The Quinn-Weinstein Proposal

Now let us discuss the recent proposal of Quinn and Weinstein[2] for eliminating fermion species doubling on the lattice. The Quinn-Weinstein proposal is based on a Hamiltonian formulation in which the gauge has been fixed according to the condition $A_0 = 0$. In the case of a non-Abelian gauge theory, the Quinn-Weinstein Hamiltonian is quite complicated, involving line integrals of the gauge field over all possible paths between the lattice points on which the fermion fields reside. In order to avoid these complications, we discuss here the Abelian theory, which is much simpler.

One can think of the Hamiltonian formulation of lattice gauge theory as a Lagrangian formulation in which time is continuous. Thus, the time components of the loop momenta range from $-\infty$ to ∞ and the time components of the vertices and the terms in the propagators that depend on the time components of momentum are give by the continuum expressions.

For the terms in the fermion propagators that depend on the spatial components of momentum, Quinn and Weinstein suggest a modified version of the DWY derivative function. In momentum space, this function reproduces the linear behavior of the continuum derivative at the origin and also crosses zero at $k_\mu = \pm\pi/a$ with a slope that tends to $-\infty$ in the continuum limit. It is very similar in form to the 'smeared' DWY derivative function that we discussed in conjunction with Rabin's procedure in Section 3.1.

In the Quinn-Weinstein Hamiltonian, the spatial components of the fermion-photon vertices couple only to a transverse photon. Quinn and Weinstein assert that the forms of these spatial vertices can be chosen independently of the form of the fermion propagator. That is, they claim that the interactions need not necessarily derive from the simplest gauge invariant combinations of fermion bilinears and link variables that one obtains from the lattice generalization of minimal substitution. In particular, Quinn and Weinstein choose for the spatial vertices the nearest neighbor form, i.e., in $O(e)$,

$$V_\mu^{(1)}(k,l) = -e\gamma_\mu \cos(k_\mu + \tfrac{1}{2}l_\mu)a. \tag{11}$$

This choice is motivated by the desire to eliminate the infinite vertex coefficient for k_μ near $\pm\pi/a$ that one usually encounters in the DWY formulation (see Section 3.1). In the absence of an infinite vertex coefficient for k_μ near $\pm\pi/a$, one would expect the infinite coefficient in the propagator denominator to suppress the extra fermion species.

Since, in the Quinn-Weinstein scheme, the current to which the transverse photon couples is itself transverse, the Hamiltonian is manifestly invariant with respect to the residual set of gauge transformations that respect the condition $A_0 = 0$. However, the Quinn-Weinstein vertices do not satisfy the usual vector Ward identities that derive from the lattice generalization of minimal substitution. For example, the Feynman identity, Eq. 7, does not hold. Recall that it is these Ward identities that allow one to see that the actual degree of divergence of the vacuum polarization is not as high as the superficial degree of divergence that one obtains by simple power-counting. For example, by application of the Ward identities it follows that the actual degree of divergence of the vacuum polarization is only logarithmic, not quadratic, in four dimensions and that the vacuum polarization is convergent, not logarithmically divergent, in two dimensions. In the Quinn-Weinstein formulation one might expect, then, that the vacuum polarization would show the superficial degree of divergence that one obtains by simple power counting.

We have investigated the leading divergence of the vacuum polarization for the Quinn-Weinstein formulation by direct calculation. In two dimensions, because of a special property of the Hamiltonian formulation, the vacuum polarization actually turns out to be finite, rather than logarithmically divergent. The time components, of course, reproduce the continuum result. For the spatial components, the term proportional to δ_{ij} appears with a coefficient that is twice the continuum one. However, this is of no consequence, since in two dimensions there is no transverse photon to couple to the spatial components of the current. In four dimensions, we have computed the most divergent part of the spatial components of the vacuum polarization and find that it is given by

$$\Pi_{ij} \sim \delta_{ij} a^{-2} \tag{12}$$

in the limit $a \to 0$.

One can interpret this divergent quantity as a transverse photon mass renormalization. It could, in principle be eliminated by adding to the Hamiltonian a Pauli-Villars fermion, whose mass would tend to infinity in the continuum limit. Of course, this would defeat the original intent of the Quinn-Weinstein proposal by explicitly breaking chiral invariance. Alternatively, one could introduce a photon mass counterterm, which would be adjusted in the limit $a \to 0$ so as to keep the photon massless. (Presumably, one would need a similar counterterm in order to deal with a logarithmic divergence in the light-by-light scattering graph.) In the case of numerical simulations, one could conceivably implement such a program for an Abelian theory. It is difficult to see how a criterion for adjusting the counterterms could be set up in the non-Abelian case. In fact, the usual continuum renormalization program would probably break down in the presence of an explicit gluon mass.

3.3 The Rebbi Proposal

Finally, let us discuss the recent proposal of Rebbi[3)] for putting fermions on the lattice. The basis for Rebbi's proposal is that, in the rationalized fermion propagator

$$\frac{1}{\not k + m} = \frac{-\not k + m}{k^2 + m^2}, \tag{13}$$

one need not use the same lattice derivative operator for the numerator and denominator. In particular, Rebbi chooses for the numerator the second-nearest neighbor derivative and for the denominator the nearest neighbor Laplacian:

$$\frac{-\not k + m}{k^2 + m^2} \longrightarrow \frac{\sum_\mu i\gamma_\mu \nabla_\mu^\pm + m_{1/2}}{-\sum_\mu \nabla_\mu^+ \nabla_\mu^- + m_0^2}. \tag{14}$$

For reasons which will become apparent later, we have generalized the propagator by allowing for the possibility of different masses in the numerator and denominator. We have denoted the mass in the spinor-like numerator with a subscript 1/2 and the mass in the scalar-like denominator with a subscript 0. Since the nearest neighbor Laplacian in the denominator has a zero-crossing only at $k = 0$ (see Fig. 1), one might hope that Rebbi's formulation would not exhibit spectrum doubling.

Couplings to the gauge field can be introduced by the lattice generalization of minimal substitution. For the numerator derivative,

$$\nabla_\mu^\pm \psi(x) \longrightarrow \nabla_\mu \psi(x) = \frac{1}{2a}[U_\mu^\dagger(x)\psi(x + a_\mu) - U_\mu(x - a_\mu)\psi(x - a_\mu)]. \tag{15a}$$

Here $U_\mu(x)$ is the usual lattice gauge field link operator that starts at site x and ends at site $x + a_\mu$. For the denominator Laplacian,

$$\nabla_\mu^+ \nabla_\mu^- \psi(x) \longrightarrow (\Delta - e \sum_{\mu\nu} \sigma_{\mu\nu} F_{\mu\nu})\psi(x), \tag{15b}$$

where

$$\sigma_{\mu\nu} = (i/4)[\gamma_\mu, \gamma_\nu], \tag{15c}$$

$$\Delta\psi(x) = \frac{1}{a^2}\sum_\mu [U_\mu^\dagger(x)\psi(x + a_\mu) + U_\mu(x - a_\mu)\psi(x - a_\mu) - 2\psi(x)], \tag{15d}$$

and $eF_{\mu\nu}$ is given by any expression with the correct continuum limit. We take the second-nearest neighbor form

$$eF_{\mu\nu} = -i[\nabla_\mu, \nabla_\nu]. \tag{15e}$$

In order to preserve chiral symmetry, we must have a vanishing mass $m_{1/2}$ in the spinor numerator of Rebbi's propagator (Eq. (14)). On the other hand, the mass m_0 in the scalar denominator is on the same footing with respect to chiral symmetry as the Laplacian, and it does not break chiral invariance. Thus, if we start with a massless bare theory ($m_{1/2} = m_0 = 0$), there is no symmetry to prevent quantum corrections from dynamically generating a mass m_0.

In order to test for this possibility, we have calculated the leading behavior of the $O(e^2)$ fermion self-energy in the limit of vanishing external momentum p (see Fig. 7). In the continuum theory, the leading behavior of the self-energy with external legs is given by

$$\sim \frac{\delta m}{p^2}, \tag{16}$$

where δm is the logarithmically divergent fermion mass correction, which vanishes in the limit of zero bare fermion mass. That is, the fermion mass is protected from dynamical generation by the chiral symmetry. The two powers of $1/p$ come from the external fermion legs, as shown in Fig. 7(b). In the case of the Rebbi propagator, the leading contributions to the fermion self-energy with external legs come from the graphs shown in Fig. 8(a). The leading behavior of these graphs is given by

$$\sim \not{p}\frac{1}{p^2}\delta m^2 \frac{1}{p^2} \sim \frac{\delta m^2}{p^3}, \tag{17a}$$

where

$$\delta m^2 \sim 1/a^2. \tag{17b}$$

We can interpret the leading behavior as being due to a quadratically divergent correction to the scalar mass m_0. (See Fig. 8(b).) As a check, we have verified that if one replaces the denominator of Eq. (14) with the second-nearest neighbor form, so that one has, in effect, the 'naive' derivative, then δm^2 vanishes.

Our calculation is similar to one carried out by Longhitano and Svetitsky[15] in connection with a second-order fermion formulation. They considered forms for $eF_{\mu\nu}$ other than the second-nearest neighbor one that we use. Based on their work, we would expect alternate forms for $eF_{\mu\nu}$ to change the coefficient of the mass correction, but not the general conclusion that there is dynamical mass generation. If one *could* find a form for $eF_{\mu\nu}$ that produced no mass generation in $O(e^2)$, then, in the absence of a protective chiral symmetry, it would almost certainly generate mass corrections in higher orders.

We conclude that, in approaching the continuum limit in the Rebbi formalism, one would need to adjust the bare scalar mass m_0 in order to maintain a zero physical mass. Furthermore, since the quantity $eF_{\mu\nu}$ is a gauge invariant, the couplings to the gauge field contained within it could, in general, be renormalized by the quantum corrections in such a way that they do not 'track' with the gauge couplings contained in Δ. Thus, the strength of the couplings contained in $eF_{\mu\nu}$ would also need to be adjusted as the continuum limit is taken. This complicated approach to the continuum would make the Rebbi formulation a rather awkward basis for lattice simulations.

4. Conclusions

We present here a table which summarizes our results for the three fermion derivative proposals that we have considered in this paper. It is clear from these results that one can learn much about the behavior of various lattice fermion schemes by subjecting them to some simple perturbative analyses. Such perturbative tests might be useful minimal criteria for judging the utility of future lattice fermion proposals.

Table 2. Summary of Results

Derivative	Calculation	Result
DWY (SLAC) (Lagrangian)	Schwinger model (all orders)	spectrum doubling anomaly$=0$ $\langle\bar{\psi}\psi\rangle=0$
Quinn-Weinstein	Π_{ij} in 4-d	$\sim\delta_{ij}a^{-2}$ gauge field mass term
Rebbi	fermion self-energy	dynamically generated fermion mass

After this material was presented at the DPF meeting, we obtained some new results with regard to the Rebbi proposal.[16] A detailed examination of the vacuum polarization in two dimensions reveals that fermion doubling actually arises in a subtle way. This phenonmenon occurs because the expressions for fermion loops in terms of the complete fermion propagator with external legs contain inverse powers of the free fermion propagator. As a result, the second-nearest neighbor numerator in Rebbi's propagator can appear in the denominator of the loop expressions, and thereby cause doubling.

We wish to acknowledge helpful discussions with G. P. Lepage and D. K. Sinclair. One of us (E. K.) wishes to thank the Argonne high energy theory group for its hospitality.

5. REFERENCES

1. Drell, S. D., Weinstein, M. and Yankielowicz, S., Phys. Rev. D 14, 487 (1976); 14, 1627 (1976).

2. Quinn, H. R. and Weinstein, M., Phys. Rev. D 34, 2440 (1986).

3. Rebbi, C., *Chiral-invariant Regularization of Fermions on the Lattice*, Brookhaven preprint BUHEP-86-6.

4. Wilson, K. G., Phys. Rev. D 10, 2445 (1974); in *New Phenomena in Subnuclear Physics*, edited by A. Zichichi (Plenum, New York, 1977).

5. Kogut, J. and Susskind, L., Phys. Rev. D 11, 395 (1975).

6. Nielsen, H. B. and Ninomiya, M., Nucl. Phys. B185, 20 (1981); Nucl. Phys. B193, 173 (1981).

7. Nielsen, H. B. and Ninomiya, M., Phys. Lett. 105B, 219 (1981).

8. Karsten, L. H. and Smit, J., Nucl. Phys. B183, 103 (1981).

9. Ninomiya, M. and Tan, C.-I, Phys. Rev. Lett. 53, 1611 (1984); in *Q.C.D. and Beyond*, edited by J. Tran Thanh Van (Kim Hup Lee, Singapore, 1985).

10. Rabin, J. M., Phys. Rev. D24, 3218 (1981).

11. Karsten, L. H. and Smit, J., Nucl. Phys. B144, 536 (1978).

12. Karsten, L. H. and Smit, J., Phys. Lett. 85B,100 (1979).

13. Bodwin, G. T. and Kovacs, E. V., *Lattice Fermions in the Schwinger Model*, Argonne preprint ANL-HEP-PR-86-96, October, 1986. To be published in Phys. Rev. D.

14. Schwinger, J., Phys. Rev. 128, 2425 (1962).

15. Longhitano, A. C. and Svetitsky, B., Phys. Lett. 126B,259 (1983).

16. Bodwin, G. T. and Kovacs, E. V., in preparation.

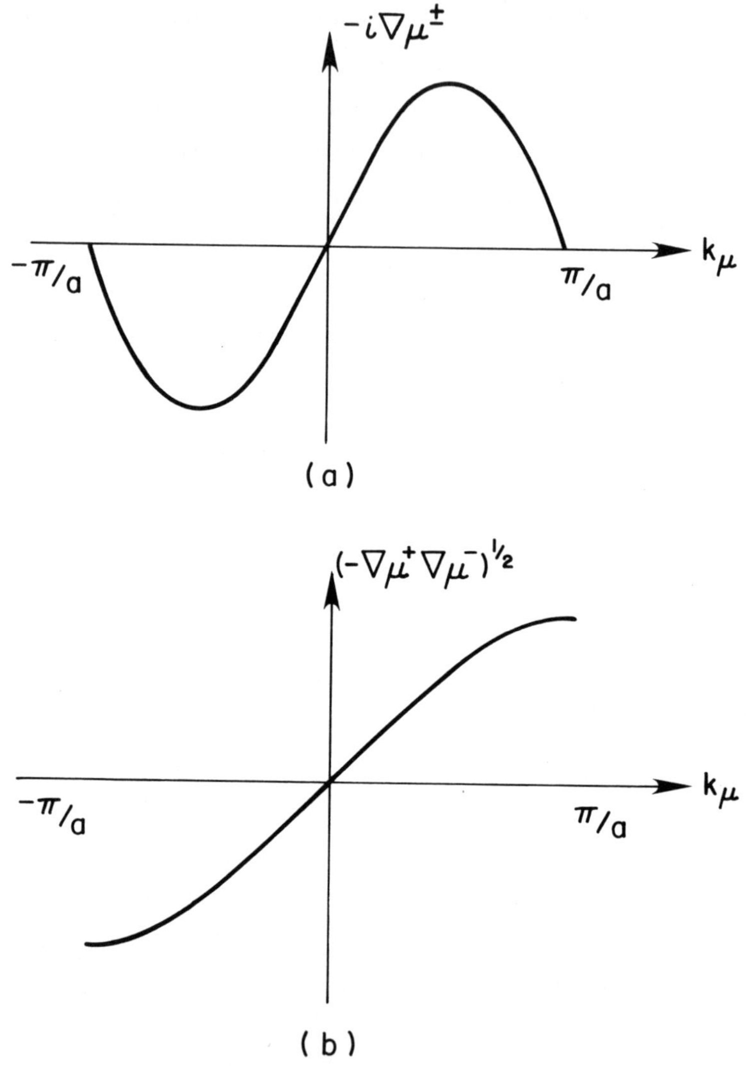

Fig. 1 The momentum space representations of the operators (a) $-i\nabla_\mu^\pm$ and (b) $(-\nabla_\mu^+\nabla_\mu^-)^{1/2}$.

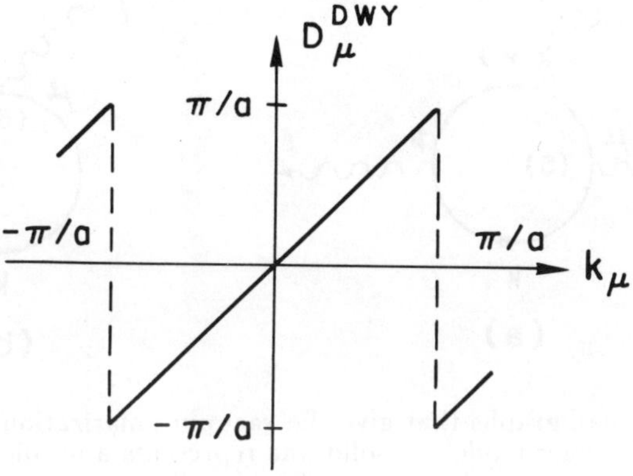

Fig. 2 The DWY derivative function in momentum space.

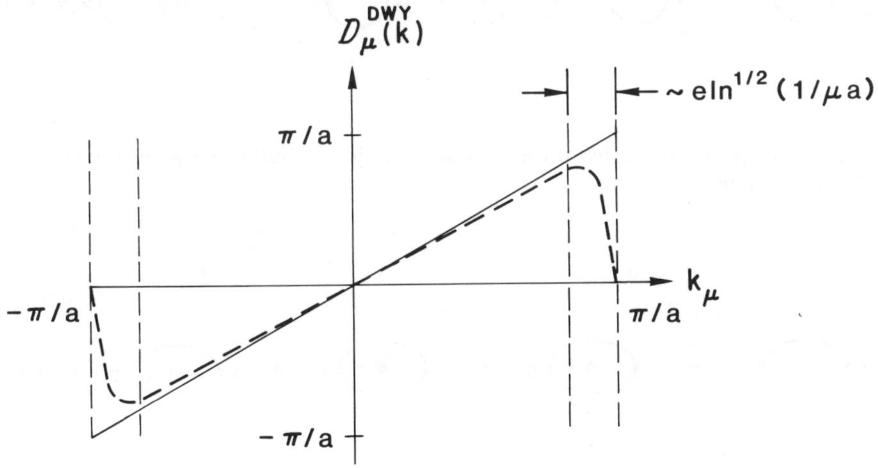

Fig. 3 The 'smeared' DWY derivative function in momentum space.

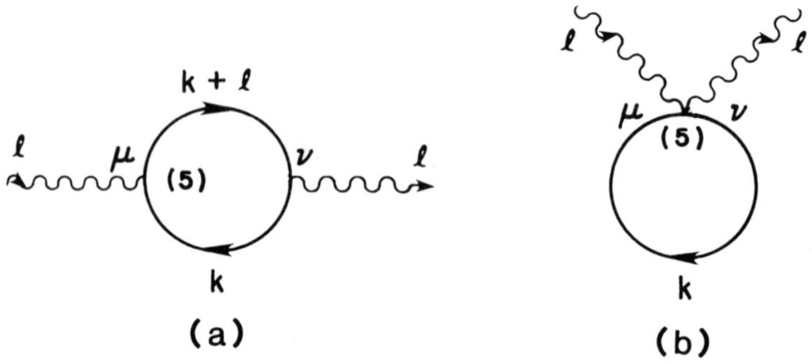

Fig. 4 The Feynman graphs that give the vacuum polarization and the anomaly in the lattice Schwinger model. A solid line represents a fermion propagator and a wavy line represents a photon propagator. (a) The non-seagull graph. (b) The seagull graph. The numeral 5 indicates that in the anomaly graphs there is a factor γ_5 to the right of one of the vertices.

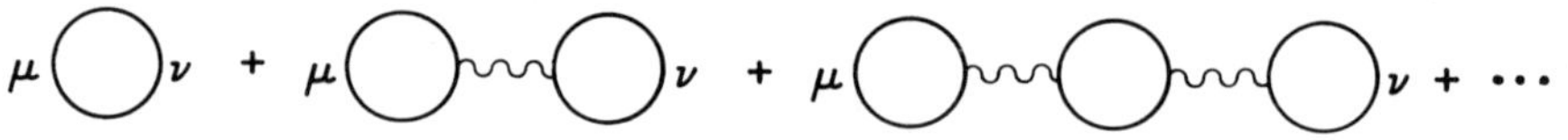

Fig. 5 The geometric series of vacuum polarization bubbles that gives the complete photon propagator.

Fig. 6 The expansion of $\langle\overline{\psi}(x)\psi(x)\overline{\psi}(0)\psi(0)\rangle$ according to the number of complete photon propagators. A wavy line with a blob on it represents a complete photon propagator.

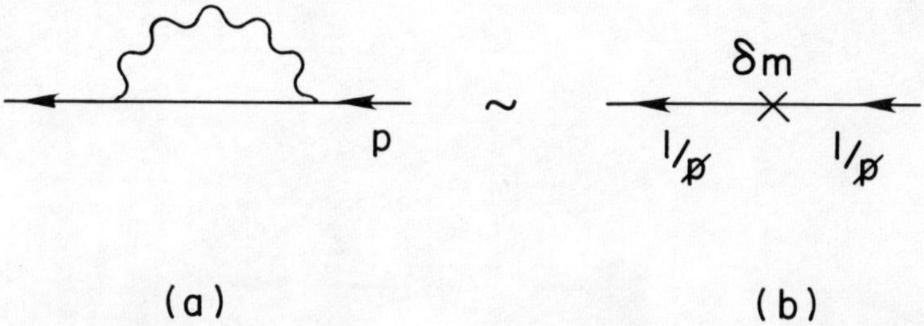

Fig. 7 (a) The $O(e^2)$ fermion self-energy graph for the continuum theory. (b) The leading behavior of that graph in the limit of vanishing external momentum p.

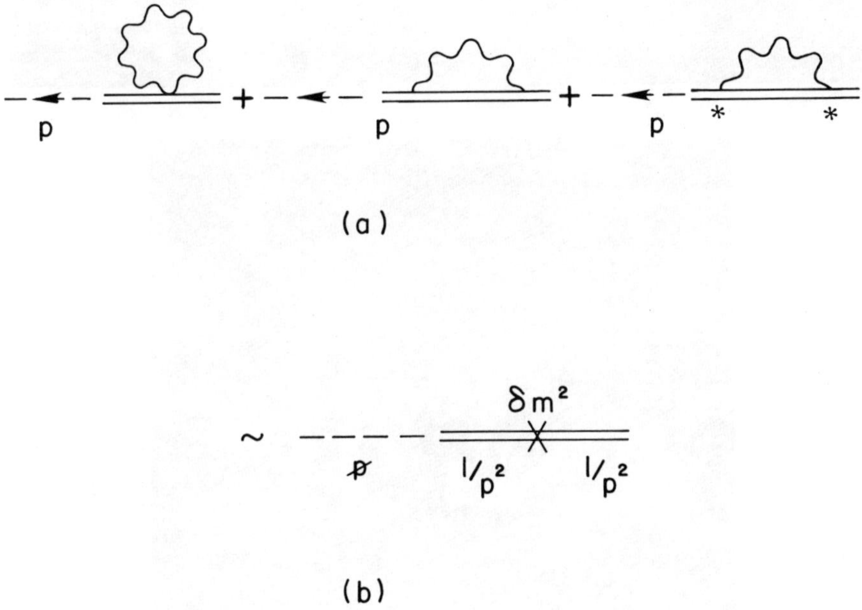

Fig. 8 (a) The graphs in the Rebbi formulation that give the leading contributions to the $O(e^2)$ fermion self-energy in the limit of vanishing external momentum p. A dashed line represents the spinor-like numerator of the Rebbi propagator and a double line represents the scalar-like denominator. (b) The leading behavior of these graphs in the limit of vanishing p.

GRAND BALLROOM

VI. Hadron Physics: Experiment

Conveners — D. Hitlin and T. Kirk

EXCLUSIVE REACTIONS AND DIMENSIONAL SCALING: NEW MEASUREMENTS AND NEW UNDERSTANDING* †

A.S. Carroll, D.S. Barton, G. Bunce, S. Gushue, Y.I. Makdisi
Brookhaven National Laboratory
Upton, New York 11973 USA

B. Baller, G.C. Blazey, H. Courant, G.Y. Fang, K.J. Heller,
S. Heppelmann, M.L. Marshak, E.A. Peterson, M.A. Shupe, D.S. Wahl
University of Minnesota
Minneapolis, Minnesota 55455 USA

M. Emond, J.J. Russell†
Southeastern Massachusetts University
North Dartmouth, Massachusetts 02747 USA

ABSTRACT

A wide range of exclusive scattering cross sections exhibit a power law energy dependence at fixed (large) scattering angle. From lepton-lepton scattering (s^{-2}) through baryon-baryon scattering (s^{-10}), this power exponent increases in step with the number of valence partons in the initial and final state. This simple result may be understood if these scattering processes are dimensionless processes between initial and final states made up of only valence partons. For large-angle hadron scattering, this is interpreted to indicate that each valence quark of each hadron undergoes a hard scatter and that at impact the valence quarks occupy a region of space small compared to 1 Fermi, a region with linear dimension scaling inversely with square root(s). This interpretation also suggests that all of the basic quark interactions are hard and that amplitudes can be calculated by perturbative QCD. When such amplitudes are folded with SU(6) wave functions for the initial and final-state hadrons, one obtains predictions for cross sections.

This experiment has measured 11 meson-baryon and 2 baryon-baryon cross sections at 90 degrees. When interpreted in terms of underlying hard quark-quark scattering diagrams, the process of quark interchange between the incident hadrons clearly dominates over other graphs such as pure gluon exchange and quark annihilation/creation. While there is no theoretical prediction for this dominance yet, there is an indication of underlying simplicity.

In a continuing program at the Brookhaven AGS we have initial indication of a strikingly large cross section for proton-

* Work performed under the auspices of the U.S. Dept. of Energy.
† Work performed under the auspices of the Nat. Science Foundation

proton elastic scattering inside nuclei at 90° and incident momentum of 10 GeV/c ($-t = (9\ \mathrm{GeV/c})^2$). This result suggests that those protons which scatter elastically to large angles have a smaller cross section for additional interactions (less shadowing) while traversing a nucleus than those undergoing normal small momentum transfer collisions, an effect predicted by the above interpretation.

1. INTRODUCTION TO EXCLUSIVE SCATTERING

First, since they are not terribly fashionable at the moment, let me define exclusive reactions (a term originally used by Feynman in about 1969).[1] They are reactions in which all the final state particles are identified and measured. The most frequently measured exclusives are the two-body reactions such as:

$$\begin{aligned} pp &\to pp \\ \pi^- p &\to p\rho^- \\ \pi^+ p &\to K^+\Sigma^+ \end{aligned} \tag{1}$$

Exclusive reactions are to be distinguished from the more common inclusive reactions where only some (usually one) of the final state particles is identified without regard to the other particles as in reaction (2).

$$\pi^- p \to \pi^+ X \tag{2}$$

To orient you to the kinematic region where our measurements were made, let us look at Fig. 1 where $\pi^- p$ elastic scattering is displayed over the entire angular range from 0° to 180°. Our data points were recorded at the minimum of the differential cross section curve corresponding to the largest possible momentum transfer.[2] For the 10 GeV/c pion beam from the AGS at $\theta_{cm} = 90°$, we have $s = 20\ \mathrm{GeV}^2$, $p_\perp = 2.1$ GeV/c and $-t = 9\ (\mathrm{GeV/c})^2$. The reason we made these measurements at a relatively low momentum is that these cross sections fall extremely rapidly (s^{-8}). Since the cross sections are only about a nanobarn, it is difficult to extend the measurements much higher in energy. This

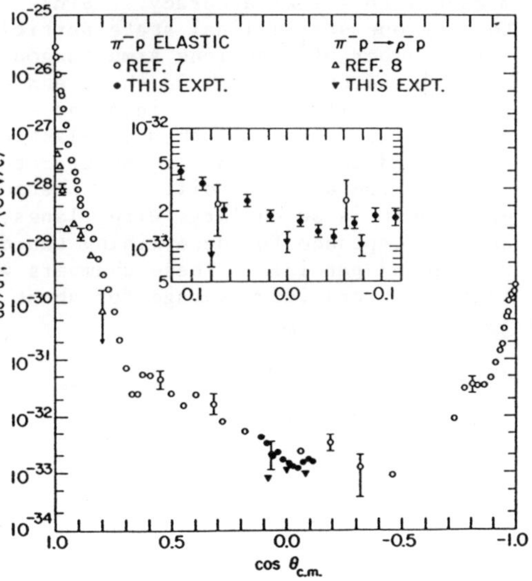

Fig. 1.

is why people measure inclusive or jet cross section at collider energies instead of exclusive cross sections. We collected 1200 π^-p elastic events which is about a factor of 20 more than previous experiments in this kinematic region.

Also on the curves are some values of the $\pi^-p \to p\rho^-$ cross section. These values, which were published about a year ago, show that there are other exclusive cross sections with values comparable to the elastic cross sections at $\theta_{cm} = 90°$.[2]

2. MEASUREMENTS

The first requirement of any good experimental program is a group of physicists interested in carrying out the experiments. The title page provides the names of this group from the University of Minnesota, Southeastern Massachusetts University and Brookhaven National Laboratory.

The apparatus shown in Fig. 2 is situated in the C1 0° beam line at the AGS and consists of three principal parts.[2] There is a beam and target system: two differential Cerenkov counters to identify incident particles, hodoscopes to measure the beam direction and momentum, counters to monitor the incident flux and a meter-long liquid hydrogen target. Second, the key to the experiments is the magnetic spectrometer arm which measured angle and momentum of the stable outgoing particle such as the K^+ or p in reaction (1). First, maxtrix logic between hodoscopes and drift chambers measured the momentum to ~ 10% for the trigger. Then the off-line analysis determines the momentum to ± 1/2% accuracy. Since the trigger requirements apply to only one stable final state particle, we are able to measure a large number of reactions simultaneously, some of which involved an unstable particles such as the ρ and the Δ. Previous measurements in this kinematic region all involved angle-angle correlations which necessarily restricted them to elastic reactions. Particle types are identified in the magnetic spectrometer with two threshold Cerenkov counters operating at different pressures. The third element of the experiment is seven large wire planes, which gives us large solid angle acceptance for determining the directions of decay products of the second particle. These chambers were not involved in the trigger logic, but provide coverage for about 90% of the ρ decays.

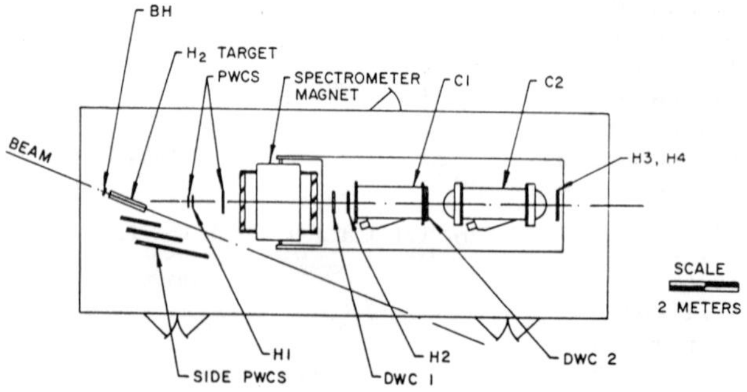

Fig. 2

Since our apparatus provides 3 constraints to elastic scattering, we can use them to determine the resolutions of our apparatus. Figure 3 shows our 1200, $\pi^- p \to p\pi^-$, events. The open histogram gives the MM^2 distribution for single tracks events based on the magnetic spectrometer. Adding the side chamber array, there is a cleanly separated elastic peak. With the elastic events removed there is a steeply falling inclusive spectrum with a shoulder at the ρ mass. We are able to measure the density matrix elements of ρ^- decay to $\pi^-\pi^\circ$. Figure 4 shows a similar distribution for an incident π^+ and a π^+ in the magnetic spectrometer. For this channel, we have 299 $\pi^+ p$ elastic events. These distributions can be resolved into an elastic peak, an inclusive background, and the Δ^+ (1232) resonance. By observing the angular distributions of the Δ^+ decay, we find that there are 72 $\pm$ 20 $\pi^+ n$ and 96 $\pm$ 30 $p\pi^\circ$ decays, consistent with the expected 1:2 ratio.

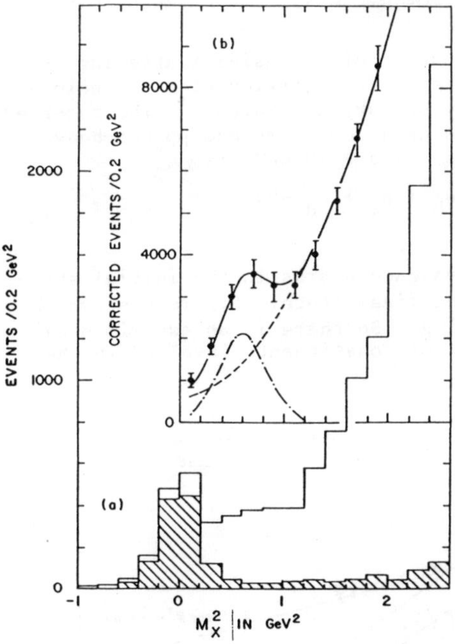

Fig. 3

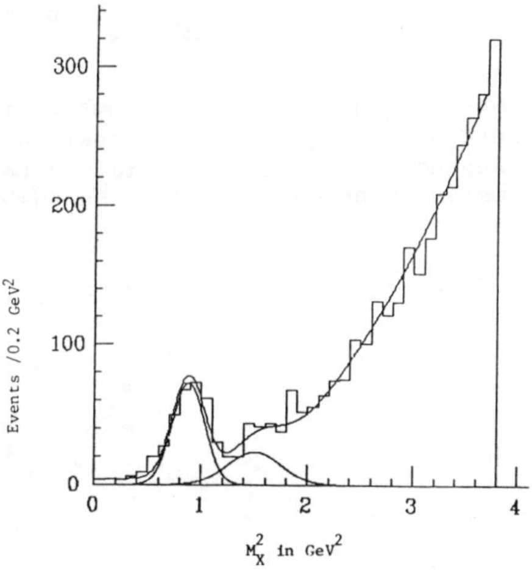

Fig. 4

A final figure on the apparatus performance (Fig. 5) shows the z vertex distribution at the downstream end target for events with nearly elastic pp kinematics. We see those from the hydrogen followed by a 3" blank space, and then an enhancement corresponding to the 1/32" thick aluminum end cap. This figure shows we have 0.2" z resolution, and a small, but significant sample of nearly elastic proton scattering from aluminum.

Fig. 5.

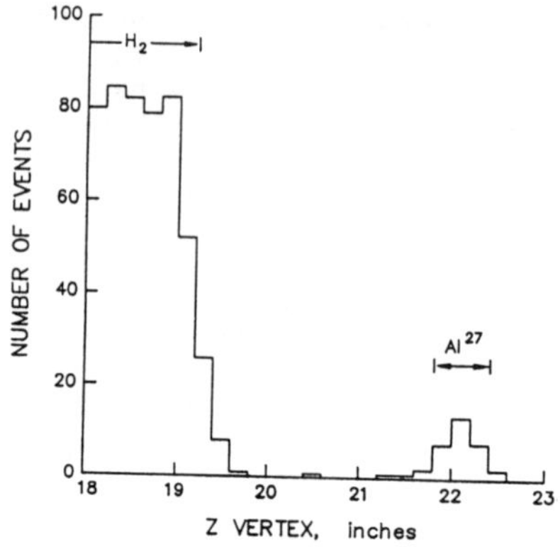

3. INTERPRETATION

Perhaps the most tantalizing aspect of exclusive scattering at large angles is the onset at remarkably low energies of a dimensional scaling law originally described by Brodsky and Farrar.[2] The observed center-of-mass energy squared (s) dependence in πp and pp two-body exclusive reactions at fixed angles for $s > 10\ \mathrm{GeV}^2$ is

$$\frac{d\sigma}{dt} \propto s^{-(n_a + n_b + n_c + n_d - 2)} \qquad (3)$$

where n_a and n_b are the number of valence quarks in the initial state, and n_c and n_d are the number in the final state. Figure 6 shows this dependence for $\pi^- p$ elastic scattering. So there is an experimental measurement which counts the number of constituents involved in the scattering!

Fig. 6

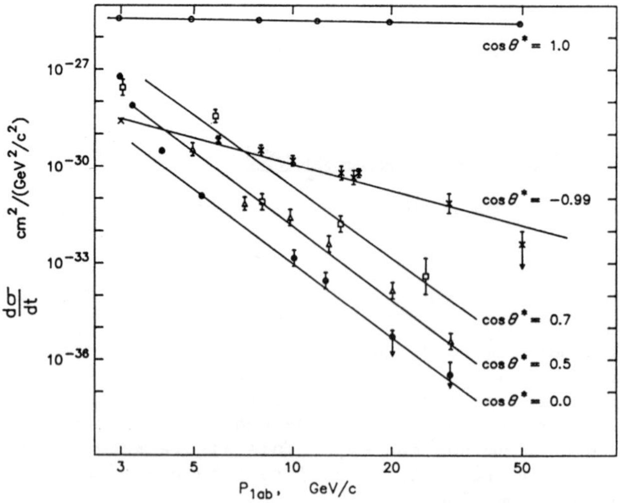

At the point of interaction, initial and final state quarks have a characteristic average distance of closest approach, r. For the Bodsky-Farrar hyposthesis, this impact distance will scale to a smaller value as the energy increases as given by Eq. (4).

$$r^2 \alpha \frac{1}{s} \quad . \tag{4}$$

This dimensional scaling argument holds for any vector field theory such as QED or QCD.

If such a scaling is true, then these exclusive processes would be good testing ground of pQCD. Even though the experimental evidence for such scaling is good, alternative suggestions for this behavior have been put forward. For example, the primary collision might take place between two $x \approx 1$ quarks and the other quarks near $x \simeq 0$ are turned by soft processes.

Mueller at Columbia and Brodsky at Stanford put forward a suggestion a few years ago to directly test the scaling hypothesis.[4] They said, if the hadrons are small at the moment of interaction, then they should be anomalously small for some distance before and after the collision. If the hadrons are small in size, then their absorption cross section in nuclear matter should be small compared to hadrons which have undergone small angle collision.

By surrounding the struck proton with nuclear matter, i.e., placing it inside a nucleus, one can measure the absorption of the incoming and outgoing hadrons in comparison to those from free protons. Nuclear radii are somewhat less than the distances it takes a proton or pion to re-expand to its normal one Fermi size in a 10 GeV/c collision. The normal attentuation length for hadrons in nuclear matter is ~ 2 Fermi. For an interaction inside an Al^{27} nucleus, this corresponds to a reduction in the free pp cross section by 0.05.

Using the data from the 1/32" thick endcap of the aluminum vacuum vessel of our hydrogen target, we can measure the ratio of pp elastic cross section for protons bound inside aluminum and those in hydrogen. The only experimental problem is that protons inside the aluminum nucleus have Fermi momentum of ~ 250 MeV/c. This smears out the elastic kinematics considerably from our ±5 mrad experimental resolution to about ±50 mrad. The event selection for both the hydrogen and aluminum is the same. Events with only a single track in both its spectrometer and side arrays are required to have a transverse momentum of greater than 1.7 GeV/c. Then the cosines of the angle θ between the observed side track and the direction predicted from the spectrometer track by momentum conservation are plotted. As shown in Fig. 7 there are clearly defined elastic signals (denoted by the shaded peaks) above the inelastic backgrounds. The ratio of these peaks is given by

$$\frac{d\sigma/dt\ pp \to pp\ (\text{aluminum})}{d\sigma/dt\ pp \to pp\ (\text{hydrogen})} \simeq .30 \pm .10$$

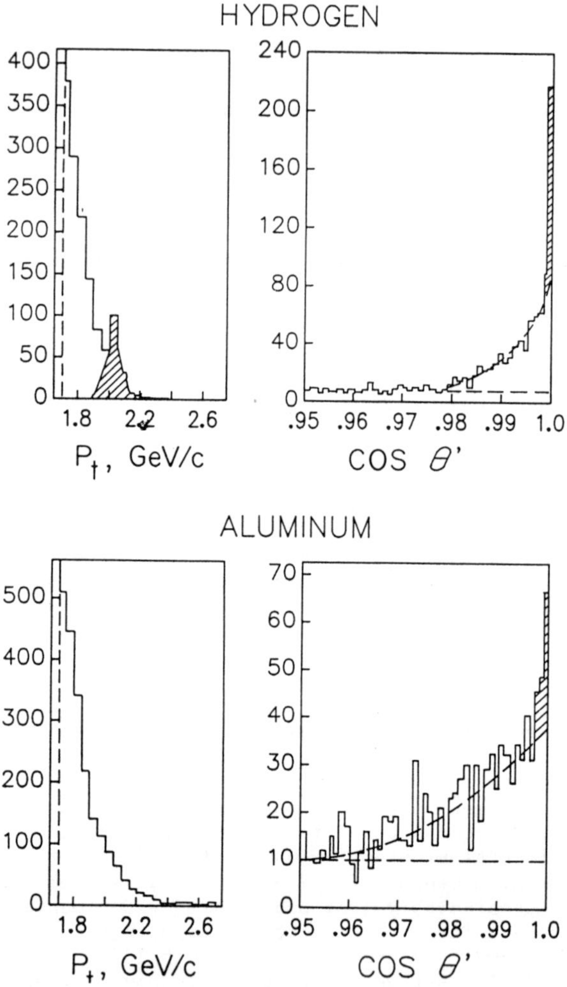

Fig. 7

Fortunately there is a measurement at Berkeley of the same quantity at an incident momentum of 1.5 GeV/c.[5] There the ratio is .06, so that one can see the increase in the attenuation length without resorting to calculations involving detailed nuclear models. Roughly speaking, the above ratio implies a mean attenuation length in aluminum of 5 Fermi at 10 GeV/c.

Clearly, this is an interesting effect and needs to be studied with better statistics, and with different nuclei and momenta. We are now in the middle of doing this more complete experiment with nuclear targets of significant thickness.[6]

Now that there is considerable optimism for success in applying pQCD approach to exclusive reactions, let us look at measurements of the 11 meson-baryon and 2 baryon-baryon cross sections at $\theta_{cm} = 90°$. These are displayed in Fig. 8. First we can compare to other measurements of elastic scattering at 10 GeV/c. We are in good agreement with the well-measured pp elastic of Allaby et al. and the π^-p elastic data of Owen et al.[8] Our π^+p and K^+p cross sections are about a factor of 3 higher than Baglin et al., and we have not been able to find the reason for the discrepancy.[9] It serves as warning that with the high beam intensities required, steep energy dependence, and small signals that it is difficult to get properly normalized cross sections. By measuring all of our cross sections simultaneously in the same apparatus, we eliminate many of the systematic errors between reactions.

The inequality of the π^+p and π^-p cross sections and the large $\pi^+\Delta^+$ production are in disagreement with the present formulation of the Massive Quark Model of Nardulli, Preparata and Soffer.[10]

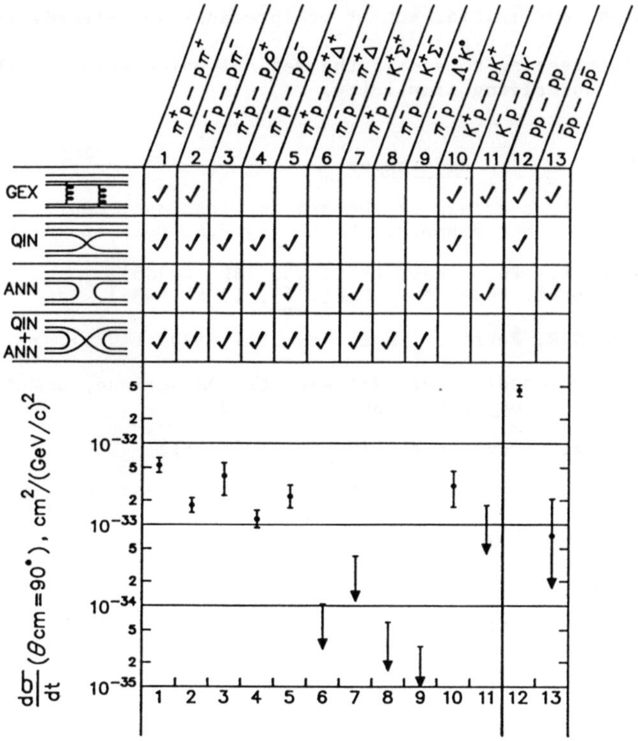

Fig. 8

In a lowest order of pQCD approach, Farrar has shown that there are nine graphs which contribute to the reaction measured in this experiment, and it is convenient to group them into four sets: gluon exchange (GEX), quark interchange (QIN), annihilation (ANN) and combination of interchange and annihilation (CIA).[11] There are two reactions (6 and 2) which only have contributions from the CIA graphs as shown in Fig. 8. Since these reactions are both small, we can assume these CIA graphs are small. Fitting to the cross section assuming that the pairs of graphs contributing to each set are the same, we find that only the QIN and GEX are significant with the QIN being three times the GEX.

Since the nine graphs above only contribute to the helicity=0 portion of the exclusive ρ production, we use our density matrix element analysis to project out this portion of the ρ production. The difficulty is that the fit to the other cross sections predicts a much larger helicity = 0 ρ production than is observed. The ability of the first order pQCD to contribute any sort of polarization phenomena is very limited since the helicity of the quarks is strictly conserved. In fact, we see a strong $\rho_{1,-1}$, density matrix element which is forbidden by pQCD,[2] and predicted by MQM.[10] We hope to shed some light on this matter by studying energy dependence of these density matrix elements.

In conclusion, our initial set of measurements has already set significant limits on the types of amplitudes which can contribute, and it will be interesting to see if theoretical calculations will indicate the same underlying simplicity.

References

1. R.P. Feynman in Proceedings of the 3rd International Conf. on High Energy Collisions, Stony Brook. Edited by J.A. Cole, p.237 (1969).

2. G.C. Blazey et al. Phys. Rev. Lett. 55, 1820 (1985).
   S. Heppelmann et al. Phys. Rev. Lett. 55, 1824 (1985).

3. S. Brodsky and G.R. Farrar. Phys. Rev. D11, 1309 (1975).

4. A.H. Mueller. Proceedings of XVII Recontre de Moriond, Les Arcs, France (1982). Editor Tran Trahn Van, p. 73.

5. I. Tanihata et al., Phys. Lett. 100B, 121 (1981).

6. D.S. Barton et al. Proposal for AGS Exp. 834, (1986). Unpublished.

7. J.V. Allaby et al. Phys. Lett. 25B, 156 (1967).

8. D.P. Owen et al. Phys. Rev. 181, 1794 (1969).

9. C. Baglin et al. Phys. Lett. 47B, 85 (1973).

10. G. Nardulli, G. Preparata and J. Soffer. Nuovo Cim. 83A, 361 (1980), Phys. Rev. D31, 626 (1985).

11. G.R. Farrar. Phys. Rev. Lett. 53, 28 (1984).

Charm Photoproduction Results from the Fermilab Tagged Photon Spectrometer

J. C. Anjos,[3] J. A. Appel,[5] S. B. Bracker,[8] T. E. Browder,[1] L. M. Cremaldi,[4] J. R. Elliott,[4,a] C. Escobar,[7] P. Estabrooks,[2] M. C. Gibney,[4] G. F. Hartner,[8] P. E. Karchin,[1,b] B. R. Kumar,[8] M. J. Losty,[6] G. J. Luste,[8] P. M. Mantsch,[5] J. F. Martin,[8] S. McHugh,[1] S. R. Menary,[8] R. J. Morrison,[1] T. Nash,[5] U. Nauenberg,[4] P. Ong,[8] J. Pinfold,[2] G. Punkar,[1] M. V. Purohit,[5] J. R. Raab,[1] A. F. S. Santoro,[3] J. S. Sidhu,[2] K. Sliwa,[5] M. D. Sokoloff,[5] M. H. G. Souza,[3] W. J. Spalding,[5] M. E. Streetman,[5] A. B. Stundžia,[8] M. S. Witherell[1]

[1]University of California, Santa Barbara, California, USA
[2]Carleton University, Ottawa, Ontario, Canada
[3]Centro Brasileiro de Pesquisas Físicas, Rio de Janeiro, Brasil
[4]University of Colorado, Boulder, Colorado, USA
[5]Fermi National Accelerator Laboratory,[*] Batavia, Illinois, USA
[6]National Research Council, Ottawa, Ontario, Canada
[7]Universidad de São Paolo, São Paolo, Brasil
[8]University of Toronto, Toronto, Ontario, Canada

Invited talk presented at the January 1987 meeting of the DPF by Milind V. Purohit

Abstract

We report results from a preliminary analysis of approximately 30 million reconstructed events (30% of the data) from E-691, a charm photoproduction experiment conducted at the Fermilab Tagged Photon Laboratory in 1985. We present ratios of antiparticles to particles, p_T^2 distributions, x_F distributions, the total charm production cross-section and its energy dependence and the fraction of D^0s from D^*s.

---

[a]Now at Electro Magnetic Applications, Inc., Denver, CO, USA.
[b]Now at Yale University, New Haven, CT, USA.
[*]Operated by the Universities Research Association, Inc. under contract with the United States Department of Energy.

INTRODUCTION

Until recently, charm hadro- and photo-production experiments have typically reconstructed 10–100 events with a charm decay in any one mode. These numbers severely limit studies of production. E-691, a charm photoproduction experiment conducted at the Fermilab Tagged Photon Lab has reconstructed 30% of its 100 million hadronic events so far, already giving us $\sim$ 3000 fully reconstructed charm decays. In this talk we focus on the production mechanism physics that we have studied using this sample (see also Ref. 1).

We will begin with a short description of the photon beam and the spectrometer, paying particular attention to the flux, triggers and vertex detector. We then explain the procedure for extracting charm signals. Next we briefly review the current status of hadro- and photo-production of charm experiments along with a general discussion of effects expected from present-day models. Finally, we present results from E-691 and comment on comparisons with models and other experiments.

BEAM AND SPECTROMETER

A 260 GeV electron beam with a momentum bite of 3.4% impinged on a 20% radiation length copper radiator. The resulting bremsstrahlung photons were incident on a 5 cm long Be target. Electrons which radiated photons in the 80-230 GeV range were bent by a magnet and entered a system of shower counters which measured their energy. Scalers counted the number of electrons showering in each individual counter and another scaler counted the total number of showers (at all times as well as when the data acquisition was live). These were used as our main flux monitors. A Monte Carlo reproduces the shape of the photon spectrum well.

Immediately downstream of the target were 9 planes of $50\mu m$ pitch silicon microstrip detectors (SMDs), which improved both tracking efficiency and resolutions. The SMDs were also crucial in reducing the non-charm background (by a factor of $\sim$ 100 to $\sim$ 300 depending on the decay mode) by identifying, event-by-event, a downstream vertex. The transverse resolution of vertices was $\approx$ 20-30μm. For 60 GeV D mesons, the resolution in the beam direction was $\approx$ 300-400μm.

Following the SMDs was a system of 35 drift chamber planes, 2 magnets (M1 and M2), 2 threshold gas Čerenkov counters and electromagnetic and hadronic calorimeters, followed by a steel wall and muon identification scintillation counters. Earlier versions of the SMDs and the downstream detector have been described in detail elsewhere.[2–4] The mass resolutions for the D^0 and D^+ lie between 5 and 10 MeV/c^2. The Čerenkov counters separate pions from kaons in the momentum range 6 – 38 GeV/c.

It is important for studies of the charm production dynamics to have an unbiased trigger while, at the same time, reducing the high trigger rate of non-charmed events. We chose to run with two triggers: a total hadronic trigger which accepted all hadronic events and with which we collected 10% of our data sample and a fast analog global transverse energy trigger. The total hadronic trigger merely required an interaction in the target and at least 40 GeV in the calorimeters. The E_T trigger required a minimum calorimetrically measured E_T in addition; it became 50% efficient at an E_T around 2.5 GeV, keeping ~ 80% of the charm events and only ~ 30% of non-charmed events.

CHARM SIGNALS

While E-691 has obtained clean high-statistics signals in several charm decay modes, we will concentrate here on the three with the highest statistics: $D^0 \to K^-\pi^+$, $D^+ \to K^-\pi^+\pi^+$ and $D^0 \to K^-\pi^+\pi^+\pi^-$. The signals in these three modes are 1370±52, 1118±47 and 432±27 events respectively. In order to reduce the background under the signal in the last mode we require that the D^0 be from the decay of a D^{*+}. Throughout this talk charge-conjugate decays are implied in any discussion. In Table 1 below, we summarize the important cuts applied to events to extract our signals (see also Ref. 5). Shown in the first column is the effect of each cut in reducing the sample size for the $D^0 \to K^-\pi^+$ mode.

Figure 1 is a picture of an E-691 charm decay in the vertex region. Seen clearly are the primary vertex and the two secondary vertices from charm decay. While most charm events do not have both secondary vertices reconstructed, the SMDs are very useful in reducing non-charm background for single reconstructed secondary vertices. In the $D^0 \to K^-\pi^+$ mode, for instance, a cut on the vertex separation of $\Delta z/\sigma(\Delta z) > 3$ gives approximately equal amounts of signal to background in the signal region. When the cut is raised to $\Delta z/\sigma(\Delta z) > 12$, the ratio of signal to background is ~6 while the signal has been depleted by only a factor of 2. Figures 2, 3 and 4 are the mass plots for the three decay modes.

Fraction Accepted in $D^0 \to K^-\pi^+$	$D^0 \to K^-\pi^+$	$D^+ \to K^-\pi^+\pi^+$	$D^0 \to K^-\pi^+\pi^+\pi^-$
$> 1/2$	Geometric Acceptance	Same	Same
$\sim 4/5$	Tracks must traverse SMDs & M1	Same	Same
$< 3/4$	Joint Čerenkov prob. > 0.2	Same	Same Also slow pion prob. > 0.5
$\sim 1/3$	Pri. vtx. $\chi^2/DF < 3.0$	Same	Same
	Sec. vtx. $\chi^2/DF < 3.5$	Same	Same
	Separation $\Delta z/\sigma(\Delta z) > 8.0$	$> 10.$	$> 5.$

Table 1. The charm selection cuts and the fraction by which each cut reduces the remaining signal sample size for the first mode, $D^0 \to K^-\pi^+$.

CURRENT STATUS OF THEORY AND EXPERIMENT

In the vector meson dominance model (VMD)[6–8] photoproduction of open charm is related to production with an incident J/ψ beam. This latter mechanism is unmeasured and there is an additional uncertainty from the Q^2 dependence of the photon-ψ coupling. However, this model does predict that the total cross-section for charm production should reach a constant value when the c.m. energy is large compared to threshold. In practice this region would be $E_\gamma >100$ GeV. The tree-level diagram for photon-gluon fusion (PGF)[9–13] describes the production of a $c\bar{c}$ pair, but it does not conserve colour and non-perturbative effects and the hadronization of the quarks are additional unknowns. However this model, unlike VMD, predicts more than the total cross-section: it predicts its energy dependence and the detailed differential cross-section.

There are several reviews of hadroproduction[14–17] and photoproduction[8, 18] experiments. From these it is clear that the mean p_T^2 in hadroproduced charm is approximately 1 GeV^2, a result that agrees well with the fusion model if the mean parton k_T^2 is assumed to be ~0.65 GeV^2. When hadroproduction experiments parameterize their x_F distributions with the form

$$\frac{dN}{dx_F} \sim (1 - x_F)^n \tag{1}$$

they find values of n which vary from 1 to 7. There is some evidence that leading charmed mesons have flatter x_F distributions. Fusion model calculations tend to predict softer distributions which fit the data well below x_F=0.3.

Total cross-sections are hard to compare because it is not obvious how to compare data from different nuclear targets. When cross-sections for hadroproduction of lighter mesons are parameterized as having an A^α dependence, it is known that α has a strong x_F dependence, varying from 0.75 at x_F=0 to 0.45 at x_F=1. Fermilab experiment E-613, a neutrino sensitive beam dump experiment, obtained the value α=0.75±0.05 for charm production.[19] However, comparisons of LEBC-EHS data with E595[14] or ACCMOR[17] lead to somewhat larger values of α (0.8–1). This problem may be circumvented altogether by comparing experiments using the same nuclear target. In the case of proton targets, LEBC data[20–22] indicate cross-sections of the order of 20μb at $\sqrt{s}$=27 GeV and ~30μb at $\sqrt{s}$=39 GeV with a ratio of $1.7^{+0.6}_{-0.5}$. ISR data[23, 24] indicate a much larger charm production cross-section (several hundred μb) at $\sqrt{s}$=60 GeV based on their observations of Λ_c^+. Fusion model calculations are in good agreement with the energy rise seen by LEBC (although the ratio has large errors), but there is considerable uncertainty regarding the total cross-section itself. Older calculations[21] indicate that the observed cross-sections are higher by a factor of ≈2–3 compared to fusion model calculations, while a more recent calculation[25] indicates that the theoretical uncertainties are larger than believed earlier and may in fact cover the measured cross-sections, except possibly for ISR data.

Prominent among the photoproduction experiments are E-87, the SLAC Hybrid photon experiment, WA58 and a ψ production experiment, E-401. Two major muon experiments, the EMC and BFP, also provide data from virtual photons which can be extrapolated to Q^2=0 for comparisons with production from real photons. Important results from these experiments are summarized below:

Experiment	Ref.	Major result
E-87	8	$< p_T > \sim 1/2$ GeV for D*
SLAC Hybrid	26	$\sigma_{TOT}=(62\pm 8^{+15}_{-10})\times 10^{-3}\mu$b at E_γ=20 GeV $\sigma(\overline{D}\Lambda_c^+ X)/\sigma_{TOT}=(71\pm 11\pm 6)\%$
WA58	27	$\sigma_{TOT}=(0.23\pm .06)\mu$b (assuming A^1 dependence) at 20 GeV $< E_\gamma <$ 70 GeV $\sigma(\overline{D}\Lambda_c^+ X)/\sigma_{TOT}=(28\pm 23)\%$ $\sigma(\overline{D}DX)/\sigma_{TOT}=(70\pm 21)\%$
E-401	8	$\gamma N \rightarrow \psi N$ (elastic) σ(200 GeV)/σ(100 GeV)=1.56±0.20 (statistical errors only)
EMC	28	$\mu\mu$ and $\mu\mu\mu$ events from virtual photons. Good agreement with PGF for Q^2, ν, p_T and $\Delta\phi$ distributions without correction for overall normalization using $xG(x)=3(1-x)^5$ σ(200 GeV)/σ(100 GeV)=1.53±0.20 (statistical errors only)
BFP	29	Comments as for EMC σ(200 GeV)/σ(100 GeV)=1.44±0.20 (statistical errors only)

Table 2. Results from major photoproduction experiments.

As we can see, there are three experiments with data on the rise of the total charm cross-section, if we assume that the E-401 result holds for open charm production. There is no good information on the p_T^2 and x_F distributions of charmed mesons, or the total cross-section. There is no information on the A-dependence of open charm production, although the A-dependence of J/ψ production has been measured[30] as α=0.94±.02±.03 for incoherent production. It would appear that associated production decreases dramatically from 20 GeV (SLAC experiment) to the somewhat higher energy range (20-70 GeV) of WA58. Whether this trend continues is a very interesting question as it throws light on how the $c\overline{c}$ pair is hadronized. Hadronization when the quark couples to a diquark in the target nucleus would favour lower values of x_F for particles as opposed to antiparticles.

E-691 RESULTS

We have obtained background subtracted and acceptance corrected p_T^2 and x_F distributions for the three modes described earlier. In Figures 5 and 6 we show these distributions for the $D^0 \to K^-\pi^+$ mode, the other distributions being similar. We fit the p_T^2 distribution to the form $dN/dp_T^2 \sim \exp(-ap_T^2)$ and the x_F distribution to the form (1). Notice that we have cut off the x_F distribution below x_F=0.2, thereby limiting ourselves to a region where the detector acceptance is well understood. We have investigated the systematic effects arising from the bias due to the E_T trigger, due to smearing of kinematic quantities, due to Fermi motion,[31] due to vertexing and other cuts and due to uncertainty in the acceptance correction calculation. For the x_F distributions, we have also investigated the effects due to the ~8% smearing of the photon energy by the tagging system. For these studies we have used a Monte Carlo based on the PGF model followed by the Lund Monte Carlo for hadronization.[32] The Monte Carlo gives a good representation of the detector and associated multiplicities. The table below summarizes our results. The systematic errors on the ratio of antiparticles to particles are small compared to the present statistical errors. It is interesting to note that the Monte Carlo, with its present default parameters (yet to be varied for a detailed study), yields a $< p_T^2 >$=1.35 GeV^2 and 2.5 for the power of (1-x_F) in the x_F distribution.

Mode	Number of $\overline{D}/D$	$< p_T^2 >$ (GeV^2)	n, where $d\sigma/dx_F \sim (1-x_F)^n$ x_F >0.2
$D^0 \to K^-\pi^+$	0.99±.07	1.37±.06±.08	2.7±.2±.4
$D^+ \to K^-\pi^+\pi^+$	1.03±.07	1.29±.09±.08	2.6±.2±.4
$D^0 \to K^-\pi^+\pi^+\pi^-$	1.07±.12	1.34±.13±.08	2.8±.3±.6
Combined	1.02±.05	1.34±.05±.08	2.7±.1±.6

Table 3. Summary of E-691 results for the ratio of antiparticles to particles, $< p_T^2 >$, and x_F distributions.

While the entire sample of $D^0 \to K^-\pi^+\pi^+\pi^-$ candidates comes from D^{*+} decays, we have examined the $D^0 \to K^-\pi^+$ sample to determine that (31.3±2.4)% of the D^0s arise from D^{*+} decays (after correcting for acceptance). This is consistent with the expectation of (27±3)% of D^0s being from this source if one assumes a 3:1 production ratio for D*s to Ds and the 1986 PDG branching ratios.[33]

The total charm production cross-section has been measured in the region $x_F > 0.2$. In order to check our flux measurement, we have determined the total hadronic cross-section *per nucleus* on Be to be (824±78) μb, consistent with the expected value of (861±40) μb.[34] Using our Monte Carlo for acceptance corrections, and the data for trigger efficiencies, we determine the total cross-section for D^0s and D^+s to be 1.70±.16 μb and 0.81±.09 μb *per Be nucleus* in the range $x_F > 0.2$ with statistical and some of the systematic errors only. Extrapolating to all x_F involves dependence on theoretical models; so does the conversion to $c\bar{c}$ production (because of events containing both a D^0 and a D^- and the missing cross-section from D_s^+, Λ_c^+ etc. production). We have made these extrapolations using the Monte Carlo described earlier to get

$$\sigma_{TOT}(c\bar{c}) = (3.17 \pm .22 \pm .8)\mu\text{b on Be nuclei.}$$

at our energies (≈170 GeV). While it is not known what the A-dependence should be, it is reasonable that α lies between 2/3 and 1. This leads to a difference of a factor of two when the *per Be* cross-section is extrapolated to a *per nucleon* cross-section!

Finally, we have studied the rise of the total charm production cross-section with energy. Figure 7 shows the uncorrected energy distribution of charm events, to be contrasted with the falling bremsstrahlung photon spectrum. The detector acceptance does not change appreciably with energy, but is an important effect that we correct. The resulting rise is fit to a linear increase with energy (Figure 8). The ratio of the cross-section at 200 GeV to that at 100 GeV was determined by combining the three decay modes weighted by their relative production cross-sections. We obtain

$$\frac{\sigma(E_\gamma = 200GeV)}{\sigma(E_\gamma = 100GeV)} = 1.83 \pm .29 \pm .37$$

This rise is consistent with the three values from other experiments quoted in Table 2. Calculations using the PGF model show a strong dependence on the power of (1-x) in the gluon structure function, going from 1.6 with n=5 to 2.0 with n=9 and a scaling distribution. There is a milder dependence on the assumptions about m_c (the mass of the charmed quark) and the QCD constant Λ.

CONCLUSIONS

A preliminary analysis of 30% of the data sample from E-691 has yielded information about the photoproduction of charm, including x_F distibutions and the ratio of D*s to Ds. The $< p_T^2 >$ of D mesons is determined to be 1.34±.05±.08 GeV^2. In our (mostly forward) acceptance region, we find that the ratio of antiparticles to particles is 1.02±.05. This indicates that there is very little associated production at our energies. The total cross-section for charm production is (3.17±.22±.8)μb *per Be nucleus*, with model dependent assumptions. The ratio of cross-sections at 200 GeV and 100 GeV is determined to be 1.83±.29±.37, consistent with previous experiments and the PGF model. We expect to continue studies of photoproduction of charm, concentrating in particular on improving our understanding of the detector acceptance and extracting the gluon structure function from the sample of events where both charmed particles are fully reconstructed.

ACKNOWLEDGEMENTS

We gratefully acknowledge the assistance of the staff of Fermilab and of all the participating institutions. This research was supported by the U. S. Department of Energy, by the Natural Science and Engineering Research Council of Canada through the Institute of Particle Physics, by the National Research Council of Canada and by the Brazilian Conselho Nacional de Desenvolvimento Científico e Tecnológico.

REFERENCES

1. Anjos, J. C. et al., in the "Proceedings of the XXIII International Conference on High Energy Physics", Berkeley, CA, 1986.
2. Karchin, P., et al., IEEE Trans. on Nucl. Sci. NS-32, 612 (1985).
3. Appel, J. A. et al., Nucl. Instr. and Meth., A243, 361 (1986) and references therein.
4. Kumar, B. R. in the Proceedings of the Workshop on Vertex Detectors, Erice, Italy, 1986. Published by Plenum Press, 1987.
5. Anjos, J. C. et al., Phys. Rev. Lett., 58, 311 (1987).

6. Feynman, R. P., Photon Hadron Interactions. Reading, Mass: Benjamin (1972).

7. Roy, D. P., Phys. Lett. 96B, 189 (1980).

8. Holmes, S. D., Lee, W. Y. and Wiss, J. E., Ann. Rev. Nucl. Part. Sci., 35, 397 (1985).

9. Shifman, M. A., Vainstein, A. I. and Zakharov, V. I., Phys. Lett. 65B, 255 (1976).

10. Babcock, J., Sivers, D. and Wolfram, S., Phys. Rev. D18, 162 (1978).

11. Fritsch, H. and Streng, K. H., Phys. Lett. 72B, 385 (1978).

12. Jones, L. M. and Wyld, H. W., Phys. Rev. D17, 759 (1978).

13. Gluck, M. and Reya, E. Phys. Lett. 79B, 453 (1978).

14. Ritchie, J. L., in the Proceedings of the Snowmass workshop, 1984.

15. MacDermott, M., in the Proceedings of the Twentieth Rencontre de Moriond (1985), edited by J. Tran Tranh Van.

16. Leedom, I. D., Fermilab TM-1396 (1986).

17. Reucroft, S., VI International Conference on Physics in Collision, Chicago (1986).

18. Nash, T. in *Proccedings of the International Symposium on Lepton and Photon Interactions at High Energies, Ithaca, New York, 1983*, ed. D. Cassel et al., Cornell University, Ithaca, 1983.

19. Duffy, M. E. et al., Phys. Rev. Lett. 55, 1816 (1985).

20. Aguilar-Benitez, M. et al., Phys. Lett. 135 237, (1984).

21. Goshaw, A. T., in the Proceedings of the XVII International Symposium on Multiparticle Dynamics, Seewinkel, Austria, 1986.

22. Ammar, R. et al., LEBC-MPS collaboration, CERN/EP 86-122, 1986.

23. Giboni, K. L. et al., Phys. Lett. 85B, 437 (1979).

24. Drijard, D. et al., Phys. Lett. 85B, 452 (1979).

25. Ellis, R. K. in the "Proceedings of the XXI Rencontre de Moriond," 1986. (Fermilab-Conf-86/35-T).

26. Abe, K. et al., Phys. Rev. D33, 1 (1986).

27. Adamovich, M. I. et al., CERN/EP 86-89 (1986).

28. Aubert, J. J. et al., Nucl. Phys. B213, 31 (1983).

29. Clark, A. R. et al., Phys. Rev. Lett. 45, 682 (1980).

30. Sokoloff, M. D. et al., Phys. Rev. Lett., 57, 3003 (1986).

31. Bodek, A. and Ritchie, J. L., Phys. Rev. D23, 1070 (1981).

32. Sjöstrand, T., LU TP 85-10, University of Lund, Sweden.

33. Particle Data Group, Phys. Lett., 170B, 1 (1986).

34. Morrison, R. J., private communication. The number comes from data on cross-sections measured in various experiments performed by Morrison et al. on targets other than Be, but the A-dependence is very accurately known.

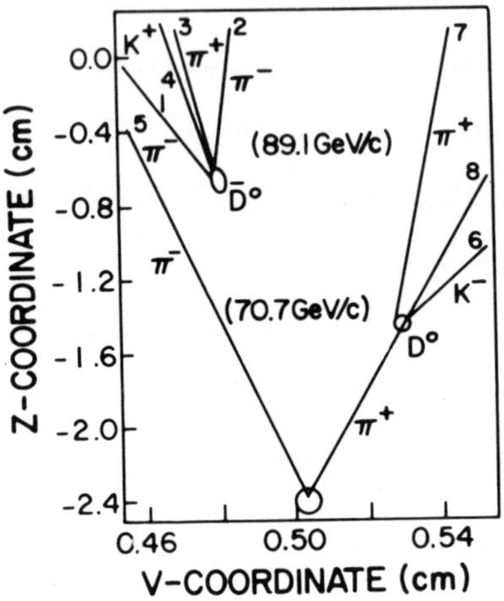

Figure 1. An E-691 charm event: magnified view of the vertex.

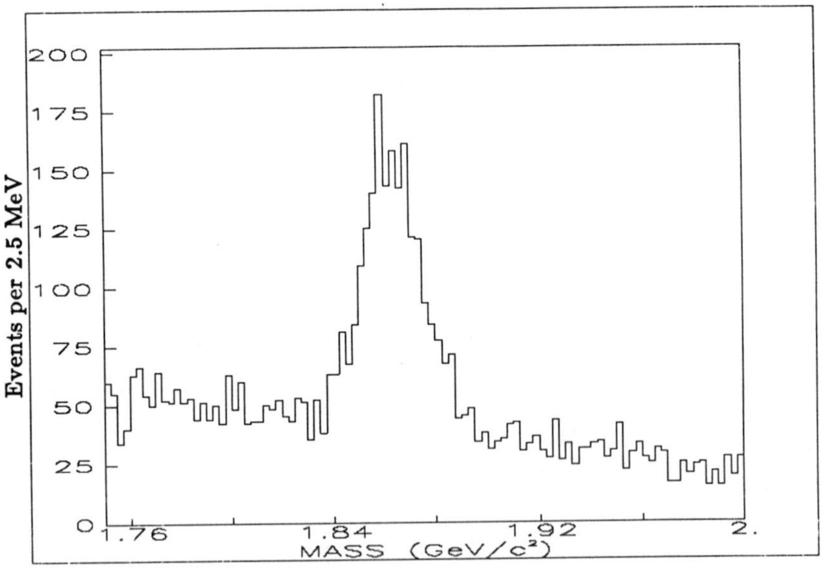

Figure 2. The $D^0 \rightarrow K^-\pi^+$ mass plot.

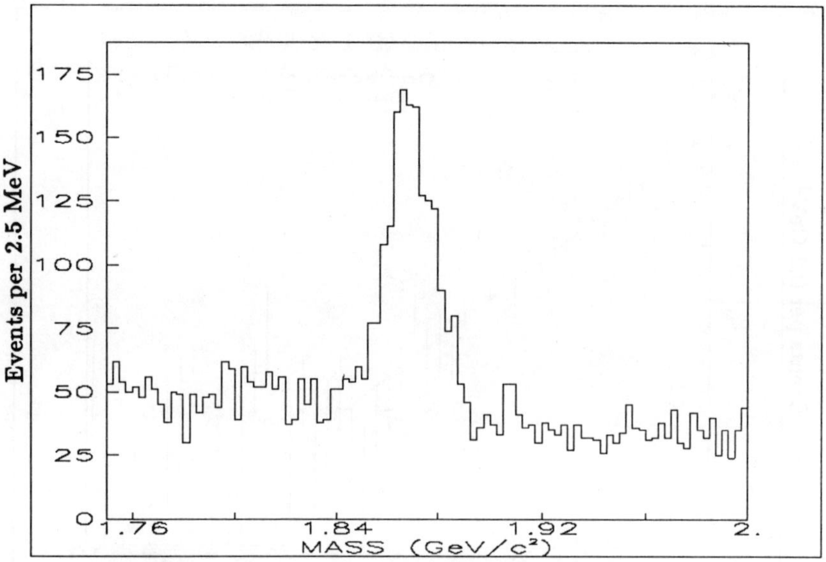

Figure 3. The $D^+ \to K^-\pi^+\pi^+$ mass plot.

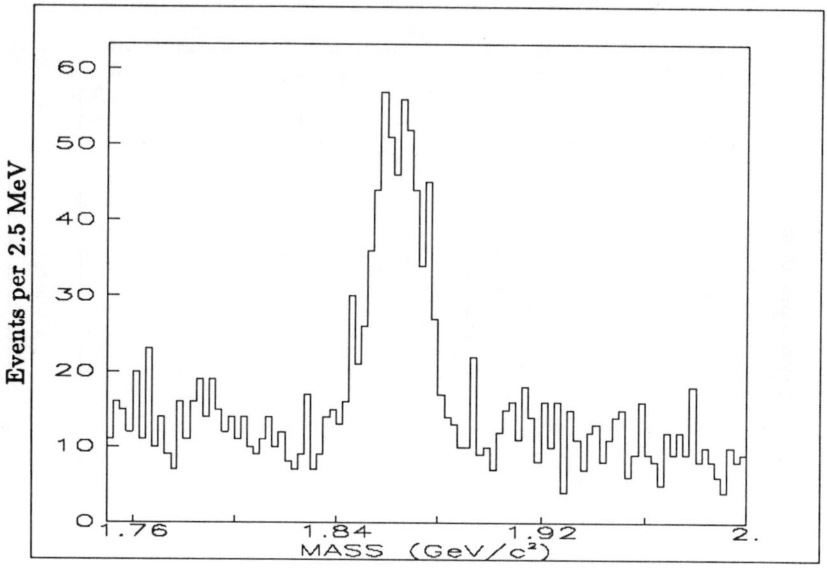

Figure 4. The $D^0 \to K^-\pi^+\pi^+\pi^-$ mass plot.

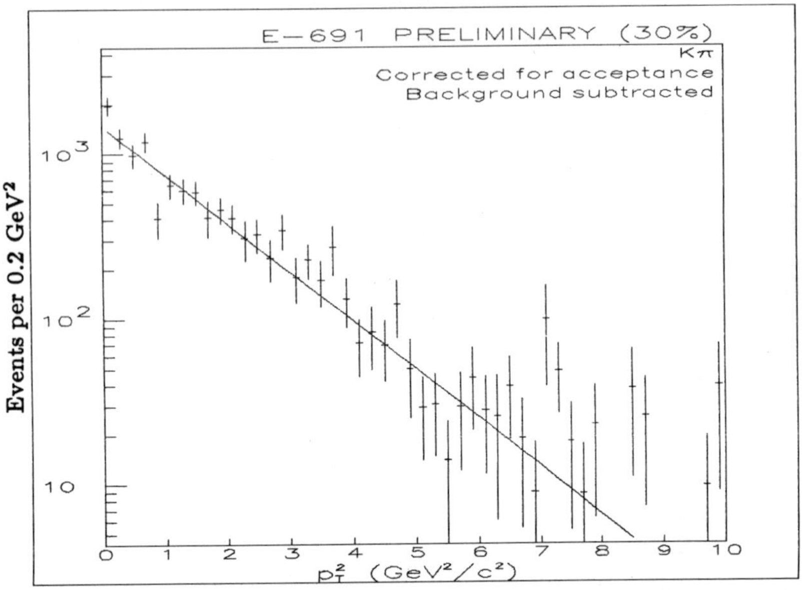

Figure 5. The p_T^2 distribution for the $D^0 \to K^-\pi^+$ mode. The solid line is the fit described in the text.

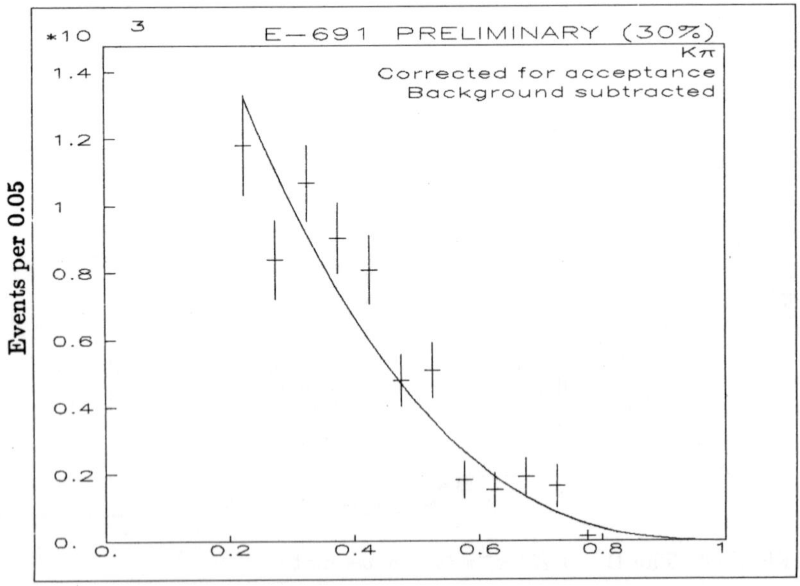

Figure 6. The x_F distribution for the $D^0 \to K^-\pi^+$ mode for $x_F > 0.2$. The solid line is the fit described in the text.

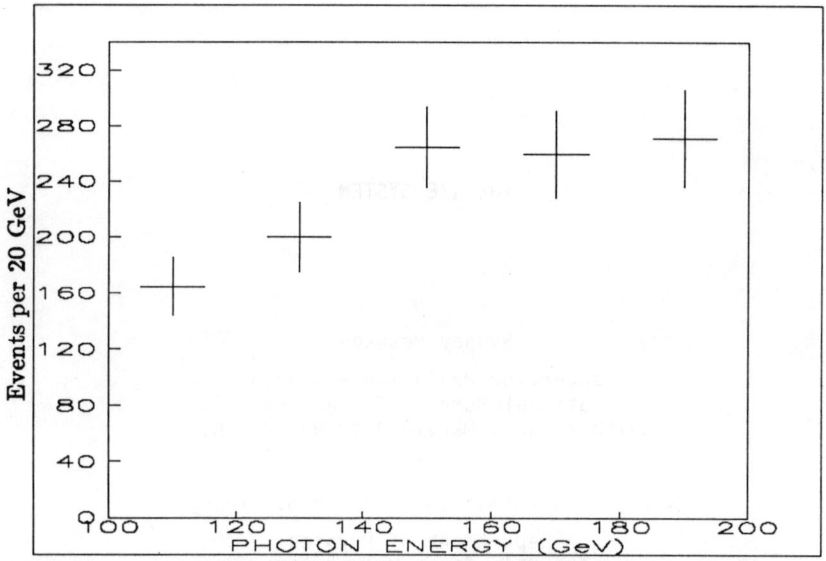

Figure 7. The background subtracted energy distribution of charm events (not corrected for acceptance).

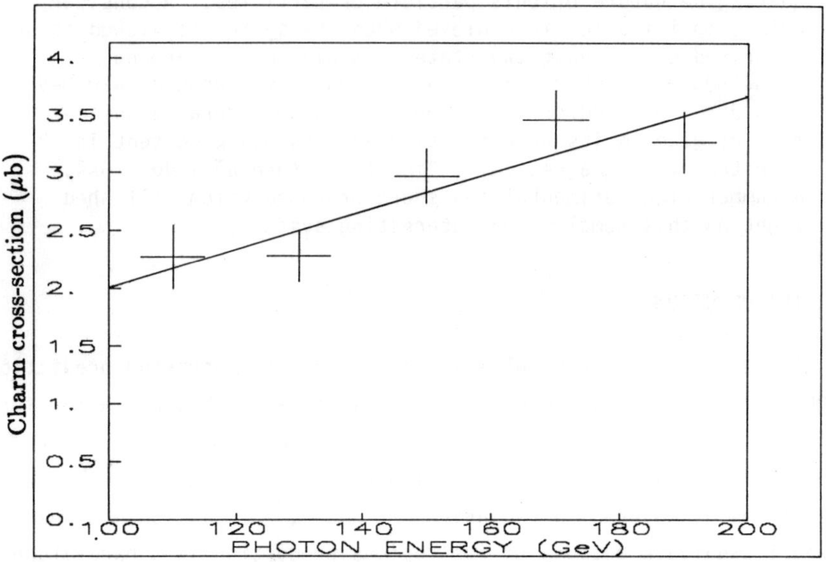

Figure 8. The rise of the total charm cross-section with energy. The solid line is the fit described in the text.

THE ι/E SYSTEM

Sydney Meshkov

Center for Radiation Research
National Bureau of Standards
Gaithersburg, Maryland 20899, U.S.A.

and

William F. Palmer and Stephen S. Pinsky

Department of Physics
The Ohio State University
Columbus, Ohio 43210

ABSTRACT

Recent data on the ι/E system is examined to elucidate the quark/glue nature of this particle or particles. A considerable simplification is achieved when the system is viewed as comprised of at least two states: a high mass ι around 1460-1500 MeV with $J^{PC} = 0^{-+}$ and a low mass E around 1420 MeV with $J^{PC} = 0^{-+}$ and/or 1^{++}. The ι has a considerable admixture of glue in its wave function and its quark content is near the electromagnetically "inert" mixture $u\bar{u} + d\bar{d} - 5s\bar{s}$. A number of experimental tests are proposed which will shed light on this complex and interesting system.

1. INTRODUCTION

The existence of glueballs is one of the most striking predictions of QCD.[1] Identifying these bound states of gauge bosons, a new form of matter unique to self interacting gauge theories, is a current experimental question of paramount importance. From the time the $\iota(1460)$ was discovered in radiative J/ψ decay,[2] it has been a leading glueball candidate because of its copious production in a hard gluon channel. A discussion of the properties of the ι with the ultimate goal of deciding whether or not it is a glueball is the main purpose of this paper. Before embarking on this analysis, it is useful to place the ι in its proper context. Hence, we start with a brief review of glueball spectroscopy , discussing level ordering and mass estimates, followed by a brief overview of recent changes in the properties of other 0^{-+} resonances with which the ι could conceivably mix.

1.1 Glueball Spectroscopy

1.1.1 Level order. This subject has been reviewed often,[1a,1b] so we only present the lowest levels for glueballs made from two gluons moving with a relative orbital angular momentum L. Note the preponderance of low lying 0^{++}, 2^{++}, and 0^{-+} states.

L = 2	2^{++},	0^{++}	...	4^{++}
L = 1	0^{-+},	$\boxed{1^{-+}}$,	2^{-+}	
L = 0	0^{++},	2^{++}		

Fig. 1. 2g glueball levels. $\boxed{}$ denotes an oddball, a glueball with quantum numbers that do not occur in $q\bar{q}$, L mesons.

1.1.2 Masses. The two main questions are: 1) Where is the ground state?; 2) What sets the scale for the excited states? Lattice Gauge Theories are our best current hope for answering these questions. After years of confusion, we are down to the only firm result — a Mass Gap (0^{++}) exists. The two most recent results for $m(0^{++})$ are both in essential agreement. De Grand[3] finds $m/\sqrt{K} = 3.10 \pm 0.15$; $m(0^{++}) = 1240 \pm 60$ MeV for string tension, $\sqrt{K} = 400$ MeV. Patel et al.,[4] found $m(0^{++}) = 1200 - 1400$ MeV for $\sqrt{K} = 420$ MeV. They also have looked for, but could not find evidence for a 2^{++} glueball. An earlier calculation by de Forcrand et al.,[5] obtains $m(0^{++}) = 1050 - 1100$ MeV, but also yields a 2^{++} state such that $m(2^{++})/m(0^{++}) \simeq 2.9$. Berg et al.,[6] a year ago found $m(0^{++}) \simeq m(2^{++})$. Au et al.,[7] recently found an extra 0^{++} state at 991 MeV in a coupled channel K-matrix $\pi\pi$ and $K\bar{K}$ analysis that they put forth as a glueball candidate.

The fact that $m(0^{++}) \simeq 1200 - 1400$ MeV, does not give us great confidence that we can have a low lying 0^{-+} glueball at 1460 MeV, especially since it comes in at the first excited level in Fig. 1.

1.2 The 0^{-+} System

Inasmuch as ι(1460) has $J^{PC} = 0^{-+}$ it has long been thought that it might mix with the other pseudoscalar resonances, η(549), η'(960), η(1275), E(1420), η(1550), and η(1800). Early calculations considered the mixings among η, η', and ι, while a more recent treatment included the η(1275) and E(1420) as well. An early "Rosner analysis"[8] of J/ψ involving η and η' in which the flavor content of these states is:

$$|\eta\rangle = X_\eta|N\rangle + Y_\eta|S\rangle + Z_\eta|G\rangle \text{ and } |\eta'\rangle = X_{\eta'}|N\rangle + Y_{\eta'}|S\rangle + Z_{\eta'}|G\rangle$$

where $|N\rangle = |(u\bar{u} + d\bar{d})/\sqrt{2}\rangle$, $|S\rangle = |s\bar{s}\rangle$ and $|G\rangle = |\text{Glueball}\rangle$ ($|G\rangle$ also includes radially excited quarks) gave; $X^2_\eta + Y^2_\eta = 1.1 \pm 0.2$, which implies that $|\eta\rangle$ is a pure $q\bar{q}$ state, and $X^2_{\eta'} + Y^2_{\eta'} = 0.65 \pm 0.18$,[9] implying lots of mixing of $|\eta'\rangle$ with $|G\rangle$. More recent "Rosner Analyses," that have incorporated the new values of $\Gamma(\pi^0 \to \gamma\gamma)$, $\Gamma(\eta \to \gamma\gamma)$, and $\Gamma(\eta' \to \gamma\gamma)$ and have led to an 8 - 1 mixing angle $\theta_P = -19° \pm 1°$,[10-13] give startlingly different results. Seiden[14] finds $Z^2_{\eta'} = 0.06 \pm 0.36$, and the TPC/2γ report of Aihara et al.,[15] gives $Z^2_{\eta'} = 0.11 \pm 0.17$. These new values for $Z^2_{\eta'}$ imply that there is no great need for mixing glue or radially excited quarks into η'! This lack of mixing should simplify the analysis of the properties of higher mass 0^{-+} states, because it implies a decoupling of the η(548) and η'(960) from them.

2. ι/E SYSTEM.

2.1 Overview.

As the quantity, quality, and variety of data relevant to the ι has increased, the picture to be interpreted has, however, become increasingly confused. There seem to be at least two or three states in the 1400-1500 MeV region, some with experimentally determined $J^{PC} = 0^{-+}$, others with $J^{PC} = 1^{++}$.[16-18]

Radiative J/ψ decay to $K\bar{K}\pi$ shows a wide, $\Gamma \approx 100$ MeV, bump at 1460 MeV with $J^{PC} = 0^{-+}$.[16,17,19,20] (See Fig. 2.) Almost all experiments that involve only hadrons produce a sharp bump at 1420 MeV with a width of 40 MeV. These include $\pi^- p \rightarrow Xn$,[18,21-24] $(\pi^+/p)p \rightarrow (\pi^+/p)Xp$,[25] and $J/\psi \rightarrow \omega X$ and ϕX,[26] where $X = K\bar{K}\pi$ or $\eta\pi\pi$, and $\bar{p}p \rightarrow K^+ K_S \pi^- Y$.[27] Spin-parity analyses of the experiments are contradictory; some yield 0^{-+}, others 1^{++}. Two photon annihilation experiments confuse the picture further: nothing is seen in $\gamma\gamma$ (real on-shell photons) in the $K\bar{K}\pi$ final state, indicating no 0^{-+} production,[28] whereas a strong peak in $\gamma\gamma^*$ (γ^* is a virtual photon) is observed at 1420 MeV for the same final state, implying production of a 1^{++} state.[29]

Our primary goal is to determine whether, in the complicated ι/E 1400 - 1500 MeV interval, there is evidence for a 0^{-+} state that is primarily glue. Our analysis is simplified if we divide the data into two distinct parts, a low mass region around 1420 MeV, called the "E", and a higher mass region from 1460- 1500 MeV, called "ι". The E region is primarily of meson spectroscopy interest. A thorough description and understanding of this region clearly also requires a better understanding of the 1^{++} nonet (see discussion later). The ι region, corresponding to the higher mass part of the broad bump seen in radiative J/ψ decay, and almost certainly 0^{-+}, is the glueball candidate region. It is where the action is! The ι and E regions may overlap, and the fundamental underlying resonance structures may actually be at somewhat different locations than where they naively appear to be in the data.

2.2 Data Features

1) The reaction $\iota/E \rightarrow \gamma\gamma$ has not been seen.[28] The limit is $\Gamma(\iota/E \rightarrow \gamma\gamma)\, B(\iota/E \rightarrow K\bar{K}\pi) < 1.6$ keV. Neither ι nor E couples strongly to two photons, in contrast with η', for which $\Gamma(\eta' \rightarrow \gamma\gamma) = 4.3 \pm 0.1$ keV; a recent TPC/2γ measurement gives $\Gamma(\eta' \rightarrow \gamma\gamma) = 4.5 \pm 0.3 \pm 0.7$ keV.

2) The reaction $\iota/E \rightarrow \gamma\gamma^*$ shows a sharp peak at 1420 MeV in the same experiment that does not see a $\gamma\gamma$ rate. Presumably this peak is a 1^{++} resonance.[29]

3) The decay $J/\psi \rightarrow \gamma\iota/E \rightarrow \gamma\gamma\rho$ yields a wide ι/E bump, centered at $M = 1420 \pm 15 \pm 20$ MeV, $\Gamma = 133 \pm 5.5 \pm 3$ MeV.[16,17,30] (See Fig. 3.)

The partial width $\Gamma(\iota/E \to \gamma\rho) = 1.9 \pm 0.7$ extracted from this data is a relatively large number. Our interpretation of this data is that there is a center section of the invariant mass distribution around 1420 MeV that is consistent with the 40 MeV seen in the $\gamma\gamma^*$ and many of the hadronic experiments. The upper tail of this distribution could be due to the ι bump, and the lower tail could be associated with the $\eta(1275)$ and/or D(1280). The $\gamma\phi$ final state has not been seen, with a limit of the same order as the $\gamma\rho$ rate.

4) The radiative decay $J/\psi \to \gamma(\eta\pi\pi)$ shows a possible peak at 1390 MeV, with a strong $\delta \to \eta\pi$ signal.[16,20,30] There might be a slight peak at 1280 MeV. There is no sign of any $\eta\pi\pi$ enhancement in the ι/E region.

5) The decay $J/\psi \to \gamma K\bar{K}\pi$, which started it all,[2,16,17,20] shows a broad ι/E peak centered at 1460 MeV and will be examined, in detail, below.

6) There is no clear ι signal in purely hadronic J/ψ decay.[26] The decays $J/\psi \to \omega K\bar{K}\pi$ and $J/\psi \to \phi K\bar{K}\pi$ go preferentially to ωE (determined experimentally to be 1^{++}) and ϕD respectively, with the $K\bar{K}\pi$ coming from the $K^*\bar{K}$ mode. We call D the mass region around 1275 MeV, though the events could be associated with the $\eta(1275)$ as well. The E is not made together with the ϕ $(B(J/\psi \to \omega E)/B(J/\psi \to \phi E) > 6)$, nor is the D made together with ω. This behavior indicates that the D and E 1^{++} states may not be "ideally" mixed but may be mixtures of strange and nonstrange quarks, as are the η and η' states.

7) The reactions $\pi^- p \to (K\bar{K}\pi)n$ and $(\pi^+/p)p \to (\pi^+/p)(K\bar{K}\pi$ or $\eta\pi\pi)p$ each have an E bump for which a Dalitz plot analysis gives $J^{PC} = 1^{++}$.[24,25]

8) The first reaction considered in 7 has been measured in a different experiment,[23] but a partial wave analysis gives $J^{PC} = 0^{-+}$ for the E, as does $\bar{p}p \to (K\bar{K}\pi)Y$.[27]

9) The reaction $\pi^- p \to (\eta\pi\pi)n$ shows an E peak, with a partial wave analysis giving $J^{PC} = 0^{-+}$.[21,22]

Obviously, all of this data cannot be reconciled! Items 8 and 9 are inconsistent with item 7.

2.3. ι REGION

2.3.1 $\iota \to \gamma\gamma$. A "classic" glueball signal has been $\Gamma(G \to \gamma\gamma) = 0$. It is therefore important to ask why ι is not seen in $\gamma\gamma$ production.

1) ι could be a 1^{++} state, and the $\gamma\gamma$ decay would be forbidden by Yang's Theorem. This is unlikely because it disagrees with the 0^{-+} assignment experimentally determined in this mass region from the decay $J/\psi \to \gamma K\bar{K}\pi$.[19]

2) ι could be a nearly pure glueball state. Glueballs have $\Gamma(G \to \gamma\gamma) \to 0$ because gluons carry no electrical charge. This description seems unlikely for two reasons.

a) $\iota \to K\bar{K}\pi$ with a large width, $\Gamma \sim 50$ Mev , assuming that to be the width of the high mass region. A state that couples this strongly to quarks has large quark admixtures, by unitarity, i.e., mixing quark states with glue increases the width of a state.

b) It is possible to construct a pure quark state that does not couple to $\gamma\gamma$ and thus mocks up the $\gamma\gamma$ behavior of pure glue. The $\gamma\gamma$ decay amplitude of a state composed of quarks measures the square of the quark flavor charge weighted by the amount of each flavor in the state wave function. Assuming that the radial dependences of each quark flavor are identical, an electromagnetically "inert" state has the quark composition:[31]

$$\text{"Inert" } \iota = (u\bar{u} + d\bar{d} - 5s\bar{s}) \,/\, \sqrt{27} \tag{1}$$

Such a state is close to a pure $s\bar{s}$ state. The $\gamma\gamma$ decay amplitude may be written as: $A(\iota \to \gamma\gamma) \sim [(2/3)^2 + (1/3)^2 - 5(1/3)^2] = 0$. Therefore, a vanishing $\gamma\gamma$ decay width for a state does not necessarily imply that the state that does not decay is a glueball. Any state that is a mixture of glueball and an inert flavor state will also not couple to $\gamma\gamma$. The same composition also assures that ι will not couple to $\gamma\gamma^*$.

We assume that the ι is such a linear combination, i.e.,

$$|\iota\rangle \sim \alpha|G\rangle + \beta/\sqrt{27}\,\{|u\bar{u}\rangle + |d\bar{d}\rangle - 5|s\bar{s}\rangle\} \;. \tag{2}$$

The relative amount of glueball to quark content, α/β, must be determined from other data. Such a linear combination is not in conflict with any two photon data. It can couple to $\gamma\rho$ and to $\gamma\phi$. States orthogonal to ι will couple more strongly to $\gamma\gamma$ than does ι; $\eta'(960)$ and $\eta(1275)$ should show this effect, coupling to $\gamma\gamma$.

The current experimental bound on $\Gamma(\iota \rightarrow \gamma\gamma)$ is given by $\Gamma(\iota/E \rightarrow \gamma\gamma)B(\iota/E \rightarrow K\bar{K}\pi) < 1.6$ keV.[28] It is not necessary to have an exactly "inert" quark flavor mixture to satisfy the bound , $\Gamma(\iota/E \rightarrow \gamma\gamma) < 1.6$ keV (for $B(\iota/E \rightarrow K\bar{K}\pi) = 1$). For an arbitrary mixture of nonstrange and strange quarks specified by a mixing angle, θ, we may take

$$|\iota\rangle \sim \alpha|G\rangle + \beta\{\cos\theta|(u\bar{u} + d\bar{d})/\sqrt{2}\rangle - \sin\theta|s\bar{s}\rangle\}. \qquad (3)$$

Such a state decays to $\gamma\gamma$ with a rate proportional to $[\beta^2/(\alpha^2 + \beta^2)] \times [1 - (\sqrt{2}/5)\tan\theta]^2$. The limit of 1.6 keV implies two solutions: $\theta = 62.3°$, with a quark content $\sim (u\bar{u} + d\bar{d} - 2.7s\bar{s})$; and $\theta = 79.1°$, with a quark content $\sim (u\bar{u} + d\bar{d} - 7.3s\bar{s})$. The inert state corresponds to $\theta = 74°$, with a quark content $\sim (u\bar{u} + d\bar{d} - 5s\bar{s})$.

2.3.2. <u>$\iota \rightarrow \gamma\rho$</u>.[16,17,30] As noted in section 2.1, item 3, $\iota/E \rightarrow \gamma\rho$ has a very large width, 1.9 MeV. If only half of the 1.9 MeV width is attributed to ι, then the vector dominance model would predict that $\Gamma(\iota \rightarrow \gamma\gamma) > 7.5$ keV, assuming only that ι is approximately an SU(3) singlet. This value is in clear contradiction to the 1.6 keV bound discussed above. This contradiction is solved by using Eq. (3), for we may then have a small $\gamma\gamma$ rate together with a large $\gamma\rho$ rate.

2.3.3 <u>Strong Decays</u>. The decays $J/\psi \rightarrow \omega K\bar{K}\pi$ and $J/\psi \rightarrow \phi K\bar{K}\pi$ are not expected to be strong sources of glueball states, because there is no hard gluon channel as in $J/\psi \rightarrow \gamma K\bar{K}\pi$. The data for both hadronic decays do not show a visible ι bump in the $K\bar{K}\pi$ channel.[26] The E is seen in conjunction with the ω,[26] suggesting that the quark content of the E is greater than that of the ι.

2.3.4 $J/\psi \rightarrow \gamma(K\bar{K}\pi)$.[2,16,17,19,20] The data on $J/\psi \rightarrow \gamma(K\bar{K}\pi)$ show a large rate for the ι/E system, in contrast to the hadronic decays described above. The radiative process is fundamentally different from purely hadronic processes. Here the production of the ι goes via its glue content. Therefore, the large rate is consistent with the conclusion that ι has a substantial amount of glue. However, for a system of several states, as we have assumed, the radiative branching fractions, $B(J/\psi \rightarrow \gamma E)$ and $B(J/\psi \rightarrow \gamma\iota)$ must be determined individually, requiring at least a 2 - resonance fit. Any quantitative conclusions about the glue content of these states await such a reevaluation.

2.3.5 ι Decay Modes. If the quark content of ι is mostly strange quarks (similar to the inert mixture), then only $K^*\bar{K}$ isobars should be seen in the $K\bar{K}\pi$ decay mode; if the quark content of ι is mostly light quarks then both $K^*\bar{K}$ and $\delta\pi$ should be seen. Assuming that δ is made of light quarks, observation of the $\delta\pi$ mode would let us conclude that ι is primarily made of light quarks (a conclusion not valid if the δ is a $K\bar{K}$ molecule).

2.3.6 $\eta\pi\pi$ Mode. As noted earlier, ι is not seen in $J/\psi \rightarrow \gamma(\eta\pi\pi)$. This three body channel has contributions from the two body channels, $\delta\pi$ and $\eta\varepsilon$. Both of these two body channels can only be produced via the light quark content of the ι, and are therefore suppressed in our model (the inert ι is mostly strange quark). (While the η has an $s\bar{s}$ content, the quark diagram for this production is disconnected and therefore carries a suppression factor.) It has also been suggested by two of us,[32] extrapolating from an experiment in the $\eta(1275)$ region, where $\delta\pi$ and $\eta\varepsilon$ amplitudes interfere destructively,[21] that a similar interference occurs in the ι region.

2.3.7 ι Hadronic Production - π^-p, π^+p, $\bar{p}p$, pp. All of the hadron production processes for the ι are expected to be suppressed because of the limited amount of light quark in the ι. The ι could be produced here via its strange quark content by popping strange quarks from the vacuum, but this process carries its own OZI suppression factor. We expect substantially more ι to be produced in a K^-p reaction,[33] where the strange quark in the K^- can enhance the production process.

3. LOW MASS REGION - E(1420)

3.1 Hypotheses for J^{PC} and Quark/Glue Content

3.1.1 E is mainly a quark state. It is strongly produced in hadronic experiments, is seen in $\gamma\gamma^*$, and it is (we have argued) seen in $\gamma\rho$.

3.1.2 $J^{PC} = 0^{-+}$. Partial wave analyses in $\pi^- p \rightarrow \eta\pi\pi n$, $\pi^- p \rightarrow K\bar{K}\pi n$, and $\bar{p}p \rightarrow K\bar{K}\pi Y$ give 0^{-+} for E(1420). This assignment is contradicted by:

a) Its observation in $\gamma\gamma^*$, but not in $\gamma\gamma$, indicating 1^{++};

b) Dalitz plot analyses in $(\pi^+/p)p \rightarrow (\pi^+/p)(K\bar{K}\pi$ or $\eta\pi\pi)p$ and $\pi^- p \rightarrow K\bar{K}\pi n$ that also give 1^{++};

c) $J/\psi \rightarrow \omega E \rightarrow \omega K^* \bar{K}$ where the E is found to be 1^{++}.

3.1.3 $J^{PC} = 1^{++}$. 1^{++} spectroscopy is a mess. This makes any attempt to correlate the production and decay properties of a 1^{++} E(1420) with other members of a 1^{++} nonet particularly difficult. The 1^{++} $I = 0$ states are the D(1280) and the E(1420). A major problem in deciding on the quark flavor constitution of these states is the continuing saga of the mass of the $I = 1$, A_1 meson. The old mass was ~ 1100 MeV. The PDG lists ~ 1275 MeV.[34] A new value of ~ 1046 ± 11 MeV derived from τ decay was reported by Soding[35] at the 1987 Aspen Winter Conference. If we use the PDG value, $m(A_1) \simeq m(D)$, implying ideal mixing for a 1^{++} nonet. Then $D = (u\bar{u} + d\bar{d})/\sqrt{2}$ and $E = s\bar{s}$. On the other hand, if $m(A_1) = 1046$ MeV, the 1^{++} nonet may be analogous to the 0^{-+} nonet. The D and E, with a low-lying $I = 1$ A_1, may be non-ideal mixtures of $(u\bar{u} + d\bar{d})/\sqrt{2}$ and $s\bar{s}$ (or of SU(3) octet and singlet) as are the $\eta(548)$ and $\eta'(960)$ with the low-lying $I = 1$ pion.

A further complication in trying to achieve a consistent picture of the 1^{++} nonet lies in the discrepancies of the decay widths of the D, E, and A_1 resonances. The values for the D and E widths are relatively narrow: $\Gamma(D) = 25 \pm 3$ MeV; $\Gamma(E) = 56 \pm 3$ MeV, whereas $\Gamma(A_1) \simeq 300 - 500$ MeV.

The experimental results listed in 4.1.2. indicate that all of the experiments quoted support a 1^{++} E(1420) assignment except for the partial wave analyses for $\pi^- p \rightarrow \eta\pi\pi n$, $\pi^- p \rightarrow K\bar{K}\pi n$, and $\bar{p}p \rightarrow K\bar{K}\pi Y$. It is possible that some of the experiments are wrong. It is also possible that there are two states in the E(1420) region, with 0^{-+} and 1^{++} quantum numbers. Nearby states with different quantum numbers are common in baryon spectroscopy. Some examples are:

1) N(1520) $3/2^-$; N(1535) $1/2^-$:
2) N(1650) $1/2^-$; N(1675) $5/2^-$; N(1680) $5/2^+$:
3) N(1700) $3/2^-$; N(1710) $1/2^+$; N(1720) $3/2^+$.

4. SUGGESTED EXPERIMENTS AND ANALYSES

1) Reanalyze the data for $J/\psi \rightarrow \gamma(K\bar{K}\pi)$ to determine whether there are two distinct states in the ι/E region.

2) Reanalyze the $\gamma\rho$ bump in $J/\psi \rightarrow \gamma(\gamma\rho)$ to see if it is a superposition of a 40 MeV wide E, the tail of a high mass ι, and the low mass D(1280) and η(1275).

3) Obtain a value for (or a better limit on) $\iota/E \rightarrow \gamma\phi$.

4) Find ι/E in the hadronic production from $K^- p$ scattering.

5) Identify the two body isobar channels $\delta\pi$ and $K^*\bar{K}$ in $K\bar{K}\pi$ in both the E and ι regions.

5. CONCLUSIONS

1) The ι/E system is composed of at least two states, a high mass ι (1460 - 1500 MeV) with $J^{PC} = 0^{-+}$ and a low mass E($\simeq$ 1420 MeV) with $J^{PC} = 1^{++}$ and/or 0^{-+}.

2) The ι appears to be a state with a lot of glue and a quark structure resembling that of an inert quark state, i.e.,

$$|\iota\rangle \sim \alpha|G\rangle + (\beta/\sqrt{27}) \{|u\bar{u}\rangle + |d\bar{d}\rangle - 5|s\bar{s}\rangle\} , \tag{4}$$

with α/β undetermined but macroscopic.

3) We must analyze $J/\psi \rightarrow \gamma(K\bar{K}\pi)$ and $J/\psi \rightarrow \gamma(\gamma\rho)$ allowing for two (or more) states in order to fully understand the ι/E system.

5. ACKNOWLEDGMENTS.

We would like to express our appreciation for helpful conversations with T. Burnett, D. Caldwell, S.U. Chung, P. Fishbane, R. Gupta, D. Hitlin, and R. Partridge. This work is supported in part by the Department of Energy under contract DE-AC02-76ER01545.

6. REFERENCES.

1. Some recent reviews with extensive references to earlier work are: a) Meshkov, S., "Glueball and Meson Spectroscopy", Proc. Aspen Winter Particle Physics Conf., Aspen, CO, Jan. 1986; b) Fishbane, P.M., and Meshkov, S., Comm. Nucl. Part. Phys. 13, 325 (1984); c) Cooper, S., Proc. XXIII Intl. Conf. on High Energy Physics, Berkeley, CA, July, 1986, SLAC-PUB-4139, Nov. 1986.
2. Scharre, D. et al., Phys. Lett. 97B, 329(1980).
3. De Grand, T.A., Colorado preprint COLO-HEP-138, Dec. 1986.
4. Patel, A. et al., Phys. Rev. Lett. 57, 1288 (1986).
5. de Forcrand, Ph. et al., CERN-TH.4167/85, April 1985.
6. Berg, B. et al., Phys. Rev. Lett. 57, 400 (1986).
7. Au, K.L. et al., Rutherford preprint RAL-86-076 (1986).
8. Rosner, J.L., Phys. Rev. D27, 1101 (1983).
9. Baltrusaitis, R.M. et al., Phys. Rev. D32, 2883 (1985).
10. Rosner, J.L., Proc. Int. Symp. on Lepton Photon Interactions, Kyoto, Japan, 1985.
11. Kolanoski, H., Proc. Int. Symp. on Lepton Photon Interactions, Kyoto, Japan, 1985.
12. Donoghue, J. et al., Phys. Rev. Lett. 55, 2766 (1985).
13. Kauffman, R. and Gilman, F.J., SLAC preprint, Dec. 1986.
14. Seiden, A., VIIth Intl. Workshop on Photon-Photon Collisions, College de France, Paris, April 1986.
15. Aihara, H. et al., Santa Barbara preprint UCSB-HEP 86-12 (1986).
16. Hitlin, D., Proc. Aspen Winter Particle Physics Conf., Aspen, CO, Jan. 1986, and references cited therein.
17. Augustin, J.E. et al., Orsay preprint LAL 85/27, July 1985.

18. Chung, S.U., Proc. Aspen Winter Particle Physics Conf., Aspen, CO, Jan. 1986.
19. Edwards, C. et al., Phys. Rev. D49, 259 (1982).
20. Mallik, U., XXIth Recontre de Moriond, Les Arcs, France, March 1986, SLAC-PUB-3946.
21. Stanton, N. et al., Phys. Rev. Lett. 42, 346 (1979).
22. Ando, A. et al., Phys. Rev. Lett. 57, 1296 (1986).
23. Chung, S.U. et al., Phys. Rev. Lett. 55, 779 (1985).
24. Dionisi, C. et al., Nucl. Phys. B169, 1 (1980).
25. Armstrong, T. et al., Phys. Lett 146B, 273 (1984).
26. Becker, G.T. et al., SLAC preprint SLAC-PUB-4149, Dec. 1986.
27. Reeves, D. F. et al., Phys. Rev. D34, 1960 (1986).
28. Aihara, H. et al., Phys. Rev. Lett. 57, 51 (1986).
29. Aihara, H. et al., Phys. Rev. Lett. 57, 2500 (1986).
30. Richman, J.D., Caltech Ph.D. Thesis, CALT-68-1231.
31. This combination has also been obtained by D. Caldwell and M. Chanowitz, private communication, D. Caldwell.
32. Palmer, W.F. and Pinsky, S.S., Phys. Rev. D27, 2219 (1983).
33. Bityukov, S.I. et al., Sov. J. Nucl. Phys. 39, 375 (1984).
34. Particle Data Group, Phys. Lett. 170B, 1 (1986).
35. Soding, P., Aspen Winter Particle Physics Conf., Aspen, CO, Jan. 1987.

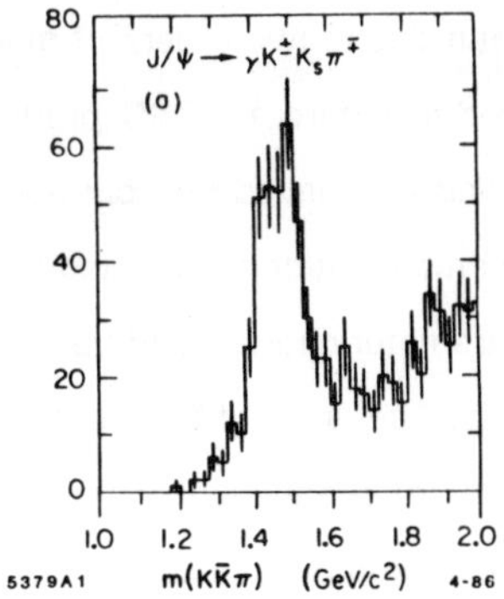

Fig. 2. The $K^{\pm}K_S\pi^{\mp}$ spectrum recoiling against a γ.

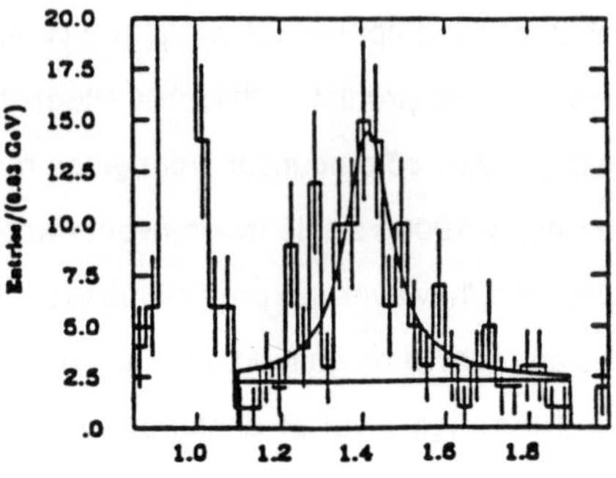

Fig. 3. Mass spectrum of the recoil system $\pi^+\pi^-\gamma_{low}$.

PRODUCTION OF SAME-SIGN DIMUONS BY 0-800 GEV NEUTRINOS AND ANTINEUTRINOS

F. Merritt, M. Oreglia, H. Schellman, B. Schumm
University of Chicago and Enrico Fermi Institute, Chicago, IL 60637

K. Bachmann, R. Bernstein, R. Blair, C. Foudas, W. Lefmann, W. Leung, S. Mishra, P. Quintas, F. Sciulli, M. Shaevitz, W. H. Smith
Columbia University, New York, NY 10027

F. Borcherding, H. E. Fisk, M. J. Lamm, W. Marsh, K. W. Merritt, D. Yovanovitch
Fermilab, Batavia, IL 60510

A. Bodek, H. Budd, W. Sakumoto
University of Rochester, Rochester, NY 14627

Presented by M. Oreglia
DPF meeting, 14-17 January 1987, Salt Lake City, UT

ABSTRACT

We present a preliminary analysis of 102 neutrino produced $\mu^-\mu^-$ and 15 antineutrino produced $\mu^+\mu^+$ same-sign dimuon events, with visible energy in the range 30 - 500 GeV. These dimuons are part of a larger sample of about 800,000 charged-current events measured by the CCFR collaboration at Fermilab. This level of same-sign dimuon production exceeds the calculated background level by about 1 standard deviation; we do not regard this as significant evidence for anomalous dimuon production.

HISTORICAL SIGNIFICANCE OF SAME-SIGN DIMUONS

From the first report of a same-sign dimuon (SSD) signal in 1973 [1)] to the two high-density dectector experiments published in 1986 [2,3)], the experimentally observed prompt rate of $(\nu\to\mu^-\mu^-)/(\nu\to\mu^-)$ has been one to two orders of magnitude higher than predicted [4c)]; this situation is depicted in Figure 1. SSD production is expected by $c\bar{c}$ production from gluon radiation [4)]. This is believed to occur about once for every 10000 single muon events when the muon momentum is restricted to exceed 9 GeV/c, however the predictions are subject to parameter choices and vary as much as an order of magnitude. The experiments are difficult because of the low overall rate and the high level of backgrounds stemming from a) production of muons in the hadron shower; b) loss of one muon in a trimuon event; and, c) accidental overlay of two charged-current events (especially in high-flux beams.) We report here on a new higher energy experiment , FNAL E744, which is less sensitive to these problems.

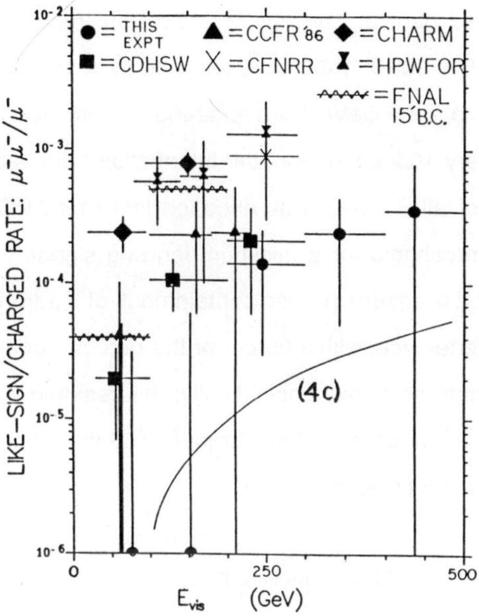

Fig. 1. SSD rate normalized to 1μ events for p_μ > 9 GeV. Solid curve is typical theoretical estimate (ref. 4). Results of this experiment are solid circles.

THE NEUTRINO BEAM

The commissioning of Fermilab's Tevatron included 8 months of fixed target running in 1985. For the neutrino area, a wide-band beam was selected. Secondary pions and kaons of both signs were focussed towards the detectors by a triplet of quadrupole magnets; the mesons decayed in a 350 m long region, yielding a beam of roughly 75% muon neutrinos and 25% antineutrinos. This beam composition resulted in a neutrino : antineutrino interaction ratio of about 9 : 1. Figure 2 shows the neutrino and antineutrino spectra which were observed at the detector and which resulted from the exposure of $5 \cdot 10^{17}$ 800 GeV protons incident on the beryllium target.

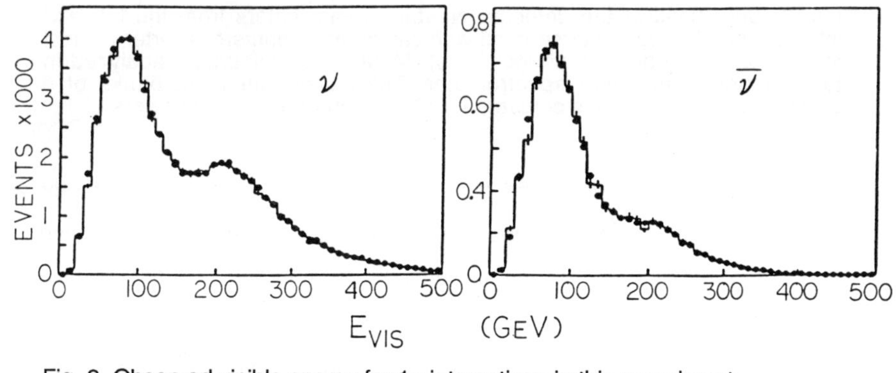

Fig. 2. Observed visible energy for 1μ interactions in this experiment. Histogram is data, dots are Monte Carlo after flux correction.

THE CCFR EXPERIMENT

The CCFR detector [5], shown in Figure 3, consists of a 690 ton iron-liquid scintillator calorimeter of dimensions 305 X 305 X 1650 cm^3, followed by a 1143 cm long toroidal magnetic spectrometer having a 2.4 GeV/c transverse-momentum kick. A scintillation counter was located after every 10.2 cm of iron; the information from each of its phototubes, as well as the sum of all 4 tubes, was encoded in 11-bit ADCs. Additionally, if the sum exceeded the threshhold for a minimum ionizing signal, the counter time was recorded in a TDC. To ensure proper containment of hadronic showers, events were rejected if their vertex was within 25 cm of the detector edges transversely, or within 100 cm of iron from the toroid longitudinally; this resulted in a fiducial volume of about 400 tons. The average density of the target is 0.53 that of solid iron; the average pion interaction length is 1.8 times that of iron.

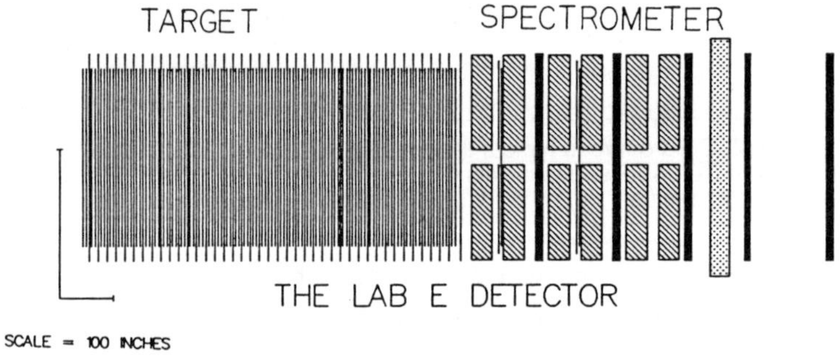

Fig. 3. Schematic of the detector. Neutrino beam enters from the left and interacts in iron target interspersed with calorimeter counters (shorter vertical lines) and drift chambers (longer lines). Muons are momentum analyzed in toroid magnet (hatched) spectrometer. Thick black lines are banks of 5 chambers; stippled area is concrete for absorbtion of delta ray showers.

For the higher-intensity and higher-energy Tevatron run, the detector was upgraded by replacing spark chambers with drift chambers and their associated TDC readout. Between the toroid magnet sections there are four banks of five consecutive drift chambers with no material between chambers; these regions, free of multiple-scattering, permit precision track timing with an accuracy of 5 nanoseconds.

The upgrade permits the detector to read out 32 events per 1 millisecond spill (there were three such "pings" in a 60 second cycle), and provides track timing with 5 nanosecond accuracy on calorimeter counter timer as well as drift chamber tracks. For multi-track events, drift chamber views in the horizontal and vertical are matched by using the information provided by two chambers tilted at approximately 10 degrees to the vertical. The calorimeter had an RMS resolution of $\Delta E_{HADRON} = 0.89\,(E_{HADRON})^{-0.5}$. Muon momentum reconstruction is dominated by the multiple scattering in iron, giving an RMS momentum resolution of $\Delta p = 0.11 \cdot p$ for essentially all momenta in the range 3-500 GeV/c.

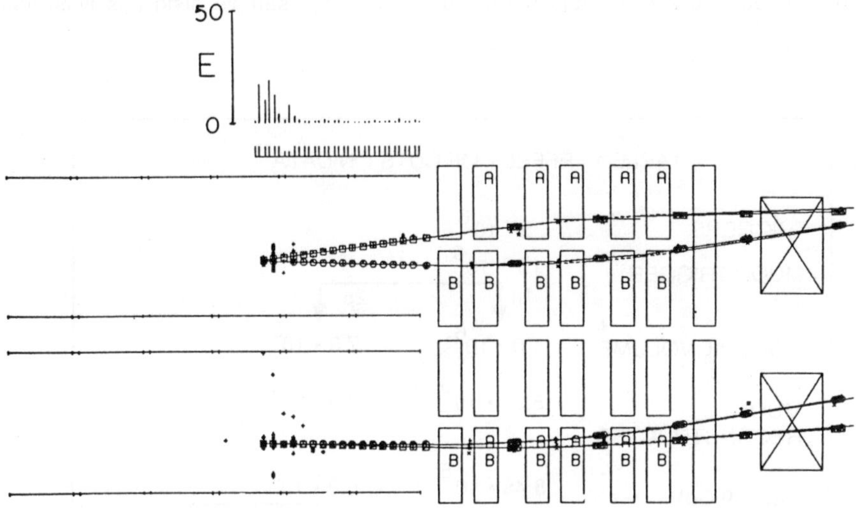

Fig. 4. Reconstruction of $\mu^+\mu^+$ event. Upper view is vertical, lower is horizontal. Histogram shows pulse height in calorimeter counters. Muon momenta are 45.4 and 32.9 GeV/c at the vertex; hadron energy is 70.3 GeV. Time differences between tracking and counter discriminator are 1.2 and 2.5 nanoseconds.

EXTRACTION OF THE SAME-SIGN EVENTS

The reconstruction of a SSD event is shown in Figure 4. The charged current events used in the SSD search had to be well contained within the detector and the muons had to traverse the good-field region of the toroid magnet. For 3% of the events the muon track momentum reconstruction had a χ^2/D.F. greater than 10. This was generally a tracking error due to delta-ray production in the toroids, and these events were corrected by physicists. The muon momenta were also required to exceed 9 GeV/c; muons of less momentum at the vertex were not likely to penetrate the target and first toroid magnet. Of the $3 \cdot 10^6$ triggers, 129 SSD candidates remain after these cuts. The efficiency of this process is now under study, but is believed to exceed 95%. The 129 event SSD sample includes trimuon events in which one muon ranges out before reaching the toroid system. Such events were eliminated if there was an in-time track longer than 9 chambers (3.1 GeV of energy loss) from the vertex. This reduced the SSD sample to 117 events. Shorter rangeouts can be hidden by the hadron shower, but are a calculable background. The SSD sample history is illustrated in Table 1.

TABLE 1: EFFECT OF CUTS ON DATA

	1μ	2μ	2μ O.S.	2μ S.S.
TRIGGERS	$3 \cdot 10^6$			
MUON TRIGGER	$1.7 \cdot 10^6$			
FIDUCIAL VOLUME	$1.0 \cdot 10^6$	$7.3 \cdot 10^3$		
MUON RECONSTRUCTED	$8.54 \cdot 10^5$	$6.4 \cdot 10^3$		
$P\mu_1 > 9$ GeV	$8.45 \cdot 10^5$	$6.3 \cdot 10^3$		
$P\mu_2 > 9$ GeV			$1.8 \cdot 10^3$	129
ELIMINATE 3μ (E > 3.1 GeV)				114
RESURRECT CHI-SQUARED FAILURES				117
				$\mu^-\mu^-$: 102; $\mu^+\mu^+$: 15

FLUX MODELING

It is necessary to know the neutrino flux for the purpose of background calculations. Flux calculations have been performed using a first order modeling of meson production and beam elements. Since it is not possible to monitor the wide-band flux directly, we use the single muon charged current events to correct the first-order calculations. This consists of reweighting the incident flux in 50 cm radial and 20 GeV visible-energy bins; the corrections are typically 5%, and are bounded by 20%. Figure 5 shows the comparison between data and flux-corrected neutrino scattering Monte Carlo for the variable $y = E_{hadron}/E_{\nu}$; the excellent agreement between data and Monte Carlo (which is based on the structure functions obtained in an earlier CCFR run[6]) is a good indicator that muon momentum and hadron energy are reconstructed correctly.

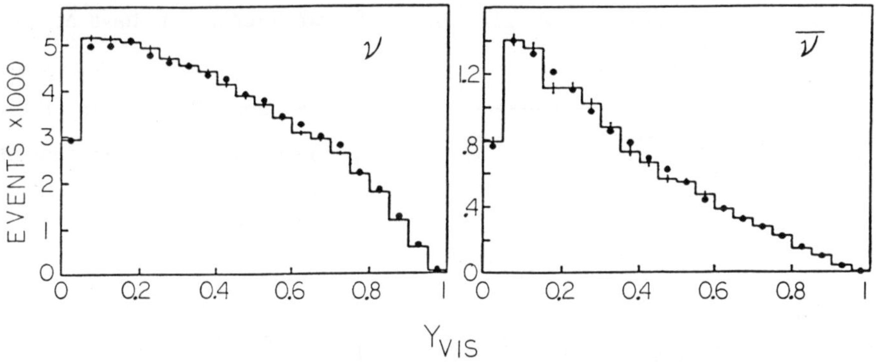

Fig. 5. Inelasticity variable $y = E_{HAD}/E_{VIS}$ for (a) neutrino and (b) antineutrino 1μ events (not corrected for detector acceptance). Histogram is data; dots are Monte Carlo calculation using CCFR structure functions from previous (lower energy) experiments.

BACKGROUND FROM ACCIDENTALS

Although accidentally coincident 1μ events, or overlapping events, have been a problem (at the 15% level) for earlier experiments, the timing information provided by the upgraded CCFR detector reduced this background to a negligible level. There are two handles on overlapping events: the time difference between the two tracks, and the distance of closest transverse approach (DCA) between the tracks. The timing was accomplished by fitting for each event a time offset for the straight track in the 5 drift chamber clusters located in the gaps between toroid magnet segments, and separately from the TDCs on the calorimeter counters; this provided track-calorimeter relative timing with an RMS accuracy of 5 nanoseconds.

The DCA was measured with an accuracy of 4 cm. The longitudinal position of the DCA had to be within about 250 cm of the event vertex so that nearly parallel tracks would not be problematic. Figure 6 demonstrates how the accidental cuts were determined. Good SSD events were required to have $\Delta t < 28$ ns (1.5 RF buckets) and DCA < 15 cm. Since the DCA distribution is independent of Δt, one can compare 6a and 6b to conclude that the accidentals background to SSD is $20 \cdot 7 / 329 = 0.5$ events.

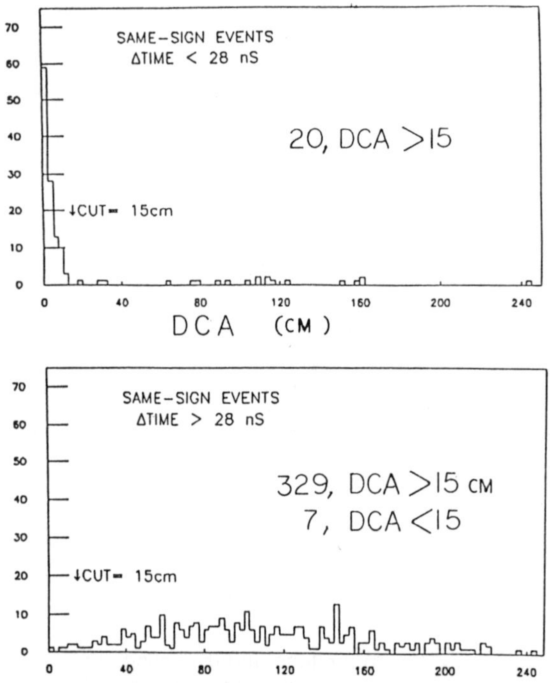

Fig. 6. DCA distribution for (a) in-time and (b) out-of-time events.

BACKGROUND FROM TRIMUONS

Trimuon events, though occurring at an extremely low level, can mimic SSD events if one muon is hidden in the hadronic shower. We calculate the spectrum for hadronically produced trimuons by using the data and parameterization of Anderson, et al.[7]. In doing this we assume that trimuons produced by neutrinos have the same spectrum as those produced by pions and protons. The trimuon Monte Carlo parameterization is then scaled so that the level predicted for observable trimuons agrees with the 53 $\mu^-\mu^-\mu^+$ observed in this experiment. Finally, cutting on the criterion for a hidden muon (E<3.1 GeV), we calculate 4.1 ν and 0.3 $\bar{\nu}$ background events; given that the level is small and that the analysis is still underway, we attribute a 100% error to these numbers.

<u>BACKGROUND FROM MESON DECAY</u>

Muons from decay of pions and kaons account for most of the SSD observed; the mechanism is shown in Figure 7. For the purpose of calculating this background, there are two very different sources: (1) direct production of mesons at the W vertex, and (2) meson production by the hadron shower in iron. The former mechanism contributes about 70% of the background and will be termed *vertex* production; the latter contributes about 30% and will be called *shower* production.

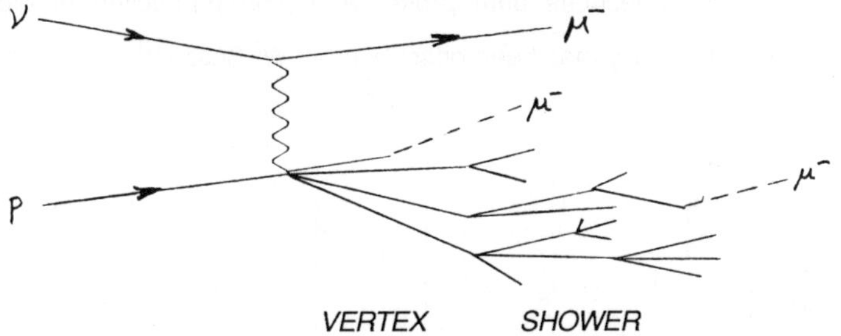

Fig. 7. Nomenclature for muons produced in the hadron shower.

Vertex production has been well parameterized by BEBC measurements of νp, νd, and νNe scattering [8]; experiments by the Fermilab 15 foot chamber and muon scattering studies by the EMC collaboration augment the BEBC data. We use the LUND Monte Carlo [9] with the BEBC/EMC choice of parameters to describe the *vertex* contribution. It is not absolutely clear as to how the *vertex* process should depend on the atomic number (A) of the target; BEBC reports no observable difference between scattering off hydrogen and neon, while EMC reports a 7% change between carbon and copper in the multiplicity of secondaries from muon scattering. Our Monte Carlo studies indicate that the effect of a 7% multiplicity increase would be to decrease the SSD background by 5%. We choose not to incorporate any non-trivial A dependence, and therefore ascribe a 5% error to the result. Other uncertainties in the *vertex* process arise from the choice of LUND parameters, and the functional form of the LUND fragmentation functions. By artificially varying parameters and shapes within the plausible bounds imposed by the data, we find a 6% uncertainty owing to the parameters (mostly the strange/u+d ratio) and 8% due to the functional forms. Added in quadrature, the three uncertainties give a total of 12%.

The *shower* process is difficult to model since it involves multiple generations in the hadron cascade; we believe it much more reliable to use measurements of muon production by pions incident on the CCFR detector. The LUND model for hadronic interactions ("LULOPT") is used to model the *shower* process; however, the LUND program is valid only for hydrogen. A model for the atomic number and x_{LAB} dependence, which decribes both proton and pion production of pions and kaons more or less universally, has been described by W. Busza[10]. We have re-weighted

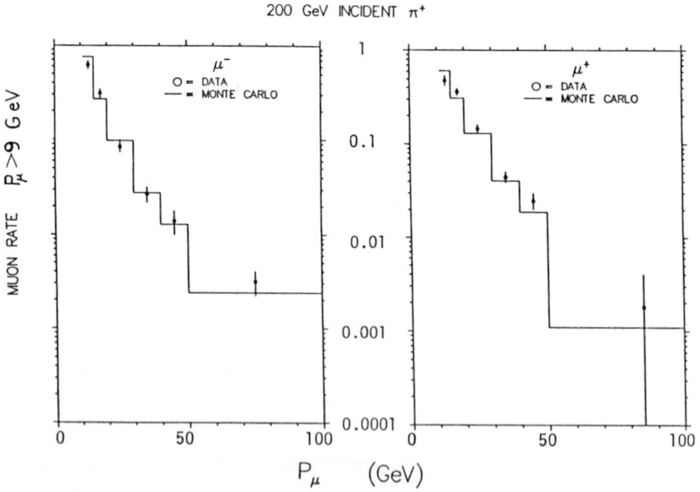

Fig. 8. Comparison of pion test beam data with shower Monte Carlo. Muon rate is number of p>9 GeV/c muons per incident pion. Monte Carlo has been boosted by a factor of 1.2 to agree with the aggregate data.

Monte Carlo secondaries by the Busza function of x_{LAB}, and have compared this to test beam muon spectra for incident π^+ and π^- of momenta 40-300 GeV/c into the CCFR detector (Figure 8). Based on this comparison, we have boosted the Monte Carlo predicted rates by 1.25+-0.35 to agree with the test beam measurements. These studies are still quite active, as we hope to incorporate more data on nuclear A and x_F dependence, and we will be taking more test beam data in 1987; the 28% uncertainty is expected to come down to 10%-15% in the near future. Uncertainty in the *vertex* process also contributes to into the *shower* uncertainty, giving a total *shower* uncertainty of 30%. Additionally, there is a 4% uncertainty on the pion and kaon absorption cross section which directly scales the predicted background rate. The total uncertainty on background SSD production is 14%.

RESULTS

The SSD data are given in Table 2 and Figure 9. For the visible energy range 30-600 GeV, data exceeds the background calculation by 13.4 ± 16.0 in the case of neutrino events, 5.1 ± 4.1 for antineutrino events, and 18.5 ± 16.5 for the total. Comparisons between data and Monte Carlo of kinematical variables indicate no unusual structure in the data. Figure 10 shows the distribution of the momentum of the lower energy muon. Here it is apparent that the SSD data are well described by the background Monte Carlo for all muon momentum cuts above 9 GeV/c; the excess of observed SSD over background is somewhat smaller for momentum cuts greater than 15 GeV/c. Referring back to Figure 1, we find that our new data and analysis are consistent with those of CDHSW [2)] (which claimed to observe a 2 standard deviation signal) or with the previous CCFR data[3)], yet the new analysis definitely gives a lower signal, and in fact no signal in the lower energy bins. We find no convincing indication of anomalous SSD production. The 90% confidence-level upper limits on a signal are:

SAME SIGN $\mu\mu / 1\mu$ $< 1.1 \cdot 10^{-4}$ for $30 < E_{vis} < 600$ GeV,

$< 5.8 \cdot 10^{-5}$ for $E_{vis} < 230$ GeV,

$< 3.6 \cdot 10^{-4}$ for $E_{vis} > 230$ GeV .

TABLE 2: COMPARISON OF DATA TO CALCULATED BACKGROUNDS

E_{vis} (GeV)	DATA	DATA - BACKGROUND
NEUTRINO:		
30-100	7	-0.4 ± 2.8
100-200	19	-4.7 ± 5.4
200-300	45	10.5 ± 8.3
300-400	22	6.5 ± 5.3
400-500	9	2.6 ± 3.2
ANTINEUTRINO:		
30-100	2	0.1 ± 1.4
100-200	8	4.3 ± 2.9
200-300	4	0.7 ± 2.1
300-400	1	0.2 ± 1.0
400-500	0	-0.2 ± 1.0

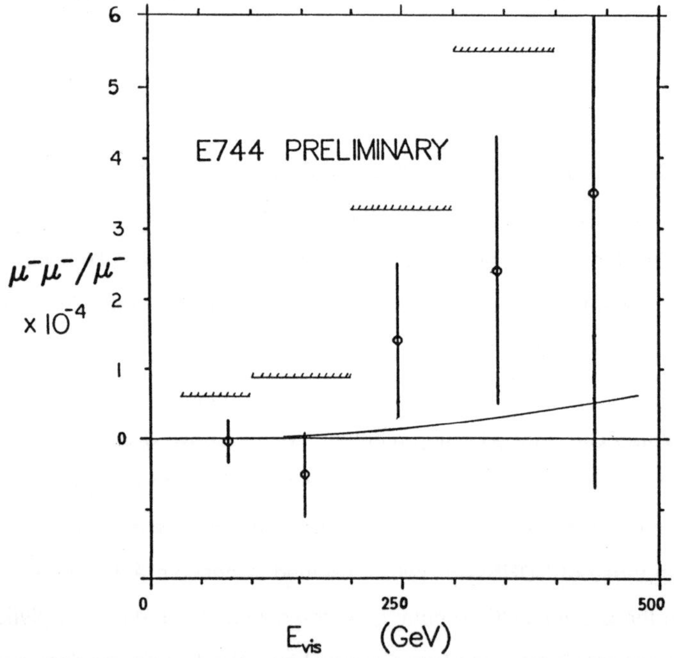

Fig. 9. SSD rate for incident neutrinos from this experiment. The solid horizontal lines are 90% C.L. upper limits. The data and theoretical curve are the same as those depicted in Fig. 1.

We expect to set stricter limits with new quadrupole triplet data which will be taken in 1987. During this new run we will also take data with particle-identified incident hadron beams in the energy range 40-100 GeV.

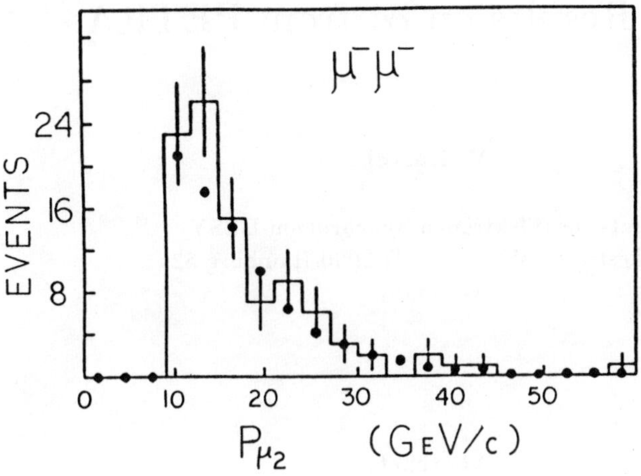

Fig. 10. Momentum of lower energy muon. Histogram is data; dots are calculated background.

REFERENCES:

1) Rubbia, C., Proc. XVII Int'l. Conf. H.E.P. IV-117, London (1974).

2) Burkhardt, H. *et al.*, Z. Phys. C31, 39 (1986).

3) Lang, K. *et al.*, Nevis Preprint 1350 (1986); to be published in Z. Phys. C.

4) a) Cudell, J. R., Halzen, F., Hikasa, K., Phys. Lett. B175, 227 (1986);
b) Barger, V., Keung, W. Y., Phillips, R. J. N., Phys. Rev. D25, 1803 (1982);
c) Young, B-L.,Walsh, T. V., Yang, T.C., Phys. Lett. 74B, 111 (1978).

5) Fisk, H. E., Sciulli, F., Ann. Rev. Nucl. Part. Sci. 32, 499 (1982).

6) MacFarlane, D. *et al.*, Z. Phys. C26, 1 (1984).

7) Anderson, K. *et al.*, Phys. Rev. Lett. 37, 799 (1976); Branson, J. G., *et al.*, Phys. Rev. Lett. 38, 1331 (1977).

8) Allasia, D. *et al.*, Z. Phys. C24, 119 (1984); Bosetti, P. *et al.*, Nucl. Phys. B209, 29 (1982); Jones, G. T. *et al.*, Z. Phys. C27, 43 (1985).

9) Sjostrand, T., Ingelman, G., CERN Pool Programs W5035/W5046.

10) Busza, W., Nucl. Phys. A418, 635 (1984).

Recent Results on α_s from PETRA.

W.Bartel

Deutsches Elektronen Synchrotron DESY
Notkestrasse 85 · D 2000 Hamburg 52
Germany

Abstract.
Some recent results on α_s are discussed. In particular the model dependence of an α_s evaluation is investigated. The strong coupling constant α_s is determined from second order QCD matrix elements combined with a phenomenological fragmentation scheme. The value of α_s depends on the fragmentation model used in the analysis. Two model independent methods are described : a determination of α_s from a measurement of the total cross section for $e^+ e^-$ annihilation into hadrons and a determination of the upper and lower limits of α_s by estimating the contribution of fragmentation. An α_s evaluated from energy energy correlations or triple energy correlations methods is found to be model dependent.

1 Introduction.

The basic idea of an α_s determination from hadronic final states in $e^+ e^-$ annihilation experiments is very simple. By the exchange of a virtual photon or Z^o, quark anti-quark pairs are produced. At sufficiently high energies an additional hard gluon may be emitted from one of the quarks, which will appear as a separate jet of hadrons in the final state. The probability for gluon emission is proportional to α_s, the strong coupling constant of QCD, as indicated in the diagrams of Fig.1. Thus two classes of hadronic final states are to be distinguished: two jet events and three jet events corresponding to two and three parton final states. The ratio of the number of two jet events to three jet events is a measure of α_s.

The actual evaluation of α_s is more involved, because second order QCD corrections are important to formulate predictions, which could be compared with data at PETRA energies; it is necessary to use second order matrix elements. Such calculations are,however, made difficult by the presence of infinities and by there being no unique prescription of treating the QCD singularities, occuring in cases of soft gluons and in case of collinear partons: thus neither theoretically nor

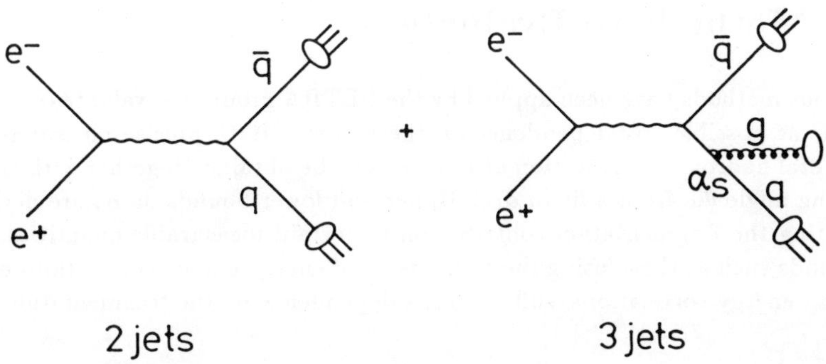

Figure 1: Basic diagrams for $e^+ e^-$ annihilation into hadrons up to first order QCD.

experimentally is it easy to define two and three jet events. Different second order matrix elements have been used giving slightly different results on α_s.

Experiments observe final state hadrons, the fragments of the initially produced partons. Therefore methods have to be developed in which QCD calculations on the parton level are implemented by a description of the hadronisation process in order to arrive at hadronic final states, which can be compared with measured data to fit α_s.

The following steps are taken to evaluate α_s.

- The expectation value for a measurable quantity $<O(\alpha_s)>$ is calculated from second order QCD matrix elements on the parton level.
- A phenomenological fragmentation model is introduced to generate hadronic final states.
- QED radiative corrections are applied.
- The response of the detector to the model events is evaluated.
- Experimentally observed distributions are compared with the QCD predictions and α_s is determined in a fitting procedure.

In the practical application of this scheme, it turns out that α_s and some of the parameters in the fragmentation models are correlated and the fitted result is model dependent. With the presently available high statistics data sample, a more accurate calculation of radiative corrections would be highly welcome, as the uncertainties introduced through the radiative corrections contribute in some cases significantly to the error on α_s.

2 Methods to Evaluate α_s.

Various methods have been applied by the PETRA groups to evaluate α_s, avoiding as far as possible any dependence on the model . With precise measurements of the total hadronic cross section at hand α_s can be obtained together with the weak mixing angle θ_W from a fit to σ_{tot}. Upper and lower bounds on α_s are derived by isolating the fragmentation contribution to certain measurable quantities. Other methods such as those using the asymmetry of energy energy correlations or triple planar energy correlations, suffer from a dependence on the fragmentation model.

2.1 Second Order Matrix Elements.

The PETRA groups have used two different second order QCD matrix elements to evaluate their data: the **ERT + AB** approach, due to Ellis,Ross and Terrano with an addition from Ali and Barreiro to treat soft gluons [1] and the **FKSS** matrix element of Fabricius,Kramer,Schierholz and Schmitt [2].

There is a basic difference in treating QCD singularities. While FKSS use the Sterman-Weinberg type of jet definition with energy flow and angular cutoff parameters corresponding to typical experimental resolutions of $\epsilon = 0.2$ and $\delta = 40^o$, the ERT + AB matrix element distinguishes one and two jets by the normalised two parton mass $y = M_{ij}^2/s$. Two jets are counted as individual jets if $y > 10^{-5}$, which is much smaller than any experimental resolution. The cut off used by FKSS would correspond to a y-cutoff between about 0.01 and 0.03.

The two matrix elements treat gluons, which are emitted at large angles with respect to the quark lines but are too soft to be accepted as a separate jet, in a different way. The ERT + AB matrix element combines the soft gluon with one of the quarks such that the mass $M_{qg} = \min$ is minimum. In the FKSS approach, soft gluons are omitted and the quark energies are rescaled. The different treatment of soft gluons accounts for about a 5% difference in α_s, ERT + AB giving lower values.

The FKSS matrix element used is not quite complete and terms of order ϵ and δ^2 have been neglected. It is estimated that α_s values would decrease by between 10% and 15% if all missing terms were added depending on the measured quantities used to determine α_s. A more complete FKSS matrix element is now available [3], although some adjustments are required, before it can be included in the presently used Mt.Carlo generators.

The ERT matrix element has a technical difficulty when applied in Mt.Carlo programs. The matrix element is not a smooth function of kinematic variables and event generation is slow and it is difficult to estimate Mt.Carlo errors.

2.2 Fragmentation Models.

Two classes of models have been used to evaluate α_s , independent fragmentation models and string fragmentation models, which are represented by Ali et al. [4] and the Lund group [5] respectively.

Independent fragmentation models assume that all partons, generated according to QCD matrix elements, fragment independently of each other, whilst the Lund string model assumes it is the massless strings between quarks and gluons which fragment into hadrons. Both models fit observed particle distributions and multiplicities fairly well but may give different results for α_s.

Independent fragmentation models have an additional uncertainty which has an impact on the evaluation of α_s. Momentum conservation after fragmentation can be enforced in different ways. Ali et al. boost all momenta in the direction of the missing momentum and then rescale all energies by a common factor. Hoyer et al. [6] first enforce local conservation of transverse momenta in each jet, then rescale longitudinal momenta by a factor which is chosen such that the ratio of the jet momentum to the momentum of the corresponding parton is the same for all partons. The Ali et al. scheme conserves the relative parton energies but not parton directions , while the Hoyer et al. scheme conserves approximately parton directions but not parton energies.

Parton shower models, like the one of Webber [7] , cannot be used in an α_s determination, since in this type of model, α_s is not a well defined quantity which can be evaluated in a fit.

2.3 Radiative Corrections.

Before evaluating α_s from measured data, radiative corrections have to be applied [8]. The errors introduced by radiative corrections are of different origin.

- The vacuum polarisation correction factor at PETRA energies is typically $\delta_{vac} = (10 \pm 0.3)\%$. The error on the correction factor is due to insufficient knowledge of the total hadronic cross section $\sigma_{tot}(e^+ e^- \rightarrow \text{hadrons})$ at lower center of mass energy and in the neighbourhood of the measured point.
- Initial state radiation reduces the center of mass energy of the annihilation process. If a photon of momentum k is emitted, the c.m. energy is reduced to $s' = s\,(1 - k/E_{beam})$. Due to experimental errors on $\sigma_{tot}(s')$, the errors introduced by radiative corrections are typically $\pm 0.9\%$ at 14 GeV and $\pm 0.5\%$ at 30 GeV on the total cross section.
- No radiative corrections of order α^4 have been calculated. They are estimated to be of the order of $(1 \pm 1)\%$ [9].

The errors introduced by radiative corrections are of the order of 1% and thus compatible with other experimental uncertainties. Errors of this magnitude have, however, a significant impact on the accuracy to which α_s can be determined. A 1% systematic change of σ_{tot} results in a 15% change of α_s. Therefore an attempt should be made to improve the accuracy of radiative corrections especially in view of the new high energy $e^+ e^-$ colliders, which come into operation soon.

2.4 α_s Derived from the Total Hadronic Cross Section.

The ratio R between the electro-weak annihilation cross section of $e^+ e^-$ into hadrons and the purely electromagnetic part of the μ pair cross section

$$R = \frac{\sigma(e^+ e^- \to \gamma, Z^o \to hadrons)}{\sigma(e^+ e^- \to \gamma \to \mu^+\mu^+)}$$

can be written in terms of the weak mixing parameter $\sin^2\theta_W$ and the QCD coupling constant α_s in second order.

$$R = 3 \cdot \sum_{N_f} \frac{\beta}{2}(3-\beta^2) \cdot \left(1 + C_1 \frac{\alpha_s}{\pi} + C_2 \cdot \left(\frac{\alpha_s}{\pi}\right)^2\right) \cdot C_{VV}$$

$$+\beta^2 \left(1 + C_1 \cdot \frac{\alpha_s}{\pi} + C_2 \cdot \left(\frac{\alpha_s}{\pi}\right)^2\right) \cdot C_{AA}$$

with

$$\alpha_s = \frac{12\pi}{(33-2N_f)\ln s/\Lambda^2 + 6\frac{153-19N_f}{33-2N_f} \cdot \ln\ln s/\Lambda^2}$$

$$C_{VV} = Q_f^2 - 2Q_f v_e v_f \cdot Re(\chi(s)) + (v_e^2 + a_e^2)v_f^2 \mid \chi(s) \mid^2$$

$$C_{VV} = (v_e^2 + a_e^2)a_f^2 \mid \chi(s) \mid^2$$

$$\chi(s) = \frac{G_F}{8 \cdot \sqrt{2}\pi\alpha} \cdot \frac{s \cdot M_Z^2}{s - M_Z^2 + iM_Z\Gamma_Z}$$

$$\lim_{\beta \to 1} C_1 = 1$$

$$C_2 = 1.99 - 0.115 \cdot N_f$$

The weak vector and axial vector charges of electrons and quarks of flavour f are denoted by v_e, v_f and a_e, a_f respectively.

The CELLO group has recently published a set of accurate results on the total hadronic cross section for 14 GeV$\leq \sqrt{s} \leq$ 46.6 GeV [10]. The statistical errors vary between 2.0% and 8.5%, the point to point errors are between 1.2% and 1.7% and the over all normalisation error is 1.7%.

The new CELLO data together with previous measurements of other groups at PEP and PETRA have been fitted to the second order QCD cross section of the electro-weak annihilation process into quark- anti-quark pairs to determine $\sin^2\theta_W$ and α_s. This method of evaluating α_s is essentially model independent,

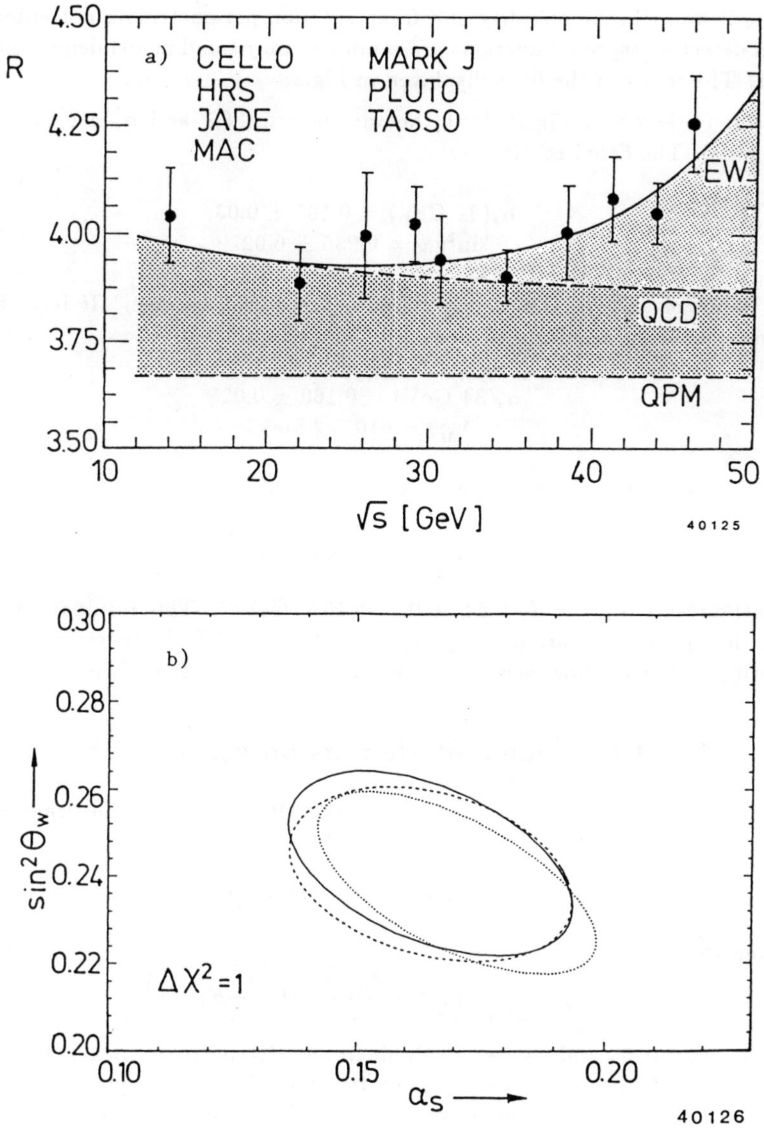

Figure 2: a) Fitting $\sin^2\theta_W$ and α_s. b) χ^2 contour plot in the $\sin^2\theta_W$,α_s plane. The dotted contours correspond to different schemes for handling normalisation errors between experiments.

as σ_{tot} is an inclusive quantity and fragmentation parameters do not enter. Only when detector response functions are computed may small model dependent terms enter. The result of the fit is displayed in Fig.2.

It is observed in Fig.2b that the variables $\sin^2\theta_W$ and α_s are only weakly correlated. The fitted result is

$$\alpha_s(34\text{ GeV}) = 0.165 \pm 0.03$$
$$\sin^2\theta_W = 0.236 \pm 0.02$$

The value of $\sin^2\theta_W$ comes out close to the world average. If it is fixed to $\sin^2\theta_W$=0.23, the result on α_s does not change significantly:

$$\alpha_s(34\text{ GeV}) = 0.169 \pm 0.025$$
$$\Lambda_{\overline{MS}} = 610^{+470}_{-340}\text{ MeV}$$

A fit has been made to the energy dependence of α_s

$$\alpha_s = \text{a} + \text{b}\ (\sqrt{s}\ \text{-}\ 34)$$

with the result $\text{b}_{exp} = (-0.64 \pm 0.73) \cdot 10^{-3}\ \text{GeV}^{-1}$. This result is compatible with the QCD expectation of $\text{b}_{QCD} = -1.3 \cdot 10^{-3}\ \text{GeV}^{-1}$. In view of the large experimental error, however, one cannot really conclude that α_s is running.

2.5 Model Independent Bounds on α_s.

For some measurable quantities $< O >$ the QCD part and the fragmentation dependent contributions F can be separated

$$< O >= C_1\alpha_s \cdot (1 + C_2\alpha_s) + F$$

leading to:

$$\alpha_s = \frac{1}{2C_2} \cdot \left(\sqrt{1 + \frac{4C_2}{C_1}(< O > -F)} - 1\right)$$

The coefficients C_1 and C_2 are calculable from QCD matrix elements. It turns out, that for some observables the fragmentation term F has a unique sign. Therefore by neglecting F, upper and lower bounds on α_s can be derived depending on whether F $>$0 or F$<$0 [11].

The JADE collaboration [12] employed the following variables for evaluating limits on α_s:

- $< 1 - T >$, where T is the event thrust variable.
- $\int_{45^\circ}^{90^\circ} A_{EEC}(\theta)d\theta$, the asymmetry of the energy energy correlation integrated over an angular region, where the influence of fragmentation is small.

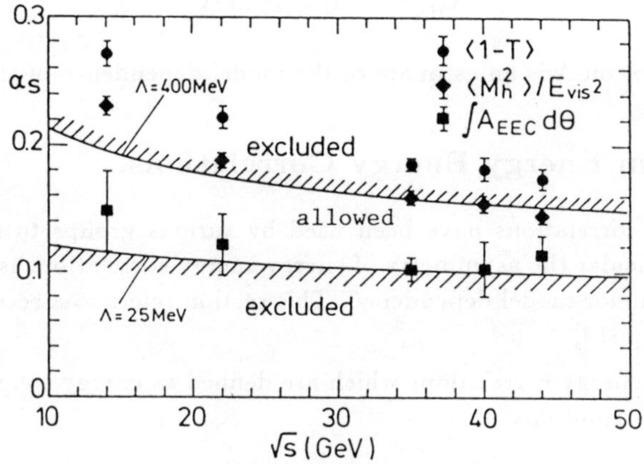

Figure 3: Limits on α_s.

- $< M_H^2 > / E_{vis}^2$,the average heavy jet mass squared normalised to the visible energy as introduced by Clavelli [13]. M_H is obtained by dividing an event into two hemispheres in such a way that the sum of the invariant masses squared of all particles in a hemisphere is a minimum, i.e. $M_H^2 + M_L^2 = \text{min}$.

For these variables the QCD coefficients calculated to second order α_s using ERT matrix elements are listed in Table 1 together with the sign of F.

Table 1

QCD parameters and sign of the fragmentation term F.

	C_1	C_2	F
$< 1 - T >$	0.334	2.88	> 0
$\int_{45^\circ}^{90^\circ} A_{EEC}(\theta) d\theta$	0.248	0.859	< 0
$< M_H^2 > / E_{vis}^2$	0.334	2.09	> 0

The limits on α_s of Fig.3 are obtained by assuming the tightest constraints which occur at a particular energy, and extrapolating in energy by using the QCD evolution of α_s given in the previous chapter.

$$25 \text{ MeV} \leq \Lambda \leq 400 \text{ MeV} \qquad \text{at } 95\% \text{ CL}$$
$$0.10 \leq \alpha_s \leq 0.15 \qquad \sqrt{s}=35 \text{ GeV}$$

The fragmentation term decreases with energy, and by fitting a $1/\sqrt{s}$ dependence one obtains values for α_s, which are only slightly model dependent. In the relation

$$< O > = C_1 \alpha_s(\sqrt{s}; \Lambda) \left\{ 1 + C_2 \alpha_s(\sqrt{s}; \Lambda) \right\} + K/\sqrt{s}$$

Λ and K were treated as free parameters in a fit to the data.

$$\Lambda_{\overline{MS}} = 150^{+90+60}_{-70-50}\ \mathrm{MeV}$$

The second error on Λ is an estimate of the model dependence of the method.

2.6 α_s from Energy Energy Correlations.

Energy energy correlations have been used by various groups to determine α_s, because in particular the asymmetry of energy energy correlations is a measurable quantity with minor model dependence. This section refers to a recent analysis of the JADE group [14].

The energy energy correlations which are defined as energy weighted two particle angular distributions

$$\frac{d\Sigma}{d\theta} = \frac{1}{N}\sum_{k=1}^{N}\sum_{i,j} x_i^k x_j^k \frac{1}{\Delta\theta}\int_{\theta-\Delta\theta/2}^{\theta+\Delta\theta/2} \delta(\Theta_{i,j}^k - \bar{\Theta})d\bar{\Theta}$$

with

$x_i = \frac{E_i}{E_{cm}}$	$x_j = \frac{E_j}{E_{cm}}$	$\Theta_{i,j} = \angle(\vec{p_i}, \vec{p_j})$
i,j	count the particles in an event	
k	counts the number of events	

The energy energy correlations are symmetric around 90° for two jet events, while they are asymmetric for three jet events. Thus the asymmetry of energy energy correlations is a measure of the three jet fraction and can be employed to determine α_s. The asymmetry is defined as

$$A(\Theta) = \frac{d\Sigma(\pi - \Theta)}{d\Theta} - \frac{d\Sigma(\Theta)}{d\Theta}$$

However, before this method can be used to evaluate α_s, events in which a hard photon is radiated in the initial state are required to be removed from the data sample, since such events also contribute to an asymmetry in the energy energy correlations. Thus events with an isolated photon with $E_\gamma > 3$ GeV have been omitted from this analysis.

Fig.4 shows results of the JADE collaboration, the measured energy energy correlation asymmetries at three center of mass energies are fitted to QCD predictions on the parton level, i.e. without fragmentation [15]. It is observed that for angles above $\Theta = 30^o$, the QCD curves fit the data well with α_s values listed in Table 2.

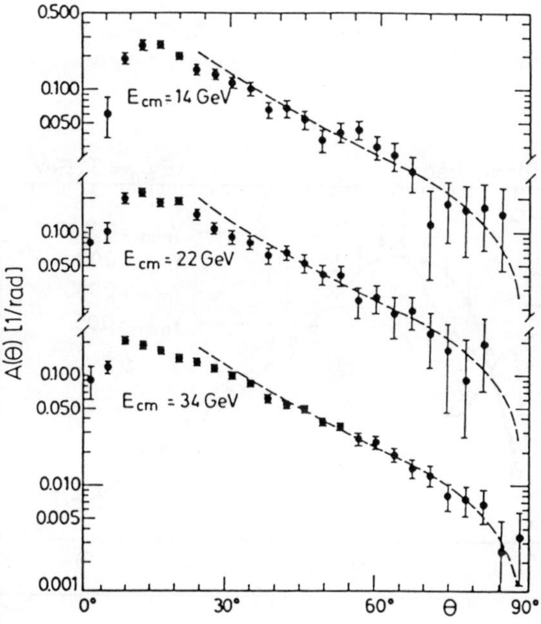

Figure 4: Fit to energy energy correlation asymmetries based on QCD predictions on the parton level.

Table 2

α_s from a fit to A(Θ) on the parton level.

E_{cm}	α_s
14 GeV	0.14 ± 0.01
22 GeV	0.13 ± 0.01
34 GeV	0.115± 0.005

At 34 GeV , where the largest data sample is available, the JADE collaboration determined α_s in a fit to the asymmetry of the energy energy correlation using the Lund model [5] to describe fragmentation. The energy energy correlation and the asymmetry are shown in Fig.5 together with the fit for various values of the cut off parameter y_{min}, α_s is found to be

$$\alpha_s(34) = 0.165 \pm 0.01 \pm 0.01.$$

An independent fragmentation model does not fit the observed energy energy correlation and thus has not been used to evaluate α_s from the asymmetry of energy energy correlations.

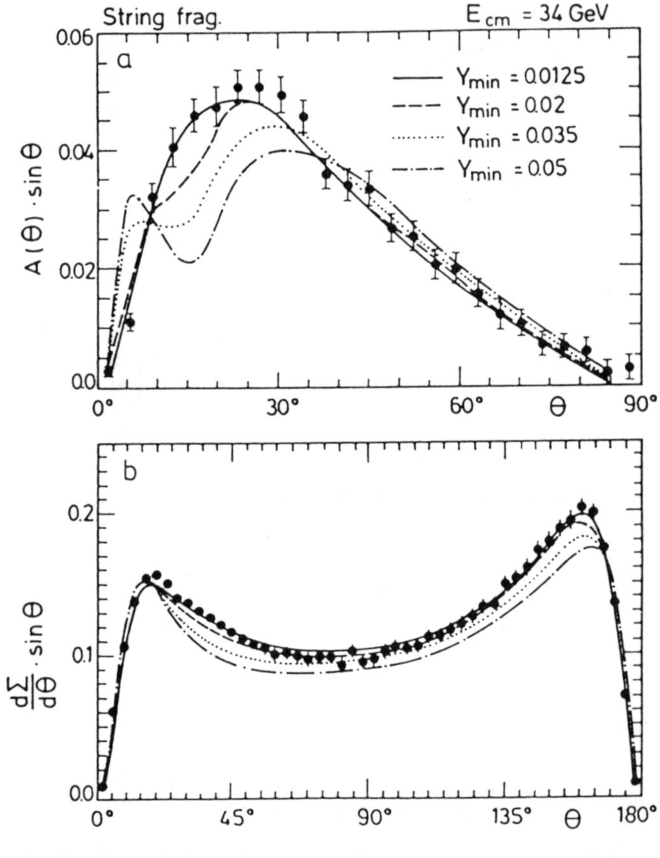

Figure 5: Fit to the asymmetry of the energy energy correlation at 34 GeV and the energy energy correlation using the Lund model .

2.7 α_s from Triple Energy Correlations.

The MARKJ group [16] used triple energy correlations (PTC) [17] to derive α_s. They are defined as

$$\frac{1}{\sigma_{tot}}\frac{d^3\Sigma}{d\chi_1 d\chi_2 d\chi_3} = \frac{1}{N d\chi_1 d\chi_2 d\chi_3}\sum_{evt}^{N}\sum_{i,j,k}\frac{E_i E_j E_k}{E_{vis}^3}$$

with

$$\chi_1 = \angle(E_j, E_k)\,;\chi_2 = \angle(E_i, E_k)\,;\chi_3 = \angle(E_j, E_i)$$

$E_{i,j,k}$ is the energy flow into solid angles i,j and k.

In order to have a measure of the three jet fraction in planar quark-quark-gloun events, the triple energy correlations are restricted to planar events by introducing the constraint

$$\mid 2\pi - \chi_1 - \chi_2 - \chi_3 \mid < 0.1\ rad$$

and defining planar triple energy correlations by

$$\frac{1}{\sigma_{tot}}\frac{d^2\Sigma}{d\chi_1 d\chi_2} = \frac{1}{N d\chi_1 d\chi_2}\sum_{evt}^{N}\sum_{i,j,k}^{plane}\frac{E_i E_j E_k}{E_{vis}^3}$$

They are a measure of planar three jet events, mathematically independent of other observables and insensitive to QCD singularities. It has been checked by Mt. Carlo calculations that radiative corrections and detector corrections are small.

The sensitivity to three jet events can be improved by integrating the planar triple energy correlation (PTC) over a region in the χ_1, χ_2 space of high three jet event density:

$$I = \int PTC(\chi_1, \chi_2) d\chi_1 d\chi_2$$

$$\left\{ \begin{array}{c} \chi_1, \chi_2 < 140^o \\ \chi_1 + \chi_2 > 220^o \end{array} \right\}$$

It has been shown that the integral I has only a 20% two jet contribution.

The integral has been calculated in second order QCD using the full ERT matrix element and two fragmentation schemes were employed, both of which fit the data equally well. The independent fragmentation model of Ali et al. [4] and the Lund string fragmentation model [5] using data between $\sqrt{s}$=22 GeV and $\sqrt{s}$=46 GeV give different results for α_s.

Model	α_s(35 GeV)
Ali et al.	0.112±0.005
Lund (string)	0.147±0.005

The results are shown in Fig.6. Averaging over the model dependence the group arrives at the following result:

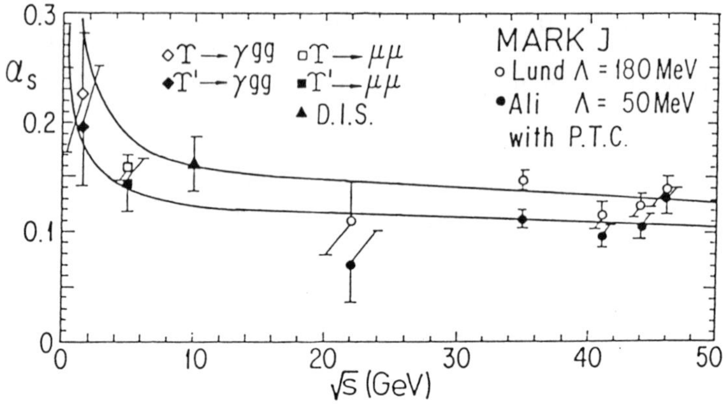

Figure 6: The measured values of α_s as a function of c.m.energy. The deep inelastic scattering data are also shown (DIS). The curves are the QCD predictions.

$$\alpha_s(44\ \mathrm{GeV})=0.121\pm0.007\pm0.010$$
$$\alpha_s(35\ \mathrm{GeV})=0.125\pm0.005\pm0.012$$
$$\Lambda = 110 \pm 30^{+70}_{-55}\ \mathrm{MeV}$$

3 Conclusions.

A compilation of second order α_s evaluations from PEP and PETRA [12] is shown in Fig.7. The results are grouped according to the type of second order matrix elements used in the evaluation, and to the fragmentation model.

- The four values of α_s marked 'neglect' are derived from a fit to the asymmetry of energy energy correlations on the parton level.
- The two model independent measurements of α_s, a fit to the total cross section and the determination of upper and lower bounds appear under the subscript 'none'.
- Assuming a $1/\sqrt{s}$ behaviour of the fragmentation term in the observables described in chapter 2.5 leads to the entry marked $1/\sqrt{s}$.
- The rest of the data points in Fig.7 are derived from fits to the asymmetry of energy energy correlations and fits to planar triple energy correlations.

Comparing the various methods used to evaluate α_s , some systematic trends are observed.

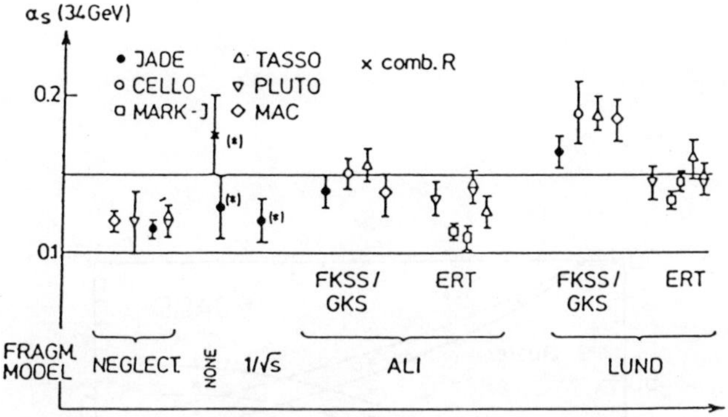

Figure 7: Compilation of α_s values derived by various methods using different 2nd order matrix elements and different fragmentation models.

- α_s values determined by fitting the second order matrix elements of FKSS are systematically 10 to 15% higher than those based on ERT matrix elements. This difference is most likely due to the approximations made by FKSS.
- Fits using the independent fragmentation model of Ali et al. are systematically low by about 15 to 20% compared to string fragmentation model calculations. This difference is probably due to different correlations between fragmentation parameters and α_s.
- The model independent evaluations of α_s agree within errors but have , however, larger errors due to the smaller sensitivity to α_s.

Although α_s has been measured over an energy range covering about 30 GeV, the accuracy is still insufficient to establish the logarithmic energy dependence predicted by QCD.

α_s has been evaluated using second order matrix elements. There are hints, that the QCD calculations used so far do not describe all features of the data correctly. The JADE collaboration [18] has compared the observed four jet rate with second order QCD predictions. As shown in Fig.8 the observed fraction of four jet events as a function of the center of mass energy is higher than the QCD prediction using second order matrix elements, while parton shower models reproduce the data better. This observation is an indication that higher order effects may play a role.

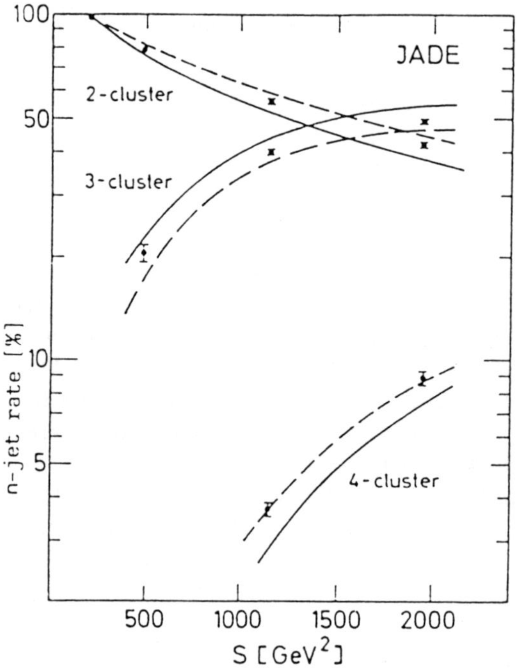

Figure 8: Fraction of n-jet events as a function of c.m. energy. 2nd order QCD (full line), parton shower model of Webber (broken line).

Acknowledgement. I greatfully acknowledge the hospitality extended to me during my stay in Salt Lake City. I thank the organisers of the meeting, in particular Professor C. DeTar, for arranging a very interesting and lively conference and D. Hitlin for giving me the opportunity to speak in his session.

References

[1] R.K. Ellis,D.Ross,A.Terrano, Phys. Lett. 45B(1980) 1226; Nucl. Phys. B187(1980) 421
A.Ali,F.Barreiro , Phys. Lett. 118B(1982)155; Nucl. Phys. B236(1984)269

[2] K.Fabricius,G.Kramer,G.Schierholz,I.Schmitt, Phys.Lett. 97B(1980)431 and Z.Phys. C11 (1982) 315.
F.Gutbrod,G.Kramer,G.Schierholz,(GKS), Z.Phys. C21(1984)235

[3] G.Kramer private communication

[4] A.Ali et al. ,Phys. Lett.,93B(1980)155
A.Ali ,Phys. Lett.,110B(1982)67

[5] B.Andersson,G.Gustafson,G.Ingelman,T.Sjöstrand, Phys. Rep. 97(1983)33 and further references therein

[6] P. Hoyer et al. Nucl. Phys. B161(1979)349

[7] G.Marchesini,B.R.Webber, Nucl. Phys. B238(1984)1
B.R.Webber, Nucl. Phys. B238(1984)492

[8] F.A.Berends,R.Kleiss, Nucl. Phys. B228 (1983) 537

[9] Y.S.Tsai , Report SLAC PUB-3129 (1983)

[10] CELLO collaboration, H.J.Behrend et al. DESY 86-133, submitted for publication

[11] R.D.Field in Proceedings of the 1083 Int. Symposium on Lepton and Photon Interactions at High Energies, Ed. D.G.Cassel and D.L.Kreinik (1983) Cornell University

[12] S.Bethke in Proceedings of the 23rd conference on High Energy Physics . Berkeley (1986)

[13] L.Clavelli,D.Wyler, Phys.Lett. 103B(1981)383

[14] JADE collaboration , W.Bartel et al. Z.Phys. C25(1984)231

[15] A.Ali,F.Barreiro, Phys.Lett. 118B(1982)155

[16] MARKJ collaboration , B.Aveda et al. MIT Technical Rep. 148(1986) to be published in Phys.Rev.Lett.

[17] F.Ciscor et al.,Phys.Rev.D31(1985)1025

[18] JADE collaboration, W.Bartel et al. Z. Phys. C33(1986)23

W/Z PRODUCTION WITH JETS IN EXPERIMENT UA1

UA1 Collaboration

Presented by

Darrel Smith

University of California
Riverside, CA 92521

ABSTRACT

The W,Z production cross-sections from the CERN $p\bar{p}$ Collider at c.m. energies 540 and 630 GeV/c are presented. The p_T distribution of W,Z and the properties of the recoil jets are compared to QCD predictions.

1. INTRODUCTION

Initial state bremsstrahlung in W production is expected to give rise to a long tail in the W transverse momentum distribution with one ore more hadronic jets produced in association with high transverse momentum W's. In the region in which QCD perturbation theory is applicable, the W transverse momentum distribution, together with the properties of the associated hadronic jets provides us with a quantitative test of QCD theory. An analysis of W production properties from the 1983 and 1984 runs appears in reference 1.

2. EVENT SELECTION AND RECONSTRUCTION

This paper describes an analysis of the W transverse momentum (p_T^W) distribution in a sample of 266 $W \rightarrow e\nu$ decays and 57 $W \rightarrow \mu\nu$ decays. The data were accumulated during the period 1982 - 1985 at center-of-mass energies $\sqrt{s}$ = 546 GeV (136 nb^{-1} for $W \rightarrow e\nu$ decays) and 630 GeV (593 nb^{-1} for $W \rightarrow$ decays). Details of the UA1 detector and of the selection criteria for the $W \rightarrow e\nu$ and $W \rightarrow \mu\nu$ events as well as the $Z^o \rightarrow e^+e^-$ and $Z^o \rightarrow \mu^+\mu^-$ events are given in reference 2. Here we briefly summarize the main selection requirements and give details of the reconstruction of hadronic jets:

i) <u>$W \to e\nu$ selection</u>. The trigger selected events containing an electromagnetic cluster with nominal transverse energy $E_T > 10$ GeV. After off-line reconstruction we require a calorimeter cluster with $E_T > 15$ GeV matching a central detector track with transverse momentum $p_T > 7$ GeV/c. The cluster must also be isolated within a cone of radius $\Delta R = (\Delta\eta^2+\Delta\phi^2)^{\frac{1}{2}} < 0.4$, where η is the pseudo-rapidity and ϕ is the azimuthal angle measured in radians. The summed $p_T(E_T)$ from extra tracks within the cone must be less than 10% of the cluster E_T and less than 3.0 GeV/c (3.2 GeV). The shower shape should be consistent with the expected shape for an isolated electromagnetic cluster. In particular, the energy measured in the associated hadronic calorimeter cells should be less than 0.6 GeV. Finally, the missing transverse energy in the fully reconstructed event must be > 15 GeV. This selection gives 266 $W \to e\nu$ candidates.

ii) <u>$W \to \mu\nu$ selection</u>. The trigger selected events with a track in the muon chambers pointing to within 150 mrad of the interaction vertex. After off-line reconstruction a good quality matching track is required in the central detector with $p_T > 15$ GeV/c. This track must be isolated within a cone of radius $\Delta R < 0.4$, with the summed E_T of calorimeter cells (excluding the expected muon deposition) < 3 GeV, and the summed p_T of extra tracks < 1 GeV/c. There must be no jet with $E_T > 10$ GeV in a larger cone of radius $\Delta R = 0.7$ in either the calorimeters or the central detector and no jet with $E_T > 10$ GeV (7.5 GeV) in the calorimeters (central detector) within ±30° of being antiparallel to the muon in the transverse plane. Finally the missing transverse energy in the event must be > 15 GeV. After appplying technical cuts to remove muons from π/K decays or cosmic rays, and validation of the muon quality by a physicist on a high resolution graphics display, we are left with 57 $W \to \mu\nu$ candidates.

iii) <u>$Z \to l^+l^-$ selection</u>. The $Z^o \to e^+e^-$ selection is similar to the W selection except the missing energy requirement is replaced by a second isolated e.m. cluster with E_T>15 GeV and a two cluster mass above 70 GeV/c^2 (32 events). Likewise, the $Z^o \to \mu^+\mu^-$ selection requires a second muon track with p_T>15 GeV/c (21 events).

iv) <u>Jet reconstruction</u>. Hadronic jets are identified by the standard UA1 jet-finding algorithm with a transverse energy initiator of 2.5 GeV/c, and are required to have a transverse momentum in excess of 7 GeV/c, and an axis within the pseudorapidity interval $|\eta| < 2.5$ [2,3]. An ISAJET[4] simulation has been used to make a detailed comparison between reconstructed jets and the corresponding partons. On average, for jets with large transverse momentum ($p_T^j \geq 20$ GeV/c), the jet energy is underestimated by about 20% and the jet momentum by about 15%. For lower transverse momenta the relationship between the observed and actual jet energies is less certain, depending critically on the details of jet fragmentation and the contribution to the jet energy from the underlying event. We have corrected all the reconstructed jets for these effects, using average correction factors which are functions of the jet E_T, η and ϕ. The estimated remaining systematic uncertainty on the jet energy scale is ~8%.

3. W,Z CROSS SECTIONS

A summary of σ K B (cross section times branching ratio) for W's and Z's is shown in table 1 for c.m. energies 546 and 630 GeV. They agree with the theoretical expectations[5]. The ratio

$$\sigma_W (\sqrt{s} = 630 \text{ GeV})/\sigma_W (\sqrt{s} = 546 \text{ GeV}) = 1.14 \pm 0.17$$

is in agreement with the theoretical expectation of 1.24[5].

Table 1: W,Z cross sections ($\sigma \cdot$ B) from experiment UA1

$\sigma \cdot$ B	$\sqrt{s}$ = 546 GeV	$\sqrt{s}$ = 630 GeV
$W^\pm \to e^\pm \nu$ (nb)	0.55 ± 0.08 ± (0.09)	0.60 ± 0.05 ± (0.09)
$W^\pm \to \mu^\pm \nu$ (nb)	0.56 ± 0.18 ± (0.12)	0.67 ± 0.08 ± (0.14)
$Z^o \to e^+ e^-$ (pb)	42^{+25}_{-18} ± (6)	73 ± 14 ± (11)
$Z^o \to \mu^+ \mu^-$ (pb)	100 ± 50 ± (15)	73 ± 19 ± (13)
$W^\pm \to \tau^\pm \nu$ (nb)	0.60 ± 0.12 ± (0.11)	

4. COMPARISON WITH QCD

The W transverse momentum distribution for the sample of 323 leptonic decays is shown in Fig. 1. The distribution has been corrected for detection and selection inefficiencies and for the effects of detector resolution so that it can be compared directly with theoretical predictions [5,6]. The overall correction factors vary as a function of p_T^W from ~2 at low p_T^W (5 GeV/c) to ~4 at very large p_T^W (100 GeV/c) for the electron channel, and are roughly a

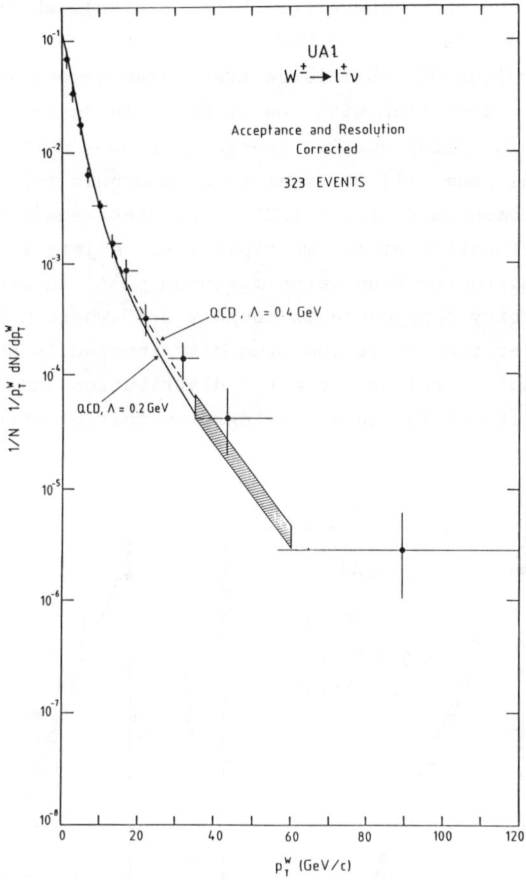

Fig. 1: The differential transverse momentum distribution of W events. The curves show the QCD predictions from reference [5]. The data are corrected for acceptance and detector resolution.

factor of four larger for the muon channel. We estimate a residual systematic uncertainty on p_T^W of ~15% of the p_T for large values of p_T^W.

The curves show the inclusive p_T^W predictions for QCD processes from reference 5; the width of the shaded band reflects the magnitude of the theoretical uncertainties. The solid curve is the inclusive p_T^W prediction using the structure functions of Duke and Owens with Λ_{QCD} = 0.2 GeV. This gives a good description of the data in the low p_T^W region, but lies systematically below the data for $p_T^W > 20$ GeV/c. Agreement can be improved by increasing the value of Λ_{QCD} =

0.4 GeV (dashed curve above the shaded region), which describes the data better out to $p_T^W \sim 40$ GeV/c.

We know from QCD that large transverse momentum W's/Z's are produced in association with one or more gluons radiated off the incoming annihilating quark. Energetic gluons emitted at large angles to the beam will be observed as hadronic jets balancing the transverse momentum of the W's/Z's. Figure 2 shows the number of W/Z events as a function of jet multiplicity. A jet, in this case, must have a reconstructed transverse momentum $p_T^{(j)}$ in excess of 5 GeV/c. The multiplicity dependence is roughly 0.3^n where n is the number of jets. The jet transverse momentum distribution is shown in figure 3. This distribution reflects the P_T^W distribution and is well described by QCD predictions for jets arising from initial-state bremsstrahlung[7].

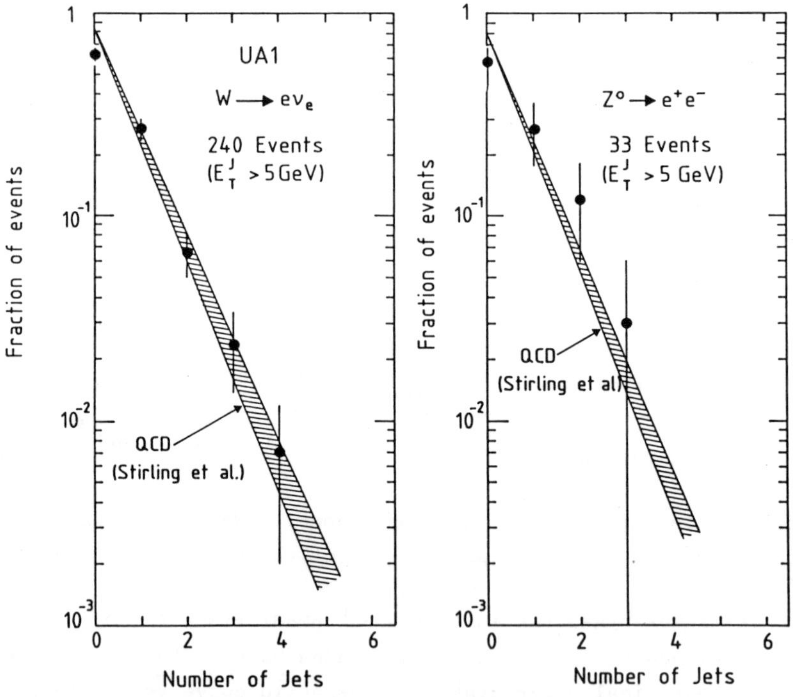

Fig. 2: The jet multiplicity distribution for $E_T^{(J)} > 5$ GeV/c associated with W and Z events. The shaded area shows the uncertainty in the QCD calculation [7,8].

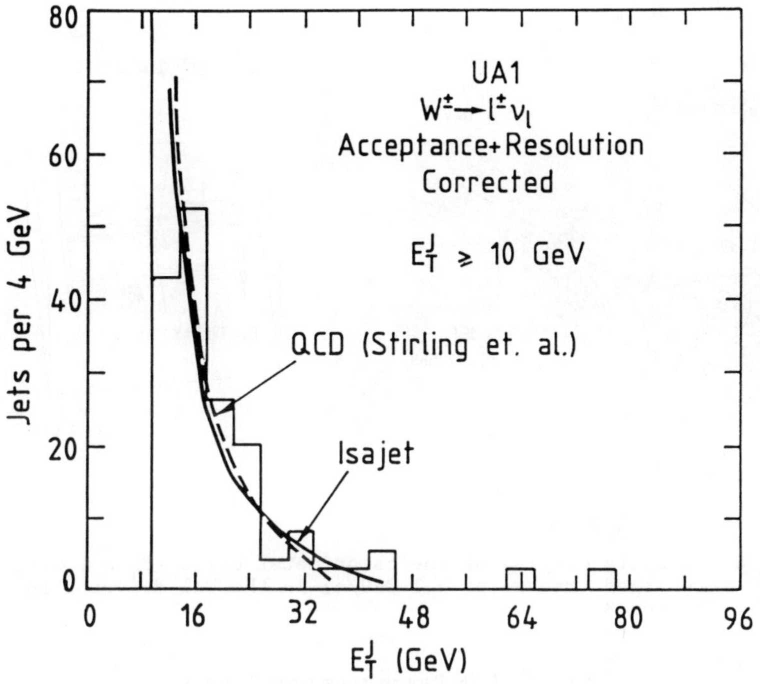

Fig. 3: The jet transverse momentum distribution for all jets produced in association with the W. The data are corrected for acceptance and detector resolution. The data agree well with ISAJET and the QCD prediction by Stirling et al.[7].

Although the jet multiplicity and E_T^{jet} distributions agree well with QCD predictions, there is a lack of W events in the measured p_T^W spectrum from ~50 GeV/c to ~80 GeV/c, followed by two events at very large transverse momentum (Table 2).

The two very high p_T^W events, labelled A and B, are shown in figures 4a and 4b. They share striking similarities in topologies and kinematics. Both events contain a W^+ candidate recoiling against an energetic dijet system. In event A(B) the W decays leptonically into a muon(electron) - neutrino system in which the neutrino (defined as missing E_T) carries a very large fraction of the transverse momentum of the W. A summary of the measured and computed parameters of the events is given in Table 2.

We have compared events A, B with events of similar topology. There are ten such W+jet+jet events (3 muon and 7 electron),

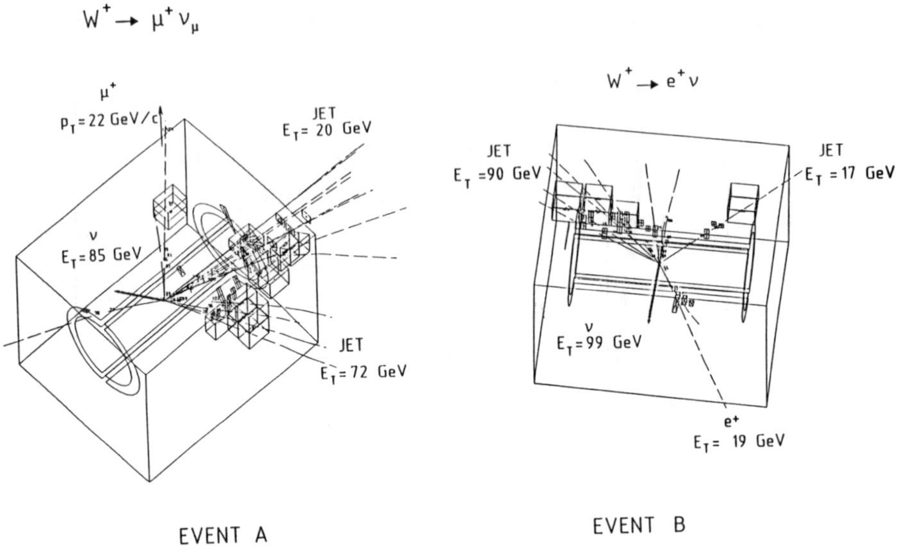

Fig. 4: Graphic display of the calorimeter cells (E_T > 1 GeV) and charged tracks (p_T > 1 GeV/c) in the UA1 detector for events A and B.

Table 2: Event Parameters for Events A and B

a.) 4 vectors x is along the beam direction.

Event A 18910/899	p_x(GeV/c)	p_y(GeV/c)	p_z(GeV/c)	E (GeV)	p_T(GeV/c)
muon	10	20	10	24	22
neutrino	-59	-53	-66	103	85
Jet 1	74	15	13	79	20
Jet 2	68	17	70	108	72
Event B 16687/1284					
electron	9	-16	-10	21	19
neutrino	-103	29	-95	143	99
Jet 1	-93	-11	90	137	90
Jet 2	17	2	17	27	17

b.) Derived kinematical parameters.

Event	P_T^W (GeV/c)	M_{jj} (GeV/c^2)	M_{Wjj}(GeV/c^2)
A	82 ± 12	82 ± 10	299 ± 28
B	105 ± 14	97 ± 12	279 ± 28

including events A and B. In figure 5 we show m(jj) v. p_T^W and compare the data with the QCD expectations. The QCD calculation has been done with the second order W+jet+jet Monte Carlo program of Ref.[8]. This program does not fragment the outgoing partons and therefore the generated events cannot be transported through the full detector simulation. We have, however, included in the calculation a simple simulation of the detector geometrical acceptance and selection losses, taking account of the effects of detector resolution on the charged lepton, jets and missing transverse energy. The resulting calculation predicts a total of 4.2 W+jet+jet events. There is an intrinsic uncertainty of about a factor 2 on this QCD prediction in the absence of a reliable estimate of the corresponding K-factor.

The contours in Fig. 5 contain respectively 68%, 95% and 99% of the expected W+jet+jet events. The experimental distribution of the 8 W+jet+jet events with $p_T^W < 30$ is consistent with QCD calculations (figure 5). In contrast, the resulting probability for observing events A and B is very small.

5. ALTERNATIVE INTERPRETATIONS

Given the difficulty of interpreting events A and B as QCD W-production, we now consider other standard processes which may

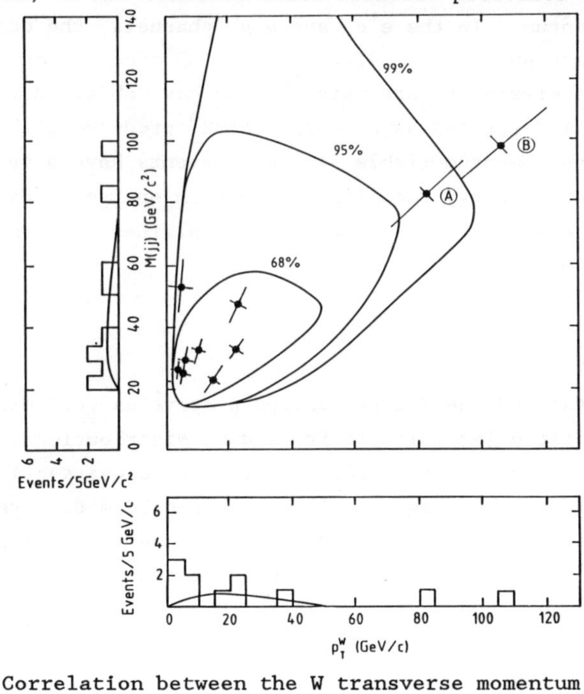

Fig. 5: Correlation between the W transverse momentum and the (jet-jet)-mass for the eight W + 2 jet events in which two hadronic jets are observed. The curves on the projections are predictions from a QCD calculation [8]. The contours contain 68%, 95%, and 99% of the predicted events.

produce an isolated energetic charged lepton together with a very large missing transverse energy and two energetic hadronic jets. For each process we require that $P_T^{l\nu} > 80$ GeV/c and $M_{l\nu jj} > 250$ GeV/c².

We have considered the following processes which can give rise to a similar topology: (1) $b\bar{b}/c\bar{c}$ production with subsequent semi-leptonic decay, (ii) tt production with $t \rightarrow Wb$ (m_t approx. 90 GeV/c², (iii) double parton scattering, and (iv) standard WW, WZ, or WH production. The standard model cross-sections for the above processes are several orders of magnitude too small to account for the observation of events A and B. The largest contribution (6×10^{-3} events) comes from process (ii) above.

We conclude that none of the above process gives a likely explanation for events A and B. Other high p_T^W events have been reported by the UA2 collaboration [9]. These events, however, have rather different properties to events A and B and can be interpreted in terms of conventional QCD production.

We have also looked for the production of Z^0's with large transverse momentum. In the e^+e^- and $\mu^+\mu^-$ channels, the observed P_T^Z distributions are entirely consistent with QCD expectations. However, in the missing E_T analysis [10], we have observed two events which could be interpreted as $\nu\nu$ decays of Z^o produced with very large transverse momentum (Table 3). Both events have a very large missing transverse energy recoiling against an energetic di-jet system. These events are also not easily explained in terms of conventional QCD.

6. CONCLUSION

We have measured the W/Z production properties including cross sections, associated jet multiplicity and E_T distributions, and the p_T^W spectrum (out to values of 100 GeV/c). The cross sections and associated jet properties agree well with standard model predictions. The data in the region $p_T^W < 40$ GeV/c are also well described by the predictions of the Standard Electroweak QCD model. However, two events with p_T^W above 80 GeV/c are unlikely according to current QCD calculations. These events have two energetic hadronic jets in association with the W.

Table 3: Event Parameters for Missing Energy $Z \rightarrow \nu\bar{\nu}$ Decay Candidates

a.) 4 vectors x is along the beam direction.

Event 13029/1289	p_x(GeV/c)	p_y(GeV/c)	p_z(GeV/c)	E (GeV)	p_T(GeV/c)
Z^o	-78	-49	56	142	75
Jet 1	-26	23	-50	66	55
Jet 2	-34	21	-8	44	22
Event 18393/1103					
Z^o	59	32	-75	137	82
Jet 1	101	-61	48	138	78
Jet 2	-20	28	32	52	43

b.) Derived kinematical parameters.

Event	M_{jj} (GeV/c)	M_{Wjj}(GeV/c^2)
13028/1289	57 ± 7	211 ± 18
18393/1103	148 ± 17	295 ± 27

References

1. G. Arnison, et al., (UA1 Collaboration), Nuovo Cimento 44, 1 (1985).

2. G. Arnison et al., (UA1 Collaboration), Phys. Lett 122B, 103 (1983), Phys. Lett. 129B, 273 (1983); G. Arnison et al., (UA1 Collaboration), Phys. Lett. 134B, 469 (1984); G. Arnison et al., (UA1 Collaboration), Phys. Lett. 166B, 484 (1986); G. Arnison et al., (UA1 Collaboration), Europhys. Lett. 1, 327 (1984).

3. G. Arnison, et al., (UA1 Collaboration), Phys. Lett. 123B, 115 (1983).

4. F.E. Paige and S.D. Protopopescu, ISAJET Version 5.20, BNL 38034 (1986).

5. G. Altarelli et al., Nucl. Phys. B246, 12 (1984). G. Altarelli et al., Z. Phys. Particles and Fields 27, 617 (1985).

6. F. Halzen, A.D. Martin and D.M. Scott, Phys. Rev. D25, 754 (1982); A. Nakamura, G. Pancheri and Y.N. Srivastava, Phys. Rev. D29, 1936 (1984).

7. S. Geer and W. Stirling, Phys. Lett. 152B, 373 (1985).

8. S.D. Ellis, R. Kleiss, and W.J. Stirling, Phys. Lett. 154B, 435 (1985).

9. P. Bagnaia et al., (UA2 Collaboration), Phys. Lett. 139B, 105 (1984); J. Appel et al., (UA2 Collaboration), Z. Phys. C Particles and Fields 30, 1 (1986).

10. C. Albajar et al., Events with Large Missing Transverse Energy at the CERN Collider: Search for the Decays of $W^\pm$ into Heavy Leptons and of Z° into Non-Interacting Particles (Paper II), submitted to Physics Letters, and CERN-EP/86-82, Nov. 18, 1986.

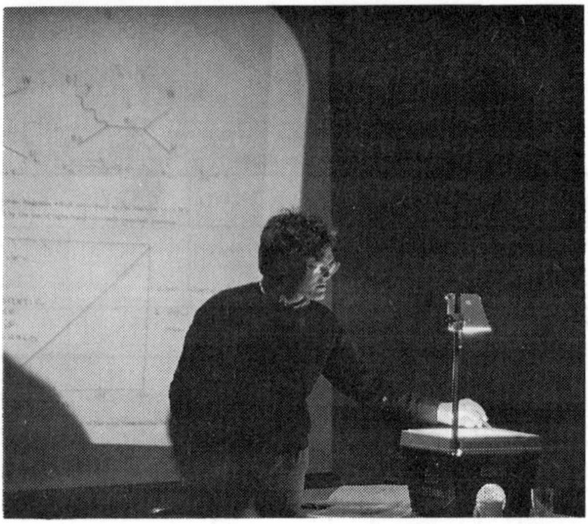

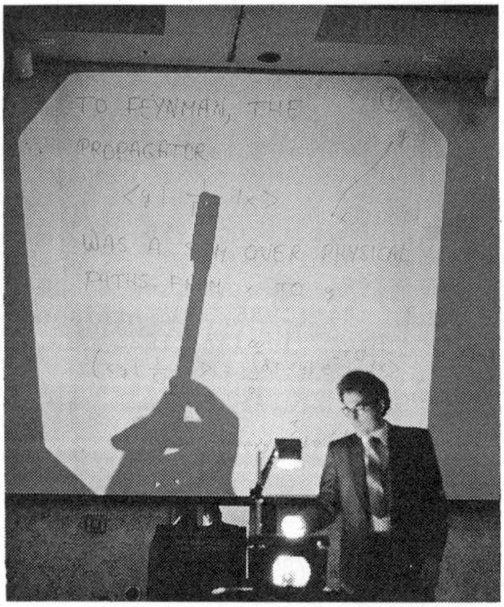
TO FEYNMAN, THE
PROPAGATOR
WAS A
OVER PHYSICAL
PATHS FROM

THE NONCOMPACT METHOD

... THEN LANGEVIN DRIVING FORCE IN 1st ORDER
DISCRETIZATION WITH "CORNELL" BILINEAR
GAUSSIAN NOISE
CAN ABSORB
INTO SHIFTS
IN PRACTICE, FIX
THEN USE
3. RESULTS

VII. Detectors and Accelerators

Conveners — M. Breidenbach and R. Talman

TRISTAN STATUS

Takayuki MATSUI
(Representing the TRISTAN e^+e^- project)[1)]

VENUS Group
National Laboratory for High Energy Physics,
Tsukuba, Ibaraki, 305,
JAPAN

ABSTRACT

Recent progress at TRISTAN is reported. The accelerator started operation at W=50 GeV in an e^+e^- colliding beam mode in Nov.'86. Four detectors, AMY, SHIP, TOPAZ and VENUS are almost ready for data taking and for producing physics output. AMY and VENUS have rolled into their respective intersection points and accumulated about $10nb^{-1}$ and $20nb^{-1}$, respectively, while TOPAZ will roll into the interaction region in coming March. Here we describe the status of the accelerator and detectors in Dec.'86, one month after the start, with emphasis on the results of test operation of AMY and VENUS.

1. INTRODUCTION

The TRISTAN e^+e^- project was started five years ago with the goal of looking for new physics in the energy region above PETRA/PEP. Even though the cross section of e^+e^- annihilation decreases as s^{-1}, the electroweak interference effects should rapidly increase at TRISTAN energy, making it a good place for testing the standard model and for searching for new physics beyond the standard model. In the interference effects, there is the possibility of finding a deviation from the standard model due to anomalous Z's. Needless to say, our activities will be directed to find a t-quark or new fermions which belong to the fourth generation. Further studies on QCD will be also done, especially aming at the determination of gluon self coupling in 4-jet events. Here we do not go into details of physics possibilities, but show the present performance of accelerator and detectors, including very preliminary results of physics signals.

Now we briefly summarize the history of the TRISTAN construction. In Table 1, we highlight the main events during the five year period between 1981 and 1986. After a great deal of preparation and design, the TRISTAN e^+e^- project was finally approved in Apr.'81 as a five year project. Ground breaking took place in Nov.'81; a very exciting event for high energy physicists in JAPAN. In Mar.'83, two major general purpose detectors VENUS and TOPAZ were approved by the TPAC (Tristan Program Advisory Committee) and immediately the detector construction was started. A half year after the full approvals of VENUS and TOPAZ, a relatively compact spectrometer with a high magnetic field of 3.0 Tesla, called AMY, and a plastic detector in search of heavily ionizing particles, SHIP, were approved. In Nov.'83, just two years from the ground breaking, we succeeded in the first circulation of electron beams in the Accumulation Ring (AR) and started to supply high energy gamma ray beams at AR test channels that were used mainly for tests and calibrations of carolimeters. After about two years of construction, TOPAZ and VENUS completed their superconducting solenoid magnet and iron rstructure, and successfully excited them to design fields of 1.2 Tesla and 0.75 Tesla, respectively. The construction of the high current electron linac for positron production also proceeded smoothly, and in Nov.'85 the first positron beam was successfully circulated and accelerated in AR.

By the end of 1985, all of the bending and Q magnets for the TRISTAN Main Ring (MR) were installed. The completion of the main ring vacuum system in Apr.'86 was immediately followed by the installation of 64 units of APS (Alternating Periodic Structure) RF cavities. In the same period, activities were very high in experimental halls for installing detector components. VENUS first rolled into the interaction region and thus the MR vacuum pipe was fully connected. Overall tests of TRISTAN MR were started in Oct.'86; in less than one month we achieved simultaneous e^+ and e^- beams in the MR at W=50 GeV. The successful beam collision was then clearly confirmed in November by a large angle Bhabha event detected by VENUS at W=48 GeV and, in fact, together with a few more large angle events, VENUS soon detected the first hadronic event at the same energy. In late Nov. the AMY also rolled into the interaction region and subsequently detected its first Bhabha event in December. In this way, both the accelerator and the detectors confirmed their potential capabilities to start an initial physics run at the highest energy in e^+e^- collider physics.

2. THE TRISTAN ACCELERATOR

Fig.1 shows a schematic layout of the TRISTAN accelarator[3]. Electrons in a 2nsec wide pulse from an electron gun are accelerated up to 2.5 GeV by a 390m long linac which has 30MW peak power. Positrons produced by a 200MeV high current linac which has a maximum current of 10A are preaccelerated up to 200 MeV and injected to the main linac. Electrons and positrons of 2.5 GeV are accumulated in the AR which is a storage accelerator of 377m in circumference and are then accelerated up to 6.0-8.0 GeV for MR injection. In Dec.'86 the typical current at the AR was about 20mA at the 2.5 GeV accumulation stage and there was no serious drop during acceleration. The energy of the AR and, correspondingly, the injection energy for the MR was limited to 6.5 GeV by the power of the AR RF system. The TRISTAN MR has a total circumference of about 3km. The bending radius is about 250m and there are 4 long straight sections of about 200m each in which about 500m of length will be used for RF sections. Nominal beam energy of the MR is 30 GeV and in this case the scyncrotron radiation loss per turn is 290 MeV which is almost the same value at LEP.

In Dec.'86 the energy of the MR was set to 48 GeV to reduce the load on the RF system for a continuous physics test run. The maximum current during this period in the MR was 1mA/bunch, resulting in a total current of 4mA in a 2-bunch on 2-bunch colliding beam mode. The current was limited at the injection stage of 6.5 GeV. In table 2 we list the maximum design current and the luminosity that is ideally limited by the beam-beam tune shift, but we expect that a practical limit will happen at about 5mA/bunch at TRISTAN. For mini-β optics, we are preparing superconducting Q-magnets to realize the β-function listed in the table, but at the moment we only have a backup optics and $\beta x^*/\beta y^*$ are 1.6/0.1m, respectively. In the Dec.'86 run, we used βx^* = 2.5m in order to reduce backgrounds at the detectors. So the estimated luminosity for the present current and optics is about $5X10^{29}cm^{-2}s^{-1}$; this value is consistent with rates observed in the VENUS luminosity monitor. Fig.2 shows operating experience in terms of beam current. The lifetime of the beam for our typical data taking run was about 50min. On the other hand it took about 20min to reestablish colliding beams in the MR, so the data taking time was about 70% of the total operational time in the normal physics runs. The main source for the lifetime limitation at that time was the vacuum, which was $8X10^{-9}torr$; the vacuum will be improved soon by aging and replacing some vacuum pumps.

3. FOUR DETECTORS AT TRISTAN

3.1 SHIP

SHIP is a detector that will be used to search for heavily ionizing particles by looking at tracks in CR-39 plate and UG5 glass. The cut American CR-39 sheets were sent by sea to avoid contamination by cosmic rays and arrived at KEK in Nov.'86. Two plates of aluminum-coated UG5 glass were mounted inside of the vacuum pipe in late Sep.'86. For the initial exposure, SHIP installed a 1cm thick module of CR-39 (multi-layers) and 11 thinner modules (double-layers) in the detector frame to take a background map around the interaction region. Also a luminosity monitor was checked out in this period. The thinner modules may be replaced by the multi-layers to get more sensitivities at the beginning of running in the spring of '87.

3.2 TOPAZ

Fig.3 shows schematic views of the TOPAZ detector. A main feature of TOPAZ is the choice of a time projection chamber (TPC) as a central tracking and particle identification device. The TPC has dimensions of 260cm in diameter and 300cm in axial length and is operated with Ar(90%) + CH4(10%) at 4-atmospheres. Each end plate of the TPC is divided in 8 sectors, and signals from sense-wires (175 channels/sector) and pads (512channels/sector) are strobed into CCD's in 100nsec intervals. The TPC is placed inside a superconducting solenoid magnet that provides a 1.0 Tesla over the volume of the TPC.

An inner drift chamber (IDC) with delay line read out for z information is installed between the TPC and a beam pipe. TOF counters, segmented in 64 units in φ, surround the TPC. The IDC and TOF counters provide the main charged particle track triggers. Outside of the magnet, there is a barrel calorimeter which consists of about 4000 of DF6 lead glass blocks. Outside of the magnet and also just in front of the lead glass, there are drift tube chambers used for finding entrance positions of particles on the lead glass array. Similar chambers are also used in the endcap region for the same purpose, and sandwich counters of lead and gas counters are installed behind them as an endcap calorimeter. All detector components including the barrel μ-chambers were installed and most electronics channels have been instrumented. Fig.4 shows a typical result of cosmic ray test. Basically all of the detector components are working and almost ready for data taking. TOPAZ will roll into the intersection ion Mar.'87 and start to take data next spring.

3.3 AMY

Fig.5 shows schematic views of the AMY detector. A main feature of AMY is a relatively compact detector with a high magnetic field of 3.0T and a X-ray detector for electron identifications. AMY has a conventional drift chamber (CDC) with about 9000 anode wires as a central tracking device. The chamber consists of 25 axial and 15 slant layers and is operated with a mixtuer of Ar/CO2/CH4 at atmospheric pressure. An inner chamber, which consists of 4 layers of straw chambers, are placed inside the CDC. Radially outside of the CDC is a cylindrical lead and proportinal tube sandwich shower counter. The proportional tubes have cathode strips and pads along anode wires, so three dimensional shower reconstruction is possible. A trigger counter system is inside the magnet coil and μ-chamber and counter system is outside of the magnet as shown in Fig.6. Almost all components of the detector are installed, except the X-ray detector. In the Dec. run, AMY had about half of its CDC electronics but enough shower counters electronics to detect Bhabha events. Now almost 100% channels are connected for all detectors. Fig.6 shows a typical Bhabha event, and the shower pattern which has a shower maximum at about 6Xo shows a typical electromagnetic showers of 25 GeV. The position reconstruction of the shower by cathode informations was done using a centroid method, and the shower center was well determined to within about 4mm. AMY accumulated 6 Bhabha events in this period, and the integrated luminositiy calculated from these events is about $10nb^{-1}$. The X-ray chamber is still under construction and will be installed in late this year.

3.4 VENUS

Fig.7 shows a schematic view of the VENUS detector. The detector is characterized by the requirement that the detector would have to be ready at the first stage of the TRISTAN, and so it is a large general purpose detector using relatively conventional techniques. VENUS also has a superconducting solenoid magnet that produces 0.75 Tesla over the tracking region. A bobbinless structure and CFRP cryostat realized a very thin magnet for a particle penetration, resulting in 0.5Xo in the radiation thickness. VENUS has a conventional drift chamber (CDC) with about 7000 anode wires as a central tracking device. The CDC is operated with a mixtuer of Ar/CO2/CH4 (89/10/1) and has 20 axial and 9 slant layers. An inner drift chamber with cathode readout for z information is placed around the beam pipe. About 30cm outside of the outer layer of the CDC, 3 layers of drift tubes (ODT) are installed to help the CDC to get a better momentum resolution for charged particle tracks. Ninety six units of TOF counters are placed between the ODT and the magnet cryostat. Outside of the magnet, VENUS has a barrel electromagnetic calorimeter with a semi-tower geometry that points at the interaction point. The barrel calorimeter consists of more than 5000 lead glass blocks of DF6. Outside of the magnet but just front of the lead glass, we have 3-layers of streamer chambers to use for finding entrance positions of particles on the lead glass array. Liquid Argon calorimeters (LA) are prepared as endcap calorimeters. The LA is not installed yet but the cooldown test has been successfully done and so the endcap will be installed on Feb.'87. The μ-chamber system was partially installed in the bottom region of the detector. Full installation of the barrel section will be finished in Mar.'87.

VENUS rolled into the interaction region in Sep.'86. In parallel with final build up of the accelerator for operations, we instrumented all electronics channels for CDC, Barrel LG, TOF and Inner CH and tested them by cosmic rays. After the success of beam acceleration and storage in e^+e^- mode, we started trigger studies for the purpose of detecting Bhabha and hadronic events. One such trigger is a total energy trigger in the barrel LG in which a energy threshold was set to 3.0 GeV. The other trigger was a track trigger that made use of the inner 7 layers the CDC. On each beam crossing, the hit patterns of the CDC are compared with a lookup table that is filled with desired patterns. We started using this track finder trigger with a minimum requirement of 2 tracks with a minimum Pt of 200 MeV/c, aiming to collect μ-pair events. However the resulting trigger rate from background triggers by beam - gas or beam -beam pipe events was more than 10 Hz. Thus, for the early running we had to put more stringent requirements as follows:

Esum(Barrel LG) > 3 GeV.
OR
TOF and Track Finder > 4 tracks. Pt > 1 GeV/c.

Now the total energy trigger rate was about 1.5 Hz and the track finder trigger rate was considerably improved about 2.5 Hz in a typical run, resulting in a total trigger rate of about 4 Hz under this condition. In Dec.'86 we started a test operation for physics and accumulated about $20nb^{-1}$ at W = 48 GeV. We collected a total of 280k triggers which include background triggers during the trigger studies at very early stage of the run. Here we show our results from quick analysis made especially for Bhabha's and hadronic events.
We applied the following criteria to select Bhabha events in the barrel region:

- SELECTION CRITERIA

- E(cluster) $\geq$ 1/3 E(beam)
- N(cluster) $\geq$ 2
- Opening Angle (X-Y plane) > 90°

Finally, by eye scanning, we identified 11 Bhabha events in this period. Fig.8 shows a typical Bhabha event. Fig.9 shows a plot of these 11 events in $cos(\theta)$ bins. The solid line shows a pure QED curve that is normarized to the first data bin. Statistics are very poor at the moment but the distribution looks reasonable.
In the same data sample, we also looked for hadronic events. In this case, we used the following criteria:

- QUICK HADRON SELECTION CRITERIA

  - E(LG) > 3 GeV

  AND

  - N(Track from origin) $\geq$ 3
    $|Rmin| < 5cm$, $|Zmin| < 20cm$
    OR
  - N(Long) $\geq$ 2
    $|Rmin| < 5cm$, $|Zmin| < 20cm$
    N(Hit on X-Y) > 10, N(Hit on R-Z) $\geq$ 4,
    Pt > 0.1GeV/c

Finally 5 events emerged as candidates for hadronic events after eye scan that required N(long) $\geq$ 5. For these 5 events we calculated visible energies and longitudinal momentum balances which are shown in Fig.10. Four of the five events are well scattered around the center of mass with a good balance of momentum. One has a small visible energy which is less than half of the center of mass energy and so this event is most likely from a two photon process. Thus we conclude that we had 4 hadronic events arising from single-photon annihilation. Their thrust distribution is shown in Fig.11. The solid line is a result calculated by LUND53 Monte Carlo and our 4 events shows a reasonable distribution in this plot.

4. FUTURE PLAN

During This New Year's day shut down, we replaced 16 klystrons of 800KW power by 1MW klystrons to allow for stable operation at W=50 GeV. In Jan.-Feb.'87 we continue taking data at W=50 GeV aiming to accumulate as many hadronic events as possible. In Feb.-Mar. the accelerator will be stopped to allow for the installation of more RF; up to 65% level of the design total, and the vacuum system will be also improved. During this period, VENUS will install liquid argon endcaps and μ-chambers and TOPAZ will roll into the interaction region. In May.-Jul. we will have a physics run at W=50 GeV or higher energy. During this period we hope to accumulate $5pb^{-1}$ data for useful physics. In the summer shut down we will add remaining RF cavities allowing us to reach W=56 GeV in the fall run. From Sep.'87 we will start a physics run between W=50 GeV and W=56 GeV. Superconducting Q-magnets for mini-β optics are under preparation and will be installed by the end of '88.

To go to higher energies we are also preparing superconducting RF cavities. A prototype cavity has been successfuly tested and showed a good performance of 4.5MV/m acceleration gradient in 5-cell cavities. The superconducting cavities will be installed starting in autumm '88 and the energy of the TRISTAN will go up to 65 GeV.

Of course we need more detailed studies to achive the design performance levels of the accelerator and the detectors. However, the present status shows that we have a good chance to explore a new physics in e^+e^- collisions with facilities described above.

REFERENCES

1) NISHIKAWA, T., Proceedings of the 1985 International Symposium on Lepton and Photon Interactions at High Energies. P.791(1985).

VENUS collaboration proposal. TRISTAN-EXP-001, KEK.

TOPAZ collaboration proposal. TRISTAN-EXP-002, KEK.

AMY collaboration proposal. TRISTAN-EXP-003, KEK.

SHIP collaboration proposal. TRISTAN-EXP-004, KEK.

TABLE 1 MAIN EVENTS OF TRISTAN

1981-APR.	GOVERNMENT APPROVAL OF TRISTAN PROJECT
-OCT.	CALL FOR LETTERS OF INTENT OF TRISTAN EXP.
-NOV.	GROUND BREAKING.
1983-MAR.	APPROVALS OF VENUS AND TOPAZ.
-NOV.	FIRST BEAM CIRCULATION e^- IN THE AR.
-DEC.	APPROVAL OF AMY.
1985-MAR.	APPROVAL OF SHIP.
-OCT.	TOPAZ MAGNET COMPLETED: 1.2T.
-NOV.	e^+ INJECTION AND ACCELERATION IN AR.
-DEC.	FULL INSTALLATION OF MAGNETS IN MR.
1986-MAR.	VENUS MAGNET COMPLETED: 0.75T.
-APR.	VACUUM SYSTEM COMPLETED.
-AUG.	INSTALLATION OF 64 UNITS OF RF CAVITIES.
-SEP.	VENUS ROLLED INTO IR.
-OCT.	OVERALL TEST OF MR.
-NOV.	BEAM COLLISION IN MR (W=50.0 GeV). FIRST BHABHA EVENT AT VENUS. FIRST HADRONIC EVENT AT VENUS.
-DEC.	AMY ROLLED INTO IR. AMY MAGNET COMPLETED: 3.0T. FIRST BHABHA EVENT AT AMY. TEST PHYSICS RUN. (W=48 GeV).

TABLE 2 GENERAL PARAMETERS OF TRISTAN

BEAM ENERGY	25 - 30	GeV
INJECTION ENERGY	6 - 8	GeV
CIRCUMFERENCE	3018	m
AVERAGE RADIUS OF CURVED SECTION	346.7	m
BENDING RADIUS	246.5	m
LONG STRAIGHT SECTION	194.4(x4)	m
TOTAL LENGTH OF RF SECTION	509.4	m
RF FREQUENCY	508.6	MHz
REVOLUTION FREQUENCY	99.3	KHz
RADIATION LOSS / TURN	290(30GeV)	MeV
RF PEAK VOLTAGE	400	MV
MAX. CURRENT	11	mA/Bunch
NUMBER OF EXP. INTERSECTIONS	4	
MAX. DESIGN LUMINOSITY	$8X10^{31}$	$cm^{-2}s^{-1}$
BETA-FUNCTION AT COLLISION POINT $\beta x^*/\beta y^*$		
MINI-BETA OPTICS	0.8/0.05	m
BACKUP OPTICS	1.6/0.1	m
LENGTH OF EXPERIMENTAL INSERTION		
MINI-BETA OPTICS	5.4	m
BACKUP OPTICS	9.0	m

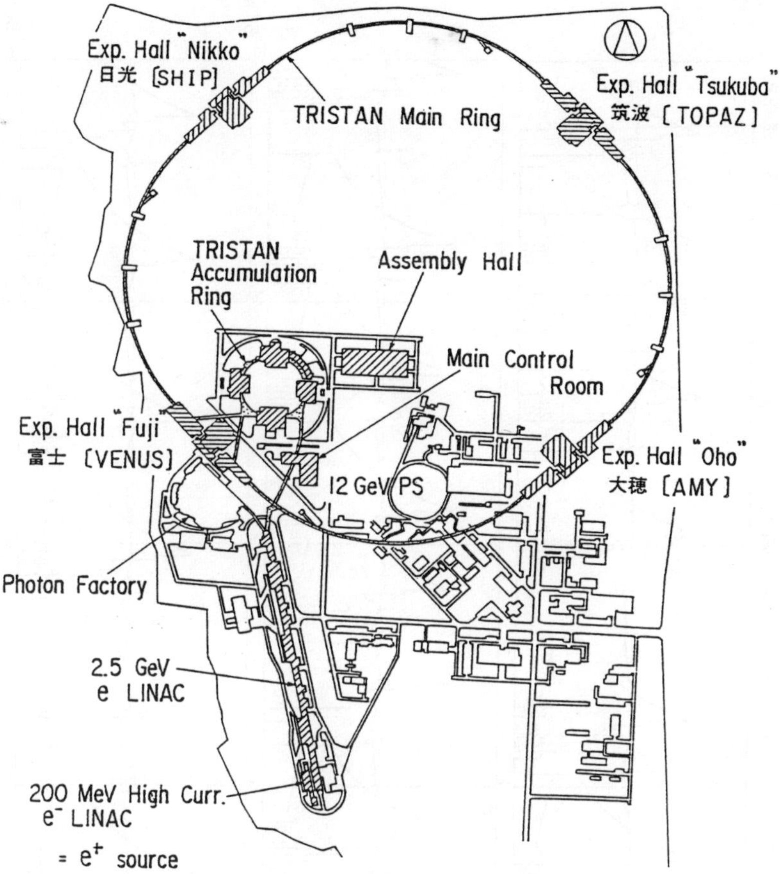

Fig.]. Schematic layout of TRISTAN.

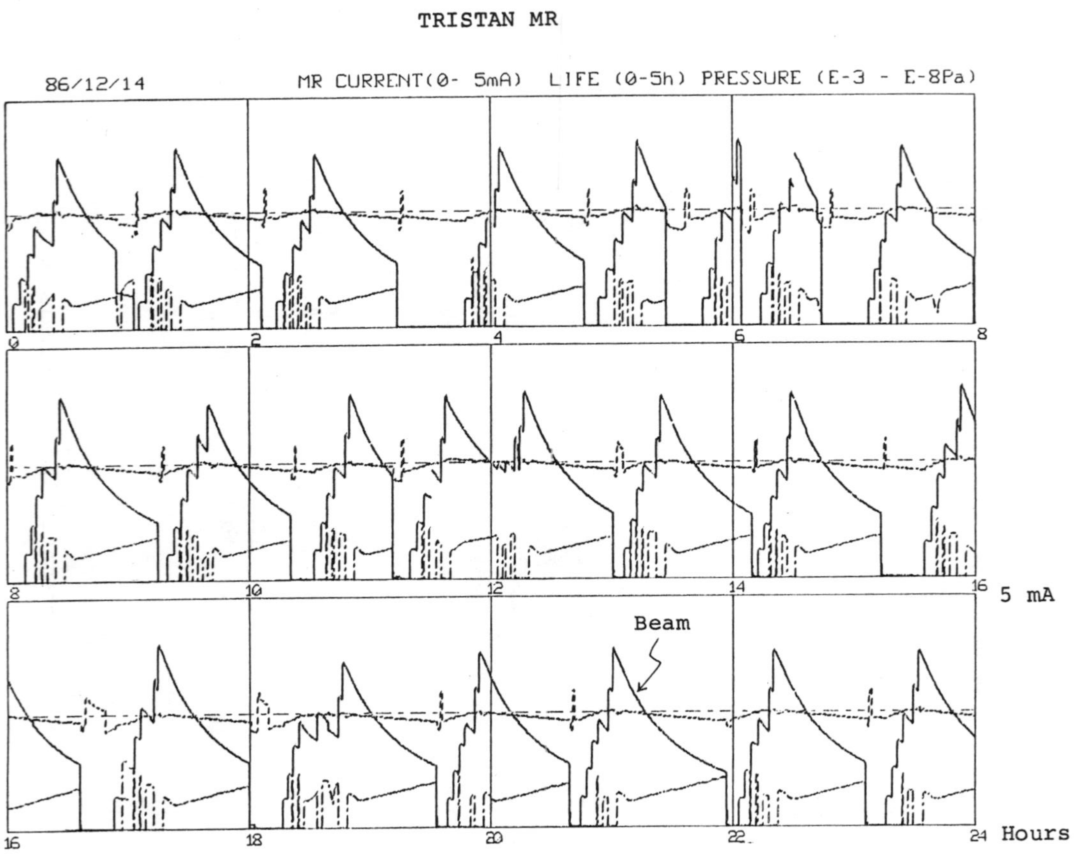

Fig.2. Typical beam performance in Nov. and Dec.'86.

TOPAZ

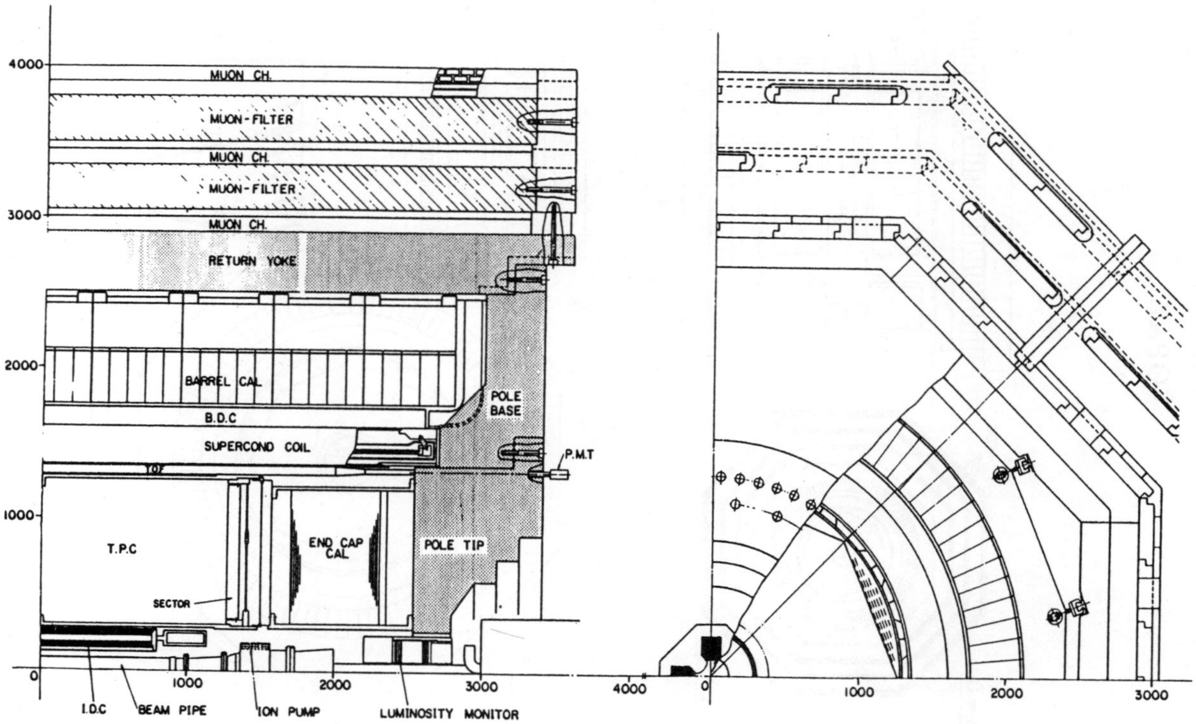

Fig.3. Schematic view of the TOPAZ detector.

TOPAZ

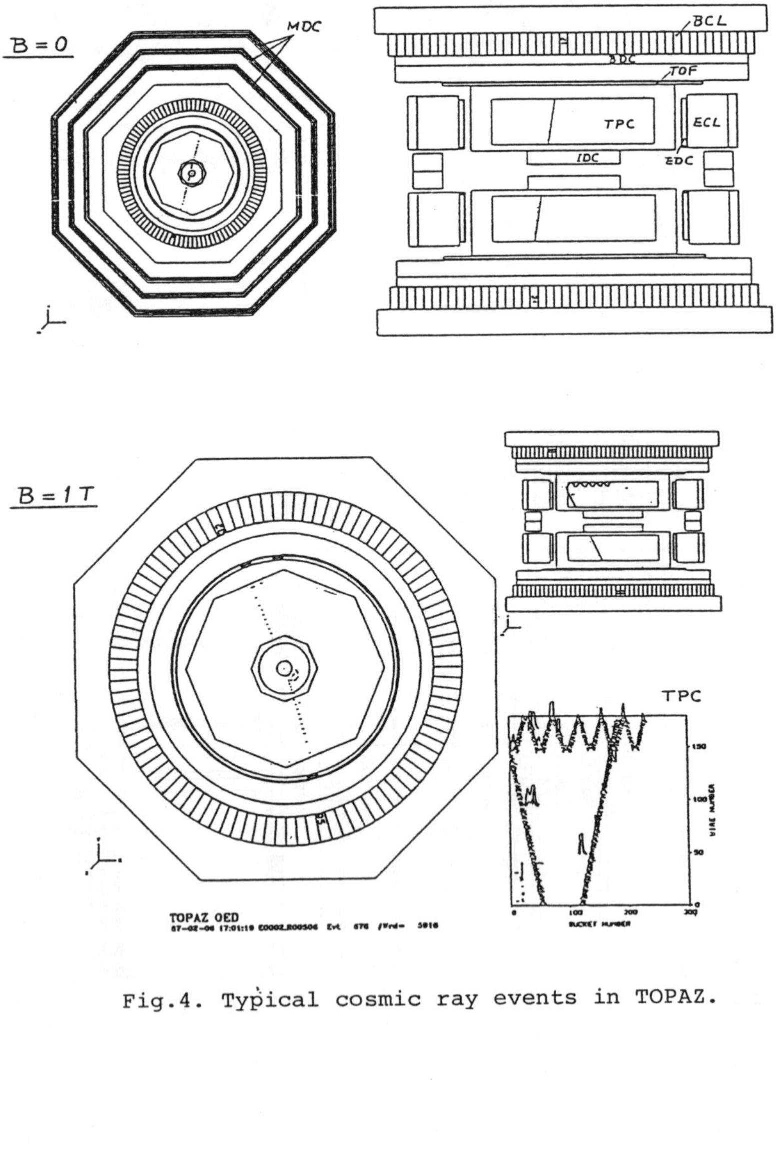

Fig.4. Typical cosmic ray events in TOPAZ.

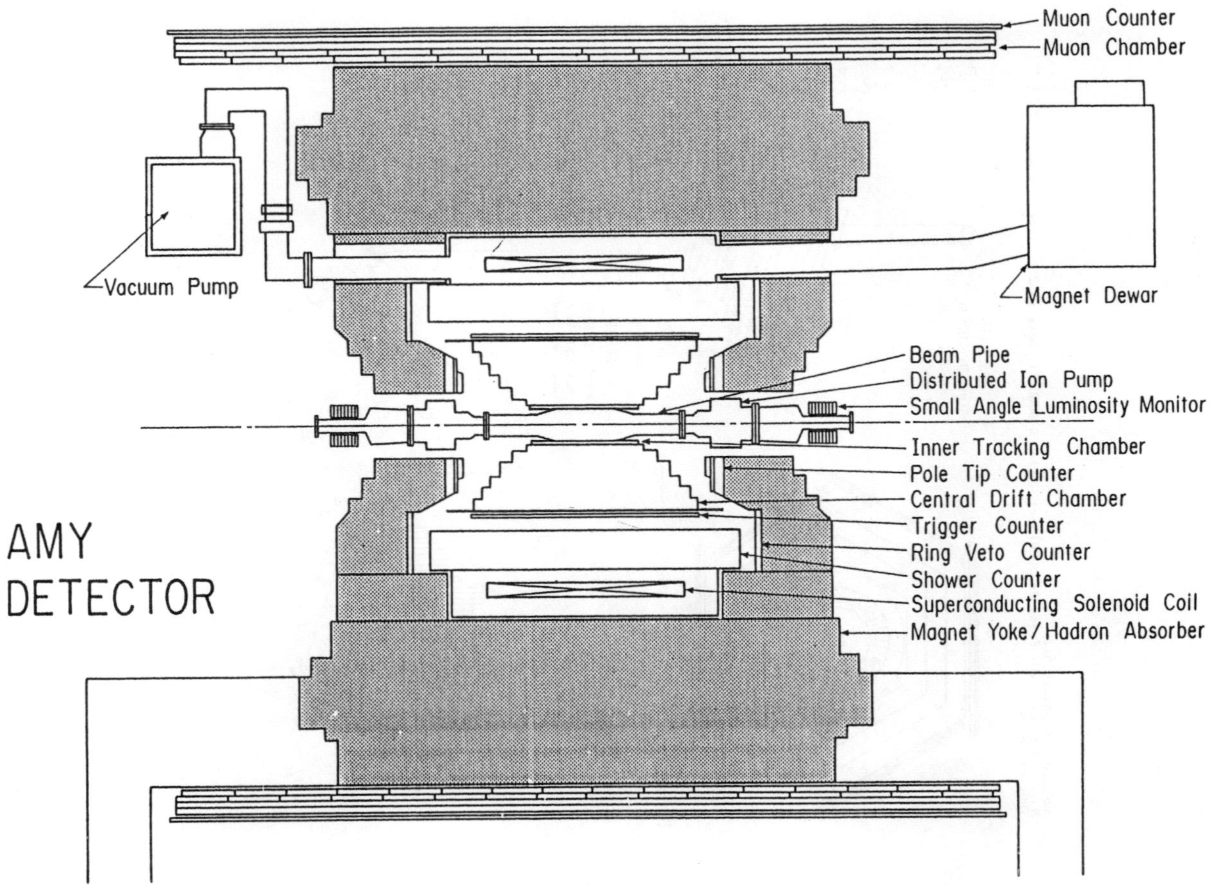

Fig.5. Schematic view of the AMY detector.

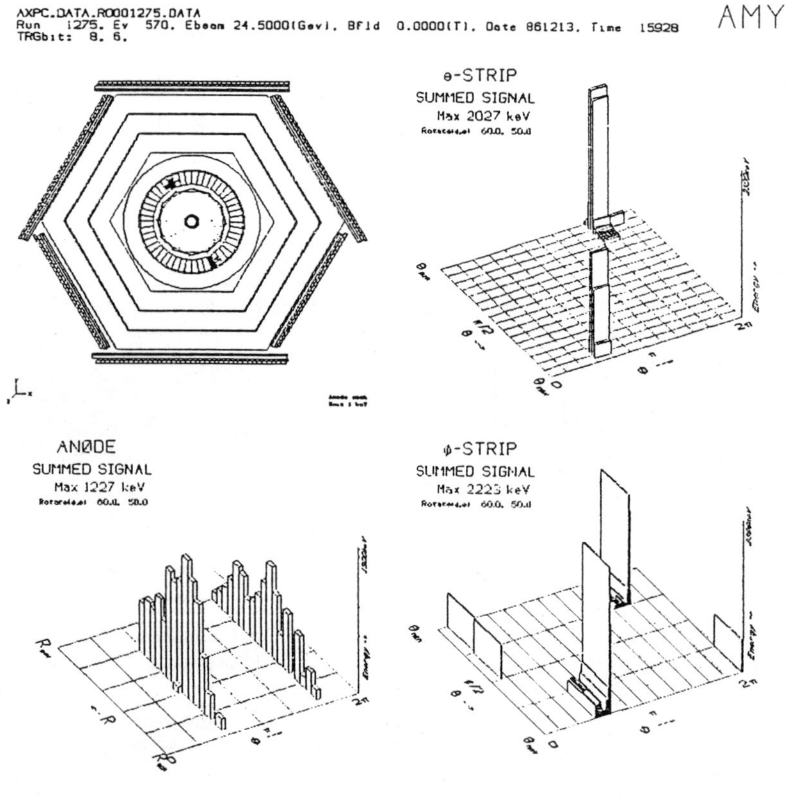

Fig.6. A typical Bhabha event in AMY.

VENUS

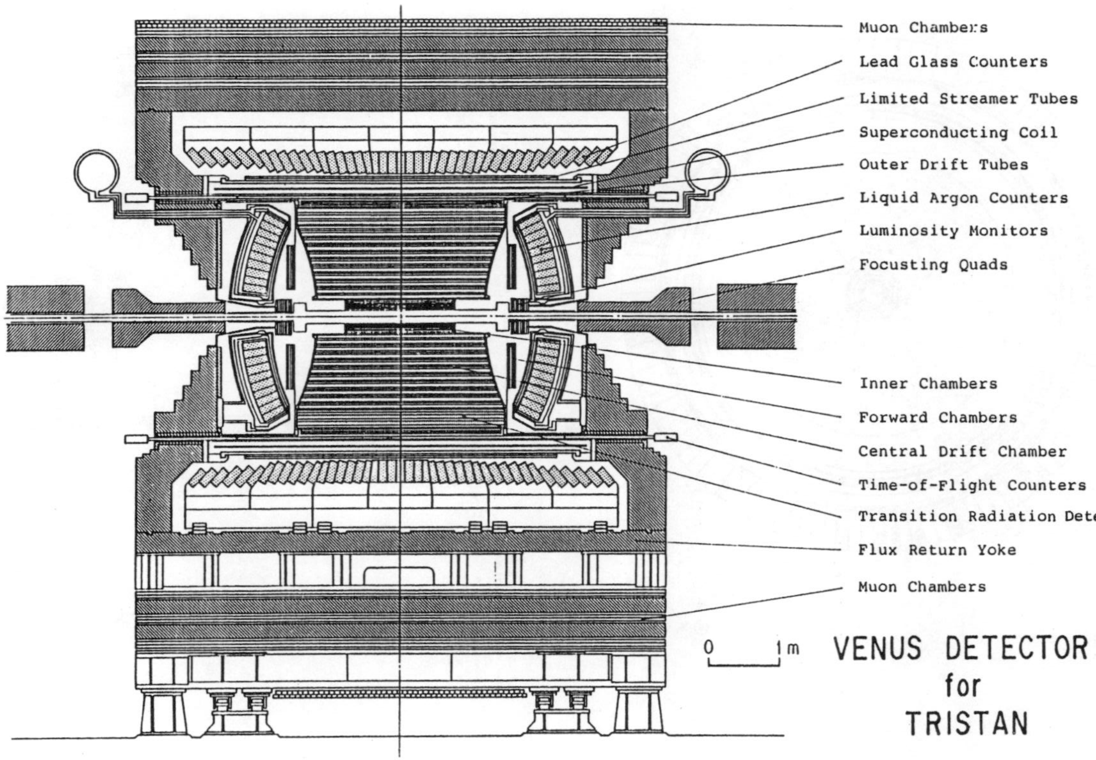

Fig.7. Schematic view of the VENUS detector.

VENUS

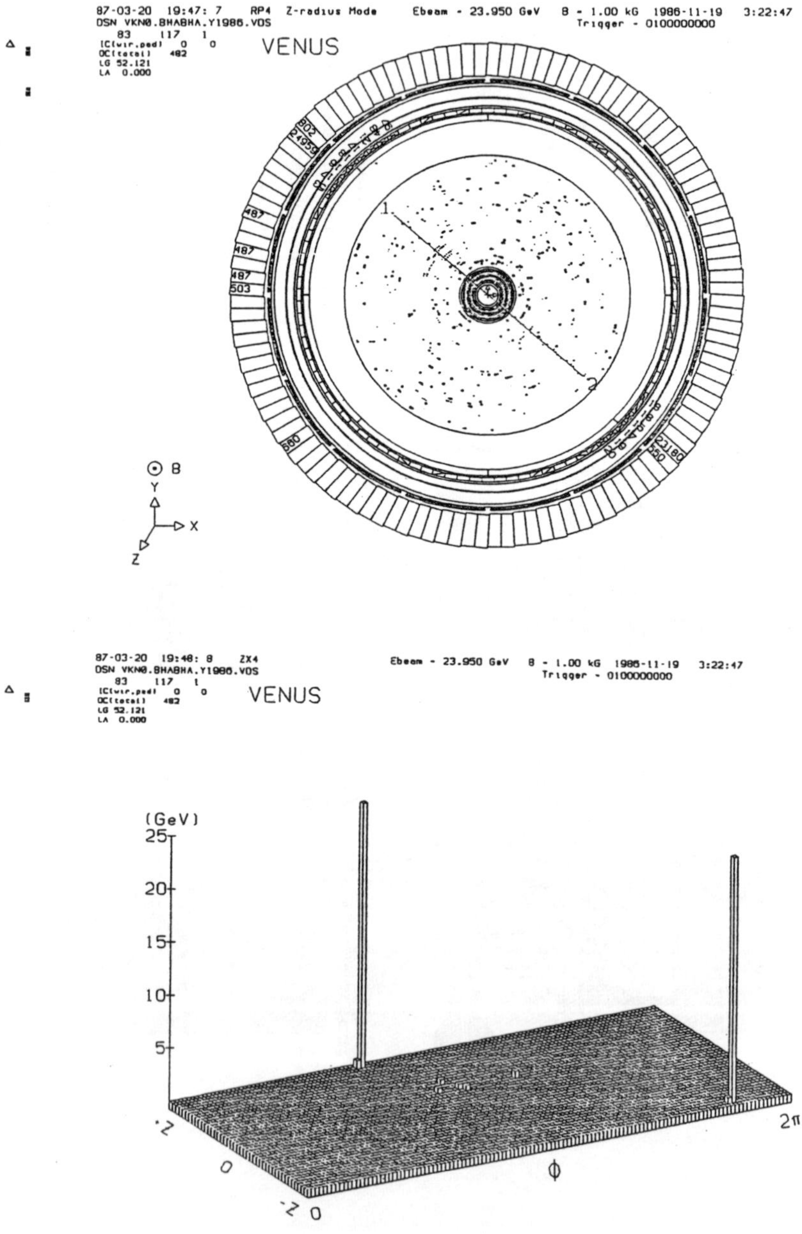

Fig.8. A typical Bhabha event in VENUS.

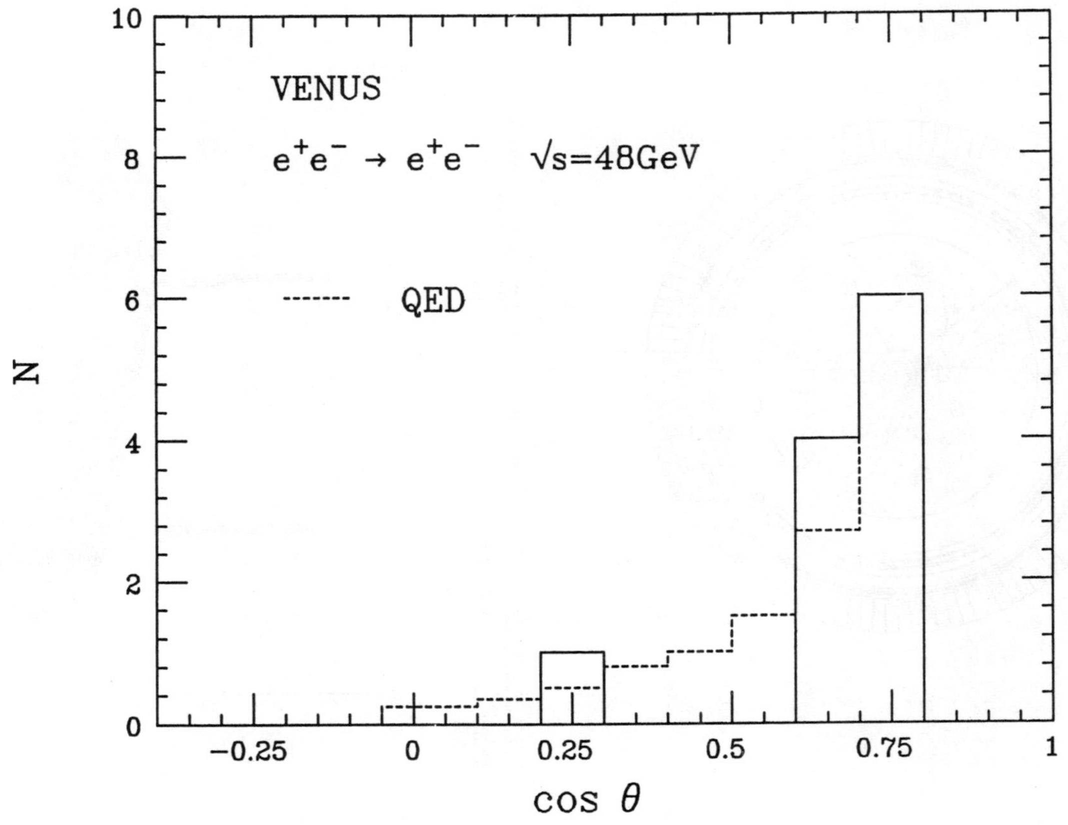

Fig.9. Angular distribution of Bhabha events.

VENUS

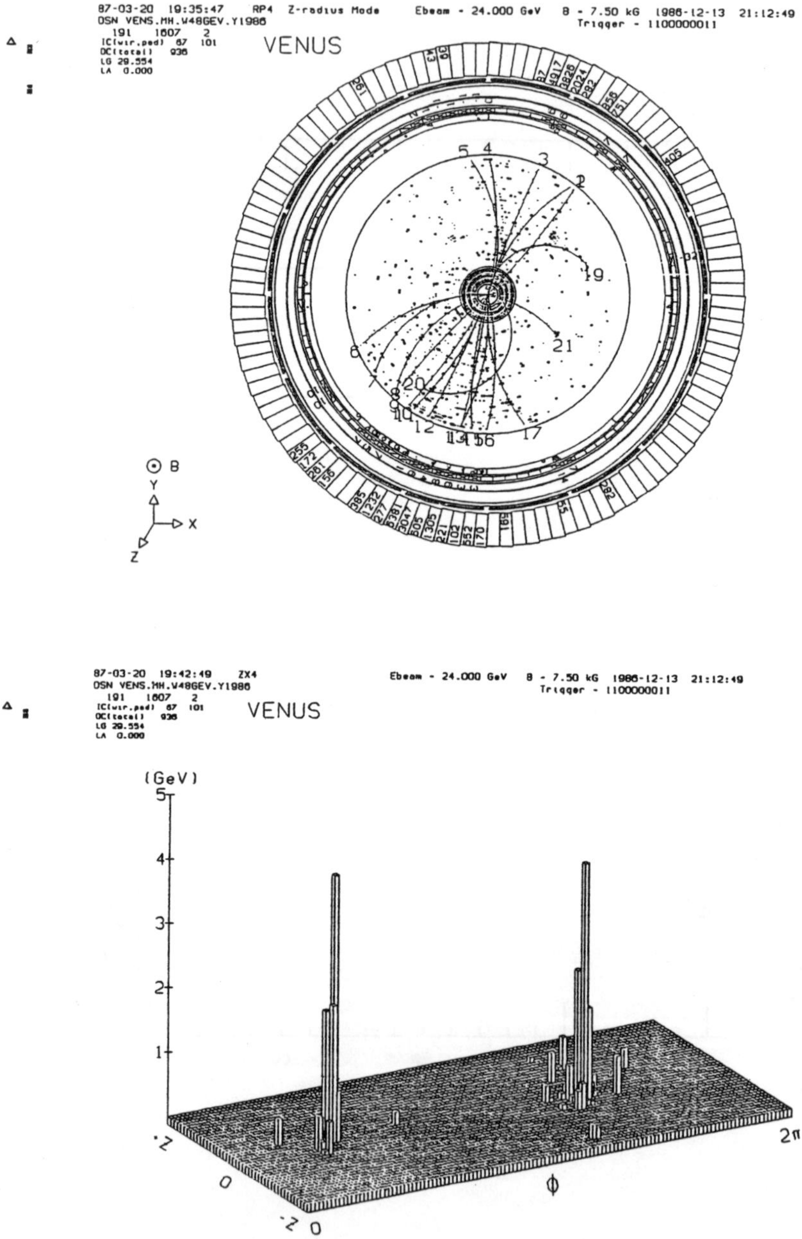

Fig.]0. A typical hadronic event in VENUS.

VENUS

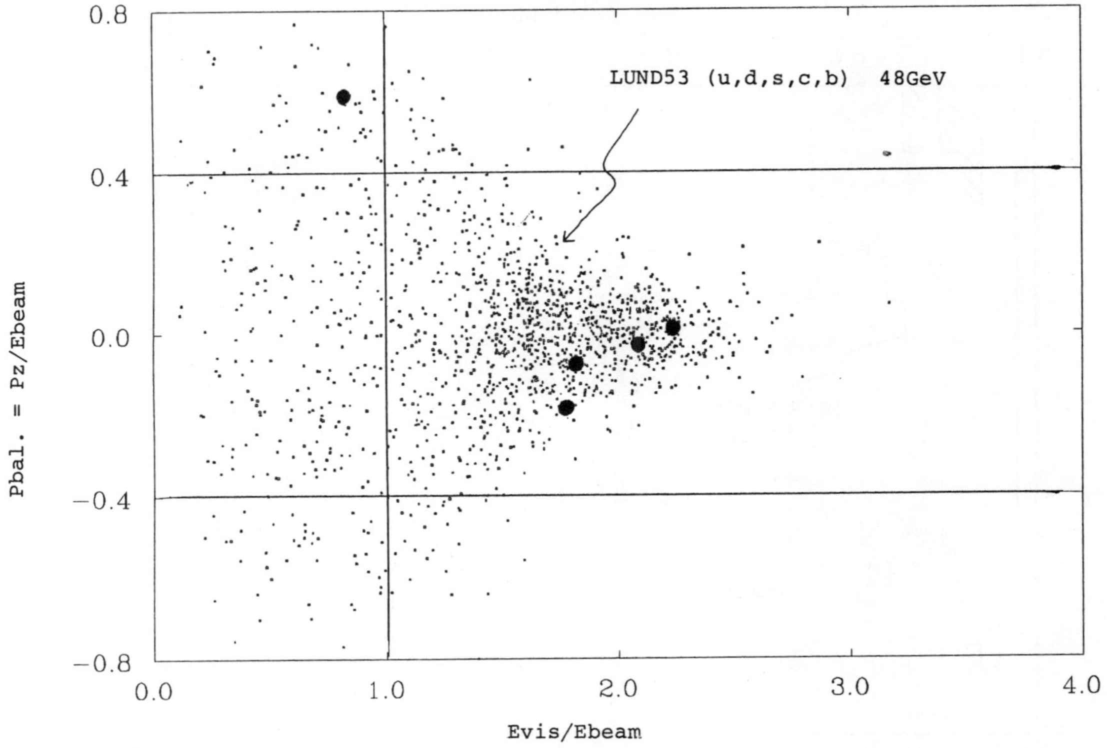

Fig.11. Visible energies and Pz balances of hadronic events.

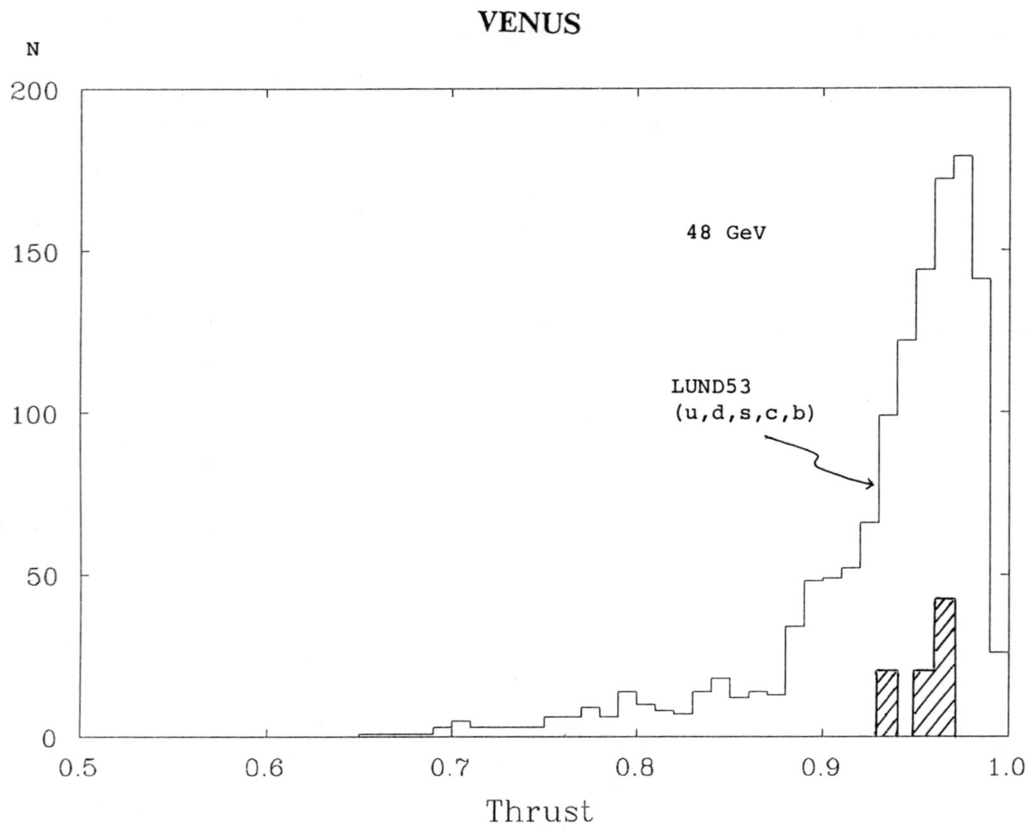

Fig.]2. Thrust distribution of hadronic events.

Probing Electroweak Symmetry Breaking at the SSC: A No-Lose Corollary *

Michael S. Chanowitz

Lawrence Berkeley Laboratory
University of California
Berkeley, California 94720, U.S.A.

Abstract

Low energy theorems are derived for scattering of longitudinally polarized W and Z's, providing the basis for an estimate of the observable signal at the SSC if electroweak symmetry breaking is due to new physics at the TeV scale.

1. INTRODUCTION

This is not a general review of SSC physics but focuses on a particular issue—electroweak symmetry breaking—which is central to the motivation for constructing the SSC. There are other excellent reasons for the SSC: the possibility of completely unanticipated discoveries that could be more important than any of the things we are able to imagine and the theorists' wish list of possible new physics such as additional gauge bosons, further generations of matter, supersymmetry, *etc.*... The energy and luminosity of the SSC will make it a tremendous instrument for these searches, which are in most cases easier than the rather difficult physics of electroweak symmetry breaking I will discuss. But while these other topics are exciting possibilities, electroweak symmetry breaking is a certainty: we know that the W and Z bosons have masses and that the photon is massless, but we do not know why. This is the outstanding open question that must be answered to complete the highly successful unified theory of the weak and electromagnetic interactions.[1] I will argue that with the proposed energy, $\sqrt{s}$ =40 TeV, and luminosity, $\mathcal{L} = 10^{33}$ cm.$^{-2}$ s.$^{-1}$, the SSC is certain to see the manifestations of the new physics responsible for electroweak symmetry breaking.[2] The SSC is strategically placed to see this new physics: I would not be prepared to make the above statement for a machine with one tenth the luminosity, $\sqrt{s}$, $\mathcal{L} = 40$, 10^{32} or for one with half the energy, $\sqrt{s}$, $\mathcal{L} = 20$, 10^{33}.

*This work was supported by the Director, Office of Energy Research, Office of High Energy and Nuclear Physics, Division of High Energy Physics of the U.S. Department of Energy under Contract DE-AC03-76SF00098.

Within the general framework of spontaneous symmetry breaking—the only known way of constructing a sensible broken gauge theory—we believe that the W and Z masses must result from new, unknown particles that interact by a new unknown force. We know neither the mass scale M_{SB} of the new particles ("SB" for symmetry breaking) nor the strength λ_{SB} of the new force. With the SSC we will be able to determine whether the new force is a strong interaction

$$\frac{\lambda_{SB}}{8\pi} \cong 0(1) \tag{1.1}$$

or a weak interaction

$$\frac{\lambda_{SB}}{8\pi} \ll 1. \tag{1.2}$$

In the first case I will argue that the new physics lies above 1 TeV. In this case it is also likely that the new spectrum begins below 2 TeV and will be directly observable. In the second, weak coupling case the new particles are much lighter than the 1 TeV scale, *i.e.*, a few hundred GeV or below, and they will be copiously produced at the SSC. These statements taken together are what I call the "No-Lose Corollary".

The basic physical point is that the longitudinal polarization modes of the W and Z, denoted W_L and Z_L, are actually degrees of freedom that originate, by the Higgs mechanism, in the symmetry breaking sector. They are essentially the "pions" of the symmetry breaking sector, and like the pions of hadron physics they obey low energy theorems characteristic of the scattering of Goldstone bosons.[3] If there are no other light (compared to 1 TeV) particles in the symmetry breaking sector, the low energy scattering amplitudes depend only on the known parameters G_F and $\rho = (M_W/M_Z \cos\theta_W)^2$ and not at all on the unknown physics of the symmetry breaking sector, denoted by its lagrangian $\mathcal{L}_{SB}$.[4] For example, one of the low energy amplitudes is

$$\mathcal{M}(W_L^+ W_L^- \to Z_L Z_L) = \sqrt{2} G_F \frac{s}{\rho}. \tag{1.3}$$

The low energy theorem provides the correlation between the mass scale M_{SB} and the interaction strength λ_{SB}. Unitarity requires that the amplitude cannot be proportional to s for arbitrarily large s, and the most likely scenario, discussed below, is that the growth in s is cut off at the mass scale M_{SB}. This observation allows us to correlate the strong coupling regime, eq. (1.1), with the mass domain $M_{SB} \gtrsim 1$ TeV (*c.f.* section 3).

For the SSC the crucial observation is that strong W_L, Z_L scattering is observable at the SSC by virtue of increased yields of gauge boson pairs produced with the WW fusion mechanism[5] shown in figure 1. At the design energy and luminosity these extra gauge boson pairs will be observable above the background sources of gauge boson pairs that are present whether $\mathcal{L}_{SB}$ is a strong or weak coupling theory. The conclusion is the

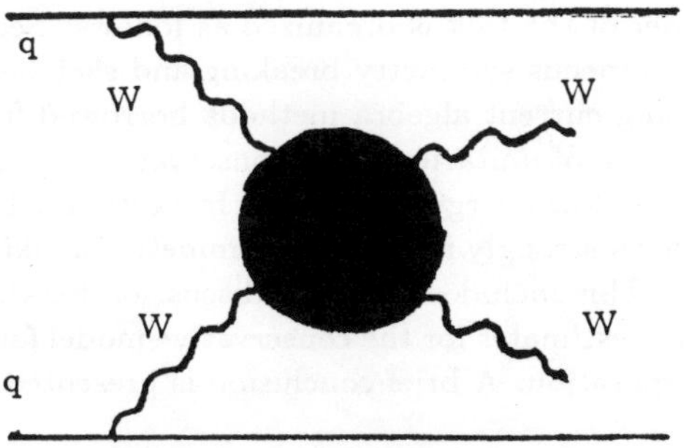

Figure 1: Production of W pairs by WW fusion.

No-Lose Corollary:

Either there are light ($\ll$ 1 TeV) particles from $\mathcal{L}_{SB}$ that can be produced and studied directly

and/or

Excess WW, WZ, ZZ production is observable, signaling strongly-coupled $\mathcal{L}_{SB}$ with $M_{SB} \gtrsim$ 1 TeV.

For the strong coupling case, if as in hadron physics resonances occur when the partial wave amplitudes are 0(1), then probably $M_{SB} \lesssim$ 2 TeV and the low-lying spectrum of $\mathcal{L}_{SB}$ is (just) visible.

In the weak coupling case there is one known exception that should be mentioned: if $\mathcal{L}_{SB}$ is given by the minimal Higgs model and if the mass happens to lie in the interval $2m_t < m_H < 2M_W$, then although 10^6 Higgs would be produced in an SSC year, it is not now known how to detect them in their $H \rightarrow \bar{t}t$ decay mode.[6] For now the only known way of discovering the Higgs boson in this mass range is to build a $\sqrt{s} = 300$ GeV e^+e^- collider with a luminosity of $\sim 10^{32}$ cm.$^{-2}$ s.$^{-1}$. Even in this rather unlikely scenario, the SSC would contribute to our understanding of symmetry breaking by verifying the absence of the excess gauge boson pairs associated with new strong interactions. (Strong coupling models might have light scalars that would approximately mimic a light Higgs boson.) In general the absence of additional gauge boson pairs from WW fusion would be our cue for a redoubled search of the sub-TeV mass scale.

The remainder of the talk is organized as follows: Section 2 reviews relevant aspects of spontaneous symmetry breaking and sketches a proof of the low energy theorems using current algebra methods borrowed from hadron physics. Section 3 is a discussion of unitarity and a conservative strong interaction model that is inspired by the low energy theorems.[7] In Section 4 I will discuss the experimental signals for a strongly interacting symmetry breaking sector that could be seen at the SSC. This includes a careful discussion for the 1 TeV Higgs case and some preliminary estimates for the conservative model for which more careful estimates are in preparation. A brief conclusion is presented in Section 5.

2. SPONTANEOUS SYMMETRY BREAKING AND LOW ENERGY THEOREMS

In order to implement spontaneous symmetry breaking, the lagrangian of the symmetry breaking sector, $\mathcal{L}_{SB}$, must possess a global symmetry group G—analogous to the flavor symmetry of QCD—which breaks by asymmetry of the vacuum to a smaller group H,

$$G \rightarrow H. \tag{2.1}$$

Gauge invariance requires that G include the electroweak $SU(2)_L \times U(1)_Y$ and that H include the unbroken electromagnetic U(1). For each broken generator of G there is a massless goldstone boson in the spectrum of $\mathcal{L}_{SB}$. Three of these couple to the weak currents and are denoted $w^{\pm}$, z. Others, if any, are denoted by $\{\phi_i\}$. Including the electroweak gauge interactions, the goldstone triplet $w^{\pm}$, z become longitudinal gauge boson modes $W_L^{\pm}$, Z_L, and the $\{\phi_i\}$ acquire small masses $0(gM_{SB})$, becoming "pseudo-goldstone" bosons.

As an example, for two flavor QCD with massless quarks the global symmetry G is $SU(2)_L \times SU(2)_R$. After spontaneous symmetry breaking the surviving invariance group is $H = SU(2)_{L+R}$ which is just the isospin group. There are three broken generators, corresponding to the axial generators $SU(2)_{L-R}$, so that three massless goldstone bosons emerge, $\pi^{\pm}$ and π^0. If there were no other symmetry breaking physics, $\mathcal{L}_{SB}$, $\pi^{\pm}$ and π^0 would indeed become longitudinal modes of $W^{\pm}$ and Z, which would however have masses of order 40 MeV rather than $\sim$ 100 GeV.

The statement that the longitudinal modes $W_L^{\pm}$, Z_L are identified with the goldstone bosons $w^{\pm}$, z is given a precise meaning by the equivalence theorem, proved to all orders in ref. (7):

$$M\left(W_L(p_1)W_L(p_2)\ldots\right)_U = M\left(w(p_1)w(p_2)\ldots\right)_R + O\left(\frac{M_W}{E_i}\right). \tag{2.2}$$

In eq. (2.2) the left side is an S-matrix element involving longitudinal modes in the U or unitary gauge while the right side is the corresponding goldstone boson amplitude in an R or renormalizeable gauge. As indicated in eq. (2.2), the equivalence holds at energies large compared to the W and Z masses. We can use the equivalence theorem to translate statements about goldstone boson scattering amplitudes into statements about scattering of longitudinally polarized W's and Z's.

As an immediate application, consider the case[7] in which the global symmetry G includes $SU(2)_L \times SU(2)_R$ and H includes an $SU(2)_{L+R}$. For such theories $\rho = 1$ up to electroweak corrections and we may immediately apply the pion low energy theorems derived from current algebra for just this case. For instance, just as for pions we have

$$M(\pi^+\pi^- \to \pi^0\pi^0) = \frac{s}{F_\pi^2} \qquad s \ll 1 \text{ GeV}^2 \tag{2.3}$$

for $\mathcal{L}_{SB}$ with no other particles that are light compared to M_{SB} we would have

$$M(w^+w^- \to zz) = \frac{s}{v^2} \qquad s \ll M_{SB}^2 \tag{2.4}$$

where $v = 0.25$ TeV is the familiar vacuum expectation value, $M_W = \frac{1}{2}gv$. With the equivalence theorem this becomes a statement about the scattering of W_L and Z_L in an intermediate energy domain:

$$M(W_L^+W_L^- \to Z_LZ_L) = \frac{s}{v^2} \qquad M_W^2 \ll s \ll M_{SB}^2. \tag{2.5}$$

Eq. (2.5) and eq. (1.3) are equal up to small corrections, $0(M_W^2/M_{SB}^2)$, for $\rho = 1$.

The assumptions used above, $G \supset SU(2)_L \times SU(2)_R$ and $H \supset SU(2)_{L+R}$, are sufficient to guarantee $\rho = 1$ to all orders in λ_{SB} but they are not known to be necessary conditions. We are therefore motivated to derive the low energy theorems for all candidate groups G and H and for all values of ρ. This has been done by three different methods:[4] a perturbative power counting analysis, the method of effective lagrangians, and current algebra. I will sketch the current algebra derivation below. Along with the low energy theorems for general values of ρ, the derivation establishes a kind of converse to the result quoted above: we find that if $\rho = 1$ then the goldstone boson sector consisting of $w^\pm$, z possesses an effective $SU(2)_{L+R}$ symmetry ("custodial" SU(2)) in the low energy domain $s \ll M_{SB}^2$.

Briefly the derivation is as follows. The global symmetry G must be at least as large as the gauge group, $G \supset SU(2)_L \times U(1)_Y$, so in particular we have the $SU(2)_L$ charge algebra

$$[L_a, L_b] = i\varepsilon_{abc}L_c \tag{2.6}$$

where the corresponding local currents L_a^μ can generally be expanded in terms of the goldstone triplet $w^\pm$, z as

$$L_a^\mu = \frac{1}{2}r_a\varepsilon_{abc}w_b\partial^\mu w_c - \frac{1}{2}f_a\partial^\mu w_a + \ldots \tag{2.7}$$

with terms involving heavy fields omitted. Since $H \supset U(1)_{EM}$ we have $f_1 = f_2$ and $r_1 = r_2$. The f_a are analogues of the PCAC constant and determine the gauge boson masses,

$$M_W = \frac{1}{2} g f_1, \tag{2.8}$$

$$\rho = (f_1/f_3)^2. \tag{2.9}$$

It is straightforward to show that the $SU(2)_L$ algebra requires

$$r_1 = \frac{1}{\sqrt{\rho}}, \tag{2.10}$$

$$r_3 = 2 - \frac{1}{\rho}, \tag{2.11}$$

so that the parameters r_a and f_a in eq. (2.7) are completely determined in terms of G_F and ρ. In particular, $\rho = 1$ implies $f_1 = f_2 = f_3$ and $r_1 = r_2 = r_3 = 1$ which means that the goldstone boson contributions to L_a^μ are the difference of $SU(2)$ vector and axial vector currents. The existence of this vector $SU(2)$ triplet of currents establishes the converse alluded to above.

The rest of the derivation is much like the usual current algebra derivation[3] except that we do not assume an $SU(2)_{L+R}$ isospin invariance. Consequently pole terms which are forbidden by G-parity in the pion case are not forbidden here. Assuming that $w^\pm$, z saturate these pole terms we find goldstone boson low energy theorems such as

$$M(w^+w^- \to zz) = \frac{s}{f_1^2}\frac{1}{\rho} \qquad s \ll M_{SB}^2 \tag{2.12}$$

which using (2.8) reduces to (2.5) for the case $\rho = 1$. By the equivalence theorem we have then

$$M(W_L^+W_L^- \to Z_L Z_L) = \frac{s}{v^2}\frac{1}{\rho} \qquad M_W^2 \ll s \ll M_{SB}^2 \tag{2.13}$$

with $v = f_1 \cong 2M_W/g$. The other two independent amplitudes are

$$M(W_L^+W_L^- \to W_L^+W_L^-) = -\frac{u}{v^2}\left(4 - \frac{3}{\rho}\right), \tag{2.14}$$

$$M(Z_L Z_L \to Z_L Z_L) = 0, \tag{2.15}$$

and by crossing we have also

$$M(W_L^\pm Z_L \to W_L^\pm Z_L) = \frac{t}{v^2}\frac{1}{\rho}, \tag{2.16}$$

$$M(W_L^+W_L^+ \to W_L^+W_L^+) = M(W_L^-W_L^- \to W_L^-W_L^-) = -\frac{s}{v^2}\left(4 - \frac{3}{\rho}\right). \tag{2.17}$$

Like (2.13), eqs. (2.14 – 2.17) are valid in the intermediate domain $M_W^2 \ll E_i^2 \ll M_{SB}^2$.

3. UNITARITY AND A CONSERVATIVE MODEL

The linear growth in s, t, u of the amplitudes (2.13 – 2.17) cannot continue indefinitely or unitarity would be violated. For instance the $W_L W_L \rightarrow Z_L Z_L$ amplitude (2.13) is pure s-wave. If we adopt the low energy amplitude (2.13) as a model of the absolute value of the scattering amplitude, then the $J = 0$ partial wave amplitude is

$$\left|a_0(W_L^+ W_L^- \rightarrow Z_L Z_L)\right| = \frac{s}{16\pi v^2} \tag{3.1}$$

where here and hereafter we set $\rho = 1$. Unitarity requires $|a_0| \leq 1$ so we see that the growth of a_0 must be cut off at a scale Λ with

$$\Lambda \leq 4\sqrt{\pi} v = 1.8 \text{ TeV}. \tag{3.2}$$

At the cutoff $\sqrt{s} = \Lambda$ the order of magnitude of the amplitude is

$$|a_0(\Lambda)| = \frac{\Lambda^2}{16\pi v^2}. \tag{3.3}$$

For $\Lambda \lesssim v \cong \frac{1}{4}$ TeV we have $|a_0(\Lambda)| \ll 1$ indicating a weakly interacting theory for the symmetry breaking dynamics $\mathcal{L}_{SB}$, while for $\Lambda \gtrsim 1$ TeV we have $|a_0(\Lambda)| \cong 0(1)$, the hallmark of a strong interaction theory. Though there is one counterexample mentioned below, the most likely dynamics is that the cutoff scale Λ is of the order of M_{SB}, the mass scale of the new quanta. Then for $\Lambda \cong 0(M_{SB})$ eq. (3.3) establishes the relationship mentioned in the introduction between the mass scale of the new quanta and the strength of the new interactions: weak coupling for $M_{SB} \ll 1$ TeV and strong coupling for $M_{SB} \gtrsim 0(1)$ TeV.

A weak coupling example is provided by the standard Higgs model[1] with a light Higgs boson, $m_H \ll 1$ TeV, which can be treated perturbatively. Then $a_0(W_L W_L \rightarrow Z_L Z_L)$ is given in tree approximation by (where I neglect M_W^2/s)

$$a_0 = \frac{-s}{16\pi v^2} \frac{m_H^2}{s - m_H^2}. \tag{3.4}$$

For $s \ll m_H^2$ this agrees with the low energy theorem (3.2) while for $s \gg m_H^2$ it saturates at the constant value $m_H^2/16\pi v^2$. Comparing with (3.3) we see that m_H indeed provides the scale for Λ.

A strong interaction example is provided by hadron physics. For the $J = I = 0$ partial wave, the low energy theorem[3] gives

$$a_{00}(\pi\pi \rightarrow \pi\pi) = \frac{s}{16\pi F_\pi^2} \tag{3.5}$$

with $F_\pi = 92$ MeV. Eq. (3.5) saturates unitarity at $4\sqrt{\pi}F_\pi = 650$ MeV which is indeed the order of the hadron mass scale. The a_{11} and a_{02} amplitudes saturate at 1100 and 1600 MeV.

The two generic possibilities are illustrated in fig. (2). For weak coupling the partial wave amplitudes saturate at values small compared to 1 giving rise to narrow resonances at masses well below 1 TeV. For strong coupling they saturate the unitarity limit with broad resonances in the TeV range.

There is one known model,[8] whose real physical significance is not clear, which has behavior different from figure (2), namely the 0(2N) Higgs model solved to leading order for $N \to \infty$ and then evaluated at $N = 2$ which corresponds to the standard model. (This is only a little worse than the large N_{color} limit for QCD which approximates $3 \cong \infty$; I admit to uneasiness with both approximations.) In that model, which is a strong coupling model, the low energy theorems are of course valid and there is indeed a slow (logarithmic) saturation of partial wave unitarity at the TeV scale, but there are no discernible resonances in the TeV region.

The 0(2N) model or the possibility (which cannot be definitively excluded by the heuristic estimates given above) that resonance structure might be deferred to 2 TeV or above both motivate a conservative model for strong interactions that Mary Gaillard and I have considered.[7] In this model we represent the absolute

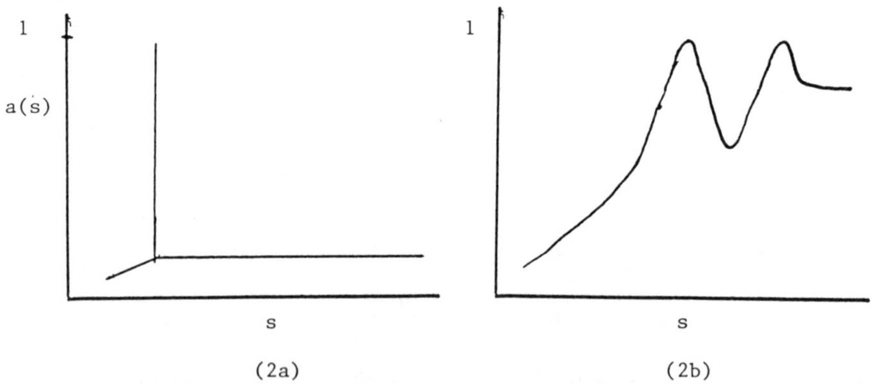

Figure 2. Typical behavior of partial wave amplitudes. Fig. (2a) corresponds to weak coupling — narrow resonance(s) much lighter than 1 TeV and saturation well below the unitarity limit — while fig. (2b) represents strong coupling — broad resonances at the TeV scale and saturation at the order of the unitarity limit.

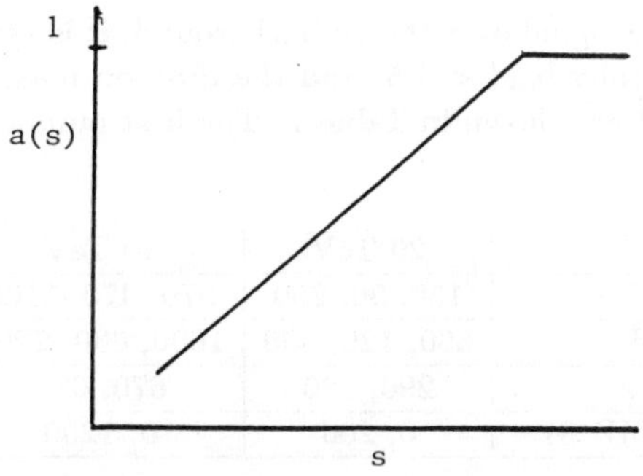

Figure 3. The conservative strong interaction model discussed in the text.

values of the partial wave amplitudes by the low energy theorems up to the energy at which unitarity is saturated and set them equal to one for higher energies, as shown in figure (3). The model is conservative in three respects:

- it neglects possible (or, I would say, likely) resonance structure, underestimating the yield for the 1 TeV Higgs boson by $\sim 50\%$.
- it neglects higher partial waves which surely begin to contribute as the lowest waves saturate.
- it correctly represents the order of magnitude seen in $\pi\pi$ data.

We discuss the experimental implications in the next section.

4. EXPERIMENTAL SIGNALS FOR STRONG INTERACTION MODELS

We consider what might actually be observed at the SSC if $\mathcal{L}_{SB}$ is a strongly interacting theory. The generic strong interaction signal is longitudinally polarized W, Z pairs produced by WW fusion, fig. (1). It will help to measure the polarization of the gauge bosons[9] (if statistics is sufficient) but I will *not* assume polarization information in the results given here. Then the irreducible background is from $\bar{q}q \to WW$, WZ, ZZ. Since $\mathcal{M}(\bar{q}q \to WW) = 0(g^2)$ while $\mathcal{M}(qq \to qqWW) = 0(g^2\lambda_{SB})$, we expect a discernible signal if and only if $\mathcal{L}_{SB}$ is strongly interacting, $\lambda_{SB} = 0(1)$. The signal may occur in W^+W^- and ZZ, as for the standard Higgs boson, but also more generally in $W^\pm Z$ and even W^+W^+ and W^-W^- ("I" = 1 and 2 channels).

In order to see the signal over the $\bar{q}q$ background, it is essential to cut on the W, Z rapidity, generally $|y_W| < 1.5$, and the diboson mass, say $M_{WW} > 1$ TeV. Yields from ref. (7) are shown in Table 1. The first number denotes the $\bar{q}q$

	20 TeV	40 TeV
ZZ	150, 90, 250	370, 470, 1100
W^+W^-	660, 120, 500	1600, 630, 2200
$W^\pm Z$	290, 120	670, 670
$W^+W^+ + W^-W^-$	0, 200	0, 1200

Table 1: yields in events per 10^4pb^{-1} for 20 and 40 TeV pp colliders, taken from ref. (7). Cuts are $M_{VV} > 1.0$ TeV and $|y_V| < 1.5$ except for $W^+W^+ + W^-W^-$ for which the rapidity cut is relaxed to $|y_V| < 4.0$. In each entry the first number is the $\bar{q}q$ annihilation background, the second is the conservative strong interaction model, and the third is the 1 TeV Higgs boson.

background, the second is the conservative model, and the third where present is for the 1 TeV Higgs boson. The units are events per $10^4 pb^{-1}$, *i.e.*, 10^7 sec. of running at $\mathcal{L} = 10^{33}$ cm.$^{-2}$ s.$^{-1}$. For like-charged W pairs the rapidity cut is relaxed to $|y_W| < 4$ since there is no $\bar{q}q$ background, the leading background from single gluon exchange being perhaps ~ 5 times smaller than the $\bar{q}q$ backgrounds in the other channels.[10]

Notice the uncharacteristically sharp dependence on the machine energy, 20 TeV versus 40 TeV. This is simply because the signal is at the edge of phase space. Not only the signal but also the signal:background ratio suffer at lower energy. The greater sensitivity of the signal than the background probably reflects the four body phase space of the signal, $qq \to qqWW$, compared to the two body phase space of the background $\bar{q}q \to WW$. To compensate for lower energy, luminosity would need to be increased beyond what would be needed just to equal the signal of a higher energy machine.

Table 1 is chiefly a theoretical exercise since it does not necessarily correspond to experimentally implementable signals. We need to consider how the gauge bosons decay and are detected. This has been done with some care for the 1 TeV Higgs but not yet for the conservative model.

The cleanest channel, $ZZ \to e^+e^-/\mu^+\mu^- + e^+e^-/\mu^+\mu^-$, has a small branching ratio, $B = 3.6 \cdot 10^{-3}$. For the 1 TeV Higgs and the cuts of Table 1 this gives a signal of 4 events over a background of 1 for $\sqrt{s} = 40$ TeV. A more promising leptonic channel is[7,11] $ZZ \to e^+e^-/\mu^+\mu^- + \bar{\nu}\nu$ with $B \sim .02$, analogous to observing $W \to e/\mu + \nu$. The signal is defined by 1) $Z \to e^+e^-/\mu^+\mu^-$ at large p_T with central rapidity, 2) large missing p_T, and 3) no hot jet activity in order to veto the background from $W+$ jet. The latter especially must be studied with a Monte

Carlo but is likely to be both clean and efficient. For the 1 TeV Higgs, Cahn and I found[11] that cuts of $|y_{e^+e^-}| < 1.5$ and $p_T(e^+e^-) > .45$ TeV give a signal of 43 events over a $\bar{q}q$ background of 7 for 40 TeV (16 standard deviations) compared to 7 over 3 for 20 TeV. The $\bar{q}q$ backgrounds are under adequate theoretical control and will in any case be measured at the SSC.

A second channel that has been well studied[12,10] is $H \to WW \to e\bar{\nu}/\mu\bar{\nu} + u\bar{d}/c\bar{s}$. Though the QCD background from Wjj is formidable, Gunion and Soldate have identified cuts which allow the signal to emerge. For the 1 TeV Higgs my extrapolation of their result for $m_H = 0.8$ TeV gives a signal of ~ 300 events over a background of ~ 500 at $\sqrt{s} = 40$ TeV (or $\sim 13\ \sigma$).

For the conservative strong interaction model defined in Section 4, comparable calculations are not yet completed. Except for $ZZ \to e^+e^-/\mu^+\mu^- + e^+e^-/\mu^+\mu^-$ (with a signal of 2 events over a background of 1), I can now offer only some crude guesses. It is however clear that the conservative model will be even more sensitive to machine energy than the 1 TeV Higgs since the signal of the former is smeared over larger values of M_{WW}.

For $ZZ \to e^+e^-/\mu^+\mu^- + \bar{\nu}\nu$ and $WW \to e\bar{\nu}/\mu\bar{\nu} + u\bar{d}/c\bar{s}$ I expect the yields to be roughly half of those quoted above for the 1 TeV Higgs boson. Signals comparable to $ZZ \to e^+e^-/\mu^+\mu^- + \bar{\nu}\nu$ are also expected in $WZ \to \ell\bar{\nu} + e^+e^-/\mu^+\mu^-$ and in $W^+W^+ + W^-W^- \to \mu\mu\nu\nu$. If the electron charge can also be measured the signal for the like-charged W channel is multiplied by a factor 4. The channel $WW \to \ell\bar{\nu} + \bar{\ell}\nu$ may also provide a useable signal without determining the lepton charges.

5. CONCLUSION

With the SSC design parameters, $\sqrt{s} = 40$ TeV and $\mathcal{L} = 10^{33}\text{cm.}^{-2}\text{s.}^{-1}$, we are assured of the capability to see the signal of a strongly interacting symmetry breaking sector. The SSC is strategically positioned in that it is close to being the minimal machine about which this statement can be made. If we do see signs of TeV scale, strong interaction, symmetry breaking physics, then (hard though it may be to imagine now) we are certainly going to think about a next generation facility that will allow more detailed studies.

Because the SSC has the capability to see the strong interaction signal, we will learn something even by not seeing the signal. We then learn that the physics of symmetry breaking involves quanta that are below 1 TeV and not above. If we have not already found them we will know where we need to look. Large numbers of the new quanta would be produced at the SSC, so that detailed studies are likely to be possible in most physics scenarios.

Light quanta—*e.g.*, pseudo goldstone bosons—can occur in TeV-scale strong interaction models. Thus without detailed studies of their properties, the discovery of light Higgs candidates may not immediately and unambiguously tell us whether $\mathcal{L}_{SB}$ involves TeV scale strong interactions. Therefore even if light scalars are discovered, say at SLC or LEP, we will need to look for the strong interaction signal at the SSC in order to clearly understand the nature of the symmetry breaking physics.

Finally it is important to keep in mind, as discussed in the introduction, the physics I have not discussed. It is potentially very rich and the SSC probes it with great facility. I have concentrated just on what must be there—electroweak symmetry breaking—and have argued that at the design energy and luminosity the SSC is sure to illuminate the underlying physics.

References

1. S. Weinberg, Phys. Rev. Lett. 19 (1967) 1264; A. Salam in "Elementary Particle Physics" (ed. N. Svartholm, Almqvist and Wiksells, Stockholm, 1968) p. 367.

2. M. Chanowitz, LBL-21973, to be published in the Proceedings of the XXIII International Conference on High Energy Physics, Berkeley, 1986.

3. S. Weinberg, Phys. Rev. Lett. 17 (1966) 616.

4. M. Chanowitz, H. Georgi and M. Golden, Phys. Rev. Lett. 57 (1986) 2344 and manuscript in preparation.

5. R. Cahn and S. Dawson, Phys. Lett. 136B (1984) 196; E 138B (1984) 464. See also D. Jones and S. Petcov, Phys. Lett. 84B (1979) 440; and Z. Hioki, S. Midorikawa, and H. Nishiura, Prog. Theor. Phys. 69 (1983) 1484. For the effective W approximation see also M. Chanowitz and M. Gaillard, Phys. Lett. 136B (1984) 196; S. Dawson, Nucl. Phys. B29 (1985) 42; G. Kane, W. Repko, and W. Rolnick, Phys. Lett. 148B (1984) 367.

6. B. Cox and F. Gilman, Proc. Snowmass 1984 (ed. R. Donaldson) p. 87.

7. M. Chanowitz and M. Gaillard, Nucl. Phys. B261 (1985) 379.

8. M. Einhorn, Nucl. Phys. B246 (1984) 75; R. Casalbuoni, D. Dominici, and R. Gatto, Phys. Lett. 147B (1984) 419.

9. M. Duncan, G. Kane, and W. Repko, Nucl. Phys. B272 (1986) 517.

10. J. Gunion *et al.*, Theoretical Report of W/Z/Higgs SSC Working Group, UCB-86-39, to be published in Proc. Snowmass 1986.

11. M. Chanowitz and R. Cahn, Phys. Rev. Lett. 56 (1986) 1327.

12. J. Gunion and M. Soldate, Phys. Rev. D34 (1986) 826.

STATUS OF THE TEVATRON I COLLIDER

Philip S. Martin

Fermi National Accelerator Laboratory*
Batavia, Illinois 60510

ABSTRACT

The Tevatron I Collider at Fermilab is being commissioned following the last major construction for the Collider program, the BO overpass and the DO experimental hall. The present status of the accelerator, the antiproton source, and the Collider experiments is reviewed.

1. BRIEF HISTORICAL REVIEW OF THE TEVATRON

The installation of the Tevatron was basically completed in early 1983, and beam studies began soon thereafter. A 400 GeV fixed target run, using the Tevatron began in October, 1983. In March, 1984, there was a brief shutdown to install the low-beta quadrupoles at the BO straight section, followed by an 800 GeV fixed-target run. That run was terminated by a series of magnet failures, traced to lack of restraint on the leads on the upstream end of half of the dipoles. During the ensuing shutdown to fix the magnets, from July through November, 1984, the Antiproton Source, the extraction line from the Main Ring to the Source, and the DO overpass were constructed.

A second 800 GeV fixed target run took place during the first 8 months of 1985. During this run, particularly during the summer months, there was much activity in the Accelerator Division and the Antiproton Source, preparing for the Collider test run which took place in September and October. Much of the Antiproton Source work could be done parasitically to the fixed-target run, as could many of the Main Ring studies. The Tevatron studies on the other hand required dedicated use of the machine, and there were frequent study periods to enable that effort to proceed.

*Operated by Universities Research Association, Inc. under contract with the U.S. Department of Energy

During the Collider test, it was clear almost from the start that the intensity of antiprotons would be low. The Antiproton Source was essentially complete, but the systems had not been sufficiently optimized, and as a result the stacking rate was about two orders of magnitude below the design value. It appeared the intensity of antiprotons in the Main Ring would not be sufficient for the rf feedback loops to work. The decision was made to use "pilot" protons, and have no feedback whatsoever on the antiprotons. This necessitated a number of changes in the configuration of the Main Ring rf and in the ramp profile, but had no affect on the Tevatron operation. The Collider test was successful, in that the accelerator systems operated at some level of efficiency, and collisions at 1.6 TeV in the center of mass were seen by the experimenters at B0. The luminosity was about six orders of magnitude below the design value.

2. 1986 SHUTDOWN

The primary purpose for the shutdown, which lasted from mid-October, 1985, until August, 1986, was to complete the construction required for the Tevatron I program: the B0 overpass and the D0 experimental hall. But these were not the only jobs. Since this was the last lengthy shutdown scheduled to take place for many years, a number of smaller projects were also undertaken.

The major effort during the shutdown was the B0 overpass, which required major civil construction and the removal and replacement of approximately 10% of the Main Ring magnets. Four new power supplies were required to provided the vertical bending and horizontal orbit correction through the overpass region.

In comparison to the B0 overpass, the construction of the D0 experimental hall was much less traumatic. A relatively small section of the Main Ring and Tevatron had to be disturbed. The Tevatron components, which consist primarily of the extraction four-bump (conventional) dipoles and extraction septa, had to be installed on girders which extend to the center of the collision hall. The girders are supported by posts, for stability reasons, and can be withdrawn to make way for the low-beta quadrupoles once those are required. The completion of the collision hall and erection of the shield wall were the critical path items which determined the accelerator start-up at the end of the construction period.

2.1 Tevatron

During the 1985 run, a series of tests were performed to identify, sector by sector, the weakest magnets, from the standpoint of quench currents. Based on those tests, and correlations between those tests and the quench data from the Magnet Test Facility, nineteen components were replaced. In addition to these, another forty components were replaced for one reason or another, as shown in Table 1.

BREAKDOWN OF TEVATRON COMPONENT REPLACEMENTS
JULY 1, 1985 THROUGH JULY 15, 1986

Reason for replacement	Number
Low quench current	19
Hipot failure	8
Open correction elements	2
Instrumentation/BPM problems	3
Leak	12
Matching to another replaced magnet	7
For E-710	2
Failed during operation	2
Damaged during shutdown	2

TABLE 1

The experience during the two fixed-target runs had shown that the heater firing units, an integral part of the quench protection system, were responsible for a large fraction of downtime. They tended to discharge spontaneously, (causing quenches,) or to fail to recharge. The shutdown provided the opportunity to rebuild all of the units. The new units have proven much more reliable.

2.2 Main Ring

The Collider test run had shown that the procedure for producing intense bunches, the coalescing procedure, did not work reliably. To improve the coalescing, two things were done to the rf systems. First, the counterphasing control for controlling the h=1113 rf was changed to allow turning off sixteen of the eighteen cavities before the remaining two are counterphased. This provides a factor of nine improvement in the control over the final rf voltage. Second, the h=53 system which rotates the ensemble of bunches so that they can be recaptured in a single h=1113 bucket, was increased by a factor of two. Of larger impact on the Main Ring was the attempt to reduce the impedance of the machine. Signals in the 1.6 GHz region were observed during the coalescing, as the bunches were being reduced to a small momentum spread. Similar signals were observed during other debunching studies. These signals are attributed to a microwave instability, which depends on the machine impedance at the appropriate harmonic number, the Z/n of the machine. To reduce the impedance, inserts were made and installed in nearly every interface in the Main Ring. Installing the inserts, which smooth the discontinuity at the interface, required opening the interface, and in most cases moving one of the adjacent magnets.

2.3 Booster

The Booster was the recipient of a new controls system, the same transformation that the Linac and Main Ring had undergone the previous year. It also has a new beam position monitor system. Still in progress is a transition-jump system to help reduce the beam emittance increase which occurs while crossing the transition energy.

2.4 Linac

The Linac was used by the Neutron Therapy Facility for patient treatment during most of the time the other machines were down. There was time, however, to realign many of the drift tanks.

2.5 Eight GeV Line

With the large vertical dispersion introduced by the overpasses, it became necessary to redesign the injection into Main Ring from Booster. The entire line was replaced, with simpler optics and better instrumentation.

2.6 Antiproton Source

2.6.1 AP-4 Line. The AP-4 Line, which connects the Booster and the Antiproton Source, was commissioned during the shutdown. This allowed studies to proceed in the Source even though the Main Ring was torn up. This line was used extensively, well into the start-up of the Main Ring and Tevatron, except for the period the Booster was down for the controls system changeover.

2.6.2 Debuncher. The Debuncher effort during the shutdown was concentrated on optimizing the aperture. This includes study of orbit control around injection and extraction devices, and installation of movable stochastic cooling tanks. The beam lifetime was measured to be two hours; the momentum aperture is approximately 4%.

2.6.3 Accumulator. The primary goal for the Accumulator during the shutdown was to increase the aperture and to improve the impedance. The stochastic cooling tanks, which should be the limiting aperture, were removed so the remainder of the ring could be explored. Many elements were placed on motorized stands so they could be remotely aligned to provide the maximum aperture. Dampers and harmonic correctors were added.

2.6.4. Target Station. A number of changes were made in the antiproton target station. The pulsed magnet now has full remote handling capability. A new lithium lens and transformer were installed. The lithium lens is now being run at a lower gradient, which required repositioning the target. Copper targets replace the tungsten-rhenium targets which clearly showed indications of damage when examined during the shutdown. The copper target will reduce the antiproton yield by approximately 20%.

2.7 Utilities

During the fixed-target runs, several failures occurred which resulted from the increased voltage stress due to higher harmonics on the power feeders. One failure, a primary-to-secondary short in a Tevatron power supply transformer, damaged several Tevatron magnets and required several weeks to repair. To reduce the harmonic voltages, the Main Ring and Tevatron feeders have been split. The added capacitance of the new feeder has shifted the harmonic frequencies, and a "harmonic filter" has been installed to reduce the voltages at the new frequencies. The filter that is now in place is temporary, and allows operation with only six Tevatron power supplies, which is appropriate for the collider run. The final filter is awaiting delivery of capacitors; only when that filter is in place can all twelve supplies be energized, as is desirable for fixed-target operation.

3. RECOMMISSIONING

The recommissioning of the accelerators has taken several months. It began with the Booster, the Main Ring and the Antiproton Source. The Tevatron was complete somewhat later, due to the period required for leak-checking, dehydration, and cooldown. The completion of the Tevatron work, together with access requirements of the various experiments, meant that the accelerator had to be shut down frequently during this time. The recommissioning was interrupted for an extended period beginning in mid-December to allow CDF to move into the collision hall at BO.

The Booster came up fairly rapidly after the shutdown for replacement of its controls system. It has recently accelerated over 3E12 protons per batch. The new beam position monitor system is now essentially fully operational. The transition-jump system is in test stage.

The commissioning of the 8 GeV line and the recommissioning of the Main Ring began in August, 1986. The DO collision hall was not yet complete; consequently, the beam studies were only allowed on weekends, and were restricted to low intensity. The new 8 GeV line behaved well with two exceptions. The vertical aperture of the last horizontal dipole before the injection Lambertson magnet was too small. This magnet was replaced during the December shutdown. Second, the new injection method required a vertical four-bump across the AO straight section. The magnets that were installed did not have sufficient field quality over the necessary aperture, and have since been replaced.

The Main Ring had two obstacles which had to be removed before circulating beam could be established. Accelerated beam followed quickly. The efficiency was poor, however, and a systematic set of aperture scans had to be done. This led to moving, and in some case replacing magnets or other components. As the intensity increased, so did the losses at the B0 straight section. These losses were of serious concern for the Collider Detector at Fermilab (CDF). A limit of 50 rads per week was established as the goal, to reduce radiation damage to the detector to an acceptable level when it moved into the collision hall.

The reduction of these losses was accomplished through a number of steps. The largest gains were made by increasing the vertical aperture of the Main Ring in the region immediately upstream of B0. The progress is best shown by plotting the losses per proton injected into the Main Ring, Figure 1. Not shown in this figure is another factor of ten reduction in the weekly radiation dose from the week of 10/8 due to A48-2 being replaced. (The normalization to injected protons was not available prior to 10/15, but the number of protons should have been comparable for the two weeks.)

The reinstallation and leak-checking of the Tevatron proceeded roughly on schedule. Each sector was checked out and powered as soon as it was cold. Part of the power tests involved ramping each sector as high as possible in current and identifying (by installing special instrumentation) which component was quenching. This provided a list of the lowest quench current magnet in each sector. Tevatron operation was reestablished at 800 GeV; ramping and storage tests showed that 875 GeV was possible, and operation was shifted to that energy.

The Antiproton Source, using the AP-4 line, was able to run as soon as the Booster came up, and was unaffected by the work in the Main Ring tunnel. One of the major problems encountered in the Source during this period was a rash of kicker cable failures. Operationally, there have been some problems with the Debuncher extraction kicker and with the Accumulator stochastic cooling systems drifting out of phase.

By mid-November, the Antiproton Source was ready to begin stacking antiprotons. The stacking rate was low, roughly the same as during the October, 1985, test. Then a vacuum problem developed in B-sector, and the two-day shutdown planned for that week had to be extended to replace the bad component. At the same time, one of the low quench current magnets was replaced. Following that repair, operation resumed with the Tevatron now at 900 GeV. Within a few days antiprotons were injected into the Main Ring and accelerated, this time without pilot protons, to 150 GeV and injected into the Tevatron. The antiproton stacking rate improved roughly threefold, partly through improvements in the Source, and partly through a 50% increase in the Main Ring intensity. Having the B0 loss problems thoroughly under control finally permitted working on increasing the intensity.

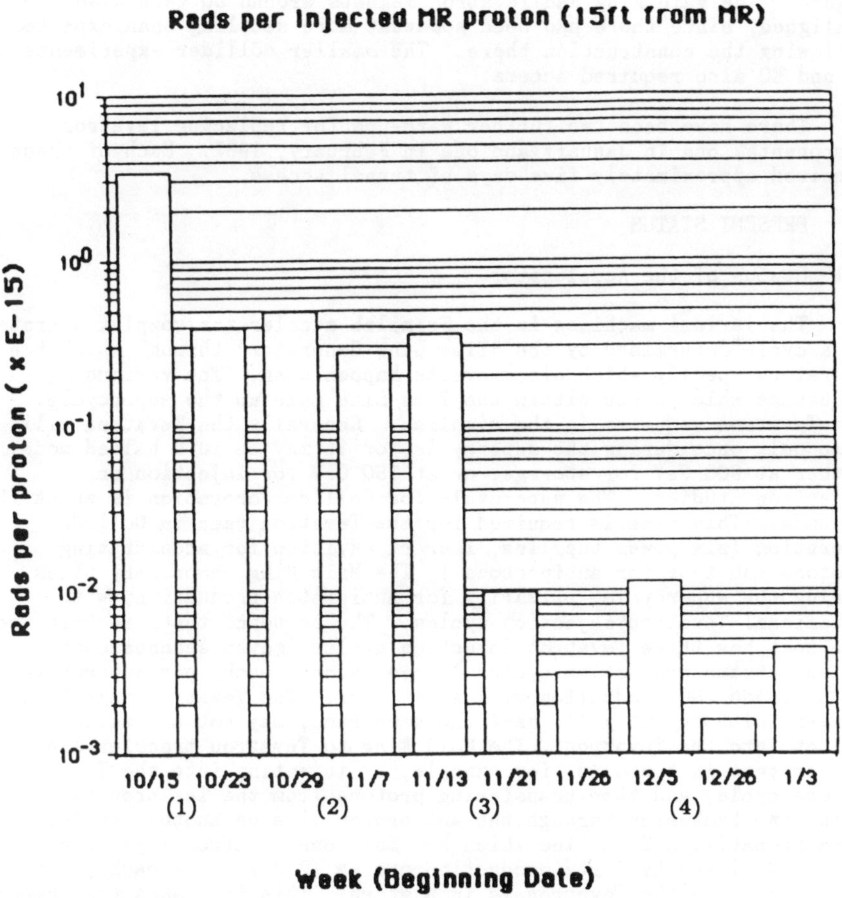

(1) Mini-tube replaced
(2) A48-3 B1 magnet replaced with B2
    (larger vertical aperture)
(3) A48-2 and A48-3 replaced with B3
(4) Collision hall- 6" pipe replaced with 12"

FIGURE 1

The CDF roll-in began in mid- December, a few days earlier than scheduled due to another Tevatron magnet problem. During this period, a number of other tasks were undertaken. Minor vacuum problems existed on either side of D0, so those regions were warmed up and a number of components replaced, including another low quench current magnet. The Main Ring and Tevatron magnets around D0 were also realigned, since there had been somewhat more settling than expected following the construction there. The smaller collider experiments at C0 and E0 also required access.

There have been two further warm-ups for replacing Tevatron components, one in January and one in February, 1987. Each of these required approximately five days of tunnel access.

4. PRESENT STATUS

4.1 Status of the Accelerator

The various machines in the Fermilab accelerator complex operate on a cycle determined by the "Time Line Generator" through which the operators specify which clock events happen when. The various sequences which occur within the Time Line make up the supercycle. The Tevatron sequence is the simplist. Generally the Tevatron will ramp only once during the supercycle, or it may be in a halted mode, either at 900 GeV for storage, or at 150 GeV for injection or injection studies. The supercycle for Collider operation is about 120 seconds. This time is required for the Tevatron ramp in Collider operation (six power supplies, four rf cavities for accelerating protons and four for antiprotons.) The Main Ring ramps many times during the supercycle, primarily for antiproton production, with occasional Tevatron injection cycles. The sequence that has been used the most has three Tevatron injection cycles (seven seconds each), and 33 antiproton production cycles (three seconds each) per supercycle. This yields 990 production cycles per hour. The Tevatron injection cycles, although they all have the same ramp, may not all actually inject into the Tevatron. The Main Ring to Tevatron transfer line for antiprotons is tuned up, for example, by injecting into the Tevatron on one cycle, and then transfering protons from the Tevatron to the Main Ring backwards through the antiproton line on another cycle. More recently, a Time Line which has only one Tevatron injection cycle, followed by forty production cycles (2.8 seconds each), has been used once the Tevatron is in a store. This increases the number of production cycles to 1210 per hour. The cycle time for the production cycles can be reduced further; 2.4 seconds may be possible, but it is unlikely that the design goal of 2 seconds can be achieved without major hardware modifications. In order extract onto the production target more frequently than every 2.4 seconds, multiple batches could be accelerated in the Main Ring, and then extracted one at a time at whatever rate the Source can tolerate.

ANTIPROTON TRANSFERS

	January 1987	February 1987
Number of attempts	44	23
Number of sessions	17	16
Pilot stores	3	0
Full stores	9	15
Main Ring only	0	2
Successful stores/ attempt	27%	71%
Successful stores/ session	53%	94%
Antiprotons extracted from Accumulator	4E11	5E11
Antiprotons extracted/ transfer	1E10	2.3E10
Hours of store	20	100
Average length of store (hours)	2.2	6.6
Number of days	14	11
Integrated luminosity (inverse nanobarns)	.04	.30

TABLE 2

Since the CDF roll-in, the stacking rate in the Antiproton Source has increased approximately three-fold, reaching a level of approximately 9E9 antiprotons per hour. (CERN's best stacking rate is 7.5E9.) Colliding beams operation, while not yet routine, is moving in that direction. Transfers from the Source to the Main Ring and Tevatron are scheduled for three intervals per day, although to date no more than two have been made. The first attempts at producing collisions involved putting only one bunch of antiprotons, together with three bunches of protons, into the Tevatron. As the numbers of antiprotons in the Source grew, so did our appetites. Transfers of two bunches of antiprotons have been successful. Recently, attempts have been made to transfer three bunches of antiprotons, but only two of them survived. Table 2 presents a summary of the attempts at antiproton transfers in January and February, 1987. The duration and initial luminosities (when available) of the 24 successful transfers are shown in Figure 2. The reasons for the ends of the stores are catalogued in Table 3.

REASONS FOR ENDING STORES

Reason	January	February	Total
Deliberate	3	5	8
Correction Elements	2	2	4
BO Quadrupoles	0	3	3
BLM/BPM	1	1	2
Power Supplies	1	1	2
Quench Protection	0	1	1
Vacuum	0	1	1
RF	0	1	1
Cryogenics	1	0	1
Kickers	1	0	1
Total	9	15	24

TABLE 3

The emittances of the beams continue to be a problem; they are roughly a factor of two larger than the design value, and result in a substantial loss of luminosity. The operation of the accelerator complex as a whole is summarized in Table 4, which presents the present status, along with the design parameters, the achievements of the October, 1985, test, and the goals for this running period. Table 5 presents a more detailed picture of the performance of the Antiproton Source.

The luminosities attained are still well below the goal for this run, but are approaching the level at which they are indeed useful for

TEV I OPERATING PARAMETERS

	DESIGN	OCT 85	PRESENT STATUS*	MAR 87
ENERGY (GEV)	1000	800	900	900
PROTONS/BUNCH	6E10	2E10	3-4E10	4E10
ANTIPROTONS/BUNCH	6E10	FEW E6	1E9	1E10
NUMBER OF BUNCHES	3X3	1X1	3X2	3X3
EMITTANCE/BUNCH LENGTH REDUCTION FACTORS	1	.6	.3	.9
LUMINOSITY AT CDF	1E30	FEW E24	5E27	1E29
ANTIPROTON ACCUMULATION RATE (PER HOUR)	10E10	1E9	8E9	1.5E10
MR INTENSITY	2E12	1E12	1.2E12	1.5E12
MR CYCLES PER HOUR	1800	720	1210	1200
ACCUM. TIME/STORE	2 HRS	---	9 HRS	5-6 HRS
ANTIPROTONS EXTRACTED	6E10	FEW E8	2.5E10	2.7E10
MR TRANSFER/TRANSMISSION				
ANTIPROTONS	1	---	.25	.75
MR COALESCING EFFICIENCY				
ANTIPROTONS	1	VERY SMALL	.50	.50
PROTONS	1	.2-.4	.80	.80
TEVATRON TRANSFER, ACCELERATION AND LOW BETA EFFICIENCY				
ANTIPROTONS	1	.5	.70	.95
PROTONS	1	.8	.95	.95

* AS OF FEBRUARY 15, 1987

TABLE 4

ANTIPROTON SOURCE MISSING FACTORS

OPERATION	DESIGN REPORT	FEB 87	MISSING FACTOR FEB 87	MISSING FACTOR OCT 85
MR INTENSITY	2E12	1.2E12	1.7	2
TARGET TO DEBUNCHER	7E7	12.4E6	3.4	10
AFTER BUNCH ROTATION-WITHIN 0.2% DP/P	7E7	10.3E6	1.2	1.1
ONTO ACCUMULATOR INJECTION ORBIT	7E7	8.6E6	1.2	1.1
STACKING ORBIT	6E7	7.8E6	1.1	1.8
CORE	6E7	6.7E6	1.0	1.4
MR CYCLES/HR	1800	1210	1.5	2.5
OVERALL STACKING RATE	10E10/HR	.8E10/HR	12.5	150

TABLE 5

the experimenters. The major problems remaining, in no particular order, are

1) Improving stacking rate to allow more frequent transfers of higher intensity

2) Improving Main Ring acceleration efficiency, both for protons and antiprotons, and reducing the loss of antiprotons during the low beta squeeze

3) Improving coalescing, particularly for antiprotons

4) Improving transverse emittances.

In addition to improvements to these aspects directly related to the beams, another problem is the length of time it takes to complete a transfer. The steps involved are primarily the checkout of the Accumulator to Main Ring and the Main Ring to Tevatron transfer lines, both of which are done using protons in the reverse direction. During this time, antiproton stacking is turned off. The entire transfer process has typically taken four to eight hours. Very recently, that time has been less than two hours. As operators are trained and better procedures are developed, it is hoped the time will decrease further, and that three transfers per day can be accomplished.

4.2 Status of the Experiments

4.2.1 <u>BO - CDF</u>. The Collider Detector at Fermilab, (CDF), which is generally well-known and hence will not be described here, has been installed in the collision hall, and is essentially complete. Problems with the purging system related to gas safety have been solved, and the central tracking chamber is now operational, although a portion does not yet have high voltage, due to a distribution problem. Main Ring beam losses prevent full use of the muon chambers. Attempts will be made to solve these problems during a week-long shutdown in late February. CDF has requested a running period beginning with two months of check-out, followed by a run yielding an integrated luminosity of 100 inverse nanobarns.

4.2.2 <u>CO - E735</u>. Experiment E735, "Search for a Deconfined Quark Gluon Phase of Strongly Interacting Matter..." is a collaboration of Duke, Fermilab, Iowa State, LLL, Notre Dame, Purdue and Wisconsin. The experiment occupies a small section in the Main Ring/Tevatron tunnel plus the spectrometer room constructed in the 1970s for the internal target experiments. It features non-magnetic tracking over a large solid angle, with a magnetic spectrometer over a smaller angular region. The spectrometer magnet has been powered during machine operation, with no ill effects on the beams. Much of the experiment is in place, with the notable exception of the central tracking chamber. It is hoped to have that chamber installed before the end of this running period in order to gain some preliminary experience with the complete detector, in preparation for a physics run during the next colliding beams run.

4.2.3 <u>DO - E713</u>. Experiment E713, a "Search for Heavily Ionizing Particles", is an effort from the University of California consisting of glass strips placed inside a specially constructed thin-walled Tevatron beam pipe, surrounded by layers of CR-39 and Rodyne. The plastic material will be replaced at intervals during the run by fresh material. The plastic and glass will be etched to search for tracks produced by heavily ionizing particles.

4.2.4 D0 - E740. Experiment E740, commonly known simply as "D0", is presently measuring backgrounds in the D0 collision hall, while the major components of the detector, which include liquid argon-uranium calorimentry, non-magnetic central tracking and transition radiation detectors, and magnetizied toroids and muon tracking chambers, are in their early fabrication stages. The completed detector is expected to be ready for data-taking in 1990, with the possibility of earlier runs with substantial subsystems of the detector in 1988 and 1989.

4.2.5 E0 - E710. Experiment E710, an experiment to measure the proton-antiproton total cross section and forward diffraction elastic scattering, is a collaboration of Bologna, Cornell, Fermilab, Maryland, George Mason and Northwestern. It consists of a scintillator array surround the interaction region, along with four "Roman pot" drift chamber assemblies, two within the long straight section and two within the accelerator lattice at D47 and E14. The experiment should be able to measure the differential elastic scattering cross section down to the Coulomb interference region, t = -.004 (GeV/c)**2.

In addition to these experiments which will be observing proton-antiproton collisions, there are two other experiments which will be taking data before Fermilab switches over to fixed-target operation. Both are related to the Superconducting Supercollider. E777, located at A17 in the Tevatron tunnel, is a collaboration of the SSC-Central Design Group, LBL and Fermilab. It uses a controlled leak rate source to spoil the beam pipe vacuum; beam loss rates and neutron fluxes in the vicinity of the spoiled vacuum will be studied as a function of the gas pressure. A flying tungsten wire is also available. The measurements will be compared with theoretical models to aid in extrapolation to SSC energies.

E778 is "An Experimental Study of the SSC Magnet Aperture Criterion" to be conducted by a collaboration of the SSC Central Design Group, Cornell, LBL, SLAC and Fermilab. They intend to measure the effects of magnetic non-linearities and how these contribute to the "smear" of the betatron motion of the particles. This requires measurements of the smear as a function of amplitude, first for the "bare" Tevatron, for which the smear is expected to be small, and then with non-linearities introduced through sextupole and skew sextupole elements. They have been approved for ten shifts, with subsequent time to be allocated following a review of their progress. Subsequent time would be used to complete the above studies, and to examine correlations between the smear and beam liftime, the effect of smear on injection, and the effects of tune modulation on particle stability.

Finally, there is another experiment which will affect accelerator operations for the Collider. E760, a study of charmonium states produced in proton-antiproton collision, is a collaboration of Fermilab and three Italian universities, Ferrara, Genova and Torino. It will utilize a hydrogen gas jet target, and detectors, located in the Accumulator ring. It will operate during the fixed-target physics program, and requires learning how to decelerate the stored antiproton beam. The civil construction for E760 was completed during the 1986 shutdown.

5. SUMMARY

The Tevatron I Collider at Fermilab is rapidly moving out of the commissioning phase into the operational phase. There has been great progress since the test run in 1985; the Accelerator Division and the experiments, present and future, look forward to continued progress towards reaching the design parameters. The era of construction is behind us, the era of exploring physics at 2 TeV is upon us!

6. ACKNOWLEDGEMENTS

I would like to thank the following people who supplied material for this paper and for the oral presentation in Salt Lake City: J. Cooper (CDF), G. Dugan (Antiproton Source), D. Finley (Accelerator), P. Grannis (DO), F. Nezrick (Experiments Support Dept.), J. Peoples (E760), and many others for smaller contributions.

Status of HERA and HERA-Experiments

W.Bartel

Deutsches Elektronen Synchrotron DESY
Notkestrasse 85 · D 2000 Hamburg 52
Germany

Abstract

The physics case for the ep-collider HERA is briefly reviewed and some experimental considerations from the nature of the machine 30 GeV electrons colliding with 820 GeV protons are discussed. Calorimetry is of crucial importance to any high energy e-p experiment. Two experiments have been proposed for HERA , H1 and ZEUS,these intend to use iron/liquid Argon and depleted Uranium/scintillator calorimeters respectively. The use of different calorimetric techniques reflects the difference in the physics emphasis of the two groups. While H1 is optimised for lepton identification, ZEUS will be better at investigating charged current reactions. The status of the HERA machine and the HERA experiments as of January 1987 is reviewed.

1 Introduction.

The standard SU(3)×SU(2)×U(1) model is supported by experiments and there is no experimental result of any statistical significance ,which contradicts the predictions of this model. However in order to be acceptable as a theory of elementary processes the standard model will have to be embedded into a more general scheme of fundamental interactions. It is thus expected that new physics will occur at centre of mass energies above those presently available i. e. above some hundred GeV.

World wide there are four accelerator projects under construction which will allow experiments in the 100 GeV energy regime.

Table 1
Machines Under Construction.

Project	Particles	Lab.	E_{cm}	Year
Tevatron	$\bar{p}p$	FNAL	1.8 TeV	1987
HERA	e p	DESY	320 GeV	1990
LEP II	$e^+ e^-$	CERN	200 GeV	1995
IHEP/UNK	pp	Serpukhov	2.2 TeV	?

Besides the projects listed in Table 1 discucssions on future machines like the SSC,the LHC and very high energy linear colliders have begun.

At first site the number of projects looks disturbing and it is legitemate to ask whether one large machine would be sufficient to unravel the new physics expected ,especially in view of the high cost involved and the limited financial resources available. There are , however, good reasons to try many different reactions to study the physics beyond the standard model, because until now there is no hint as to which direction to go. In this sense the machines listed in Table 1 allow a complementary approach. In e^+e^- collisions primarily neutral currents will be examined, $e^{\pm}$ p collisions study neutral and charged current interactions in the space like region, while pp and $\bar{p}p$ reactions probe the interactions between quarks and gluons and thus QCD at short distances. Therefore all projected storage rings and accelerators will test different aspects of physics.

2 Physics Case for HERA.

Much literature exists concerning the significance of high energy ep collisions [1]. In this brief report only a few items can be touched upon.

2.1 Structure Functions.

At HERA, where 30 GeV electrons collide with 820 GeV protons at a centre of mass energy of 314 GeV, the main objective will be measurements of the proton structure functions. According to the diagram of Fig.1 the current j couples to one of the proton constituents and kicks it out of the proton. Thus the final state will contain two jets ,the current jet and the proton jet, containing the fragments of the struck quark and the proton fragments respectively. Regarding final state leptons two classes of events have to be distinguished, neutral current events due to Z^o and photon exchange and charged current events due to $W^{\pm}$ exchange. While the incoming lepton is conserved in neutral current scattering, it is converted into a

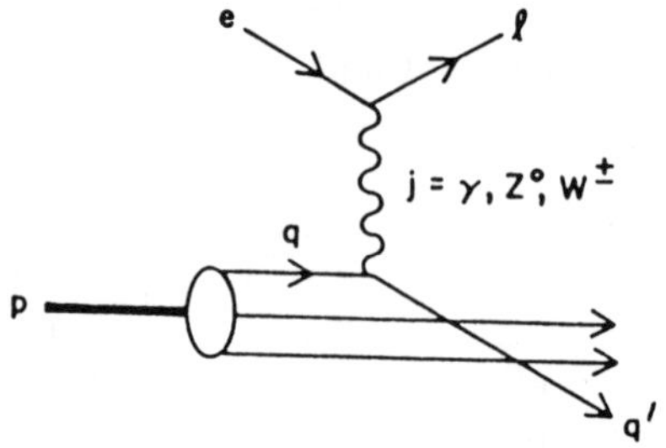

Figure 1: Basic diagram for lepton proton scattering.

neutrino in charged current reactions.

$$\begin{array}{ll} \text{NC} & \text{ep} \rightarrow \text{e X} \\ \text{CC} & \text{ep} \rightarrow \nu \text{ X} \end{array}$$

One of the advantages of HERA over previous experiments at fixed target machines measuring structure functions is that neutral current and charged current structure functions are determined by one detector using identical normalisation. Thus most of the uncertainties which enter when different experiments are combined combined.

HERA will extend the accessible Q^2 range up to Q^2=4·10^4 GeV^2. This value has to compared with the present limit of Q^2=300 GeV^2. It is assumed that HERA will accumulate about 200 pb^{-1} in its first two years of operation. Assuming this to be the case the expected counting rates in bins of Q^2 and y are shown in Fig.2 a and b for neutral current and charged current reactions.

A precise measurement of structure functions opens up a wide area of physics investigations.

- The QCD evolution of the coupling between quarks and gluons can be studied.
- Structure functions are sensitive to substructure of quarks and leptons.
- A comparison of structure functions measured with polarised electrons and positrons is sensitive to new currents.
- Structure functions are an important input to $\overline{p}p$ and pp scattering.

2.1.1 QCD Evolution.

When a quark is kicked out of a proton with a sufficiently high momentum, then with high probability a hard gluon is also emitted . In this case two jets may be observed, a quark jet and a gluon jet. The relevant diagram is shown in Fig.3.

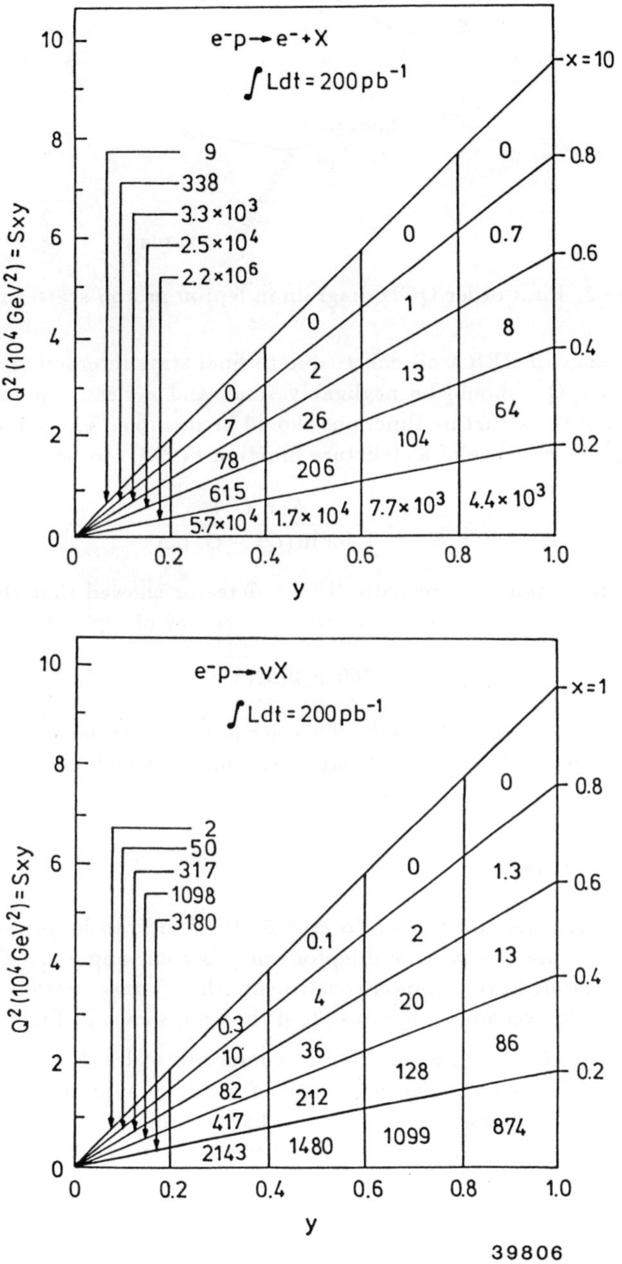

Figure 2: Counting rates for 200 pb^{-1} in CC and NC scattering.

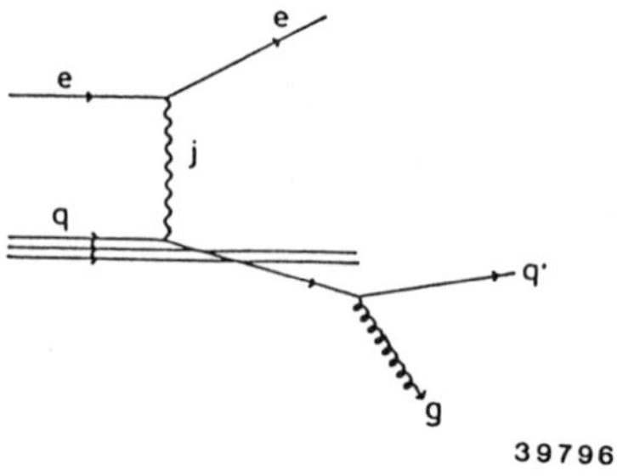

Figure 3: First order QCD diagram in lepton proton scattering.

In the Q^2 range of HERA all effects due to final state interactions, which are proportional to $1/Q^2$, should be negligably small and a clean separation of the QCD evolution of the structure functions shoold be possible. The standard model predicts the Q^2 dependence of a structure function $F(x,Q^2)$ to be

$$F(x,Q^2) \sim \frac{F(x,Q_o^2)}{1 + c\ln(Q^2 - Q_0^2)/\Lambda^2}$$

A Mt. Carlo simulation of a realistic HERA detector showed that the scale parameter of QCD could be determined with an accuracy of typically

$$\Lambda = 200 \pm 40MeV$$

from a data set based on 200 pb^{-1}. This accuracy has to be compared to errors of about $\pm$ 100 MeV for present day's experiments which have an additional uncertainty due to higher twist contributions.

2.1.2 Substructure

With HERA spacial distances down to $\Delta \approx 3 \cdot 10^{-18}$ cm can be probed.At these distances substructure of quarks and leptons may become apparent. Especially if electrons and protons have common constituents the ep cross section will receive a contribution from exchange diagrams [2] of the type shown in Fig.4

HERA experiments will be sensitive to composite scales $\Lambda_{comp} \approx 7$ TeV. At present the best values for this parameter come from $e^+ e^-$ annihilation into μ pairs and quark anti quark pairs, where the point like coupling of quarks and leptons is deduced from the energy dependence of the cross sections [3]. The limits are $\Lambda_{comp} = 2 \div 3$ TeV. Also model dependent values of $\Lambda_{comp} = 20 \div 30$ TeV are quoted [4]. They are derived from limits on the decay $K^+ \rightarrow \pi^+ \mu$ e.

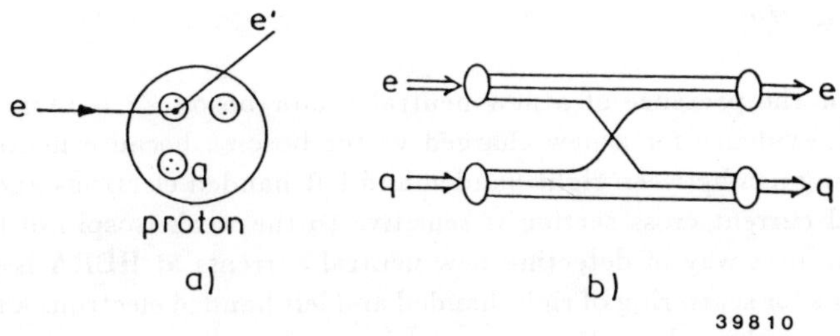

Figure 4: a) Scattering off quark constituents. b) Exchange diagram.

2.2 New Currents.

Charged current interactions at present energies appear to be left handed. It has been speculated, that this left right asymmetry, which cannotbe explained within the framework of the standard model, may be a low energy phenomenon and that the theory is in principle symmetric, with right handed vector bosons which are heavier than the left handed $W^{\pm}$.

The present mass limits are of the order of 300 GeV and come from the study of non leptonic decays and an analysis of the electron momentum spectrum in the decay $\mu \to e\nu\nu$ [5]. A model dependent analysis of the mass difference $\delta m(K^o_L$-$K^o_S)$ suggests a much higher mass M_R of the right handed W with 10 TeV < M_R <100 TeV [6]

At HERA a model independent mass limit of 500 GeV can be obtained by searching for neutral current events in the reaction

$$e^-_R p \to \nu X$$

where right handed electrons are scattered off protons. If this cross section were non zero one would have unambiguous proof of the existence of charged right handed currents.

The appearance of a right handed W is a special case of a more general scheme with two charged vector bosons W_1 and W_2, which mix to form the physical particles W and $\tilde{W}$.

$$W = W_1 \cos\chi - W_2 \sin\chi$$
$$\tilde{W} = W_1 \sin\chi + W_2 \cos\chi$$

The sensitivity of experiments to a second vector boson W depends on the value of the mixing angle χ and the coupling strength of the bosons to quarks and leptons. In the extreme,but improbable ,situation without mixing HERA experiments will be sensitive to a mass of M_2=800 GeV if the second W couples with the same strength as the standard one.

2.3 New Z^o.

Evidence for the presence of a new neutral vector boson $\tilde{Z}^o$ is more difficult to obtain than evidence for a new charged vector boson , because neutral currents do not distinguish between right handed and left handed electrons and positrons. The neutral current cross section is sensitive to the weak isospin of leptons and quarks. The best way of detecting new neutral currents at HERA is to compare cross sections for scattering of right handed and left handed electrons and positrons and to measure the polarisation asymmetries.

$$A_{RL}^{-+} = \frac{d\sigma(e_R^-) - d\sigma(e_L^+)}{d\sigma(e_R^-) + d\sigma(e_L^+)}$$

$$A_{LR}^{--} = \frac{d\sigma(e_L^-) - d\sigma(e_R^-)}{d\sigma(e_L^-) + d\sigma(e_R^-)}$$

$$A_{RR}^{-+} = \frac{d\sigma(e_R^-) - d\sigma(e_R^+)}{d\sigma(e_R^-) + d\sigma(e_R^+)}$$

An additional neutral gauge boson will result in a polarisation asymmetry which is different from that predicted by the standard model.

$$\delta(A) = \mid A(\tilde{Z}^o, Z^o) - A(Z^o) \mid$$

with

$$Z^o = Z_1 \cos\chi - Z_2 \sin\chi$$
$$\tilde{Z}^o = Z_1 \sin\chi + Z_2 \cos\chi$$

Thus a consistent set of deviations $\delta(A)$ would unambigously signal the presence of a heavier $\tilde{Z}^o$. The size of the effect depends on the masses of Z_1 and Z_2,the mixing angle and the coupling strengths. In a specific broken E_6 superstring model, $\delta(A)$ is expected to be of the order of 10% [7] .With an accuracy of about 2% HERA experiments will be sensitive up to $\tilde{Z}^o$ masses of 300 GeV for mixing angles around 2°. The present mass limits for a second Z boson are of the order of 100 GeV [8].

2.4 Particle Production.

The main contribution to heavy quark production in ep collisions can be attributed to current-gluon fusion diagrams as shown in Fig.5 and in particular to photon gluon fusion processes.

The cross section for this type of process

$$\sigma(ep \rightarrow Q\bar{Q}X) \sim M_Q^{-4}$$

decreases steeply with the heavy quark mass, but despite this, for a quark mass M_Q=40 GeV and a luminosity of 200 pb^{-1} about 300 events would be observed.

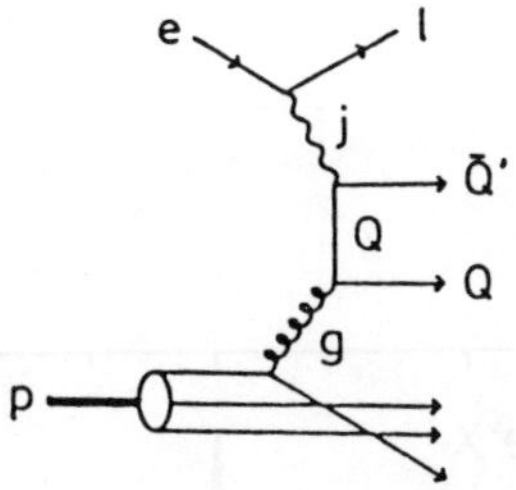

Figure 5: Diagram for current gluon fusion processes.

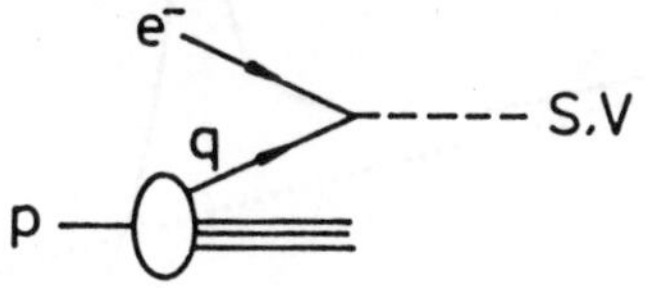

Figure 6: Diagram for lepto-quark production.

An ep machine is the ideal instrument for directly producing lepto-quarks as shown in Fig.6. Cross sections are large and unique signals will show up. The cross section for generating lepto-quarks of mass 200 GeV at HERA, calculated using the model of Ref.[9], is shown as a function of the scaling variable x for electrons and positrons in Fig.7.

2.5 Summary.

Some aspects of the physics motivations for building the ep collider HERA have been briefly mentioned.

- An ep collider at a centre of mass energy of about 300 GeV makes a valuable contribution to the world wide efforts to observe and undestand the physics beyond the standard model.
- The Q^2 range in which structure functions can be measured will be enlarged by two orders of magnitude.
- HERA experiments will be sensitive to a possible substructure of quarks and leptons, to new currents and to new particles. The structure of charged and neutral currents can be investigated.

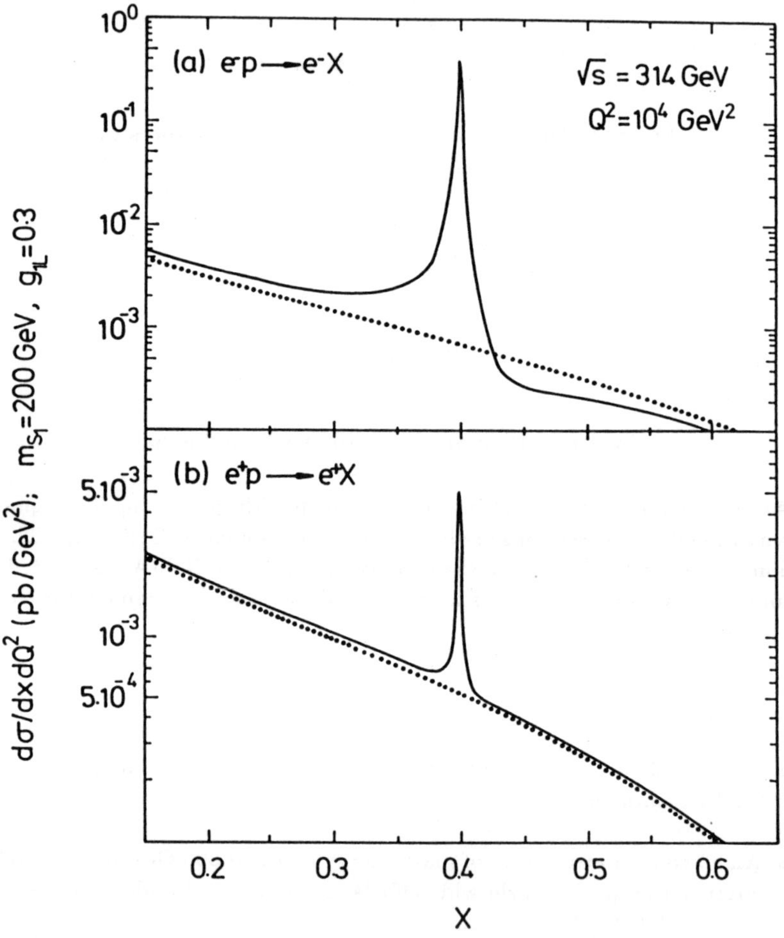

Figure 7: Lepto quark cross section for m_S=200 GeV.

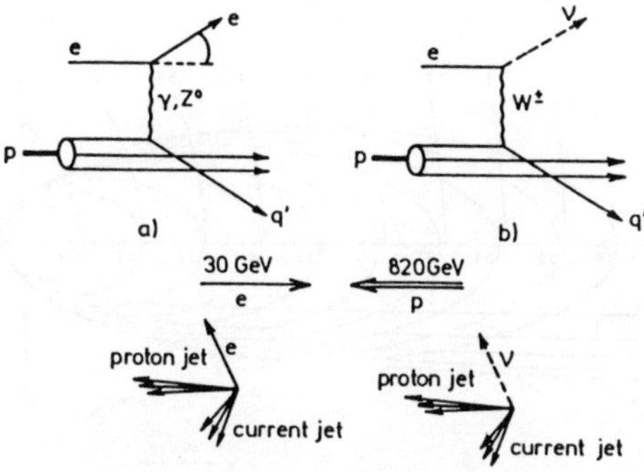

Figure 8: Kinematics NC and CC events.

- The possibility of colliding linearely polarised electrons and positrons with protons allows HERA to investigate broad areas of physics.

3 Experimenting at HERA.

Experiments at HERA are influenced by the large energy difference of the two beams. When 30 GeV electrons collide with 820 GeV protons, the centre of mass moves with the proton beam(forward direction) and all secondary particles are boosted forward. Therefore, in order to match the ep kinematics, all HERA detectors have to be asymmetric with respect to the interaction point. This has consequences for the energy resolution and granularity of calorimeters and the design of tracking systems.

3.1 ep Kinematics.

Two event classes ,sketched in Fig.8, have to be distinguished: neutral current events with a final state electron plus hadron jets and charged current events with a neutrino replacing the electron. The kinematics of neutral current ep interactions can be reconstructed from the measured angle and energy of the final state electron as well as from the measured angle and energy of the current jet. Charged current events do not allow this redundancy because only the hadron jet is seen in the detector.

To describe the kinematics of an ep-reaction three independent invariants are used , Q^2 the momentum transfer squared,x the parton momentum fraction and y the scaled energy of the current- parton interaction. In order to get a feeling for energies and angles of final states at HERA the kinematics are mapped out in

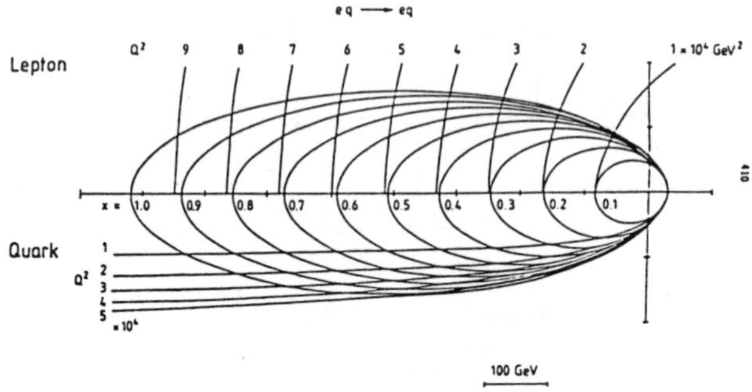

Figure 9: ep-Kinematics , plot of p_L vs. $p_\perp$.

Fig.9. In the upper half of the figure a plot of the longitudinal versus transverse momenta of final state leptons is shown and in the lower half the equivalent plot final state current jets. Lines of constant Q^2 and x are indicated. The proton beam enters from the left and one clearly sees the effect of the forward Lorentz boost. Typically, HERA detectors will have to cope with energies of 50 GeV around 90° scattering angle and $200 \div 300$ GeV in the forward region.

The density of lines in Fig.9 translates directly into the energy and angular resolution functions necessary for HERA detectors to measure Q^2 and x with the same relative accuracy across the Q^2 and x range. The impact of the detector resolution on the physics performance can be seen in figures 10 to 11. The figures show the limits imposed on structure function measurements by statistics and detector resolution. The statistical limit assumes at least 100 events per bin for 100 pb^{-1}. The limits set by the absolute energy calibration error δE allow a maximum relative change of the structure function F_2 of $\Delta F_2/F_2 = 10\%$.

The plots tell us that the electromagnetic as well as the hadronic calorimeters need good energy and angular resolution. The energy resolution is not the only important factor the absolute energy calibration has also to be sufficiently well known in order to fully profit from a good resolution.These constraints are highly demanding and are a challenge to the groups building calorimeters for HERA experiments.

3.2 Calorimetry.

As calorimetry is so important at HERA, the different philosophies followed in the calorimeter design by H1 and ZEUS are described here.

Two physics processes , which have a major impact on the energy resolution of

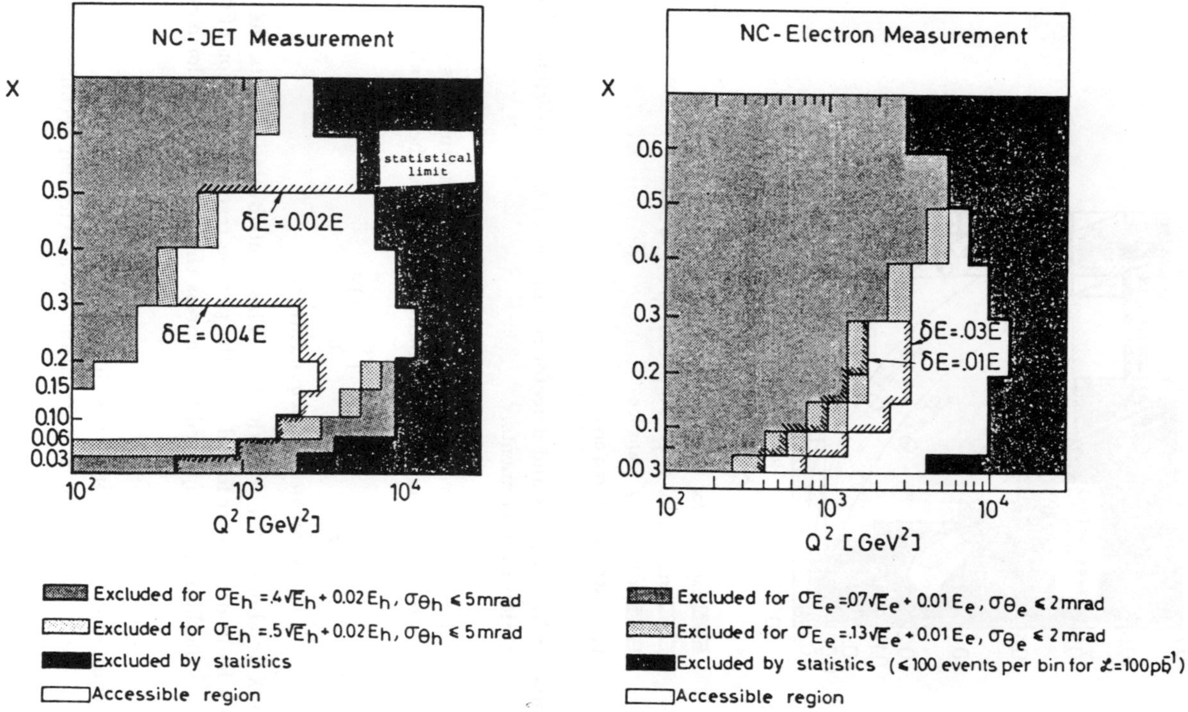

Figure 10: Limits set by statistics and detector resolution to the accessible region for NC structure function measurements.

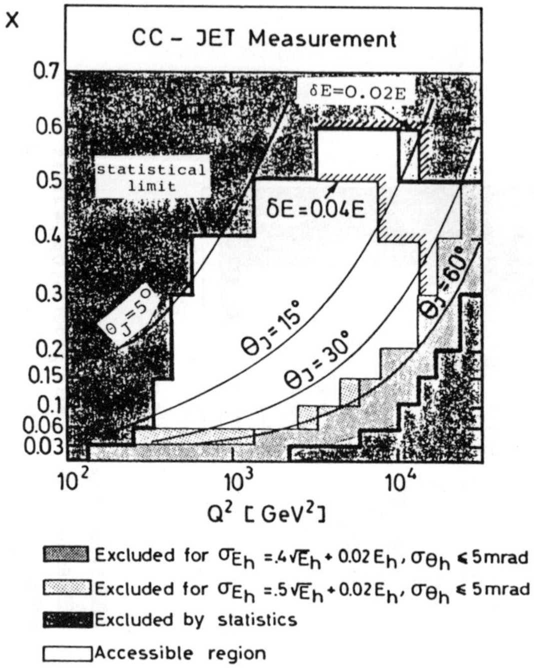

Figure 11: Limits set by statistics and detector resolution to the accessible region for CC structure function measurements.

a sampling hadron calorimeter have been identified , shower fluctuations and the difference in response to electrmagnetic and hadronic showers the so called e/h ratio.

$$\frac{\sigma_E}{E} = \frac{c}{\sqrt{E}} \oplus X \cdot (\frac{e}{h} - 1)$$

$$\oplus\ means\ add\ in\ quadrature$$

The coefficient X is about 8.2% for e/h=1.4. This contribution reflects the π^o fluctuations in a hadronic shower folded with the photon response function of the calorimeter. Two methods are employed to suppress this part of the calorimeter resolution.

- Use of a compensating calorimeter with e/h=1.
- Event by event correction for π^o fluctuations by a weighting technique.

3.2.1 Compensating calorimeter.

The ZEUS collaboration is designing a compensating calorimeter using depleted Uranium and scintillator. In order to achieve equal response for electrons and hadrons the response to electrons and photons has to be decreased and the response to hadrons increased. The compensation which is observed in a DU scintillator calorimeter is the result of a complicated interplay of several reactions.

- Low energy photons and electrons are absorbed in the high Z material of the absorber plates thus reducing the sensitivity to electromagnetic showers.
- Fission products increase the response to hadrons.
- Neutrons from nuclear cascades are registered in the scintillator through recoiling protons.
- Fast neutrons are moderated in the scintillator and initiate further fission processes.
- Photons from fission and spallation reactions are registered in the scintillator via knock on compton electrons.

The ZEUS collaboration has made test measurements in a beam containing hadrons and electrons.In Fig.12 a lead scintillator sampling calorimeter is compared with a DU calorimeter. The lead calorimeter shows a clear difference in the response to electrons and hadrons, while the DU calorimeter is nearly compensating. The e/h ratio is found to lie between 1.08 and 1.13. In the meantime the group has optimized the absorber and scintillator thicknesses and with 3 mm DU and 2.5 mm scintillator they achieve compensation between 3 GeV and 9 GeV beam energy with a resolution of $\sigma(E)/E = 34\%/\sqrt{E}$. More details are given in Ref. [10]

3.2.2 Weighting technique.

In a sampling calorimeter large localised energy deposits stem in general from π^o fluctuations. These fluctuations can be suppressed by giving them a smaller weight when forming the energy sum over all calorimeter bins contributing to the shower.

$$E^{corr} = \sum_i E_i^{obs} - c \cdot \frac{\sum_i (E_i^{obs})^2}{\sqrt{E_{tot}}}$$

This technique of course requires a calorimeter set up which samples the shower as a function of depth as does the liquid Argon calorimeter proposed by the H1 collaboration. The weighting factors c have to be measured in a test beam prior to installation. Based on the experience of the CDHS group at CERN a sampling frequency of about 1λ is sufficient [11]. The results of this group obtained with an iron-scintillator calorimeter are shown in Fig.13.

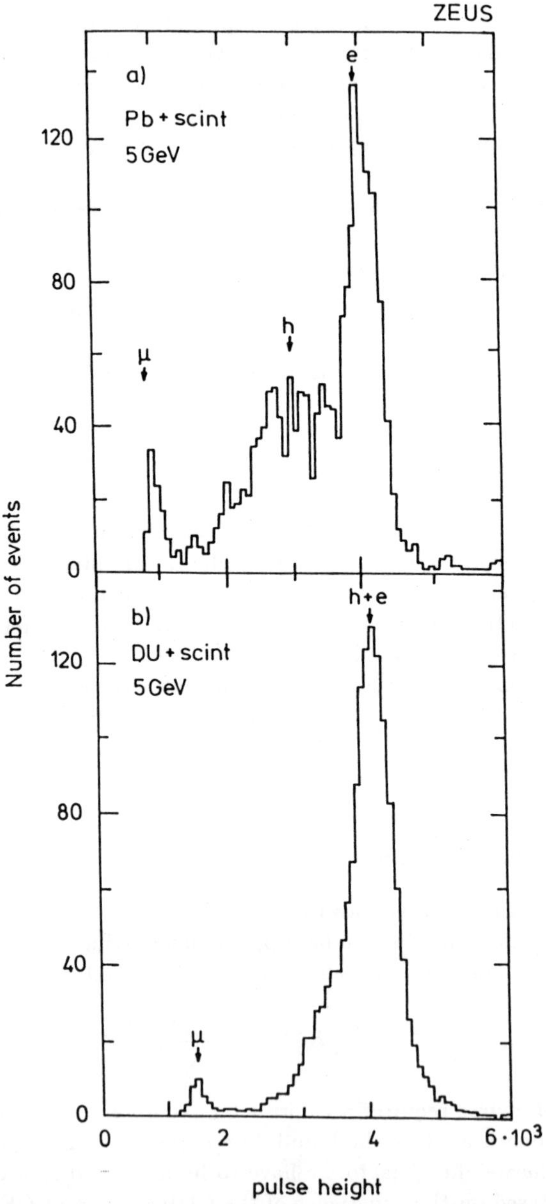

Figure 12: a) 5 mm lead 5 mm scintillator calorimeter exposed to 7 GeV electrons and pions. b) 3.2 mm DU 5 mm scintillator calorimeter exposed to 7 GeV electrons and pions.

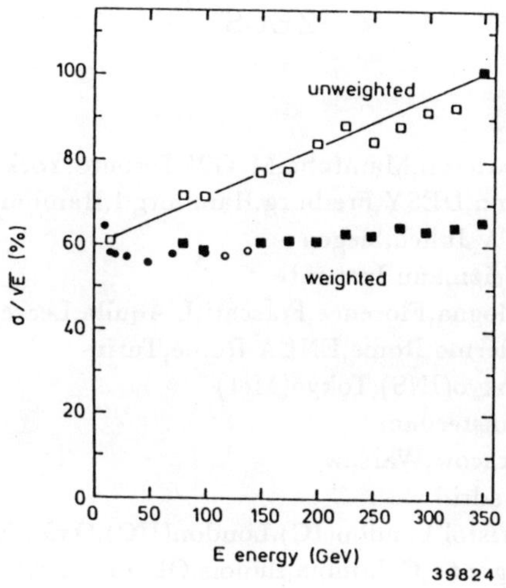

Figure 13: Weighting technique applied to a Fe-scintillator calorimeter with 5 samplings in depth.

4 HERA Experiments.

Two large international collaborations have formed to run the experiments H1 and ZEUS at HERA.

H1

France	Ecole Polytechnique,Orsay,Paris,Saclay
Germany	Aachen I,Aachen III,DESY,Dortmund,Hamburg I,Hamburg II, MPI-Munich,Wuppertal,Berlin-Zeuthen(DDR)
Italy	Rome
Poland	Cracow
Switzerland	Zurich
UK	Glasgow,Lancaster,Liverpool,Manchester,RAL
USA	Davis,Houston,Northeastern
USSR	ITEP,Lebedev Institut

ZEUS

Canada	Carleton,Manitoba,McGill,Toronto,York
Germany	Bonn,DESY,Freiburg,Hamburg I,Hamburg II, KFA Julich,Siegen
Israel	Weizmann Institute
Italy	Bologna,Florence,Frascati,L'Aquila,Lecce,Milan,Padua Palermo,Rome,ENEA Rome,Turin
Japan	Tokyo(INS),Tokyo(Met)
Holland	Amsterdam
Poland	Cracow,Warsaw
Spain	Madrid
UK	Bristol,London(IC),London(UC),Oxford,RAL
USA	Argonne,Columbia,Illinois,Ohio,Pennsylvania State, Princeton,Virrginia,Wisconsin

The physics emphasis of the two experiments is different and they can be regarded as being complemantary.

The **H1** detector, which is shown in Fig.14 and 15, is optimized for lepton identification with a large diameter coil housing the tracking device and the calorimeter thus minimising the amount of material in front of the calorimeter. The calorimeter is of the liquid Argon type with lead absorbers in the electromagnetic section and steel absorber plates in the hadronic part allowing a fine granularity, which aids the electron pion separation. It is backed up by an instrumented return yoke, which also serves as a muon filter. The large homogeneous magnetic field and the relatively low density calorimeter support muon identification because track matching between the central detector and the muon filter is easier when multiple scattering is small and the curvature introduced by the magnetic field is large. The muon detection system is complemented by a toroid of magnetised iron with appropriate tracking chambers in the forward direction.

The **ZEUS** detector, which is shown in Fig.16 and 17, is optimised for the investigation of charged current reactions that is for calorimetric energy resolution. Hence the use of a DU-scintillator sandwich with photo multiplier read out. The field free room for the phototubes is provided between a small superconducting coil, which houses the central tracking chamber, and the return yoke. As in H1, the main hadron calorimeter is backed up by an instrumented iron yoke, which also serves as a muon spectrometer. In order to improve the momentum resolution for muons, the iron in the barrel section is magnetised. As in H1 the muon detection system is complemented by a forward toroid. Some more details are listed in Table 2 where H1 and ZEUS are compared.

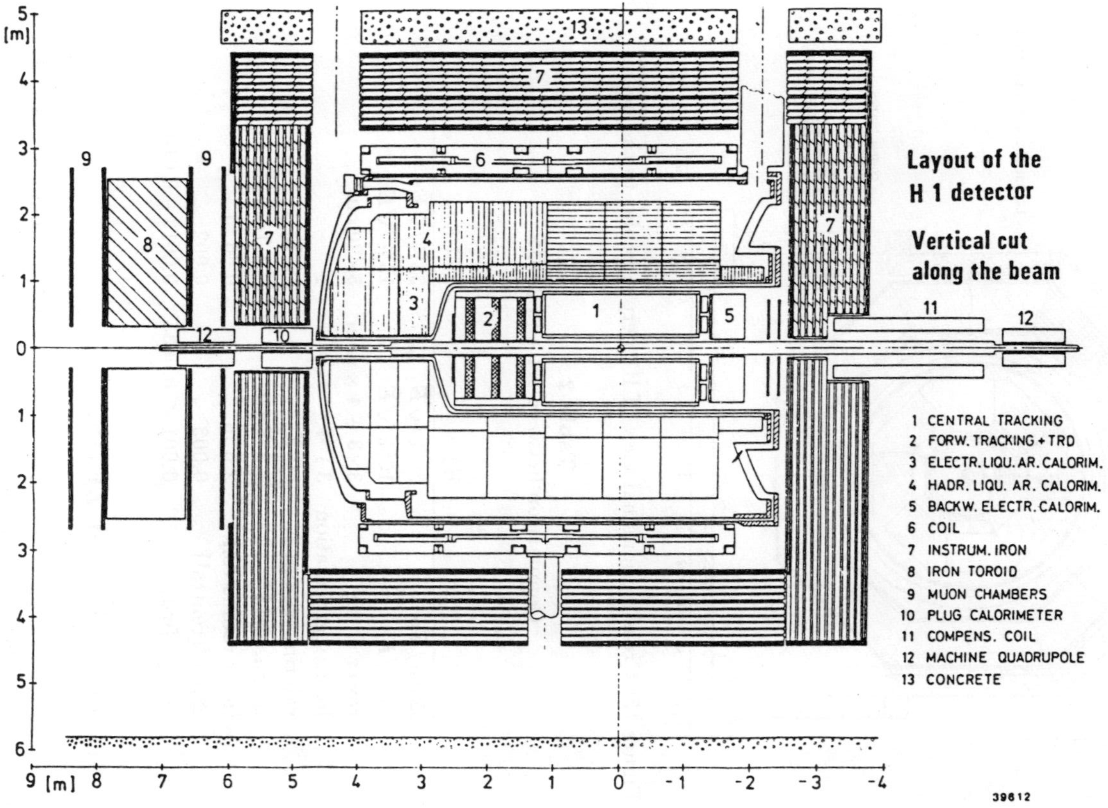

Figure 14: Cross sectional view of H1 along the beam line.

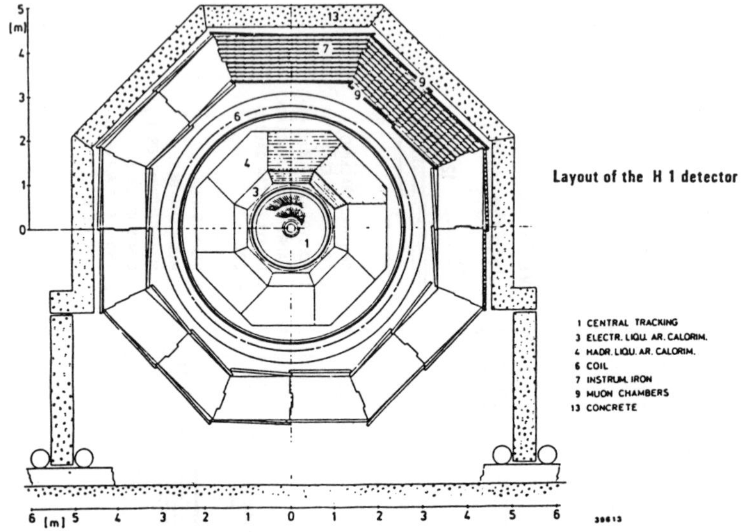

Figure 15: Cross sectionional view of H1 perpendicular to the beam.

Table 2

Comparison between H1 and ZEUS.

	H1	ZEUS
$\Delta E_H/E_H$	$55\%/\sqrt{E}$	$35\%/\sqrt{E}$
$\Delta E_e/E_e$	$10\%/\sqrt{E}$	$15\%/\sqrt{E}$
tower el.mg.	$3\times3 \div 4\times 4\text{cm}^2$	$5\times20\ \text{cm}^2$
long. segmentation	$3\div 4$	1
calorim. channels	60 K	13 K
magn. field	1.2 Tesla	1.8 Tesla
inner radius coil	2.6 m	0.86 m
$\Delta p/p^2$(central)	0.003	0.002
$\Delta p/p^2$(fwd)	$0.001 \div 0.0035$	$0.0013\div 0.010$

4.1 Status of H1 and ZEUS.

Both experiments have been approved by the scientific committees of DESY and accepted by the DESY board of directors. The finances of the first stages of the experiments are in good shape and contracts between participating laboratories have either already been signed or will be signed soon. Many large components have been ordered in industrie or are out for tenders. Both groups pursue an

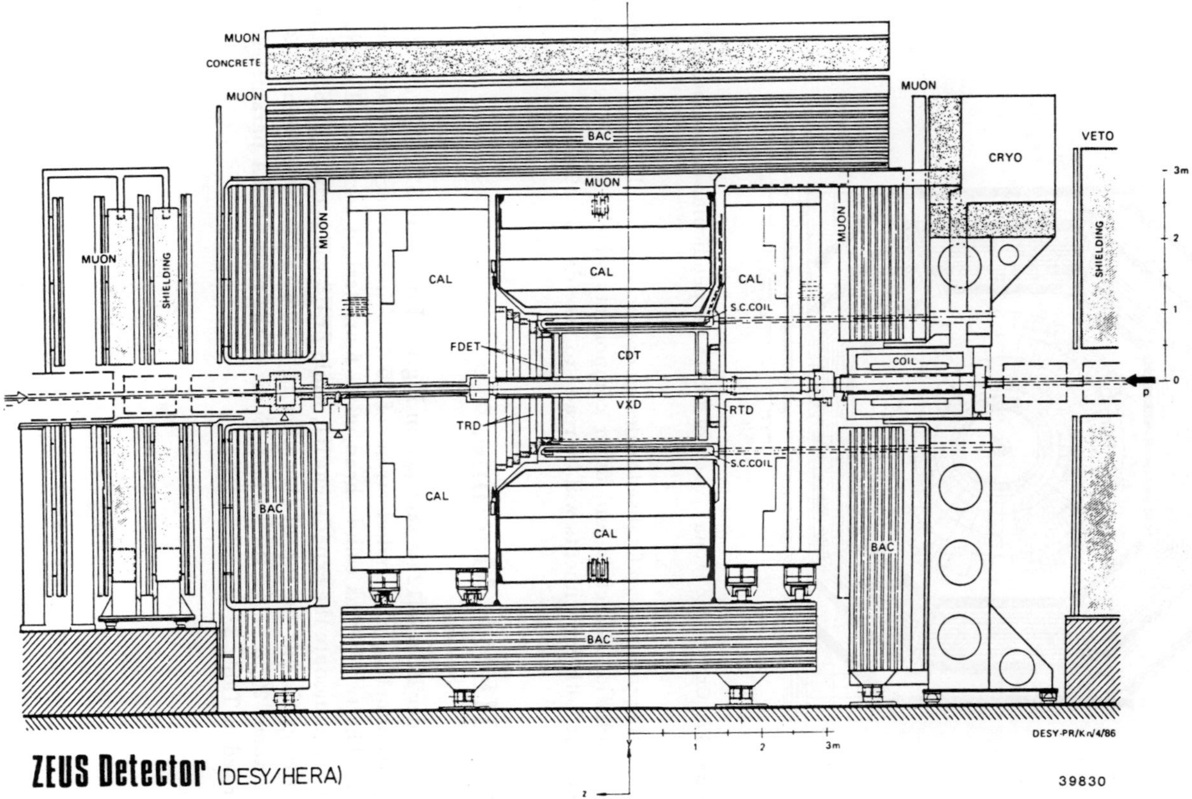

Figure 16: Cross sectional view of ZEUS along the beam line.

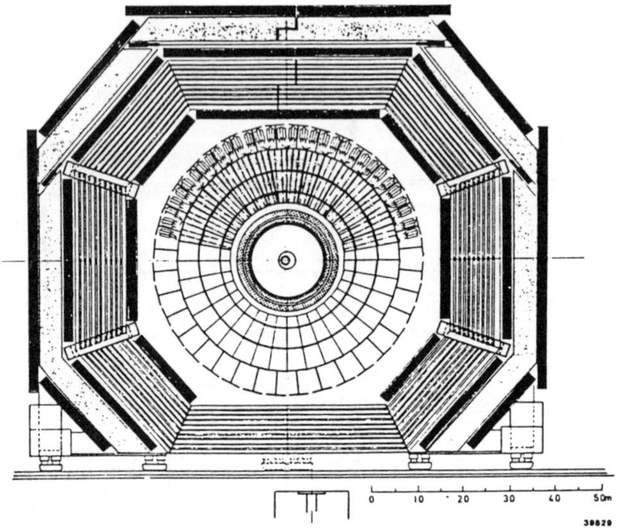

Figure 17: Cross sectional view of ZEUS perpendicular to the beam.

extinsive testing program to finalise the design of tracking chambers, calorimeters, transition radiation detectors and other components which need R & D. It is expected that both experiments will be ready to take data in 1990.

5 The HERA Machine.

The ep - collider HERA [12] is sketched in Fig.18. Protons of 820 GeV are stored in a superconducting ring and electrons of 30 GeV in a storage ring with warm magnets. The two beams collide head on in four straight sections. The North hall will accommodate the experiment H1, while ZEUS will occupy the South hall. The two rings sit one on top of the other in a 10.4 m diameter tunnel 6.3 km long and about 20 m below the surface. Some of the main HERA parameters are summarised in Table 3.

Table 3
HERA parameters.

	p-ring		e-ring	units
Nominal energy	820		30	GeV
c.m. energy		314		GeV
Luminosity		$1.5 \cdot 10^{31}$		cm^{-2} sec^{-1}
Polarisation time		25		min
No. of interaction pnts.		4		
Crossing angle		0		
Free space for exp.		± 5.5		m
Circumference		6336		m
Bending radius	588		608	m
Magnetic field	4.65		0.165	Tesla
Energy range	$300 \div 820$		$10 \div 30$	GeV
Injection energy	40		14	GeV
Circulating current	163		58	mA
Total no. of particles	2.110^{13}		0.810^{13}	
No of bunches		210		
Time between X-ing		96		ns
Beta fct. β_x^*/β_y^*	10/1.0		2/0.7	m
Beam size at x-ing σ_x	0.29		0.26	mm
Beam size at x-ing σ_y	0.07		0.07	mm
Beam size at x-ing σ_z	110		8	mm
Energy loss / turn	$1.4 \cdot 10^{-10}$		127	MeV
Max. circumf. volt	0.2/2.4		260	MV
Total RF power	1		13.2	MW
RF frequency	52.033/208.13		499.667	MHz
Filling time	20		15	min
Tune shift Q_x	0.001		0.016	
Tune shift Q_y	0.00042		0.025	

In autumn 1987 all major **civil engeneering** will be completed. The tunnel is almost 3/4 excavated. Two of the experimental halls are ready and the other two will be completed during the summer. The transfer tunnels between PETRA and HERA are almost ready and magnet installation has started.

5.1 Injection.

Electrons, positrons and protons are pre accelerated in linacs before they are brought up to 7 GeV in synchrotrons, the electrons in DESY II and the protons

a)

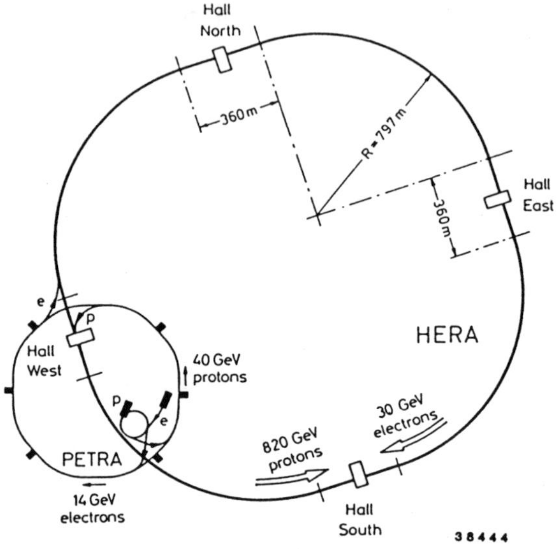

b)

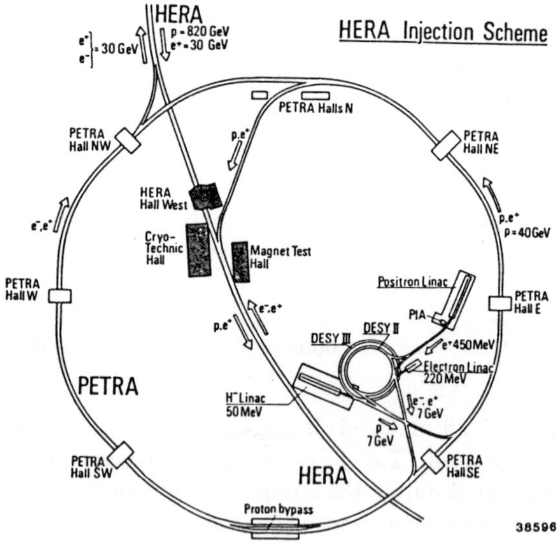

Figure 18: a)Sketch of the HERA machine.b)Injection scheme.

in DESY III. These synchrotrons inject particles into the storage ring PETRA, where electrons or positrons are accelerated to 14 GeV and protons to 40 GeV before they are injected into HERA. A schematic of the injection system is shown in Fig.18 b.

The DESY II machine is a new accelerator, which is also used to fill the e^+ e^- storage ring DORIS now operating in the Υ region, while the proton synchrotron DESY III uses the magnets of the old DESY machine. PETRA no longer exists as an e^+ e^- storage ring for physics. It is now being converted into an injector for HERA.

DESY II is already running and tuning of the machine is in progress. DESY III is in the assembly stage as is the proton linac. The proton negative ion source is operational. It is planned to bring positrons through the proton transfer line to the HERA injection kicker magnet in the spring of this year and electrons to the electron kicker magnet this summer.

5.2 Components.

The electron ring will be assembled in modular form, each module consisting of a 9 m long bending magnet with a quadrupole and sextupole on the same support structure. The first production modules are already at DESY and will be installed in February. The installation work will be completed in April 1988.

The warm **electron RF system**, which allows a maximum energy of 27 GeV, has already been built. It is that part of the PETRA system, which is no longer needed for the injector mode of operation. To reach 30 GeV 8 cryostats with a total of 64 superconducting cavity cells are needed. A prototype will be tested at PETRA this summer.

The **superconducting quadrupoles** , which have been developed at the Saclay laboratories, reach a 40% higher gradient than the design value. Orders have been placed with industry and the first 9 quads will be completed by October 1987 . The production line will be finished one year later.

Until now four prototype **superconducting dipoles** have been tested. They are 9 m long and have a design in which the super conductor is held together with aluminium brackets, which are then bracketed with iron collars. Both the aluminium and the iron are cold, as shown in the cross sectional view of Fig.19 The diploes are out for tender and will be ordered from industry. The field which has been achieved with the prototype magnets exceeds the design value so that mass production will start with a comfortable safety margin.

The **refrigeration plant** has been installed at DESY and acceptance tests have started. These include a 1000 hrs uninterrupted performance test. In summer the refrigerators will be handed over to DESY.

The **He transfer line** is out for tender.

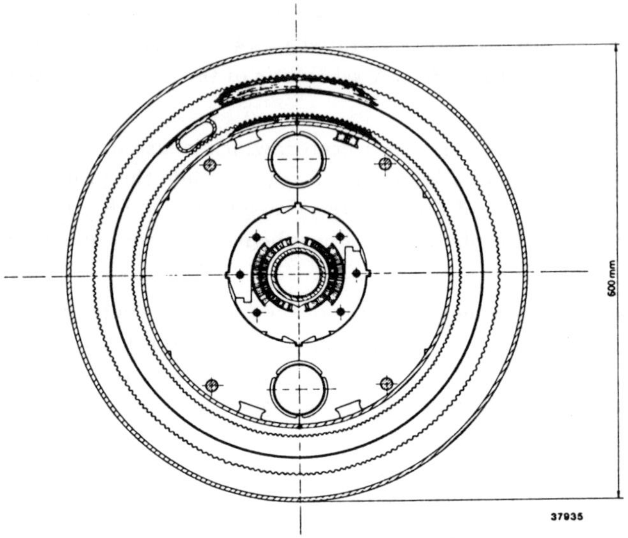

Figure 19: Super conducting HERA dipole.

The **hall for field measurements** is ready and the installation of transfer lines has started.

To **summarise** all components tested so far meet their specifications and component production is well under way so that the first physics runs may start mid 1990.

5.3 International collaboration.

A project like HERA is too big to be built by a single laboratory like DESY. Therefore DESY has asked for international help and several countries and laboratories are now contributing to the construction of HERA.

Canada	p beam transport, p-RF
China	staff
France	sc quadrupoles
Holland	sc correction magnets
Israel	sc current leads
Italyl	1/2 sc dipoles
Poland	staff and components
Switzerland	R& D transfer line, prototype
UK	technical help
USA	test of sc cable, cryo components

This type of international collaboration is a novelty and it has proven succesful so far. Like in other laboratories physicists, who were doing $e^+ e^-$ experiments so far, joined the machine group temporarily for the construction phase and may go back to an experiment group later on.

Acknowledgement. I greatfully acknowledge the hospitality extended to me during my stay in Salt Lake City. I thank the organisers of the meeting, in particular Professors C. DeTar, for arranging a very interesting and lively conference. I thank M. Breidenbach for giving me the opportunity to speak in his session.

References

[1] Physics with ep Colliders in view of HERA, DESY-HERA 81-18 (1981)
Experimentation at HERA (Workshop NIKHEF) DESY-HERA 83-20 (1983)
G.Altarelli,B.Mele,R.Rückl CERN TH3932 (1984)
J.A.Bagger,M.E.Peskin SLAC-PUB-3447 (1984)
R.J.Cashmore, Exotic Phenomena, Phys.Rep. (1985)

[2] E.J. Eichten,K.D. Lane,M.E. Peskin Phys. Rev. Lett. 50(1983) 811

[3] B. Naroska DESY 86-113(1986)

[4] I.Bars,M.J.Bowick Phys. Lett. 138B(1984) 159

[5] J. Carr et al. Phys. Rev. Lett. 51 (1983) 627, 1222(E)
D.P. Stoker et al. LBL-18935 (1985)

[6] G.Ecker, W. Grimus Z. Phys.C30 (1986) 293

[7] I.Cornet,R.Rückl Z. Phys.C30 (1986) 293

[8] P.Langacker,R.W.Robinett,J.L.Rosner Phys. Rev. D30(1984)1470

[9] W.Buchmüller,R.Rückl,D.Wyler DESY 86-150 (1986)

[10] B.Anders et al. DESY 86-105 (1986)

[11] C.W.Fabjan,T.Ludlam Ann.Rev.Nucl.Part.Sci.32(1982)335

[12] HERA proposal DESY-HERA 81-10 (1981)

1) DARK MATTER SEARCHES
2) NEUTRINO MASS EXPERIMENTS
3) SOLAR NEUTRINO DIRECT DETECTION
VIA NEUTRAL CURRENT INTERACTIONS
COLLABORATE WITH ZARAGOZA GROUP
SO I CAN RETIRE IN EUROPE (YOU ARE
AWARE!)

10 CM

LIST OF PARTICIPANTS

Abachi, Shahriar	Purdue
Abolins, Maris A.	Michigan State
Acharya, Raghunath	Arizona State
Adler, Stephen L.	Inst. for Advanced Study
Alexander, James P.	SLAC
Allen, Roland E.	Texas A&M
Alvarez, Orlando	U. C. Berkeley
Amiri, Farhang	Weber State
Anderson, Scott B.	Colorado State
Arnowitt, Richard L	Texas A & M
Aurenche, Patrick J.P.	LAPP
Avignone III, Frank	Univ. of S. Carolina
Baer, Howard A.	Argonne
Bagger, Jonathan A.	Harvard
Balke, Brian K.E.	U. C. Berkeley
Ball, Austin H.	Univ. of Maryland
Ball, James S.	Univ. of Utah
Baltrusaitis, Rose Mary	Univ. of Utah
Barish, Barry C.	Caltech
Bartel, Wulfrin	DESY
Bauer, Gerry	Harvard
Bennett, Bradley F.	
Berley, David	NSF
Bernard, Claude	U. C. Los Angeles
Betounes, David E.	Univ. of S. Mississippi
Blucher, Edward C.	Cornell
Bodwin, Geoffrey T.	Argonne
Bonner, Billy E.	Rice
Breidenbach, Martin	SLAC
Bromberg, Carl M.	Michigan State
Brown, Stanley G.	Phys. Rev. D and Letters
Bunce, Gerry M.	BNL
Burnett, T. H.	Univ. of Washington
Burnstein, Ray A.	Illinois Inst. of Tech.
Cady, D. Robert	LANL
Cahill, Kevin E.	Univ. of New Mexico
Campbell, William B.	Univ. of Nebr.-Lincoln
Carbon, Steve L.	Univ. of Central Florida
Carroll, Alan S.	BNL
Cason, Neal M.	Univ. of Notre Dame
Caticha, Ariel	Univ. of Utah
Cavalli-Sforza, Matteo	U. C. Santa Cruz
Chanowitz, Michael	LBL
Charlton, Gordon R.	DOE
Chau, Ling-Lie	U. C. Davis
Chen, Chu-Xin	Univ. of Utah
Chen, Hua	Univ. of Utah
Chen, Min	MIT
Cheng, Hai Yang	Indiana
Cherry, Michael L.	Univ. of Pennsylvania
Chien, Chih-Yung	Johns Hopkins
Ciampa, David P.	Univ. of Michigan

Cline, David	Univ. of Wisconsin and UCLA
Cohn, Hans O.	Oak Ridge Nat'l Lab
Collins, John	Illinois Inst. of Tech.
Cook, Steve	Univ. of Utah
Cooper, Martin D.	LANL
Cooper, Ron G.	Univ. of Utah
Coteus, Paul W.	Univ. of Colorado
Cremaldi, Lucien	Univ. of Colorado
Creutz, Michael	BNL
Dalitz, Richard H.	Oxford
Das, Bhanu P.	Colorado State
Dawson, Bruce	Univ. of Utah
DeGrand, Thomas A.	Univ. of Colorado
DeTar, Carleton E.	Univ. of Utah
Dombeck, Thomas	LANL
Domokos, Gabor	Johns Hopkins
Dorfan, Jonathan	SLAC
Durkin, L. Stanley	Ohio State
Eastaugh, Alexander G.	SUNY Stony Brook
Einsweiler, K.	CERN
Elbert, Jerome W.	Univ. of Utah
Elbert, Paul	Middle Georgia College
Ensign, Philip W.	Univ. of Colorado
Fabbrichesi, Marco E.	Univ. of Virginia
Ferbel, Thomas	Univ. of Rochester
Fick, Brian E.	Univ. of Utah
Fisher, John C.	
Flippen, Leroy	
Forinash III, Kyle	Indiana U. Southeast
Franklin, Jerrold	Temple
Franzini, Paolo	Columbia Univ.
Frederiksen, Soren G.	Univ. of Ottawa
Froidevaux, Daniel	LAL Orsay
Fry, Alan	Univ. of Utah
Fuchs, Norman H.	Purdue
Gaisser, Thomas K.	Bartol
Gan, Kock K.	SLAC
Gettner, Marvin	Northeastern
Giddings, Steve B.	Princeton
Gilbert, Gordon	Mesa College
Gilchriese, Murdock	Cornell
Gingrich, Douglas M.	Univ. of Toronto
Gittelman, Bernard	Cornell
Goldman, Terry J.	LANL
Gomez, Ricardo	Calif. Inst. of Tech.
Goss, Dave	Nebraska Wesleyan
Gravel, Pierre	CMR St-Jean
Gray, Richard	Univ. of Washington
Green, Kevin	Univ. of Utah
Groom, Donald E.	LBL
Gunion, John F.	U. C. at Davis
Gustafson, H. Richard	Univ. of Michigan
Haas, Peter M.	Ohio State
Hagelin, John S.	Maharishi Int. Univ.

Haines, Todd	U. C. Irvine
Haller, Kurt	Univ. of Connecticut
Halverson, Peter G.	Univ. of Arizona
Hancock, Arthur D.	EG&G
Hearty, Christopher	Univ. of Washington
Heitowit, Ezra D.	Univ. Res. Assoc.
Herold, Wolf-Dietrich	SIN
Highland, Virgil L.	Temple
Hill, Christopher	Fermilab
Hitlin, David	Caltech
Holzscheiter, Michael	LANL
Hou, George Wei-Shu	Univ. of Pittsburgh
Innes, Walter R.	SLAC
Isenberg, Martin	Univ. of Utah
Jacob, Richard J.	Arizona State
Ji, Cheung R.	CUNY Brooklyn
Johns, Kenneth	Univ. of Minnesota
Jones, Lawrence W.	Univ. of Michigan
Jones, Lorella M.	Univ. of Illinois
Jones, Stanley T.	Univ. of Alabama
Kaarsberg, Tina M.	SUNY Stony Brook
Kalelkar, Mohan S.	Rutgers
Kane, Gordon L.	Univ. of Michigan
Kang, Ju Hwan	Univ. of New Mexico
Karatas, Debra	Argonne/IIT
Karlen, Dean A.	SLAC
Kaufman, William A.	Univ. of Colorado
Keung, Wai-Yee	Univ. of Illinois
Kirk, Thomas B.W.	Fermilab
Klein, Spencer R.	SLAC
Klinkhamer, Frans R.	LBL
Kliwer, James K.	Univ. of Nevada, Reno
Klopfenstein, Christopher	LBL
Knapp, David A.	LLL
Koetke, Donald D.	Valparaiso Univ.
Kolb, Edward W.	Fermilab
Kovacs, Eve V.	Univ. Ill. at Chicago
Kovesi-Domokos, Susan	Johns Hopkins
Kozanecki, Witold	SLAC
Krakauer, Daniel A.	Univ. of Maryland
Lach, Joseph T.	Fermilab
Lander, Richard L.	U. C. Davis
Lang, Karol	Stanford
Lassila, Kenneth E.	Iowa State
Lau, Kwong H.	Univ. of Houston
Lee, Wen-Piao	U. C. Irvine
Lee, Yong Y.	BNL
Lee-Franzini, Juliet	SUNY Stony Brook
Lepage, G. Peter	Cornell
Lichtenberg, Don B.	Indiana
Liebing, David	Univ. of Utah
Lind, V. Gordon	Utah State
Lingle, Charles P.	Univ. of Utah
Lipps, Frederick W.	Univ. of Houston

Littenberg, Laurence S.	BNL
Lobkowicz, Frederick	Univ. of Rochester
Loew, Gregory A.	SLAC
Loh, Eugene C.	Univ. of Utah
London, David	Univ. of B. C.
Loveless, Richard John	Univ. of Wisconsin
Lovelock, Dale M.	SUNY Stony Brook
Lu, Adolph	U. C. Santa Barbara
Luth, Vera	SLAC
Ma, Ernest	Univ. of Hawaii
MacKenzie, Paul	Fermilab
Mandula, Jeffrey E.	DOE / Washington Univ.
Manweiler, Robert W.	Valparaiso Univ.
Marciano, William	BNL
Marks, Monica H.	Johns Hopkins
Martin, Philip	Fermilab
Marx, Jay	LBL
Masten, Jeffrey P.	
Matsui, Takayuki	KEK
Matthews, James M.	Univ. of Michigan
McCarthy, Robert L.	Stony Brook
McClain, Bruce S.	Univ. of Utah
McDonald, Keith Leon	
McDonald, Kirk T.	Princeton
McDonough, James E.	Temple Univ.
McHugh, Sean F.	U.C. Santa Barbara
Meacham, Brian	Univ. of Utah
Meadows, Brian T.	Univ. of Cincinnati
Meshkov, Sydney	NBS
Meurice, Yannick L.	Argonne
Migneron, Roger	Univ. of W. Ontario
Milner, E. Cas	LANL
Minten, Adolf	CERN
Month, Mel	BNL
Morfin, Jorge G.	Fermilab
Morris, Thomas W.	BNL
Mountain, Raymond J.	Univ. of Notre Dame
Murtagh, Michael J.	BNL
Nandi, Satyanarayan	Oklahoma State
Narain, Kumar S.	Rutherford Lab.
Nash, Thomas	Fermilab
Nauenberg, Uriel	Univ. of Colorado
Neal, Rufus	Cambridge Univ. Press
Nelson, Donald J.	
Nelson, Phil	Harvard
Nemeschansky, Dennis D.	Univ. of S. California
Nguyen, Ai G.	Michigan State
Nitz, David F.	Univ. of Michigan
Nordstrom, Dennis	Physical Review D
O'Halloran, Thomas A.	Univ. of Illinois
Oh, Benedict Y.	Penn State
Ohshima, Takayoshi	Univ. of Tokyo
Olness, Fredrick I.	Illinois Inst. of Tech.
Oreglia, Mark J.	Univ. of Chicago

Padley, Paul	Univ. of Toronto
Pakvasa, Sandip	Univ. of Hawaii
Palmer, William F.	Ohio State
Pan, C.Y.	Utah State
Parsa, Zohreh	BNL
Parsley, Scott L.	East River Elec. Power Corp
Pedigo, R. Daryl	Austin Community College
Perez-Mendez, Victor	LBL
Perger, Warren F.	Colorado State
Petcher, Donald N.	NIKHEF - Holland
Piilonen, Leo E.	LANL
Pollard, Ken	Univ. of Utah
Pondrom, Lee	Univ. of Wisconsin
Pope, Bernard G.	Michigan State
Potter, Douglas	Carnegie-Mellon
Potter, Michael E.	U. C. Irvine
Prasad, Sudhakar	Univ. of New Mexico
Purohit, Milind V.	Fermilab
Raab, Johannes R.	U.C. Santa Barbara
Ramond, Pierre	Univ. of Florida
Randa, James P.	Nat. Bureau of Standards
Rankin, Patrica	SLAC
Reeder, Don D.	Univ. of Wisconsin
Reno, M. Hall	Fermilab
Repond, Jose O.	Argonne
Riles, John K.	SLAC
Rohm, Ryan	Caltech
Rosen, S. Peter	LANL
Rosenblum, Arnold	Utah State
Rosenzweig, Carl	Syracuse
Rosner, Jonathan L.	Univ. of Chicago
Roszkowski, Leszek	U. C. Davis
Rubin, Howard A.	Illinois Inst. of Tech.
Ruchti, Randal C.	Univ. of Notre Dame
Sandberg, Vern	LANL
Sanielevici, Sergui E.	BNL
Sarmiento, Miguel	Univ. of Notre Dame
Scanio, Joseph J.G.	Univ. of Cincinnati
Scarlatella, Michele	U. C. Santa Cruz
Schramm, David	Univ. of Chicago
Schultz, David	SLAC
Schwarz, Andreas S.	U. C. Santa Cruz
Schwarz, John H.	Caltech
Schwitters, Roy	Harvard
Seidel, Sally C.	Univ. of Michigan
Shaevitz, Michael H.	Columbia
Shapero, Donald C.	Natl. Research Council
Shapiro, Anatole M.	U.S. DOE
Shih, Chia C.	Univ. of Tennessee
Shirvel, Stanislaus	Univ. of B. C.
Shubert, Richard	Cal. State Fullerton
Siegel, Steven J.	Idaho State
Siemann, Robert	Cornell
Simhony, Menahem	The Hebrew Univ.

Sinclair, Daniel	Univ. of Michigan
Skorski, Roman	Univ. of Alabama
Sliwa, Christopher Jan	Fermilab
Smith, Darrel W.	U. C. Riverside
Smith, Gerald A.	Penn. State
Snow, Gregory R.	Univ. of Michigan
Sokoloff, Michael D.	Fermilab
Sokolsky, Pierre	Univ. of Utah
Soldate, Mark A.	Fermilab
Somalwar, Sunil V.	Univ. of Chicago
Sommers, Paul D.	Univ. of Utah
Soni, Amarjit	UCLA
Soper, Davison E.	Univ. of Oregon
Soulliere, Mark J.	Penn. State
Staebler, Gary M.	VPI
Steele, Tom	Univ. of W. Ontario
Steger, Herbert	Univ. of Michigan
Stein-Schabes, Jaime	Fermilab
Stephens, Christopher R.	Univ. of Utah
Stickland, David P.	Princeton
Stone, Sheldon	Cornell
Straub, P. Bruce	Univ. of Washington
Strovink, Mark W.	U. C. Berkeley
Sulak, Lawrence R.	Boston Univ.
Summers, Don	Univ. of Wisconsin
Suranyi, Peter	Univ. of Cincinnati
Svobod, Robert	U. C. Irvine
Swamy, P. Narayana	SIU Edwardsville
Swartz, Morris L.	SLAC
Talman, Richard	Cornell
Tangherlini, Frank R.	College of the Holy Cross
Thews, Robert L.	Univ. of Arizona
Thomas, Scott	Utah State
Thorndike, Edward H.	Univ. of Rochester
Thornton, Greg	Univ. of Michigan
Ticho, Harold K.	U. C. San Diego
Toothacker, William S.	Penn. State
Torre, Charles G.	Univ. of Utah
Tosa, Yasunari	Univ. of Colorado
Toussaint, W. Douglas	U. C. San Diego
Trigg, George L.	Physical Review Letters
Tschirhart, Robert S.	Univ. of Michigan
Turcotte, Stephen	Univ. of Utah
Tye, Henry S.H.	Cornell
Voyvodic, Louis	Department of Energy
Wang, Z.	Utah State
Wasserbaech, Steven	SLAC
Weinstein, Alan J.	U. C. Santa Cruz
Westgard, James B.	Indiana State
Wienke, Bruce R.	LANL
Wijewardhana, L.C.R.	Yale
Wilkerson, John F.	LANL
Wilkinson, Carol A.	LANL
Willenbrock, Scott S. D.	Univ. of Wisconsin

Winstein, Bruce	Univ. of Chicago
Witherell, Michael S.	U. C. Santa Barbara
Witten, Edward	Princeton
Wojcicki, Stanley G.	SSC CDG
Word, Gary B.	Vanderbilt
Wu, Yongshi	Univ. of Utah
Xu, Yao-Hui	BYU
Yamin, Peter	BNL
Yu, Feng	Univ. of Utah
Zdarko, Richard W.	SLAC

INDEX OF TITLES AND ABSTRACTS FOR INVITED TALKS AND CONTRIBUTED PAPERS

* Those marked with an asterix are included in these Proceedings

Abachi, Shahriar (Purdue)
"Strange Baryon Production in e^+e^- Annihilations at 29 GeV"
"Production of Charmed Baryons in e^+e^- Annihilations at 29 GeV"
Alexander, James P. (SLAC)
"Design, Prototype Studies, and Construction of a Vertex Chamber for the Mark II at SLC"
Allen, Roland E. (Texas A&M)
"Nature as an Ordered Phase"
Alvarez, Orlando (U. C. Berkeley)
"Geometry of the Operator Product Expansion"
Anderson, Scott B. (Colorado State)
"Exact Solution of a Non-trivial, Non-renormalizable Quantum Field Theory in N Dimensions"
Argus Collaboration (see D. Gingrich and P. Padley)
Arnowitt, Richard L (Texas A & M)
*"Prospects for Observation of Supersymmetry at the Tevatron
ASP Collaboration (see C. Hearty)
Aurenche, Patrick J.P. (LAPP)
*"Photons in Hard Collisions: An Optimized QCD Approach"
Avignone III, Frank (Univ. of S. Carolina)
*"Experimental Evidence Suggestive of Neutrinoless $\beta\beta$-Decay with a Near Massless Particle"
Baer, Howard A. (Argonne)
"The Search for Supersymmetric Particles at Hadron Colliders"
Bagger, Jonathan A. (Harvard)
*"Thirring Strings"
Balke, Brian K.E. (U. C. Berkeley)
"Precise Measurement of the Asymmetry Parameter δ in Muon Decay"
Ball, James S. (Univ. of Utah)
"Superconductivity: A Testing Ground for Confinement"
Bartel, Wulfrin (DESY)
*"Recent Results on α_s from PETRA"
*"Status of HERA and HERA-Experiments"
Bauer, G. (Harvard)
*"W and Z Properties in UA1"
Bernard, Claude (U. C. Los Angeles)
"Weak Interactions on the Lattice"
Betounes, David E. (Univ. of S. Mississippi)
"Gauge Invariant Lagrangians on the Jet Bundle J^1 (E plus C)"
Blucher, Edward C. (Cornell)
"Charm Production in B Meson Decay"
Bodwin, Geoffrey T. (see E. Kovacs)
Cady, Donald R. (LANL)
"Observation of 100 TeV Gamma Rays from Cygnus X-3"
Cahill, Kevin E. (Univ. of New Mexico)
"Noncompact Simulations of SU(2) in 3 dimensions"
Campbell, William B. (Univ. of Nebr.-Lincoln)
"What is the Charge of the Charged Vacuum"

Carrol, Alan S. (BNL)
*"Exclusive Reactions and Dimensional Scaling: New Measurements and New Understanding"
Cason, Neal M. (Univ. of Notre Dame)
"Study of the $K_sK_s\pi^o$ System Produced in the Reaction $\pi^- p \rightarrow nK_sK_s\pi^o$ at 23 GeV/c"
Cavalli-Sforza, Matteo (SLAC)
"Results and Developments on Cerenkov Ring Imaging for the SLD Detector"
CERN PS183 Collaboration (see M Soulliere)
CERN UA1 Collaboration (see D.W. Smith and D. Summers)
CERN UA2 Collaboration (see K. Einsweiler)
CERN UA6 Collaboration (see G. R. Snow)
Chanowitz, Michael (LBL)
"Probing Electroweak Symmetry Breaking at the SSC: A No Lose Corollary"
Chau, Ling-Lie (U. C. Davis)
*"Analysis of Exclusive Two-Body Decays of Charm Mesons Using the Quark-Diagram Scheme"
*"Charmless Beauty Decays and CP Non-Invariance"
*"Ways to Measure $|V_{ub}|$ and $|V_{ub}/V_{cb}|$ from the Quark Diagram Scheme"
"Linear Systems and Conservation Laws of Gravitational Fields in Four Plus Extended Super Space"
Chen, Chu-Xin (Univ. of Utah)
"A Phenomenological Lattice Model of the High Temperature Phase Transition in Quantum Chromodynamics"
Chen, Min (MIT)
*"Low Thrust Hadronic Events with Isolated μ from PETRA
Cherry, Michael L. (Univ. of Pennsylvania)
"Diurnal Variations in the ^{8}B Solar Neutrino Flux"
CLEO Collaboration (see R.S. Galik)
Cline, David (Univ. of Wisconsin and UCLA)
*"Missing Transverse Energy Events and a Search for Additional Neutrino Families and a 4th Generation Lepton"
Collins, John (Illinois Inst. of Tech.)
*"Evaluation of Perturbative QCD in Monte Carlo Event Generators"
Coteus, Paul W. (Univ. of Colorado)
"Production of the Charmed Strange Baryon Ξ c+ by Incident Neutrons"
Cremaldi, Lucien (Univ. of Colorado)
"Measurement of the Λ_c^+ Lifetime"
CUSB Collaboration (see J. Lee-Franzini)
DeGrand, Thomas A. (Univ. of Colorado)
"Mass and Wave Function of the Lightest Scalar Glueball"
DESY Argus Collaboration (see D. Gingrich and P. Padley)
DeTar, Carleton E. (Univ. of Utah)
"Static Screening Lengths in the Quark-Gluon Plasma"
Durkin, L. Stanley (Ohio State University)
*"Elastic Scattering of Muon Neutrinos at B.N.L.
Einsweiler, K. (CERN)
"Recent Electro-Weak Resuts from UA2 at the CERN $\bar{p}p$ Collider"
"The UA2 Experiment at ACOL"

Elias, V. (Univ. of W. Ontario)
"$\langle q \bar{q} \rangle$ and Higher-Dimensional-Condensate Contributions to the Nonperturbative Quark Mass"
Ensign, Philip W. (Univ. of Colorado)
"The Supercurrent and the Adler-Bardeen Theorem in Coupled Supersymmetric Yang-Mills Theories"
Fabbrichesi, Marco E. (Univ. of Virginia)
"Dynamical Mass Generation in BRS Invariant Theories"
Fermilab D-Zero (see C. Klopfenstein)
Fermilab E400 (see P. Coteus)
Fermilab E531 (see S.G. Frederiksen)
Fermilab E605 (see Richard Gray)
Fermilab E691 (see L. Cremaldi, S. McHugh, M. Purohit, J.R. Raab, C. J. Sliwa, and M. Sokoloff)
Fermilab E744 (see M.J. Oreglia)
Frederiksen, Soren G. (Univ. of Ottawa)
"A Study of the Production of Charm Particles by Neutrinos"
Froidevaux, Daniel (LAL Orsay)
"Hadron Collider Results and Prospects"
Gan, Kock K. (SLAC)
"Study of τ Decay with Multiple Neutral Mesons in the Final States"
Giddings, Steve B. (Princeton)
*"The Equivalence of Polyakov and Light Cone Bosonic String Theory"
Gilchriese, Murdock (U. C. Berkeley)
"SSC Detectors"
Gingrich, Douglas M. (Univ. of Toronto)
"Inclusive D_s Meson Production from B Meson Decay"
"D and D* Production in e^+e^- Annihiliation"
Gravel, Pierre (CMR St-Jean)
"Charge Computations and Masses in Multiconnected Kaluza-Klein Theories"
Gray, Richard (Univ. of Washington)
"Production Mechanisms of the Upsilon"
Gunion, John F. (U. C. at Davis)
"Probing the W-Z-Higgs Sector of Electroweak Gauge Theories at Hadron Colliders"
"Analytic and Intuitive Understanding of Nonperturbative Enhancements in Heavy Quark Production"
Haas, Peter M. (Cornell)
"An Upper Limit on D^o - $\bar{D}^o$ Mixing"
Hagelin, John S. (Maharishi Int. Univ.)
"A High Energy Neutrino Signature from Supersymmetric Relics"
Haller, Kurt (Univ. of Connecticut)
"Gauge Theories in the Temporal Gauge"
Hearty, Christopher (Univ. of Washington)
"New Results from ASP on Single - Photon Production in e^+e^- Annihilation at 29 GeV"
Holzscheiter, Michael (LANL)
"The Gravitational Acceleration of the Antiproton"

Hou, George Wei-Shu (Univ. of Pittsburgh)
"Implications of a Heavy Top Quark and a Fourth Generation on the Decays $B \to K\ell^+\ell^-$, $K\nu\bar{\nu}$"
*"Charmless Rare B Decays and a Fourth Generation"
IMB Collaboration (see S. Seidel and G. Thornton)
Ji, C.R. (CUNY Brooklyn)
"Exclusive Heavy Meson Production in e^+e^- Annihilation"
Johns, Kenneth (Univ. of Minnesota)
"Search for Underground Muons Associated with X-Ray Binaries by Soudan I"
Jones, Stanley T. (Univ. of Alabama)
"Fixed Angle Asymptotic Behavior of the Type I Superstring Amplitude"
Kaarsberg, Tina M. (SUNY Stony Brook)
"Measurement of the Branching Ratio for $T" \to \mu\mu$"
Kalelkar, Mohan S. (Rutgers)
"Multiparticle Production from 200 GeV/c p, π^+, and K^+ Interactions on Au, Ag, and Mg"
Kang, Ju Hwan (Univ. of New Mexico)
"Search for Neutrino Oscillations in Line E at LAMPF"
Karlen, Dean A. (SLAC)
"Low q^2 Radiative Bhabha Scattering at High Energies"
Kaufman, William A. (Univ. of Colorado)
"Subleading QCD Corrections to the $K^o - \bar{K}$ System
Klein, Spencer R. (SLAC)
"Strange Baryon Production in e^+e^- Collisions at an E_{cm} of 29 GeV"
Klinkhamer, Frans R. (LBL)
"Contact Interactions for Closed Superstring Field Theory"
Kliwer, James K. (Univ. of Nevada, Reno)
"Moeller Scattering and the Measurement of Beta-Ray Longitudinal Polarization"
Klopfenstein, Christopher (LBL)
"Test Results from a Precision Drift Chamber Vertex Detector Prototype Using Dimethyl Ether"
Kolb, Edward W. (Fermilab)
"Neutrino Masses and Oscillations"
Kovacs, Eve (Univ. of Illinois/Chicago)
*"Unorthodox Lattice Fermion Derivatives and Their Shortcomings"
Kozanecki, Witold (SLAC)
"SLC Status"
Krakauer, Daniel A. (Univ. of Maryland)
"Recent Results from Electron Neutrino - Electron Elastic Scattering at LAMPF"
Lach, Joseph T. (Fermilab)
"Measurement of the Form Factors in Σ^- β Decay"
LAMPF E764 Collaboration (see J. Kang)
Lang, Karol (SLAC)
"An Electron Beam Dump Search for Light, Short-Lived Particles"
Lassila, K. (Iowa State)
"Electro and Photo Reaction to Study the EMC Effect"

Lee-Franzini, Juliet (SUNY Stony Brook)
"Observation of $\Upsilon'' \to \chi_b' \gamma \to (\Upsilon')\Upsilon\gamma\gamma \to \mu^+\mu^-(e^+e^-)\gamma\gamma$ and the Hadronic Widths of the χ_b' States"
LEP L3 Collaboration (see D. Stickland)
Lepage, G. Peter (Cornell)
"Methods and Algorithms of Lattice Gauge Theory"
London, David (Univ. of B. C.)
"Effects of Z^0 Missing with an E_6 Gauge Boson at e^+e^- Colliders"
Lovelock, Dale M. (SUNY Stony Brook)
"Fine Structure of the ψ_b' States"
Luth, Vera (SLAC)
*"Lifetimes of Heavy Leptons and Hadrons"
Ma, Ernest (Univ. of Hawaii)
"New Improved Supersymmetric Left-Right Model"
Marciano, William J. (BNL)
*"Electroweak Radiative Corrections and Neutral Current Phenomenology"
*"Charged Current Unitarity and Extra Neutral Gauge Bosons"
Martin, Philip (Fermilab)
*"Status of the Tevatron I Collider"
Matsui, Takayuki (KEK)
*"Tristan Status"
McDonald, Keith Leon
"Cosmic Radiation Hydromagnetism"
McDonald, Kirk T. (Princeton)
"J/Ψ Longitudinal Polarization in 252-GeV πN Interactions"
"The Synchrotron-Cherenkov Effect"
"Experiments on the Nonlinear QED of Intense Laser Fields"
McDonough, J. (Temple)
"New Upper Limit for the Decay $\pi^o \to 3\gamma$"
McHugh, Sean F. (U.C. Santa Barbara)
"Measurements of D_s^+ and Cabibbo-supressed D^+ Decays"
Meshkov, Sydney (NBS)
*"The ι/E System"
Meurice, Yannick L. (Argonne)
"From Points To Gauge Fields"
Minten, Adolf (CERN)
*"LEP Status"
Nandi, Satyanarayan (Oklahoma State)
"Higgs Boson Production at Heavy Z-Resonance"
Narain, K.S. (Rutherford Lab.)
"Left/Right Asymmetric Orbifolds"
Nelson, Phil (Harvard)
*"A New Topological Term in 2D Field Theory"
Nemeschansky, Dennis (Univ. of S. California)
"Multiloop Amplitudes on Orbifolds"
Nguyen, Ai G. (Michigan State)
"Inclusive Charm Cross Sections in 800 GeV/c p-p Interactions"
Oh, Benedict Y. (Penn State)
"Hadroproduction of Prompt Photons in the Second Order of α Strong"
"Detection of W Decay into $e\nu$ and Jets"
Ohshima, Takayoshi (Univ. of Tokyo)
"An Upper Limit for the Mass of Electron Anti-Neutrino"

Olness, Fredrick I. (Illinois Inst. of Tech.)
"The Effective Vector Boson Method for High Energy Collisions"
Oreglia, Mark J. (Univ. of Chicago)
*"Production of Same-Sign Dimuons by 0-800 GeV Neutrinos and Antineutrinos"
Padley, Paul (Univ. of Toronto)
"Lifetime Measurements at ARGUS"
"Observation of the Charmed Baryon Σ_c in e^+e^- Annihilation at 10 GeV CM Energy"
Perez-Mendez, Victor (LBL)
"Position Sensing Detection of Charged Particles in Hydrogenated Amorphous Silicon Layers"
Perger, Warren F. (Colorado State)
"Theory of Parity Nonconservation in Atoms due to Neutral Weak Current"
Piilonen, Leo E. (LANL)
"Status of the MEGA Experiment"
Pope, Bernard G. (Michigan State)
"A Study of Massive Electron Pairs and Associated Particles Produced at the CERN ISR"
Prasad, Sudhakar (Univ. of New Mexico)
"Ratios of Wilson Loops in Three Dimensions"
Purohit, Milind V. (Fermilab)
*"Charm Photoproduction Results from the Fermilab Tagged Photon Spectrometer"
Raab, Johannes R. (U.C. Santa Barbara)
"Measurement of the D^o, D^+, D_s^+ Lifetime"
Rankin, Patricia (SLAC)
"A Search for Short - lived Axions in a 20GeV Photo-production Experiment"
Repond, Jose O. (Argonne)
"Observation of η Meson Production in Tau Lepton Decay"
Riles, John K. (SLAC)
"Upper Limit on $D^o \rightarrow e^{\pm}\mu^{\mp}$ in e^+e^- Annihilation at 29 GeV"
Rohm, Ryan (Caltech)
*"Exotic Spheres as Topological Defects in String Field Theory"
Rosner, Jonathan L. (Univ. of Chicago)
*"Quark Mixing and CP Violation"
Roszkowski, L. (U. C. Davis)
"Limits on the Masses and Couplings of the Z′ and the Higgs Bosons and Z′ Decays in an E_6 Superstring Based Model"
Ruchti, Randal C. (Univ. of Notre Dame)
"A Study of Charm Production in 200 GeV/c π^-Be Interactions"
"Development of a Scintillating Fiber Target (SFT) for Fixed Target Applications"
Sanielevici, Sergui E. (BNL)
"Flavor Number Dependence of the Chiral Phase Transition in QCD"
Scanio, Joseph J.G. (Univ. of Cincinnati)
"Higgs Physics at Future e^+e^- Colliders"
Scarlatella, Michele (U. C. Santa Cruz)
"Observation of Decays of J/ψ to Vector Meson Pairs from the Mark III"

Schultz, David (SLAC)
"A Study of Strangeonium States Observed in K^-p Interactions at 11 GeV/c"
Schwarz, Andreas S. (U. C. Santa Cruz)
"A Silicon Strip Vertex Detector for the Mark II Experiment at the SLC"
Schwarz, John H. (Caltech)
*"Superstrings - A Progress Report"
Seidel, Sally C. (Univ. of Michigan)
"Recent Results from the IMB Nucleon Decay Experiment"
Shafi, Q. (Bartol)
"Magnetic Monopoles"
Shubert, Richard (Cal. State Fullerton)
"Particles of the Vacuum"
Siemann, Robert (Cornell)
*"Advanced Accelerator Research"
Simhony, Menahem (The Hebrew Univ.)
"Inexpanding Phenomena Causing the Hubble Redshift and the Diversity of the Hubble Constant"
"'Grand Unification' of Quantum Physics and Relativity in the Electron-Positron Lattice (Epola) Model of Space"
"Experiments Proving the Massive Existence of Bound Electrons and Positrons in the Vacuum Space"
Skorski, Roman (Univ. of Alabama)
"The Sun"
SLAC HRS Collaboration (see S. Abachi, J. Repond, and R. Tschirhart)
SLAC Hybrid Facility Photon Collaboration (see P. Rankin)
SLAC Mark II Collaboration (see J.P. Alexander, K.K. Gan, S.R. Klein, K. Riles, A.S. Schwarz, A.J. Weinstein)
SLAC Mark III Collaboration (see M. Scarlatella and S. Wasserbaech)
SLAC SLD (see M. Cavalli-Sforza)
Sliwa, Christopher Jan (Fermilab)
"Measurement of the Semileptonic Charm Decay $D^o \rightarrow K^-e^+\nu$"
Smith, Darrel W. (U. C. Riverside)
*"W/Z Production with Jets in Experiment UA1"
Snow, Gregory R. (Univ. of Michigan)
"Neutral Pion, Eta, Direct Photon and J/Ψ Production in $\bar{p}$ p and pp Interactions at $\sqrt{s} = 24.3$ GeV"
Sokoloff, Michael D. (Fermilab)
"A Dependence of J/Ψ Photoproduction"
Sokolsky, Pierre (Univ. of Utah)
*"Study of Ultra High Energy Cosmic Rays with the Fly's Eye"
Soldate, Mark A. (Fermilab)
"Partial Wave Unitarity and Closed-String Amplitudes"
Somalwar, Sunil V. (Univ. of Chicago)
"Monopole Detection by Faraday Induction: A General Study"
Soper, Davison E. (Univ. of Oregon)
"Statistical Analysis in New Particle Searches"
"The Effective W Approximation"
Soulliere, Mark J. (Penn. State)
"Precision Measurement of Charged Two-Meson Channels in Antiproton-Proton Annihilation at LEAR."

Staebler, Gary M. (VPI)
"Gauge Theory of the Microscopic Dynamics of Superfluid He4"
"Supersymmetric Non-linear Sigma Models with Torsion in 3+1 Dimensions"
Steele, T. (Univ. of W. Ontario)
"$<\bar{q}q>$ and Higher Dimensional-Condensate Contributions to the Non-Perturbative Quark Mass"
Steger, Herbert (Univ. of Michigan)
"Nonlinear Realization of Heavy Fermions"
Stein-Schabes, Jaime (Fermilab)
*"Inflation in Tolman-Bondi Cosmological Models"
Stephens, Christopher R. (Univ. of Utah)
"Odd Statistics in Odd Dimensions"
Stickland, David P. (Princeton)
"Status of BGO Calorimetry in L3"
Stone, Sheldon (Cornell)
*"e^+e^- Collider Results"
Straub, P. Bruce (Univ. of Washington)
"High p_T Particle Production from Beryllium and Tungsten"
Summers, Don (Univ. of Wisconsin, Madison)
*"Beauty Physics at UA1"
Swamy, P. Narayana (SIU Edwardsville)
"Constraints of Dynamical Flavor Symmetry Breaking in Quantum Chromodynamics: Vafa-Witten Theorem"
Tangherlini, Frank R. (College of the Holy Cross)
"Phenomenological Expressions for Hyperon Magnetic Moments"
Thorndike, Edward H. (Univ. of Rochester)
*"The Mystery of the Missing D's"
Thornton, Greg (Univ. of Michigan)
*"Underground Search for Muons Correlated with Cygnus X-3"
Toothacker, William S. (Penn. State)
"Characteristics of Leading Particle Production from pA and $\bar{p}$A Interactions at 100 GeV/c"
Tosa, Yasunari (Univ. of Colorado)
"New Anomaly-Free Theories Beyond Superstrings?"
Toussaint, W. Douglas (U. C. San Diego)
*"Supercomputations in QCD"
Tschirhart, Robert S. (Univ. of Michigan)
"Measurement of the Cabibbo Suppressed Decay of the τ Lepton"
Tye, Henry (Cornell)
*"Construction of Four Dimensional Fermionic String Models"
Wang, Z. (Utah State)
"A Very Weakly Coupled Graviphoton from Kaluza-Klein Supergravity"
Wasserbaech, Steven (SLAC)
*"Results on the Charmed Meson Decays from Mark III"
Weinstein, Alan J. (U. C. Santa Cruz)
"The Central Drift Chamber for the Mark II Experiment at the SLC"
Westgard, James B. (Indiana State)
"Electromagnetism and Conformal Geometry"
Wilkerson, John F. (LANL)
*"Limits on the Electron Anti-Neutrino Mass"

Willenbrock, Scott S. D. (Univ. of Wisconsin)
*"Pair Production of Longitudinal W and Z Bosons and the Goldstone Boson Equivalence Theorem"
Winstein, Bruce (Univ. of Chicago)
*"CP and Other Tests of the Standard Model"
Witherell, Michael S. (U. C. Santa Barbara)
*"Heavy Quark Physics"
Witten, Edward (Princeton)
*"String Theory in 1987"
Wojcicki, Stanley G. (SSC CDG)
"SSC Status"
Word, Gary B. (Vanderbilt)
"The Charm Content in Semileptonic B Meson Decay"
Wu, Yongshi (Univ. of Utah)
"Superstring Cosmology at Late Times and Time Variation of Fundamental Coupling Constants"
Zdarko, Richard W. (SLAC)
"Results from Studies of Limited Streamer Tubes with External Electrode Readout"

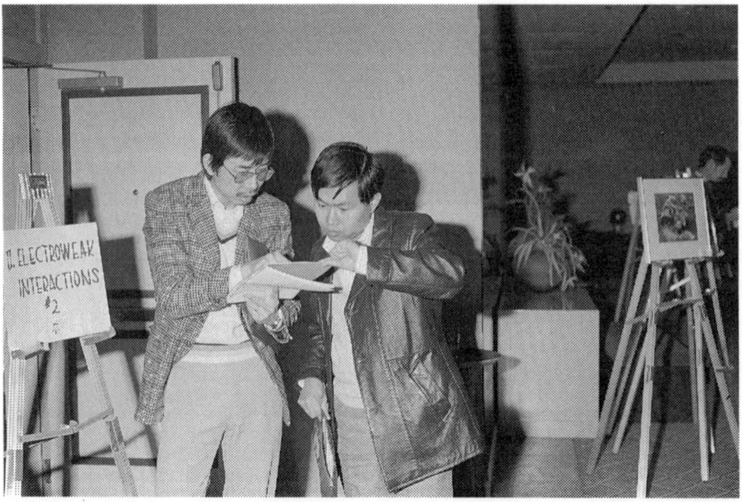
II. ELECTROWEAK
INTERACTIONS
#2

AUTHOR INDEX

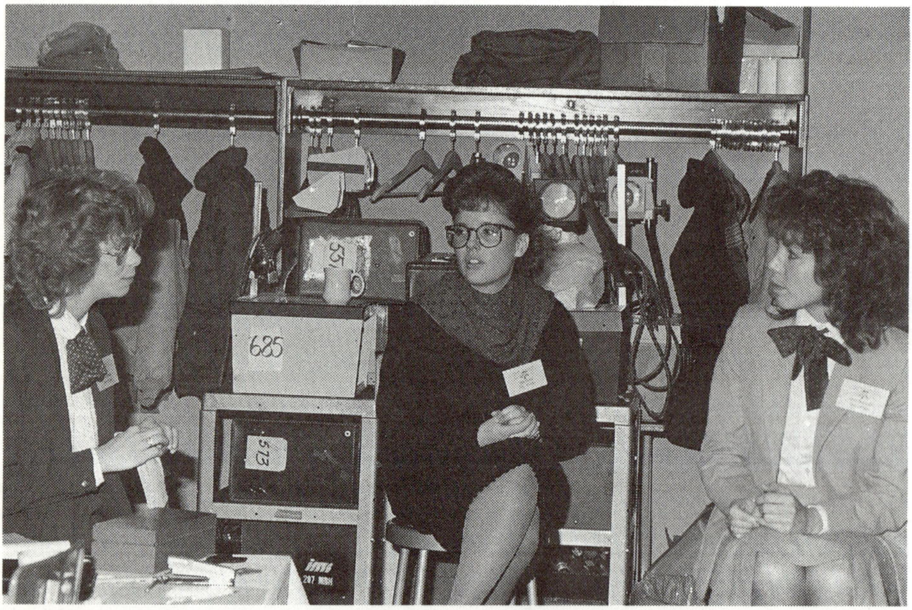
685
513